Plant Stress Physiology, 2nd Edition

_____

Plant Stress Physiology, 2nd Edition

Edited by

Sergey Shabala

School of Land and Food
University of Tasmania
Australia

CABI is a trading name of CAB International

CABI
Nosworthy Way
Wallingford
Oxfordshire OX10 8DE
UK

CABI
745 Atlantic Avenue
8th Floor
Boston, MA 02111
USA

Tel: +44 (0)1491 832111
Fax: +44 (0)1491 833508
E-mail: info@cabi.org
Website: www.cabi.org

Tel: +1 (617)682-9015
E-mail: cabi-nao@cabi.org

A catalogue record for this book is available from the British Library, London, UK.

Library of Congress Cataloging-in-Publication Data

Names: Shabala, Sergey, editor.
Title: Plant stress physiology / Sergey Shabala, editor.
Description: Boston, MA : CABI, 2017. | "2nd Edition." | Includes
    bibliographical references and index.
Identifiers: LCCN 2016031184 (print) | LCCN 2016047000 (ebook) | ISBN
    9781780647296 (hbk : alk. paper) | ISBN 9781780647302 (ePDF) | ISBN
    9781780647319 (ePub)
Subjects: LCSH: Plant physiology. | Plants--Effect of stress on.
Classification: LCC QK754 .P585 2017 (print) | LCC QK754 (ebook) | DDC
    581.7--dc23
LC record available at https://lccn.loc.gov/2016031184

ISBN-13: 978 1 78064 729 6

Commissioning editor: Rachael Russell
Editorial assistant: Emma McCann
Production editor: Tracy Head

Typeset by SPi, Pondicherry, India.
Printed and bound in the UK by CPI Group (UK) Ltd, Croydon, CR0 4YY, UK.

Contents

Contributors

Suleyman I. Allakhverdiev, Institute of Plant Physiology, Russian Academy of Sciences, Moscow, Russia; Institute of Basic Biological Problems, Russian Academy of Sciences, Pushchino, Russia; Department of Plant Physiology, Moscow State University, Moscow, Russia; E-mail: suleyman. allakhverdiev@gmail.com

Joanne Bentley, Department of Molecular and Cell Biology, University of Cape Town, Private Bag X3, Rondebosch 7701, South Africa; E-mail: joanne.bentley@uct.ac.za

Gerald A. Berkowitz, Department of Plant Science and Landscape Architecture, University of Connecticut, Storrs, CT 06269, USA; E-mail: gerald.berkowitz@uconn.edu

Keren Cooper, Department of Molecular and Cell Biology, University of Cape Town, Private Bag X3, Rondebosch 7701, South Africa; E-mail: Keren.Cooper@uct.ac.za

Halford J.W. Dace, Department of Molecular and Cell Biology, University of Cape Town, Private Bag X3, Rondebosch 7701, South Africa; E-mail: halford.dace@uct.ac.za

Emmanuel Delhaize, CSIRO Agriculture and Food, GPO Box 1600, Canberra ACT 2601, Australia; E-mail: manny.delhaize@csiro.au

Vadim Demidchik, Department of Planet Cell Biology and Bioengineering, Biological Faculty, Belarusian State University, Minsk, Belarus; Russian Academy of Sciences, Komarov Botanical Institute, St Petersburg, Russia; E-mail: Dzemidchyk@bsu.by

Jill M. Farrant, Department of Molecular and Cell Biology, University of Cape Town, Private Bag X3, Rondebosch 7701, South Africa; E-mail: jill.farrant@uct.ac.za

Stephen R. Grattan, Department of Land, Air and Water Resources, University of California, Davis, CA 95616, USA; E-mail: srgrattan@ucdavis.edu

Lawrence V. Gusta, Department of Plant Sciences, University of Saskatchewan, Saskatoon, Canada S7N 5A8; E-mail: larry.gusta@usask.edu

Amelia Hilgart, Department of Molecular and Cell Biology, University of Cape Town, Private Bag X3, Rondebosch 7701, South Africa; E-mail: HLGAME001@myuct.ac.za

Anjana Jajoo, School of Life Sciences, Devi Ahilya University, Indore, India; E-mail: anjanajajoo@hotmail.com

Marcel A.K. Jansen, School of Biological, Earth and Environmental Sciences, University College Cork, Ireland; E-mail: m.jansen@ucc.ie

Andre Läuchli[†], formerly of Department of Land, Air and Water Resources, University of California, Davis, CA 95616, USA

Yi Ma, Department of Plant Science and Landscape Architecture, University of Connecticut, Storrs, CT 06269, USA; E-mail: yi.ma@uconn.edu

Lakshmi Praba Manavalan, Division of Plant Sciences, University of Missouri, Columbia, Missouri, 65211, USA; E-mail: manavalanl@missouri.edu

Rana Munns, CSIRO Agriculture, Canberra ACT 2601; School of Plant Biology and ARC Centre of Excellence in Plant Energy Biology, University of Western Australia, Crawley WA 6009, Australia; E-mail: rana.munns@sciro.au

Henry T. Nguyen, Division of Plant Sciences, University of Missouri, Columbia, Missouri, 65211, USA; E-mail: nguyenhenry@missouri.edu

Pierdomenico Perata, PlantLab, Institute of Life Sciences, Scuola Superiore Sant'Anna, Pisa, Italy; E-mail: p.perata@sssup.it

Paula Pongrac, The James Hutton Institute, Invergowrie, Dundee DD2 5DA, UK; E-mail: paula.pongrac@hutton.ac.uk

Geert Potters, Antwerp Maritime Academy, 2030 Antwerp, Belgium, and Department of Bioscience Engineering, University of Antwerp, 2020 Antwerp, Belgium; E-mail: geert.potters@uantwerpen.be

Chiara Pucciariello, PlantLab, Institute of Life Sciences, Scuola Superiore Sant'Anna, Pisa, Italy; E-mail: c.pucciariello@sssup.it

Eric Ruelland, Institute of Ecology and Environmental Sciences of Paris, Centre National de la Recherche Scientifique, CNRS UMR7618, Université Paris-Est, France; E-mail: eric.ruelland@upmc.fr

Ryan, Peter R. CSIRO Agriculture and Food, GPO Box 1600, Canberra, ACT 2601, Australia; E-mail: Peter.Ryan@csiro.au

Sergey Shabala, School of Land and Food, University of Tasmania, Private Bag 54, Hobart, Tasmania, Australia; E-mail: sergey.shabala@utas.edu.au

Philip J. White, The James Hutton Institute, Invergowrie, Dundee DD2 5DA, United Kingdom; E-mail: Philip.White@hutton.ac.uk

Ian R. Willick, Department of Plant Sciences, University of Saskatchewan, Saskatoon S7N 5A8, Canada; E-mail: irw412@mail.usask.ca

Michael Wisniewski, United States Department of Agriculture, Agricultural Research Service (USDA-ARS), Appalachian Fruit Research Station, Kearneysville, WV 25430, USA; E-mail: Michael.Wisniewski@ARS.USDA.GOV

Stress: The Way of Life

Marcel A.K. Jansen[1]* and Geert Potters[2,3]
[1]*School of Biological, Earth and Environmental Sciences, University College Cork, Cork, Ireland; [2]Antwerp Maritime Academy, 2030 Antwerp, Belgium; [3]Department of Bioscience Engineering, University of Antwerp, 2020 Antwerp, Belgium*

If terminology is not corrected, then what is said cannot be followed.
If what is said cannot be followed, then work cannot be accomplished.
Confucius, Analects 13, 3
(Waley, 1938, transl.)

Abstract

Stress is a common aspect in the life cycle of a plant. However, the principles of 'plant stress' remain poorly defined, and key terminology can be interpreted in more than one way. This ambiguity can potentially impede scientific progress. Several theoretical frameworks detailing the principles of plant stress have been conceived, based on analogies with human stress, mechanics and thermodynamic states. Particularly important are the concepts of eustress and distress (good and bad stress), and how these are affected by stressor dose and duration. Using these concepts can help place experimental data in a conceptual framework, and avoid generation of seemingly contradictory data.

What is Stress?

One of the originators of the concept of stress, Selye (1973), wrote that 'everybody knows what stress is and nobody knows what it is'. The public perception of stress is informed by the human experience of stress, which places a strong emphasis on disruptions in psychological and/or emotional processes with consequences for physical well-being. Clearly such a concept does not apply easily to plants, and a plant-specific stress concept is required. Unfortunately, there is no widely accepted, concise definition of plant stress that is broad enough to accommodate a broad range of 'responses to stressors', without including every physiological change in an organism. Plant stress is considered to be one of the most important topics in plant biology, and is referred to in large numbers of publications, discussed at conferences and subject to intense exchanges of ideas between researchers and agronomists; however, the concept itself is usually passed on as a 'black box' term. Wang *et al.* (2003) state that 'Abiotic stress is the primary cause of crop loss worldwide, reducing average yields for most major crops by more than 50%.' However, in the absence of an accepted definition of stress, conclusions about the impact of

---

* Corresponding author: m.jansen@ucc.ie

stress seem futile. Textbooks hardly ever allow for a deeper insight into the nature of stress. A short survey of published handbooks shows widely diverging descriptions of plant stress. Buchanan *et al.* (2000) define stresses as 'external conditions that adversely affect growth, development or productivity'. Smith *et al.* (2010) refer to 'stressful environments' which are 'environments that are less than optimal for plant growth' while Taiz and Zeiger (2010) refer to stress as 'a disadvantageous influence exerted on a plant by external abiotic or biotic factor(s), such as infection, or heat, water and anoxia'. What these descriptions of plant stress have in common is an agronomic perspective; stress decreases plant production and hence world food production. Another common aspect of these descriptions is the strong focus on environmental factors as a cause for stress. However, the link between environmental conditions and plant responses is, typically, very loose. Treatment with a chilling temperature of 10°C causes a rapid decrease in the photosynthetic rate of tropical species such as sorghum, but has only minor effects on a species such as ryegrass (Taylor and Rowley, 1971). Is a temperature of 10°C a stress? Similarly, a soil contaminated with heavy metals may be considered non-optimal for plant growth, but are the metallophyte species growing on such a substrate stressed? These examples demonstrate that in lieu of a concept of stress that focuses on environmental factors, actual impacts on specific organisms need to be considered (Hopkins, 1999). Larcher (1980) referred to stress as 'Changes in physiology that occur when species are exposed to extraordinary unfavourable conditions', and Leclerc (2003) describes stress as 'deviations from normal average conditions of the plant'. In the latter concept, the causal environmental factor responsible for plant stress is variously named the 'external constraint', 'stress factor' or 'stressor'. Conversely, the result of exposure to a stressor is named 'the state of stress', 'stress response' or simply 'stress', and this state may refer to molecular, cellular, tissue or organismal aspects of a plant's physiology (Leclerc, 2003). Unfortunately, in the literature the term 'stress' is used arbitrarily to refer either to the actual environmental stressor, or to the plant response, while some papers completely fail to clarify how terms are used, and this hampers the exchange of information within the scientific community.

From Stress to Eustress and Distress

Defining stress as a deviation from 'normal average conditions of the plant' does, of course, not change the problem of defining stress. Every change in an environmental factor influences plant metabolism, growth and development (Gaspar *et al.*, 2002). In fact, the plant environment fluctuates regularly and predictably under daily and seasonal cycles, emphasizing that physiological plasticity is essential for survival of sessile plants (Gaspar *et al.*, 2002). For example, a tree will develop sun leaves in exposed parts of the crown and shade leaves in more sheltered parts. These two types of leaves will differ with respect to morphology, pigment composition and photosynthetic rates (Lichtenthaler *et al.*, 2007). Such adjustments of physiology optimize performance and are central to the survival of sessile plants in their dynamic environment. Nevertheless, few authors would define such environmental plasticity as stress (Gaspar *et al.*, 2002). Plants will also adjust their metabolism to optimize performance when exposed to hostile conditions such as drought, heat, salinity or other stressors. However, in this scenario the plant response is more complex, and distinctions can be made between the destructive influences of stressor exposure and adjustments of metabolism to achieve a new optimized state. An early proponent of this more analytical analysis of stress responses was the endocrinologist Hans Selye. Selye developed the concept of the 'general adaptation[1] syndrome', which provides a framework for analysing stress responses. This framework consists of two dimensions: (i) the severity of the stressor; and (ii) the dynamic response of the organism over time. Regarding the dose of the stressor, Selye (1964) and Lichtenthaler (1988) defined a positive, adaptive stress triggered by low levels of a stressor as 'eustress', and a negative stress caused by high levels of a stressor as 'distress'. Eustress drives the adjustments of metabolism that result in a new optimized state under the new environmental conditions, while distress refers to destructive influences. It is the balance between eustress and distress that determines whether a plant will thrive or perish. Thus, stress may be defined as a state in which increasing demands made upon a plant lead to an initial destabilization of functions, followed by either normalization and improved tolerance, or permanent damage or death (Gaspar *et al.*, 2002).

In practical terms this may mean that a scientist who analyses responses to a stressor may find under mild stress conditions (eustress-prevalent) extensive readjustment of metabolism, including specific changes in gene expression, induction of repair and protection responses, accumulation of phytochemicals and antioxidants, and morphological and developmental adjustments. Conversely, under severe stress conditions (distress-prevalent) emphasis is more likely on physiological destabilization, including oxidative damage, DNA damage, inactivation of photosynthesis and cell death. In a study of plant UV-B responses, Casati and Walbot (2004) showed only limited overlap in induced gene expression between plants exposed to 8-h low UV-B or high UV-B. Similarly, Claeys *et al.* (2014) used transcript analysis to show that expression of a range of stress markers occurs under severe stress conditions (induced by NaCl, H_2O_2, sorbitol and mannitol), but was absent under the mild stress conditions which are associated with eustress. These authors argue the importance of measuring morphological parameters to monitor mild stress responses. This is what was done by Brodführer (1955), who showed that low doses of UV-B triggered a rather different morphological response from high levels of UV-B. These studies demonstrate that low and high doses of a potential stressor can trigger different plant responses. Thus, (common) experiments whereby plants are exposed to a single dose of a stressor do not necessarily highlight the full extent of plant stress responses, and dose-response curves are an important aspect of stress physiology (Lichtenthaler, 1988).

The second fundamental consideration in the 'general adaptation syndrome' concerns the dynamics of plant responses to a particular stressor. Selye (1936) and Lichtenthaler (1998) recognized an alarm phase, which occurs when an organism is first exposed to a particular stressor. If the plant survives the initial stress (i.e. eustress-dominant), a subsequent resistance phase is dominated by adjustment of metabolism to cope with the stressor. Long-term exposure to

a stressor might trigger an exhaustion phase, leading to plant death. Thus, single time point stress experiments do not necessarily highlight the full extent of the readjustment of plant metabolism, and response kinetics are vital to fully understand the complexity of plant stress biology. This point is highlighted in a detailed gene-expression study by Kilian *et al.* (2007), who showed that exposure to stressors such as cold, drought and UV-B initially induced expression of a relatively high number of generic genes. At later time points, more stressor specific genes were found to be expressed. Thus, the character of the stressor-induced response changes with time. In this context, it should also be mentioned that different stress responses will also depend on whether plants have been primed, are newly exposed to a stressor, or are already acclimated.

Other Stress Concepts

Levitt (1980) developed a concept of plant stress based on similarities with material properties of solid objects. This concept refers to forces (stressors) that can deform (strain) a body, and this strain can be reversible (elastic), irreversible (plastic) or lead to a breakage. Levitt (1980) described a mild stress as an elastic strain on the plant. Elastic deformations represent reversible readjustments that optimize metabolism under the new environmental conditions. Just as in material science, the imbalance is not large enough to introduce lasting effects. However, if the impact of the stressor becomes more substantial, irreversible change becomes inevitable, which in its most severe form represents cell death.

A rather distinct stress concept was developed by Strasser (Tsimilli-Michael *et al.*, 1996), whose thermodynamic state-change concept focuses on how stressors cause suboptimality; that is, a lack of adjustment between physiological state and the environment, or a disruption of physiological homeostasis. Plant stress is thus a temporary, non-optimal state before a plant either reaches a new thermodynamically equilibrium, or dies (Tsimilli-Michael *et al.*, 1996). Given that changes in environmental parameters are ubiquitous, Strasser refers to 'perpetual state changes in plants approaching harmony with their environment'.

Chasing Equilibrium

The concepts of distress and eustress (Selye, 1964; Larcher, 1980; Lichtenthaler, 1988), elastic and plastic stress (Levitt, 1980) or thermodynamic state-change (Tsimilli-Michael *et al.*, 1996) are all about driving readjustments of metabolism, which optimize physiological performance under a particular set of environmental conditions. Mild stress-inducing conditions (Fig. 1) cause an initial destabilization of functions, followed by a physiological adjustment (Gaspar *et al.*, 2002). This process may involve changes in gene expression, physiology and accumulation of metabolites, but also a more comprehensive reorientation of growth. The latter type of adjustment can literally be visualized by studying plant morphological responses. Low doses of a stressor can induce adaptive, phenotypic alterations, known as stress-induced morphogenic responses (SIMR) (Potters *et al.*, 2007). The resulting phenotype is characterized by active redistribution of growth, including: (i) inhibition of elongation; (ii) alteration in cell differentiation status; and (iii) localized stimulation of cell division leading to lateral growth and a more bushy phenotype (Potters *et al.*, 2007). The mechanism underlying this redistribution of growth includes well-directed changes in plant cell redox metabolism and hormone activities. From an ecological viewpoint, this can be interpreted as an avoidance reaction which may contribute to tolerance (Potters *et al.*, 2007).

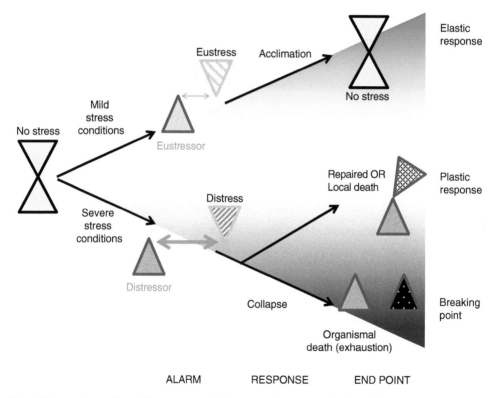

Fig. 1. Schematic overview of the concepts of distress and eustress, elastic and plastic stress and thermodynamic state-change. Exposure to mild stress-inducing conditions leads to an imbalance between environmental conditions (eustressor) and physiology (eustress) in the initial, alarm phase. In turn, this drives acclimation, and ultimately optimizes physiological performance under the new environmental conditions (i.e. an elastic response). Exposure to high stress-inducing conditions leads to a major imbalance between environmental conditions (distressor) and physiology (distress). The plant's metabolism cannot cope and collapses (breaking point), leading to death. If, however, damage repair mechanisms can be activated, a plastic response can be envisaged.

Thus eustress, elastic stress and state changes are all similarly linked to the concept of acclimation. Stress acclimation is defined as induction of reversible, non-heritable, physiological or biochemical responses that lead to increased tolerance; that is, an increase in the range of environmental factors that a species can survive. Stress acclimation is not to be confused with stress adaptation, which is a much slower, evolutionary process. Some acclimation responses are generic, such as increased total antioxidant neutralizing capacity (Mittler, 2002), SIMR (Potters et al., 2007) or stress-induced flowering (Wada and Takeno, 2010), which can all be induced by widely different stressors. Other acclimation responses are highly specific, and only protect against one specific stressor.

Exposure to high levels of a stressor (Fig. 1) leads to distress (destabilization of cellular functions). In the terminology of Strasser, the thermodynamic equilibrium is disrupted to such an extent that the plant's metabolism cannot reach a new optimal state, but rather collapses, leading to the death of the plant. A breaking point has been reached in the perspective by Levitt (1980). In metabolic terms, these conditions are associated with generic responses such as high levels of reactive oxygen species (ROS) production, cellular disruption, membrane damage and possibly cell death. Repair processes can either restore some cellular functions, or at least avoid more than local death (Fig. 1), and this captures the concept of plastic responses proposed by Levitt (1980). A biological example of a plastic response would be necrotic spots, which may arise due to a hypersensitive response following a biotic attack.

Conclusion

There is too much variation in the way in which plant stress researchers use and understand terminology such as 'stress', 'stressor', 'acclimation' and 'adaptation'. This causes ambiguity, and impedes scientific progress. Moreover, there is a lack of recognition that plant stress responses comprise a mixture of eustress and distress, and that this mixture depends on the dose of the stressor, as well as on exposure kinetics. Thus, without appropriate calibration of stress conditions, contradictory data can be produced that are of limited use for the understanding of plant stress responses. Selye (1964), Levitt (1980), Lichtenthaler (1988) and Tsimilli-Michael et al. (1996) have provided theoretical frameworks defining stress, and these frameworks can be used to place molecular, biochemical or physiological data in the appropriate context. The theoretical stress frameworks have demonstrated that, in the plant world, stress is more than just a clinical condition. Rather, stress conditions are important drivers that help a plant to perceive the outside environment, to harmonize itself with it and thus to optimize growth and development.

Acknowledgement

MAKJ acknowledges the support of WoB.

Note

[1] Strictly speaking, and in retrospect, the proper term should have been 'general acclimation syndrome', as (physiological) acclimation is not to be confused with stress adaptation, which is a much slower, evolutionary process that leads to increased tolerance.

References

Brodführer, U. (1955) Der Einfluss einer abgestuften Dosierung von ultravioletter Sonnenstrahlung auf das Wachstum der Pflanzen. Planta 45, 1–56.

Buchanan, B.B., Gruissem, W. and Jones, R.L. (2000) Biochemistry & Molecular Biology of Plants, Vol. 40. American Society of Plant Physiologists, Rockville, Maryland.

Casati, P. and Walbot, V. (2004) Rapid transcriptome responses of maize (Zea mays) to UV-B in irradiated and shielded tissues. Genome Biology 5, R16–R16.

Claeys, H., Van Landeghem, S., Dubois, M., Maleux, K. and Inzé, D. (2014) What is stress? Dose-response effects in commonly used in vitro stress assays. *Plant Physiology* 165, 519–527.

Gaspar, T., Franck, T., Bisbis, B., Kevers, C., Jouve, L., Hausman, J.F. and Dommes, J. (2002) Concepts in plant stress physiology. Application to plant tissue cultures. *Plant Growth Regulation* 37, 263–285.

Hopkins, W.G. (1999) *Introduction to Plant Physiology*, 2nd edn. Wiley, New York.

Kilian, J., Whitehead, D., Horak, J., Wanke, D., Weinl, S., Batistic, O., D'Angelo, C., Bornberg-Bauer, E., Kudla, J. and Harter, K. (2007) The AtGenExpress global stress expression data set: protocols, evaluation and model data analysis of UV-B light, drought and cold stress responses. *The Plant Journal* 50, 347–363.

Larcher, W. (1980) *Physiological Plant Ecology*, 2nd edn. Springer, Berlin/Heidelberg/New York.

Leclerc, J.-C. (2003) *Plant Ecophysiology*. Science Publishers, Enfield, CT/Plymouth, UK.

Levitt, J. (1980) Responses of plants to environmental stresses: chilling, freezing and high temperature stresses. In: *Physiological Ecology*, Vol. 1. Academic Press, New York, pp. 347–447.

Lichtenthaler, H.K. (1988) In vivo chlorophyll fluorescence as a tool for stress detection in plants. In: Lichtenthaler, H.K. (ed.) *Applications of Chlorophyll Fluorescence in Photosynthesis Research, Stress Physiology, Hydrobiology and Remote Sensing*. Springer, Dordrecht, The Netherlands, pp. 129–142.

Lichtenthaler, H.K. (1998) The stress concept in plants: an introduction. *Annals of the New York Academy of Sciences* 851, 187–198.

Lichtenthaler, H.K., Ač, A., Marek, M.V., Kalina, J. and Urban, O. (2007) Differences in pigment composition, photosynthetic rates and chlorophyll fluorescence images of sun and shade leaves of four tree species. *Plant Physiology and Biochemistry* 45, 577–588.

Mittler, R. (2002) Oxidative stress, antioxidants and stress tolerance. *Trends in Plant Science* 7, 405–410.

Potters, G., Pasternak, T.P., Guisez, Y., Palme, K.J. and Jansen, M.A.K. (2007) Stress-induced morphogenic responses: growing out of trouble? *Trends in Plant Science* 12, 98–105.

Selye, H. (1936) A syndrome produced by diverse nocuous agents. *Nature* 138, 32.

Selye, H. (1964) *From Dream to Discovery*. McGraw-Hill, New York.

Selye, H. (1973) The evolution of the stress concept: the originator of the concept traces its development from the discovery in 1936 of the alarm reaction to modern therapeutic applications of syntoxic and catatoxic hormones. *American Scientist* 61, 692–699.

Smith, A.M., Coupland, G., Dolan, L., Harberd, N., Jones, J., Martin, C., Sablowski, R. and Amey, A. (2010) *Plant Biology*. Garland Science, New York/Abingdon, UK.

Taiz, L. and Zeiger, E. (2010) *Plant Physiology*, 5th edn. Sinauer Associates, Sunderland, Massachusetts.

Taylor, A.O. and Rowley, J.A. (1971) Plants under climatic stress I. Low temperature, high light effects on photosynthesis. *Plant Physiology* 47, 713–718.

Tsimilli-Michael, M., Kruger, G.H.J. and Strasser, R.J. (1996) About the perpetual state changes in plants approaching harmony with their environment. *Archives des Sciences* 49, 173–203.

Wada, K.C. and Takeno, K. (2010) Stress-induced flowering. *Plant Signaling & Behavior* 5, 944–947.

Waley, A. (transl.) (1938) *The Analects of Confucius*. Random House, New York.

Wang, W., Vinocur, B. and Altman, A. (2003) Plant responses to drought, salinity and extreme temperatures: towards genetic engineering for stress tolerance. *Planta* 218, 1–14.

1 Drought Tolerance in Crops: Physiology to Genomics

Lakshmi Praba Manavalan and Henry T. Nguyen*

Division of Plant Sciences, University of Missouri, Columbia, USA

Abstract

More frequent and severe drought combined with high temperatures have been recognized as a potential impact of global warming on agriculture. Improving crop yield under water stress is the goal of agricultural researchers worldwide. Direct selection for yield under drought has been the major breeding strategy and was successful in some crops. Drought modifies the structure and function of plants. An understanding of the impact, mechanisms and traits underlying drought tolerance is essential to develop drought-tolerant cultivars. Identification and evaluation of key physiological traits would aid and strengthen molecular breeding and genetic engineering programmes in targeting and delivering traits that improve water use and/or drought tolerance of crops. There is an overlap between different osmotic stresses and the selection of appropriate drought evaluation methods. The benefits of genetic engineering have been realized in crop improvement for quality traits, and several promising genes have emerged in the last decade as candidates for drought tolerance. Combining the physiological traits that would sustain yield under drought, and incorporating elite quantitative trait loci (QTL) and genes underlying these traits into high-yielding cultivars, would be a successful strategy.

1.1 Introduction

Food production worldwide is affected by periodic droughts. Drought is an extended abnormal dry period that occurs in a region consistently receiving a below-average rainfall. Globally, agriculture is the biggest consumer of water, accounting for almost 70% of all withdrawals, and up to 95% in developing countries (FAO, 2007). Out of 1474 million hectares (ha) of cultivated land in the world, 86% comes under rain-fed cultivation (Kumar, 2005). Drought is classified into three major categories (Dai, 2010): (i) agricultural drought; (ii) meteorological drought; and (iii) hydrological drought. Meteorological drought is a period with less-than-average precipitation, and is often associated with above-normal temperatures that precede and cause other types of drought. Meteorological drought is caused by constant changes in large-scale atmospheric circulation patterns such as high pressure. Agricultural drought is a period with below-average precipitation, less frequent rain events or above-normal evaporation, resulting in reduced crop production and plant growth. Hydrological drought occurs when there is a reduced supply of water or water levels from river streams and other water storage structures such as aquifers, lakes or reservoirs fall below long-term mean levels. A lack of rainfall triggers agricultural and hydrological droughts; but other factors, including

---

* Corresponding author: nguyenhenry@missouri.edu

high temperature, poor irrigation management and external factors like overgrazing and erosion, also cause drought. The proportion of the land surface globally in extreme drought is predicted to increase from 1% to 3% at present to 30% by the 2090s. The number of extreme drought events per 100 years and mean drought duration are likely to increase by factors of two and six, respectively, by the 2090s (Burke *et al.*, 2006). According to the World Bank, drought is the world's most expensive disaster, destroying the economic livelihood and food source for those dependent on the agricultural sector. Much effort is being made by agricultural researchers around the globe to reduce water use by crops to address the challenges that especially affect farmers in drought-prone environments across the developing world.

1.2 Global Impact of Drought On Crop Production

Global climate change, in the form of increasing temperature and fluctuating soil moisture conditions including drought and floods, is projected to decrease the yield of food crops over the next 50 years (Leakey *et al.*, 2006). By 2025, around 1800 million people will be living in countries or regions with absolute water scarcity, and two-thirds of the world population could be under stress conditions (FAO Water; www.fao.org/nr/water/issues/scarcity.html). It has been reported that 2000–2009 was the warmest decade since the 1880s (http://www.nasa.gov/topics/earth/features/temp-analysis-2009.html). Over the past 10 years, large-scale periodic regional drought and a general drying trend over the southern hemisphere have reduced global terrestrial net primary production (Zhao and Running, 2010).

The United States Department of Agriculture classified risk in agriculture into seven categories. Among them, yield risk is the most common and has a direct impact on agriculture, which is mostly influenced by water supply and temperature (Motha and Menzie, 2007). Drought has an extensive impact on agriculture as it disrupts cropping programmes, reduces breeding stock, and reduces the assets and farm

inputs. The National Climatic Data Center (NCDC) of the US Department of Commerce estimated that, from 1980 to 2010, a combination of drought and heatwaves caused a total loss of around US$1825 million (http://www.ncdc.noaa.gov/img/reports/billion/state2010.pdf). It has been reported that about 50% of world rice production is affected by drought (Bouman *et al.*, 2005). Most of the 160 million ha of maize grown globally is rain-fed. The average annual yield losses to drought are around 15% of potential yield on a global basis and, as temperatures rise and rainfall patterns change, additional losses of maize grain may approach 10 million t/year, currently worth almost US$5 billion (Edmeades, 2008). Total wheat production in the wider drought-affected regions of the Middle East and Central Asia is currently estimated to have declined by at least 22% in 2009 compared to 2008 (de Carbonnel, 2009). Drought stress after flowering is one of the most common and serious environmental limitations to yield in pearl millet, resulting in 50% yield loss (Mahalakshmi *et al.*, 1987). Drought is the most damaging abiotic stress to soybean production and, in the USA, dry-land soybeans yield approximately 60–70% less than irrigated systems (Egli, 2008).

Other than the socio-economic impacts, nearly every plant physiological process is directly or indirectly affected by water deficit. Cell enlargement depends on the level of cell turgor; photosynthesis is directly inhibited by insufficient water; and stomatal control of transpiration and CO_2 absorption rely on the water status of guard cells (Giménez *et al.*, 2005). The adverse effect of drought on plant structure and function such as xylem embolism, reduced carbohydrate pool size, leaf and fine root production, on the ability of plants to resist pathogen attacks, the impacts on soil microbial dynamics, decomposition and nutrient-supply processes, and shifting competitive abilities between plant species cannot be underestimated (Ciais *et al.*, 2005). While the negative impact of drought on crop yields is obvious, the adverse effects on crop quality are less recognized. Severe drought can result in a loss in food quality in terms of feed value, starch and lipid concentration, or physical/sensory traits (Wang and Frei, 2011). Crop improvement through conventional breeding and modern biotechnology both offer potential for substantial progress (Cominelli and Tonelli, 2010).

1.3 Drought Resistance Mechanisms

Understanding the concept and components of drought resistance is a key factor for improving drought tolerance of crops. Drought resistance mechanisms for different crops have been extensively reviewed and summarized from crop physiology, plant breeding and molecular perspectives (Nguyen *et al.*, 1997; Turner *et al.*, 2001; Manavalan *et al.*, 2009; Shao *et al.*, 2009; Mittler and Blumwald, 2010; Todaka *et al.*, 2015). Drought resistance can be classified broadly into three categories (Taiz and Zeiger, 2002): (i) desiccation postponement (the ability to maintain tissue hydration or drought tolerance at high water potential); (ii) desiccation tolerance (the ability to function while dehydrated or drought tolerance at low water potential); and (iii) drought escape where the plants avoid drought by completing life cycles before the onset of a dry period to sustain some reproduction. These drought resistance mechanisms vary with the geographical area, based on soil and climatic conditions. For example, tolerance to extreme drought conditions (air < 0% relative humidity) exhibited by desert-adapted resurrection plants such as *Craterostigma plantagineum* (Bartels *et al.*, 1990) and *Tortula ruralis* (Oliver and Bewley, 1997), is achieved by limiting their metabolic functions. In contrast, most cultivated plants cannot withstand a water deficit less than 85% of relative humidity during the vegetative period (Bartels and Salamini, 2001), and these plants adapt to drought by either dehydration avoidance or dehydration tolerance mechanisms to maintain biological functions. Dehydration avoidance (a plant's capacity to sustain high water status by water uptake or a reduction of water loss in dry conditions) is achieved through the development of a large and deep root system to acquire water from the soil, as well as through the closure of stomata or a non-permeable leaf cuticle to reduce transpiration. Physiological traits such as leaf osmotic adjustment, proportion/quantity of ABA, chlorophyll, proline and soluble sugars; and toxic removal mechanisms such as peroxidase or superoxide dismutase activity contribute to dehydration tolerance (Luo, 2010).

1.4 Physiological Traits Affecting Crop Response to Drought

Effects of water deficit at the whole-plant level are manifested by effects on plant phenology, growth and development, source–sink relations and plant reproduction processes. An understanding of the various physiological traits controlling/regulating crop responses to drought is required for identifying natural genetic variation for drought tolerance. These traits can be broadly classified as shoot- and root-related traits.

1.4.1 Phenology

Plant developmental traits such as early vigour or phenology may be particularly significant in water-limiting conditions (Cairns *et al.*, 2009). Faster phenological development is particularly useful in drought situations where late season drought is prominent. The early planting system of soybeans in the USA is an example where short season cultivars are planted during March–April. The early maturing cultivars start flowering in late April and set pods in late May, thus completing the reproductive stage before the period of possible drought during July–August (Heatherly and Elmore, 2004). Seed size and early seedling vigour were found to be associated with drought tolerance in pearl millet (Manga and Yadav, 1995), wheat (Rebetzke and Richards, 1999), sorghum (Harris, 1996), cotton (Basal *et al.*, 1995), and rice (Cui *et al.*, 2008). While plants could potentially escape and avoid drought, theory and previous findings suggest that there are likely to be trade-offs between these strategies. For instance, early flowering is an adaptive strategy under drought conditions in *Brassica*, where a trade-off between drought avoidance and escape indicates that selection for drought escape through earlier flowering is more important than phenotypic plasticity (Franks, 2011).

1.4.2 Root system architecture

The role of the plant root system is to uptake water and nutrients from the soil through its highly responsive and plastic morphology, which allows the plant to adjust and exploit the varying soil physical and chemical properties (Armengaud et al., 2009). An increased depth and density of roots is considered a major mechanism for improving water uptake under drought conditions (Turner, 1986). Extensive information is available on the value of root traits in relation to drought avoidance in crops (Courtois et al., 2009; Hochholdinger and Tuberosa, 2009; Hodge, 2009; Maurel et al., 2010; Yamaguchi and Sharp, 2010). In addition, the alteration of root hydraulic conductance by different anatomical and biochemical traits provides the plants the ability to regulate plant water use for the critical crop stages (Vadez, 2014).

Screening for root architectural traits is one of the major bottlenecks in root research due to the difficulties associated with separation of a whole root system from the soil and the huge amount of time and labour requirements for field evaluation. In addition, most of the destructive analysis is done at the end of the experiment and hence monitoring root system development over a crop's life cycle is not feasible. Several encouraging assays using root observation chambers (Singh et al., 2010), soil-less media (Bengough et al., 2004; Manschadi et al., 2008; Manavalan et al., 2010), image-based phenotyping platforms (Hund et al., 2009; Iyer-Pascuzzi et al., 2010) and tools to analyse the images such as Root-Flow (van der Weele et al., 2003), EZ-RHIZO (Armengaud et al., 2009) and RootTrace (French et al., 2009) are showing exciting opportunities to understand the root traits and apply them in crop improvement.

1.4.3   Leaf water potential

Leaf water potential (LWP) is recognized as an index for whole-plant water status (Turner, 1982). When irrigated normally, plants transpire and create a negative LWP, which results in the uptake of water. Under water deficit conditions, LWP becomes more and more negative with no water to fill the xylem, resulting in cavitations and leading to the loss of turgor and wilting of plants. Significant differences between cultivars for LWP and its relation with plant performance and yield have been documented in rice (Jongdee et al., 2002), wheat (Winter et al., 1988), maize (Cary and Fisher, 1971), sorghum (Jones and Turner, 1978), soybean (Djekoun and Planchon, 1991), sunflower (Boyer, 1968), cotton (Grimes and Yamada, 1982) and other crops. LWP can be measured in the field/ greenhouse by using a pressure chamber apparatus. Another method is thermocouple psychrometry which is advantageous because of the flexibility to use small sample size and easy measurement of the components of water potential such as turgor pressure and osmotic potential (Boyer, 2010).

1.4.4   Leaf relative water content

Relative water content (RWC) is closely related to cell size and may strongly reflect the balance between water supply to the leaf and transpiration (Fischer and Wood, 1979). Estimation of leaf RWC is quite simple and certainly applicable to a large number of plants. It has been suggested that plant water status, rather than plant function, controls crop performance under drought. Therefore, those genotypes that can maintain higher LWP and RWC are drought resistant simply because of their superior internal water status (Kamoshita et al., 2008). Genotypic differences that exist for RWC under drought are well documented in rice (Courtois et al., 2000), wheat (Schonfeld et al., 1988), soybean (Carter and Patterson, 1985; James et al., 2008) and other crops. A positive relationship was observed between grain yield and RWC measured during the reproductive stage in wheat, where the high-yield selections maintained a significantly higher RWC than the low-yield selections (Tahara et al., 1990). However, studies suggest that differences in RWC among cultivars are highly influenced by plant maturity, adaptation and severity of stress, and hence it may be used as a secondary selection trait (Lafitte, 2002). Boyer et al. (2008) found that osmotic adjustment causes discrepancies in RWC values (10–15% lesser values than the actual plant water status) and suggested possible alterations. If maturity differences can be feasibly monitored, with a suitable rehydration procedure, RWC should serve as a practical and reliable indicator of drought resistance in mass selection programmes.

1.4.5   Stomatal conductance

Stomata, the specialized cells performing gas exchange in plants, account for water loss through transpiration. Transpiration rate is influenced by the diffusion resistance, provided by the stomatal pores and by the humidity gradient between the leaf's internal air spaces and ambient air. Stomata close when LWP decreases (Brodribb and Holbrook, 2003). The plant hormone ABA is involved in stomatal closure and appears to trigger stomatal closure even before significant decline in water potential occurs (Zhang and Davies, 1989; Liu et al., 2003). When leaf RWC falls to around 70%, photosynthesis in most species becomes irreversibly depressed (Lawlor and Cornic, 2002), and thus the resistance of the photosynthetic apparatus to desiccation is also a potential trigger for stomatal closure. Plants exhibit adaptation to drought by stomatal closure (Vignes et al., 1986; Muchow and Sinclair, 1989; Blum, 1996; Price et al., 2002; Reynolds et al., 2005). It is argued that stomatal conductance is not a desirable trait as it affects productivity under non-stress conditions. For regions that depend entirely on rain-fed agriculture, and for dry-land conditions where a conservative response to drying soils is needed, this trait should not be neglected. Successful application of stomatal regulation using partial root-zone drying (Loveys et al., 2000) and regulated deficit irrigation (RDI) in maize (Kang et al., 2000) and horticultural crops (Costa et al., 2007) showed that reducing stomatal conductance by partial irrigation resulted in improved water use efficiency and productivity equivalent to non-stress conditions. Through genetic engineering and mutant approaches, several genes were manipulated to regulate stomatal closure downstream of ABA production (Schroeder et al., 2001). Although non-invasive instruments are available for rapid measurement of stomatal conductance, they have drawbacks such as the control of humidity. Some of the improvements for detecting water stress-induced stomatal closure as a guide to irrigation scheduling are the use of infrared thermometry (Idso et al., 1981) or thermography (Jones, 2004).

1.4.6   Anatomical modifications to reduce water loss (sunken stomata/ glaucousness/epicuticular wax/leaf pubescence)

Plants such as *Nerium oleander*, *Ficus* spp., and modified leaves of certain plants (pine needles), avoid drought by sunken stomata, which is an anatomical adaptation. In these species, stomata are sunken below the epidermal plane (Fig. 1.1A). The guard cells are located in a depression, creating a more humid microclimate in the boundary layer. Air in the depression is slightly protected from wind, and any molecule of water that escapes from the stoma may remain in the depression long enough to actually bounce back into the leaf rather than evaporate (Fig. 1.1A, insert). Glaucousness is the waxy covering of the plant cuticle that renders a dull-white or bluish-green cast referred to as bloom in crops such as sorghum and wheat (Fig. 1.1B). Genotypes with low cuticular transpiration rates can conserve RWC in water-deficient conditions. Glossy leaf trait was found to be associated with seedling stage drought tolerance in sorghum (Maiti et al., 1984). A positive association between water use efficiency and glaucousness was reported in wheat (Richards et al., 1986), peas (Sánchez et al., 2001) and maize (Ristic and Jenks, 2002). Leaf pubescence density is considered as an adaptive trait for drought tolerance in soybean (Fig. 1.1C). Pubescent hairs reflect excess radiation and reduce epidermal conductance. A significant negative correlation between epidermal conductance and water use efficiency in soybean (Hufstetler et al., 2007) supports the importance of these traits.

1.4.7   Cell membrane stability

One of the cellular components that is intensively affected by water stress is the cell membrane. During a water deficit, membrane permeability increases, leading to disruption of the cell membrane and to the efflux of electrolytes. The measurement of ion leakage and further estimation of membrane stability have been used as criteria for selection for drought resistance in wheat (Blum and Ebercon, 1981) and rice (Tripathy et al., 2000). A positive association between cell membrane stability (CMS)

Fig. 1.1. Anatomical, morphological and genetic modifications in plants exhibiting drought tolerance. (A) Sunken stomata in *Nerium oleander*; insert: anatomy of sunken stomata. (B) Glaucousness in sorghum (Center for Plant Environmental Stress Physiology, Purdue University). (C) Pubescence in soybean. (D) Rice NIL (centre) with QTL for improved basal root thickness (Chamarark and Nguyen). Parents (IR62266 and CT9993) on both sides. (E) Transgenic rice with *sNAC1* gene (right) exhibiting greater drought tolerance than control (left) in field. Courtesy of Dr Lizhong Xiong, China. (F) Transgenic soybean with *AtDREB1* gene (right) exhibiting vegetative stage drought tolerance and control (left) in the greenhouse (Guttikonda *et al.*, 2014). NIL, near isogenic lines; QTL, quantitative trait loci.

and high phospholipid content was observed in drought-tolerant maize cultivars (Premachandra *et al.*, 1991). However, the ion leakage method is time consuming and needs many replications to capture genotypic differences.

1.4.8 Oxidative damage and reactive oxygen species scavenging indicators

Reactive oxygen species (ROS) are produced as by-products of various metabolic pathways localized in different cellular compartments. In plants, ROS are produced continuously, predominantly in chloroplasts, mitochondria and peroxisomes. The equilibrium between production and scavenging of ROS may be perturbed by a number of adverse abiotic stress factors, including drought (Apel and Hirt, 2004). Abnormal increase in ROS leads to irreversible damage to the cellular membrane and photosynthesis, and ultimately to cell death. ROS molecules such as hydrogen peroxide, superoxide and singlet oxygen are detoxified by

non-enzymatic antioxidants such as ascorbate and glutathione (GSH), as well as tocopherol, flavonoids, alkaloids and carotenoids. Enzymatic ROS-scavenging mechanisms in plants include superoxide dismutase (SOD), ascorbate peroxidase (APX), glutathione peroxidase (GPX) and catalase (CAT) (Vernoux *et al.*, 2002). Several methods are available for quantification of ROS in plants (Verslues *et al.*, 2006). The role of ROS in stomatal regulation in *Arabidopsis* (Pei *et al.*, 2000), auxin signalling and gravitropism in maize roots (Joo *et al.*, 2005), and maintenance of root elongation in maize under water deficit conditions (Zhu *et al.*, 2007) indicate that ROS may perform as positive signalling molecules to regulate the response of plant growth to water stress.

1.4.9 Osmotic adjustment

Osmotic adjustment (OA) is defined as the active accumulation of organic solutes intracellularly in response to an increasing water deficit. OA is considered a useful trait because it provides a means for maintaining cellular turgor when tissue water potential declines. OA has been shown to maintain stomatal conductance and photosynthesis at lower water potentials, delayed leaf senescence and death, reduced flower abortion, improved root growth and increased water extraction from the soil as water deficit develops (Turner *et al.*, 2001). Consistent differences in OA exist among cultivars which can be associated with plant production under drought stress in *Brassica* (Kumar *et al.*, 1984), sorghum (Ludlow *et al.*, 1990), wheat (Blum *et al.*, 1999) and maize (Chimenti *et al.*, 2006). Beneficial effects of OA on root growth under water deficit conditions clearly show the value of this trait under water-limiting conditions.

1.4.10   Canopy temperature

Canopy temperature is considered as a sister/surrogate trait in relation to stomatal conductance, as they are directly related. Plants with high stomatal conductance transpire more and thus maintain a cooler canopy temperature. Canopy temperature and its depression relative to ambient air temperature indicate how much transpiration cools the leaves under the hot and humid climate that is typically associated with drought stress. Canopy temperature, measured with an infrared thermometer, has been used as a secondary trait to evaluate cultivars for drought tolerance in rice (Garrity and O'Toole, 1995), wheat (Reynolds *et al.*, 2007), sorghum (Blum, 1988) and maize (Araus *et al.*, 2012). New remote-sensing tools based on the use of thermal imaging to estimate plant water status at field level are gaining in importance. Use of thermography has been proposed for high-throughput phenotyping of tropical maize adaptation in water stress (Zia *et al.*, 2012).

1.4.11   Chlorophyll fluorescence and reflection indices

Drought affects the photosynthetic activity of leaves as a consequence of altered chlorophyll *a* fluorescence kinetics. The analysis of changes in chlorophyll fluorescence kinetics provides detailed information on the structure and function of the photosynthetic apparatus, especially photosystem II (Strasser *et al.*, 1995). Measurement of chlorophyll fluorescence was used as a non-destructive measure of drought avoidance in wheat (Araus *et al.*, 1998), barley (Oukarroum *et al.*, 2007), rice (Pieters and El Souki, 2005) and maize (Earl and Davis, 2003). Use of spectral reflection indices and imaging for crop monitoring would allow us to detect stress at an early stage (Virlet *et al.*, 2015).

1.4.12   Effective use of water

Water use efficiency (WUE) has been used widely as a breeding target in water-saving agriculture (Condon *et al.*, 2004). High WUE, which is the ratio between the amount of dry matter produced per unit of water applied, could contribute to crop productivity under drought. The positive association between WUE and total biomass yield in a drought environment suggests that improvement of the WUE of a crop plant should result in superior yield performance if a high harvest index can be maintained (Wright *et al.*, 1996). Carbon-isotope analysis has been used as a tool in selection for improved WUE in breeding programmes for C_3 species (Farquhar and Richards, 1984). A surrogate method to determine WUE is the measurement of ash content and potassium concentration (Tsialtas *et al.*, 2002). Another relevant trait related to WUE is transpiration efficiency (TE), which is the assimilation or dry matter accumulation per unit of transpiration (Fischer, 1981). TE is under genetic control and excludes the amount of water lost by soil evaporation, and hence should be considered as a potential trait. However, under most dry-land situations where crops depend on unpredictable seasonal rainfall, the maximization of soil moisture use is a crucial component of drought resistance (avoidance), which is generally expressed in lower WUE (Blum, 2005). Blum (2009) suggested the term effective use of water (EUW), which implies maximal soil moisture extraction for transpiration (constitutive root traits) which also involves reduced non-stomatal transpiration and minimal water loss by soil evaporation (shoot adaptive traits).

1.5 Overlap Between Different Osmotic Stresses

The osmotic stresses (drought, salinity and cold) are often inter-related and result in the disruption of homeostasis and ion distribution leading to the arrest of shoot growth, cell division and expansion. Recent molecular and genetic studies have revealed that the signalling pathways induced in response to these osmotic stresses involve many signal components such as histidine kinases (HK), mitogen-activated protein kinases (MAPK), Ca^{2+}-mediated salt overly sensitive signal transduction (SOS), calcium-dependent protein kinases (CDPK) and phospholipid signalling. These signalling compounds in turn trigger transcription factors such as dehydration-responsive (DREB) and ABA responsive elements (ABRE), and zinc finger proteins (ZFP) (Wang et al., 2003). These transcription factors activate the synthesis and differential accumulation of osmoprotectants, late embryogenesis abundant proteins, antifreeze proteins, chaperones (heat shock proteins) and ROS-scavenging molecules (aldose/aldehyde reductases, thio redoxins, peroxiredoxins), which are common among these stresses (Bartels and Sunkar, 2005). Overexpression of dehydration-responsive DREB1/CBF also increased the tolerance of transgenic plants to freezing, drought and salt stresses, suggesting that the system is important for the development of stress tolerance in plants (Seki et al., 2003). Recent studies indicate that cross-talk exists among cytokinins (CK), ABA and osmotic stress-signalling pathways, and that CK signalling and CK metabolism may play crucial roles not only in plant growth and development but also in osmotic stress signalling (Tran et al., 2010).

1.6 Drought Evaluation Methods

In order to comprehend the balance between the different drought-tolerance traits and their values to plants, it is critical that drought evaluation studies include measurement of both plant growth condition (soil water status) and plant responses including tissue water status and its regulators such as leaf area and stomatal conductance (Jones, 2007). With the transgenic and mutant approaches to identify/characterize gene function under stress, the most important requirement is reliable and repeatable drought evaluation methods. Specific physiological and biochemical conditions had to be met to test these plants in growth chambers, greenhouses or in field conditions (Boyer, 2010). Technologies such as automated plant phenotyping platforms to study the plant responses to soil water deficit under controlled conditions (Granier et al., 2006), automated rotating lysimeter systems (Lazarovitch et al., 2006), non-destructive measurement of plant water status over time using portable nuclear magnetic resonance equipment (Capitani et al., 2009) and other precision equipment to quantify plant water use should be exploited. Rain-out shelter facilities provide a useful measure to evaluate germplasm in field conditions with precise control over irrigation. The emerging field of phenomics focuses on the characterization of the whole-plant phenotype. The Plant Accelerator (formerly the Australian Plant Phenomics Facility) is a world-leading plant growth and analysis facility based at the Waite Campus of the University of Adelaide. This facility utilizes digital imaging technologies, high capacity computing and robotics, which allow the dissection of traits that contribute to drought and salinity tolerance for large populations of plants (Furbank, 2009; Rajendran et al., 2009; Berger et al., 2010; Harris et al., 2010; Munns et al., 2010). Plant phenotyping has become an increasingly important tool to quantify the link between the genotype and the environment, and exciting new discoveries are paving the way to experimentally explore the entire genotype–environment matrix for individual factors and their interactions (Pieruschka and Lawson, 2015).

1.7 Molecular Breeding for Drought Tolerance

Molecular breeding approaches through identification of QTL and marker-assisted selection offer an opportunity for significant improvements in the drought tolerance of crops. To gain momentum in the progress of breeding for drought tolerance, two sets of cutting-edge tools are being actively utilized. One involves the use of molecular markers to better understand the genetic basis of drought tolerance and to select more efficiently for this trait. For traits that are difficult to phenotype at a large scale, such as root traits and OA, molecular tagging with specific markers will facilitate the efficient identification of genes controlling these traits. The other tool, known as participatory plant breeding, offers a more active role to farmers, who make important contributions to selection for better drought tolerance as they closely observe plant performance. Identification of genetic hotspots in chromosomes through genome mapping across crop species will aid in prioritizing the set of genes to be used for crop improvement under drought. The combination of the above two approaches was employed successfully for breeding drought-tolerant rice (Steele *et al.*, 2006) and sorghum (Harris *et al.*, 2007). In rice, some of the many QTLs for roots are common across different genetic backgrounds (Li *et al.*, 2005). Marker-assisted back crossing has been used to introgress these QTLs into inbred hybrid parental lines for the subsequent production of improved hybrids (Hash *et al.*, 2000). The introgression of QTLs for yield under stress is proving to be a more successful strategy in pearl millet (Serraj *et al.*, 2005). Introgression of QTL for basal root thickness (BRT) improved the BRT by 25–40% in 18 near-isogenic lines of rice (Fig. 1.1D; Chamarark and Nguyen, unpublished data). High-density genetic maps and confirmed QTLs/genes, which are screened across various environments and across genetic backgrounds, are the most important criteria for developing drought-resistant soybeans through marker-assisted selection. The discovery of major QTL(s) representing 47% of the average rice yield under stress (Bernier *et al.*, 2007), 25% of variance in grain yield in pearl millet (Bidinger *et al.*, 2007), a major constitutive QTL for root and agronomic traits in maize (Landi *et al.*, 2010), QTL explaining 66% of phenotypic variance for deep rooting in rice (Uga *et al.*, 2011) and novel root QTL identified from the soybean interspecific population (Manavalan *et al.*, 2015) shows promising opportunities to improving drought resistance through molecular breeding.

QTL mapping using bi-parental populations has limitations because of limited allelic diversity and genomic resolution. A multi-parent advanced generation inter-cross populations (MAGIC) strategy has been proposed to interrogate multiple alleles and to provide increased recombination and mapping resolution (Bandillo *et al.*, 2013). The increased recombination in MAGIC populations can lead to novel rearrangements of alleles and greater genotypic diversity. The limitation with multi-parental populations is mapping resolution limitations as it depends on meiotic events. In contrast, the genome-wide association study (GWAS) approach provides opportunities to explore the tremendous allelic diversity existing in natural germplasm (Deshmukh *et al.*, 2014). Recently GWAS was used successfully to identify loci associated with carbon-isotope discrimination in maize (Farfan *et al.*, 2015) and soybean (Dhanapal *et al.*, 2015).

1.8 Biotechnological Approaches to Improving Drought Tolerance

1.8.1 Genomics

Novel strategies for gene discovery based on germplasm screening and functional genomic research is needed for developing drought-resistant crops. Genomics research results will supply information on the biology of traits, especially for complex quantitative traits such as drought (Varshney *et al.*, 2005). Plant functional genomics has emerged as a new and rapidly evolving scientific discipline to study the functions of genes. Gene expression profiling through microarrays has been used successfully to identify genes regulating drought resistance in crops (Zinselmeier *et al.*, 2002; Hazen *et al.*, 2005; Wang *et al.*, 2011). Most of the drought-responsive genes identified from transcriptomics are classified into ABA-dependent, ABA-independent and DREB2A/ubiquitination-related mechanisms (Ahuja *et al.*, 2010). Genes associated with the production of osmolytes (raffinose family oligosaccharides, sucrose, trehalose, mannitol and sorbitol), amino acids (proline) and amines (glycine betaine and polyamines) are differentially expressed in response to drought stress (Umezawa *et al.*, 2006).

1.8.2 Proteomics

Proteomics, the systematic analysis of (differentially) expressed proteins, is a tool for the identification of proteins involved in cellular processes (Jacobs *et al.*, 2000). Proteomics provides information on the amount of the gene products, their isoforms and which post-transcriptional modifications regulate protein activation. Several drought-responsive proteins have been identified by proteomics in different plant tissues (Salekdeh *et al.*, 2002; Hajheidari *et al.*, 2007; Alvarez *et al.*, 2008; Liu and Bennett, 2011; Sengupta *et al.*, 2011).

1.8.3 Metabolomics

Plant metabolism is highly altered in response to drought, and downstream transcript-level changes lead to the alteration in quality and quantity of various metabolites (Shulaev *et al.*, 2008). Metabolic profiling can give an instantaneous snapshot of the physiology and biochemical changes in the cell. Recent developments in the field of metabolomics with techniques such as mass spectrometry (MS), liquid chromatography mass spectrometry (LCMS), gas chromatography mass spectrometry (GCMS) and nuclear magnetic resonance spectroscopy (NMR) makes such comprehensive analysis possible, resulting in the identification of natural variants and phenotypic mutants (Keurentjes, 2009). Metabolomics is becoming a key tool in comprehensively understanding the cellular response to abiotic stress, in addition to techniques currently employed in genomics-assisted selection for plant improvement (Ruan and Teixeira da Silva, 2011). Current approaches are targeted to identify metabolic biomarkers that can predict the phenotypes associated with drought tolerance. Emerging techniques such as Transcriptome-To-Metabolome™ (TTM™) biosimulations (Phelix and Feltus, 2015) are interesting: the simulated results on metabolites of 30 primary and secondary metabolic pathways in rice (*Oryza sativa*) were used as the biomarkers to predict whether the transcriptome was from a plant that had been under drought conditions. The authors identified three metabolic markers including trehalose using this approach.

In addition to gene transcripts, proteins and metabolites, small RNAs (miRNAs, siRNAs) are reported to be involved in adaptive responses to abiotic stresses (Sunkar *et al.*, 2007). The next challenge is to incorporate these genes into the genetic backgrounds of elite cultivars and hybrids and to evaluate their performance under field conditions.

1.8.4 Genetic engineering

The applications of genetic engineering of food crops have already led to examples of improved drought tolerance and increased yield under drought. Much genetic engineering work has been carried out in *Arabidopsis* with the dehydration-responsive element-binding (DREB) transcription factors of the ABA-dependent pathway IV that was first associated with improved cold tolerance (Gilmour *et al.*, 2004). Several transcription factors belonging to the MYB family were reported to play an important role in both stomatal and non-stomatal responses by regulation of stomatal numbers and sizes, and metabolic components (Saibo *et al.*, 2009). Many genes related to drought tolerance have been tested and characterized in the model dicot plant, *Arabidopsis thaliana*. Since the scope of this chapter is on drought resistance of crop plants, we will exclude the work on *Arabidopsis* and focus on the studies that are pertinent to crop plants. The transgenic/mutant approach resulted in the identification of several drought tolerance-related genes in crops. However, careful assessment of physiological mechanisms and meaningful slow progressive drought experiments is essential to confirm their function in drought tolerance. A comprehensive summary of genetic engineering for drought resistance in crops with known physiological mechanisms is presented in Table 1.1. Several studies showed promising drought tolerance in field screenings (Hu *et al.*, 2006, 2008 (Fig. 1.1E); Wang *et al.*, 2005, 2009; Nelson *et al.*, 2007; Xiao *et al.*, 2007; Oh *et al.*, 2009; Jeong *et al.*, 2010). Transgenic crops, with quantifiable drought-tolerance mechanisms were reported in controlled environmental studies in crops, including alfalfa (Zhang *et al.*, 2005), peanut (Bhatnagar-Mathur *et al.*, 2007), potato (Stiller *et al.*, 2008), rice (Manavalan *et al.*, 2012) and soybean (Guttikonda *et al.*, 2014; Fig. 1.1F). Surprisingly, only a few genetic engineering studies identified genes that showed a role in root growth. Considering the importance of roots

Table 1.1. Genetic engineering for drought resistance in crops with known physiological mechanisms.

Trait	Crop	Gene & promoter	Gene family	Function	Reference
Root architecture	Tobacco	*HDG11*- constitutive promoter *CaMV35S*	Homeodomain (HD)-START transcription factor	Overexpression of gene conferred drought tolerance associated with improved root architecture and reduced leaf stomatal density	Yu et al., 2008
	Rice	*OsNAC10*- constitutive promoter *GOS2*	NAM (no apical meristem) ATAF1-2 and CUC2 (Cup-Shaped Cotyledon) (NAC domain)	Overexpression of gene enlarged roots and increased yield by 25–42% under field drought conditions	Jeong et al., 2010
	Tobacco	*CKX1*- constitutive *WRKY6* promoter	Cytokinin oxidase/dehydrogenase	Root specific overexpression resulted in improved root biomass up to 60% and high survival rate after drought	Werner et al., 2010
	Rice	*DRO1*-	Deeper rooting 1-QTL	Under upland conditions, Dro1-NIL showed deeper rooting and maintained better grain filling under drought	Uga et al., 2013
Relative water content	Rice	*RWC3* - stress-inducible promoter *SWPA2*	Aquaporin	Compared to the wild-type (WT) plant, the transgenic rice exhibited higher root osmotic hydraulic conductivity, LWP and relative cumulative transpiration at the end of 10 h PEG treatment	Lian et al., 2004
	Tobacco	*TaPP2Ac-1*- constitutive promoter *CaMV35S*	Catalytic subunit (c) of protein phosphatase 2A	Maintenance of RWC	Xu et al., 2007
Osmolyte accumulation/ turgor maintenance	Soybean	*p5cr*-inducible heat shock promoter *IHSP*	Δ1-pyrroline-5-carboxylate reductase	Accumulation of proline was positively associated with maintenance of RWC and lesser degree of damage to phytosystem II	De Ronde et al., 2004
	Potato	*TPS1*- drought-inducible promoter *StDS2*	Trehalose phosphate synthase	Maintenance of better water status and delayed wilting under drought	Stiller et al., 2008
	Cotton	*TsVP*- constitutive promoter *CaMV35S* / *AVP1*- constitutive promoter *CaMV35S*	Vacuolar H$^+$-PPase gene from *Thellungiella halophila* / Vacuolar H$^+$-PPase gene from *Arabidopsis thaliana*	Enhanced drought tolerance and root biomass associated with higher solute content such as soluble sugars and free amino acids, and improved cotton fibre yield	Lv et al., 2009; Pasapula et al., 2011; Wei et al., 2011
	Maize	*TsVP and BetA*- constitutive promoter *ZmUbi*	Vacuolar H$^+$-PPase gene from *Thellungiella halophila* and BetA, encoding choline dehydrogenase from *Escherichia coli*	Retention of high RWC, less anthesis silking interval and higher yield under drought	

Continued

Table 1.1. Continued.

Trait	Crop	Gene & promoter	Gene family	Function	Reference
Membrane stabilization	Tobacco	*GmERF3*- constitutive promoter *CaMV35S*	AP2/ERF transcription factor	Enhanced tolerance to drought and dehydration through accumulation of proline and soluble sugars	Zhang *et al.*, 2009
	Rice	*HVA1*- constitutive promoter *rice Act1*	Group 3 LEA protein	Maintenance of favourable water status and CMS leading to better recovery after drought	Babu *et al.*, 2004
	Maize	*betA*- constitutive promoter *CaMV35S*	Glycine betaine	Increased grain yield under drought due to protection of the integrity of the cell membrane and of the activity of enzymes	Quan *et al.*, 2004
	Maize	*ZmPIS*- constitutive promoter *ZmUbi*	phosphatidylinositol synthase	Overexpression of the phosphatidylinositol synthase gene conferred drought-stress tolerance by altering membrane lipid composition and increasing ABA synthesis	Liu *et al.*, 2013
Water use efficiency	Peanut	*AtDREB1A*-stress inducible promoter *RD29A*	Transcription factor	Increased TE under water stress	Bhatnagar-Mathur *et al.*, 2007
	Rice	*HARDY*- constitutive promoter *CaMV35S*	AP2/ERF-like transcription factor	Improved WUE and drought resistance by increase in adaptive root biomass under drought	Karaba *et al.*, 2007
	Tomato	*NCED*- super promoter *rbcS3C*	9-cis-epoxy-carotenoid dioxygenase	Increased accumulation ABA and improved transpiration efficiency	Thompson *et al.*, 2007
	Tobacco	*TaCRT*- constitutive promoter *CaMV35S*	Calcium binding protein	Better water status, WUE and membrane stability	Jia *et al.*, 2008
Stomatal regulation	Rice	*SNAC1*- constitutive promoter *CaMV35S*	NAC domain	22–34% higher seed setting than WT at reproductive stage	Hu *et al.*, 2006
	Rice	*DST*- constitutive promoter *CaMV35S*	Zinc finger protein	Loss of DST function increases stomatal closure and reduces stomatal density, consequently resulting in enhanced drought and salt tolerance in rice	Huang *et al.*, 2009
	Canola	*FTA,B*-RD29A and shoot specific HPR promoter	Farnesyl transferase	Downregulation through RNAi resulted in reduced stomatal conductance, increased yield under drought in field	Wang *et al.*, 2005, 2009
	Potato	*StMYB1R-1*- constitutive promoter *CaMV35S*	MYB transcription factor	Transgenic plants exhibited reduced rates of water loss and more rapid stomatal closing than WT plants under drought stress	Shin *et al.*, 2011

Category	Crop	Gene/promoter	Gene product	Effect	Reference
	Rice	ZmSQS1- constitutive promoter ZmUbi	Squalene synthase	RNAi mediated downregulation of squalene synthase-reduced stomatal conductance and conserved more moisture that led to increased grain yield	Manavalan et al., 2012
Photochemical efficiency	Tobacco	Ipt-senescence associated SARK promoter	Isopentenyl transferase	Delayed senescence and photosynthesis protection through overproduction of cytokinins	Rivero et al., 2007
		HvCBF4- constitutive maize ubiquitin1 promoter	Barley C-repeat DREB factor 3	Maintenance of photochemical efficiency	Oh et al., 2007
	Maize	CspB-constitutive rice Act-1 promoter	Cold shock protein	High photosynthsis, increased yield under drought	Castiglioni et al., 2008
	Maize	ZmPLC1- constitutive maize ubiquitin promoter	Phosphatidylinositol-specific phospholipase C	High RWC, OA, photosynthesis, yield	Wang et al., 2008
	Rice	DSM2- constitutive promoter CaMV35S	β-carotene hydroxylase	Drought and oxidative stress resistance by increased synthesis of xanthophylls and abscisic acid	Du et al., 2010
Antioxidants	Rice	AP37- constitutive promoter OsCc1	Transcription factor encoding Oryza sativa cytochrome c gene	Increased expression of antioxidant genes and maintenance of cellular homeostasis resulting in increased grain yield under field drought	Oh et al., 2009
	Potato	codA, SOD, APX- stress-inducible SWPA2	Choline oxidase, superoxide dismutase and ascorbate peroxide	Protection from ROS	Ahmad et al., 2010
	Rice	OsSIK1- constitutive promoter CaMV35S	Receptor-like protein kinase	Reduced water loss through regulation of stomatal density and improved antioxidant activity	Ouyang et al., 2010

CMS, cell membrane stability; DREB, dehydration-responsive element-binding; DST, drought and salt tolerance–zinc finger protein; LWP, leaf water potential; OA, osmotic adjustment; PEG, polyethylene glycol; ROS, reactive oxygen species; RWC, relative water content; TE, transpiration efficiency; WT, wild type; WUE, water use efficiency.

in drought avoidance and water uptake, more emphasis should be given to develop transgenic crop plants with improved root architecture. Although it is being argued that contribution of a single gene to a complex trait such as drought tolerance is questionable, recent evidence indicates that incorporation of a single gene does impart measurable drought tolerance to crops (Castiglioni et al., 2008).

Another approach to improve drought tolerance by gene manipulation is by knocking down genes. Targeting induced local lesions in genomes (TILLING) is a powerful reverse genetics approach for functional genomics studies. Mutations induced in stress-related genes are being identified using this approach (Guo et al., 2015). Since the technology is non-transgenic, and mutations are stably inherited, the alleles identified by TILLING can easily be incorporated into traditional breeding programmes.

1.9 Conclusion and Perspectives

The United Nations General Assembly declared the period from 2005 to 2015 the International Decade for Action, 'Water for Life', which officially started on World Water Day (22 March 2005) (http://www.un.org/waterforlifedecade/background.shtml). The Resolution stated that the main goal of the decade should be a greater focus on water-related issues at all levels and on the implementation of water-related programmes to achieve water-related goals that were internationally agreed upon. Identification of large-effect QTLs controlling yield under drought, targeting genes from those specific regions and pyramiding of a few genes in those regions and introgressing these genes into elite germplasm will be an effective strategy. For example, the maize hybrids AQUAmax® from Dupont Pioneer (Johnston, Iowa, USA) and

Agrisure Artesian™ from Syngenta (Greensbro, North Carolina, USA) were developed using this strategy. Also, recent advances in candidate gene approaches and genetic engineering of crops have shown promising improvements in drought tolerance of crops such as rice, maize, canola and soybean. As another possible strategy, given the wealth of genomics information and examples in model crop species, stacking of genes involved in a particular pathway related to dehydration tolerance should be considered. Recent results from Ramu et al. (2016) are encouraging, finding that pyramiding drought-adaptive traits by simultaneous expression of genes regulating drought-adaptive mechanisms resulted in abiotic stress tolerance in groundnut.

Production of crops with several genetically engineered traits may seem a logical step to follow and, as more locally adapted transgenic varieties become available and accepted on the market, these will become the source material into which novel genes will be further incorporated. However, difficulties associated with multiple gene inserts such as trait silencing due to similarities between transgene cassettes or epistatic interactions between transgenes should be addressed (Dhlamini et al., 2005). The recent availability of genome editing tools such as clustered regularly interspaced short palindromic repeat (CRISPR) are opening new avenues to efficiently introduce targeted modifications in the genome. These will allow study of the functional aspects of various components of the genome in diverse plants and offer potential avenues for production of abiotic stress-tolerant crop plants (Jain, 2015). Such novel approaches to simplify and improve the process of precise introduction of single or multiple genes into crop varieties should be explored to enhance drought tolerance and subsequent crop improvement.

References

Ahmad, R., Kim, Y.H., Kim, M.D., Kwon, S.Y., Cho, K., Lee, H.S. and Kwak, S.S. (2010) Simultaneous expression of choline oxidase, superoxide dismutase and ascorbate peroxidase in potato plant chloroplasts provides synergistically enhanced protection against various abiotic stresses. *Physiologia Plantarum* 138, 520–533.
Ahuja, I., de Vos, R.C.H., Bones, A.M. and Hall, R.D. (2010) Plant molecular stress responses face climate change. *Trends in Plant Science* 15, 664–674.

Alvarez, S., Marsh, E.L., Schroeder, S.G. and Schachtman, D.P. (2008) Metabolomic and proteomic changes in the xylem sap of maize under drought. *Plant, Cell and Environment* 31, 325–340.

Apel, K. and Hirt, H. (2004) Reactive oxygen species: metabolism, oxidative stress, and signal transduction. *Annual Review of Plant Biology* 55, 373–399.

Araus, J.L., Amaro, T., Voltas, J., Nakkoul, H. and Nachit, M.M. (1998) Chlorophyll fluorescence as a selection criterion for grain yield in durum wheat under Mediterranean conditions. *Field Crops Research* 55, 209–223.

Araus, J.L., Serret, M.D. and Edmeades, G.O. (2012) Phenotyping maize for adaptation to drought. *Frontiers in Physiology* 3, 305. Available at: http://doi.org/10.3389/fphys.2012.00305.

Armengaud, P., Zambaux, K., Hills, A., Sulpice, R., Pattison, R.J., Blatt, M.R. and Amtmann, A. (2009) EZ-Rhizo: integrated software for the fast and accurate measurement of root system architecture. *Plant Journal* 57, 945–956.

Babu, R.C., Zhang, J., Blum, A., Ho, T.H.D., Wu, R. and Nguyen, H.T. (2004) HVA1, a LEA gene from barley confers dehydration tolerance in transgenic rice (*Oryza sativa* L.) via cell membrane protection. *Plant Science* 166, 855–862.

Bandillo, N., Raghavan, C., Muyco, P.A., Sevilla, M.A.L., Lobina, I.T., Dilla-Ermita, C.J. *et al.* (2013) Multi-parent advanced generation inter-cross (MAGIC) populations in rice: progress and potential for genetics research and breeding. *Rice* 6, 1–15.

Bartels, D. and Salamini, F. (2001) Desiccation tolerance in the resurrection plant *Craterostigma plantagineum*. A contribution to the study of drought tolerance at the molecular level. *Journal of Plant Physiology* 127, 1346–1353.

Bartels, D. and Sunkar, R. (2005) Drought and salt tolerance in plants. *Critical Reviews in Plant Sciences* 24, 23–58.

Bartels, D., Schneider, K., Terstappen, G., Piatkowski, D. and Salamini, F. (1990) Molecular cloning of abscisic acid-modulated genes which are induced during desiccation of the resurrection plant *Craterostigma plantagineum*. *Planta* 181, 27–34.

Basal, H., Smith, C.W., Thaxton, P.S. and Hemphill, J.K. (1995) Seedling drought tolerance in upland cotton. *Crop Science* 45, 766–771.

Bengough, A.G., Gordon, D.C., Al-Menaie, H., Ellis, R.P., Allan, D., Keith, R., Thomas, W.T.B. and Forster, B.P. (2004) Gel observation chamber for rapid screening of root traits in cereal seedlings. *Plant and Soil* 262, 63–70.

Berger, B., Parent, B. and Tester, M.A. (2010) High-throughput shoot imaging to study drought responses. *Journal of Experimental Botany* 61, 3519–3528.

Bernier, J., Kumar, A., Venuprasad, R., Spaner, D. and Atlin, G. (2007) A large-effect QTL for grain yield under reproductive-stage drought stress in upland rice. *Crop Science* 47, 505–517.

Bhatnagar-Mathur, P., Devi, J.M., Reddy, D.S., Lavanya, M., Vadez, V., Serraj, R., Yamagunchi-Shinozaki, K. and Sharma, K.K. (2007) Stress-inducible expression of at DREB1A in transgenic peanut (*Arachis hypogia* L.) increase transpiration efficiency under water-limiting conditions. *Plant Cell Reports* 26, 2071–2082.

Bidinger, F.R., Nepolean, T., Hash, C.T., Yadav, R.S. and Howarth, C.J. (2007) Quantitative trait loci for grain yield in pearl millet under variable post flowering moisture conditions. *Crop Science* 47, 969–980.

Blum, A. (1988) *Plant Breeding for Stress Environments*. CRC Press, Boca Raton, Florida.

Blum, A. (1996) Crop response to drought and the interpretation of adaptation. *Plant Growth Regulation* 20, 135–148.

Blum, A. (2005) Drought resistance, water-use efficiency, and yield potential—are they compatible, dissonant, or mutually exclusive? *Australian Journal of Agricultural Research* 56, 1159–1168.

Blum, A. (2009) Effective use of water (EUW) and not water use efficiency (WUE) is the target of crop yield improvement under drought stress. *Field Crops Research* 112, 119–123.

Blum, A. and Ebercon, A. (1981) Cell membrane stability as a measure of drought and heat tolerance in wheat. *Crop Science* 21, 43–47.

Blum, A., Zhang, J. and Nguyen, H.T. (1999) Consistent differences among wheat cultivars in osmotic adjustment and their relationship to plant production. *Field Crops Research* 64, 287–291.

Bouman, B.A.M., Peng, S., Castañeda, A.R. and Visperas, R.M. (2005) Yield and water use of irrigated tropical aerobic rice systems. *Agricultural Water Management* 74, 87–105.

Boyer, J.S. (1968) Relationship of water potential to growth of leaves. *Journal of Plant Physiology* 43, 1056–1062.

Boyer, J.S. (2010) Drought decision-making. *Journal of Experimental Botany* 61, 3493–3497.

Boyer, J.S., James, R.A., Munns, R., Condon, A.G. and Passioura, J.B. (2008) Osmotic adjustment leads to anomalously low estimates of relative water content in wheat and barley. *Functional Plant Biology* 35, 1172–1182.

Brodribb, T.J. and Holbrook, N.M. (2003) Stomatal closure during leaf dehydration, correlation with other leaf physiological traits. *Plant Physiology* 132, 2166–2173.

Burke, E.J., Brown, S.J. and Christidis, N. (2006) Modeling the recent evolution of global drought and projections for the twenty-first century with the Hadley centre climate model. *Journal Hydrometeorology* 7, 1113–1125.

Cairns, J.E., Namuco, O.S., Torres, R., Simborio, F.A., Courtois, B., Aquino, G.A. and Johnson, D.E. (2009) Investigating early vigour in upland rice (*Oryza sativa* L.): part II. Identification of QTLs controlling early vigour under greenhouse and field conditions. *Field Crops Research* 113, 207–217.

Capitani, D., Brilli, F., Mannania, L., Proietti, N. and Loreto, F. (2009) In situ investigation of leaf weater status by portable unilateral NMR. *Plant Physiology* 149, 1638–1647.

Carter Jr, T.E. and Patterson, R.P. (1985) Use of relative water content as a selection tool for drought tolerance in soybean. In: *Agronomy Abstracts*. ASA, Madison, Wisconsin, p. 77.

Cary, J.W. and Fisher, H.D. (1971) Plant water potential gradients measured in the field by freezing point. *Journal of Plant Physiology* 24, 397–402.

Castiglioni, P., Warner, D., Bensen, R.J., Anstrom, D.C., Harrison, J., Stoecker, M., Abad, M., Kumar, G., Salvador, S., Ordine, R.D., Navarro, S., Back, S., Fernandes, M., Targolli, J., Dasgupta, S., Bonin, C., Luethy, M. and Heard, J.E. (2008) Bacterial RNA chaperones confer abiotic stress tolerance in plants and improved grain yield in maize under water-limited conditions. *Plant Physiology* 147, 446–455.

Chimenti, C.A., Marcantonio, M. and Hall, A.J. (2006) Divergent selection for osmotic adjustment results in improved drought tolerance in maize (*Zea mays* L.) in both early growth and flowering phases. *Field Crops Research* 95, 305–315.

Ciais, P., Reichstein, M., Viovy1, N., Granier, A., Oge, J., Allard, V., Aubinet, M., Buchmann, M. *et al.* (2005) Europe-wide reduction in primary productivity caused by the heat and drought in 2003. *Nature* 237, 529–533.

Cominelli, E. and Tonelli, C. (2010) Transgenic crops coping with water scarcity. *New Biotechnology* 27, 473–477.

Condon, A.G., Richards, R.A., Rebetzke, G.J. and Farquhar, G.D. (2004) Breeding for high water-use efficiency. *Journal of Experimental Botany* 55, 2447–2460.

Costa, J.M., Ortuño1, M.F. and Chaves, M.M. (2007) Deficit irrigation as a strategy to save water: physiology and potential application to horticulture. *Journal of Integrative Plant Biology* 49, 1421–1434.

Courtois, B., McLaren, G., Sinha, P.K., Prasad, K., Yadav, R. and Shen, L. (2000) Mapping QTLs associated with drought avoidance in upland rice. *Molecular Breeding* 6, 55–66.

Courtois, B., Ahmadi, N., Perin, C., Luquet, D. and Guiderdoni, E. (2009) The rice root system: from QTLs to genes to alleles. In: Serraj, R. *et al.* (eds) *Drought Frontiers in Rice: Crop Improvement for Increased Rainfed Production*. World Scientific Publishing, Singapore, pp. 171–188.

Cui, J.H., Xing, Y., Yu, S., Xu, C. and Peng, S. (2008) Mapping QTLs for seedling characteristics under different water supply conditions in rice (*Oryza sativa*). *Journal of Plant Physiology* 132, 53–68.

Dai, A. (2010) Drought under global warming: a review. *Wiley Interdisciplinary Reviews: Climate Change* 2(1), 45–65. doi: 10.1002/wcc.81.

de Carbonnel, E. (2009) Catastrophic fall in 2009 global food production. Available at: http://www.globalresearch.ca/index.php?context=va&aid=12252, accessed 28 December 2010.

De Ronde, J.A., Cress, W.A., Kruger, G.H.J., Strasser, R.J. and Van Staden, J. (2004) Photosynthetic response of transgenic soybean plants, containing an *Arabidopsis P5CR* gene, during heat and drought stress. *Journal of Plant Physiology* 161, 1211–1224.

Deshmukh, R., Sonah, H., Patil, G., Chen, W., Prince, S. *et al.* (2014) Integrating omic approaches for abiotic stress tolerance in soybean. *Frontiers Plant Science* 5, 244. doi: 10.3389/fpls.2014.00244.

Dhanapal, A.P., Ray, J.D., Singh, S.K., Hoyos-Villegas, V., Smith, J.R. *et al.* (2015) Genome-wide association study (GWAS) of carbon isotope ratio ($\delta^{13}C$) in diverse soybean [*Glycine max* (L.) Merr.] genotypes. *Theoretical and Applied Genetics* 128(1), 73–91.

Dhlamini, Z., Spillane, C., Moss, J.P., Ruane, J., Urquia, N. and Sonnino, A. (2005) Analysis of the FAO-BioDeC data on genetically modified (GM) crop varieties. *Status of Research and Application of Crop Biotechnologies in Developing Countries*, pp. 19–42. FAO, Rome.

Djekoun, A. and Planchon, C. (1991) Tolerance to low leaf water potential in soybean genotypes. *Euphytica* 55, 247–253.

Du, H., Wang, N., Cui, F., Li, X., Xiao, J. and Xiong, L. (2010) Characterization of the β-carotene hydroxylase gene DSM2 conferring drought and oxidative stress resistance by increasing xanthophylls and abscisic acid synthesis in rice. *Plant Physiology* 154, 1304–1318.

Earl, H.J. and Davis, R.F. (2003) Effect of drought stress on leaf and whole canopy radiation use efficiency and yield of maize. *Agronomy Journal* 95, 688–696.

Edmeades, G.O., (2008) Drought tolerance in maize: an emerging reality. Companion document to Executive Summary, ISAAA Briefs 39. In: James, C. (ed.) *Global Status of Commercialized. Biotech/GM Crops: 2008*. ISAAA, Ithaca, New York.

Egli, D.B. (2008) Soybean yield trends from 1972 to 2003 in mid-western USA. *Field Crops Research* 106, 53–59.

FAO (2007) *Water at a glance*. Available at: http://www.fao.org/nr/water/docs/wateraglance.pdf, accessed 21 July 2016.

Farfan, I.D.B., De La Fuente, G.N., Murray, S.C., Isakeit, T., Huang, P.C. *et al.* (2015) Genome wide association study for drought, aflatoxin resistance, and important agronomic traits of maize hybrids in the sub-tropics. *PLoS ONE* 10(2), e0117737. doi: 10.1371/journal.pone.0117737.

Farquhar, G.D. and Richards, R.A. (1984) Isotopic composition of plant carbon correlates with water-use efficiency of wheat genotypes. *Australian Journal of Plant Physiology* 11, 539–552.

Fischer, R.A. (1981) Optimizing the use of water and nitrogen through breeding of crops. *Plant and Soil* 58, 249–278.

Fischer, R.A. and Wood, J.T. (1979) Drought resistance in spring wheat cultivars. III. Yield association with morpho-physiological traits. *Australian Journal of Agricultural Research* 30, 1001–1020.

Franks, S.J. (2011) Plasticity and evolution in drought avoidance and escape in the annual plant *Brassica rapa*. *New Phytologist* 190, 249–257.

French, A., Ubeda-Tomas, S., Holman, T.J., Bennett, M.J. and Pridmore, T. (2009) High-throughput quantification of root growth using a novel image analysis tool. *Plant Physiology* 150, 1784–1795.

Furbank, R.T. (2009) Plant phenomics: from gene to form and function. *Functional Plant Biology* 36, 5–6.

Garrity, D.P. and O'Toole, J.C. (1995) Selection for reproductive stage drought avoidance in rice using infra-red thermometry. *Agronomy Journal* 87, 773–779.

Gilmour, S.J., Fowler, S.G. and Thomashow, M.F. (2004) Arabidopsis transcriptional activators CBF1, CBF2, and CBF3 have matching functional activities. *Plant Molecular Biology* 54, 767–781.

Giménez, C., Gallardo, M. and Thompson, R.B. (2005) Plant water relations. In: Hillel, D. (ed.) *Encyclopedia of Soils in the Environment*. Vol. 3. Elsevier, Oxford, pp. 231–238.

Granier, C., Aguirrezábal, L., Chenu, K., Cookson, S.J., Dauzat, M., Hamard, P. *et al.* (2006) PHENOPSIS, an automated platform for reproducible phenotyping of plant responses to soil water deficit in *Arabidopsis thaliana* permitted the identification of an accession with low sensitivity to soil water deficit. *New Phytologist* 169, 623–635.

Grimes, D.W. and Yamada, H. (1982) Relation of cotton growth and yield to minimum leaf water potential. *Crop Science* 22, 134–139.

Guo, Y., Abernathy, B., Zeng, Y. and Ozias-Akins, P. (2015) TILLING by sequencing to identify induced mutations in stress resistance genes of peanut (*Arachis hypogaea*). *BMC Genomics* 16(1), 157. Available at: http://doi.org/10.1186/s12864-015-1348-0.

Guttikonda, S.K., Valliyodan, B., Neelakandan, A.K., Tran, L.S., Kumar, R., Quach, T.N. *et al.* (2014) Overexpression of AtDREB1D transcription factor improves drought tolerance in soybean. *Molecular Biology Reports* 41, 7995–8008. doi: 10.1007/s11033-014-3695-3.

Hajheidari, M., Eivazi, A., Buchanan, B.B., Wong, J.H., Majidi, I. and Salekdeh, G.H. (2007) Proteomics uncovers a role for redox in drought tolerance in wheat. *Journal of Proteome Research* 6, 1451–1460.

Harris, B.N., Sadras, V.O. and Tester, M.A. (2010) A water-centred framework to assess the effects of salinity on the growth and yield of wheat and barley. *Plant and Soil* 336, 377–389.

Harris, D. (1996) The effects of manure, genotype, seed priming, depth and date of sowing on the emergence and early growth of *Sorghum bicolor* (L.) Moench in semi-arid Botswana. *Soil and Tillage Research* 40, 73–88.

Harris, K., Subudhi, P.K., Borrell, A., Jordan, D., Rosenow, D. and Nguyen, H. (2007) Sorghum stay-green QTL individually reduce post-flowering drought-induced leaf senescence. *Journal of Experimental Botany* 58, 327–338.

Hash, C.T., Yadav, R.S., Cavan, G.P., Howarth, C.J., Liu, H., Qi, X., Sharma, A., Kolesnikova-Allen, M.A., Bidinger, F.R. and Witcombe, J.R. (2000) Marker-assisted backcrossing to improve drought tolerance in pearl millet. In: Ribaut, J.M. and Poland, D. (eds) *Molecular Approaches for the Genetic Improvement of Cereals for Stable Production in Water-Limited Environments*. A Strategic Planning Workshop held at International Maize and Wheat Improvement Center (CIMMYT), El Batan, Mexico, D.F., pp. 114–119.

Hazen, S.P., Pathan, M.S., Sanchez, A., Baxter, I., Dunn, M., Estes, B., Chang, H.S., Zhu, T., Kreps, J.A. and Nguyen, H.T. (2005) Expression profiling of rice segregating for drought tolerance QTLs using a rice genome array. *Functional Integrative Genomics* 5, 104–116.

Heatherly, L.G. and Elmore, R.W. (2004) Managing inputs for peak production. In: Specht, J.E. and Boerma, H.R. (eds) *Soybeans: Improvement, Production, and Uses. Agronomy Monographs*, 3rd edn. No. 16, ASA-CSSA-SSSA, Madison, Wisconsin, pp. 451–536.

Hochholdinger, F. and Tuberosa, R. (2009) Genetic and genomic dissection of maize root development and architecture. *Current Opinions in Plant Biology* 12, 172–177.

Hodge, A. (2009) Root decisions. *Plant Cell Environment* 32, 628–640.

Hu, H., Dai, M., Yao, J., Xiao, B., Li, X., Zhang, Q. *et al.* (2006) Overexpressing a NAM, ATAF, and CUC (NAC) transcription factor enhances drought resistance and salt tolerance in rice. *Proceedings of the National Academy of Sciences USA* 103, 12987–12992.

Hu, H., You, J., Fang, Y., Zhu, X., Qi, Z. and Xiong, L. (2008) Characterization of transcription factor gene SNAC2 conferring cold and salt tolerance in rice. *Plant Molecular Biology* 67, 169–181.

Huang, X.Y., Chao, D.Y., Gao, J.P., Zhu, M.Z., Shi, M. and Lin, H.X. (2009) A previously unknown zinc finger protein, DST, regulates drought and salt tolerance in rice via stomatal aperture control. *Genes & Development* 23, 1805–1818.

Hufstetler, E.V., Boerma, H.R., Carter, T.E. and Earl, H.G. (2007) Genotypic variation for three physiological traits affecting drought tolerance in soybean. *Crop Science* 47, 25–35.

Hund, A., Traschel, S. and Stamp, P. (2009) Growth of axile and lateral roots of maize: I development of a phenotyping platform. *Plant and Soil* 325, 335–349.

Idso, S.B., Jackson, R.D., Pinter, P.J., Reginato, R.J. and Hatfield, J.L. (1981) Normalizing the stress-degree-day parameter for environmental variability. *Agricultural Meteorology* 24, 45–55.

Iyer-Pascuzzi, A.S., Symonova, O., Mileyko, Y., Hao, Y., Belcher, H., Harer, J., Weitz, J.S. and Benfey, P.N. (2010) Imaging and analysis platform for automatic phenotyping and trait ranking of plant root systems. *Plant Physiology* 152, 1148–1157.

Jacobs, D.I., van der Heijden, R. and Verpoorte, R. (2000) Proteomics in plant biotechnology and secondary metabolism research. *Phytochemical Analysis* 11, 277–287.

Jain, M. (2015) Function genomics of abiotic stress tolerance in plants: a CRISPR approach. *Frontiers in Plant Science* 6, 375. Available at: http://doi.org/10.3389/fpls.2015.00375.

James, A.T., Lawn, R.J. and Cooper, M. (2008) Genotypic variation for drought stress response traits in soybean. I. Variation in soybean and wild *Glycine* spp. for epidermal conductance, osmotic potential, and relative water content. *Australian Journal of Agricultural Research* 59, 656–669.

Jeong, J.S., Kim, Y.S., Baek, K.H., Jung, H., Ha, S.H., Do Choi, Y. *et al.* (2010) Root-specific expression of OsNAC10 improves drought tolerance and grain yield in rice under field drought conditions. *Plant Physiology* 153, 185–197.

Jia, X.Y., Xu, C.Y., Jing, R.L., Li, R.Z., Mao, X.G., Wang, J.P. and Chang, X.P. (2008) Molecular cloning and characterization of wheat calreticulin (CRT) gene involved in drought-stressed responses. *Journal of Experimental Botany* 59, 739–751.

Jones, H.G. (2004) Application of thermal imaging and infrared sensing in plant physiology and ecophysiology. *Advances in Botanical Research* 41, 107–163.

Jones, H.G. (2007) Monitoring plant and soil water status: established and novel methods revisited and their relevance to studies of drought tolerance. *Journal of Experimental Botany* 58, 119–130.

Jones, M.M. and Turner, N.C. (1978) Osmotic adjustment in leaves of sorghum in response to water deficits. *Plant Physiology* 61, 122–126.

Jongdee, B., Fukai, S. and Cooper, M. (2002) Leaf water potential and osmotic adjustment as physiological traits to improve drought tolerance in rice. *Field Crops Research* 76, 153–164.

Joo, J.H., Yoo, H.J., Hwang, I., Lee, J.S., Nam, K.H. and Bae, Y.S. (2005) Auxin-induced reactive oxygen species production requires the activation of phosphatidylinositol 3-kinase. *FEBS Letters* 579, 1243–1248.

Kamoshita, A., Babu, R.C., Boopathi, N.M. and Fukai, S. (2008) Phenotypic and genotypic analysis of drought-resistance traits for development of rice cultivars adapted to rainfed environments. *Field Crops Research* 109, 1–23.

Kang, S., Shi, W. and Zhang, J. (2000) An improved water-use efficiency for maize grown under regulated deficit irrigation. *Field Crops Research* 67, 207–214.

Karaba, A., Dixit, S., Greco, R., Aharoni, A., Trijatmiko, K.R., Marsch-Martinez, N., Krishnan, A., Nataraja, K.N., Udayakumar, M. and Pereira, A. (2007) Improvement of water use efficiency in rice by expression of HARDY, an *Arabidopsis* drought and salt tolerance gene. *Proceedings of the National Academy of Sciences USA* 104, 15270–15275.

Keurentjes, J.J.B. (2009) Genetical metabolomics: closing in on phenotypes. *Current Opinion in Plant Biology* 12, 223–230.

Kumar, A., Singh, P., Singh, D.P., Singh, H. and Sharma, H.C. (1984) Differences in osmoregulation in *Brassica* species. *Annals of Botany* 54, 537–541.

Kumar, D. (2005) Breeding for drought resistance. In: Ashraf, P.J.C. and Harris, J. (eds) *Abiotic Stresses*. The Howarth Press Inc., New York, pp. 145–175.

Lafitte, R. (2002) Relationship between leaf relative water content during reproductive stage water deficit and grain formation in rice. *Field Crops Research* 76, 165–174.

Landi, P., Giuliani, S., Salvi, S., Ferri, M., Tuberosa, R. and Sanguineti, M.C. (2010) Characterization of root-yield-1.06, a major constitutive QTL for root and agronomic traits in maize across water regimes. *Journal of Experimental Botany* 61, 3553–3562.

Lawlor, D.W. and Cornic, G. (2002) Photosynthetic carbon assimilation and associated metabolism in relation to water deficits in higher plants. *Plant Cell Environment* 25, 275–294.

Lazarovitch, N., Ben-Gal, A. and Shani, U. (2006) An automated rotating lysimeter system for greenhouse evapo-transpiration studies. *Vadose Zone Journal* 5, 801–804.

Leakey, A.D.B., Uribelarrea, M., Ainsworth, E.A., Naidu, S.L., Rogers, A., Ort, D.R. and Long, S.P. (2006) Photo-synthesis, productivity and yield of *Zea mays* are not affected by open-air elevation of CO_2 concentration in the absence of drought. *Plant Physiology* 140, 779–790.

Li, Z., Mu, P., Li, C., Zhang, H., Li, Z., Gao, Y. and Wang, X. (2005) QTL mapping of root traits in a doubled haploid population from a cross between upland and lowland *japonica* rice in three environments. *Theoretical and Applied Genetics* 110, 1244–1252.

Lian, H.L., Yu, X., Ye, Q. *et al.* (2004) The role of aquaporin RWC3 in drought avoidance in rice. *Plant Cell Physiology* 45, 481–489.

Liu, F., Andersen, M.N. and Jensen, C.R. (2003) Loss of pod set caused by drought stress is associated with water status and ABA content of reproductive structures in soybean. *Functional Plant Biology* 30, 271–280.

Liu, J.X. and Bennett, J. (2011) Reversible and irreversible drought induced changes in the anther proteome of rice (*Oryza sativa* L.) genotypes IR64 and Moroberekkan. *Molecular Plant* 4, 59–69.

Liu, X., Zhai, S., Zhao, Y., Sun, B., Liu, C., Yang, A. and Zhang, J. (2013) Overexpression of the phosphatidylinosi-tol synthase gene (ZmPIS) conferring drought stress tolerance by altering membrane lipid composition and increasing ABA synthesis in maize. *Plant Cell Environment* 36, 1037–1055.

Loveys, B.R., Stoll, M., Dry, P.R. and McCarthy, M.G. (2000) Using plant physiology to improve the water use effi-ciency of horticultural crops. *Acta Horticulturae* 537, 187–197.

Ludlow, M.M., Santamaria, J.M. and Fukai, S. (1990) Contribution of osmotic adjustment to grain yield in *Sorghum bicolor* (L.) Moench under water-limited conditions. II. Water stress after anthesis. *Australian Journal of Plant Physiology* 41, 67–78.

Luo, L. (2010) Breeding for water-saving and drought-resistance rice in china. *Journal of Experimental Botany* 61, 3509–3517.

Lv, S.L., Lian, L.J., Tao, P.L., Li, Z.X., Zhang, K.W. and Zhang, J.R. (2009) Overexpression of *Thellungiella halophila* H^+–PPase (TsVP) in cotton enhances drought stress resistance of plants. *Planta* 229, 899–910.

Mahalakshmi, V., Bidinger, F.R. and Raju, D.S. (1987) Effects of timing of stress in pearl millet [*Pennisetum amer-canum* (L.) Leeke]. *Field Crops Research* 15, 327–339.

Maiti, R.K., Prasada-Rao, K.E., Raju, P.S. and House, L.R. (1984) The glossy trait in sorghum: its characteristics and significance in crop improvement. *Field Crops Research* 9, 279–289.

Manavalan, L.P., Guttikonda, S.K., Tran, L.P. and Nguyen, H.T. (2009) Physiological and molecular approaches to improve drought resistance in soybean. *Plant Cell Physiology* 50, 1260–1276.

Manavalan, L.P., Guttikonda, S.K., Nguyen, V.T., Shannon, J.G. and Nguyen, H.T. (2010) Evaluation of diverse soybean germplasm for root growth and architecture. *Plant and Soil* 330, 503–514.

Manavalan, L.P., Chen, X., Clarke, J., Salmeron, J. and Nguyen, H.T (2012) RNAi-mediated disruption of squalene synthase improves drought tolerance and yield in rice. *Journal of Experimental Botany* 63, 163–175.

Manavalan, L.P., Prince, S.J., Musket, T.A., Chaky, J., Deshmukh, R., Vuong, T.D. *et al.* (2015) Identification of novel QTL governing root architectural traits in an interspecific soybean population. *PLoS ONE* 10(3), e0120490. doi:10.1371/journal.pone.0120490.

Manga, V.K. and Yadav, O.P. (1995) Effect of seed size on developmental traits and ability to tolerate drought in pearl millet. *Journal of Arid Environments* 29, 169–172.

Manschadi, A.M., Hammer, G.L., Christopher, J.T. and deVoil, P. (2008) Genotypic variation in seedling root architectural traits and implications for drought adaptation in wheat (*Triticum aestivum* L.). *Plant and Soil* 303, 115–129.

Maurel, C., Simonneau, T. and Sutka, M. (2010) The significance of roots as hydraulic rheostats. *Journal of Experi-mental Botany* 61, 3191–3198.

Mittler, R. and Blumwald, E. (2010) Genetic engineering for modern agriculture: challenges and perspectives. *Annual Review of Plant Biology* 61, 443–462.

Motha, R.P. and Menzie, K.L. (2007) Meteorological information for evaluating agrometeorological risk and un-certainty for agricultural marketing systems. *WMO Bulletin* 56, 30–33.

Muchow, R.C. and Sinclair, T.R. (1989) Epidermal conductance, stomatal density and stomatal size among geno-types of *Sorghum bicolor* (L.) Moench. *Plant Cell Environment* 12, 425–431.

Munns, R., James, R.A., Sirault, X.R.R., Furbank, R.T. and Jones, H.G. (2010) New phenotyping methods for screening wheat and barley for beneficial responses to water deficit. *Journal of Experimental Botany* 61, 3499–3507.

Nelson, D.E., Repetti, P.P., Adams, T.R., Creelman, R.A., Wu, J., Warner, D.C. *et al.* (2007) Plant nuclear factor Y (NF-Y) B subunits confer drought tolerance and lead to improved corn yields on water-limited acres. *Proceedings of the National Academy of Sciences USA* 104, 16450–16455.

Nguyen, H.T., Babu, R.C. and Blum, A. (1997) Breeding for drought resistance in rice: physiology and molecular genetics considerations. *Crop Science* 37, 1426–1434.

Oh, S.J., Kwon, C.W., Choi, D.W., Song, S.I. and Kim, J.K. (2007) Expression of barley HvCBF4 enhances tolerance to abiotic stress in transgenic rice. *Plant Biotechnology Journal* 5, 646–656.

Oh, S.J., Kim, Y.S., Kwon, C., Park, H.K., Jeong, J.S. and Kim, J.K. (2009) Overexpression of the transcription factor AP37 in rice improves grain yield under drought conditions. *Plant Physiology* 2, 191–200.

Oliver, M.J. and Bewley, J.D. (1997) Desiccation-tolerance of plant tissues: a mechanistic overview. *Horticultural Reviews* 18, 171–121.

Oukarroum, A., Madidi, S.E., Schansker, G. and Strasser, R.J. (2007) Probing the responses of barley cultivars (*Hordeum vulgare* L.) by chlorophyll *a* fluorescence OLKJIP under drought stress and re-watering. *Environmental and Experimental Botany* 60, 438–446.

Ouyang, S.Q., Liu, Y.F., Liu, P., Lei, G., He, S.J., Ma, B., Zhang, W.K., Zhang, J.S. and Chen, S.Y. (2010) Receptor-like kinase OsSIK1 improves drought and salt stress tolerance in rice (*Oryza sativa*) plants. *Plant Journal* 62, 316–329.

Pasapula, V., Shen, G., Kuppu, S., Paez-Valencia, J., Mendoza, M., Hou, P., Chen, J., Qiu, X., Zhu, L., Zhang, X., Auld, D., Blumwald, E., Zhang, H., Gaxiola, R. and Payton, P. (2011) Expression of an *Arabidopsis vacuolar* H^+–pyrophosphatase gene AVP1in cotton improves drought- and salt tolerance and increases fibre yield in the field conditions. *Plant Biotechnology Journal* 9, 88–99.

Pei, Z.-M., Murata, Y., Benning, G., Thomine, S., Klusener, B., Allen, G.J., Grill, E. and Schroeder, J.I. (2000) Calcium channels activated by hydrogen peroxide mediate abscisic signaling in guard cells. *Nature* 406, 731–734.

Phelix, C.F. and Feltus, F.A. (2015) Plant stress biomarkers from biosimulations: the Transcriptome-To-Metabolome™ (TTM™) technology – effects of drought stress on rice. *Plant Biology* 17, 63–73. doi: 10.1111/plb.12221.

Pieruschka, R. and Lawson, T. (2015) Preface on special issue on phenotyping in plants. *Journal of Experimental Botany* 66(18), 5385–5387.

Pieters, A.J. and El Souki, S. (2005) Effects of drought during grain filling on PS II activity in rice. *Journal of Plant Physiology* 162, 903–911.

Premachandra, G.S., Saneoka, H., Kanaya, M. and Ogata, S. (1991) Cell membrane stability and leaf surface wax content as affected by increasing water deficits in maize. *Journal of Experimental Botany* 42, 167–171.

Price, A.H., Cairns, J.E., Horton, P., Jones, H.G. and Griffiths, G. (2002) Linking drought-resistance mechanisms to drought avoidance in upland rice using a QTL approach: progress and new opportunities to integrate stomatal and mesophyll responses. *Journal of Experimental Biology* 53, 989–1004.

Quan, R., Shang, M., Zhang, H., Zhao, Y. and Zhang, J. (2004) Engineering of enhanced glycine betaine synthesis improves drought tolerance in maize. *Plant Biotechnology Journal* 2, 477–486.

Rajendran, K., Tester, M.A. and Roy, S.J. (2009) Quantifying the three main components of salinity tolerance in cereals. *Plant, Cell and Environment* 32, 237–249.

Ramu, V.S., Swetha, T.N., Sheela, S.H., Babitha, C.K., Rohini, S., Reddy, M.K., Tuteja, N., Reddy, C.P., Prasad, T.G. and Udayakumar, M. (2016) Simultaneous expression of regulatory genes associated with specific drought-adaptive traits improves drought adaptation in peanut. *Plant Biotechnology Journal* 14, 1008–1020. doi:10.1111/pbi.12461.

Rebetzke, G.J. and Richards, R.A. (1999) Genetic improvement of early vigour in wheat. *Australian Journal of Agricultural Research* 50, 291–302.

Reynolds, M.P., Mujeeb-Kazi, A. and Sawkins, M. (2005) Prospects for utilising plant-adaptive mechanisms to improve wheat and other crops in drought- and salinity-prone environments. *Annals of Applied Biology* 146, 239–259.

Reynolds, M.P., Saint Pierre, C., Saad Abu, S.I., Vargas, M. and Condon, A.G. (2007) Evaluating potential genetic gains in wheat associated with stress-adaptive trait expression in elite genetic resources under drought and heat stress. *Crop Science* 47,172–189.

Richards, R.A., Rawson, H.M. and Johnson, D.A. (1986) Glaucousness in wheat: its development and effect on water-use efficiency, gas exchange and photosynthetic tissue temperatures. *Australian Journal of Plant Physiology* 13, 465–473.

Ristic, Z. and Jenks, M.A. (2002) Leaf cuticle and water loss in maize lines differing in dehydration avoidance. *Journal of Plant Physiology* 159, 645–651.

Rivero, R.M., Kojima, M., Gepstein, A., Sakakibara, H., Mittler, R., Gepstein, S. and Blumwald, E. (2007) Delayed leaf senescence induces extreme drought tolerance in a flowering plant. *Proceedings of the National Academy of Sciences USA* 104, 19631–19636.

Ruan, C.J. and Teixeira da Silva, J.A. (2011) Metabolomics: creating new potentials for unraveling the mechanisms in response to salt and drought stress and for the biotechnological improvement of xero-halophytes. *Critical Reviews in Biotechnology* 31, 153–169.

Saibo, N.J.M., Lourenço, T. and Oliveira, M.M. (2009) Transcription factors and regulation of photosynthetic and related metabolism under environmental stresses. *Annals of Botany* 103, 609–62.

Salekdeh, G.H., Siopongco, J., Wade, L.J., Ghareyazie, B. and Bennett, J. (2002) Proteomic analysis of rice leaves during drought stress and recovery. *Proteomics* 2, 1131–1145.

Sánchez, F.J., Manzanares, M., de Andrés, E.F., Tenorio, J.L. and Ayerbe, L. (2001) Residual transpiration rate, epicuticular wax load and leaf colour of pea plants in drought conditions. Influence on harvest index and canopy temperature. *European Journal of Agronomy* 15, 57–70.

Schonfeld, M.A., Johnson, R.C., Carver, B.F. and Mornhinweg, D.W. (1988) Water relations in winter wheat as drought resistance indicators. *Crop Science* 28, 526–531.

Schroeder, J.I., Kwak, J.M. and Allen, G.J. (2001) Guard cell abscisic acid signalling and engineering drought hardiness in plants. *Nature* 410, 327–330.

Seki, M., Kamei, A., Yamaguchi-Shinozakiz, K. and Shinozaki, K. (2003) Molecular responses to drought, salinity and frost: common and different paths for plant protection. *Current Opinion in Biotechnology* 14, 194–199.

Sengupta, D., Kannan, M. and Reddy, A.R. (2011) A root proteomics-based insight reveals dynamic regulation of root proteins under progressive drought stress and recovery in *Vigna radiata* (L.). *Planta* 233, 1111–1127. doi: 10.1007/s00425-011-1365-4.

Serraj, R., Hash, C.T., Rizvi, S.M.H., Sharma, A., Yadav, R.S. and Bidinger, F.R. (2005) Recent advances in marker-assisted selection for drought tolerance in pearl millet. *Plant Production Science* 8, 334–337.

Shao, H., Chu, L., Jaleel, C.A., Manivannan, P., Panneerselvam, R. and Shao, M. (2009) Understanding water deficit stress-induced changes in the basic metabolism of higher plants – biotechnologically and sustainably improving agriculture and the ecoenvironment in arid regions of the globe. *Critical Reviews in Biotechnology* 29, 131–151.

Shin, D., Moon, S., Han, S., Kim, B., Park, S., Lee, S., Yoon, H., Lee, H., Kwon, H., Baek, D., Yi, B.Y. and Byun, M. (2011) Expression of StMYB1R-1, a novel potato single MYB-like domain transcription factor, increases drought tolerance. *Plant Physiology* 155, 421–432.

Shulaev, V., Cortes, D., Miller, G. and Mittler, R. (2008) Metabolomics for plant stress response. *Journal of Plant Physiology* 132, 199–208.

Singh, V., van Oosterom, E.J., Jordan, D.R., Messina, C.D., Cooper, M. and Hammer, G.L. (2010) Morphological and architectural development of root systems in sorghum and maize. *Plant and Soil* 333, 287–299.

Steele, K.A., Price, A.H., Shashidar, H.E. and Witcombe, J.R. (2006) Marker-assisted selection to introgress rice QTLs controlling root traits into an Indian upland rice variety. *Theoretical and Applied Genetics* 112, 208–221.

Stiller, I., Dulai, S., Kondrák, M., Tarnai, R., Szabó, L., Toldi, O. and Bánfalvi, Z. (2008) Effects of drought on water content and photosynthetic parameters in potato plants expressing the trehalose-6-phosphate synthase gene of *Saccharomyces cerevisiae*. *Planta* 227, 299–308.

Strasser, R.J., Srivastava, A. and Govindjee, G. (1995) Polyphasic chlorophyll *a* fluorescence transient in plants and cyanobacteria. *Photochemistry and Photobiology* 61, 32–42.

Sunkar, R., Chinnusamy, V., Zhu, J. and Zhu, J.K. (2007) Small RNAs as big players in plant abiotic stress responses and nutrient deprivation. *Trends in Plant Science* 12, 301–309.

Tahara, M., Carver, B.F., Johnson, R.C. and Smith, E.L. (1990) Relationship between relative water-content during reproductive development and winter wheat grain yield. *Euphytica* 49, 255–262.

Taiz, L. and Zeiger, E. (2002) *Plant Physiology*, 3rd edn. Sinauer Associates, Sunderland, Massachusetts.

Thompson, A.J., Andrews, J., Mulholland, B.J., McKee, J.M.T., Hilton, H.W., Black, C.R. and Taylor, I.B. (2007) Overproduction of abscisic acid in tomato increases transpiration efficiency and root hydraulic conductivity and influences leaf expansion. *Plant Physiology* 143, 1905–1917.

Todaka, D., Shinozaki, K. and Yamaguchi-Shinozaki, K. (2015) Recent advances in the dissection of drought-stress regulatory networks and strategies for development of drought-tolerant transgenic rice plants. *Frontiers in Plant Science* 6, 84. doi.org/10.3389/fpls.2015.00084.

Tran, L.P.S., Shinozaki, K. and Yamaguchi-Shinozaki, K. (2010) Role of cytokinin responsive two-component system in ABA and osmotic stress signaling. *Plant Signaling Behavior* 5, 148–150.

Tripathy, J.N., Zhang, J., Robin, S., Nguyen, T.T. and Nguyen, H.T. (2000) QTLs for cell-membrane stability mapped in rice (*Oryza sativa* L.) under drought stress. *Theoretical and Applied Generics* 100, 1197–1202.

Tsialtas, J.T., Kassioumi, M. and Veresoglou, D.S. (2002) Evaluating leaf ash content and potassium concentration as surrogates of carbon isotope discrimination in grassland species. *Journal of Agronomy and Crop Science* 188, 168–175.

Turner, N.C. (1982) The role of shoot characteristics in drought tolerance of crop plants. In: *Drought Tolerance in Crop with Emphasis on Rice*. IRRI, Los Banos, Manila, pp. 115–134.

Turner, N.C. (1986) Adaptation to water deficits: a changing perspective. *Australian Journal of Plant Physiology* 13, 175–90.

Turner, N.C., Wright, G.C. and Siddique, K.H.M. (2001) Adaptation of grain legumes (pulses) to water limited environments. *Advances in Agronomy* 71, 193–231.

Uga, Y., Okuno, K. and Yano, M. (2011) Dro1, a major QTL involved in deep rooting of rice under upland field conditions. *Journal of Experimental Botany* 62(8), 2485–2494. doi: 10.1093/jxb/erq429.

Uga, Y., Sugimoto, K., Ogawa, S., Rane, J. *et al.* (2013) Control of root system architecture by DEEPER ROOTING 1 increases rice yield under drought conditions. *Nature Genetics* 45, 1097–1102.

Umezawa, T., Fujita, M., Fujita, Y., Yamaguchi-Shinozaki, K. and Shinozaki, K. (2006) Engineering drought tolerance in plants: discovering and tailoring genes unlock the future. *Current Opinion in Biotechnology* 17, 113–122.

Vadez, V. (2014) Root hydraulics: the forgotten side of roots in drought adaptation. *Field Crops Research* 165, 15–24.

van der Weele, C.M., Jiang, H.S., Palaniappan, K.K., Ivanov, V.B., Palaniappan, K. and Baskin, T.I. (2003) A new algorithm for computational image analysis of deformable motion at high spatial and temporal resolution applied to root growth: roughly uniform elongation in the meristem and also, after an abrupt acceleration, in the elongation zone. *Journal of Plant Physiology* 132, 1138–1148.

Varshney, R.K., Graner, A. and Sorrells, M.E. (2005) Genomics assisted breeding for crop improvement. *Trends in Plant Science* 10(12), 621–630.

Vernoux, T., Sánchez-Fernández, R. and May, M. (2002) Glutathione biosynthesis in plants. In: Inze, D. and Montagu, M.V. (eds) *Oxidative Stress in Plants*. Taylor & Francis, London, pp. 297–311.

Verslues, P.E., Agarwal, M., Katiyar-Agarwal, S., Zhu, J. and Zhu, J.K. (2006) Methods and concepts in quantifying resistance to drought, salt and freezing, abiotic stresses that affect plant water status. *Plant Journal* 45, 523–539.

Vignes, D., Djekoun, A. and Planchon, C. (1986) Responses de different genotypes de soja au deficit hydrique. *Canadian Journal of Plant Science* 66, 247–255.

Virlet, N., Costes, E., Martinez, S., Kelner, J.J. and Regnard, J.L. (2015) Multispectral airborne imagery in the field reveals genetic determinisms of morphological and transpiration traits of an apple tree hybrid population in response to water deficit. *Journal of Experimental Botany* 66, 5453–5465.

Wang, C.R., Yang, A.F., Yue, G.D., Gao, Q., Yin, H.Y. and Zhang, J.R. (2008) Enhanced expression of phospholipase C 1 (ZmPLC1) improves drought tolerance in transgenic maize. *Planta* 227, 1127–1140.

Wang, D., Pan, Y., Zhao, X., Zhu, L., Fu, B. and Li, Z. (2011) Genome-wide temporal-spatial gene expression profiling of drought responsiveness in rice. *BMC Genomics* 12, 149.

Wang, W., Vinocur, B. and Altman, A. (2003) Plant responses to drought, salinity and extreme temperatures: towards genetic engineering for stress tolerance. *Planta* 218, 1–14.

Wang, Y. and Frei, M. (2011) Stressed food – the impact of abiotic environmental stresses on crop quality. *Agriculture, Ecosystems and Environment* 141, 271–286.

Wang, Y., Ying, J., Kuzma, M., Chalifoux, M., Sample, A., McArthur, C., Uchacz, T., Sarvas, C., Wan, J., Dennis, D.T., McCourt, P. and Huang, Y. (2005) Molecular tailoring of farnesylation for plant drought tolerance and yield protection. *Plant Journal* 4, 413–424.

Wang, Y., Beaith, M., Chalifoux, M., Ying, J., Uchacz, T., Sarvas, C., Griffiths, R., Kuzma, M., Wan, J. and Huang, Y. (2009) Shoot-specific down-regulation of protein farnesyltransferase (α-subunit) for yield protection against drought in canola. *Molecular Plant* 2, 191–200.

Wei, A., He, C., Li, B., Li, N. and Zhang, J. (2011) The pyramid of transgenes TsVP and BetA effectively enhances the drought tolerance of maize plants. *Plant Biotechnology Journal* 9, 216–229.

Werner, T., Nehnevajova, E., Köllmer, I., Novak, O., Strnad, M., Krämer, U. and Schmülling, T. (2010) Root-specific reduction of cytokinin causes enhanced root growth, drought tolerance, and leaf mineral enrichment in *Arabidopsis* and tobacco. *Plant Cell* 22, 3905–3920.

Winter, S.R., Musick, J.T. and Porter, K.B. (1988) Evaluation of screening techniques for breeding drought resistant winter wheat. *Crop Science* 512–516.

Wright, G.C., Nageswara Rao, R.C. and Basu, M.S. (1996) A physiological approach to the understanding of genotype by environment interactions – a case study on improvement of drought adaptation in groundnut.

In: Cooper, M. and Hammer, G.L. (eds) *Plant Adaptation and Crop Improvement*. CAB International, Wallingford, UK, pp. 365–381.

Xiao, B., Huang, Y., Tang, N. and Xiong, L. (2007) Over-expression of a LEA gene in rice improves drought resistance under the field conditions. *Theoretical and Applied Genetics* 115, 35–46.

Xu, C., Jing, R., Mao, X., Jia, X. and Chang, X. (2007) A wheat (*Triticum aestivum*) protein phosphatase 2A catalytic subunit gene provides enhanced drought tolerance in tobacco. *Annals of Botany* 99, 439–450.

Yamaguchi, M. and Sharp, R.E. (2010) Complexity and coordination of root growth at low water potentials: recent advances from transcriptomic and proteomic analyses. *Plant Cell and Environment* 33, 590–603.

Yu, H., Chen, X., Hong, Y.Y., Wang, Y., Xu, P., Ke, S.D., Liu, H.Y., Zhu, J.K., Oliver, D.J. and Xiang, C.B. (2008) Activated expression of an *Arabidopsis* HD-START protein confers drought tolerance with improved root system and reduced stomatal density. *Plant Cell* 20, 1134–1151.

Zhang, G.Y., Chen, M., Li, L.C., Xu, Z.S., Chen, X.P., Guo, J.M. and Ma, Y.Z. (2009) Overexpression of the soybean GmERF3 gene, an AP2/ERF type transcription factor for increased tolerances to salt, drought, and diseases in transgenic tobacco. *Journal of Experimental Botany* 60, 3781–3796.

Zhang, J. and Davies, W.J. (1989) Sequential response of whole plant water relations to prolonged soil drying and the involvement of zylem sap ABA I the regulation of stomatal behaviour of sunflower plants. *New Phytologist* 113, 167–174.

Zhang, J.-Y., Corey, D., Blancaflor, B., Sledge, K., Sumner, W. and Wang, Z. (2005) Overexpression of WXP1, a putative *Medicago truncatula* AP2 domain-containing transcription factor gene, increases cuticular wax accumulation and enhances drought tolerance in transgenic alfalfa (*Medicago sativa*). *Plant Journal* 42, 689–707.

Zhao, M. and Running, S.W. (2010) Drought-induced reduction in global terrestrial net primary production from 2000 through 2009. *Science* 329, 940–943.

Zhu, J., Alvarez, S., Marsh, E.L., Lenoble, M.E., Cho, I.J., Sivaguru, M., Chen, S., Nguyen, H.T., Wu, Y., Schachtman, D.P. and Sharp, R.E. (2007) Cell wall proteome in the maize primary root elongation zone. II. Region-specific changes in water soluble and lightly ionically bound proteins under water deficit. *Plant Physiology* 145, 1533–1548.

Zia, S., Romano, G., Spreer, W., Sanchez, C., Cairns, J., Araus, J.L. and Müller, J. (2012) Infrared thermal imaging as a rapid tool for identifying water stress tolerant maize genotypes of different phenology. *Journal of Agronomy and Crop Science* 199, 75–84.

Zinselmeier, C., Sun, Y.J., Helentjaris, T., Beatty, M., Yang, S., Smith, H. and Habben, J. (2002) The use of gene expression profiling to dissect the stress sensitivity of reproductive development in maize. *Field Crops Research* 75, 111–121.

2 Salinity Stress: Physiological Constraints and Adaptive Mechanisms

Sergey Shabala[1],* and Rana Munns[2,3]

[1]*School of Land and Food, University of Tasmania, Hobart, Australia;*
[2]*CSIRO Agriculture, Canberra, Australia;* [3]*School of Plant Biology
and ARC Centre of Excellence in Plant Energy Biology,
University of Western Australia, Crawley, Australia*

Abstract

A significant part of the world's land area is salt-affected, including areas in which crops and pastures are grown for food and forage. Growth and yield of most crops is reduced by salinity, and only halophytes are able to handle large amounts of salt without penalty. This chapter summarizes our current knowledge of physiological mechanisms conferring plant adaptive responses to salinity. The classification of saline soils is given with causes of primary and secondary types of salinity. Major physiological constraints are then summarized, and physiological and genetic diversity of plant responses to salinity are presented. Key physiological and anatomical mechanisms conferring salinity tolerance in plants are then analysed in detail, with emphasis on how salt uptake, transport and accumulation in tissues within the plant are controlled. This chapter shows that plants have evolved numerous mechanisms to prevent accumulation of toxic Na^+ concentrations in leaves, and to regulate concentrations of Na^+, K^+ and Cl^- within the various cell compartments. This ability is complemented by mechanisms enabling efficient osmotic adjustment and maintenance of cell turgor, as well as mechanisms of coping with oxidative stress imposed by salinity. A more complete physiological and genetic understanding of these processes will enable targeted breeding for new salt-tolerant plants for the future.

2.1 Introduction: Salinity as an Issue

A significant proportion of the world's land area is salt-affected, either by salinity or sodicity. Most of this salinity is natural; however, cultivation has caused a significant amount of land to become saline because of land clearing or irrigation (Szabolcs, 1989). Irrigation schemes have also caused extensive salinization and land degradation in many countries (Szabolcs, 1989; Ghassemi *et al.*, 1995). While irrigated land represents only 15% of total cultivated land, because it yields twice as much as rain-fed land, it produces one-third of the world's food (Munns, 2005). Thus, it is highly unlikely that irrigation will be abandoned as 'unsustainable practice' in the near future. On the contrary, the shortage of suitable agricultural land and a need to meet the 2050 challenge of feeding 9 billion people will most likely increase the proportion of irrigated land. Thus, the salinity issue will remain one of the key threats to global food production in the 21st century.

2.1.1 Classification of saline soils

A soil is defined as saline when the electrical conductivity of the saturation paste extract (EC_e)

---

* Corresponding author: sergey.shabala@utas.edu.au

exceeds 4 dS m^{-1} (Richards, 1954), equivalent to 40 mM NaCl. This definition of salinity derives from the EC_e that substantially reduces the yield of most crops. Soils are rarely saturated, and the salt concentration experienced by roots can be several times that in the saturation extract (Rengasamy, 2002), so a soil with an EC_e of 4 dS m^{-1} will have a salt concentration of 80–100 mM NaCl most of the time, and substantially reduce the yield of most crop and pasture species.

The salts that give rise to salinity come mainly from weathering of rocks, or from aerial deposition of ocean aerosols via wind or rain (Rengasamy, 2002). The main salt of saline soils is NaCl, but sometimes there are also significant concentrations of Ca^{2+}, Mg^{2+}, SO_4^{2-} and CO_3^{2-}. Seawater intrusion on to low-lying coastal land can also deposit large amounts of salt. Salinity can occur naturally (primary salinity), or as a result of human activities (secondary salinity).

2.1.2   Primary salinity

Natural salinity occurs where the rainfall is low and the salt remains in the subsoil. Salt can move in and out of the root zone with seasonal rainfall, giving rise to the term 'transient salinity' (Rengasamy, 2002). Neither agronomic nor engineering practices can address this form of salinity, and genetic improvement provides the only way to increase productivity on these lands.

Sodicity occurs in soils in which Na^+ makes up a high percentage of the cations bound to clay particles. This causes loss of soil structure, and the soil becomes waterlogged when wet, and hard as it dries. Sodic soils affect plant growth because roots cannot penetrate layers that are hypoxic or hard (Qadir and Schubert, 2002). A soil is defined as sodic if the exchangeable sodium percentage (ESP) is greater than 15. This definition replaces the earlier one of Richards (1954) in which 'sodic' was synonymous with 'alkali'. Most sodic soils have a pH above 7, the pH depending on whether they were formed from carbonate, bicarbonate, chloride or sulfate salts (Szabolcs, 1989). Some sodic soils are also saline, or have serious chemical constraints such as very high pH, boron toxicity, aluminium toxicity and micronutrient deficiency (Qadir and Schubert, 2002; Rengasamy, 2002).

2.1.3   Secondary salinity

The two main causes of secondary salinity are irrigation and the clearing of land.

Irrigation systems are prone to salinization and waterlogging, due to brackish irrigation water, or to excessive leaching and subsequent rising water tables (Rhoades et al., 1992). The amount of salt removed from the soil by crops is negligible, so salt accumulates in the root zone. This accumulation can be managed by leaching and drainage. Salt-tolerant crops need only half the leaching requirement of salt-sensitive crops (Rhoades et al., 1992), so improving the salt tolerance of crops can lessen the costs of irrigation, both in the need to import fresh water and to dispose of saline water. On a global scale, it has been estimated that every minute 3 hectares (ha) of currently arable land become unproductive due to secondary-induced salinization (Zhu et al., 2005), and that between 10 and 20 mega hectares (Mha) of irrigated land deteriorates to zero productivity each year (Choukr-Allah, 1996; Hamdy, 1996).

Salinity can also be caused by dryland agriculture (Halvorson, 1990). The cause of this 'dryland salinity' is the clearing of land for dryland agriculture, and replacement of native perennial vegetation by annual crops. The replacement of perennial deep-rooted native vegetation by shallow-rooted annual crops or pastures results in wetter subsoils and accompanying larger drainage beyond the reach of shallow roots, leading eventually to rising water tables. Clearing of land in moderate rainfall zones (350–600 mm) causes water tables to rise and form saline seeps. Annual crops do not use all the rainfall, and allow more rainwater to escape their roots than the original perennial vegetation. This 'dryland salinity' can be managed by crop rotations. The introduction of deep-rooted perennial species such as lucerne (alfalfa) can lower the water tables (Dunin et al., 2001), but salt tolerance will be required not only for the 'de-watering' species, but also for the annual crops to follow, as some salt will be left in the soil when the water table is lowered.

2.1.4   Economic impacts of salinity

Soil salinity has a significant impact on food production in many countries of the world, including the USA, Australia, China, India and Pakistan, resulting in annual losses of an estimated US$27.3 billion to the agricultural sector (Qadir et al., 2014).

There are no easy solutions to the problem of soil salinity. Where salinization of the soil is caused by irrigation or land clearing, the problem can be tackled by altered agronomic methods, including crop rotations or drainage. Where salinity is naturally present in the soil, or is caused by seawater intrusions exacerbated by global warming and seawater rise, production of staple crops can be lifted only by breeding of more salt-tolerant varieties.

Development of new halophytic crops such as quinoa is an alternative way of maintaining global food production, as is the use of salt-tolerant forage grasses for grazing animals on highly saline land.

Difficult decisions have to be made on the economics of developing new irrigation schemes, when water is available, to take in the environmental and social impacts of these decisions. Many irrigation systems are short-lived as soil salinity increases, even though the irrigation water is initially low in salts (Szabolcs, 1989). Other difficult decisions are how to treat land that has become salinized due to irrigation or land clearing. Should these be drained, or revegetated by trees or other perennial species with the aim of lowering water tables, or planted with halophytic species with the aim of a sustainable yield, or should they be abandoned?

2.2 Physiological Constraints Imposed by Salinity

2.2.1 Osmotic stress

A salt concentration in the soil of 4 dSm^{-1} or 40 mM NaCl has an osmotic pressure of about 0.2 MPa, which affects the ability of plants to take up water. The osmotic pressure can be calculated from the concentration of the salt solution by the equation $\pi = cRT$, where π is osmotic pressure (expressed in MPa), c is concentration (osmoles per litre of water), R is the universal gas constant and T is absolute temperature in degrees Kelvin ($RT = 2.48$ at 25°C, in litre-MPa per mole). This osmotic effect has a flow-on effect, via internal signals, to reduce the rate of cell expansion in growing tissues, and the degree of stomatal aperture in leaves. The reduction in stomatal conductance of CO_2 limits the rate of photosynthesis which, together with the slower formation of photosynthetic leaf area, reduces the flow of assimilates to the meristematic and growing tissues of the plant, both leaves and roots (although leaves are often more affected than roots) (Munns and Sharp, 1993).

The decreased rate of leaf growth after an increase in soil salinity is primarily due to the osmotic effect of the salt around the roots. A sudden increase in soil salinity causes leaf cells to lose water, but this loss of cell volume and turgor is transient (Passioura and Munns, 2000). Within hours, cells regain their original volume and turgor owing to osmotic adjustment but, despite this, cell elongation rates are reduced (Passioura and Munns, 2000; Fricke and Peters, 2002). Over days, reductions in cell elongation and also cell division lead to slower leaf appearance and smaller final size. Cell dimensions change, with more reduction in area than depth, so leaves are smaller and thicker.

For a moderate salinity stress, an inhibition of lateral shoot development becomes apparent over weeks, and over months there are effects on reproductive development, such as early flowering or a reduced number of florets. During this time, a number of older leaves may die. However, production of younger leaves continues. All these changes in plant growth are responses to the osmotic effect of the salt, and are similar to drought responses.

The reduction in leaf development is due to the salt outside the roots. That this reduction is largely due to the osmotic effect of the salt is supported by experiments using mixed salts such as concentrated Hoagland's solution (Termaat and Munns, 1986), other single salts such as KCl and non-ionic solutes such as mannitol or polyethylene glycol (PEG) (Yeo et al., 1991). These different osmotica all have a similar qualitative effect as NaCl on leaf expansion. A similar situation occurs for roots, which might seem paradoxical as roots are more directly exposed to salt than leaves, but their control of uptake coupled with the ability to efflux excessive salts enables them to control their salt content precisely and avoid toxicity.

2.2.2 Ionic imbalance

The second major constraint imposed by salinity is Na$^+$ toxicity and ionic imbalance in the cell cytosol. Due to the similarity in physico-chemical properties between Na$^+$ and K$^+$ (i.e. ionic radius and ion hydration energy), the former competes with K$^+$ for major binding sites in key metabolic processes in the cytoplasm, such as enzymatic reactions, protein synthesis and ribosome functions (Marschner, 1995). With over 50 cytoplasmic enzymes being activated by K$^+$, the disruption to metabolism is severe, both in root and leaf tissues. Salinity also affects a plant's ability to acquire and metabolize other essential nutrients such as calcium, nitrogen, phosphorous and magnesium. Several major factors contribute to this process (Shabala and Cuin, 2008):

1. High concentrations of Na$^+$ in the soil substantially reduce the activity of many essential nutrients, making them less available for plants (as illustrated for K$^+$ in Fig. 2.1). As an example, the presence of 100 mM NaCl in the soil solution results in a nearly threefold drop in Ca^{2+} activity.
2. Na$^+$ may compete directly at uptake sites with many essential cations such as K$^+$, Mg^{2+} or NH$_4^+$ (Fig. 2.1). This may affect both low- (e.g. non-selective cation (NSCC)) and high- (e.g. HKT) affinity transporters.
3. A significant membrane depolarization occurs when positively charged Na$^+$ crosses the plasma membrane. Such depolarization makes passive uptake of many essential cations thermodynamically impossible and, at the same time, dramatically increases efflux of some of them (e.g. K$^+$ leak through depolarization-activated outward-rectifying K$^+$ channels; KOR in Fig. 2.1).
4. Increased *de novo* synthesis of various compatible solutes used for osmoprotection under saline conditions (see Section 2.4.1 for details) severely reduces the available ATP pool, making high-affinity cation uptake even more problematic. This also significantly reduces uptake of anions such as NO$_3^-$ or PO$_4^{-3}$, as their transport is energized by the H$^+$-ATPase activity.

2.2.3 Oxidative stress

Oxidative stress is defined as the toxic effect of chemically reactive oxygen species (ROS) on

biological structures. Both osmotically induced stomatal closure and accumulation of high levels of toxic Na$^+$ species in the cytosol under saline conditions impair photosynthetic machinery and reduce a plant's capacity to fully utilize light that was absorbed by photosynthetic pigments. This leads to formation of ROS in green tissues (see Chapter 3 for more details on ROS classification). The detrimental effects of ROS are a result of their ability to cause lipid peroxidation in cellular membranes, DNA damage, protein denaturation, carbohydrate oxidation, pigment breakdown and an impairment of enzymatic activity (Scandalios, 1993; Noctor and Foyer, 1998). The major sources of ROS production are cell wall peroxidize and amine oxidase, plasma membrane NADPH oxidase, and intracellular oxidases and peroxidizes in mitochondria, chloroplasts and peroxisomes (see Chapter 3).

In general, ROS production under saline conditions is attributed to disruption of two key metabolic processes: photosynthesis and respiration. Plants respond to salinity stress by decreasing stomatal conductance to minimize the water loss. As a result, the amount of absorbed light exceeds the demand for photosynthesis (Ozgur et al., 2013). This affects the rate of electron transport through photosystems and results in an increased ROS (mainly O$_2^{\bullet-}$ and ^{1}O$_2$) production (Allakhverdiev et al., 2002). Mitochondrial respiration is another major source of salt-induced ROS production. Over-reduction of the ubiquinone pool during salt-stress allows the electrons to leak from complexes I and III of the mitochondrial electron transport chain to molecular oxygen resulting in O$_2^{\bullet-}$ production (Noctor et al., 2007; Miller et al., 2010). Naturally salt-tolerant halophyte species increase (by several-fold; Ottow et al., 2005) activity of the alternative oxidase (AOX) to remove electrons from the ubiquinone pool and transfer them to oxygen to form H$_2$O, for ROS detoxification (Millar et al., 2011). Constitutively overexpressing alternative oxidase (*Ataox1a*) in *Arabidopsis* decreased ROS formation and improved growth by 30–40% during salt stress (Smith et al., 2009).

Importantly, ROS production under saline conditions occurs not only in leaf but also in root tissues (Luna et al., 2000; Mittler, 2002; Miller et al., 2008). In *Arabidopsis*, root exposure to 100 mM NaCl resulted in 2.5- to 3-fold increase in hydroxyl radical generation, whereas

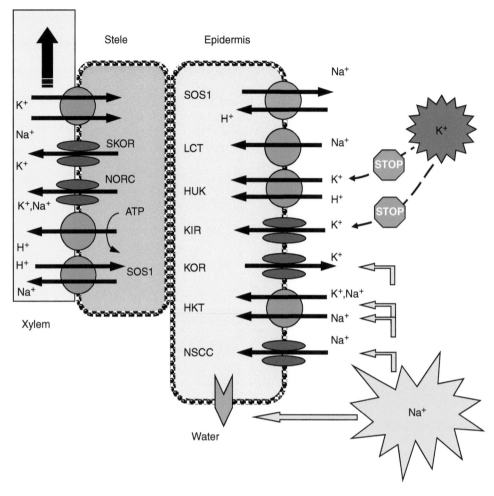

Fig. 2.1. Major pathways of radial Na$^+$ and K$^+$ uptake and xylem loading in plant roots under saline conditions.

250 mM NaCl triggered a 4- to 4.5-fold HO$^\bullet$ increase (Demidchik *et al.*, 2010). ROS accumulation in roots occurs very rapidly (within minutes) and has an immediate and very significant impact on cell metabolism. First, ROS directly activate Ca^{2+}-permeable plasma-membrane channels triggering a rapid Ca^{2+} uptake (Demidchik *et al.*, 2003). The resultant elevation in cytosolic Ca^{2+} activates NADPH oxidase and causes further increase in [Ca^{2+}]$_{cyt}$ via positive feedback mechanisms (Lecourieux *et al.*, 2002) (Fig. 2.2). ROS are also known to activate a certain class of K$^+$-permeable NSCC channels (Demidchik *et al.*, 2003), resulting in a massive K$^+$ leak from the cytosol and a rapid decline in the cytosolic K$^+$ pool (Shabala *et al.*, 2006).

Taken together, sustained elevation in cytosolic Ca^{2+} and depletion of the K$^+$ pool activate caspase-like proteases and trigger programmed cell death (Fig. 2.2; Shabala, 2009). Importantly, salinity-induced programmed cell death (PCD) in plant roots is observed within 1 h after stress onset. This is comparable with the impact of the osmotic component of salinity stress.

2.2.4 Impact on plant growth

Salts in the soil water inhibit plant growth for two reasons:

1. The presence of salt in the soil solution reduces the ability of the plant to take up water,

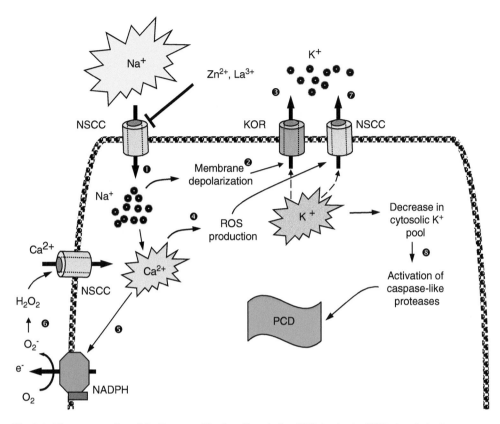

Fig. 2.2. The proposed model of ion-specific signalling during PCD in plants. KOR, depolarization-activated outward-rectifying K⁺ channel; NSCC, non-selective cation channel; PCD, programmed cell death; ROS, reactive oxygen species. Reproduced from Shabala (2009).

and this leads to reductions in the growth rate. This is referred to as the osmotic or water-deficit effect of salinity.

2. If excessive amounts of salt enter the plant in the transpiration stream there will be injury to cells in the transpiring leaves and this may cause further reductions in growth. This is called the salt-specific or ion-excess effect of salinity (Greenway and Munns, 1980). The definition of salt tolerance is usually the per cent biomass production in saline soil relative to plants in non-saline soil, after growth for an extended period of time. For slow-growing, long-lived or uncultivated species it is often difficult to assess the reduction in biomass production, so per cent survival is often used.

The salt in the soil solution (the 'osmotic stress') reduces leaf growth, and to a lesser extent root growth, and decreases stomatal conductance and thereby photosynthesis (Munns, 1993). The cellular and metabolic processes involved are in common to drought-affected plants (Munns, 2002). The rate at which new leaves are produced depends largely on the water potential of the soil solution, in the same way as for a drought-stressed plant. Salts themselves do not build up in the growing tissues at concentrations that inhibit growth: meristematic tissues are fed largely in the phloem from which salt is effectively excluded, and rapidly elongating cells can accommodate the salt that arrives in the xylem within their expanding vacuoles. So, the salt taken up by the plant does not directly inhibit the growth of new leaves.

The salt within the plant enhances the senescence of old leaves. Continued transport of salt into transpiring leaves over a long period of time eventually results in very high Na⁺ and Cl⁻ concentrations, and they die. The rate of leaf death is crucial for the survival of the plant. If new leaves are continually produced at a rate greater than that at which old leaves die, then there are enough photosynthesizing leaves for the plant to produce flowers and seeds, although in reduced numbers. However, if the rate of leaf death exceeds the rate at which new leaves are produced, then the plant may not survive to produce seed. For an annual plant there is a race against time to initiate flowers and form seeds, while the leaf area is still adequate to supply the necessary photosynthate. For perennial species, there is an opportunity to enter a state of dormancy, and thus survive the stress.

The two responses give rise to a two-phase growth response to salinity over time (Munns, 1993; Munns *et al.*, 1995). The first phase of growth reduction is quickly apparent, and is due to the salt outside the roots. Then there is a second phase of growth reduction, which takes time to develop, and results from internal injury. The initial growth reduction is due to the osmotic effect of the salt outside the roots, and what distinguishes a salt-sensitive plant from a more tolerant one is the inability to prevent salt from reaching toxic levels in the transpiring leaves over time.

2.2.5 Disturbance to photosynthesis

The most dramatic and readily measurable whole-plant response to salinity is a decrease in stomatal aperture. Stomatal responses are undoubtedly induced by the osmotic effect of the salt outside the roots. Salinity affects stomatal conductance quickly: firstly and transiently owing to perturbed water relations and shortly afterwards owing to the local synthesis of ABA (Fricke *et al.*, 2004). A short-lived increase in ABA was detected in the photosynthetic tissues within 10 min of the addition of 100 mM NaCl to barley (Fricke *et al.*, 2006); the rapidity of the increase suggesting *in situ* synthesis of ABA rather than transport from the roots. However, a new reduced rate of transpiration stabilized within hours while ABA tissue levels returned to control concentrations (Fricke *et al.*, 2004, 2006). This stomatal response is probably regulated by root signals in common with plants in a drying soil (Davies *et al.*, 2005), as evidenced by stomatal closure in salt-treated plants whose

water status was kept high by applying a balance pressure (Termaat *et al.*, 1985).

Rates of photosynthesis per unit leaf area in salt-treated plants are often unchanged, even though stomatal conductance is reduced (James *et al.*, 2002). This paradox is explained by the changes in cell anatomy described in Section 2.5 that give rise to smaller, thicker leaves and result in a higher chloroplast density per unit leaf area. When photosynthesis is expressed on a unit chlorophyll basis, rather than a leaf area basis, a reduction due to salinity can usually be measured. In any case, the reduction in leaf area due to salinity means that photosynthesis per plant is always reduced.

A large part of plant resistance to salinity, as to drought and other abiotic stresses, is the ability to dissipate excess radiation (Chaves *et al.*, 2009). The effects of water stress on photosynthesis, be it due to low soil moisture or to high soil salinity, are either direct (such as the diffusion limitations through the stomata and the mesophyll, and the alterations in photosynthetic metabolism) or secondary, such as the oxidative stress arising from the superimposition of multiple stresses (Chaves *et al.*, 2009).

At high salinity, salts can build up in leaves to excessive levels. Exactly how they exert their toxicity is not known. They may build up in the apoplast and dehydrate the mesophyll cells, they may build up in the cytoplasm and inhibit enzymes involved in carbohydrate metabolism, or they may build up in the chloroplast and affect photosynthetic processes. It is most likely that the damage is caused by Na^+ rather than by Cl^-, but this is not known for sure (Munns and Tester, 2008). A toxic effect of Na^+ could be directly on the photosystems or on pH homeostasis due to H^+-coupled Na^+ efflux mechanisms.

2.2.6 Metabolic disturbances

Cause–effect relationships between photosynthesis and growth rate can be difficult to untangle. It is always difficult to know whether a reduced rate of photosynthesis is the cause of a growth reduction, or the result. With the onset of salinity stress, a reduced rate of photosynthesis is certainly not the sole cause of a growth reduction because of the rapidity of the change in leaf expansion rates described earlier (Cramer and Bowman, 1991; Passioura and Munns, 2000; Fricke *et al.*, 2004), and also because of the increase in stored carbohydrate, which indicates unused assimilate (Munns *et al.*, 2000). However, with time, feedback inhibition from sink to source may fine-tune the rate of photosynthesis to match the reduced demand arising from growth inhibition (Paul and Foyer, 2001). Reduced leaf expansion resulting in a build-up of unused photosynthate in growing tissues may generate feedback signals to downregulate photosynthesis.

The carbon balance of a plant during a period of water stress, be it drought or salinity, depends on the degree and velocity of photosynthesis decline during water depletion (Chaves *et al.*, 2009). If the stress develops slowly, and photosynthesis continues with little decline, then carbohydrate metabolism is diverted from growth processes to pools of soluble carbohydrates which provide osmotic adjustment, and to storage compounds (starch, fructans) which provide a source of new growth if the stress is relieved by rain or irrigation. However, if the stress is severe and old leaves senesce prematurely, then the carbon balance becomes negative and the plant will die.

Costs of osmotic adjustment also affect the carbon balance of a plant. As explained below (Section 2.4.1), it is not just diversion of photosynthate from growth to soluble carbohydrate pools that is necessary for osmotic adjustment. ATP is needed for the synthesis of specific compatible solutes, and for the uptake and compartmentation of Na^+ and Cl^-. Halophytes and the more salt-tolerant crop species use ions rather than organic compounds for the bulk of their osmotic adjustment but, even so, there are costs (Section 2.4.1). This requires a significant diversion of respiratory energy.

Metabolic disturbances such as those described above are adaptive mechanisms that are beneficial to the plant, at least in the long term, but there can also be disturbances that are costly and potentially fatal. These arise if Na^+ concentrations in metabolic compartments are not controlled, and they rise to toxic levels. The following sections discuss the concept of Na^+ and Cl^- toxicity, and summarize the experimental evidence for optimal concentrations of Na^+ and Cl^- in the various cell compartments, and the likely concentration at which they would be toxic.

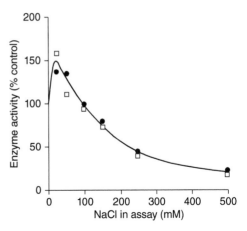

Fig 2.3. Enzymes from halophytes are not intrinsically salt tolerant. NaCl effects on the activity of malate dehydrogenase from saltbush, *Atriplex spongiosa* (closed symbols), and bean, *Phaseolus vulgaris* (open symbols), confirm comparable sensitivity. (Based on Greenway and Osmond, 1972. Figure reproduced, with permission, from Plants in Action: http://plantsinaction.science.uq.edu.au/edition1/?q=content/17-3-1-devices-manage-leaf-salt, accessed 10 July 2016.)

Halophytes avoid Na^+ and Cl^- toxicity by sequestering these ions in vacuoles as the concentration in the leaf rises with time. Enzymes extracted from halophytes are no more tolerant of salt *in vitro* than the corresponding enzymes from non-halophytes, indicating that compartmentation of Na^+ and Cl^- in the vacuole is an essential mechanism of adaptation to high soil salinity (summarized in Flowers *et al.*, 2015). For example, enzymes extracted from the halophytes *Atriplex spongiosa* or *Suaeda maritima* were just as sensitive to NaCl in their assay media as were enzymes extracted from beans or peas (Greenway and Osmond, 1972). This is shown for malate dehydrogenase in Fig. 2.3.

2.2.7 Na⁺ toxicity

Na^+ must be partitioned within cells so that concentrations in the cytoplasm and its component organelles are kept low, and do not rise above a given level when the amount in the cells or tissues rises with time. This critical concentration for the cytosolic component of the cytoplasm is possibly in the order of 10–30 mM (Munns and Tester, 2008), but for some organelles it may be higher.

Na⁺ toxicity is a widespread concept, although rarely defined. Na⁺ is often referred to as a toxic ion, or at least a potentially toxic ion as we know that some C_4 species cannot grow without it, and that many C_3 species benefit from low concentrations of Na⁺ in the soil (reviewed by Kronzucker *et al.*, 2013). Curiously, little is known about exactly why Na⁺ is toxic, and at what concentration it becomes toxic. Cheeseman (2013) has pointed out the complex relations between proteins and electrolyte concentrations in the cytosol, in that protein structure, ionic strength and the stability of protein complexes both in membranes and in the cytosol, depend so much on charge density as well as on individual ion properties.

Exactly where in the cell Na⁺ exerts its apparently toxic effect is not known. It is presumably on some enzyme or metabolic process in the cytosol or the mitochondria. As explained below, chloroplasts are not particularly sensitive to Na⁺. And exactly what is the concentration at which it becomes toxic – what might this threshold be? This question is not easy to answer, particularly as we do not know what is 'normal' or optimal in the cytoplasm and its three main components: the cytosol, chloroplasts and mitochondria. The literature for halophytes relevant to this question was published by Flowers *et al.* (2015), and for non-halophytes is summarized briefly below.

Cytosol

Direct measurement of the Na⁺ concentrations in the cytosolic compartment is difficult as the volume is so small, just a thin layer of 1–2 μM between the cell wall and the organelles. The only *in vivo* study in plants has been on cortical cells of barley roots growing in 200 mM NaCl. Use of cation-specific electrodes found activities of 2–28 mM cytosolic Na⁺ and a surprisingly low K⁺ activity of 40–60 mM (Carden *et al.*, 2003). More information is available for animal cells, where despite being bathed in fluid with about 150 mM Na⁺, cytosolic Na⁺ of heart muscle cells is less than 10 mM in the cytosol (e.g. Bers *et al.*, 2003. A rise to 25 mM signals a heart attack (Murphy and Eisner, 2009).

The other approach has been to measure cells with small vacuoles, found in the meristems of plants, and in microalgae. These data indicate concentrations for Na⁺ in the range of 20–50 mM. A study on the micro-alga *Chlorella* grown at 100 and 335 mM NaCl found only 21 mM Na⁺ (Greenway and Setter, 1979). Analyses of shoot apices and root tips found 20–50 mM Na⁺ in plants exposed to medium to high NaCl concentrations (Munns *et al.*, 1988; Lazof and Läuchli, 1991; Zhong and Läuchli, 1994; Munns and Rawson, 1999). From this we could conclude that normal Na⁺ concentrations in the cytosol are about 10 mM, and that concentrations could be 10–50 mM before toxic effects occur.

In vitro enzyme studies showed that Na⁺ starts to inhibit enzymes at concentrations over 30 mM NaCl. In some studies a stimulation of activity is apparent, with significant inhibition compared to control without NaCl not occurring until 100 mM (see Fig. 2.3 above). However, if compared to the maximum, there is a progressive inhibition of activity from 30 mM NaCl onwards which is consistent with the presumed concentration in the cytosol. Some studies have found that enzymes are active in the range 50–150 mM NaCl, but there are doubts how this relates to *in vivo* activity as the response to NaCl added *in vitro* could be altered by the substrate concentration, the preparative procedures used to extract enzymes from tissues and the protein concentration (summarized in Flowers *et al.*, 2015).

Mitochondria

Ions in plant mitochondria have not been measured, so the data for animals may be relevant. In heart cells of rats, Na⁺ in mitochondria estimated *in vivo* by fluorescence was about 10 mM, half that in the cytosol (Donoso *et al.*, 1992). Na⁺ estimated *in vivo* by a number of techniques, all with their limitations, was considered to be about 6 mM (Bers *et al.*, 2003; Murphy and Eisner, 2009). These authors consider that the mitochondrial concentration is lower than in the cytosol.

Isolated plant mitochondria exposed to a range of NaCl concentrations tolerate much higher Na⁺ concentrations than the 10 mM Na⁺ understood to be present *in vivo*, if plant cells are similar to animal cells in this regard. With mitochondria extracted from a range of cereal and vegetable species, significant inhibition of O_2 uptake occurred only above 100 mM NaCl (Flowers and Hanson, 1969; Flowers, 1974; Campbell et al., 1976). Mitochondria isolated from pea responded similarly to those from the halophyte *Suaeda maritima* (Flowers, 1974). KCl was found to have similar effects to NaCl, suggesting Cl⁻ was inhibitory, as organic osmotica such as sucrose and mannitol caused no inhibition of respiration until much higher concentrations.

Flowers and Hanson (1969) point out that the inhibition of state III respiration (ADP added) varies with the substrate being oxidized. This was investigated further by Jacoby et al. (2016), who showed that respiration could be stimulated by NaCl over the range 50–200 mM when exogenous NADH was provided as substrate and electron flow was coupled to the generation of a proton gradient across the inner membrane. Where the substrate was succinate or malate, or when phosphorylation was uncoupled by inhibitors, there was a progressive inhibition in response to added NaCl, so that respiration was reduced by 50% at 100 mM NaCl. That is, different pathways of electron transport showed divergent responses to NaCl concentrations between 0 mM and 200 mM. All pathways were inhibited by NaCl concentrations above 400 mM (Jacoby et al., 2016).

Chloroplasts

In contrast to mitochondria and the cytosol, Na⁺ and Cl⁻ concentrations seem optimal at around 100 mM. For example, chloroplasts isolated from spinach grown in 200 mM NaCl contained 165 mM Na⁺ and 117 mM Cl⁻, and chloroplasts isolated from control plants grown without added NaCl contained 96 mM Na⁺ and 100 mM Cl⁻ (Robinson et al., 1983). Similar values were found for halophytes grown at high salinity (reviewed by Flowers et al., 2015, see their Table 4). Consistent with this, O_2 evolution in isolated chloroplasts was not inhibited until NaCl concentrations over 100 mM were applied (e.g. Preston and Chritchley, 1986).

cause toxicity in species that have already excluded most of the Na⁺. Species that are considered as 'Cl⁻ sensitive', and in which genetic variation in Cl⁻ uptake correlates with salt tolerance, appear to be those that keep Na⁺ at relatively low concentrations in the leaves by withholding it within older roots, stems and petioles (e.g. soybean, citrus, grapevine). These fall within the relative category of salt sensitive, possibly because of limited ability for osmotic adjustment.

By growing plants in different salt solutions with or without Na⁺ or Cl⁻ respectively, toxicity has been associated with Na⁺ rather than Cl⁻ (e.g. in wheat: Kingsbury and Epstein, 1986). Some species exclude almost all the Na⁺ from the leaf lamina but not Cl⁻, so that Cl⁻ rises in parallel with the external solution, e.g. in beans, *Phaseolus vulgaris* (Seemann and Chritchley, 1985). The reverse never occurs. With Na⁺-excluding species, Cl⁻ may eventually rise to such high concentrations that it is toxic, but this does not mean that Cl⁻ is more toxic than Na⁺. The reason for Cl⁻ toxicity is not known, nor the concentration at which it is toxic, nor even the optimal concentration in the cytoplasmic compartments. Activities of many extracted enzymes show almost identical responses to NaCl and KCl; for example, malate dehydrogenase from bean showed the same response to KCl as to NaCl (shown in Fig. 2.3). At face value, this suggests that Cl⁻ is the toxic component of the salt effect. Where a complex of enzymes such as the ribosomal system that catalyses protein synthesis has been investigated, a toxic effect of Cl⁻ is implied. Translation of mRNA extracted from a range of species on a wheat germ translation system was optimal in 100–200 mM potassium acetate, and substitution of Na⁺ for K⁺ or Cl⁻ for acetate was inhibitory (Gibson et al., 1984). A similar finding was reported by Flowers and Dalmond (1992).

Toxicity associated with high Cl⁻ concentrations in tissues might be due to the costs of maintaining charge regulation, or compartmentation of the Cl⁻ or accompanying anions, instead of a direct effect of the ion on a protein.

2.2.8 Cl⁻ toxicity

The question of Cl⁻ toxicity is often raised. Excessively high concentrations of Cl⁻ in leaves could

2.2.9 Toxicology and tissue tolerance

The concept of toxicology rather than toxicity is relevant, as suggested by Flowers et al. (2015).

In toxicology, the length of exposure of the organism to the toxic compound, as well as any other stresses suffered by the individual, influences the outcome (Miller et al., 2000). This could be because of slow accumulation of inhibitory compounds such as ROS which become toxic at a given concentration, or it could be that the amount of energy required to combat the salt and/or the inhibitory compounds is exhausted and exceeds the demand to maintain ion concentrations in cytoplasmic compartments.

It is possible that injury to the cell and tissue is not due precisely to Na^+ rising above a threshold level of toxicity, but to an energy deficit caused by unsustainable expenditure of ATP to maintain the Na^+ level in the cytosol below the threshold, which appears to be in the order of 10–30 mM. In other words, it is possible that Na^+ does not reach a toxic level before the tissue is injured: it is the energy expended to keep it in the vacuole and out of the cytosol and mitochondria that causes the death of the cell and the tissue. In species that have very low Na^+ in their leaves, it is also the energy expended to divert assimilates from growth processes, and in the synthesis of specific metabolites.

Tissue tolerance can be considered as the ability to tolerate an elevated tissue concentration of Na^+ or Cl^-. By *elevated* we mean higher than 50–100 mM Na^+ and 100–150 mM Cl^-. It would be satisfying to have a more precise definition, such as 'the ability of a cell to tolerate Na^+ above x mM, that is, a concentration that is toxic to the cytosol'. However, as shown above, we do not know what this potentially toxic concentration is. Another definition is a relative concept that has a practical application for pre-breeding within a given species: the ability of one genotype to tolerate leaf Na^+ concentrations above that in another. This would be for a given salinity after a defined period of time, as the degree of underlying osmotic stress and the period of time of exposure are also important.

Much focus remains on Na^+ toxicity, but some workers have highlighted the possibility of high Na^+/K^+ ratio (Greenway and Munns, 1980; Maathuis and Amtmann, 1999) or low K^+ per se (Shabala and Cuin, 2008; Shabala and Pottosin, 2014) as possibly having specific adverse effects on metabolism. K^+-requiring enzymes might be particularly vulnerable. It is well established that plants growing in saline soil have lower K^+ concentrations in tissue, and sudden exposure of leaves or roots to NaCl causes K^+ efflux. Genotypic differences in K^+ retention in some species are linked to genotypic differences in salt tolerance, such as in barley (Wu et al., 2015b).

2.3 Physiological and Genetic Diversity of Plant Responses to Salinity

2.3.1 General classification

Species vary strongly in their capacity to tolerate salinity. Among the major food crops, barley, cotton, sugar beet and canola are the most tolerant; bread wheat is moderately tolerant, while rice and most legume species are sensitive. The three most widely grown cereal crops in the world are wheat, rice and maize. Wheat can be grown in salinities up to 150 mM NaCl (15 dS/m) as long as rainfall or irrigation can rescue the crop at critical stages such as pollen meiosis and fertilization. Rice is much more salt sensitive (Fig. 2.4). Maize falls in between these two species in terms of salt sensitivity. Of the important forage species of the world, lucerne (alfalfa) is by far the most tolerant (Fig. 2.4).

A comprehensive survey of the variation in salt tolerance between crop, pasture forage and horticultural species is given by Maas and Hoffman (1977), now published electronically (USDA-ARS, 2011). This survey presents for each species a threshold salinity below which there is no reduction in yield, and then a linear reduction in yield with increasing salinity (a 'bent stick' relationship). It is a very useful starting point for comparisons in tolerance between species. For example, the yield of rice ('sensitive') starts to decline at an EC_e of 3 dS m^{-1} (30 mM NaCl) compared to 6–8 dS m^{-1} (60–80 mM NaCl) for wheat ('moderately tolerant'), and the subsequent linear yield decline of rice with increasing salinity is double that of wheat.

It should be emphasized that the 'bent stick' relationships are for yield and not total biomass. The relationship between soil salinity and total biomass is linear for most cultivated species. Low salinity may not reduce grain yield even though the leaf area is reduced, as there is compensation for reduced leaf area during grain filling by remobilization of stored carbohydrate reserves.

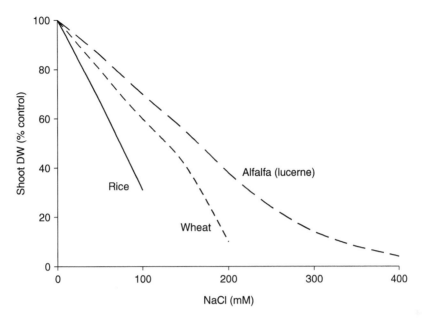

Fig. 2.4. Growth response to salinity of the world's three most important crops. Rice (*Oryza sativa*) and wheat (*Triticum aestivum*) are the world's staple food crops, and lucerne (*Medicago sativa*, alfalfa) is the most important forage species. DW, dry weight. Reproduced from Munns (2007).

Significant diversity exists between and within cultivated species, and there is potential for increasing yield in saline soil. This requires new germplasm, and more efficient techniques for identifying important genes and their performance in the field.

2.3.2 Genetic variability within halophytes

Halophytes are the native flora of saline soils, which are found in environments ranging from estuaries to remnant salt lakes in the arid deserts. Vascular halophytes are distributed among many families of flowering plants, and include some grasses, shrubs and trees (Flowers and Colmer, 2008). Well-known halophytes of the Australasian region are saltbushes (*Atriplex* species) and mangroves (diverse genera).

The term 'halophyte', meaning a salt-loving plant, reflects the observation that many plants native to saline areas require some salt to grow well: some need 10–50 mM NaCl to reach maximum growth, and a few halophytes, such as *Atriplex nummularia* (old man saltbush) grow

best at around 200 mM NaCl (Fig. 2.5). Many halophytes can grow at salinities near those of seawater or even higher.

Because NaCl is the most soluble and widespread salt, it is not surprising that all plants have evolved mechanisms to regulate its accumulation and to select against it in favour of other nutrients commonly present in low concentrations, such as K^+ and NO_3^-. In most plants, Na^+ and Cl^- are effectively excluded by roots while water is taken up from the soil (Munns, 2005). Halophytes are able to maintain this exclusion at higher salinities than their antithesis, glycophytes. For example, sea barleygrass, *Hordeum marinum*, excludes both Na^+ and Cl^- until at least 450 mM NaCl. It is also not surprising that because salinity is a common feature of arid and semi-arid lands, plants have evolved mechanisms to tolerate the low soil water potential caused by salinity, as well as by drought, and so tolerance to osmotic stress is a feature of most glycophytes and halophytes.

Curiously, the only species to have a substantial increase in growth with increasing salinity are dicotyledonous; monocotyledonous species have little or no positive growth response

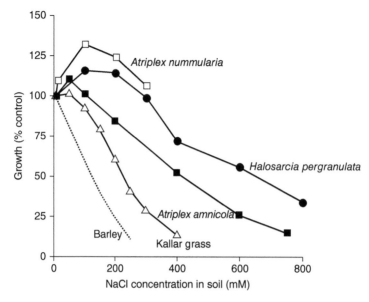

Fig. 2.5. Growth response of four halophytes compared to one of the most salt-tolerant glycophytes, barley (*Hordeum vulgare*). The halophytes shown are Kallar grass (*Leptochloa fusca* or *Diplachne fusca*), swamp saltbush (*Atriplex amnicola*), old man saltbush (*A. nummularia*) and black-seeded samphire (*Halosarcia pergranulata*). (Figure reproduced, with permission, from Chapter 17 in Plants in Action: http://plantsinaction.science.uq.edu.au/edition1/?q=content/17-3-halophytes-and-adaptation-salt, accessed 10 July 2016.)

to increasing salinity. All the same, most halophytes accumulate high concentrations of NaCl in their tissues, which in itself contributes substantially to the dry weight of the plant. For example, salts contributed about 10% of the dry weight of *A. nummularia* grown near its optimum NaCl concentration of 200 mM. The salt concentration in the leaf tissues of halophytes can be greater than 500 mM, which is much higher than the maximum concentration found in glycophyte species. Analysis of leaf sap from some halophytes shows that NaCl can be up to 1 M NaCl. As enzymes cannot function in such a high NaCl concentration, there must be special cellular features that allow metabolism to continue.

2.3.3 Genetic variability within glycophytes

Wheat (*Triticum aestivum*) is the world's most widely grown crop. It is moderately salt tolerant and drought tolerant and, with barley, is the preferred cereal in most arid and semi-arid

agricultural regions. Much of the world's wheat is produced under irrigation; however, the area sown to rain-fed wheat is very substantial and is expected to grow further as water for irrigation declines globally. As wheat is grown in many arid or semi-arid regions of the world, it is likely to encounter salinity, whether caused by irrigation, land clearing or natural processes (Munns, 2007).

Bread wheat is moderately salt-tolerant, compared to other cereals. In a field where the salinity rises to 10 dS m^{-1} (about 100 mM NaCl), rice (*Oryza sativa*) will die before maturity, while bread wheat will yield, although at a reduced rate. Even barley (*Hordeum vulgare*), the most tolerant cereal, dies after extended periods at salt concentrations higher than 250 mM NaCl (equivalent to 50% seawater). Some selected genotypes, however, are capable of handling much higher salinity levels, and some salt-tolerant barley varieties yielded as much as 50% of control when grown at 320 mM NaCl (Chen *et al.*, 2007). Durum wheat (*Triticum turgidum* ssp. *durum*) is less salt tolerant than bread wheat (Munns *et al.*, 2006). Genetic variation for ability to control

Na+ transport to leaves exists within wild wheat species (Colmer *et al.*, 2006). Two Na+ transporters originating from a wheat relative, *Nax1* and *Nax2*, are being used to increase the salt tolerance of durum wheat (Munns, 2007).

Global rice production is close to that of wheat. It is the staple food for the largest number of people on Earth. It is particularly salt sensitive, the grain yield of many varieties being reduced by half at an EC_e of only 6 dS m^{-1} (Maas and Hoffman, 1977). Rice has evolved in freshwater marshes, and is adapted to waterlogged but not saline conditions, so salinity is a major constraint to rice production in deltas with seawater intrusions. Salt-tolerant cultivars include the traditional tall *indica* landraces Pokkali and Nona Bokra, which come from estuarine areas in India (Munns, 2007). Genes from these and other rice genotypes are being used to breed more salt-tolerant rice cultivars (Zeng *et al.*, 2004).

2.3.4 Correlation between salinity and drought tolerance

Mechanisms controlling leaf and root growth are in common to drought and salinity; they are due to factors associated with water stress. This is supported by the evidence that Na+ and Cl− are below toxic concentrations in the growing cells themselves, in leaves (Hu and Schmidhalter, 1998; Fricke, 2004) and roots (Jeschke, 1984; Jeschke *et al.*, 1986).

In cereals, the major effect of salinity on total leaf area is a reduction in the number of tillers; in dicotyledonous species, the major effect is the dramatic curtailing of the size of individual leaves or the numbers of branches. Curiously, shoot growth is more sensitive than root growth, a phenomenon that also occurs in drying soils and for which there is as yet no mechanistic explanation (see the following section). The benefit is that a reduction in leaf area development relative to root growth would decrease the water use by the plant, thus allowing it to conserve soil moisture and prevent an escalation in the salt concentration in the soil.

When roots sense a soil water deficit, root cells change in growth rate and differentiation, and the root system architecture changes in the degree of branching or rate of branch root elongation. Although root elongation rate may decrease, the decrease is less than of shoot growth and, in some cases, when the water deficit is relatively mild, the root mass can increase. In *Arabidopsis*, the proportion of roots as a fraction of whole-plant biomass increased in drying soil by 30% (Hummel *et al.*, 2010). In grapevines in the field, roots in drying soil continue to grow into deeper wetter layers, whereas the roots of irrigated plants proliferate in the topsoil (Lovisolo *et al.*, 2010). Roots in dry soil also transmit a signal to the shoot which may be abscisic acid (ABA) or a more complex suite of hormones (Davies *et al.*, 2005; Lovisolo *et al.*, 2010).

When roots sense a saline soil, the responses are similar. ABA is not the only signal, at least in species other than grapevine, as cytokinins, auxins and the ethylene precursor ACC are also implicated (Pérez-Alfocea *et al.*, 2010). Hormonal signals may coordinate assimilate production and use in the competing sinks of roots, leaves and lateral shoots. The hormonal regulation of source–sink relations during the osmotic phase of salinity, the phase when growth rate and development is reduced and before ions build up to toxic levels in leaves, affects the whole-plant energy balance and is critical to delaying the accumulation of ions to toxic levels (Pérez-Alfocea *et al.*, 2010). Productivity in saline as well as in dry soil depends on maintaining a high growth rate of young leaves, while at the same time delaying the senescence of older leaves.

Hormonal control of cell division and differentiation is clear from the appearance of leaves, which are smaller in area but often thicker, indicating that cell size and shape has changed (James *et al.*, 2002). Leaves from salt-treated plants have a higher weight/area ratio, which means that their transpiration efficiency is higher (more carbon fixed per water lost), a feature that is common in plants adapted to dry and to saline soil.

2.4 Key Physiological Mechanisms Conferring Salinity Tolerance in Plants

2.4.1 Osmotic adjustment

As mentioned in the earlier sections, salinity inhibits plant growth by imposing osmotic stress and decreasing cell turgor. To maintain the normal growth, plants must therefore readjust to increased external osmolality. This can be done by accumulating a variety of molecules in cytoplasm to counteract the external osmotic pressure. Three major avenues are available for organisms (Shabala and Shabala, 2011): (i) plants can accumulate a range of organic osmolytes (so-called *compatible solutes*) by increasing their uptake from external media; (ii) osmotic adjustment can be achieved by *de novo* synthesis of compatible solutes; and (iii) plants can rely on inorganic rather than organic osmolytes and increase accumulation of Na^+, Cl^- and K^+ for osmotic adjustment purposes. Each of these options has its own advantages and disadvantages.

Uptake of organic osmolytes is the most preferred option. Four major classes of osmolytes are usually distinguished (Delauney and Verma, 1993): sugars, polyoles, amino acids and quaternary ammonium compounds. All these are small water-soluble molecules that may be accumulated in cells at high concentrations without affecting metabolic reactions inside the cell. According to the classical view, accumulation of these non-toxic (thus *compatible*) osmotically active solutes will result in an increase in cellular osmolarity leading to the influx of water into, or at least reduced efflux from cells, thus providing the turgor necessary for cell expansion (Shabala and Shabala, 2011). However, in most cases the concentration of these organic osmolytes in soils is extremely low and, thus, this option is simply not practically viable.

Instead of relying on a limited pool of organic osmolytes in the soil, most plants increase their *de novo* production. As a result, the amounts of organic osmolytes may increase manyfold when plants are confronted by saline media (Bohnert *et al.*, 1995; Sakamoto and Murata, 2000). Osmolyte accumulation has long been emphasized as a selection criterion in traditional crop breeding programmes (Ludlow and Muchow, 1990; Zhang *et al.*, 1999), and numerous genetic engineering attempts have been made to manipulate the biosynthetic pathways of compatible solutes in order to enhance osmotic stress tolerance (reviewed in Bohnert *et al.*, 1995; Bajaj *et al.*, 1999; Bohnert and Shen, 1999; Shabala and Shabala, 2011).

There are, however, several major hurdles that make this strategy not very competitive:

(i) *de novo* synthesis of compatible solutes comes at extremely high cost. Typically, between 30 and 80 moles of ATP must be spent to synthesize 1 mole of the different compatible solutes (Raven, 1985; Oren, 1999). Such high carbon cost results in substantial yield penalties, and genotypes of high osmotic adjustment capacity under saline conditions usually have a relatively low yield potential under non-stressed conditions (Muñoz-Mayor *et al.*, 2008). The trade-off is unavoidable and explains the failure of most transgenic plants with increased osmolyte production to perform well under field conditions (Bajaj *et al.*, 1999; Bohnert and Shen, 1999); (ii) the biosynthesis of compatible solutes is a rather slow process, operating in a timescale of hours and days. At the same time, many organisms may experience much faster fluctuations in media osmolality and, hence, require much quicker ways for osmotic adjustment; and (iii) the concentrations of organic osmolytes are very low (even being increased manyfold under saline conditions), and usually do not exceed several mM (Igarashi *et al.*, 1997; Shen *et al.*, 1997), which is far too low to be accountable for osmotic adjustment of whole cells or tissues. Therefore, it is currently accepted that *de novo* synthesis of organic osmolytes may be involved in osmotic adjustment only in small cell compartments such as cytosol or chloroplasts, but cannot generate cell turgor required to maintain tissue growth under saline conditions.

A viable alternative to the energetically expensive and rather slow process of osmolyte biosynthesis is osmotic adjustment by means of inorganic ions. Direct experiments using the single-cell pressure probe technique have shown that 90% of the turgor recovery in *Arabidopsis* root epidermis was achieved within 40 min by mean of increased uptake of K^+, Na^+ and Cl^- ions (Shabala and Lew, 2002). Indeed, water retention within the cell may be achieved equally well by increased concentration of both organic and inorganic molecules. However, accumulation of these ions should not interfere significantly with cell metabolism. K^+ seems to be well suited for this role; however, under stress conditions, the electrochemical gradients favour potassium loss but not uptake. Two other ions, Na^+ and Cl^-, are very abundant in the external media under saline conditions and can therefore be used as cheap osmotica to maintain normal cell turgor.

Unfortunately, they are both toxic at high concentrations and can cause severe disruptions to cell metabolism. With a few possible exceptions for extremophilic bacteria (Oren, 1999), enzymatic activity appears to be substantially inhibited by elevated cytosolic Na^+ levels, regardless whether in halophyte or glycophyte species. Thus, efficient sequestration in the vacuole is absolutely essential. Even so, pumping one mole of Na^+ against the electrochemical gradient into the vacuole takes only 3.5 mol of ATP which is only one-tenth of the amount required to produce one mole of organic osmolyte (Shabala and Shabala, 2011). Thus, the energetic benefits are obvious.

2.4.2 Sodium exclusion from uptake by roots

Several pathways for Na^+ uptake across the plasma membrane have been identified using electrophysiological (patch-clamp technique) and molecular genetics approaches. These include (Fig. 2.1):

1. NSCC (Amtmann and Sanders, 1999; Tyerman and Skerrett, 1999; Demidchik et al., 2002). These may be either voltage independent (so-called VIC channels), or weakly voltage-dependent (Demidchik et al., 2002). The NSCC channels also differ dramatically in their sensitivity to extracellular Ca^{2+} (Davenport and Tester, 2000). NSCC are believed to be the major route for Na^+ uptake into the root, and are gated by numerous factors such as cytosolic or external Ca^{2+} and pH levels, cyclic nucleotides, ATP, glutamate, G-proteins, ROS and mechanical tension on the membrane (Demidchik and Maathuis, 2007). The precise molecular identity of NSCC channels remains obscure.
2. The high-affinity potassium transporter, HKT1 (Maathuis and Amtmann, 1999; Laurie et al., 2002). The HKT family can be divided into two distinct subfamilies based on their transport selectivity. Transporters in subfamily one are highly selective to Na^+ over K^+, while those in subfamily two are selective to K^+ or transport both ions (Platten et al., 2006). Comparing alignments of selected Trk/Ktr/HKT protein sequences from bacteria and plants with that in *Dionaea muscipula* (Venus flytrap) gland cells (used to take up Na^+ from digested sodium-rich

insects), the highly selective Na^+ channel was identified that belongs to the HKT subclass 1, with closest homology to HKT1.1 and HKT1.2 from eucalyptus (Bohm et al., 2016a). When the corresponding flytrap gene was expressed in *Xenopus* oocytes, the recorded Na^+ dependent currents carried the hallmark features of a sodium channel. The essential feature for this strict Na^+ selectivity was the presence of the serine residue in their first pore loop. In contrast, K^+-permeable Trk/Ktr/HKTs, such as the wheat, barley, and rice orthologues belonging to subfamily 2, frequently have a glycine in the respective position (Bohm et al., 2016b).

HKT-type transporters, many of which seem to be sodium-specific, may be regulators of K^+ homeostasis in the presence of Na^+ (Rus et al., 2001).
3. The low-affinity cation transporter, LCT1 (Amtmann et al., 2001). However, so far this transporter has been found exclusively in wheat species and, thus, is unlikely represent a major pathway for Na^+ entry into the root.
4. A bypass flow, resulting from Na^+ 'leakage' into the root via the apoplast. This uptake usually takes place in the mature root zone, where the integrity of the Caparian strip in the root endodermis is ruptured by protruding lateral roots. The extent of the contribution of this apoplastic flow to total Na^+ influx into the plant varies dramatically between species. It is believed, for example, that a bypass flow represents the major pathway of Na^+ uptake in rice (Yadav et al., 1996; Yeo et al., 1999) but is about tenfold smaller in wheat (Garcia et al., 1997). There also appears to be not much control over apoplastic Na^+ uptake except the factors affecting transpiration, such as stomatal closure with the onset of darkness or extreme water deficit.

Although the relative contribution of each of these pathways varies with species and growth conditions, the consensus is that the unidirectional influx of Na^+ in most glycophytes is thermodynamically passive and appears to be quite poorly controlled (Tester and Davenport, 2003). There appears also to be little (if any) difference in unidirectional Na^+ uptake among genotypes contrasting in their salinity tolerance (Davenport et al., 1997; Chen et al., 2007; Cuin et al., 2011). Also, the amount of Na^+ entering the root is much bigger than the amount of Na^+ eventually accumulated in plant tissues. Both these factors imply that a major bulk (up to 95–97%; Munns, 2002) of the Na^+ entering the root is extruded back into the rhizosphere.

Thermodynamically, Na^+ extrusion from the cytosol to the external medium under saline conditions is an active, energy-consuming process. In animals and microorganisms, such extrusion is mediated by specialized Na^+-pumps, energized directly by the hydrolysis of ATP (Gimmler, 2000). Such Na^+ pumps, however, have been not found in higher plants (Garciade-blás et al., 2001), and active Na^+ extrusion from plant roots is believed to be mediated by plasma membrane Na^+/H^+ exchangers (Fig. 2.1) fuelled by the existence of sharp H^+ gradients at both sides of the plasma membrane (Zhu, 2003; Apse and Blumwald, 2007). In Arabidopsis, a Na^+/H^+ antiporter function has been attributed to the SOS1 gene (Shi et al., 2000; Qiu et al., 2003), and experimental evidence for the presence of SOS1-homologues has been shown for other species, both glycophytes (Mullan et al., 2007) and halophytes (Chen et al., 2010). Overexpression of SOS1 has been found to reduce Na^+ accumulation and improve salinity tolerance in transgenic Arabidopsis (Shi et al., 2003), and comparative transcript analyses revealed higher levels of basal as well as salt-induced SOS1 expression in Thellungiella (a salt-tolerant relative) compared with Arabidopsis species (Oh et al., 2010). Wheat varieties with superior salinity tolerance also had functionally higher SOS1-like Na^+/H^+ exchanger activity (Cuin et al., 2011). All these observations point out the most essential role of plasma membrane Na^+/H^+ exchangers as components of plant salinity-tolerance mechanisms.

2.4.3 Intracellular Na^+ sequestration

As commented in Section 2.4.1, efficient Na^+ sequestration in the vacuole is energetically the most favourable and efficient way to achieve osmotic adjustment under saline conditions. There appears to be little if any difference in Na^+ sensitivity of major cytosolic enzymes between halophyte and glycophyte species (Flowers and Colmer, 2008), and it is absolutely essential that plants maintain their cytosol Na^+ content at non-toxic levels.

The compartmentation of Na^+ into the vacuoles provides an efficient mechanism to achieve this aim.

The transport of Na^+ into the vacuoles is mediated by a tonoplast Na^+/H^+ antiporter that is driven by the electrochemical gradient of protons generated by the vacuolar H^+-translocating enzymes, the H^+-ATPase and the H^+-PPiase (Blumwald, 2000; Zhang and Blumwald, 2001). Such exchangers are encoded by NHX genes and were found to be present and operate in both root and leaf cells. Arabidopsis plants bearing a knockout in gene NHX1 show impaired leaf expansion (Apse et al., 2003), and expression of AtNHX1 in an nhx1 yeast mutant suppressed its NaCl sensitivity and enhanced intracellular compartmentalization of Na^+ (Gaxiola et al., 1999; Quintero et al., 2000). The Arabidopsis AtNHX1 protein was the first Na^+/H^+ exchanger identified in plants (Gaxiola et al., 1999). Since then, the number of homologous NHX transporters identified has grown dramatically, and DNA sequences encoding NHX proteins have been identified from >60 plant species (Pardo et al., 2006). In total, six isoforms of AtNHX are known in Arabidopsis. AtNHX3 and AtNHX4 transcript are found exclusively in flowers and roots (Yokoi et al., 2002; Aharon et al., 2003). AtNHX5 and AtNHX6 are believed to be localized in endosomes and are involved in growth and development, salinity tolerance and vesicle trafficking (Bassil et al., 2011a; Qiu, 2012). Recently, NHX1 and NHX2 were found to transport K^+ into the vacuole giving them a new role in K^+ homeostasis in Arabidopsis (Bassil et al., 2011b; Barragán et al., 2012).

Tonoplast antiporters are constitutively expressed in halophytes (Barkla et al., 1995; Glenn et al., 1999), whereas they must be activated by NaCl in salt-tolerant glycophytes (e.g. Garbarino and Dupont, 1988), while in salt-sensitive plants their expression levels are extremely low and not salt-inducible (Apse et al., 1999; Zhang and Blumwald, 2001). Overexpression of NHX antiporters has been used to improve salt tolerance in several plant species (Apse et al., 1999; Zhang and Blumwald, 2001; Zhang et al., 2001; Li et al., 2007), with all transformed plants showing improved plant survival and increased shoot growth over the control lines under salt stress. Interestingly, it is osmotic rather than ion-specific stress that activates NHX activity at the vacuole (Pardo et al., 2006). A possible role of ABA, a drought- and osmotic stress-related hormone, was suggested (Quintero *et al.*, 2000), along with the transient shifts of intracellular and apoplastic pH (Pardo *et al.*, 2006).

The activity of the vacuolar primary H^+-pumps, that provide the driving force for the operation of the vacuolar Na^+/H^+ antiporters, is regulated by sodium, and increases in tonoplast H^+-ATPase activity in response to NaCl have been reported for many plant species (Shabala and Mackay, 2011). In contrast, vacuolar H^+-PPiase appears to be inhibited by increased NaCl concentrations (Golldack and Dietz, 2001), presumably as a result of competition of Na^+ for K^+ binding sites in the enzyme (Rea and Poole, 1993). Thus, overexpressing H^+-PP_iase activity resulted in improved salinity and osmotic stress tolerance in several species (Bao *et al.*, 2009; Liu *et al.*, 2010). However, more recent studies (Bose *et al.*, 2015) have revealed that salinity tolerance in the halophyte species is not related to the constitutively higher AHA transcript levels in the root epidermis, but to a plant's ability to rapidly up-regulate the PM H^+-ATPase upon salinity treatment.

Efficient Na^+ sequestration relies not only on transport across the tonoplast, but also on retention of ions within vacuoles. Indeed, given the four- to fivefold concentration gradient between the vacuole and the cytosol, Na^+ may easily leak back, unless some efficient mechanisms are in place to prevent this process (Shabala and Mackay, 2011). Thus, tonoplast cation channels, through which Na^+ may potentially leak back to the cytoplasm, should be tightly controlled. This has been shown recently by direct patch-clamp experiments on slow (SV) and fast (FV) channels in vacuoles from *Chenopodium quinoa* mesophyll cells (Bonales-Alatorre *et al.*, 2013). It was shown that old leaves that possess only a few salt bladders and rely almost exclusively on vacuolar Na^+ sequestration had intrinsically much lower density of the FV current compared with young leaves that rely heavily on salt sequestration in epidermal bladder cells. This density has further decreased about twofold in plants grown under high salinity. Also much smaller (up to sevenfold) was SV channel density in old leaves. These results provided strong supporting evidence that negative control of SV and FV tonoplast channels activity in old leaves reduces Na^+ leak thus enabling efficient sequestration of Na^+ in their vacuoles, enabling optimal photosynthetic performance and conferring salinity tolerance. A follow-up study (Pottosin *et al.*, 2014) revealed that the above down-regulation of vacuolar SV currents in quinoa mesophyll cell vacuoles is a result of the blocking effect of choline. Choline is one of the known 'compatible solutes', and onset of salt-stress rapidly (in minutes) induces several-fold increase in the levels of choline in leaf tissues (Summers and Weretilnyk, 1993).

2.4.4 Potassium retention in the cytosol

With Na^+ toxicity occurring as a result of its competition with K^+ for enzyme activation and protein biosynthesis (see previous section), it is likely that it is not the absolute quantity of Na^+ per se, but rather the cytosolic K^+/Na^+ ratio that determines cell metabolic competence and, ultimately, the ability of a plant to survive in saline environments (Shabala and Cuin, 2008). Thus, plant salinity tolerance may be achieved not only by cytosolic Na^+ exclusion but also by efficient cytosolic K^+ retention. A strong positive correlation between shoot K^+ concentration and a genotype's salinity tolerance was reported for a wide range of plant species (Cuin *et al.*, 2003, 2010; Chen *et al.*, 2005, 2007; Colmer *et al.*, 2006). It may be at least partly due to the difference in salt sensitivity between bread and durum wheat being in enhanced K^+/Na^+ discrimination, a feature controlled by a *Kna1* locus on chromosome 4D (Gorham *et al.*, 1991; Dvořák *et al.*, 1994; Dubcovsky *et al.*, 1996). While *Arabidopsis* plants showed a progressive decline in leaf K^+ content with increasing salinity, their salt-tolerant relative *Thellungiella halophila* was capable of even increasing mesophyll K^+ content under saline conditions (Volkov *et al.*, 2004).

Potassium uptake and retention in plant cells is mediated by a large number of various transporters (75 transport genes belonging to seven different families in *Arabidopsis*; Véry and Sentenac, 2002; Shabala, 2003). These transporters display high specificity of expression, both at the tissue level and as a function of the stress per se. However, it appears that under saline conditions only two of these, namely depolarization-activated outward-rectifying K^+ (GORK) channel, and NSCC channel, play the major role in maintaining the optimal cytosolic K^+ homeostasis and controlling salinity-induced K^+ leak (Fig. 2.2).

When plants are grown under normal conditions, most of the K^+ uptake is mediated by inward-rectifying K^+-selective (abbreviated KIR) channels (such as AKT or KAT in *Arabidopsis*). Under saline conditions, however, the root plasma membrane is strongly depolarized as a consequence of a massive influx of positively charged Na^+ ions (Fig. 2.2). The observed shift in membrane potential is as big as 50–70 mV, depending on the species and the actual salt concentration. This makes K^+ uptake through KIR channels thermodynamically impossible, and plants should rely exclusively on K^+ uptake via high-affinity transport systems. Such uptake is much less efficient; also, in plants grown under 'luxury' K^+ supply (i.e. using K^+ fertilizers), the expression levels of these high-affinity transport systems are rather low, and it takes a long time to get them expressed to a level sufficient to meet plant demands for reduced K^+ availability. Even more importantly, the observed depolarization not only makes K^+ uptake more problematic but also causes a massive K^+ efflux through GORK channels. Taken together, these two factors result in a massive depletion in the cytosolic K^+ pool, with high metabolic cost to plants (discussed below).

Another factor contributing to massive K^+ leak from the cytosol under saline condition may be ROS production. Several types of K^+-permeable channels, including NSCC (Demidchik et al., 2003) and GORK (Demidchik et al., 2010) are activated by ROS, providing an additional pathway (independent of membrane depolarization) for K^+ leak from the cell (Fig. 2.2). The physiological consequences of such a leak are at least threefold:

1. To enable normal metabolic activity in the cytosol, plants try to replenish the cytosolic K^+ pool by drawing available K^+ from vacuoles (Cuin et al., 2003). This results in the loss of the turgor and immediate growth penalties. If the process takes too long, the vacuolar K^+ pool is also depleted, and the cell collapses.
2. To prevent a depolarization-induced K^+ leak, plants have to restore (otherwise depolarized) membrane potential by more active H^+ pumping. Indeed, a strong positive correlation exists between a plant's ability to maintain negative membrane potential in root cells, and differential salt tolerance, at least in some species (Chen et al., 2007). Moreover, salt-tolerant genotypes appear to have an intrinsically higher (two- to threefold) rate of H^+-pumping, even under control conditions. However, the latter comes at high metabolic (e.g. ATP) cost. Thus, this strategy may be very efficient under saline conditions but will result in yield penalties under control (non-saline) conditions.
3. Depletion of the cytosolic K^+ pool may also result in activation of caspase-like proteins and trigger PCD, a cell's 'suicide programme' (Fig. 2.2). It is assumed that PCD helps plants to survive under stress (Apel and Hirt, 2004) by providing a mechanical 'shield' from stress factors and signal stress to surviving cells. However, if this process is extended to a large number of cells, the root's ability to take up nutrients and water will be severely compromised, impacting shoot growth and development. Thus, such strategy may have a physiological advantage only if used in the very short term.

While most of the earlier studies were focused on the essential nature of K^+ retention in roots (e.g. Chen et al., 2005, 2007; Cuin et al., 2008, 2011; Smethurst et al., 2008), the role of K^+ retention in the leaf mesophyll remained essentially unexplored. Recently we have shown (Wu et al., 2015b) that the ability of leaf mesophyll to retain K^+ represents an important and essentially overlooked component of a salinity-tolerance mechanism, at least in barley. The strong positive correlation between mesophyll K^+ retention ability under saline conditions (quantified by the magnitude of NaCl-induced K^+ efflux from mesophyll) and the overall salinity tolerance (relative fresh weight and/or survival or damage under salinity stress) was found. Contrary to previous reports for barley roots (Chen et al., 2007), K^+ retention in mesophyll was not associated with an increased H^+-pumping in tolerant varieties but instead correlated negatively with this trait. These findings are explained by the fact that increased H^+ extrusion may be needed to charge balance the activity and provide the driving force for the high-affinity HAK/KUP K+ transporters required to restore cytosolic K^+ homeostasis in salt-sensitive genotypes (Wu et al., 2015b).

2.4.5 Tissue-specific Na⁺ sequestration

Salt concentrations are always lowest in the youngest tissue, especially when it is still growing and cells are enlarging in size. This is because growing tissues are fed by the phloem, which has very low salt concentrations.

The leaf blades are the most vulnerable tissue in a plant growing in saline soil, as the transpiration stream ends there, and with limited capacity to export the salt the leaf blades may accumulate salt to potentially toxic levels. Many species can retrieve Na⁺ from the xylem as it flows through a leaf base towards the leaf tip or through an elongated stem towards the shoot apex. In dicotyledonous species, Na⁺ can be retrieved and stored in leaf petioles; and in monocotyledonous species, Na⁺ can be retrieved and stored in leaf sheaths (Wolf *et al.*, 1990; Davenport *et al.*, 2005). When comparing fully grown leaves on a plant, salt concentrations are lower in the younger than in the older leaves, because they have been transpiring for shorter periods of time, and in most species are lower in leaf blades than in leaf bases, that is, petioles (dicotyledonous species) or sheaths (monocotyledonous species), because of retrieval of salts from the xylem as the transpiration stream flows through the leaf bases.

When the stem has elongated, Na⁺ can be retrieved from the xylem flowing to the leaves, and stem internodes become a substantial sump for ions in the transpiration stream (Wolf *et al.*, 1991). In barley plants at the reproductive stage, concentrations of Na⁺ and Cl⁻ in the stem internodes increased over time and significantly reduced the amount of Na⁺ flowing into the leaf above them (Fig. 2.6).

An interesting observation was made recently while studying cell- and organelle-specific distribution of Na⁺ in roots of several wheat varieties contrasting in salinity stress tolerance (Wu *et al.*, 2015a), using fluorescence CoroNa green dye. It was shown that salinity stress tolerance correlated positively with vacuolar Na⁺ sequestration ability in the mature root zone but not in the root apex. Moreover, contrary to all expectations, cytosolic Na⁺ levels in root meristem were significantly higher in the salt-tolerant than in the sensitive group. It was suggested (Wu *et al.*, 2015a) that meristem cells may play a role of the 'salt sensor', triggering

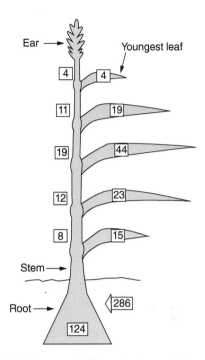

Fig. 2.6. Increases in Na⁺ content over a 7-day period in different parts of a barley plant grown in 100 mM NaCl. The numbers are μmol Na⁺. The total amount of Na⁺ taken up by the plant over 7 days was 286 μmol, the net amount deposited in the roots was 124 μmol (43%), and the remaining amount was deposited in the stem and leaves as shown. (Based on figure in Wolf *et al.*, 1991. Reproduced, with permission, from Plants in Action: http://plantsinaction.science.uq.edu.au/edition1/?q=content/17-2-1-annual-plants, accessed 10 July 2016.)

the cascade of signalling pathways and leading to adaptive plant responses, both at transcriptional and functional levels.

2.4.6 Control of xylem loading

Among the multiple physiological mechanisms contributing to plant salinity tolerance, reducing Na⁺ loading into the xylem is often named as one of the most crucial features (Tester and Davenport, 2003; Munns and Tester, 2008). This can be achieved by either minimization of Na⁺ entry to the xylem from the root symplast, or by maximization of retrieval back out from the xylem before it reaches sensitive tissues in the shoot. The underlying transport mechanisms involved in Na$^+$ loading into the xylem remain highly controversial (Shabala, 2007). At least two possible mechanisms have been suggested (Fig. 2.1): (i) passive loading mediated by Na$^+$-permeable ion channels located at the xylem–parenchyma interface; and (ii) active Na$^+$ loading mediated by SOS1 Na$^+$/H$^+$ exchangers (Wegner and Raschke, 1994; Wegner and De Boer, 1997; Köhler and Raschke, 2000; De Boer and Volkov, 2003; Munns and Tester, 2008). The biggest element of uncertainty is what are the actual concentrations of Na$^+$ in the xylem sap itself and in the parenchyma's cell cytosol, and what is the membrane potential of the xylem parenchyma.

The actual Na$^+$ concentration in the xylem depends on the species, as well as on the severity of the salt stress, and the transpiration rate at the time of collection. However, there is a general consensus that Na$^+$ xylem content is typically between 10 to 30 mM (Flowers and Colmer, 2008; Shabala et al., 2010; Shabala and Mackay, 2011). Getting the membrane potential values for xylem parenchyma cells is a more challenging task, but both conventional measurements using isolated root stele (Shabala et al., 2010; Wegner et al., 2011) and in planta measurements using a multifunctional xylem pressure probe (Wegner et al., 2011) suggest that xylem parenchyma cells are usually slightly hyperpolarized, compared with root epidermis, and have membrane potential (MP) values of around −120 mV to −140 mV (data for maize and barley). Once plants are hit by the salt stress, a sequential depolarization of all cells along the radial root axis occurs, and parenchyma cells are depolarized by 40–60 mV, under typical experimental conditions (Wegner et al., 2011).

In the light of the above, one can assume that a 'typical' glycophyte species has about 20 mM Na$^+$ in the xylem sap at low transpiration rates as would occur at high salinity, and a parenchyma cell MP of about −70 mV to –80 mV. A simple calculation of the Nernst potential for these parameters will indicate that plants must have more than 300 mM Na$^+$ in the parenchyma's cell cytosol to get Na$^+$ loaded into the xylem passively. This is highly unlikely. As mentioned before, most published reports using either X-ray microanalysis techniques or ion-selective electrodes agree that cytosolic Na$^+$ values do not exceed 30 mM (Tester and Davenport, 2003, and references within). Hence, active loading is most likely to be the key mechanism responsible for

control of xylem Na$^+$ level in plants under saline conditions, unless the apoplastic route (see above) dominates.

The most likely candidate for this active loading is the Na$^+$/H$^+$ antiporter SOS1 (Fig. 2.1). Such antiporters were shown to be preferentially expressed at the xylem–symplast boundary of roots (indicated by promoter-GUS fusions) (Shi et al., 2002). The driving force for this active Na$^+$ loading will be an H$^+$ gradient maintained by H$^+$-translocating ATPases that are localized to the endodermis. Addition of NaCl to the halophyte Atriplex nummularia increased mRNA levels of such an ATPase (Niu et al., 1993), and a T-DNA knockout of an endodermal H$^+$-ATPase in Arabidopsis increased the salt sensitivity and decreased the ability of plants to maintain low shoot Na$^+$ in salinized conditions (Vitart et al., 2001).

The fact that Na$^+$ loading into the xylem appears to be an active process that requires metabolic energy poses an interesting question: why do plants apparently 'waste' the energy to get Na$^+$ pumped into the xylem and then try to minimize its accumulation in the shoot? It is worth remembering that Na$^+$ can be used as a cheap osmoticum to maintain cell turgor (and, ultimately, tissue growth), assuming it can be efficiently sequestered in the cell vacuoles by the tonoplast NHX Na$^+$/H$^+$ exchanger. Thus, the conclusion about the importance of restriction of Na$^+$ loading into the xylem cannot be possibly generalized to all cases. It was shown, for example, that exclusion of Na$^+$ loading into the xylem plays no major role in barley salt tolerance in barley (Shabala et al., 2010); moreover, tolerant barley varieties had higher xylem Na$^+$ content compared with salt-sensitive genotypes. The same is true for halophyte species (reviewed in Shabala and Mackay, 2011).

There are two important caveats here, however. The elevated xylem Na$^+$ loading is only acceptable when plants have a superior ability to safely sequester in the leaf vacuoles. Second, xylem Na$^+$ loading must also be accompanied by the concurrent loading of sufficient amounts of K$^+$ into the xylem, to maintain the optimal xylem K/Na ratio (Shabala et al., 2010). Here, the role of passive transport systems (i.e. SKOR and NORC channels; Fig. 2.1) becomes crucial. The SKOR channels are known to be preferentially expressed in xylem parenchyma tissue and be activated by membrane depolarization in a time-dependent manner at membrane potentials slightly positive of the Nernst potential of K^+ (Wegner and Raschke, 1994; Gaymard *et al.*, 1998). The role of SKOR channels in K^+ xylem loading in barley was confirmed recently in direct electrophysiological experiments (Shabala *et al.*, 2010). NORC channels are much less selective and can also permeate other cations such as Na^+ and Ca^{2+} (Wegner and Raschke, 1994). Thus, if cytosolic Na^+ content is much higher (e.g. above 100 mM as was suggested in some works, such as Flowers and Colmer, 2008), and xylem Na^+ content is, on the contrary, at the lower end, NORC channels may be responsible for the passive Na^+ loading into the xylem. It is reasonable to assume that this role may be most significant immediately after salt exposure, when xylem Na+ content is still low, and a progressive Na^+ build-up in root cortical cells takes place.

Further insights into mechanisms of xylem Na^+ loading have been gained by conducting a series of electrophysiological experiments from the xylem parenchyma tissue in response to H_2O_2 and ABA, both of them associated with salinity stress signalling (Zhu *et al.*, 2016a). These results suggested that NADPH oxidase-mediated apoplastic H_2O_2 production acts upstream of the xylem Na^+ loading and is causally related to ROS-inducible Ca^{2+} uptake systems in the root stelar tissue. It was also found (Zhu *et al.*, 2016a) that ABA regulates (directly or indirectly) the process of Na^+ retrieval from the xylem and the significant reduction of Na^+ and K^+ fluxes induced by bumetanide (a known blocker of mammalian chloride cation co-transporter (CCC); Starremans *et al.*, 2003). These latter findings lead to the conclusion that the above CCC plays a major role in xylem Na^+ loading in barley (Zhu *et al.*, 2016a). AtCCC is preferentially expressed in the root and shoot vasculature at the xylem–symplast boundary (Colmenero-Flores *et al.*, 2007) and was found to catalyse the coordinated symport of K^+, Na^+ and Cl^- into the xylem in *A. thaliana*.

2.4.7 Na$^+$ retrieval from the xylem

Another strategy to reduce Na^+ accumulation in the shoot may be the retrieval of Na^+ that has entered the xylem before it reaches the bulk of the shoot (Tester and Davenport, 2003). The most likely 'checkpoints' for this process include mature zones of root and shoot vasculature, and the base of the shoot.

Little is known about the thermodynamics of Na^+ loading and unloading of the phloem at the molecular level. HKT transporters have been proposed to function in both processes (Berthomieu *et al.*, 2003). HKT transporters can operate in two transport modes, either as a high-affinity sodium–potassium symporter (in the presence of low potassium and sodium concentrations), or as a low-affinity sodium–sodium (co)-transporter (when the sodium–potassium concentration ratio in the external solution is high) (Rubio *et al.*, 1995; Gassmann *et al.*, 1996). *Arabidopsis* hkt1;1 mutants are salt sensitive compared to wild type and hyperaccumulate Na^+ in the shoot, but show hypoaccumulation of Na^+ in the root (Berthomieu *et al.*, 2003).

HKT group 1 genes are a family of uptake transporters that are Na^+-selective and transport Na^+ across the plasma membrane from the apoplast into cells such as xylem parenchyma cells lining the xylem vessels. The rice Na^+ transporter OsHKT1;5 localizes to the plasma membrane and is expressed in the xylem tissues in roots (Ren *et al.*, 2005). Genetic variation in rates of Na^+ accumulation in leaves correlate highly with salt tolerance of rice, and allelic variation of *OsHKT1;5* is largely responsible for this variation (Negrão *et al.*, 2013; Platten *et al.*, 2013). Conformational 3D modelling of the rice HKT transporters OsHKT1;4 and OsHKT1;5 was presented by Cotsaftis *et al.* (2012). They suggested that Na^+ accumulation in the leaf blade is controlled by regulation of complex gene expression, alternate splicing and the protein structure of transporters under salt stress. Recently, expression of *OsHKT1;4* was studied in rice organs and tissues at different stages of plant development (Suzuki *et al.*, 2016). OsHKT1;4 was expressed in leaf sheaths but not in the roots of young rice plants. At the reproductive stage, when the stem has elongated, the gene was expressed most strongly in the peduncle (the uppermost internode) as well as in the upper nodes and the flag leaf. The phenotype was investigated with RNA interference gene silencing which showed that only in the reproductive stage did it have an effect on tissue accumulation of Na^+ (Suzuki *et al.*, 2016).

The *Arabidopsis* equivalent is AtHKT1;1 (Davenport *et al.*, 2007). Targeted overexpression of the *Arabidopsis* HKT1;5 homologue, AtHKT1;1, in *Arabidopsis* and rice has been shown to increase Na^+ exclusion from the shoot (Møller *et al.*, 2009; Plett *et al.*, 2010).

In bread wheat, orthologues of the two rice genes occur on all three genomes (Huang *et al.*, 2008), but the only physiologically active member of the gene family is on the D genome (Byrt *et al.*, 2007), which is present in bread wheat but not in durum wheat. *TaHKT1;5-D* (known previously as the locus *Kna1*) retrieves Na^+ from the xylem and reduces its flux to leaves (Byrt *et al.*, 2014). Transport and expression results indicated that *TaHKT1;5-D* encodes a Na^+-selective transporter localized on the plasma membrane in the wheat root stele (Byrt *et al.*, 2014). RNA interference-induced silencing decreased the expression of *TaHKT1;5-D* in transgenic bread wheat lines, which led to an increase in the Na^+ concentration in the leaves.

Durum wheat lacks any active physiologically active HKT group 1 genes, but two loci that reduced Na^+ accumulation in the shoot, named *Nax1* and *Nax2*, were discovered in an unusual durum wheat (James *et al.*, 2006). Both *Nax1* and *Nax2* could unload Na^+ from the xylem in the root, while *Nax1* could also unload Na^+ from the xylem at the leaf base (the sheath) so leading to a high Na^+ ratio between the sheath and the blade (James *et al.*, 2006). *Nax1* and *Nax2* lines also had higher rates of K^+ transport from the root to the shoot, resulting in an enhanced discrimination of K^+ over Na^+ (James *et al.*, 2006). The candidate gene for *Nax1* was identified as *TmHKT1;4* (Huang *et al.*, 2006), and the candidate gene for *Nax2* as *TmHKT1;5* (Byrt *et al.*, 2007). Later work showed that it was localized on the plasma membrane of cells surrounding the xylem (Munns *et al.*, 2012). These loci are not present in modern durum and, when crossed into an elite Australian durum cultivar, were found to confer a yield benefit of 25% on saline soil in a farmer's field (Munns *et al.*, 2012).

Nax2 was also found to benefit plants in sodic soil (Genc *et al.*, 2016). Three cereal species differing in Na^+ exclusion were grown in a potting mix made sodic with Na^+ humate, and leaf nutrient profiles and pot yield were determined. Na^+-excluding bread wheat, and durum wheat with the *Nax2* gene had higher yield than Na^+ accumulating barley and durum wheat without the *Nax2* gene (Genc *et al.*, 2016).

However, it was shown recently that *Nax* loci also affect activity and expression levels of the SOS1-like Na^+/H^+ exchanger in both root cortical and stelar tissues. The non-invasive ion flux measuring MIFE technique was used to measure the magnitude of Na^+ efflux from stelar tissues of salt-treated parental Tamaroi line and Nax lines (Zhu *et al.*, 2016b). The measured net Na^+ efflux decreased in the sequence Tamaroi > $Nax1 = Nax2 > Nax1:Nax2$, was sensitive to amiloride (a known inhibitor of the Na^+/H^+ exchanger) and was mirrored by net H^+ flux changes. Also lower (six- to tenfold) were *TdSOS1* transcript levels in *Nax* lines compared with Tamaroi. These findings led to the conclusion that *Nax* loci confer two highly complementary mechanisms, both of which contribute towards reducing the xylem Na^+ content. One of these mechanisms enhances the retrieval of Na^+ back into the root stele via HKT1;4 or HKT1;5, while the other reduces the rate of Na^+ loading into the xylem via SOS1. Thus, the question of which process – reducing xylem Na^+ loading or recirculation of Na^+ back to the roots by the phloem – plays the bigger role in controlling shoot ion content remains to be answered in the future. Regardless of this answer, however, there is little doubt that such duality plays an important adaptive role with greater versatility for responding to a changing environment and controlling Na^+ delivery to the shoot.

2.4.8 Sodium removal from the shoot via the phloem

Sodium removal from the shoot is mediated by recirculation of Na^+ back to the roots by the phloem (Munns *et al.*, 1988; Blom-Zandstra *et al.*, 1998). A strong correlation between the extent of this recirculation and plant salinity tolerance has been reported in several species (Matsushita and Matoh, 1991; Pérez-Alfocea *et al.*, 2000). Yet, as shown below, Na^+ fluxes in the phloem are much less than in the xylem, indicating that retranslocation in the phloem has limited capacity to regulate Na^+ concentrations in leaves.

Little is known about the thermodynamics of Na^+ loading and unloading of the phloem at a molecular level. HKT transporters have been proposed to function in both processes (Berthomieu *et al.*, 2003). *Arabidopsis hkt1;1* mutants are salt sensitive compared to wild type, and hyperaccumulate Na^+ in the shoot, but show hypoaccumulation of Na^+ in the root (Berthomieu *et al.*, 2003). Based on RNA *in situ* hybridization and promoter fusion, Berthomieu *et al.* (2003) concluded that AtHKT1;1 was located predominantly in the phloem. This was challenged by Sunarpi *et al.* (2005), who demonstrated localization of the protein to the plasma membrane of xylem parenchyma cells in the shoot, raising a possibility for the role of AtHKT1;1 in withdrawal of Na^+ from the xylem in the shoot. This view was unequivocally supported by Davenport *et al.* (2007), who concluded that AtHKT1;1 is not involved in Na^+ recirculation via the phloem and is not responsible for Na+ influx into roots. Instead, the role of AtHKT1;1 in controlling both retrieval of Na^+ from the xylem and root vacuolar loading was endorsed (Davenport *et al.*, 2007). An answer to the question of how Na^+ is removed from the shoot and recirculated to roots is still to come.

An important question that has not been answered concerns the fate of the Na^+ remobilized in the phloem. It is a general consensus that Na^+ moves to the root – but is this the final destination? Does it stay in root vacuoles or is it effluxed to the soil?

If the sieve tubes carrying Na^+ exported from a leaf are connected exclusively with downwards movement of phloem sap, Na^+ could be exported without risk to the young leaves. However, if the sieve tubes are connected to younger leaves or the shoot apex, export in the phloem could be directed to growing leaves and meristematic regions of the shoot where it could quickly reach toxic levels. As shown by C^{14} urea feeding studies in lupin, old leaves feed their photosynthate to the root, young leaves feed the shoot apex and mid-position leaves feed both shoot apex and root (Layzell *et al.*, 1981). Limited export of salt via the phloem ensures that it is not redirected to growing tissues of the shoot.

The more salt-tolerant species have low rates of export in the phloem. The salt-tolerant species, barley, exports only about 10% of the Na^+ that enters leaves in the xylem (Wolf *et al.*, 1990). Salt-sensitive species appear to have a higher rate of phloem export than salt-tolerant species. For white lupin, an extremely sensitive

plant, phloem export was estimated to be 30% of the import for plants growing in 25–40 mM NaCl, and 50% of import for those growing in 100 mM NaCl (Munns *et al.*, 1988; Jeschke *et al.*, 1992). The higher the concentration of Na^+ and Cl^- in the phloem, the more likely the meristematic and reproductive tissues are to receive a load of salt. In lupin, it is very likely that the increased phloem export of Na^+ may contribute to its sensitivity to saline soil (Jeschke *et al.*, 1992).

It thus remains critical for a plant growing in saline soil to control the loading of Na^+ in the xylem in the roots, and to have a back-up mechanism to retrieve Na^+ from the xylem in the upper part of the roots or at the leaf base or stem base. Export in the phloem is not the solution.

2.4.9 Oxidative stress tolerance

The main line of defence that protect cells against oxidative injury are various *antioxidant components* – a number of enzymes and low molecular weight compounds capable of quenching ROS without themselves undergoing conversion to a destructive radical (Scandalios, 1993). Both enzymatic and non-enzymatic components contribute to detoxication of ROS species (Mittler *et al.*, 2004). This includes (Mittler, 2002): (i) superoxide dismutase (SOD; found in all cellular compartments); (ii) the water–water cycle (in chloroplasts); (iii) the ascorbate–glutathione cycle (in chloroplasts, mitochondria, cytosol and apoplastic space); (iv) glutathione peroxidase (GPX); and (v) catalase (CAT). Both (iv) and (v) are found in peroxisomes. The major antioxidant enzymes are SOD, ascorbate peroxidase (APX), CAT and GPX (Noctor and Foyer, 1998), while most important non-enzymatic ROS scavenging compounds are ascorbate and glutathione (Colville and Smirnoff, 2008). The existing balance between these, together with sequestration of metal ions, is important in preventing the formation of the highly toxic hydroxyl radical (Mittler, 2002).

While traditionally ROS are considered to be toxic by-products of aerobic metabolism (Mittler, 2002), it has become apparent that plants actively produce ROS as signalling molecules to control numerous physiological processes such as defence responses and cell death, cross tolerance, gravitropism, stomatal aperture, cell expansion and polar growth, hormone action, and leaf and flower development. In many cases, the production of ROS is genetically programmed, and superoxide and H_2O_2 are used as second messengers (Foyer and Noctor, 2005), and a new concept of 'oxidative signalling' instead of 'oxidative stress' was proposed. The 'positive' role of ROS has been reported both at the physiological (e.g. regulation of ion channels activity; Foreman et al., 2003) and genetic (e.g. control of gene expression; Mittler et al., 2004) levels. Thus, it appears that ROS have pleiotropic effects in plants, as they do in animals. When ROS are produced in a controlled manner within specific compartments, they have key roles in plant metabolism. When they are produced in excess, the resultant uncontrolled oxidation leads to cellular damage and eventual death. To prevent damage, yet allow beneficial functions of ROS to continue, the antioxidant defences must keep active oxygen under control. Thus, plants are facing a really challenging dilemma to delicately balance ROS production and scavenging. This is achieved by orchestrated regulation of a large network of genes (termed the 'ROS gene network'; Mittler et al., 2004) and efficient control of ion transporters and cytosolic ionic homeostasis.

The link between antioxidant activity and salinity stress tolerance seems to be highly complicated. While dozens of papers claim a positive association between antioxidant production in plant tissues and plant salinity tolerance, the equal amount showed no, or even negative, correlation between these two traits (reviewed in Bose et al., 2014). It was argued in this context that truly salt-tolerant species possessing efficient mechanisms for Na^+ exclusion from the cytosol may not require high level of antioxidant activity, as they simply do not allow excessive ROS production in the first instance (Bose et al., 2014). Thus, it appears that increased antioxidant activity should be treated as a damage control mechanism rather than a trait directly conferring salinity stress tolerance.

Despite the above controversy, a comparative analysis between glycophyte and halophyte species revealed generally higher level of antioxidant enzymes and, specifically, SOD in halophytes. As the major role of SOD is in a rapid conversion of O_2- to H_2O_2, it was suggested (Bose et al., 2014) that higher SOD levels are essential

for the rapid induction of 'H_2O_2 signatures' under stress conditions, to trigger a range of adaptive responses at both transcriptional and post-translational levels. It should also be added that hydroxyl radical, the most harmful of all ROS species, cannot be scavenged by means of enzymatic antioxidants, and the only way to prevent its detrimental effects on membrane structures and transport proteins is via control of non-enzymatic antioxidants. Halophytes species appear to be highly superior in this component compared with their glycophyte counterparts (Bose et al., 2014).

2.5 Anatomical Adaptation

There is no single anatomical feature that will make halophytes or salt-tolerant glycophytes strikingly different from their salt-sensitive counterparts. Still, several anatomical features are considered to be essential in plant adaptation to highly saline environment. While most of them are related to efficient Na^+ sequestration or secretion, a few other physiological benefits must be also considered.

2.5.1 Leaf succulence

Leaf succulency is a term used to describe thickening of leaf tissues and the resultant increase in the volume of leaf sap (Shabala and Mackay, 2011). The physiological rationale beyond this phenomenon is a significant increase in the cell (and, hence, vacuole) size leading to the possibility of more efficient intracellular Na^+ sequestration in this organelle. In glycophytes, leaf succulence is typically achieved by increasing the size of mesophyll cells (Gorham et al., 1985). Given the importance of Na^+ removal from the cytosol to enable optimal cell metabolism, plants capable of rapidly increasing leaf thickness are considered to have an adaptive advantage. Accordingly, a relative increase in leaf thickness was suggested for use as a physiological marker in crop breeding for salinity tolerance. In halophytes, especially those having no salt glands, leaf succulence is a constitutive feature which often results in a high degree of specialization between leaf tissue cells. This is further illustrated in Fig. 2.7, depicting a

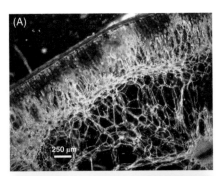

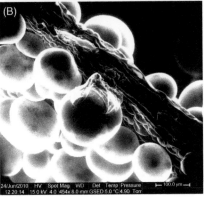

Fig. 2.7. Anatomical adaptation to salinity in plant leaves. A – leaf succulence shown in a cross-section of the leaf lamina of a halophyte species, *Carpobrotus rossii* (photo courtesy of Adam Pirie). B – numerous salt bladders on both abaxial and adaxial surfaces of *Atriplex halimus* leaves (photo courtesy of Alex Mackay).

cross-section over the leaf lamina of the halophyte species *Carpobrotus rossii*. As one can see, photosynthetically competent mesophyll cells are restricted to a relatively small (*c.*10%) area while the major bulk of the leaf is made of spongy parenchyma cells with very few chloroplasts. Being very large in size (typically between 300 and 500 µm; Fig. 2.7), these cells present a vast possibility for Na^+ sequestration in the leaf lamina. The specific ionic mechanisms enabling preferential Na^+ loading into spongy parenchyma cells but not into mesophyll remain to be investigated.

2.5.2 Salt glands and bladders

Salt bladders and glands are arguably the most remarkable anatomical feature related to salinity.

Three major types of gland or bladder structures are known (Agarie *et al.*, 2007; Shabala and Mackay, 2011): (i) two-celled excretory structures found in most of the graminoids; (ii) multi-cellular structures found in some graminoids and several dicotyledonous families; and (iii) the epidermal bladder cells (EBC) characteristic of the Chenopodioideae, Oxalidaceae and Mesembryanthemaceae. The main function of salt glands and bladders is the elimination or sequestration of excess salt from metabolically active tissues (Agarie *et al.*, 2007; Flowers and Colmer, 2008). It was reported that plants containing salt glands are capable of excreting three to four times the amount of Na^+ present in leaves daily (Marcum and Murdoch, 1992; Marcum *et al.*, 2007). However, the bladder function appears to be far more advanced rather than being merely a storage space for toxic ion species. Being rather dense structures (Fig. 2.7), salt bladders may also play an important role as a secondary epidermis to reduce water loss and prevent excessive UV damage. Other functions postulated for EBC include storage of water and various metabolic compounds such as malate, flavinoids, cysteine, pinitol, inositol and calcium oxalate crystals (Agarie *et al.*, 2007; Jou *et al.*, 2007). From this point of view, salt bladders appear to be more multitask compared with glands whose function is merely in secretion of excessive Na^+ and Cl^-.

Salt bladder density can vary dramatically depending upon leaf surface (adaxial versus abaxial), leaf age and plant species. The multilayer bladder structures are well reported. As a rule of thumb, the bladder density is higher in young, developing leaves and becomes less while the leaf matures (Bonales-Alatorre *et al.*, 2013). Once filled to its full capacity, the bladder is ruptured and accumulated salts are released. As the leaf surface is heavily cutinized, the possibility of reabsorption is minimal, and most Na^+ is then washed from the leaf surface by rain.

The mechanism of ion loading into salt bladders is still an area of conjecture. Both apoplastic and symplastic pathways have been suggested, and several specialized cell types (such as stalk cells) have been reported (Shabala and Mackay, 2011). Recently, a new model has been suggested based on the structural and functional similarities with the solute-accumulating plant systems studied (Shabala *et al.*, 2014) that describes major transport systems contributing to salt loading into EBC. According to this model, the stalk cell (SC) functions as a solute 'traffic controller' that regulates salt loading into EBC, comparable to companion cells (CC) loading sugars into the phloem (De Schepper et al., 2013). The movement of salt across the SC and across the cytoplasmic layer of the bladder cell into the huge central vacuole can be considered like kidney cells epithelia (Gattineni and Baum, 2015) that possess a physiologically similar role. This process is mediated by transporters organized in a polar manner. From a thermodynamic view, Na^+ and Cl^- transport from the EC into the EBC vacuole should be energetically an uphill process, requiring coupling to the electrochemical potential gradient of the ions in question. This role is attributed to the plasma membrane and vacuole proton-coupled sodium transporters encoded by SOS1- and NHX1-type genes (Kronzucker and Britto, 2011; Yamaguchi et al., 2013), accordingly. The downhill Na^+ transport is mediated by HKT-type genes encoding for Na^+ channel (Almeida et al., 2013). Chloride loading into EBC is mediated by ClC-type proton-coupled carriers and SLAC/SLAH-type channels (Plett et al., 2010; Roelfsema et al., 2012). Vacuolar Na^+ and Cl^- deposition is energized by the V-ATPase and V-PPase H^+ pumps (Shabala, 2013) and is osmotically adjusted by cytosolic potassium and organic osmolytes (Fig. 2.8).

2.5.3   Changes in the root anatomy

Root growth is usually less affected than leaf growth, and root elongation rate recovers remarkably well after exposure to NaCl or other osmotica (Munns, 2002). In maize, recovery from a moderate stress of up to 0.4 MPa of mannitol, KCl or NaCl (i.e. an osmotic shock that does not cause plasmolysis) is complete within 1 h, and recovery from NaCl concentrations as high as 150 mM can occur within 1 day (Munns, 2002). In contrast to leaves, these recoveries of growth rate of roots take place despite turgor not being fully restored (Frensch and Hsiao, 1994). This indicates different changes in cell wall properties compared with leaves, but the mechanism is unknown. With time, reduced initiation of new seminal or lateral roots probably occurs, but little is known about this.

Previous studies of other plant species growing in salty conditions that surrounded the entire root systems have found that vascular root tissues continue to differentiate despite slowing of seminal (axile) tip elongation. In detailed anatomical studies of cotton roots growing in vermiculite with 150 or 200 mM NaCl, Reinhardt and Rost (1995a) showed that, as axile root (tap root) elongation slowed, xylem vessel differentiation proceeded such that mature vessels were found closer to the tips of salt-stressed plants than control plants without exposure to salt. They also found that branch roots emerged about three times closer to the tip when the tap elongation was severely restricted (over ten times slower than the control) in 200 mM NaCl compared to the control (Reinhardt and Rost, 1995b).

The adaptive advantage of increased branch root growth when the seminal or axile tip is slowed could be the maintenance to total root length for access to water and nutrients, as was found by Rahnama et al. (2011), where axile roots of durum wheat were shortened by the salt but the concomitant increase in branch roots compensated, maintaining root length. Since branch roots are generally thinner than axile roots, more root length can be generated per unit carbon in branch roots, perhaps explaining the shift from axile elongation.

Studies in cotton and other species have indicated that the reduction rate of root elongation, and the increase in proportion of branch roots, is due to the osmotic stress and not to salt toxicity (reviewed in Rahnama et al., 2011).

2.5.4   Control over leaf transpiration

Rates of photosynthesis per unit leaf area in salt-treated plants are often unchanged, even though stomatal conductance is reduced (James et al., 2002). This paradox is explained by the changes in cell anatomy described above, which give rise to smaller, thicker leaves and result in a higher chloroplast density per unit leaf area. In a detailed study with durum wheat, the stress-induced reduction in stomatal conductance was seen as soon as the leaf emerged. Photosynthetic rate, in contrast, was not reduced when the leaf emerged (James et al., 2002). The high photosynthetic rate per unit leaf area was caused by

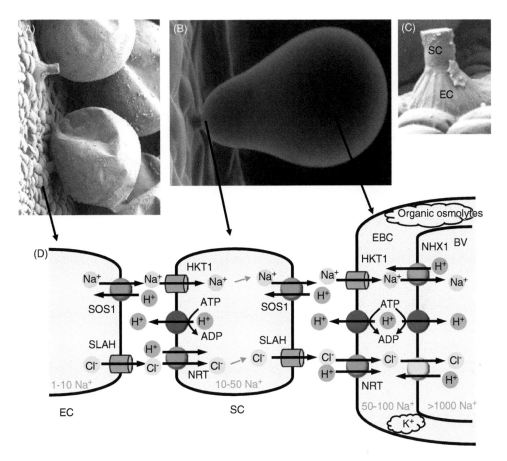

Fig. 2.8. Anatomy of the bladder cell complex and predicted transport system mediating salt sequestration in epidermal bladder cells (EBC) (reproduced from Shabala *et al*. (2014), with permission of Elsevier). (A) SEM image of quinoa leaf surface showing several (deflated) EBC. (B) A fully turgid EBC on the surface of a young quinoa leaf. (C) A close-up image of the epidermal cell–stalk cell (EC–SC) complex shown in panel (A). Both EBC and SC can be easily removed by a gentle brushing, suggesting an apoplastic connection within the EC–SC–EBC complex. (D) Given the pronounced concentration gradient for NaCl between bladder cells and the common ECs (numbers in red), questions about the molecular mechanism of cellular Na^+ and Cl^- transport and its regulation within the functional epidermis–bladder complex arise. Based on thermodynamic considerations and preliminary experimental data, an orchestrated operation of several transport systems is postulated. These include: SOS1 (Na^+/H^+ plasma membrane exchanger), HKT1 (high-affinity transporter operating as Na^+ channel), SLAH (Cl-permeable anion channel), NRT (Cl^-/H^+ co-transporter) and NHX1 (tonoplast-based Na^+/H^+ exchanger).

the changed leaf and cell morphology, so that leaves with smaller cells of different dimensions, giving a higher chloroplast density per unit leaf area, were able to compensate for the lower stomatal conductance and maintain their photosynthetic rate (on a unit leaf area basis). This results in a high transpiration efficiency for the leaf, because the water lost per

unit of carbon fixed is less: an adaptation to dry or saline soil.

When photosynthesis is expressed on a unit chlorophyll basis, rather than on a leaf area basis, a reduction due to salinity can usually be measured. In any case, the reduction in leaf area due to salinity means that photosynthesis per plant is always reduced.

2.6   Putative Mechanisms of Salt Stress Sensing and Early Signalling

While the molecular identity and functional expression of Na⁺ transport systems mediating Na⁺ exclusion from the cytosol has been studied in detail, far less is known about the mechanisms by which plants sense high Na⁺ levels in the soil and the rapid signalling events that optimize plant performance under saline conditions (Maathuis, 2014; Shabala *et al.*, 2015). Semantically, the term sensor implies the component that is *the first* to detect changes in the external environment. This implicates a biophysical nature for the sensor (operating within the milliseconds range) and attributes its location to the plasma membrane. Physiologically, however, detrimental effects of salinity became evident at a much slower (hours to days) timescale. Hence, from the practical point of view, these are the early signalling events that confer genotypic difference and determine adaptive potential of plants to a saline environment. In this context, several possible candidates should be considered (summarized in Table 2.1).

It was also suggested (Shabala *et al.*, 2015) that many (if not all) of these sensing systems operate *in parallel* (although at very different timescales) and are integrated via common components, namely changes in cytosolic free Ca^{2+} and H_2O_2. Taken together, such a sophisticated

Table 2.1. Putative mechanisms of salt stress sensing and early signalling in plants.

Putative sensory mechanism	Candidate genes/proteins	Details/comments	References
Na⁺ transport systems	SOS1 Na⁺/H⁺ antiporter	SOS1 protein has 10–12 transmembrane domains and a 700 amino acids-long tail that is predicted to reside in the cytoplasm and potentially sense Na⁺. However, no direct experimental support for this hypothesis has been provided	Zhu, 2003
Na⁺ transport systems	Na⁺/Ca²⁺ exchangers from NCX family	The existence of a putative NCX-like (*AtNCL*) gene was suggested for *Arabidopsis*. It encodes a protein with 9–11 transmembrane domains and a large intracellular hydrophilic loop. The protein has some critical amino acids conserved in NCX (such as threonine, serine or asparagine), which form the Na⁺ site sensing domain on the intracellular side	Wang *et al.*, 2012
Proteins with regulatory Na⁺ binding sites	Cation-H⁺ exchangers from CHX family	No Na⁺-selective ion channels have yet been identified in plants (Hedrich, 2012). However, plants may contain other proteins that have regulatory Na⁺ binding sites such as those found in mammals. Possible candidates should possess the side chains of an aspartate and a histidine, which are located across from one another in a cytosolic loop and form a DxR/KxxH motif	Maathuis, 2014
NSCC/NADPH oxidases tandem	NOX genes (RbohA-J)	RbohD and RbohF are critical to salt stress responses and may operate as a salt sensor in plants in tandem with Ca²⁺-permeable NSCC channels. The model assumes that NSCC are located in the immediate proximity of the NADPH oxidase (e.g. form a microdomain in a lipid raft) and thus are capable of providing a very rapid activation of NADPH oxidase and a concurrent increase in ROS accumulation in the apoplastic space	Shabala *et al.*, 2015

Continued

Table 2.1. Continued.

Putative sensory mechanism	Candidate genes/proteins	Details/comments	References
Mechanosensory channels and transporters	Transporters from MSL (MscS-like), MCA and Piezo family; RLK (receptor-like kinases)	Mechano-sensory channels convert a mechanical force into electric trans-membrane variation within milliseconds and are, therefore, the most likely candidates for sensing changes in the external media imposed by a sudden change in the solution osmolality	Monshausen and Haswell, 2013; Kurusu et al., 2013
Cyclic nucleotide receptors	CNGC family	20 non-selective cation channels belong to the CNGC family in *Arabidopsis*. These channels are permeable to Ca^{2+} and, in a tandem with NADPH oxidase, may compose a self-amplifying functional unit shaping Ca^{2+} 'signatures' and activating gene network-mediated plant adaptive responses to salinity	Julkowska and Testerink, 2015; Talke et al., 2003; Maathuis and Sanders, 2001
Purino-receptors	P2X receptors	Extracellular purines such ATP and ADP function as signalling molecules under saline conditions. Salt-induced eATP may be sensed by the P2X purino-receptors in the plasma membrane and then translated into downstream signals, such as H_2O_2 and cytosolic Ca^{2+}, which are required for the upregulation of genes linked to K^+/Na^+ homeostasis and plasma membrane repair	Demidchik et al., 2009; Tanaka et al., 2010; Sun et al., 2012
Annexins	ANN1	AtANN1 is a plasma membrane protein capable of forming Ca^{2+}-permeable channels in planar lipid bilayers. While AtANN1 does not contribute to root Na^+ uptake, it is a component of the cytosolic Ca^{2+} signal generated upon salt treatment and is suggested to be a key component in root cell adaptation to salinity	Lee et al., 2004; Laohavisit et al., 2013
H^+-ATPase/ GORK tandem	AHA; GORK	The plant plasma membrane H^+-ATPase is stimulated by potassium bound to the proton pump at a site involving Asp(617) in the cytoplasmic phosphorylation domain. Massive GORK-mediated K^+ efflux induced by salinity results in rapid (within minutes) twofold decreases in cytosolic K^+ concentration. Assuming GORK channels are located in close proximity to H^+-ATPase, a reduction in cytosolic K^+ concentration in the H^+-ATPase microdomain may rapidly activate H^+-pump by releasing inhibition at Glu(184), as described above	Buch-Pedersen et al., 2006; Shabala et al., 2006

sensing mechanism may provide plants with a robust system of decoding information about the specific nature and severity of the salt stress, converting it into stress-specific Ca^{2+} and H_2O_2 'signatures'.

2.7 Conclusions

Secondary salinity imposes both an environmental and a social threat, as salts are moved towards the soil surface by rising water tables, and enter creeks or rivers which feed irrigation storage bodies or town water supplies. Continued cultivation of these soils, without consideration of the changes to water tables, may have a further negative impact, and revegetation or changing agricultural practices may be necessary to halt the salinization process. Lack of remediation or controls on further use of irrigation water, or further clearing of vegetated land, may cause irreversible damage to soils and water catchments.

With natural salinity, where rainfall is low and water tables are not rising, continued cultivation is important to maintain global food supplies. The mechanisms by which plants cope with salt is a subject of increasing importance as agriculture moves into more marginal lands, and new genetic sources of salt tolerance need to be found.

This chapter has shown that plants have evolved numerous mechanisms to prevent accumulation of toxic Na^+ levels in cell cytosol. This ability has to be complemented by mechanisms enabling osmotic adjustment and maintenance of the cell turgor, as well as efficient mechanisms of coping with ROS stress imposed by salinity. A more complete physiological and genetic understanding of these processes will enable breeders to produce new salt-tolerant plants for the future.

References

Agarie, S., Shimoda, T., Shimizu, Y., Baumann, K., Sunagawa, H., Kondo, A., Ueno, O., Nakahara, T., Nose, A. and Cushman, J.C. (2007) Salt tolerance, salt accumulation, and ionic homeostasis in an epidermal bladder-cell-less mutant of the common ice plant *Mesembryanthemum crystallinum*. *Journal of Experimental Botany* 58, 1957–1967.

Aharon, G.S., Apse, M.P., Duan, S.L., Hua, X.J. and Blumwald, E. (2003) Characterization of a family of vacuolar Na^+/H^+ antiporters in *Arabidopsis thaliana*. *Plant and Soil* 253, 245–256.

Allakhverdiev, S.I., Nishiyama, Y., Miyairi, S., Yamamoto, H., Inagaki, N., Kanesaki, Y. and Murata, N. (2002) Salt stress inhibits the repair of photodamaged Photosystem II by suppressing the transcription and translation of *psbA* genes in *Synechocystis*. *Plant Physiology* 130, 1443–1453.

Almeida, P., Katschnig, D. and de Boer, A.H. (2013) HKT transporters – state of the art. *International Journal of Molecular Science* 14, 20359–20385.

Amtmann, A. and Sanders, D. (1999) Mechanisms of Na^+ uptake by plant cells. *Advances in Botanical Research* 29, 75–112.

Amtmann, A., Fischer, M., Marsh, E.L., Stefanovic, A., Sanders, D. and Schachtman, D.P. (2001) The wheat cDNA LCT1 generates hypersensitivity to sodium in a salt-sensitive yeast strain. *Plant Physiology* 126, 1061–1071.

Apel, K. and Hirt, H. (2004) Reactive oxygen species: metabolism, oxidative stress, and signal transduction. *Annual Review of Plant Biology* 55, 373–399.

Apse, M.P. and Blumwald, E. (2007) Na^+ transport in plants. *FEBS Letters* 581, 2247–2254.

Apse, M.P., Aharon, G.S., Snedden, W.A. and Blumwald, E. (1999) Salt tolerance conferred by overexpression of a vacuolar Na^+/H^+ antiport in *Arabidopsis*. *Science* 285, 1256–1258.

Apse, M.P., Sottosanto, J.B. and Blumwald, E. (2003) Vacuolar cation/H^+ exchange, ion homeostasis, and leaf development are altered in a T-DNA insertional mutant of AtNHX1, the *Arabidopsis* vacuolar Na^+/H^+ antiporter. *Plant Journal* 36, 229–239.

Bajaj, S., Targolli, J., Liu, L.F., Ho, T.H.D. and Wu, R. (1999) Transgenic approaches to increase dehydration-stress tolerance in plants. *Molecular Breeding* 5, 493–503.

Bao, A.K., Wang, S.M., Wu, G.Q., Xi, J.J., Zhang, J.L. and Wang, C.M. (2009) Overexpression of the *Arabidopsis* H^+-PPase enhanced resistance to salt and drought stress in transgenic alfalfa (*Medicago sativa* L.). *Plant Science* 176, 232–240.

Barkla, B.J., Zingarelli, L., Blumwald, E. and Smith, J.A.C. (1995) Tonoplast Na^+/H^+ antiport activity and its energization by the vacuolar H^+-atpase in the halophytic plant *Mesembryanthemum crystallinum* L. *Plant Physiology* 109, 549–556.

Barragán, V., Leidi, E.O., Andrés, Z., Rubio, L., De Luca, A., Fernandez, J.A., Cubero, B. and Pardo, J.M. (2012) Ion exchangers NHX1 and NHX2 mediate active potassium uptake into vacuoles to regulate cell turgor and stomatal function in *Arabidopsis*. *The Plant Cell* 24, 1127–1142.

Bassil, E., Ohto, M.A., Esumi, T., Tajima, H., Zhu, Z., Cagnac, O., Belmonte, M., Peleg, Z., Yamaguchi, T. and Blumwald, E. (2011a) The *Arabidopsis* intracellular Na^+/H^+ antiporters NHX5 and NHX6 are endosome associated and necessary for plant growth and development. *The Plant Cell* 23, 224–239.

Bassil, E., Tajima, H., Liang, Y.C., Ohto, M., Ushijima, K., Nakano, R., Esumi, T., Coku, A., Belmonte, M. and Blumwald, E. (2011b) The *Arabidopsis* Na⁺/H⁺ antiporters NHX1 and NHX2 control vacuolar pH and K⁺ homeostasis to regulate growth, flower development, and reproduction. *The Plant Cell* 23, 3482–3497.

Bers, D.M., Barry, W.H. and Despa, S. (2003) Intracellular Na⁺ regulation in cardiac myocytes. *Cardiovascular Research* 57, 897–912.

Berthomieu, P., Conéjéro, G., Nublat, A., Brackenbury, W.J., Lambert, C., Savio, C., Uozumi, N., Oiki, S., Yamada, K., Cellier, F., Gosti, F., Simonneau, T., Essah, P.A., Tester, M., Véry, A.A., Sentenac, H. and Casse, F. (2003) Functional analysis of AtHKT1 in *Arabidopsis* shows that Na⁺ recirculation by the phloem is crucial for salt tolerance. *EMBO Journal* 22, 2004–2014.

Blom-Zandstra, M., Vogelzang, S.A. and Veen, B.W. (1998) Sodium fluxes in sweet pepper exposed to varying sodium concentrations. *Journal of Experimental Botany* 49, 1863–1868.

Blumwald, E. (2000) Sodium transport and salt tolerance in plants. *Current Opinion in Cell Biology* 12, 431–434.

Bohm, J., Scherzer, S., Krol, E., Kreuzer, I., von Meyer, K., Lorey, C., Mueller, T.D., Shabala, L., Monte, I., Solano, R., Al-Rasheid, K.A.S., Rennenberg, H., Shabala, S., Neher, E. and Hedrich, R. (2016a) The Venus flytrap *Dionaea muscipula* counts prey-induced action potentials to induce sodium uptake. *Current Biology* 26, 286–295.

Bohm, J., Scherzer, S., Shabala, S., Krol, E., Neher, E., Mueller, T.D. and Hedrich, R. (2016b) Venus flytrap HKT1-type channel provides for prey sodium uptake into carnivorous plant without conflicting with electrical excitability. *Molecular Plant* 9, 428–436.

Bohnert, H.J. and Shen, B. (1999) Transformation and compatible solutes. *Scientia Horticulturae* 78, 237–260.

Bohnert, H.J., Nelson, D.E. and Jensen, R.G. (1995) Adaptation to environmental stresses. *The Plant Cell* 7, 1099–1111.

Bonales-Alatorre, E., Shabala, S., Chen, Z.H. and Pottosin, I. (2013) Reduced tonoplast fast-activating and slow-activating channel activity is essential for conferring salinity tolerance in a facultative halophyte, quinoa. *Plant Physiology* 162, 940–952.

Bose, J., Rodrigo-Moreno, A. and Shabala, S. (2014) ROS homeostasis in halophytes in the context of salinity stress tolerance. *Journal of Experimental Botany* 65, 1241–1257.

Bose, J., Rodrigo-Moreno, A., Lai, D.W., Xie, Y.J., Shen, W.B. and Shabala, S. (2015) Rapid regulation of the plasma membrane H⁺-ATPase activity is essential to salinity tolerance in two halophyte species, *Atriplex lentiformis* and *Chenopodium quinoa*. *Annals of Botany* 115, 481–494.

Buch-Pedersen, M.J., Rudashevskaya, E.L., Berner, T.S., Venema, K. and Palmgren, M.G. (2006) Potassium as an intrinsic uncoupler of the plasma membrane H⁺-ATPase. *The Journal of Biological Chemistry* 281, 38285–38292.

Byrt, C.S., Platten, J.D., Spielmeyer, W., James, R.A., Lagudah, E.S., Dennis, E.S. and Munns, R. (2007) HKT1;5-like cation transporters linked to Na⁺ exclusion loci in wheat, *Nax2* and *Kna1*. *Plant Physiology* 143, 1918–1928.

Byrt, C.S., Xu, B., Krishnan M, Lightfoot, D.J., Athman, A., Jacobs, A.K., Watson-Haigh, N.S., Plett, D., Munns, R., Tester, M. and Gilliham, M. (2014) The Na⁺ transporter, TaHKT1;5-D, limits shoot Na⁺ accumulation in bread wheat. *The Plant Journal* 80, 516–526.

Campbell, L.C., Raison, J.K. and Brady, C.J. (1976) Response of plant-mitochondria to media of high solute content. *Journal of Bioenergetics and Biomembranes* 8, 121–129.

Carden, D.E., Walker, D.J., Flowers, T.J. and Miller, A.J. (2003) Single-cell measurements of the contributions of cytosolic Na⁺ and K⁺ to salt tolerance. *Plant Physiology* 131, 676–683.

Chaves, M.M., Flexas, J. and Pinheiro, C. (2009) Photosynthesis under drought and salt stress: regulation mechanisms from whole plant to cell. *Annals of Botany* 103, 551–560.

Cheeseman, J.M. (2013) The integration of activity in saline environments: problems and perspectives. *Functional Plant Biology* 40, 759–774.

Chen, J.A., Xiao, Q.A., Wu, F.H., Dong, X.J., He, J.X., Pei, Z.M. and Zheng, H.L. (2010) Nitric oxide enhances salt secretion and Na⁺ sequestration in a mangrove plant, *Avicennia marina*, through increasing the expression of H⁺-ATPase and Na⁺/H⁺ antiporter under high salinity. *Tree Physiology* 30, 1570–1585.

Chen, Z., Newman, I., Zhou, M., Mendham, N., Zhang, G. and Shabala, S. (2005) Screening plants for salt tolerance by measuring K⁺ flux: a case study for barley. *Plant Cell and Environment* 28, 1230–1246.

Chen, Z.H., Pottosin, II, Cuin, T.A., Fuglsang, A.T., Tester, M., Jha, D., Zepeda-Jazo, I., Zhou, M.X., Palmgren, M.G., Newman, I.A. and Shabala, S. (2007) Root plasma membrane transporters controlling K⁺/Na⁺ homeostasis in salt-stressed barley. *Plant Physiology* 145, 1714–1725.

Choukr-Allah, R. (1996) The potential of halophytes in the development and rehabilitation of arid and semi-arid zones. In: Choukr-Allah, R., Malcolm, C.V. and Hamdy, A. (eds) *Halophytes and Biosaline Agriculture*. Marcel Dekker, New York, pp. 3–13.

Colmenero-Flores, J.M., Martinez, G., Gamba, G., Vazquez, N., Iglesias, D.J., Brumos, J. and Talon, M. (2007) Identification and functional characterization of cation-chloride cotransporters in plants. *Plant Journal* 50, 278–292.

Colmer, T.D., Flowers, T.J. and Munns, R. (2006) Use of wild relatives to improve salt tolerance in wheat. *Journal of Experimental Botany* 57, 1059–1078.

Colville, L. and Smirnoff, N. (2008) Antioxidant status, peroxidase activity, and PR protein transcript levels in ascorbate-deficient *Arabidopsis thaliana vtc* mutants. *Journal of Experimental Botany* 59, 3857–3868.

Cotsaftis, O., Plett, D., Shirley, N., Tester, M. and Hrmova, M. (2012) A two-staged model of Na^+ exclusion in rice explained by 3D modelling of HKT transporters and alternative splicing. *PloS ONE* 7, e39865.

Cramer, G.R. and Bowman, D.C. (1991) Kinetics of maize leaf elongation. I. Increased yield threshold limits short-term, steady-state elongation rates after exposure to salinity. *Journal of Experimental Botany* 42, 1417–1426.

Cuin, T.A., Miller, A.J., Laurie, S.A. and Leigh, R.A. (2003) Potassium activities in cell compartments of salt-grown barley leaves. *Journal of Experimental Botany* 54, 657–661.

Cuin, T.A., Betts, S.A., Chalmandrier, R. and Shabala, S. (2008) A root's ability to retain K^+ correlates with salt tolerance in wheat. *Journal of Experimental Botany* 59, 2697–2706.

Cuin, T.A., Parsons, D. and Shabala, S. (2010) Wheat cultivars can be screened for NaCl salinity tolerance by measuring leaf chlorophyll content and shoot sap potassium. *Functional Plant Biology* 37, 656–664.

Cuin, T.A., Bose, J., Stefano, G., Jha, D., Tester, M., Mancuso, S. and Shabala, S. (2011) Assessing the role of root plasma membrane and tonoplast Na^+/H^+ exchangers in salinity tolerance in wheat: *in planta* quantification methods. *Plant Cell and Environment* 34, 947–961.

Davenport, R., James, R.A., Zakrisson-Plogander, A., Tester, M. and Munns, R. (2005) Control of sodium transport in durum wheat. *Plant Physiology* 137, 807–818.

Davenport, R.J. and Tester, M. (2000) A weakly voltage-dependent, nonselective cation channel mediates toxic sodium influx in wheat. *Plant Physiology* 122, 823–834.

Davenport, R.J., Reid, R.J. and Smith, F.A. (1997) Sodium-calcium interactions in two wheat species differing in salinity tolerance. *Physiologia Plantarum* 99, 323–327.

Davenport, R.J., Muñoz-Mayor, A., Jha, D., Essah, P.A., Rus, A. and Tester, M. (2007) The Na^+ transporter AtHKT1;1 controls retrieval of Na^+ from the xylem in *Arabidopsis*. *Plant Cell and Environment* 30, 497–507.

Davies, W.J., Kudoyarova, G. and Hartung, W. (2005) Long-distance ABA signaling and its relation to other signaling pathways in the detection of soil drying and the mediation of the plant's response to drought. *Journal of Plant Growth Regulators* 24, 285–295.

De Boer, A.H. and Volkov, V. (2003) Logistics of water and salt transport through the plant: structure and functioning of the xylem. *Plant Cell and Environment* 26, 87–101.

Delauney, A.J. and Verma, D.P.S. (1993) Proline biosynthesis and osmoregulation in plants. *Plant Journal* 4, 215–223.

Demidchik, V. and Maathuis, F.J.M. (2007) Physiological roles of nonselective cation channels in plants: from salt stress to signalling and development. *New Phytologist* 175, 387–404.

Demidchik, V., Davenport, R.J. and Tester, M. (2002) Nonselective cation channels in plants. *Annual Review of Plant Biology* 53, 67–107.

Demidchik, V., Shabala, S.N., Coutts, K.B., Tester, M.A. and Davies, J.M. (2003) Free oxygen radicals regulate plasma membrane Ca^{2+} and K^+- permeable channels in plant root cells. *Journal of Cell Science* 116, 81–88.

Demidchik, V., Shang, Z.L., Shin, R., Thompson, E., Rubio, L. *et al.* (2009) Plant extracellular ATP signalling by plasma membrane NADPH oxidase and Ca^{2+} channels. *Plant Journal* 58, 903–913.

Demidchik, V., Cuin, T.A., Svistunenko, D., Smith, S.J., Miller, A.J., Shabala, S., Sokolik, A. and Yurin, V. (2010) *Arabidopsis* root K^+-efflux conductance activated by hydroxyl radicals: single-channel properties, genetic basis and involvement in stress-induced cell death. *Journal of Cell Science* 123, 1468–1479.

De Schepper, V., De Swaef, T., Bauweraerts, I. and Steppe, K. (2013) Phloem transport: a review of mechanisms and controls. *Journal of Experimental Botany* 64, 4839–4850.

Donoso, P., Mill, J.G., Oneill, S.C. and Eisner, D.A. (1992) Fluorescence measurements of cytoplasmic and mitochondrial sodium concentration in rat ventricular myocytes. *Journal of Physiology-London* 448, 493–509.

Dubcovsky, J., Maria, G.S., Epstein, E., Luo, M.C. and Dvořák, J. (1996) Mapping of the K^+/Na^+ discrimination locus Kna1 in wheat. *Theoretical and Applied Genetics* 92, 448–454.

Dunin, F.X., Smith, C.J., Zegelin, S.J., Leuning, R., Denmead, O.T. and Poss, R. (2001) Water balance changes in a crop sequence with lucerne. *Australian Journal of Agricultural Research* 52, 247–261.

Dvořák, J., Noaman, M.M., Goyal, S. and Gorham, J. (1994) Enhancement of the salt tolerance of *Triticum turgidum* L. by the Kna1 locus transferred from the *Triticum aestivum* L. chromosome 4D by homoeologous recombination. *Theoretical and Applied Genetics* 87, 872–877.

Flowers, T.J. (1974) Salt tolerance in *Suaeda maritima* (L.) Dum. A comparison of mitochondria isolated from green tissues of *Suaeda* and *Pisum*. *Journal of Experimental Botany* 25, 101–110.

Flowers, T.J. and Colmer, T.D. (2008) Salinity tolerance in halophytes. *New Phytologist* 179, 945–963.

Flowers, T.J. and Dalmond, D. (1992) Protein synthesis in halophytes: the influence of potassium, sodium and magnesium in vitro. *Plant and Soil* 146, 153–161.

Flowers, T.J. and Hanson, J.B. (1969) The effect of reduced water potential on soybean mitochondria. *Plant Physiology* 44, 939–945.

Flowers, T.J., Colmer, T.D. and Munns, R. (2015) Sodium chloride toxicity and the cellular basis of salt tolerance in halophytes. *Annals of Botany* 115, 419–431.

Foreman, J., Demidchik, V., Bothwell, J.H.F., Mylona, P., Miedema, H., Torres, M.A., Linstead, P., Costa, S., Brownlee, C., Jones, J.D.G., Davies, J.M. and Dolan, L. (2003) Reactive oxygen species produced by NADPH oxidase regulate plant cell growth. *Nature* 422, 442–446.

Foyer, C.H. and Noctor, G. (2005) Oxidant and antioxidant signalling in plants: a re-evaluation of the concept of oxidative stress in a physiological context. *Plant Cell and Environment* 28, 1056–1071.

Frensch, J. and Hsiao, T.C. (1994) Transient responses of cell turgor and growth of maize roots as affected by changes in water potential. *Plant Physiology* 104, 247–54.

Fricke, W. (2004) Rapid and tissue-specific accumulation of solutes in the growth zone of barley leaves in response to salinity. *Planta* 219, 515–525.

Fricke, W. and Peters, W.S. (2002) The biophysics of leaf growth in salt-stressed barley. A study at the cell level. *Plant Physiology* 129, 374–388.

Fricke, W., Akhiyarova, G., Veselov, D. and Kudoyarova, G. (2004) Rapid and tissue-specific changes in ABA and in growth rate response to salinity in barley leaves. *Journal of Experimental Botany* 55, 1115–1123.

Fricke, W., Akhiyarova, G., Wei, W., Alexandersson, E., Miller, A., Kjellbom, P.O., Richardson, A., Wojciechowski, T., Schreiber, L., Veselov, D., Kudoyarova, G. and Volkov, V. (2006) The short-term growth response to salt of the developing barley leaf. *Journal of Experimental Botany* 57, 1079–1095.

Garbarino, J. and Dupont, F.M. (1988) NaCl induces a Na$^+$/H$^+$ antiport in tonoplast vesicles from barley roots. *Plant Physiology* 86, 231–236.

Garcia, A., Rizzo, C.A., Ud-Din, J., Bartos, S.L., Senadhira, D., Flowers, T.J. and Yeo, A.R. (1997) Sodium and potassium transport to the xylem are inherited independently in rice, and the mechanism of sodium: potassium selectivity differs between rice and wheat. *Plant Cell and Environment* 20, 1167–1174.

Garciadeblás, B., Benito, B. and Rodríguez-Navarro, A. (2001) Plant cells express several stress calcium ATPases but apparently no sodium ATPase. *Plant and Soil* 235, 181–192.

Gassmann, W., Rubio, F. and Schroeder, J.I. (1996) Alkali cation selectivity of the wheat root high-affinity potassium transporter HKT1. *Plant Journal* 10, 869–882.

Gattineni, J. and Baum, M. (2015) Developmental changes in renal tubular transport – an overview. *Pediatric Nephrology* 30, 2085–2098.

Gaxiola, R.A., Rao, R., Sherman, A., Grisafi, P., Alper, S.L. and Fink, G.R. (1999) The *Arabidopsis thaliana* proton transporters, AtNhx1 and Avp1, can function in cation detoxification in yeast. *Proceedings of the National Academy of Sciences USA* 96, 1480–1485.

Gaymard, F., Pilot, G., Lacombe, B., Bouchez, D., Bruneau, D., Boucherez, J., Michaux-Ferriere, N., Thibaud, J.B. and Sentenac, H. (1998) Identification and disruption of a plant shaker-like outward channel involved in K$^+$ release into the xylem sap. *Cell* 94, 647–655.

Genc, Y., Oldach, K., Taylor, J. and Lyons, G.H. (2016) Uncoupling of sodium and chloride to assist breeding for salinity tolerance in crops. *New Phytologist* 210, 145–156.

Ghassemi, F., Jakeman, A.J. and Nix, H.A. (1995) *Salinisation of Land and Water Resources: Human Causes, Extent, Management and Case Studies.* UNSW Press, Sydney, Australia, and CAB International, Wallingford, UK.

Gibson, T.S., Speirs, J. and Brady, C.J. (1984) Salt-tolerance in plants. II. *In vivo* translation of m-RNAs from salt-tolerant and salt-sensitive plants on wheat germ ribosomes. Responses to ions and compatible organic solutes. *Plant, Cell and Environment* 7, 579–587.

Gimmler, H. (2000) Primary sodium plasma membrane ATPases in salt-tolerant algae: facts and fictions. *Journal of Experimental Botany* 51, 1171–1178.

Glenn, E.P., Brown, J.J. and Blumwald, E. (1999) Salt tolerance and crop potential of halophytes. *Critical Reviews in Plant Sciences* 18, 227–255.

Golldack, D. and Dietz, K.J. (2001) Salt-induced expression of the vacuolar H$^+$-ATPase in the common ice plant is developmentally controlled and tissue specific. *Plant Physiology* 125, 1643–1654.

Gorham, J., Jones, R.G.W. and McDonnell, E. (1985) Some mechanisms of salt tolerance in crop plants. *Plant and Soil* 89, 15–40.

Gorham, J., Bristol, A., Young, E.M. and Jones, R.G.W. (1991) The presence of the enhanced K/Na discrimination trait in diploid *Triticum* species. *Theoretical and Applied Genetics* 82, 729–736.

Greenway, H. and Munns, R. (1980) Mechanisms of salt tolerance in nonhalophytes. *Annual Review of Plant Physiology* 31, 149–190.

Greenway, H. and Osmond, C.B. (1972) Salt responses of enzymes from species differing in salt tolerance. *Plant Physiology* 49, 256–259.

Greenway, H. and Setter, T. (1979) Na⁺, Cl⁻ and K⁺ concentrations in *Chlorella emersonii* exposed to 100 and 335 mM NaCl. *Australian Journal of Plant Physiology* 6, 61–67.

Halvorson, A.D. (1990) Management of dryland saline seeps. In: Tanji, K.K. (ed.) *Agricultural Salinity Assessment and Management*. ACSE Manuals and Reports on Engineering Practice, No. 71. American Society of Civil Engineers, New York, pp. 372–392.

Hamdy, A. (1996) Saline irrigation: assessment and management techniques. In: Choukr-Allah, R., Malcolm, C.V. and Hamdy, A. (eds) *Halophytes and Biosaline Agriculture*. Marcel Dekker, New York, pp. 147–180.

Hedrich, R. (2012) Ion channels in plants. *Physiological Reviews* 92, 1777–1811.

Hu, Y. and Schmidhalter, U. (1998) Spatial distributions and net deposition rates of mineral elements in the elongating wheat (*Triticum aestivum* L.) leaf under saline soil conditions. *Planta* 204, 212–219.

Huang, S., Spielmeyer, W., Lagudah, E.S., James, R.A., Platten, J.D., Dennis, E.S. and Munns, R. (2006) A sodium transporter (HKT7) is a candidate for *Nax1*, a gene for salt tolerance in durum wheat. *Plant Physiology* 142, 1718–1727.

Huang, S., Spielmeyer, W., Lagudah, E.S. and Munns, R. (2008) Comparative mapping of HKT genes in wheat, barley and rice, key determinants of Na⁺ transport and salt tolerance. *Journal of Experimental Botany* 59, 927–937.

Hummel, I., Pantin, F., Sulpice, R., Piques, M., Rolland, G., Dauzat, M., Christophe, A., Pervent, M., Bouteille, M., Stitt, M., Gibon, Y. and Muller, B. (2010) *Arabidopsis* plants acclimate to water deficit at low cost through changes of carbon usage: an integrated perspective using growth, metabolite, enzyme, and gene expression analysis. *Plant Physiology* 154, 357–372.

Igarashi, Y., Yoshiba, Y., Sanada, Y., Yamaguchi-Shinozaki, K., Wada, K. and Shinozaki, K. (1997) Characterization of the gene for Δ(1)-pyrroline-5-carboxylate synthetase and correlation between the expression of the gene and salt tolerance in *Oryza sativa* L. *Plant Molecular Biology* 33, 857–865.

Jacoby, R.P., Che-Othman, M.H., Millar, A.H. and Taylor, N.L. (2016) Analysis of the sodium chloride-dependent respiratory kinetics of wheat mitochondria reveals differential effects on phosphorylating and non-phosphorylating electron transport pathways, *Plant Cell and Environment* 39, 823–833.

James, R.A., Rivelli, A.R., Munns, R. and von Caemmerer, S. (2002) Factors affecting CO₂ assimilation, leaf injury and growth in salt-stressed durum wheat. *Functional Plant Biology* 29, 1393–1403.

James, R.A., Davenport, R.J. and Munns, R. (2006) Physiological characterisation of two genes for Na⁺ exclusion in durum wheat: *Nax1* and *Nax2*. *Plant Physiology* 142, 1537–1547.

Jeschke, W.D. (1984) K⁺-Na⁺ exchange at cellular membranes, intracellular compartmentation of cations, and salt tolerance. In: Staples, R.C. (ed.) *Salinity Tolerance in Plants: Strategies for Crop Improvement*. Wiley, New York, pp. 37–66.

Jeschke, W.D., Aslam, Z. and Greenway, H. (1986) Effects of NaCl on ion relations and carbohydrate status of roots and on osmotic regulation of roots and shoots of *Atriplex amnicola*. *Plant, Cell and Environment* 9, 559–569.

Jeschke, W.D., Wolf, O. and Hartung, W. (1992) Effect of NaCl salinity on flows and partitioning of C, N, and mineral ions in whole plants of white lupin, *Lupinus albus* L. *Journal of Experimental Botany* 43, 777–788.

Jou, Y., Wang, Y.L. and Yen, H.C.E. (2007) Vacuolar acidity, protein profile, and crystal composition of epidermal bladder cells of the halophyte *Mesembryanthemum crystallinum*. *Functional Plant Biology* 34, 353–359.

Julkowska, M.M. and Testerink, C. (2015) Tuning plant signaling and growth to survive salt. *Trends in Plant Science* 20, 586–594.

Kingsbury, R.A. and Epstein, E. (1986) Salt sensitivity in wheat a case for specific ion toxicity. *Plant Physiology* 80, 651–654.

Köhler, B. and Raschke, K. (2000) The delivery of salts to the xylem. Three types of anion conductance in the plasmalemma of the xylem parenchyma of roots of barley. *Plant Physiology* 122, 243–254.

Kronzucker, H. and Britto, D.T. (2011) Sodium transport in plants: a critical review. *New Phytologist* 189, 54–81.

Kronzucker, H.J., Coskun, D., Schulze, L.M., Wong, J.R. and Britto, D.T. (2013) Sodium as a nutrient and toxicant. *Plant and Soil* 369, 1–23.

Kurusu, T., Kuchitsu, K., Nakano, M., Nakayama, Y. and Iida, H. (2013) Plant mechanosensing and Ca²⁺ transport. *Trends in Plant Science* 18, 227–233.

Laohavisit, A., Richards, S.L., Shabala, L., Chen, C., Colaco, R. *et al.* (2013) Salinity-induced calcium signaling and root adaptation in Arabidopsis require the calcium regulatory protein Annexin1. *Plant Physiology* 163, 253–262.

Laurie, S., Feeney, K.A., Maathuis, F.J.M., Heard, P.J., Brown, S.J. and Leigh, R.A. (2002) A role for HKT1 in sodium uptake by wheat roots. *Plant Journal* 32, 139–149.

Layzell, D.B., Pate, J.S., Atkins, C.A. and Canvin, D.T. (1981) Partitioning of carbon and nitrogen and the nutrition of root and shoot apex in a nodulated legume. *Plant Physiology* 67, 30–36.

Lazof, D. and Läuchli, A. (1991) The nutritional status of the apical meristem of *Lactuca sativa* as affected by NaCl salinization: an electron probe microanalytic study. *Planta* 184, 334–342.

Lecourieux, D., Mazars, C., Pauly, N., Ranjeva, R. and Pugin, A. (2002) Analysis and effects of cytosolic free calcium increases in response to elicitors in *Nicotiana plumbaginifolia* cells. *The Plant Cell* 14, 2627–2641.

Lee, S., Lee, E.J., Yang, E.J., Lee, J.E., Park, A.R., Song, W.H. and Park, O.K. (2004) Proteomic identification of annexins, calcium-dependent membrane binding proteins that mediate osmotic stress and abscisic acid signal transduction in Arabidopsis. *The Plant Cell* 16, 1378–1391.

Li, J.Y., Jiang, A.L. and Zhang, W. (2007) Salt stress-induced programmed cell death in rice root tip cells. *Journal of Integrative Plant Biology* 49, 481–486.

Liu, S.P., Zheng, L.Q., Xue, Y.H., Zhang, Q.A., Wang, L. and Shou, H.X. (2010) Overexpression of OsVP1 and OsNHX1 increases tolerance to drought and salinity in rice. *Journal of Plant Biology* 53, 444–452.

Lovisolo, C., Perrone, I., Carra, A., Ferrandino, A., Flexus, J., Medrano, H. and Schubert, A. (2010) Drought-induced changes in development and function of grapevine (*Vitis* spp.) organs in their hydraulic and non-hydraulic interactions at the whole-plant level: a physiological and molecular update. *Functional Plant Biology* 37, 98–116.

Ludlow, M.M. and Muchow, R.C. (1990) A critical evaluation of traits for improving crop yields in water-limited environments. *Advances in Agronomy* 43, 107–153.

Luna, C., Seffino, L.G., Arias, C. and Taleisnik, E. (2000) Oxidative stress indicators as selection tools for salt tolerance in *Chloris gayana*. *Plant Breeding* 119, 341–345.

Maas, E.V. and Hoffman, G.J. (1977) Crop salt tolerance – current assessment. *Journal of the Irrigation and Drainage Division American Society of Civil Engineering* 103, 115–134.

Maathuis, F.J.M. (2014) Sodium in plants: perception, signalling, and regulation of sodium fluxes. *Journal of Experimental Botany* 65, 849–858.

Maathuis, F.J.M. and Amtmann, A. (1999) K^+ nutrition and Na^+ toxicity: the basis of cellular K^+/Na^+ ratios. *Annals of Botany* 84, 123–133.

Maathuis, F.J.M. and Sanders, D. (2001) Sodium uptake in Arabidopsis roots is regulated by cyclic nucleotides. *Plant Physiology* 127, 1617–1625.

Marcum, K.B. and Murdoch, C.L. (1992) Salt tolerance of the coastal salt-marsh grass, *Sporobolus virginicus* (L.) Kunth. *New Phytologist* 120, 281–288.

Marcum, K.B., Yensen, N.P. and Leake, J.E. (2007) Genotypic variation in salinity tolerance of *Distichlis spicata* turf ecotypes. *Australian Journal of Experimental Agriculture* 47, 1506–1511.

Marschner, H. (1995) *The Mineral Nutrition of Higher Plants.* Academic Press, London.

Matsushita, N. and Matoh, T. (1991) Characterization of Na^+ exclusion mechanisms of salt-tolerant reed plants in comparison with salt-sensitive rice plants. *Physiologia Plantarum* 83, 170–176.

Millar, A.H., Whelan, J., Soole, K.L. and Day, D.A. (2011) Organization and regulation of mitochondrial respiration in plants. *Annual Review of Plant Biology* 62, 79–104.

Miller, F., Schlosser, P. and Janszen, D. (2000) Haber's rule: a special case in a family of curves relating concentration and duration of exposure to a fixed level of response for a given endpoint. *Toxicology in Vitro* 149, 21–24.

Miller, G., Shulaev, V. and Mittler, R. (2008) Reactive oxygen signaling and abiotic stress. *Physiologia Plantarum* 133, 481–489.

Miller, G.A.D., Suzuki, N., Ciftci-Yilmaz, S. and Mittler, R.O.N. (2010) Reactive oxygen species homeostasis and signalling during drought and salinity stresses. *Plant, Cell & Environment* 33, 453–467.

Mittler, R. (2002) Oxidative stress, antioxidants and stress tolerance. *Trends in Plant Science* 7, 405–410.

Mittler, R., Vanderauwera, S., Gollery, M. and Van Breusegem, F. (2004) Reactive oxygen gene network of plants. *Trends in Plant Science* 9, 490–498.

Møller, I.S., Gilliham, M., Jha, D., Mayo, G.M., Roy, S.J., Coates, J.C., Haseloff, J. and Tester, M. (2009) Shoot Na^+ exclusion and increased salinity tolerance engineered by cell type-specific alteration of Na^+ transport in Arabidopsis. *The Plant Cell* 21, 2163–2178.

Monshausen, G.B. and Haswell, E.S. (2013) A force of nature: molecular mechanisms of mechanoperception in plants. *Journal of Experimental Botany* 64, 4663–4680.

Mullan, D.J., Colmer, T.D. and Francki, M.G. (2007) *Arabidopsis*–rice–wheat gene orthologues for Na^+ transport and transcript analysis in wheat–*L. elongatum* aneuploids under salt stress. *Molecular Genetics and Genomics* 277, 199–212.

Munns, R. (1993) Physiological processes limiting plant growth in saline soil: some dogmas and hypotheses. *Plant Cell and Environment* 16, 15–24.

Munns, R. (2002) Comparative physiology of salt and water stress. *Plant Cell and Environment* 25, 239–250.

Munns, R. (2005) Genes and salt tolerance: bringing them together. *New Phytologist* 167, 645–663.

Munns, R. (2007) Utilising genetic resources to enhance productivity of salt-prone land. *CAB Reviews: Perspectives in Agriculture, Veterinary Science, Nutrition and Natural Resources* 2, No. 009. doi: 10.1079/PAVSNNR20072009.

Munns, R. and Rawson, H.M. (1999) Effect of salinity on salt accumulation and reproductive development in the apical meristem of wheat and barley. *Australian Journal of Plant Physiology* 26, 459–464.

Munns, R. and Sharp, R.E. (1993) Involvement of abscisic acid in controlling plant growth in soils of low water potential. *Australian Journal of Plant Physiology* 20, 425–437.

Munns, R. and Tester, M. (2008) Mechanisms of salinity tolerance. *Annual Review of Plant Biology* 59, 651–681.

Munns, R., Tonnet, M.L., Shennan, C. and Gardner, P.A. (1988) Effect of high external NaCl concentrations on ion transport within the shoot of *Lupinus albus*. II. Ions in phloem sap. *Plant, Cell & Environment* 11, 291–300.

Munns, R., Schachtman, D.P. and Condon, A.G. (1995) The significance of a two-phase growth response to salinity in wheat and barley. *Australian Journal of Plant Physiology* 22, 561–569.

Munns, R., Guo, J., Passioura, J.B. and Cramer, G.R. (2000) Leaf water status controls day-time but not daily rates of leaf expansion in salt-treated barley. *Australian Journal of Plant Physiology* 27, 949–957.

Munns, R., James, R.A. and Läuchli, A. (2006) Approaches to increasing the salt tolerance of wheat and other cereals. *Journal of Experimental Botany* 57, 1025–1043.

Munns, R., James, R.A., Xu, B., Athman, A., Conn, S.J., Jordans, C., Byrt, C.S., Hare, R.A., Tyerman, S.D., Tester, M., Plett, D. and Gilliham, M. (2012) Wheat grain yield on saline soils is improved by an ancestral Na$^+$ transporter gene. *Nature Biotechnology* 30, 360–364.

Muñoz-Mayor, A., Pineda, B., Garcia-Abellan, J.O., Garcia-Sogo, B., Moyano, E., Atares, A., Vicente-Agullo, F., Serrano, R., Moreno, V. and Bolarin, M.C. (2008) The HAL1 function on Na$^+$ homeostasis is maintained over time in salt-treated transgenic tomato plants, but the high reduction of Na$^+$ in leaf is not associated with salt tolerance. *Physiologia Plantarum* 133, 288–297.

Murphy, E. and Eisner, D.A. (2009) Regulation of intracellular and mitochondrial sodium in health and disease. *Circulation Research* 104, 292–303.

Negrão, S., Almadanim, C.M., Pires, I.S., Abreu, I.A., Maroco, J., Courtois, B., Gregorio, G.B., McNally, K.L. and Oliveira, M.M. (2013) New allelic variants found in key rice salt-tolerance genes: an association study. *Plant Biotechnology Journal* 11, 87–100.

Niu, X.M., Narasimhan, M.L., Salzman, R.A., Bressan, R.A. and Hasegawa, P.M. (1993) NaCl regulation of plasma membrane H$^+$-ATPase gene expression in a glycophyte and a halophyte. *Plant Physiology* 103, 713–718.

Noctor, G. and Foyer, C.H. (1998) Ascorbate and glutathione: keeping active oxygen under control. *Annual Review of Plant Physiology and Plant Molecular Biology* 49, 249–279.

Noctor, G., De Paepe, R. and Foyer, C.H. (2007) Mitochondrial redox biology and homeostasis in plants. *Trends in Plant Science* 12, 125–134.

Oh, D.H., Dassanayake, M., Haas, J.S., Kropornika, A., Wright, C., d'Urzo, M.P., Hong, H., Ali, S., Hernandez, A., Lambert, G.M., Inan, G., Galbraith, D.W., Bressan, R.A., Yun, D.J., Zhu, J.K., Cheeseman, J.M. and Bohnert, H.J. (2010) Genome structures and halophyte-specific gene expression of the extremophile *Thellungiella parvula* in comparison with *Thellungiella salsuginea* (*Thellungiella halophila*) and *Arabidopsis*. *Plant Physiology* 154, 1040–1052.

Oren, A. (1999) Bioenergetic aspects of halophilism. *Microbiology and Molecular Biology Reviews* 63, 334–348.

Ottow, E.A., Brinker, M., Teichmann, T., Fritz, E., Kaiser, W., Brosché, M., Kangasjärvi, J., Jiang, X. and Polle, A. (2005) *Populus euphratica* displays apoplastic sodium accumulation, osmotic adjustment by decreases in calcium and soluble carbohydrates, and develops leaf succulence under salt stress. *Plant Physiology* 139, 1762–1772.

Ozgur, R., Uzilday, B., Sekmen, A.H. and Turkan, I. (2013) Reactive oxygen species regulation and antioxidant defence in halophytes. *Functional Plant Biology* 40, 832–847.

Pardo, J.M., Cubero, B., Leidi, E.O. and Quintero, F.J. (2006) Alkali cation exchangers: roles in cellular homeostasis and stress tolerance. *Journal of Experimental Botany* 57, 1181–1199.

Passioura, J.B. and Munns, R. (2000) Rapid environmental changes that affect leaf water status induce transient surges or pauses in leaf expansion rate. *Australian Journal of Plant Physiology* 27, 941–948.

Paul, M.J. and Foyer, C.H. (2001) Sink regulation of photosynthesis. *Journal of Experimental Botany* 52, 1383–1400.

Pérez-Alfocea, F., Balibrea, M.E., Alarcon, J.J. and Bolarin, M.C. (2000) Composition of xylem and phloem exudates in relation to the salt-tolerance of domestic and wild tomato species. *Journal of Plant Physiology* 156, 367–374.

Pérez-Alfocea, F., Albacete, A., Ghanem, M.E. and Dodd, I.C. (2010) Hormonal regulation of source–sink relations to maintain crop productivity under salinity: a case study of root-to-shoot signalling in tomato. *Functional Plant Biology* 37, 592–603.

Platten, J.D., Cotsaftis, O., Berthomieu, P., Bohnert, H., Davenport, R.J., Fairbairn, D.J., Horie, T., Leigh, R.A., Lin, H.X., Luan, S., Maser, P., Pantoja, O., Rodríguez-Navarro, A., Schachtman, D.P., Schroeder, J.I., Sentenac, H., Uozumi, N., Véry, A.A., Zhu, J.K., Dennis, E.S. and Tester, M. (2006) Nomenclature for HKT transporters, key determinants of plant salinity tolerance. *Trends in Plant Science* 11, 372–374.

Platten, J.D., Egdane, J.A. and Ismail, A.M. (2013) Salinity tolerance, Na$^+$ exclusion and allele mining of HKT1;5 in *Oryza sativa* and *O. glaberrima*: many sources, many genes, one mechanism? *BMC Plant Biology* 13, 32.

Plett, D., Toubia, J., Garnett, T., Tester, M., Kaiser, B.N. and Baumann, U. (2010) Dichotomy in the NRT gene families of dicots and grass species. *PLoS ONE* 5, dx.doi.org/10.1371/journal.pone.0015289.

Pottosin, I., Velarde-Buendia, A.M., Bose, J., Zepeda-Jazo, I., Shabala, S. and Dobrovinskaya, O. (2014) Cross-talk between reactive oxygen species and polyamines in regulation of ion transport across the plasma membrane: implications for plant adaptive responses. *Journal of Experimental Botany* 65, 1271–1283.

Preston, C. and Critchley, C. (1986) Differential effects of K$^+$ and Na$^+$ on oxygen evolution activity of photosynthetic membranes from two halophytes and spinach. *Australian Journal of Plant Physiology* 13, 491–498.

Qadir, M. and Schubert, S. (2002) Degradation processes and nutrient constraints in sodic soils. *Land Degradation and Development* 13, 275–294.

Qadir, M., Quillerou, E., Nangia, V., Murtaza, G., Singh, M., Thomas, R.J., Drechsel, P and Noble, A.D. (2014) Economics of salt-induced land degradation and restoration. *Natural Resources Forum* 38, 282–295.

Qiu, Q.-S. (2012) Plant and yeast NHX antiporters: roles in membrane trafficking. *Journal of Integrative Plant Biology* 54, 66–72.

Qiu, Q.S., Barkla, B.J., Vera-Estrella, R., Zhu, J.K. and Schumaker, K.S. (2003) Na$^+$/H$^+$ exchange activity in the plasma membrane of *Arabidopsis*. *Plant Physiology* 132, 1041–1052.

Quintero, F.J., Blatt, M.R. and Pardo, J.M. (2000) Functional conservation between yeast and plant endosomal Na$^+$/H$^+$ antiporters. *FEBS Letters* 471, 224–228.

Rahnama, A., Munns, R., Poustini, K. and Watt, M. (2011) A screening method to identify genetic variation in root growth response to a salinity gradient. *Journal of Experimental Botany* 62, 69–77.

Raven, J.A. (1985) Regulation of pH and generation of osmolarity in vascular plants: a cost-benefit analysis in relation to efficiency of use of energy, nitrogen and water. *New Phytologist* 101, 25–77.

Rea, P.A. and Poole, R.J. (1993) Vacuolar H$^+$-translocating pyrophosphatase. *Annual Review of Plant Physiology and Plant Molecular Biology* 44, 157–180.

Reinhardt, D.H. and Rost, T.L. (1995a) On the correlation of primary root growth and tracheary element size and distance from the tip in cotton seedlings grown under salinity. *Environmental and Experimental Botany* 35, 575–588.

Reinhardt, D.H. and Rost, T.L. (1995b) Developmental changes of cotton root primary tissues induced by salinity. *International Journal of Plant Sciences* 156, 505–513.

Ren, Z.-H., Gao, J.-P., Li, L.-G., Cai, X.-L., Huang, W., Chao, D.-Y., Zhu, M.-Z., Zong-Yang, W., Luan, S. and Hong-Xuan, L. (2005) A rice quantitative trait locus for salt tolerance encodes a sodium transporter. *Nature Genetics* 37, 1141–1146.

Rengasamy, P. (2002) Transient salinity and subsoil constraints to dryland farming in Australian sodic soils: an overview. *Australian Journal of Experimental Agriculture* 42, 351–361.

Rhoades, J.D., Kandiah, A. and Mashali, A.M. (1992) The use of saline waters for crop production. FAO Irrigation and Drainage Paper 48. Available at: http://www.fao.org/docrep/T0667E/T0667E00.htm, accessed 10 July 2016.

Richards, L.A. (ed.) (1954) *Diagnosis and improvement of saline and alkali soils*. Agriculture Handbook No 60. United States Department of Agriculture, Agricultural Research Service. Available at: http://www.ars.usda.gov/Services/docs.htm?docid=10158 (accessed 10 July 2016).

Robinson, S.P., Downton, W.J.S and Millhouse, J.A. (1983) Photosynthesis and ion content of leaves and isolated chloroplasts of salt-stressed spinach. *Plant Physiology* 73, 238–242.

Roelfsema, M.R.G., Hedrich, R. and Geiger, D. (2012) Anion channels: master switches of stress responses. *Trends in Plant Science* 17, 221–229.

Rubio, F., Gassmann, W. and Schroeder, J.I. (1995) Sodium-driven potassium uptake by the plant potassium transporter HKT and mutations conferring salt tolerance. *Science* 270, 1660–1663.

Rus, A., Yokoi, S., Sharkhuu, A., Reddy, M., Lee, B.H., Matsumoto, T.K., Koiwa, H., Zhu, J.K., Bressan, R.A. and Hasegawa, P.M. (2001) AtHKT1 is a salt tolerance determinant that controls Na$^+$ entry into plant roots. *Proceedings of the National Academy of Sciences USA* 98, 14150–14155.

Sakamoto, A. and Murata, N. (2000) Genetic engineering of glycinebetaine synthesis in plants: current status and implications for enhancement of stress tolerance. *Journal of Experimental Botany* 51, 81–88.

Scandalios, J.G. (1993) Oxygen stress and superoxide dismutases. *Plant Physiology* 101, 7–12.

Seemann, J.R. and Chritchley, C. (1985) Effects of salt stress on the growth, ion content, stomatal behaviour and photosynthetic capacity of a salt-sensitive species, *Phaseolus vulgaris* L. *Planta* 164, 151–162.

Shabala, S. (2003) Regulation of potassium transport in leaves: from molecular to tissue level. *Annals of Botany* 92, 627–634.

Shabala, S. (2007) Transport from root to shoot. In: Yeo, A.R. and Flowers, T.J. (eds) *Plant Solute Transport*. Blackwell Publishing, Oxford, UK, pp. 214–234.

Shabala, S. (2009) Salinity and programmed cell death: unravelling mechanisms for ion specific signalling. *Journal of Experimental Botany* 60, 709–711.

Shabala, S. (2013) Learning from halophytes: physiological basis and strategies to improve abiotic stress tolerance in crops. *Annals of Botany* 112, 1209–1221.

Shabala, S. and Cuin, T.A. (2008) Potassium transport and plant salt tolerance. *Physiologia Plantarum* 133, 651–669.

Shabala, S. and Mackay, A. (2011) Ion transport in halophytes. *Advances in Botanical Research* 57, 151–199.

Shabala, S. and Pottosin, I. (2014) Regulation of potassium transport in plants under hostile conditions: implications for abiotic and biotic stress tolerance. *Physiologia Plantarum* 151 (3), 257–279.

Shabala, S. and Shabala, L. (2011) Ion transport and osmotic adjustment in plants and bacteria. *BioMolecular Concepts* 2, 407–419.

Shabala, S., Demidchik, V., Shabala, L., Cuin, T.A., Smith, S.J., Miller, A.J., Davies, J.M. and Newman, I.A. (2006) Extracellular Ca²⁺ ameliorates NaCl-induced K⁺ loss from *Arabidopsis* root and leaf cells by controlling plasma membrane K⁺-permeable channels. *Plant Physiology* 141, 1653–1665.

Shabala, S., Cuin, T.A., Pang, J.Y., Percey, W., Chen, Z.H., Conn, S., Eing, C. and Wegner, L.H. (2010) Xylem ionic relations and salinity tolerance in barley. *Plant Journal* 61, 839–853.

Shabala, S., Bose, J. and Hedrich, R. (2014) Salt bladders: do they matter? *Trends in Plant Science* 19, 687–691.

Shabala, S., Wu, H.H. and Bose, J. (2015) Salt stress sensing and early signalling events in plant roots: current knowledge and hypothesis. *Plant Science* 241, 109–119.

Shabala, S.N. and Lew, R.R. (2002) Turgor regulation in osmotically stressed *Arabidopsis* epidermal root cells. Direct support for the role of inorganic ion uptake as revealed by concurrent flux and cell turgor measurements. *Plant Physiology* 129, 290–299.

Shen, B., Jensen, R.G. and Bohnert, H.J. (1997) Increased resistance to oxidative stress in transgenic plants by targeting mannitol biosynthesis to chloroplasts. *Plant Physiology* 113, 1177–1183.

Shi, H.Z., Ishitani, M., Kim, C.S. and Zhu, J.K. (2000) The *Arabidopsis thaliana* salt tolerance gene SOS1 encodes a putative Na⁺/H⁺ antiporter. *Proceedings of the National Academy of Sciences USA* 97, 6896–6901.

Shi, H.Z., Quintero, F.J., Pardo, J.M. and Zhu, J.K. (2002) The putative plasma membrane Na⁺/H⁺ antiporter SOS1 controls long-distance Na⁺ transport in plants. *The Plant Cell* 14, 465–477.

Shi, H.Z., Kim, Y., Guo, Y., Stevenson, B. and Zhu, J.K. (2003) The *Arabidopsis* SOS5 locus encodes a putative cell surface adhesion protein and is required for normal cell expansion. *The Plant Cell* 15, 19–32.

Smethurst, C.F., Rix, K., Garnett, T., Auricht, G., Bayart, A., Lane, P., Wilson, S.J. and Shabala, S. (2008) Multiple traits associated with salt tolerance in lucerne: revealing the underlying cellular mechanisms. *Functional Plant Biology* 35, 640–650.

Smith, C.A., Melino, V.J., Sweetman, C. and Soole, K.L. (2009) Manipulation of alternative oxidase can influence salt tolerance in *Arabidopsis thaliana*. *Physiologia Plantarum* 137, 459–472.

Starremans, P., Kersten, F.F.J., Van Den Heuvel, L., Knoers, N. and Bindels, R.J.M. (2003) Dimeric architecture of the human bumetanide-sensitive Na-K-Cl Co-transporter. *Journal of the American Society of Nephrology* 14, 3039–3046.

Summers, P.S. and Weretilnyk, E.A. (1993) Choline synthesis in spinach in relation to salt stress. *Plant Physiology* 103, 1269–1276.

Sun, J., Zhang, X., Deng, S.R., Zhang, C.L., Wang, M.J. *et al.* (2012) Extracellular ATP signaling is mediated by H₂O₂ and cytosolic Ca²⁺ in the salt response of *Populus euphratica* cells. *PloS ONE* 7, doi:10.1371/journal.pone.0053136.

Sunarpi, Horie, T., Motoda, J., Kubo, M., Yang, H., Yoda, K., Horie, R., Chan, W.Y., Leung, H.Y., Hattori, K., Konomi, M., Osumi, M., Yamagami, M., Schroeder, J.I. and Uozumi, N. (2005) Enhanced salt tolerance mediated by AtHKT1 transporter-induced Na⁺ unloading from xylem vessels to xylem parenchyma cells. *Plant Journal* 44, 928–938.

Suzuki, K., Yamaji, N., Costa, A., Okuma, E., Kobayashi, N.I., Kashiwagi, T., Katsuhara, M., Wang, C., Tanoi, K., Murata, Y., Schroeder, J.I., Ma, J.F. and Horie, T. (2016) OsHKT1;4-mediated Na⁺ transport in stems contributes to Na⁺ exclusion from leaf blades of rice at the reproductive growth state. *BMC Plant Biology* 16, 22.

Szabolcs, I. (1989) *Salt-Affected Soils*. CRC Press, Boca Raton, Florida.

Talke, I.N., Blaudez, D., Maathuis, F.J.M. and Sanders, D. (2003) CNGCs: prime targets of plant cyclic nucleotide signalling? *Trends in Plant Science* 8, 286–293.

Tanaka, K., Gilroy, S., Jones, A.M. and Stacey, G. (2010) Extracellular ATP signaling in plants. *Trends in Cell Biology* 20, 601–608.

Termaat, A. and Munns, R. (1986) Use of concentrated macronutrient solutions to separate osmotic from NaCl-specific effects on plant growth. *Australian Journal of Plant Physiology* 13, 109–122.

Termaat, A., Passioura, J.B. and Munns, R. (1985) Shoot turgor does not limit shoot growth of NaCl-affected wheat and barley. *Plant Physiology* 77, 869–872.

Tester, M. and Davenport, R. (2003) Na⁺ tolerance and Na⁺ transport in higher plants. *Annals of Botany* 91, 503–527.

Tyerman, S.D. and Skerrett, I.M. (1999) Root ion channels and salinity. *Scientia Horticulturae* 78, 175–235.

USDA-ARS (2011) Research databases. Bibliography on salt tolerance. George E. Brown Jr Salinity Lab. US Department of Agriculture, Agricultural Research Service, Riverside, California. Available at: http://www.ars.usda.gov/Services/docs.htm?docid=8908 (accessed 10 July 2016).

Véry, A.A. and Sentenac, H. (2002) Cation channels in the *Arabidopsis* plasma membrane. *Trends in Plant Science* 7, 168–175.

Vitart, V., Baxter, I., Doerner, P. and Harper, J.F. (2001) Evidence for a role in growth and salt resistance of a plasma membrane H⁺-ATPase in the root endodermis. *Plant Journal* 27, 191–201.

Volkov, V., Wang, B., Dominy, P.J., Fricke, W. and Amtmann, A. (2004) *Thellungiella halophila*, a salt-tolerant relative of *Arabidopsis thaliana*, possesses effective mechanisms to discriminate between potassium and sodium. *Plant Cell and Environment* 27, 1–14.

Wang, P., Li, Z.W., Wei, J.S., Zhao, Z.L., Sun, D.Y. and Cui, S.J. (2012) A Na⁺/Ca²⁺ exchanger-like protein (AtNCL) involved in salt stress in Arabidopsis. *The Journal of Biological Chemistry* 287, 44062–44070.

Wegner, L.H. and De Boer, A.H. (1997) Properties of two outward-rectifying channels in root xylem parenchyma cells suggest a role in K⁺ homeostasis and long-distance signaling. *Plant Physiology* 115, 1707–1719.

Wegner, L.H. and Raschke, K. (1994) Ion channels in the xylem parenchyma of barley roots – a procedure to isolate protoplasts from this tissue and a patch-clamp exploration of salt passageways into xylem vessels. *Plant Physiology* 105, 799–813.

Wegner, L.H., Stefano, G., Shabala, L., Rossi, M., Mancuso, S. and Shabala, S. (2011) Sequential depolarization of root cortical and stelar cells induced by an acute salt shock – implications for Na⁺ and K⁺ transport into xylem vessels. *Plant Cell and Environment* 34, 859–869.

Wolf, O., Munns, R., Tonnet, M.L. and Jeschke, W.D. (1990) Concentrations and transport of solutes in xylem and phloem along the leaf axis of NaCl-treated *Hordeum vulgare*. *Journal of Experimental Botany* 41, 1133–1141.

Wolf, O., Munns, R., Tonnet, M.L. and Jeschke, W.D. (1991) The role of the stem in the partitioning of Na⁺ and K⁺ in salt-treated barley. *Journal of Experimental Botany* 42, 697–704.

Wu, H.H., Shabala, L., Liu, X.H., Azzarello, E., Zhou, M., Pandolfi, C., Chen, Z.H., Bose, J., Mancuso, S. and Shabala, S. (2015a) Linking salinity stress tolerance tissue-specific Na⁺ sequestration in wheat roots. *Frontiers in Plant Science* 6, doi: 10.3389/fpls.2015.00071.

Wu, H.H., Zhu, M., Shabala, L., Zhou, M.X. and Shabala, S. (2015b) K⁺ retention in leaf mesophyll, an overlooked component of salinity tolerance mechanism: a case study for barley. *Journal of Integrative Plant Biology* 57, 171–185.

Yadav, R., Flowers, T.J. and Yeo, A.R. (1996) The involvement of the transpirational bypass flow in sodium uptake by high- and low-sodium-transporting lines of rice developed through intravarietal selection. *Plant, Cell and Environment* 19, 329–336.

Yamaguchi, T., Hamamoto, S. and Uozumi, N. (2013) Sodium transport system in plant cells. *Frontiers in Plant Science* 4, doi: 10.3389/fpls.2013.00410.

Yeo, A.R., Lee, K.S., Izard, P., Boursier, P.J. and Flowers, T.J. (1991) Short- and long-term effects of salinity on leaf growth in rice (*Oryza sativa* L.). *Journal of Experimental Botany* 42, 881–889.

Yeo, A.R., Flowers, S.A., Rao, G., Welfare, K., Senanayake, N. and Flowers, T.J. (1999) Silicon reduces sodium uptake in rice (*Oryza sativa* L.) in saline conditions and this is accounted for by a reduction in the transpirational bypass flow. *Plant Cell and Environment* 22, 559–565.

Yokoi, S., Quintero, F.J., Cubero, B., Ruiz, M.T., Bressan, R.A., Hasegawa, P.M. and Pardo, J.M. (2002) Differential expression and function of *Arabidopsis thaliana* NHX Na⁺/H⁺ antiporters in the salt stress response. *Plant Journal* 30, 529–539.

Zeng, L.H., Kwon, T.R., Liu, X.A., Wilson, C., Grieve, C.M. and Gregorio, G.B. (2004) Genetic diversity analyzed by microsatellite markers among rice (*Oryza sativa* L.) genotypes with different adaptations to saline soils. *Plant Science* 166, 1275–1285.

Zhang, H.X. and Blumwald, E. (2001) Transgenic salt-tolerant tomato plants accumulate salt in foliage but not in fruit. *Nature Biotechnology* 19, 765–768.

Zhang, H.X., Hodson, J.N., Williams, J.P. and Blumwald, E. (2001) Engineering salt-tolerant *Brassica* plants: characterization of yield and seed oil quality in transgenic plants with increased vacuolar sodium accumulation. *Proceedings of the National Academy of Sciences USA* 98, 12832–12836.

Zhang, J.X., Nguyen, H.T. and Blum, A. (1999) Genetic analysis of osmotic adjustment in crop plants. *Journal of Experimental Botany* 50, 291–302.

Zhong, H. and Läuchli, A. (1994) Spatial distribution of solutes, K, Na, Ca and their deposition rates in the growth zone of primary cotton roots: effects of NaCl and CaCl₂. *Planta* 194, 34–41.

Zhu, J.K. (2003) Regulation of ion homeostasis under salt stress. *Current Opinion in Plant Biology* 6, 441–445.

Zhu, J.K., Bressan, R.A., Hasegawa, P.M. and Pardo, J. (2005) Salt and crops: salinity tolerance. Success Stories in Agriculture. Council for Agricultural Science and Technology. Available at: https://www2.ag.purdue.edu/hla/zhulab/Documents/Publications/2005/bohnert.pdf (accessed 17 November 2016).

Zhu, M., Zhou, M.X., Shabala, L. and Shabala, S. (2016a) Physiological and molecular mechanisms mediating xylem Na⁺ loading in barley in the context of salinity stress tolerance. *Plant, Cell and Environment* (in press). doi: 10.1111/pce.12727.

Zhu, M., Shabala, L., Cuin, T.A., Huang, X., Zhou, M., Munns, R. and Shabala, S. (2016b) *Nax* loci affect SOS1-like Na⁺/H⁺ exchanger expression and activity in wheat. *Journal of Experimental Botany* 67, 835–844.

3 Reactive Oxygen Species and Their Role in Plant Oxidative Stress

Vadim Demidchik[1,2,*]

[1]*Department of Plant Cell Biology and Bioengineering, Belarusian State University, Minsk, Belarus;* [2]*Russian Academy of Sciences, Komarov Botanical Institute, St Petersburg, Russia*

Abstract

Oxidative stress is a physiological response due to progressive accumulation under some circumstances of reactive oxygen species (ROS) and oxidized forms of biomolecules. This response is associated with damage of all components of the cell, leading to pathophysiological processes. At the mechanistic level, oxidative stress can be caused by: (i) external oxidizers (UV, O_3, halogens, gamma radiation, extreme light, some xenobiotics, etc.) and ˙OH-producing transition metals (Cu, Fe, Mn, Hg, Ni, etc.); (ii) cellular programmes inducing ROS generation as a part of a response to abiotic and biotic stresses; and (iii) ROS production for needs of normal physiology, such as programmed cell death and autophagy. The intensity and consequences of an oxidative stress depends on a biological system's ability to detoxify ROS and to repair oxidative damage. Antioxidants stop or delay the oxidation of biomolecules ameliorating oxidative stress-induced damage, and orchestrate ROS signalling. The origins of ROS generation leading to oxidative stress include electron-transport chains of chloroplasts, mitochondria and peroxisomes, NADPH oxidases, peroxidases, phospholipases, oxygenases and some other systems. These systems produce $O_2^{\cdot-}$, singlet oxygen and NO as well as oxidized forms of organic molecules, which can give other ROS and free radicals in living cells. The most dangerous and highly oxidizing ROS is ˙OH, formed via Fenton-like reactions catalysed by transition metals. ROS are sensed via specific regulatory proteins, and are the reason for altered cell signalling and gene expression. They can trigger cytosolic Ca^{2+} elevation, K^+ loss, autophagy and programmed cell death. Downstream ROS-Ca^{2+}-regulated signalling cascades include regulatory systems with one (ion channels and transcription factors), two (Ca^{2+}-activated NADPH oxidases and calmodulin) or multiple components (receptor-like and Ca^{2+}-dependent protein kinases, mitogen-activated protein kinases, etc.). Currently, research into plant oxidative stress is at the forefront of developing stress-tolerant agricultural plants and designing strategies for adapting to global climate changes.

3.1 Introduction

A physiological response (physiological state) induced by over-production and accumulation of molecules containing activated oxygen, so-called reactive oxygen species (ROS; Fig. 3.1) is known as an oxidative stress. This state is accompanied by a progressive accumulation of oxidized and sometimes inactive forms of biomolecules, including proteins, nucleic acids, polysaccharides, lipids and metabolites, leading to pathophysiological phenomena. Most abiotic and biotic stresses, which do not induce direct oxidation, trigger oxidative stress through *de novo* biosynthesis of ROS. Oxidative stress then serves as a damaging factor for these stresses. At the same time, ROS accumulation during stresses plays the role of ubiquitous stress signal, triggering adaptation and immunity reactions.

---

* dzemidchyk@bsu.by

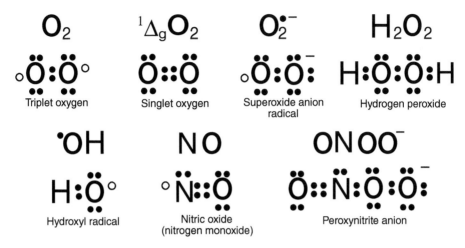

Fig. 3.1. Lewis dot structures of reactive oxygen species (ROS) and reactive nitrogen species (RNS).

The major physiological mechanisms of an oxidative stress include: (i) an imbalance between quantities of produced and detoxified ROS due to disturbance of 'normal' cell metabolism; and (ii) *de novo* biosynthesis of ROS as a constituent part of developmental processes, such as programmed cell death (PCD) or signalling and immunity response needed for defence and adaptation. These mechanisms can co-exist, because stress factors such as transition metals, ultraviolet, gamma radiation, some herbicides or ozone produce ROS directly but, at the same time, stimulate NADPH oxidases and peroxidases, producing ROS *de novo* (Zhang *et al.*, 2010; Nawkar *et al.*, 2013; Shahid *et al.*, 2014).

High concentration of ROS is one of the oldest 'stresses' on the planet for living systems (Cannio *et al.*, 2000; Dowling and Simmons, 2009). Plants have been exposed to ROS for at least 2.7 billion years, i.e. since they started producing O_2 from H_2O. The constantly increasing level of O_2 has directed species evolution and determined the biochemistry of modern organisms (Dowling and Simmons, 2009). Superoxide dismutases have been found in Archaea, the prokaryotes living in extreme environments, even in anaerobic ones (Cannio *et al.*, 2000). Thousands of redox-active enzymes and biomolecules, which react with ROS, have evolved. Primitive ancient mechanisms grew up into sophisticated systems of oxidative stress recognition and antioxidative defence. Moreover, plants

started to use ROS as signalling molecules and their biosynthesis as a regulatory tool for sensing other stresses and managing growth, development and death (Murata *et al.*, 2001; Demidchik *et al.*, 2003, 2004, 2009, 2010, Foreman *et al.*, 2003; Krishnamurthy and Rathinasabapathi, 2013), generating gravitropic response (Joo *et al.*, 2001) and a number of other processes that are not primarily related to stress or oxidation.

This chapter discusses the biochemical mechanisms and cell biology reactions underlying plant oxidative stress and ROS-induced regulation of plant functions. The chemistry of individual ROS and plasma membrane switches amplifying oxidation signals are the main focus of this review.

3.2 Reactive Oxygen Species and Free Radicals: Common and Different Features

ROS (Fig. 3.1; Table 3.1) contain one or more activated atoms of oxygen but are not necessarily radicals. Major ROS H_2O_2 is not a radical. Free radicals are any chemical species that exist independently and contain unpaired electron(s). Free radicals do not necessarily have oxygen atoms. For example, transition metals or carbon-centred radicals do not contain oxygen. However, both ROS and

Table 3.1. Reactive oxygen species and reactive nitrogen species that can exist in plants.

Reactive oxygen species		Reactive nitrogen species	
Non-radical	Free radicals	Non-radical	Free radicals
Hydrogen peroxide H_2O_2	Superoxide $O_2^{\cdot-}$	Nitrous acid HNO_2	Nitric oxide $NO^{\cdot}$
Hypobromous acid **HOB**	Hydroxyl **$^{\cdot}OH$**	Nitrosyl cation NO^+	Nitrogen dioxide $NO_2^{\cdot}$
Hypochlorous acid **HOCl**	Hydroperoxyl $HO_2^{\cdot}$	Nitroxyl anion NO^-	Nitrate radical $NO_3^{\cdot}$
Ozone O_3	Peroxyl $RO_2^{\cdot}$	Dinitrogen tetroxide N_2O_4	
Organic peroxides **ROOH**	Carbonate $CO_3^{\cdot-}$	Dinitrogen trioxide N_2O_3	
Singlet oxygen $^1\Delta_gO_2$	Alkoxyl $RO^{\cdot}$	Peroxynitrite $ONOO^-$	
Peroxynitrite $ONOO^-$	Carbon dioxide radical $CO_2^{\cdot-}$	Peroxynitrate O_2NOO^-	
Peroxynitrate O_2NOO^-	Singlet oxygen $^1\Sigma g^+O_2$	Peroxynitrous acid **ONOOH**	
Peroxynitrous acid **ONOOH**		Nitronium cation NO_2^+	
Peroxomonocarbonate $HOOCO_2^-$		Alkyl peroxynitrites **ROONO**	
		Alkyl peroxynitrates RO_2ONO	
		Nitryl chloride NO_2Cl	
		Peroxyacetyl nitrate $CH_3C(O)OONO_2$	

free radicals can induce oxidative stress through oxidation of cell compounds.

Surprisingly, the term 'oxidative stress' has a number of meanings. The first and most widely used meaning is the physiological state, response or conditions when loss of electrons (oxidation) exceeds gain of electrons (reduction) leading to chemical (oxidative) damage of cell compounds. The oxidative stress is therefore referred to long-term reduction/oxidation (redox) imbalance caused by 'deficiency' of electrons. Antioxidants donate electrons and prevent electron 'leakage' from biomolecules and their oxidation. The second meaning is a 'stress factor' (along with pathogen attack, drought, heavy metals, salinity, etc.) causing direct damage and triggering signals.

An oxidative stress usually begins from the activation of O_2. Gaseous 'normal' O_2 is called a triplet oxygen, having actually three 'triplet states' which are energetically more favourable, and correspond to the 'ground state' of the molecule with a total electron spin of $S = 1$. Excitation to the $S = 0$ state gives a singlet oxygen, which is much more reactive than O_2. The formation of a high-energy form of oxygen, singlet oxygen, can happen due to the effect of photons from visible light, but most likely from UV and gamma radiation. Some chemical reactions also give singlet oxygen. However, O_2 can also be activated via gaining high-energy electrons from redox enzymes, transition metals and some organic substances. This mechanism is more important for cell physiology than formation of singlet oxygen. Energy input during the 'activation' process makes O_2 chemically more active or 'reactive'; therefore oxidative stress can be defined as the stress caused by 'reactive oxygen species', 'reactive oxygen intermediates', 'oxygen-derived species', 'free oxygen radicals', etc.

Reactive nitrogen species (RNS) are another important class of substances involved in plant oxidative stress (Fig. 3.1; Table 3.1). RNS cause so-called 'nitrosative stress'. ROS and RNS stresses are often collectively referred to as ROS/RNS-induced stress. At present there are experimental data sufficient to demonstrate significance of nitrosative stress in plants. For example, NO can interact with metal clusters in redox enzymes, resulting in the formation of a stable metal–nitrosyl complex, oxidized species and chemical signals, leading to significant modifications of protein structure and function as well as to specific gene expression changes (Arora *et al.*, 2016). As space is limited, the nitrosative stress will not be considered here in detail.

3.3 Chemistry and Cell Biology of Reactive Oxygen Species and Pro-oxidants

The small group of chemical species that includes superoxide, hydrogen peroxide, hydroxyl radical, singlet oxygen, hypochlorous acid, peroxinitrite and transition metals are prime or early products of O_2 'activation' in most living cells (Fig. 3.1). These molecules, their production and metabolism, are mainly responsible for molecular and physiological changes associated with oxidative stress in plants (Demidchik, 2015). A number of other ROS could also be involved, such as peroxyl, alkoxyl and hydroperoxyl radicals, peroxinitrite, ozone and hypochlorous acid (Table 3.1).

In the Earth's crust, oxygen is the most abundant element (Guido, 2001). H_2O is made of atoms of hydrogen and oxygen. The latter 'occupies' approximately 89% of its mass, making oxygen the most abundant element in living organisms (by mass). Triplet oxygen (O_2) is the major atmospheric form of oxygen. Triplet oxygen is a free radical (diradical) having two unpaired electrons ($O_2^{2 \cdot}$). Unpaired electrons in O_2 have the same spin numbers or 'parallel spins', which restrict the number of O_2 targets to those that have two similar electrons with antiparallel spins. This phenomenon is called a 'spin restriction' and decreases the reactivity of O_2. O_2 is not very chemically active and is not toxic to aerobic organisms. To 'acquire' higher reactivity, O_2 requires an input of energy to remove the spin restriction. This energy comes from some chemical and biochemical reactions, electrons containing high energy leaking from electron-transport chains (ETC), ultraviolet (UV) and ionizing radiation, ultrasound cavitation and some other sources.

3.3.1 Superoxide radical is a key player in oxidative stress

Chemical properties of superoxide

Formation of superoxide from O_2 is a critical phenomenon for free radical biology of all plant species (Figs 3.1–3.4). This event is a starting point for development of oxidative stress and reactions of redox signalling and regulation of physiology. Scavenging superoxide is a major approach to stop oxidative stress at the early stage.

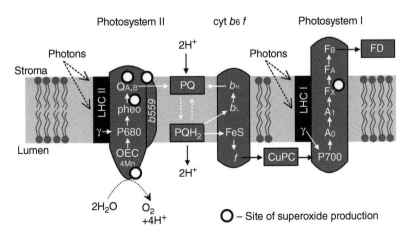

Fig. 3.2. Predicted sites of superoxide generation in photosystems I and II.

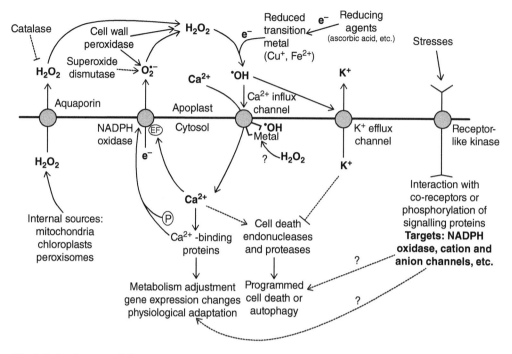

Fig. 3.3. Involvement of plasma membrane transport and signalling systems in the regulation of oxidative stress.

$$O_2 \xrightarrow{+e} O_2^{\bullet-} \xrightarrow{+e,\ 2H^+} H_2O_2 \xrightarrow{+e} {}^\bullet OH + OH^- \text{ (I.)}$$

$$Me^{+/2+} + H_2O_2 \longrightarrow Me^{2+/3+} + {}^\bullet OH + OH^- \text{ (II.)}$$

$$Me^{+/2+}\text{-}OOH \longrightarrow Me^{2+/3+}\text{-}O^- + {}^\bullet OH \xrightarrow{+H^+} Me^{2+/3+} + {}^\bullet OH + OH^- \text{ (III.)}$$

$$O_2^{\bullet-} + {}^\bullet NO \longrightarrow ONO_2^- \longrightarrow ONO_2CO_2^- \cdots\cdots\rightarrow {}^\bullet NO_2 + CO_3^{\bullet-} \text{ (IV.)}$$

Fig. 3.4. Reactions synthesizing hydroxyl radical in plants. I. Reduction of oxygen leads to formation of hydroxyl radicals. II. 'Outer-sphere electron transfer' in classical Fenton-like reaction (when transition metals do not bind covalently to H_2O_2). III. 'Inner-sphere electron transfer' which involves strong peroxide binding to a transition metal. IV. Sequential formation of reactive nitrogen species (RNS) from NO and superoxide. Peroxynitrite interacts with CO_2 and gives nitrosoperoxycarbonate, which decomposes to nitrate, CO_2, nitrogen oxide radical ($^\bullet NO_2$) and carbonate radical ($CO_3^{\bullet-}$).

In living plant cells, O_2 can lose its 'spin restriction' by accepting a single electron (Fig. 3.1), for example, via electron leakage from ETC or NADPH oxidase activity (Fig. 3.2). This reaction produces $O_2^{\bullet-}$ which is more reactive than O_2 although having just one unpaired electron. $O_2^{\bullet-}$ can be called 'superoxide', 'superoxide anion radical', 'superoxide radical anion' or 'superoxide radical'. This molecule is unstable. The half-life of $O_2^{\bullet-}$ maximally reaches a few milliseconds under biological conditions (Kavdia, 2006). This is enough to diffuse for a few micrometres; therefore $O_2^{\bullet-}$ can be considered as an autocrine or maximally paracrine agent.

Superoxide participates in a number of reactions. The most prevalent reaction will be the protonation: $O_2^{\bullet-} + H^+ = HO_2^\bullet$ (hydroperoxyl radical or perhydroxyl radical) (Fig. 3.4). $HO_2^\bullet$ is a gaseous, non-ionic and relatively stable molecule. $HO_2^\bullet$ can probably diffuse through the membrane.

$HO_2^{\cdot}$ is also a more powerful oxidant than $O_2^{\cdot-}$. This substance can oxidize tocopherol and polyunsaturated fatty acids.

Two $HO_2^{\cdot}$ molecules can produce O_2 and H_2O_2 in so-called 'superoxide dismutation', although in reality superoxide does not participate in this reaction (Fig. 3.4). The $O_2^{\cdot-}/HO_2^{\cdot}$ ratio increases from 1:1 at pH 4.8 to 10:1 at pH 5.8 and 100:1 at pH 6.8, respectively (Sawyer and Gibian, 1979; Ross, 1985). This means that $HO_2^{\cdot}$ is critically important ROS in cell compartments with lower pH, such as apoplastic space, mitochondrial intermembrane space, chloroplast thylakoid lumen. In growing cell walls, vacuoles, peroxisomes and lysosomes $HO_2^{\cdot}$ could also play crucial roles.

$O_2^{\cdot-}/HO_2^{\cdot}$ pair is a major 'origin' of an oxidative stress. It can reduce other radicals and ROS ($Cu^{+/2+}$, $Fe^{2+/3+}$, Fe-S clusters, $NO^{\cdot}$, phenoxyl radical, etc.), leading to formation of highly reactive oxidants (Sawyer and Gibian, 1979; Ross, 1985; Halliwell and Gutteridge, 2015). Central to this is $O_2^{\cdot-}$-mediated reduction of oxidized transition metals (Fig. 3.4). $O_2^{\cdot-}$ reduces Fe^{3+} and Cu^{2+} to Fe^{2+} and Cu^{+}, respectively. This is a key process, underlying oxidative stress, because it reduces transition metals Fe^{2+} and Cu^{+}, reacts with H_2O_2 and produces $^{\cdot}OH$, which is the most 'aggressive' oxidant of all ROS (Pryor and Squadrito, 1995).

Additionally, the $O_2^{\cdot-}/HO_2^{\cdot}$ pair reacts with $NO^{\cdot}$, producing two extremely reactive RNS, peroxynitrite ($ONOO^-$) and alkyl peroxinitrite (ROONO) which could be involved in plant nitrosative stress and signalling (Squadrito and Pryor, 1995; del Río et al., 2002, 2003, 2006; Fig. 3.4). Peroxynitrite may also decompose to $^{\cdot}OH$ (Pryor and Squadrito, 1995).

Detection of superoxide

Superoxide generation may vary in kinetics, duration and intensity; but it accompanies almost all major plant stresses, such as salinity, drought, heavy metals, UV, O_3, xenobiotics, hypothermia and hyperthermia. It is most likely that the first wave of $O_2^{\cdot-}$ generation develops a few seconds after the addition of a stress factor or stress hormone (Kawano et al., 1998). Quantitative measurements of $O_2^{\cdot-}$ do not make much sense because it is short-living and produced locally.

Thousands of studies reported $O_2^{\cdot-}$ production under stress conditions as increase in superoxide dismutase activity (SOD; enzyme scavenging

$O_2^{\cdot-}$). This kind of $O_2^{\cdot-}$ detection is definitely inadequate. Only a few studies have really dealt with $O_2^{\cdot-}$ biosynthesis in vivo. A number of fluorescent (dihydrorhodamine 123, MitoSOX, etc.), colorimetric (nitroblue tetrazolium) and luminescent (luminol, lucigenin, coelenterazine, etc.) probes have been used for $O_2^{\cdot-}$ detection. However, all these dyes are not specific to $O_2^{\cdot-}$ (Henderson and Chappell, 1993; Kervinen et al., 2004; Halliwell and Gutteridge, 2015). Lucigenin, lumino, nitroblue tetrazolium and dihydrorhodamine probably can catalyse the redox cycling themselves and generate $O_2^{\cdot-}$ (Halliwell and Gutteridge, 2015). However, some new data show that luminol can be used as a sensitive and specific dye for measuring superoxide in living systems (Yamazaki et al., 2011).

Electron paramagnetic resonance (EPR) spectroscopy is considered so far to be the most specific and reliable technique for detection of $O_2^{\cdot-}$ (Mojovic et al., 2005; Pospíšil, 2014; Halliwell and Gutteridge, 2015; Krieger-Liszkay et al., 2015). EPR spectroscopy is used in combination with spin traps, such as 'DMPO', 'DEPMPO' and others, providing characteristic spectra of stable oxygen-centred radical adducts of these traps (Mojovic et al., 2005; Krieger-Liszkay et al., 2015). EPR spectroscopy has been applied successfully for measurements of extracellular $O_2^{\cdot-}$ and other radicals in intact maize (Liszkay et al., 2004) and Arabidopsis roots (Renew et al., 2005; Demidchik et al., 2010) and different in vitro preparations (Mojovic et al., 2005; Krieger-Liszkay et al., 2015).

EPR techniques have significant limitations. For example, recording an EPR signal from thylakoids requires preparation of thylakoid-rich vesicles or isolated chloroplasts (Menconi et al., 1995; Krieger-Liszkay et al., 2015). This procedure causes wounding stress that triggers additional ROS generation. Other intracellular tests based on EPR spectroscopy, such as isolation of mitochondria or peroxisomes, 'suffer' similar problems.

Among fluorescent probes, hydroethidine was demonstrated to have a very high specificity to $O_2^{\cdot-}$ (Zhao et al., 2005). Hydroethidine can interact with $O_2^{\cdot-}$, forming a two-electron oxidation product (E-oxy) while other ROS, RNS and halogens cannot oxidize hydroethidine to the same product (Zhao et al., 2005). Hydroethidine has now been used successfully for the detection of $O_2^{\cdot-}$ during metal-induced oxidative stress in plant cell cultures, roots and leaves (reviewed by Sandalio *et al.*, 2008). The problem of hydroethidine is its high instability in water solutions at room temperature and under light.

Superoxide generation

An 'electron leakage' in ETCs of chloroplasts and mitochondria is probably the major source of O_2^- and contributor to oxidative stress in photosynthesizing tissues (Smirnoff, 1993; Møller, 2001; Apel and Hirt, 2004; Lesser, 2006; Møller *et al.*, 2007; Rinalducci *et al.*, 2008; Takahashi and Badger, 2011; Gupta *et al.*, 2015; Mignolet-Spruyt *et al.*, 2016). A few per cent of electrons (1–5%) are 'lost' in ETCs and some of these electrons target and activate O_2, forming O_2^-. Some moderate constitutive level of O_2^- biosynthesis in organelles is necessary for normal life and maintaining redox homeostasis, but this increases dramatically in stress conditions. Therefore stressed plants produce more O_2^- than they can detoxify (Gupta *et al.*, 2015).

Electron leakage seems to occur in several sites (Fig. 3.2) in photosystem I (Asada, 2006; Caffarri *et al.*, 2014; Takagi *et al.*, 2016) and photosystem II (Pospíšil *et al.*, 2004; Caffarri *et al.*, 2014; Pospíšil, 2014). Moreover, mitochondrial complexes I and III are also 'leaky' and produce superoxide (Møller, 2001; Hirst *et al.*, 2008; Rinalducci *et al.*, 2008). Recent data have shown that succinate dehydrogenase (mitochondrial complex II) is also a source of superoxide and other ROS in plants that contribute to ROS-controlled developmental and stress responses (Jardim-Messeder *et al.*, 2015). Mechanisms of mitochondrial ETC-catalysed production of superoxide are not discussed here due to space limitations.

Generation of superoxide in photosystems has been intensively studied since the 1970s; however, the exact structure of superoxide-producing sites is still missing (Caffarri *et al.*, 2014). Experiments with EPR spectroscopy have shown that pheophytin (pheo.), primary quinone acceptor (Q_A) and cytochrome $b559$ probably reduce O_2 to O_2^- at the electron acceptor side of photosystem II (Fig. 3.2) (Ananyev *et al.*, 1994; Cleland and Grace, 1999; Pospíšil *et al.*, 2004, 2006; Caffarri *et al.*, 2014; Pospíšil, 2014). O_2^- is also generated by one-electron oxidation of H_2O_2 at the electron donor side of photosystem II

(Chen *et al.*, 1992, 1995; Caffarri *et al.*, 2014). However, photosystem I is widely considered to be a major site for O_2^- generation in chloroplasts (Genty and Harbinson, 1996; Asada, 2006; Foyer and Noctor, 2009). In photosystem I, O_2^- is probably synthesized by the 4Fe-4S complex (clusters X) on psaA and psaB or A/B on psaC at the electron accepting (stromal) side (Asada, 1999; Caffarri *et al.*, 2014).

Antioxidant enzymes prevent oxidative stress in organelles and cytoplasm (Smirnoff, 2005; Foyer and Noctor, 2009). In chloroplasts, O_2^- undergoes superoxide dismutase-catalysed disproportionation to O_2 and H_2O_2, which is detoxified by ascorbate peroxidases (ascorbate + $H_2O_2 \rightarrow$ dehydroascorbate + H_2O) (Asada, 2006). Mitochondrial O_2^- is converted by SOD to O_2 and H_2O_2 but is mainly detoxified by glutathione peroxidases. However, in some tissues, catalase and enzymes of the so-called ascorbate–gluthatione cycle can also be involved in H_2O_2 detoxification (Foyer and Halliwell, 1976; Møller, 2001).

Recent data suggested that generation of O_2^- by peroxisomes is an important source of this ROS under biotic and abiotic stresses, such as pathogen attack, salinity, Cd^{2+} and xenobiotics (Sandalio and Romero-Puertas, 2015; del Río and López-Huertas, 2016). Major physiological functions of these organelles include the catabolism of fatty acids, amino acids, glycolate, polyamines, and some xenobiotics, detoxification of hydrogen peroxide and participation in the glyoxylate cycle (del Río *et al.*, 2006; Reumann and Weber, 2006; del Río and López-Huertas, 2016). Peroxisomes physiologically maintain high levels of superoxide generation and accumulate H_2O_2 for oxidation of organic molecules; therefore, their high ROS signal detected with fluorescent imaging may be 'misread' as an oxidative burst in the cell. Highly oxidized peroxisomes are degraded by autophagy (Shibata *et al.*, 2013). This prevents damage to plant cells.

At least two systems of O_2^- generation exist in peroxisomes: (i) it is a xanthine oxidase that is active in the cellular matrix; and (ii) it is an NADH/NADPH-dependent ETC (so-called small ETC) in the peroxisomal membrane comprising NADH:ferricyanide reductase, cytochrome b, monodehydroascorbate reductase and NADPH:-cytochrome P450 reductase and producing

O_2^- in the cytosol (Lopez-Huertas *et al.*, 2000; del Río *et al.*, 2002, 2006; Sandalio and Romero-Puertas, 2015). Stresses stimulate O_2^- generation leading to H_2O_2 accumulation in peroxisomes and cytosol by a yet unknown pathway. They also decrease the activity of peroxisomal antioxidative defence systems (del Río *et al.*, 2006). SOD is a major O_2^--scavenging system in peroxisomes (del Río *et al.*, 2002, 2006; Sandalio and Romero-Puertas, 2015). H_2O_2 produced during the HO_2^- dismutation reaction is probably detoxified through the ascorbate–glutathione cycle (Foyer and Halliwell, 1976), similar to other plant cell compartments (del Río *et al.*, 2003; Reumann and Weber, 2006; del Río, 2015). Peroxisomes could also be involved in the control of plant antioxidant levels and enzymes repairing oxidized cell components (Reumann *et al.*, 2007; del Río, 2015).

All major stresses can be divided into two cohorts by the major source of superoxide. The first cohort includes an extreme light, UV, ozone, some xenobiotics and some transition metals. These stresses directly affect the chloroplast, mitochondria or peroxisome ETCs, leading to massive generation of O_2^- and other ROS. Undoubtedly, a prime cause of oxidative stress induced by these environmental cues is ETC-mediated leakage of electrons. In the second cohort the disturbance of ETCs seems to be a secondary process, occurring in the later stages of the stress response. These stresses include drought, salinity, pathogen attack, wounding, hypothermia, hyperthermia, heavy metals, hypoxia and some others. A prime mechanism for these stresses is O_2^- generation due to increased activities of NADPH oxidases and extra- and intracellular peroxidases (Doke, 1983; Torres *et al.*, 1998, 2005; Bolwell *et al.*, 2002; Foremen *et al.*, 2003; Torres and Dangl, 2005; Bindschedler *et al.*, 2006; Fluhr, 2009; Demidchik, 2010, 2015; Chang *et al.*, 2012; Steffens *et al.*, 2013). They can still trigger generation of ROS in chloroplasts and mitochondria via decrease of general antioxidant activity, which results from an oxidative burst induced by a primary oxidative burst (induced by NADPH oxidases, peroxidases and other systems generating ROS *de novo*). In NaCl-treated photosynthesizing cells, such as the mesophyll, chloroplast-generated ROS can finally dominate total ROS production (Wu *et al.*, 2015).

Peroxidase- and NADPH oxidase-generated ROS can be distinguished by the sensitivity to the NADPH oxidase inhibitor diphenylene iodonium (DPI) (Foreman *et al.*, 2003; Fluhr, 2009). Usually, NADPH oxidase is blocked by 1–10 μM DPI while peroxidases are inhibited by levels of DPI one to two orders higher. Additionally, peroxidases are sensitive to azide and cyanide while NADPH oxidases are not inhibited by this drug (Bindschedler *et al.*, 2006; Halliwell and Gutteridge, 2015). One cautionary note should be made about the use of these pharmaceuticals. DPI is dissolved in dimethyl sulfoxide (final concentration: 1–3%), which is a strong radical scavenger affecting any ROS measurements. A proper control test with dimethyl sulfoxide should be carried out. A number of more specific and reliable inhibitors of NADPH oxidase are available (Williams and Griendling, 2007).

The function of NADPH oxidase in ROS promoting stress reactions, signalling, survival or death, is conserved among Kingdoms (Kawahara *et al.*, 2007; Fluhr, 2009; Jiang *et al.*, 2011; Lambeth and Neish, 2014). NADPH oxidases are encoded by the Respiratory Burst Oxidase Homologues (*RBOH*) gene family (Lambeth and Neish, 2014). *Arabidopsis thaliana* has ten *RBOH* genes (*AtRBOHA-J*), rice has nine *RBOHs*; they are also abundant in other plant genomes (Groom *et al.*, 1996; Torres *et al.*, 1998; Torres and Dangl, 2005; Kawahara *et al.*, 2007; Wong *et al.*, 2007; Jiang *et al.*, 2011). A reverse genetics approach and 'over-expression' of *RBOH* genes have demonstrated the relation between specific NADPH oxidase homologues (their activity and expression level) and reactions induced by different stresses (reviewed by Apel and Hirt, 2004; Torres and Dangl, 2005; Fluhr, 2009; del Río, 2015). For example, increase of *AtrbohD* and *AtrbohF* expression is required for oxidative burst induced by pathogenic *Pseudomonas syringae* or *Hyaloperonospora parasitica* (Torres *et al.*, 2002). The mechanism of this specificity is yet to be understood.

The structure–function relationships of NADPH oxidase have recently been clarified (Kawahara *et al.*, 2007; Sumimoto, 2008; Fluhr, 2009; Lambeth and Neish, 2014). The transfer of electrons in NADPH oxidase is catalysed by a C-terminal cytoplasmic superdomain that is homologous to the ferredoxin reductase. It includes the NADPH-binding and FAD-binding sites, which transfer the electron to the N-terminal six transmembrane segments containing the di-heme system (Kawahara *et al.*, 2007; Sumimoto, 2008). Di-heme reacts with O_2 producing O_2^- at the apoplastic side of the plasma membrane. NADPH oxidase is not functional in the absence of Ca^{2+}. The activation by Ca^{2+} is structurally related to the N-terminus of plant NADPH oxidase containing two Ca^{2+}-binding helix–loop–helix structural domains (EF-hands) which are similar to Ca^{2+}-binding domains in calmodulin and troponin-C. Binding of Ca^{2+} causes a conformational change and intramolecular interaction of the N-terminal Ca^{2+}-binding domain with the C-terminal superdomain, resulting in the activation of electron transfer (Bánfi *et al.*, 2004). Half-maximal activation of model NADPH oxidase (animal Nox5; *in vitro*) is caused by 1 µM Ca^{2+} (Bànfi *et al.*, 2004). Maximal stress-induced $[Ca^{2+}]_{cyt}$ increases are in the range from 0.3 to 3 µM, fitting well to this number (Demidchik *et al.*, 2003, 2009; Demidchik and Maathuis, 2007).

Sensitivity to Ca^{2+} increases after calmodulin binding to the NADPH-binding domain or phosphorylation of serine/tryptophan residues in the FAD-binding domain by protein kinase C (Kobayashi *et al.*, 2007; Tirone and Cox, 2007). Calmodulin-like domain protein kinases (CDPKs) stimulate NADPH oxidases through phosphorylation (Xing *et al.*, 2001; Wu *et al.*, 2010). For example, an increased CDPK expression causes elevation of plant NADPH oxidase activity, which is abolished by protein phosphatase 2A and unaffected by protein phosphatase 1 (Xing *et al.*, 2001). Small G-proteins (Rac/Rop GTPases) increase NADPH oxidase activity in a Ca^{2+}-dependent manner (Baxter-Burrell *et al.*, 2002; Wong *et al.*, 2007).

The plasma membrane-associated cytoplasmic kinase BIK1 (BOTRYTIS-INDUCED KINASE1) is a positive modulator of NADPH oxidase activity, which causes enzyme activation in a Ca^{2+}-independent manner (Kadota *et al.*, 2015). BIK1 is activated by pathogen-associated molecular patterns (PAMPs) and directly phosphorylates NADPH oxidase (RBOHD). This mechanism can probably underlie ROS production under pathogen attack.

Ca^{2+}-activated NADPH oxidase works in concert with ROS-activated Ca^{2+}-permeable cation channels to generate and amplify stress-induced Ca^{2+} and ROS signals (reviewed by Demidchik and Maathuis, 2007) (Fig. 3.3). Elevation of cytosolic $[Ca^{2+}]$ causes an increase in O_2^- production and vice versa, and O_2^- activates Ca^{2+} influx through ROS-activated cation channels (Demidchik and Maathuis, 2007; Demidchik *et al.*, 2009; Demidchik, 2010). This self-amplification mechanism, which is likely to be an upstream component for many stresses, may catalyse amplification and encode weak stimuli (transducing them into dramatic O_2^--Ca^{2+} alterations). For example, a marker of wounding stress, an extracellular ATP (Dark *et al.*, 2011), acts through O_2^--Ca^{2+} signal amplification cycle (Demidchik *et al.*, 2009). Initial small transient ATP-induced increase in $[Ca^{2+}]_{cyt}$ results in the production of ROS, which in turn induces massive activation of Ca^{2+}-permeable cation channels (Demidchik *et al.*, 2009).

Systems that control $[Ca^{2+}]_{cyt}$, such as Ca^{2+}-permeable non-selective cation channels (NSCCs), depolarization- and hyperpolarization-activated Ca^{2+} channels, Ca^{2+}-ATPase, Na^+/Ca^{2+} and Ca^{2+}/H^+ exchangers, cytosolic Ca^{2+}-binding proteins and endomembrane Ca^{2+} transporters are probably key regulators of the NADPH oxidase activity in plant cells. These systems are modulated by a number of regulatory enzymes, for example specialized kinases and phosphatases, as well as physical/chemical regulators (pH, hormones, etc.) (Demidchik, 2010; Hedrich, 2012). They also undergo specific expression in various physiological and pathophysiological conditions (Demidchik and Maathuis, 2007; Hedrich, 2012). Overall, this maintains fine oxidative balance and generates adequate O_2^--Ca^{2+} responses, which encode information about individual stress factors.

The spatial distribution of the O_2^--Ca^{2+} system within a cell is regulated via SCN1/AtrhoGDI1 RhoGTPase GDP dissociation inhibitor which concentrates O_2^- production by AtrbohC to specific zones of the cell (Carol *et al.*, 2005). This explains greater ROS-Ca^{2+} responses in the tips of root hairs or cell of root elongation zone, which explore new area of substrate and provide first reaction to stresses and various stimuli (Demidchik *et al.*, 2003, 2007, 2009, 2010; Foreman *et al.*, 2003). Having a four-dimensional system (X–Y–Z–time), O_2^--Ca^{2+}-mediated signals have the highest complexity and diversity, which is necessary for generation of combined response to a multitude of environmental cues and developmental stimuli (Demidchik and Maathuis, 2007; Himschoot et al., 2015).

Virtually all stress factors tested were able to produce an $O_2^{\cdot-}$ burst through this mechanism (Fluhr, 2009; Marino et al., 2012). This has been shown for a number of plant species, all organs and tissues, calluses, suspensions and protoplasts (Torres and Dangl, 2005; Demidchik et al., 2009; Fluhr, 2009; Marino et al., 2012; Dubiella et al., 2013). The analysis of the available data suggests that the activation of NADPH oxidase leading to superoxide production during stress is required for the following: (i) recognition of stress to adjust gene expression and metabolism for adaptation; (ii) triggering the autophagy and PCD needed for the stress response and adaptation on tissue and organismal level; (iii) stomatal closure during drought stress; and (iv) simultaneous 'processing' stress, immunity, defence and developmental signals (abscisic acid, ethylene, brassinosteroids, auxin, gibberellic acid, methyl jasmonate, salicylic acid, volatiles, etc.). NADPH oxidase and Ca^{2+}-permeable cation channels form a 'regulatory hub' catalysing perception, amplification, transduction and encoding of stress stimuli (Fig. 3.3).

Synthesis of peroxynitrite and nitrosoperoxycarbonate

Superoxide is also the source of peroxynitrite and inducer of nitrosative stress in plant cells. $O_2^{\cdot-}$ reacts with nitric oxide ($\cdot NO$) (Fig. 3.4). This reaction yields a non-radical and very powerful oxidant known as peroxynitrite (ONO_2^-). ONO_2^- can react with a number of organic molecules such as lipids and proteins, with reactivity similar to hydroxyl radical (Rubbo et al., 1994, 2009; Alvarez and Radi, 2003). Moreover, under biological conditions, peroxynitrite reacts with CO_2 and forms nitrosoperoxycarbonate (Veselá and Wilhelm, 2002; Medinas et al., 2007) (Fig. 3.4; reaction IV). This substance decomposes to nitrate, CO_2, nitrogen oxide radical ($\cdot NO_2$) and carbonate radical ($CO_3^{\cdot-}$) (Medinas et al., 2007). Peroxynitrite formation and involvement in the hypersensitive response to pathogen attack, as well as to other stresses, has

recently been investigated in plants (Vandelle and Delledonne, 2011).

3.3.2 Hydrogen peroxide

Properties of H_2O_2 and its role in oxidative stress

H_2O_2 (Fig. 3.1) is the most stable ROS that is critically important for plant physiology (Apel and Hirt, 2004; Foyer and Noctor, 2009; Demidchik, 2015; Halliwell and Gutteridge, 2015). H_2O_2 is a weak acid without unpaired electrons (non-radical), and it is a much more stable molecule as compared to superoxide, hydroperoxyl, NO, hydroxyl and singlet oxygen. Nevertheless, the lifetime of H_2O_2 in living tissues is still short (up to 0.1–1 s) due to activities of catalases and peroxidases (Halliwell and Gutteridge, 2015). In contrast to cytoplasm, which is a highly reduced and antioxidant-enriched alkaline compartment, an extracellular space is acidic and has lower activities of antioxidant enzymes (Hernández et al., 2001; Mhamdi et al., 2012). This provokes H_2O_2 accumulation in the apoplast and promotes a so-called cell wall oxidative burst.

The cytoplasm is a very thin layer (as the vacuole occupies up to 95% of cell volume); therefore H_2O_2 produced in organelles needs to diffuse just a few micrometres to reach the plasma membrane and then pass it through aquaporins (Dynowski et al., 2008; Bienert and Chaumont, 2014). H_2O_2 is also enzymatically produced in the apoplast by NADPH oxidases and extracellular heme-containing Class III peroxidases, which synthesize its precursor ($O_2^{\cdot-}$) (Cosio and Dunand, 2009; Demidchik, 2015).

Apart from the simplest peroxide, H_2O_2, a number of other non-radical organic peroxides (ROOR, ROOH or RO_2OH) can be produced in the cell, such as lipid peroxides. These compounds promote oxidative stress in different ways, for example in lipid peroxidation chain reactions, and are probably also involved in 'shaping' the intra- and extracellular redox signal (Møller et al., 2007; Suzuki et al., 2012).

Detection of H_2O_2

There is some value in making quantitative measurements of H_2O_2: they can provide useful information on the time course and strength of an oxidative stress. In most cases, researchers deal with testing H_2O_2 level rather than unstable oxygen-centred radicals. Classical tests on H_2O_2 include colorimetric, fluorescent and luminescent probes, reporting the presence of H_2O_2 via change of spectral characteristics or emission of light (Rhee et al., 2010; Demidchik, 2015). H_2O_2-sensitive probes can be membrane-permeant or membrane-impermeant. This helps to measure fractions of intracellular and extracellular H_2O_2. Not all probes for H_2O_2 are absolutely specific to H_2O_2 and, in fact, interact with other ROS (Rhee et al., 2010).

In response to stresses, bulk tissue H_2O_2 concentration increases from 0.01–1 µM to 0.1–10 mM depending on technique, stress factor, preparation and many other variables. Significant variation in the kinetics and amplitude of H_2O_2 accumulation has also been reported. Increase in H_2O_2 levels can be found from a few seconds to several days after the application of a stress factor. There may be a number of reasons for this difference in results:

1. H_2O_2 is produced locally in so-called hot spots, such as tips of growing root hairs (Foreman et al., 2003); meaning at organismal level the increase can be very low.
2. H_2O_2 'quantitative' tests show the cumulative effect of H_2O_2 reaction with the probe and do not demonstrate real H_2O_2 level.
3. Chemiluminescent probes are more sensitive than fluorescent probes. They report H_2O_2 faster and more accurately.
4. Chemical conditions used in tests, such as high pH values (pH 8–9), can alter the cell potency to generate H_2O_2. H_2O_2 measurements can also be affected by the structure of plant tissues. For example, the leaf cuticula is virtually impermeable to H_2O_2 probes.
5. The nature, exposure time and intensity of the imposed stress vary dramatically across studies.

More efficient methods for measuring H_2O_2 have been recently developed (Rhee et al., 2010; Michelet et al., 2013). Boronate-based fluorescent probes (the 'peroxysensor family' of H_2O_2 probes) fluoresce after the H_2O_2-induced removal of a boronate group (Miller et al., 2005; Lin et al., 2013). They are membrane-permeant and can be combined with SNAP-tag, which

targets them to different cell compartments to assess $[H_2O_2]$ spatially (Lin et al., 2013). Although these probes do not interact with other ROS (Lin et al., 2013), they react with H_2O_2 irreversibly and cannot test $[H_2O_2]$ dynamics (Dickinson et al., 2010). Another group of probes utilizes specially modified green fluorescent proteins (GFP), which include two redox-sensitive cysteine residues (Rhee et al., 2010). H_2O_2 induces formation of a disulfide bond and modifies spectral characteristics of protein fluorescence. However, this probe lacks specificity to H_2O_2.

The latest addition to fluorescent redox probes is a genetically encoded and reversible 'hydrogen peroxide sensor' (HyPer; Malinouski et al., 2011; Lukyanov and Belousov, 2014). The structure of HyPer is based on bacterial OxyR transcription factor (Kim et al., 2002) with a fluorescent protein inserted into the regulatory domain of this molecule (Belousov et al., 2006; Lukyanov and Belousov, 2014). H_2O_2 acts on the OxyR regulatory domain and causes the formation of a disulfide bond between C199 and C208, which changes the spectral properties of YFP (Kim et al., 2002; Belousov et al., 2006; Lukyanov and Belousov, 2014). This disulfide bond can be repaired by thiol–disulfide oxidoreductase in vivo; therefore HyPer can react with H_2O_2 again (Belousov et al., 2006; Lukyanov and Belousov, 2014). Formation of oxidized HyPer shows the dynamics of H_2O_2 in individual cell compartments (Malinouski et al., 2011; Lukyanov and Belousov, 2014). Another approach for detection of $[H_2O_2]$ involves peroxalate nanoparticles, which undergo a three-component chemiluminescent reaction between H_2O_2, peroxalate esters and fluorescent dyes inside the cell (D. Lee et al., 2007). This technique is sensitive and specific to H_2O_2; however, the delivery of nanoparticles and their potential intracellular redox activity could be problematic. Michelet et al. (2013) have recently developed novel EPR spectroscopy-based techniques to measure extracellular H_2O_2 content in cells of Chlamydomonas reinhardtii. These authors have used a spin-trapping assay containing 4-POBN/ethanol/Fe-EDTA. Fe-EDTA reacts with H_2O_2 forming ·OH which is trapped by 4-POBN/ethanol (giving stable adduct with strong signal).

H₂O₂ generation and physiological 'targets'

The origins of H_2O_2 generation are related to production of its precursor superoxide. Peroxidases were first proposed to be involved in the generation of H_2O_2 during biotic stress (Bolwell and Wojtaszek, 1997; Bolwell *et al.*, 1998; Bindschedler *et al.*, 2006; Camejo *et al.*, 2016). It is now widely accepted that increase in the activities of Class III peroxidases, for example Cu-containing amino oxidases, polyamine oxidases, gluthatione and ascorbate oxidases, is critically important for stress-induced H_2O_2 production (Rodríguez *et al.*, 2002, 2007; Chang *et al.*, 2009; Camejo *et al.*, 2016). Interestingly, peroxidases are probably regulated by a negative feedback mechanism; some of them are inhibited by H_2O_2 (Kitajima, 2008).

There is an opinion that an oxidative stress is directly caused by H_2O_2. It is not exactly true, because ˙OH biosynthesis is required to accomplish H_2O_2-mediated oxidative damage of biomolecules. The biosynthesis of H_2O_2 is difficult to relate directly with distinct physiological functions. This is why, in a number of studies, 0.01–10 mM H_2O_2 is added to plant cells exogenously and the induced reactions are investigated. H_2O_2 is a weak oxidant and cannot modify DNA, amino acids or lipids. H_2O_2 reacts with SH-groups, although this reaction is completely reversible. To become a reason for oxidative stress, H_2O_2 must be converted into hydroxyl radicals. Thus its main 'target' in terms of oxidative stress is transition metal-binding sites where metals can donate an electron for formation of ˙OH (Fry *et al.*, 2002; Fry, 2004). H_2O_2 has been reported to deactivate some enzymes, for example fructose biphosphate; however, this may be caused by residues of transition metals in experimental solutions (Halliwell and Gutteridge, 2015). Most proteins withstand 100 mM H_2O_2 in transition metal-free solutions (Halliwell and Gutteridge, 2015). A number of H_2O_2 'sensors' have been proposed to exist in plant cells (Apel and Hirt, 2004); however, only a few of them have been proved by experiments.

3.3.3 Hydroxyl radical

The hydroxyl radical (˙OH; Fig. 3.1) is an extremely short-living ROS, and a key cause of oxidative stress, damaging all biomolecules and triggering chain reactions of lipid peroxidation (Chen and Schopfer, 1999; Demidchik, 2015). It is also directly involved in redox signalling and oxidative stress-induced PCD (Demidchik *et al.*, 2003, 2010; Foreman *et al.*, 2003). For example, ˙OH induces the greatest activation of Ca^{2+} and K^+ channels among all known ROS, leading to Ca^{2+} influx and K^+ efflux immediately after application of a stress factor (Demidchik *et al.*, 2010). Theoretically, '*in vivo*' half-life of ˙OH does not exceed 1 ns, which allows ˙OH diffusion over 1 nm (Sies, 1993). Second-order rate constants for reactions of ˙OH with most organic molecules are extremely high. ˙OH interactions with cell biomolecules are probably diffusion-limited (Anbar and Neta, 1967; Sies, 1993). Specific ˙OH scavengers and antioxidants do not exist and the widely reported effects of mannitol, sorbitol, dimethyl sulfoxide, thiourea or others are, in fact, not due to ˙OH scavenging. These substances probably interact with ˙OH 'longer-living' precursors or chelate transition metals.

Fenton-like reactions are central to ˙OH biosynthesis (Fig. 3.4). Experiments conducted by Fenton in the 19th century aimed to establish the effect of Fe^{2+} on tartaric acid (Fenton, 1894). Surprisingly, his findings were so important for natural sciences that they have been used for more than a century. 'Fenton chemistry' and 'Fenton-like reagents' are used to refer to reactions of H_2O_2 and transition metals producing ˙OH, water and superoxide (Goldstein *et al.*, 1993; Chen and Schopfer, 1999). Although some intermediates are formed, the net 'Fenton reaction' is as follows: (i) metal reduced + H_2O_2 → metal oxidized + ˙OH + OH⁻; and (ii) metal oxidized + H_2O_2 → metal reduced + $HO_2^•$ + H+ (Koppenol, 2001). These equations were proposed by Fritz Haber (Nobel Prize winner in 1918) and his student Joseph Weiss in the 1930s and since called the Haber–Weiss cycle (Haber and Weiss, 1932). Importantly, ascorbic acid is likely to serve as a pro-oxidant reductant for iron, copper and probably some other metals and their complexes in Haber–Weiss-like cycles in plants, because its concentration in plants (including the apoplast) is extremely high (1–20 mM) (Fry *et al.*, 2002; Fry, 2004; Foyer and Noctor, 2011). This pro-oxidant activity of the ascorbate and its role in plant physiology is still poorly understood.

'OH can also be produced by homolytic bond fission of H_2O. In this case, electrons in covalent bonds are equally distributed to atoms. This requires significant input of energy by ultraviolet quanta, freezing–drying cycle, heat or ionizing radiation (Halliwell, 2006; Halliwell and Gutteridge, 2015). The generation of 'OH from H_2O_2, though, has a lower energy threshold. Therefore, under some natural conditions, 'OH can probably be directly generated from H_2O_2 (HOOH $\rightarrow$ 'OH + 'OH) and hydroperoxides (ROOH $\rightarrow$ 'RH + 'OH) by sunlight (Downes and Blunt, 1879; Halliwell, 2006). This process is called 'photo Fenton'. The physiological importance of 'OH produced by homolytic fission is still debated, apart from studies on UV stress where it is directly involved in oxidative destruction of cell components (Jain et al., 2004; Kataria et al., 2005; Halliwell, 2006).

In classical Fenton-like reactions or so-called outer-sphere electron transfer (Fig. 3.4), metals do not bind covalently to H_2O_2. However, the inner-sphere electron transfer leads to formation of a temporary covalent bond between the peroxide and metal ion (Fig. 3.4; Sawyer et al., 1993, 1996; Fridovich, 1998; Pospíšil et al., 2004). Cytochrome P450 (Sono et al., 1996), heme oxygenases (Ortiz de Montellano, 1998), bleomycin (Burger, 2000), superoxide reductases (Mathe et al., 2002) and some PSII proteins (Pospíšil et al., 2004) are examples of systems that have this mechanism of 'OH production.

Measurement of 'OH in living tissues is problematic because this substance has extremely high reactivity and a short lifetime. Several techniques for 'OH detection have been developed, including colorimetric, fluorescent, luminescent and radioactively labelled probes. Nevertheless, EPR (or ESR) spectroscopy is the only method providing high specificity and accuracy of 'OH measurement (Liszkay et al., 2003, 2004; Jain et al., 2004; Renew et al., 2005; Demidchik et al., 2010; Michelet et al., 2013; Šeršeň and Kráľová, 2013; Sosan et al., 2016).

EPR-based studies showed great sensitivity and specificity when it was used in plant in vitro preparations (Pospíšil, 2009, 2014). However, measurements that are conducted in vivo have a number of complications, such as a rapid decomposition of spin traps (POBN, DMPO, EMPO and DEPMPO are the most widely used probes)

at room temperature; reaction of spin trap with other radicals, for example with superoxide or transition metals; and noisy signal due to radical compositions of living tissues (Pou et al., 1989; Demidchik et al., 2010; Sosan et al., 2016). EPR spectroscopy cannot be used efficiently for studying inner plant tissues.

Using EPR spectroscopy Renew et al. (2005) carried out first measurements of constitutive 'OH production in a single non-stressed root of Arabidopsis thaliana in vivo. This study has clearly demonstrated that the root produces 'OH without special stimuli or pathophysiological conditions. Stresses such as salinity induce over-production of 'OH, which was also measured by EPR spectroscopy in vivo (Demidchik et al., 2010). This over-production of 'OH can be the source of severe oxidative stresses and cell death (Demidchik et al., 2010). The importance of 'OH for oxidative damage in photosynthetic apparatus is widely acknowledged (Møller et al., 2007; Šeršeň and Kráľová, 2013), but this has only recently been confirmed by direct EPR spectroscopy tests (Pospíšil et al., 2004; Šnyrychová et al., 2006; Pospíšil, 2009, 2014). Under stress conditions, 'OH is generated by both photosystems. In PSI, leakage of electrons leads to biosynthesis of superoxide/hydroperoxyl (Fig. 3.2), which forms H_2O_2, which in turn accumulates in the stroma. H_2O_2 is reduced to 'OH by free transition metals in stroma via a classical Fenton-like reaction (Fig. 3.4) or via inner-sphere electron transfer formation of H_2O_2-Fe complexes of ferredoxin (Šnyrychová et al., 2006; Fig. 3.4). In PSII, three transition metal-binding sites are probably involved in 'OH production from H_2O_2 (Pospíšil et al., 2004; Pospíšil, 2009, 2014): (i) Fenton-like reaction of H_2O_2 with free transition metals can occur in stroma; (ii) non-heme iron (the ligand is unknown) could be involved through inner-sphere electron transfer; and (iii) heme iron of cyt b_{559} could also participate in a similar reaction and form Fe-H_2O_2 complexes.

Diffusion limitation in 'OH-induced oxidation reactions suggests the presence in the cell of the specific chemical sites, which bind catalytically active transition metal cations, such as $Cu^{+/2+}$, $Fe^{2+/3+}$ or $Mn^{2+/3+}$ (Demidchik et al., 2014) (Fig. 3.3). These sites may contain pairs of cysteine and histidine residues forming a complex with transition metal (Demidchik et al., 2014; Demidchik, 2015). Demidchik et al. (2007) have demonstrated that the activation of single-channel conductances by H_2O_2 requires the delivery of H_2O_2 directly to the channel macromolecule at the extracellular side of the plasma membrane (excised patches). This strongly suggests that cation channels activated by H_2O_2 have transition metal-binding sites at the cytoplasmic face. Supporting this hypothesis, Rodrigo-Moreno et al. (2013) have recently found that Cu^{2+} acts on K^+ efflux at the cytosolic side of the plasma membrane.

3.3.4 Singlet oxygen

Formation of singlet oxygen is related to rearrangement of the electron in O_2 molecule, making this one much more reactive, rather than making a pair to one of its free radicals. The activation of O_2 in chloroplasts and mitochondria, for example through absorption of light quanta in P680 (Asada, 2006), can lead to the formation of two types of extremely reactive O_2-derived species: non-radical $^1\Delta_g O_2$ (22.4 kcal) and the more reactive free radical $^1\Sigma g^+ O_2$ (37.5 kcal) (Schweitzer and Schmidt, 2003). The term 'singlet O_2' traditionally covers both species. $^1\Sigma g^+ O_2$ can decay into $^1\Delta_g O_2$, but the significance of this process in the cell is still debated. Singlet O_2 is detoxified rapidly by beta-carotene in the PSII reaction centre, water, tocopherol, reduced plastoquinone or flavonoids (Trebst and Depka, 1997; Schweitzer and Schmidt, 2003; Asada, 2006; Kruk and Trebst, 2008; Fischer et al., 2013). It is believed that over-production of singlet oxygen takes place under stress conditions (in particular, under so-called photo-oxidative stress) (Asada, 2006; Fischer et al., 2013; Shumbe et al., 2016). This results in oxidative injuries, triggering PCD and 'retrograde' signalling between organelles (Møller et al., 2007; Fornazari et al., 2008; Przybyla et al., 2008; Fischer et al., 2013; Shumbe et al., 2016; Takagi et al., 2016).

The classical view claims that the lifetime and diffusion distance of singlet oxygen is short (3.1–3.9 µs and maximally 190 nm, respectively, accordingly to Asada, 2006). Accordingly to Krasnovsky (1998) the diffusion limit for singlet oxygen in chloroplasts is just a few nanometres. Recently, measurements of singlet O_2 with the fluorescent probe Singlet Oxygen Sensor Green (SOSG) showed that this substance can diffuse outside the chloroplast and reach apoplastic space (Flors et al., 2006; Driever et al., 2009). However, these measurements have not been repeated so far by other authors.

3.3.5 Role of transition metals in oxidative stress

The IUPAC Gold Book defines a transition metal as any element with an incomplete d sub-shell, or which can give rise to cations with an incomplete d sub-shell. Forty elements (21–30, 39–48, 71–80 and 103–112) fit this definition and can be considered as transition metals (McCleverty, 1999). Only a few transition metals have demonstrated significance for biological systems. The most important transition metals for living systems are Fe and Cu and, to a much lesser extent, Mn. Ni, Hg, Cr and Co are involved in some metabolic reactions but they are clearly less important for plant cell physiology (Bergman, 1992). Fe and Cu are the most abundant transition metals in living systems (Bergman, 1992; Outten and Theil, 2009; Robinson and Winge, 2010). Their physiological role is directly related to electron transport in redox enzymes. Ionic forms of these metals change their valence more easily as compared to the ionic form of other metals found in cells (Bergmann, 1992; Robinson and Winge, 2010). Mn and Ni show very similar properties to Cu and Fe, when coordinated by organic ligands, such as His or Cys residues; however, they almost completely lack electron transfer capacity in free ionic form.

Complexes of transition metals with S-containing amino acids, His and some other organic ligands are major redox switches in living systems (Outten and Theil, 2009; Robinson and Winge, 2010). The catalytic activity of Cu and Fe may increase several times during stress conditions (Becana and Klucas, 1992; Moran et al., 1994, 1997; Becana et al., 1998). The toxic and regulatory effects of Cu and Fe are mainly caused by $\cdot OH$ (Fig. 3.3). Cu is a well-known eco-toxicant; iron toxicity is sometimes found at low pH in oxygen-deficient reduced conditions (Bergmann, 1992). Cu^{2+} directly affects plants, causing a significant oxidative stress damaging all cell components and processes, and triggering cell death (Bergmann, 1992; Demidchik et al., 1997, 2001).

Copper is about 60 times more potent as a catalyst of the Haber–Weiss cycle and several billion times more soluble than Fe under the biological pH range, but Fe is more abundant in the cell due to its role in heme (Bergmann, 1992; Fry *et al.*, 2002; Halliwell and Gutteridge, 2015). Taking this into account, Cu as a major catalyst of ˙OH generation seems to be much more probable (Fry *et al.*, 2002; Demidchik *et al.*, 2003). Cu and Fe are typically bound in organic complexes with carbohydrates, proteins and even lipids where they maintain their catalytic activity. Moran *et al.* (1997) discovered that specific phenolic compounds are synthesized in response to stresses that chelate Fe and increase its catalytic activity, inducing ˙OH generation, DNA damage and lipid peroxidation. Another potential mechanism of Cu and Fe 'mobilization' and increase of their catalytic activities during stresses is via complexes with polyamines such as spermine, spermidine and putrescine. These substances have been shown to be synthesized in plant cells in response to stresses (Alcázar *et al.*, 2006; Moschou *et al.*, 2009; Shi and Chan, 2014) and proved to form redox-active complexes with both Cu (Guskos *et al.*, 2007) and Fe (Tadolini, 1988). Zepeda-Jazo *et al.* (2011) have recently demonstrated that polyamines potentiate effects of Cu^{2+}-containing ˙OH-generating mixtures. This suggests that polyamine–Cu complexes could be more redox-active than free Cu^{2+}.

H_2O_2-induced oxidative stress seems to be targeted on Cu- and Fe-binding centres that generate ˙OH (Demidchik *et al.*, 2007, 2014). Cu and Fe may also bind to His- or Cys-containing pockets and cause their oxidation, leading to damage (Demidchik *et al.*, 2014). There is evidence that oxidative stress would not be possible without Fe and Cu, as removal of these metals (using specific chelators or decreasing accumulation) results in termination of oxidative burst (Sayre *et al.*, 2008; Demidchik *et al.*, 2014).

3.4   Reactive Oxygen Species Sensing and Signalling in Plant Cells

Oxidative stress does not always cause plant death. Moderate oxidative stress is a reason for induction of plant resistance to stresses. A number of mechanisms of ROS sensing and signalling in plant cells have been found (Apel and Hirt, 2004; Asada, 2006; Demidchik and Maathuis, 2007; Foyer and Noctor, 2009; Sierla *et al.*, 2013; Demidchik *et al.*, 2014; Demidchik, 2015). Potential ROS receptors include: (i) receptor-like kinases (RLKs), particularly cysteine-rich receptor-like kinases (CRKs); (ii) two-component histidine kinases; (iii) redox-sensitive transcription factors, such as NPR1 or heat shock factors; (iv) ROS-sensitive phosphatases; and (v) redox-regulated ion channels (reviewed by Apel and Hirt, 2004; Demidchik and Maathuis, 2007; Osakabe *et al.*, 2013; Sierla *et al.*, 2013; Demidchik *et al.*, 2014). The role of the first four mechanisms requires investigation, while the direct involvement of ion channels in ROS sensing has recently been studied in detail.

The changes in ion activities through modulation of ion channel/receptor conductance are the fastest and most important switches of physiological and biochemical parameters in animal cells responding to external and internal factors, including oxidative stress. At least two mechanisms are involved: (i) a change in ionic composition modifying metabolic interactions, as ions are direct regulators of enzymes of osmotic pressure; and (ii) changing the electric potential difference across the membrane, which can modulate the activities of active transporters, such as H^+-ATPase, and the functioning of vesicular transport and cytoskeleton. Evidence obtained *in vitro* using recombinant protein techniques shows that plant G-proteins (modulators of signalling cascades and ion channels) are directly regulated by ROS (20 μM H_2O_2) via stimulation of alpha subunit dissociation (Wang *et al.*, 2008). Nevertheless, this observation requires confirmation by tests performed *in vivo*. The activity of several plant cation channels have been shown to be modified in the presence of high [H_2O_2], showing that ion channels can be involved in the perception of H_2O_2-encoded messages. Exogenous H_2O_2 activates Ca^{2+}-permeable NSCCs in protoplasts isolated from *Arabidopsis thaliana* guard cells (Pei *et al.*, 2000). These channels are probably involved in ABA- and jasmonate-induced stomata closure (Munemasa *et al.*, 2011). Exogenous H_2O_2 inhibits K^+ outwardly rectifying channels in guard cells and root epidermis (Demidchik *et al.*, 2003), but it does not change the activity of Ca^{2+}-, K^+-, Cl^--selective channels and NSCCs in the green alga

Nitella flexilis or in the *Arabidopsis thaliana* root plasma membrane NSCCs when applied in the whole-cell configuration (Demidchik *et al.*, 1997, 2001, 2003, 2007). Exogenous H_2O_2 stimulates anion efflux in cultured *Arabidopsis thaliana* cells, mimicking the ABA effect (Trouverie *et al.*, 2008). However, this action seems to be related to the activation of Ca^{2+} conductance, which in turn activates Cl^- currents. Endogenously applied H_2O_2 activates *Arabidopsis thaliana* root Ca^{2+}-permeable NSCCs (Demidchik *et al.*, 2003, 2007). This activation is observed only in the outside-out mode when H_2O_2 was delivered close to the plasma membrane cytoplasmic side. The direct voltage-dependent activation of *Arabidopsis thaliana* plasma membrane K^+ channel SKOR (heterologously expressed in HEK cells) by H_2O_2 has recently been discovered (Garcia-Mata *et al.*, 2010). Cys-168 residing in the S3 alpha-helix of SKOR voltage sensor complex is responsible for the sensitivity to H_2O_2.

The effect of $\cdot OH$ on ion channels has been investigated by the addition of an $\cdot OH$-generating mixture (Cu^{2+}, L-ascorbic acid and H_2O_2) to protoplasts and intact cells (Demidchik *et al.*, 2003, 2007, 2010; Foreman *et al.*, 2003). In these experiments, $\cdot OH$-activated Ca^{2+}-permeable NSCCs (Ca^{2+} influx channels) and K^+ outwardly rectifying channels (catalysing K^+ efflux) in mature root epidermal cells, elongation zone, pericycle cells, cortex and root hairs of *Arabidopsis thaliana*. $\cdot OH$-induced activation of Ca^{2+} influx and K^+ efflux has also been observed in mature and young root epidermis of maize, clover, pea, wheat and spinach (Demidchik *et al.*, 2003). $\cdot OH$-activated K^+ efflux conductance is encoded by Gork-1 while the genetic nature of $\cdot OH$-activated Ca^{2+}-permeable channels is still unclear. Both $\cdot OH$-induced Ca^{2+} and K^+ conductances could also be related to activities of annexins, as they are decreased in the KO line lacking these systems (Laohavisit *et al.*, 2012). Moreover, Ca^{2+}/K^+ transporting annexins might include functional peroxidase domain producing ROS (Laohavisit *et al.*, 2009). $\cdot OH$-activated Ca^{2+} influx channels have recently been found in the pollen tube plasma membrane where $\cdot OH$-Ca^{2+}-driven mechanism (Fig. 3.3) plays a critical role for pollen tube rupture to release sperm for fertilization.

Massive loss of K^+ from roots can be caused by $\cdot OH$-induced K^+ channel activation (Demidchik

et al., 2003). This happens in response to NaCl, pathogen elicitors and oxidants (Demidchik *et al.*, 2003, 2010, 2014). Animal cell death enzymes, endonucleases and caspases, are controlled by cytosolic K^+, which is their natural suppressor. Activation of cation channels and K^+ loss can stimulate these enzymes and induce apoptosis (Seon and Ja-Eun, 2002). This mechanism also exists in plants where generates PCD (Demidchik *et al.*, 2010; Fig. 3.3). Root cell endonucleases and proteases have been shown to be activated by K^+ loss, causing symptoms typical of PCD. K^+ channel antagonists or radical scavengers can stop or delay K^+ loss and the appearance of PCD symptoms (Demidchik *et al.*, 2010, 2014). Surprisingly, expression of the animal anti-apoptotic CED-9 gene can decrease the H_2O_2-induced K^+ efflux from leaf segments (Shabala *et al.*, 2007). This suggests that plants and animals share similar K^+-mediated mechanisms of apoptosis-like PCD.

Singlet O_2 generation can lead to inhibition of the mitochondrial inner membrane K^+ influx channels which regulate mitochondrial volume in animal cells (Duprat *et al.*, 1995; Fornazari *et al.*, 2008). This induces cytochrome *c* release and transports superoxide anion, which is produced in mitochondria during stress, to the cytosol. Whether similar reactions take place in plants is unknown. The involvement of singlet O_2 in ion channel activation could be through H_2O_2, which can accumulate after singlet oxygen detoxification in organelles and diffuse to the cytosol where it activates Ca^{2+} channels and triggers signalling cascades, for example sending ROS/Ca^{2+} messages to the nucleus. The latter is called retrograde signalling and might play an essential role in the regulation of organelle protein biosynthesis under high light and probably other environmental stresses (Fernándeza and Stranda, 2008; Chang *et al.*, 2009; Karpiński *et al.*, 2013; Kim and Apel, 2013).

ROS-induced activation of Ca^{2+} influx channels causes a transient elevation of cytosolic Ca^{2+}. This provides a link between H_2O_2 accumulation and intracellular signalling and gene expression, activating Ca^{2+}-dependent regulatory cascades through Ca^{2+}-binding proteins (CBPs). Approximately 250 CBP genes exist in the *Arabidopsis thaliana* genome (Day *et al.*, 2002; Plattner and Verkhratsky, 2015). CBPs undergo reversible interaction with Ca^{2+} that leads to their conformational change and facilitates inter-actions with a number of cell targets (Gifford et al., 2007; Dubiella et al., 2013; Plattner and Verkhratsky, 2015). Classical CBPs contain 'EF-hand' motifs providing high-affinity binding of Ca^{2+} (Plattner and Verkhratsky, 2015). The five classes of CBPs are calmodulins, calmodulin-like proteins, Ca^{2+}-dependent protein kinases (CD-PKs), calcineurin B-like proteins and NADPH oxidases. CDPKs directly transduce ROS-Ca^{2+} sig-nals to catalytic activity while the calmodulins, calmodulin-like proteins, and calcineurin B-like proteins are intermediate sensors regulating downstream systems, which in turn react with final target or other regulators. NADPH oxidas-es, as mentioned above, enhance weak Ca^{2+} sig-nals, amplifying these signals using ROS biosynthesis de novo and ROS-activated Ca^{2+} in-flux channels. Apart from calmodulins and NA-DPH oxidases that exist in animals, the other three CBP classes are only found in plants and some bacteria. These three classes of CBPs are found in all plant cell organelles and are involved in a multitude of functions (DeFalco et al., 2010).

The activity and expression level of certain CBPs both increase in the presence of elevated ROS concentrations, and in response to biotic and abiotic stresses which generate ROS (DeFal-co et al., 2010; Ranty et al., 2016). For example, H_2O_2 causes fast induction of CDPK3 in Arabi-dopsis root suspension culture protoplasts, lead-ing to change in the expression of some stress-responsive genes and activity of 28 target proteins (Mehlmer et al., 2010). Stress hor-mones (abscisic and jasmonic acids) also stimu-late CDPKs (Munemasa et al., 2011). This effect probably relies on ROS-Ca^{2+} signalling induced by these hormones. CDPKs are involved in the regulation of specific NADPH oxidases (such as AtrbohD, StRBOHB and others) via Ca^{2+}-dependent phosphorylation of these proteins (Kobayashi et al., 2007; Dubiella et al., 2013; Ranty et al., 2016). This is another amplification mechanism of ROS-Ca^{2+} signalling.

Oxidative stress can change the activity of a number of other regulatory enzymes, in particu-lar specific kinases and phosphatases such as MAP kinases and other Ser/Thr kinases and MAPK phosphatases (reviewed by van Breusegem et al., 2008; Pitzschke and Hirt, 2009; Rodriguez et al., 2010). The mechanism by which this regu-lation occurs is unclear, nor is much known about the downstream targets. In some cases, re-searchers do not know the nature of reactions catalysed by proteins involved in the redox regu-lation. For example, it was demonstrated that EXECUTER 1 and 2 control the singlet oxygen-induced retrograde signalling from chloroplasts to nucleus. They regulate stress-specific gene expression and this process is dependent upon enzymatic lipid peroxidation. However, the mechanism by which these proteins function has yet to be identified (K.P. Lee et al., 2007; Przybyla et al., 2008; Kim and Apel, 2013). It is unlikely that many kinase/phosphatase-trig-gered cascades form primary components/sen-sors of ROS-mediated signalling. They probably serve as long-term downstream metabolic and genetic adjustment elements (switches) inte-grating ROS signalling into the cellular context and helping plants to adapt or make a decision for PCD. Some kinase/phosphatase systems lie upstream of ROS production, providing a stimu-lation feedback loop (van Breusegem et al., 2008).

ROS-Ca^{2+} signalling is not restricted to cyto-plasm and may exist in the nucleus (Mazars et al., 2010), chloroplasts (Johnson et al., 1995) and mitochondria (Logan and Knight, 2003), where stress-induced Ca^{2+} and ROS transients have been measured. These reactions could play the role of regulators of genomes in the nucleus and the organelles, although their exact functions have not yet been identified (Mazars et al., 2010; Kim and Apel, 2013).

Some plant transcription factors and their regulators can potentially play the role of ROS sensors (Despres et al., 2003; Hong et al., 2013; Guo et al., 2016). TGA1 transcription has two specific Cys residues (Cys-260 and Cys-266), which enable its binding to NPR1 regulator (Despres et al., 2003). Oxidation of Cys-260 and Cys-266 leads to loss of this interaction and so regulates transcription factor binding to DNA. Heat shock transcription factors may also be in-volved in direct ROS sensing (Miller and Mittler, 2006; Hong et al., 2013; Guo et al., 2016). They regulate transcription of various defence-related genes and are active during stresses, including oxidative stress (Miller and Mittler, 2006; Guo et al., 2016). Similar systems in fungi (Hahn and Thiele, 2004; Hong et al., 2013) and animals (Ahn and Thiele, 2003) sense H_2O_2 by two Cys residues; but this mechanism has yet to be con-firmed in plants.

3.5 Effect of Oxidative Stress on Key Biomolecules

3.5.1 Lipids

Oxidative stress induces modifications of biomolecules such as proteins, polynucleic acids, carbohydrates and lipids (Sies and Cadenas, 1985; Møller et al., 2007; Farmer and Mueller, 2013). Among these, oxidation of lipids is particularly dangerous because it propagates free radicals through so-called 'chain reactions' (Anjum et al., 2015). Lipid oxidation (also known as lipid peroxidation) is widely considered as a 'hallmark' of oxidative stress (Farmer and Mueller, 2013). This irreversible damage causes complete loss of some organismal activities and often results in cell death.

The three stages of lipid peroxidation are: initiation, propagation and termination (Catalá, 2006; Farmer and Mueller, 2013; Anjum et al., 2015). Initiation of lipid peroxidation (initiation stage) is triggered by hydrogen atom abstraction from the lipid molecule. This can be caused by hydroxyl, alkoxyl and peroxyl radicals, as well as by peroxynitrite but not by hydrogen peroxide or superoxide (Halliwell and Gutteridge, 2015). H^+ is abstracted from the methylene group ($-CH_2-$) giving $-\dot{C}H-$ or lipid radical ($L\dot{\ }$), which is a carbon-centred radical. Phospholipids (the most abundant membrane lipids) are susceptible to radicals and peroxidation because the double bond in the fatty acid weakens the C-H bond and facilitates H^+ subtraction. $L\dot{\ }$ can activate O_2 and form an oxygen-centred 'lipid peroxyl radical' ($LOO\dot{\ }$), which in turn is capable of abstracting H^+ from a neighbouring fatty acid to produce a lipid hydroperoxide ($LOOH$) and a second lipid radical ($L\dot{\ }$) (Catalá, 2006). This gives rise to the propagation phase. $LOOH$ can undergo 'reductive cleavage' by reduced transition metals (mainly Fe^{2+} or Cu^+) and form lipid alkoxyl radical ($LO\dot{\ }$), which is also reactive and induces further abstraction of H^+ from neighbouring fatty acid. Another important mechanism of lipid peroxidation is via direct reaction of double bonds with singlet oxygen from the PSII reaction centre, which gives $LOOH$ (Krieger-Liszkay et al., 2008; Przybyla et al., 2008; Farmer and Mueller, 2013). Singlet oxygen is also formed in reaction of two $LOO\dot{\ }$ molecules.

In cases of severe lipid peroxidation, membranes undergo loss of integrity followed by lysis of organelles, oxidation and dysfunction of proteins, DNA and RNA (Farmer and Mueller, 2013; Anjum et al., 2015; Halliwell and Gutteridge, 2015). Terminal products of lipid peroxidation are 'aggressive' substances, such as aldehydic secondary products (malondialdehyde, 4-hydroxy-2-nonenal, 4-hydroxy-2-hexenal and acrolein), which are markers of oxidative stress (Del et al., 2005; Farmer and Mueller, 2013). They are easy to measure, for example using classical thiobarbituric acid assay for malondialdehyde (Hodges et al., 1999) or more modern sensitive mass spectrometry-based techniques that can identify individual lipid species targeted by peroxidation and to study the chemical complexity of oxidative products formed due to this process (reviewed by Shulaev and Oliver, 2006; Farmer and Mueller, 2013).

3.5.2 Proteins

Such ROS as hydroxyl radicals, singlet O_2 or peroxynitrite can oxidize any proteinogenic amino acid (Møller et al., 2007; Avery, 2011; Halliwell and Gutteridge, 2015). These modifications result in loss of a given protein-mediated function such as specific metabolic, structural, transport or regulatory activities. Protein oxidation also results in accumulation of toxic protein aggregates and, in the case of severe damage, induces PCD (Demidchik et al., 2010; Avery, 2011; Anjum et al., 2015). Major ROS-induced modifications to amino acids are summarized in Table 3.2.

Terminal products of lipid peroxidation, such as 4-hydroxynonenal and malondialdehyde, have been demonstrated to react and oxidize a number of amino acids (such as lysine or histidine) (Table 3.2). A review by Møller et al. (2007) summarized a huge amount of literature data on oxidative damages to the most important plant protein complexes, such as photosystem I, D1 protein of photosystem II, ribulose-1,5-bisphosphate carboxylase/oxygenase and SOD. With the exception of D1 protein, which is probably affected by singlet oxygen, this damage requires the presence or increase in activity of catalytically active transition metals (catalysing biosynthesis of $\dot{\ }OH$).

Table 3.2. ROS-induced modifications to amino acids (according to Berlett and Stadtman, 1997; Shacter, 2000; Stadtman and Levine, 2000; Cecarini *et al.*, 2007; Avery, 2011).

Amino acid	Oxidized form
Cysteine	Cysteine → cystine → cysteine sulfenic acid → cysteine sulfinic acid → cysteic acid
Methionine	R- and S-stereoisomers of methionine sulfoxide → methionine sulfone
Histidine	2-oxohistidine, asparagine, aspartate, 4-hydroxynonenal-histidine (HNE-His)
Glutamyl (glutamine, glutathione, glutamate)	Oxalic acid, pyruvic acid
Lysine	α-aminoadipic semialdehyde, chloramines, malondialdehyde-lysine (MDA-Lys), 4-hydroxynonenal-lysine (HNE-Lys), acrolein-lysine, carboxymethyllysine, p-hydroxyphenylacetaldehyde-lysine (pHA-Lys)
Tyrosine	p-hydroxyphenylacetaldehyde, dityrosine, nitrotyrosine, chlorotyrosines, L-3,4-dihydroxyphenylalanine (L-DOPA)
Threonine	2-amino-3-ketobutyric acid
Tryptophan	Hydroxy- and nitro-tryptophans, kynurenines
Phenylalanine	Hydroxyphenylalanines
Valine, Leucine	Hydroperoxides and hydroxides
Proline	Hydroxyproline, pyrrolidone, glutamic semialdehyde
Arginine	Glutamic semialdehyde, chloramines

The significance of protein oxidation reversibility is not well understood due to a lack of analytical tools for studying protein structure *in vivo*, and hence the existing data are mainly based on *in vitro* analyses. It is believed that most types of protein oxidation damage are irreversible, with the exception of S-containing amino acids such as Met and Cys (Shacter, 2000; Bechtold *et al.*, 2004; Møller *et al.*, 2007; Hawkins *et al.*, 2009; Onda, 2013). The oxidation of most amino acids is widely considered to be a pathophysiological phenomenon, while the oxidation of S-containing amino acids is thought to play a regulatory role (for example through protein folding).

Cys reaction with ROS is reversible. This is based on incomplete oxidation of this amino acid. 'OH and singlet oxygen induce the sequential formation of more oxidized derivatives of this amino acid: cysteine → cystine → cysteine sulfenic acid → cysteine sulfinic acid → cysteic acid, all of which (apart from the last one) are enzymatically reversible by glutaredoxin or thioredoxin systems (Biteau *et al.*, 2003; Tang *et al.*, 2004; Buchanan and Balmer, 2005; Møller *et al.*, 2007; Onda, 2013). Formation of a disulfide bond between two oxidized SH-groups is a key mechanism of protein regulation. Cys–Cys dimer can then be reduced by several electron donors in the cell.

Oxidative damage of different amino acids in the protein have different functional consequences (Shacter, 2000; Stadtman and Levine, 2000; Cecarini *et al.*, 2007; Avery, 2011; Onda, 2013; Anjum *et al.*, 2015). For example, Met is sensitive to ROS and is easily and reversibly oxidized; however, its modification typically does not affect the function of the entire protein, although it sometimes prevents oxidation of other amino acids. At the same time, oxidation of Cys results in significant regulatory consequences, although induced dysfunction has also been recorded.

The most commonly occurring oxidative protein modification after modifications to Cys and Me is the formation (insertion) of carbonyl group called carbonylation (Lounifi *et al.*, 2013; Anjum *et al.*, 2015). It requires higher 'inputs of energy' than oxidation of Cys and Met, involves most amino acids and results in severe changes in protein structure/function and pathophysiological effects (Berlett and Stadtman, 1997; Shacter, 2000; Tanou *et al.*, 2009; Lounifi *et al.*, 2013). Carbonylation usually refers to a process that forms reactive ketones or aldehydes which can be detected by 'Brady's test' with 2,4-dinitrophenylhydrazine (leading to formation of hydrazones). The oxidation of side chains of lysine, arginine, proline and threonine is considered to be a primary protein carbonylation reaction, which produces 2,4-dinitrophenylhydrazine detectable products. 'Secondary protein carbonylation' reaction occurs via the addition of aldehydes that are produced during lipid peroxidation (usually they are aggressive carbonyl species, three to nine carbons in length). Carbonylation leads to the addition of a large and reactive group into the protein chain. It has a number of dangerous effects on protein characteristics, including covalent intermolecular cross-linking, cleavage or changing the rate of protein degradation. All these significantly modify (usually inhibit) protein enzymatic and physiological activities. Elevated protein carbonylation has been found for a number of plant stresses such as salinity (Tanou et al., 2009), drought (Bartoli et al., 2004) and cadmium toxicity (Romero-Puertas et al., 2002). Protein carbonylation has been always considered as irreversible (Berlett and Stadtman, 1997) but evidence has now appeared that it can be enzymatically reversed for some transcription factors (Wong et al., 2008). Therefore, this reaction can potentially be a novel ROS signalling mechanism (Wong et al., 2008; Lounifi et al., 2013).

The analyses of Arabidopsis thaliana proteome under oxidative stress conditions have revealed that nitrosative stress (NO˙ donors) results in protein modification known as S-nitrosylation (Lindermayr et al., 2005; Lounifi et al., 2013; Anjum et al., 2015). It is likely that this reaction is induced by peroxynitrite (the product of NO˙ reaction with O_2^-) and leads to severe protein function disturbance or signalling events, by the analogy with animal cells (Lounifi et al., 2013; Halliwell and Gutteridge, 2015).

3.5.3 Carbohydrates

Although carbohydrates are the most abundant group of organic molecules in plants (and on the planet), they are much less studied in terms of oxidative damage and role in stress signalling. They mechanically support and shape plant cells (cellulose, pectin, etc.), store reduced carbon (starch, sucrose, etc.), regulate enzyme activities and osmotic pressure (low molecular weight sugars), provide non-enzymatic antioxidant defence (flavonoids, mannitol, etc.) and play other key roles. Oxidation of carbohydrates is potentially harmful for plants.

Hydroxyl radicals, which are generated by catalytically active Cu^{2+} or other transition metals, can react non-enzymatically with xyloglucans and pectins, breaking them down into parts and causing cell wall loosening (Fry et al., 2002). This may facilitate expansive cell growth and promote fruit ripening (Fry et al., 2002; Fry, 2004). This reaction is beneficial but, in stress conditions when catalytic Cu and Fe activities increase several-fold, it could have pathophysiological consequences (Becana and Klucas, 1992; Moran et al., 1994).

Mono- and disaccharides can probably act as ROS scavengers (Couée et al., 2006). Their ability to scavenge ˙OH is as follows (EPR and HPLC tests): maltose > sucrose > fructose > glucose > deoxyribose > sorbitol (Morelli et al., 2003). Nevertheless, the metabolism of products (apart from formate) synthetized in these reactions is unclear. Accumulation of some carbohydrates (for example mannitol) has been shown to correlate with increased resistance to oxidative stress in a number of species (Shen et al., 1997; Couée et al., 2006; Patel and Williamson, 2016), but the direct link between oxidative stress-induced modifications to carbohydrates and plant physiology is still missing. They probably act as structural, osmotic, nutrient and signalling agents rather than redox switches or major targets for ROS (Couée et al., 2006).

3.5.4 Polynucleic acids

Damage to DNA caused by hydroxyl and other ROS is a major reason for cancerogenesis and many other diseases in animal physiology. However, this is not relevant to agronomically important plants; therefore the physiological significance of the damage induced by ROS in polynucleic acids is insufficiently studied in plant physiology.

Oxidative DNA damage can be the reason for ageing of seed stocks and, sometimes, for death of crop plants (Britt, 1996). This is classically subdivided into three types of lesions: mismatched bases, double-strand breaks and chemically modified bases (Cooke et al., 2003; Yoshiyama et al., 2013). Hydroxyl radicals are a main damaging factor for polynucleic acids, reacting with them by addition to double bonds of nucleotide bases and by abstraction of H^+ from each of the C-H bonds of 2'-deoxyribose and methyl group of thymine. 8-oxo-7,8-dihydroguanine (8-oxoG) and 2,6-diamino-4-hydroxy-5-formamidopyrimidine (FapyG) are commonly detected products of ˙OH-induced DNA/RNA oxidation (Cooke et al., 2003; Wang et al., 2010; Yoshiyama et al., 2013).

Systems for DNA repair exist in plants, including direct repair of the damaged part of the molecule, in addition to base and nucleotide replacement (Britt, 1996; Tuteja et al., 2001; Yoshiyama et al., 2013). Protection also includes enforcement of antioxidant defence in both cytosol and organelles. Vanderauwera et al. (2011) have shown that nuclear ROS-scavenging enzymes (peroxiredoxin and glutathione) are insufficient to protect DNA during oxidative stress. They have demonstrated that catalase and cytosolic ascorbate peroxidase are important for protection of nuclear DNA in stress conditions. Moreover, ROS-induced DNA damage triggers signalling phenomena and activates specific transcription factors, inducing DNA repair (Yoshiyama et al., 2013).

3.6   A Short Overview of Defence Mechanisms

Plant antioxidants have been the subject of extensive research, and has been the focus of a number of very good recent reviews (including, most recently: Smirnoff, 2005; Dietz et al., 2006; Pitzschke et al., 2006; Santosa and Reya, 2006; Møller et al., 2007; Foyer and Noctor, 2009, 2011; Asensi-Fabado and Munne-Bosch, 2010; Gill and Tuteja, 2010; Farmer and Mueller, 2013; Zagorchev et al., 2013; del Río, 2015; Noctor et al., 2016). They will therefore be discussed briefly.

Plants evolved several lines of defence against an oxidative stress (reviewed by Alscher et al., 2002; Dietz, 2003, Mittler et al., 2004; Dietz et al., 2006; Gill and Tuteja, 2010; Mhamdi et al., 2012; Noctor et al., 2016):

1. An activation and *de novo* synthesis of antioxidants – enzymes and non-enzymatic substances directly scavenging (removing) ROS and free radicals. Major enzymatic antioxidants, which show high affinity to specific ROS, include cytosolic Cu-Zn-SOD, mitochondrial Mn-SOD, chloroplastal Fe-SOD (all SODs: superoxide + $2H^+ \rightarrow H_2O_2 + O_2$), catalases ($2H_2O_2 \rightarrow 2H_2O + O_2$), peroxidases ($R/HOOH + R-H_2 \rightarrow R + 2H_2O/ROH$),

peroxiredoxins ($ROOH \rightarrow ROH$), and thioredoxins and glutaredoxins (both: $R-S-S-R \rightarrow 2R-SH$). Non-enzymatic antioxidants are non-specific to different ROS and comprise ascorbic acid, glutathione, proline, polyamines, betaine, carotenes, some flavonoids and α-tocopherol. These are most important, but a number of other molecules act as antioxidants. There is also a group of enzymes maintaining ROS scavenger function, such as monodehydroascorbate reductase, dehydroascorbate reductase, thioredoxin reductase, glutathione reductase, glutathione S-transferases and others, which reduce oxidized antioxidants, such as ascorbic acid and glutathione. The first line of defence also includes substances that decrease catalytic activity of transition metals, such as metollothioneines (small Cys-rich proteins), phytochelatines (oligomers of glutathione), pectins and other cell wall polysaccharides and structural proteins (reviewed by Cobbett and Goldsbrough, 2002; Fry et al., 2002; Hassinen et al., 2011; Zagorchev et al., 2013).
2. Plants can probably synthetize protein isoforms and lipids which are less sensitive to oxidation. This requires an upstream signalling step to activate specific genetic/metabolic pathways (Myouga et al., 2008). This mechanism is well established for animals but it is still debated whether it is important for plants.
3. Living tissues can be defended from stresses by the layers of dead cells, which die rapidly by programmed ROS-induced mechanisms (Ca^{2+} influx/K^+ efflux; Demidchik and Maathuis, 2007; Demidchik et al., 2010, 2014). These provide a shield against infection or aggressive agents. Enhanced mycorrhization is another mechanism of this kind. Fungal hyphae can create a shield around a root, isolating it from stresses. This mechanism is proved for heavy metals (Schutzendubel and Polle, 2002).
4. Plants activate biosynthesis systems for repair of damaged components (reviewed by Møller et al., 2007; Yoshiyama et al., 2013).

3.7   Summary

The theme of ROS and oxidative stress has become one of the most important in experimental plant biology. This review has demonstrated the following key aspects of this problem:

- The oxidative damage in plant cells is mainly caused by hydroxyl radical and singlet oxygen while hydrogen peroxide, superoxide radical, nitric oxide and other ROS and RNS play the roles of intermediate molecules or induce signalling events. The presence of transition metals, such as $Cu^{+/2+}$ or $Fe^{2+/3+}$, is critical for formation of hydroxyl radical and development of oxidative stress.

- Apart from exogenous sources (UV, O_3, gamma radiation, Cu, Fe, etc.), ROS are generated by photosystems, electron-transporting mitochondrial complexes, ETC of peroxisomal membranes, xanthine oxidase of the peroxisome matrix, NADPH oxidases of the plasma membrane, peroxidases expressed in all cell compartments including cell walls, and some other less important systems.

- A number of major environmental and biotic stresses activate the NADPH oxidase and Class III peroxidases. This mechanism triggers the apoplastic oxidative burst that may result in oxidative stress and trigger signalling cascades.

- Novel genetically encoded and 'reversible' ROS probes such as HyPer provide advanced tools for testing ROS dynamics *in planta* and can shed light on unknown mechanisms of plant oxidative stress and redox signalling.

- ROS are sensed by the plasma membrane cation channels and probably by RLKs, which send Ca^{2+}- and phosphorylation-encoded messages to metabolic machinery and systems regulating gene expression.

- ROS activates K^+ efflux through K^+ efflux channels such as GORK, SKOR and probably annexins, inducing K^+ loss and subsequent PCD.

- Lipid peroxidation and degradation is caused by hydroxyl radical or singlet oxygen via interaction of these ROS with double bonds of polyunsaturated fatty acids. This can lead to a chain reaction converting functional lipids into toxic aldehydes and ketones. Reversible oxidation of S-containing amino acids by ROS plays the role of a regulatory redox switch in plant cells. Major mechanisms of irreversible modification of amino acids include 'primary' and 'secondary' carbonylation. Plant carbohydrates and nucleic acids are sensitive to 'OH but the role of their oxidative damage in physiology is still not understood.

- The complex networks of enzymatic (specific) and non-enzymatic (non-specific) antioxidants, as well as repair systems, defend plant cells against oxidative stress and orchestrate stress signalling.

- The future frontiers for research into mechanisms of oxidative stress include identification of 'ROS sensors' among the plant RLKs and delineating the mechanism of their interaction with ROS. Another key issue will be the characterization at the genetic level of Ca^{2+}-transporting cation channels which are responsible for ROS-induced Ca^{2+} influx. Future research into identification and genetic improvement of structural moieties related to oxidation-induced protein dysfunction (for example by the directed evolution approach) can help in understanding oxidative stress and enhancing stress resistance and crop productivity. A deeper analysis of regulation of ROS-producing systems; defining the significance of carbohydrate and polynucleotide damage, and the roles of transition metal metabolism and antioxidant networks; and establishing precise protein reparation mechanisms are also critically important tasks.

Acknowledgement

This study was supported by Russian Science Foundation grant #15-14-30008 to VD.

References

Ahn, S.G. and Thiele, D.J. (2003) Redox regulation of mammalian heat shock factor 1 is essential for Hsp gene activation and protection from stress. *Genes and Development* 17, 516–528.

Alcázar, R., Marco, F., Cuevas, J.C., Patron, M., Ferrando, A., Carrasco, P., Tiburcio, A.F. and Altabella, T. (2006) Involvement of polyamines in plant response to abiotic stress. *Biotechnology Letters* 28, 1867–1876.

Alscher, R.G., Erturk, N. and Heath, L.S. (2002) Role of superoxide dismutases (SODs) in controlling oxidative stress in plants. *Journal of Experimental Botany* 53, 1331–1341.

Alvarez, B. and Radi, R. (2003) Peroxynitrite reactivity with amino acids and proteins. *Amino Acids* 25, 295–311.

Ananyev, G.M., Renger, G., Wacker, U. and Klimov, V.V. (1994) The production of superoxide radicals and the superoxide dismutase activity of photosystem II. The possible involvement of cytochrome b559. *Photosynthetic Research* 41, 327–338.

Anbar, M. and Neta, P. (1967) A compilation of specific bimolecular rate constants for the reactions of hydrated electrons, hydrogen atoms and hydroxyl radicals with inorganic and organic compounds in aqueous solution. *International Journal of Applied Radiation and Isotopes* 18, 493–523.

Anjum, N.A., Sofo, A., Scopa, A., Roychoudhury, A., Gill, S.S., Iqbal, M., Lukatkin, A.S., Pereira, E., Duarte, A.C. and Ahmad, I. (2015) Lipids and proteins – major targets of oxidative modifications in abiotic stressed plants. *Environmental Science and Pollution Research International* 22, 4099–4121.

Apel, K. and Hirt, H. (2004) Reactive oxygen species: metabolism, oxidative stress, and signal transduction. *Annual Reviews of Plant Biology* 55, 373–399.

Arora, D., Jain, P., Singh, N., Kaur, H. and Bhatla, S.C. (2016) Mechanisms of nitric oxide crosstalk with reactive oxygen species scavenging enzymes during abiotic stress tolerance in plants. *Free Radical Research* 50, 291–303.

Asada, K. (1999) The water-water cycle in chloroplasts: scavenging of active oxygen species and dissipation of excess photons. *Annual Review Plant Physiology Plant Molecular Biology* 50, 601–639.

Asada, K. (2006) Production and scavenging of reactive oxygen species in chloroplasts and their functions. *Plant Physiology* 141, 391–396.

Asensi-Fabado, M.A. and Munne-Bosch, S. (2010) Vitamins in plants: occurrence, biosynthesis and antioxidant function. *Trends in Plant Science* 15, 582–592.

Avery, S.V. (2011) Molecular targets of oxidative stress. *Biochemical Journal* 434, 201–210.

Bánfi, B., Tirone, F., Durussel, I., Knisz, J., Moskwa, P., Molnár, G.Z., Krause, K.H. and Cox, J.A. (2004) Mechanism of Ca^{2+} activation of the NADPH oxidase 5 (NOX5). *Journal of Biological Chemistry* 279, 18583–18591.

Bartoli, C.G., Gomez, F., Martinez, D.E. and Guiamet, J.J. (2004) Mitochondria are the main target for oxidative damage in leaves of wheat (*Triticum aestivum* L.). *Journal of Experimental Botany* 55, 1663–1669.

Baxter-Burrell, A., Yang, Z., Springer, P.S. and Bailey-Serres, J. (2002) RopGAP4-dependent Rop GTPase rheostat control of *Arabidopsis* oxygen deprivation tolerance. *Science* 296, 2026–2028.

Becana, M. and Klucas, R.V. (1992) Transition metals in legume root nodules: iron-dependent free radical production increases during nodule senescence. *Proceedings of the National Academy of Sciences USA* 89, 8958–8962.

Becana, M., Moran, J.F. and Iturbe-Ormaetxe, I. (1998) Iron-dependent oxygen free radical generation in plants subjected to environmental stress: toxicity and antioxidant protection. *Plant and Soil* 201, 137–147.

Bechtold, U., Murphy, D.J. and Mullineaux, P.M. (2004) *Arabidopsis* peptide methionine sulfoxide reductase prevents cellular oxidative damage in long nights. *The Plant Cell* 16, 908–919.

Belousov, V.V., Fradkov, A.F., Lukyanov, K.A., Staroverov, D.B., Shakhbazov, K.S., Terskikh, A.V. and Lukyanov, S. (2006) Genetically encoded fluorescent indicator for intracellular hydrogen peroxide. *Nature Methods* 3, 281–286.

Bergmann, W. (1992) *Nutritional Disorders of Plants – Development, Visual and Analytical Diagnosis*. Fisher, Jena, Germany.

Berlett, B.S. and Stadtman, E.R. (1997) Protein oxidation in aging, disease, and oxidative stress. *Journal of Biological Chemistry* 272, 20313–20316.

Bienert, G.P. and Chaumont, F. (2014) Aquaporin-facilitated transmembrane diffusion of hydrogen peroxide. *Biochimica et Biophysica Acta* 1840, 1596–1604.

Bindschedler, L.V., Dewdney, J., Blee, K.A., Stone, J.M., Asai, T., Plotnikov, J., Denoux, C., Hayes, T., Gerrish, C., Davies, D.R., Ausubel, F.M. and Bolwell, G.P. (2006) Peroxidase-dependent apoplastic oxidative burst in *Arabidopsis* required for pathogen resistance. *Plant Journal* 47, 851–863.

Biteau, B., Labarre, J. and Toledano, M.B. (2003) ATP-dependent reduction of cysteine–sulphinic acid by *S. cerevisiae* sulphiredoxin. *Nature* 425, 980–984.

Bolwell, G.P. and Wojtaszek, P. (1997) Mechanisms for the generation of reactive oxygen species in plant defence – a broad perspective. *Physiological Molecular Plant Pathology* 51, 347–366.

Bolwell, G.P., Davies, D.R., Gerrish, C., Auh, C.K. and Murphy, T.M. (1998) Comparative biochemistry of the oxidative burst produced by rose and French bean cells reveals two distinct mechanisms. *Plant Physiology* 116, 1379–1385.

Bolwell, G.P., Bindschedler, L.V., Blee, K.A., Butt, V.S., Davies, D.R., Gardner, S.L., Gerrish, C. and Minibayeva, F. (2002) The apoplastic oxidative burst in response to biotic stress in plants: a three component system. *Journal of Experimental Botany* 53, 1367–1376.

Britt, A.B. (1996) DNA damage and repair in plants. *Annual Reviews of Plant Physiology and Plant Molecular Biology* 47, 75–100.

Buchanan, B.B. and Balmer, Y. (2005) Redox regulation: a broadening horizon. *Annual Reviews of Plant Biology* 56, 187–220.

Burger, R.M. (2000) Nature of activated bleomycin. *Structural Bonding* 97, 287–303.

Camejo, D., Guzmán-Cedeño, Á. and Moreno, A. (2016) Reactive oxygen species, essential molecules, during plant-pathogen interactions. *Plant Physiology and Biochemistry* 103, 10–23.

Cannio, R., Fiorentino, G., Morana, A., Rossi, M. and Bartolucci, S. (2000) Oxygen: friend or foe? Archaeal superoxide dismutases in the protection of intra- and extracellular oxidative stress. *Frontiers in Bioscience* 1, 768–779.

Caffarri, S., Tibiletti, T., Jennings, R.C. and Santabarbara, S. (2014) A comparison between plant photosystem I and photosystem II architecture and functioning. *Current Protein and Peptide Science* 15, 296–331.

Carol, R.J., Takeda, S., Linstead, P., Durrant, M.C., Kakesova, H., Derbyshire, P., Drea, S., Zarsky, V. and Dolan, L. (2005) A RhoGDP dissociation inhibitor spatially regulates growth in root hair cells. *Nature* 438, 1013–1016.

Catalá, A. (2006) An overview of lipid peroxidation with emphasis in outer segments of photoreceptors and the chemiluminescence assay. *International Journal of Biochemistry and Cell Biology* 38, 1482–1495.

Cecarini, V., Gee, J., Fioretti, E., Amici, M., Angeletti, M., Eleuteri, A.M. and Keller, J.N. (2007) Protein oxidation and cellular homeostasis: emphasis on metabolism. *Biochimica et Biophysica Acta* 1773, 93–104.

Chang, C.C.C., Slesak, I., Jorda, L., Sotnikov, A., Melzer, M., Miszalski, Z., Mullineaux, P.M., Parker, J.E., Karpinska, B. and Karpinski, S. (2009) *Arabidopsis* chloroplastic glutathione peroxidases play a role in cross talk between photooxidative stress and immune responses. *Plant Physiology* 150, 670–683.

Chang, M.L., Chen, N.Y., Liao, L.J., Cho, C.L. and Liu, Z.H. (2012) Effect of cadmium on peroxidase isozyme activity in roots of two *Oryza sativa* cultivars. *Botanical Studies* 53, 31–44.

Chen, G.X., Kazimir, J. and Cheniae, G.M. (1992) Photoinhibition of hydroxylamine-extracted photosystem II membranes: studies of the mechanism. *Biochemistry* 31, 11072–11083.

Chen, G.X., Blubaugh, D.J., Homann, P.H., Golbeck, J.G. and Cheniae, G.M. (1995) Superoxide contributes to the rapid inactivation of specific secondary donors of the photosystem II reaction center during photodamage of manganese-depleted photosystem II membranes. *Biochemistry* 34, 2317–2332.

Chen, S.X. and Schopfer, P. (1999) Hydroxyl-radical production in physiological reactions. A novel function of peroxidase. *European Journal of Biochemistry* 260, 726–735.

Cleland, R.E. and Grace, S.C. (1999) Voltammetric detection of superoxide production by photosystem II. *FEBS Letters* 457, 348–352.

Cobbett, C. and Goldsbrough, P. (2002) Phytochelatins and metallothioneins: roles in heavy metal detoxification and homeostasis. *Annual Review of Plant Biology* 53, 159–182.

Cooke, M.S., Evans, M.D., Dizdaroglu, M. and Lunec, J. (2003) Oxidative DNA damage: mechanisms, mutation, and disease. *FASEB Journal* 17, 1195–1214.

Cosio, C. and Dunand, C. (2009) Specific functions of individual classIII peroxidases genes. *Journal of Experimental Botany* 62, 391–408.

Couée, I., Sulmon, C., Gouesbet, G. and El Amrani, A. (2006) Involvement of soluble sugars in reactive oxygen species balance and responses to oxidative stress in plants. *Journal of Experimental Botany* 57, 449–459.

Dark, A., Demidchik, V., Richards, S.L., Shabala, S. and Davies, J.M. (2011) Release of extracellular purines from plant roots and effect on ion fluxes. *Plant Signal and Behaviour* 6, 1855–1857.

Day, I., Reddy, V., Ali, G.S. and Reddy, A.S.N. (2002) Analysis of EF-hand-containing proteins in *Arabidopsis*. *Genome Biology* 3, RESEARCH0056.

DeFalco, T.A., Bender, K.W. and Snedden, W.A. (2010) Breaking the code: Ca^{2+} sensors in plant signalling. *Biochemical Journal* 425, 27–40.

Del, R.D., Stewart, A.J. and Pellegrini, N. (2005) A review of recent studies on malondialdehyde as toxic molecule and biological marker of oxidative stress. *Nutrition, Metabolism and Cardiovascular Diseases* 15, 316–328.

del Río, L.A. (2015) ROS and RNS in plant physiology: an overview. *Journal of Experimental Botany* 66, 2827–2837.

del Río, L.A. and López-Huertas, E. (2016) ROS generation in peroxisomes and its role in cell signaling. *Plant and Cell Physiology* doi: 10.1093/pcp/pcw076. [Epub ahead of print.]

del Río, L.A., Corpas, F.J., Sandalio, L.M., Palma, J.M., Gomez, M. and Barroso, J.B. (2002) Reactive oxygen species, antioxidant systems and nitric oxide in peroxisomes. *Journal of Experimental Botany* 53, 1255–1272.

del Río, L.A., Corpas, F.J., Sandalio, L.M., Palma, J.M., Gomez, M. and Barroso, J.B. (2003) Plant peroxisomes, reactive oxygen metabolism and nitric oxide. *IUBMB Life* 55, 71–81.

del Río, L.A., Sandalio, L.M., Corpas, F.J., Palma, J.M. and Barroso, J.B. (2006) Reactive oxygen species and reactive nitrogen species in peroxisomes. Production, scavenging, and role in cell signaling. *Plant Physiology* 141, 330–335.

Demidchik, V. (2010) Reactive oxygen species, oxidative stress and plant ion channels. In: Demidchik, V. and Maathuis, F.J.M. (eds) *Ion Channels and Plant Stress Responses*. Springer, Heidelberg, pp. 207–232.

Demidchik, V. (2015) Mechanisms of oxidative stress in plants: from classical chemistry to cell biology. *Environmental and Experimental Botany* 109, 212–228.

Demidchik, V. and Maathuis, F.J.M. (2007) Physiological roles of nonselective cation channels in plants: from salt stress to signalling and development. Tansley review. *New Phytologist* 175, 387–405.

Demidchik, V., Sokolik, A. and Yurin, V. (1997) The effect of Cu^{2+} on ion transport systems of the plant cell plasmalemma. *Plant Physiology* 114, 1313–1325.

Demidchik, V., Sokolik, A. and Yurin, V. (2001) Characteristics of non-specific permeability and H^+-ATPase inhibition induced in the plasma membrane of *Nitella flexilis* by excessive Cu^{2+}. *Planta* 212, 583–590.

Demidchik, V., Shabala, S.N., Coutts, K.B., Tester, M.A. and Davies, J.M. (2003) Free oxygen radicals regulate plasma membrane Ca^{2+}- and K^+-permeable channels in plant root cells. *Journal of Cell Science* 116, 81–88.

Demidchik, V., Adobea, P. and Tester, M.A. (2004) Glutamate activates sodium and calcium currents in the plasma membrane of *Arabidopsis* root cells. *Planta* 219, 167–175.

Demidchik, V., Shabala, S. and Davies, J. (2007) Spatial variation in H_2O_2 response of *Arabidopsis thaliana* root epidermal Ca^{2+} flux and plasma membrane Ca^{2+} channels. *Plant Journal* 49, 377–386.

Demidchik, V., Shang, Z., Shin, R., Thompson, E., Rubio, L., Chivasa, S., Slabas, A.R., Glover, B.J., Schachtman, D.P., Shabala, S.N. and Davies, J.M. (2009) Plant extracellular ATP signaling by plasma membrane NADPH oxidase and Ca^{2+} channels. *Plant Journal* 58, 903–913.

Demidchik, V., Cuin, T.A., Svistunenko, D., Smith, S.J., Miller, A.J., Shabala, S., Sokolik, A. and Yurin, V. (2010) *Arabidopsis* root K^+ efflux conductance activated by hydroxyl radicals: single-channel properties, genetic basis and involvement in stress-induced cell death. *Journal of Cell Science* 123, 1468–1479.

Demidchik, V., Straltsova, D., Medvedev, S., Pozhvanov, G., Sokolik, A. and Yurin, V. (2014) Stress-induced electrolyte leakage: the role of K^+-permeable channels and involvement to programmed cell death and metabolic adjustment. *Journal of Experimental Botany* 65, 1259–1270.

Despres, C., Chubak, C., Rochon, A., Clark, R., Bethune, T., Desveaux, D. and Fobert, P.R. (2003) The *Arabidopsis* NPR1 disease resistance protein is a novel cofactor that confers redox regulation of DNA binding activity to the basic domain/leucine zipper transcription factor TGA1. *The Plant Cell* 15, 2181–2191.

Dickinson, B.C., Huynh, C. and Chang, C.J. (2010) A palette of fluorescent probes with varying emission colors for imaging hydrogen peroxide signaling in living cells. *Journal of American Chemical Society* 132, 5906–5915.

Dietz, K.J. (2003) Plant peroxiredoxins. *Annual Reviews of Plant Biology* 54, 93–107.

Dietz, K.J., Jacob, S., Oelze, M.-L., Laxa, M., Tognetti, V., Marina, S., de Miranda, N., Baier, M. and Finkemeier, I. (2006) The function of peroxiredoxins in plant organelle redox metabolism. *Journal of Experimental Botany* 57, 1697–1709.

Doke, N. (1983) Involvement of superoxide anion generation in the hypersensitive response of potato tuber tissues to infection with an incompatible race of *Phytophthora infestans* and to the hyphal wall components. *Physiological Plant Pathology* 23, 345–357.

Dowling, D.K. and Simmons, L.W. (2009) Reactive oxygen species as universal constraints in life-history evolution. *Proceedings of the Royal Society, Series B, Biological Sciences* 276, 1737–1745.

Downes, A. and Blunt, T.P. (1879) The effect of sunlight upon hydrogen peroxide. *Nature* 20, 521.

Driever, S.M., Fryer, M.J., Mullineaux, P.M. and Baker, N.R. (2009) Imaging of reactive oxygen species *in vivo*. *Methods in Molecular Biology* 479, 109–116.

Dubiella, U., Seybold, H., Durian, G., Komander, E., Lassig, R., Witte, C.P., Schulze, W.X. and Romeis, T. (2013) Calcium-dependent protein kinase/NADPH oxidase activation circuit is required for rapid defense signal propagation. *Proceedings of the National Academy of Sciences USA* 110, 8744–8749.

Duprat, F., Guillemare, E., Romey, G., Fink, M., Lesage, F. and Alzdunski, M. (1995) Susceptibility of cloned K^+ channels to reactive oxygen species. *Proceedings of the National Academy of Sciences USA* 92, 11796–11800.

Dynowski, M., Schaaf, G., Loque, D., Moran, O. and Ludewig, U. (2008) Plant plasma membrane water channels conduct the signalling molecule H_2O_2. *Biochemical Journal* 414, 53–61.

Farmer, E.E. and Mueller, M.J. (2013) ROS-mediated lipid peroxidation and RES-activated signaling. *Annual Review of Plant Biology* 64, 429–450.

Fenton, H.J.H. (1894) Oxidation of tartaric acid in presence of iron. *Journal of Chemical Society Transactions* 65, 899–911.

Fernándeza, A.P. and Stranda, A. (2008) Retrograde signaling and plant stress: plastid signals initiate cellular stress responses. *Current Opinion in Plant Biology* 11, 509–513.

Fischer, B.B., Hideg, É. and Krieger-Liszkay, A. (2013) Production, detection, and signaling of singlet oxygen in photosynthetic organisms. *Antioxidants and Redox Signaling* 18, 2145–2162.

Flors, C., Fryer, M.J., Waring, J., Reeder, B., Bechtold, U., Mullineaux, P.M., Nonell, S., Wilson, M.T. and Baker, N.R. (2006) Imaging the production of singlet oxygen in vivo using a new fluorescent sensor, singlet oxygen sensor green (R). *Journal of Experimental Botany* 57, 1725–1734.

Fluhr, R. (2009) Reactive oxygen-generating NADPH oxidases in plants. In: del Río, L.A. and Puppo, A. (eds) *Reactive Oxygen Species in Plant Signalling*. Springer, Berlin, pp. 1–23.

Foreman, J., Demidchik, V., Bothwell, J.H.F., Mylona, P., Miedema, H., Torres, M.A., Linstead, P., Costa, S., Brownlee, C., Jones, J.D.G., Davies, J.M. and Dolan, L. (2003) Reactive oxygen species produced by NADPH oxidase regulate plant cell growth. *Nature* 422, 442–446.

Fornazari, M., de Paula, J.G., Castilho, R.F. and Kowaltowski, A.J. (2008) Redox properties of the adenoside triphosphate-sensitive K$^+$ channel in brain mitochondria. *Journal of Neuroscience Research* 86, 1548–1556.

Foyer, C.H. and Halliwell, B. (1976) The presence of glutathione and glutathione reductase in chloroplasts: a proposed role in ascorbic acid metabolism. *Planta* 133, 21–25.

Foyer, C.H. and Noctor, G. (2009) Redox regulation in photosynthetic organisms: signaling, acclimation, and practical implications. *Antioxidants and Redox Signalling* 11, 861–710.

Foyer, C.H. and Noctor, G. (2011) Ascorbate and glutathione: the heart of the redox hub. *Plant Physiology* 155, 2–18.

Fridovich, I. (1998) Oxygen toxicity: a radical explanation. *Journal of Experimental Biology* 201, 1203–1209.

Fry, S.C. (2004) Primary cell wall metabolism: tracking the careers of wall polymers in living plant cells. *New Phytologist* 161, 641–675.

Fry, S.C., Miller, J.G. and Dumville, J.C. (2002) A proposed role for copper ions in cell wall loosening. *Plant and Soil* 247, 57–67.

Garcia-Mata, C., Wang, J.W., Gajdanowicz, P., Gonzalez, W., Hills, A., Donald, N., Riedelsberger, J., Amtmann, A., Dreyer, I. and Blatt, M.R. (2010) A minimal cysteine motif required to activate the SKOR K$^+$ channel of *Arabidopsis* by the reactive oxygen species H$_2$O$_2$. *Journal of Biological Chemistry* 285, 29286–29294.

Genty, B. and Harbinson, J. (1996) Regulation of light utilization for photosynthetic electron transport. In: Baker, N.R. (ed.) *Photosynthesis and the Environment*. Kluwer Academic, Dordrecht, The Netherlands, pp. 67–99.

Gifford, J.L., Walsh, M.P. and Vogel, H.J. (2007) Structure and metal-ion-binding properties of the Ca^{2+}-binding helix-loop-helix EF-hand motifs. *Biochemical Journal* 405, 199–221.

Gill, S.S. and Tuteja, N. (2010) Reactive oxygen species and antioxidant machinery in abiotic stress tolerance in crop plants. *Plant Physiology and Biochemistry* 48, 909–930.

Goldstein, S., Meyerstein, D. and Czapski, G. (1993) The Fenton reagents. *Free Radicals in Biology and Medicine* 15, 435–445.

Groom, Q.J., Torres, M.A., Fordham-Skelton, A.P., Hammond-Kosack, K.E., Robinson, N.J. and Jones, J.D. (1996) *RbohA*, a rice homologue of the mammalian gp91phox respiratory burst oxidase gene. *Plant Journal* 10, 515–522.

Guido, V. (2001) *Fundamentals of Physics and Chemistry of the Atmosphere*. Springer, Berlin.

Guo, M., Liu, J.H., Ma, X., Luo, D.X., Gong, Z.H. and Lu, M.H. (2016) The plant heat stress transcription factors (HSFs): structure, regulation, and function in response to abiotic stresses. *Frontiers in Plant Science* 7, 114.

Gupta, D.K., Palma, J.M. and Corpas, F.J. (2015) *Reactive Oxygen Species and Oxidative Damage in Plants under Stress*. Springer International Publishing, Heidelberg, Germany.

Guskos, N., Likodimos, V., Typek, J., Maryniak, M., Grech, E. and Kolodziej, B. (2007) Photoacoustic and EPR studies of two copper(II) complexes with spermine analoques. *Reviews of Advances in Material Sciences* 14, 97–103.

Haber, F. and Weiss, J. (1932) On the catalysis of hydroperoxide. *Naturwissenschaften* 20, 948–950.

Hahn, J.S. and Thiele, D.J. (2004) Activation of the *Saccharomyces cerevisiae* heat shock transcription factor under glucose starvation conditions by Snf1 protein kinase. *Journal of Biological Chemistry* 279, 5169–5176.

Halliwell, B. (2006) Reactive species and antioxidants. Redox biology is a fundamental theme of aerobic life. *Plant Physiology* 141, 312–322.

Halliwell, B. and Gutteridge, J.M.C. (2015) *Free Radicals in Biology and Medicine*. 5th edn. Oxford University Press, Oxford, UK.

Hassinen, V.H., Tervahauta, A.I., Schat, H. and Karenlampi, S.O. (2011) Plant metallothioneins – metal chelators with ROS scavenging activity? *Plant Biology* 13, 225–232.

Hawkins, C.L., Morgan, P.E. and Davies, M.J. (2009) Quantification of protein modification by oxidants. *Free Radicals Biology and Medicine* 46, 965–988.

Hedrich, R. (2012) Ion channels in plants. *Physiological Reviews* 92, 1777–1811.

Henderson, L.M. and Chappell, J.B. (1993) Dihydrorhodamine 123: a fluorescent probe for superoxide generation? *European Journal of Biochemistry* 217, 973–980.

Hernández, J.A., Ferrer, M.A., Jiménez, A., Barceló, A.R. and Sevilla, F. (2001) Antioxidant systems and $O_2^{\bullet-}/H_2O_2$ production in the apoplast of pea leaves. Its relation with salt-induced necrotic lesions in minor veins. *Plant Physiology* 127, 817–831.

Himschoot, E., Beeckman, T., Friml, J. and Vanneste, S. (2015) Calcium is an organizer of cell polarity in plants. *Biochimica at Biophysica Acta* 1853, 2168–2172.

Hirst, J., King, M.S. and Pryde, K.R. (2008) The production of reactive oxygen species by complex I. *Biochemical Society Transactions* 36, 976–980.

Hodges, D.M., DeLong, J.M., Forney, C.F. and Prange, R.K. (1999) Improving the thiobarbituric acid-reactive-substances assay for estimating lipid peroxidation in plant tissues containing anthocyanin and other interfering compounds. *Planta* 207, 604–611.

Hong, S.Y., Roze, L.V. and Linz, J.E. (2013) Oxidative stress-related transcription factors in the regulation of secondary metabolism. *Toxins (Basel)* 18, 683–702.

Jain, K., Kataria, S. and Guruprasad, K.N. (2004) Oxyradicals under UV-b stress and their quenching by antioxidants. *Indian Journal of Experimental Biology* 42, 884–892.

Jardim-Messeder, D., Caverzan, A., Rauber, R., de Souza Ferreira, E., Margis-Pinheiro, M. and Galina, A. (2015) Succinate dehydrogenase (mitochondrial complex II) is a source of reactive oxygen species in plants and regulates development and stress responses. *New Phytologist* 208, 776–789.

Jiang, F., Zhang, Y. and Dusting, G.J. (2011) NADPH oxidase-mediated redox signaling: roles in cellular stress response, stress tolerance, and tissue repair. *Pharmacological Reviews* 63, 218–242.

Johnson, C., Knight, M., Kondo, T., Masson, P., Sedbrook, J., Haley, A. and Trewavas, A. (1995) Circadian oscillations of cytosolic and chloroplastic free calcium in plants. *Science* 269, 1863–1865.

Joo, J.H., Bae, Y.S. and Lee, J.S. (2001) Role of auxin-induced reactive oxygen species in root gravitropism. *Plant Physiology* 126, 1055–1060.

Kadota, Y., Shirasu, K. and Zipfel, C. (2015) Regulation of the NADPH oxidase RBOHD during plant immunity. *Plant Cell Physiology* 56, 1472–1480.

Karpiński, S., Szechyńska-Hebda, M., Wituszyńska, W. and Burdiak, P. (2013) Light acclimation, retrograde signalling, cell death and immune defences in plants. *Plant Cell and Environment* 36, 736–744.

Kataria, S., Jain, K. and Guruprasad, K.N. (2005) Involvement of oxyradicals in promotion/inhibition of expansion growth in cucumber cotyledons. *Indian Journal of Experimental Biology* 43, 910–915.

Kavdia, M. (2006) A computational model for free radicals transport in the microcirculation. *Antioxidant and Redox Signalling* 8, 1103–1111.

Kawahara, T., Quinn, M.T. and Lambeth, J.D. (2007) Molecular evolution of the reactive oxygen-generating NADPH oxidase (Nox/Duox) family of enzymes. *BMC Evolutionary Biology* 7, 109.

Kawano, T., Sahashi, N., Takahashi, K., Uozumi, N. and Muto, S. (1998) Salicylic acid induces extracellular superoxide generation followed by an increase in cytosolic calcium ion in tobacco suspension culture: the earliest events in salicylic acid signal transduction. *Plant Cell Physiology* 39, 721–730.

Kervinen, M., Pätsi, J., Finel, V. and Hassinen, L.E. (2004) Lucigenin and coelenterazine as superoxide probes in mitochondrial and bacterial membranes. *Analytical Biochemistry* 324, 45–51.

Kim, C. and Apel, K. (2013) $^1O_2$-mediated and EXECUTER-dependent retrograde plastid-to-nucleus signaling in norflurazon-treated seedlings of *Arabidopsis thaliana*. *Molecular Plant* 6, 1580–1591.

Kim, S.O., Merchant, K., Nudelman, R., Beyer, W.F.J., Keng, T., DeAngelo, J., Hausladen, A. and Stamler, J.S. (2002) OxyR: a molecular code for redox-related signaling. *Cell* 109, 383–396.

Kitajima, S. (2008) Peroxide-mediated inactivation of two chloroplastic peroxidases, ascorbate peroxidase and 2-Cys peroxiredoxin. *Photochemistry and Photobiology* 84, 1404–1409.

Kobayashi, M., Ohura, I., Kawakita, K., Yokota, N., Fujiwara, M., Shimamoto, K., Doke, N. and Yoshioka, H. (2007) Calcium-dependent protein kinases regulate the production of reactive oxygen species by potato NADPH oxidase. *The Plant Cell* 19, 1065–1080.

Koppenol, W.H. (2001) The Haber-Weiss cycle – 70 years later. *Redox Reports* 6, 229–234.

Krasnovsky, A.A.J. (1998) Singlet molecular oxygen in photobiochemical systems: IR phosphorescence studies. *Membrane Cell Biology* 12, 665–690.

Krieger-Liszkay, A., Fufezan, C. and Trebst, A. (2008) Singlet oxygen production in photosystem II and related protection mechanism. *Photosynthesis Research* 98, 551–564.

Krieger-Liszkay, A., Trösch, M. and Krupinska, K. (2015) Generation of reactive oxygen species in thylakoids from senescing flag leaves of the barley varieties Lomerit and Carina. *Planta* 241, 1497–1508.

Krishnamurthy, A. and Rathinasabapathi, B. (2013) Oxidative stress tolerance in plants: novel interplay between auxin and reactive oxygen species signaling. *Plant Signalling and Behavior* 18, e25761.

Kruk, J. and Trebst, A. (2008) Plastoquinol as a singlet oxygen scavenger in photosystem II. *Biochimica et Biophysica Acta – Bioenergetics* 1777, 154–162.

Lambeth, J.D. and Neish, A.S. (2014) Nox enzymes and new thinking on reactive oxygen: a double-edged sword revisited. *Annual Reviews of Pathology* 9, 119–145.

Laohavisit, A., Mortimer, J.C., Demidchik, V., Coxon, K.M., Stancombe, M.A., Macpherson, N., Brownlee, C., Hofmann, A., Webb, A.A., Miedema, H., Battey, N.H. and Davies, J.M. (2009) *Zea mays* annexins modulate cytosolic free Ca^{2+} and generate a Ca^{2+}-permeable conductance. *The Plant Cell* 21, 479–493.

Laohavisit, A., Shang, Z., Rubio, L., Cuin, T.A., Véry, A.A., Wang, A., Mortimer, J.C., Macpherson, N., Coxon, K.M., Battey, N.H., Brownlee, C., Park, O.K., Sentenac, H., Shabala, S., Webb, A.A. and Davies, J.M. (2012) *Arabidopsis* annexin1 mediates the radical-activated plasma membrane Ca^{2+}- and K$^+$-permeable conductance in root cells. *The Plant Cell* 24, 1522–1533.

Lee, D., Khaja, S., Velasquez-Castano, J.C., Dasari, M., Sun, C., Petros, J., Taylor, W.R. and Murthy, N. (2007) *In vivo* imaging of hydrogen peroxide with chemiluminescent nanoparticles. *Nature Materials* 6, 765–769.

Lee, K.P., Kim, C., Landgraf, K. and Apel, K. (2007) EXECUTER1- and EXECUTER2-dependent transfer of stress-related signals from the plastid to the nucleus of *Arabidopsis thaliana*. *Proceedings of the National Academy of Sciences USA* 104, 10270–10275.

Lesser, M.P. (2006) Oxidative stress in marine environments: biochemistry and physiological ecology. *Annual Reviews of Physiology* 68, 253–278.

Lin, V.S., Dickinson, B.C. and Chang, C.J. (2013) Boronate-based fluorescent probes: imaging hydrogen peroxide in living systems. *Methods of Enzymology* 526, 19–43.

Lindermayr, C., Saalbach, G. and Durner, J. (2005) Proteomic identification of S-nitrosylated proteins in *Arabidopsis*. *Plant Physiology* 137, 921–930.

Liszkay, A., Kenk, B. and Schopfer, P. (2003) Evidence for the involvement of cell wall peroxidase in the generation of hydroxyl radicals mediating extension growth. *Planta* 217, 658–667.

Liszkay, A., van der Zalm, E. and Schopfer, P. (2004) Production of reactive oxygen intermediates (O$_2^-$, H$_2$O$_2$, and ˙OH) by maize roots and their role in wall loosening and elongation growth. *Plant Physiology* 136, 3114–3123.

Logan, D.C. and Knight, M.R. (2003) Mitochondrial and cytosolic calcium dynamics are differentially regulated in plants. *Plant Physiology* 133, 21–24.

Lopez-Huertas, E., Charlton, W.L., Johnson, B., Graham, I.A. and Baker, A. (2000) Stress induces peroxisome biogenesis genes. *EMBO Journal* 19, 6770–6777.

Lounifi, I., Arc, E., Molassiotis, A., Job, D., Rajjou, L. and Tanou, G. (2013) Interplay between protein carbonylation and nitrosylation in plants. *Proteomics* 13, 568–578.

Lukyanov, K.A. and Belousov, V.V. (2014) Genetically encoded fluorescent redox sensors. *Biochimica et Biophysica Acta* 1840, 745–756.

Malinouski, M., Zhou, Y., Belousov, V.V., Hatfield, D.L. and Gladyshev, V.N. (2011) Hydrogen peroxide probes directed to different cellular compartments. *PLoS ONE* 6, e14564.

Marino, D., Dunand, C., Puppo, A. and Pauly, N. (2012) A burst of plant NADPH oxidases. *Trends in Plant Sciences* 17, 9–15.

Mathe, C., Mattioli, T.A., Horner, O., Lombard, M., Latour, J.-M., Fontecave, M. and Niviere, V. (2002) Identification of iron-(III) peroxo species in the active site of the superoxide reductase SOR from *Desulfoarculus baarsii*. *Journal of American Chemical Society* 124, 4966–4967.

Mazars, C., Thuleau, P., Lamotte, O. and Bourque, S. (2010) Cross-talk between ROS and calcium in regulation of nuclear activities. *Molecular Plant* 3, 706–718.

McCleverty, J. (1999) *Chemistry of the First-Row Transition Metals*. Oxford University Press, Oxford, UK.

Medinas, D.B., Cerchiaro, G., Trindade, D.F. and Augusto, O. (2007) The carbonate radical and related oxidants derived from bicarbonate buffer. *IUBMB Life* 59, 255–262.

Mehlmer, N., Wurzinger, B., Stael, S., Hofmann-Rodrigues, D., Csaszar, E. and Pfister, B. (2010) The Ca^{2+}-dependent protein kinase CPK3 is required for MAPK-independent salt-stress acclimation in *Arabidopsis*. *Plant Journal* 63, 484–498.

Menconi, M., Sgherri, C.L.M., Pinzino, C. and Navari-Izzo, F. (1995) Activated oxygen production and detoxification in wheat plants subjected to a water deficit programme. *Journal of Experimental Botany* 46, 1123–1130.

Mhamdi, A., Noctor, G. and Baker, A. (2012) Plant catalases: peroxisomal redox guardians. *Archives of Biochemistry and Biophysics* 15, 181–194.

Michelet, L., Roach, T., Fischer, B.B., Bedhomme, M., Lemaire, S.D. and Krieger-Liszkay, A. (2013) Down-regulation of catalase activity allows transient accumulation of a hydrogen peroxide signal in *Chlamydomonas reinhardtii*. *Plant Cell and Environment* 36, 1204–1013.

Mignolet-Spruyt, L., Xu, E., Idänheimo, N., Hoeberichts, F.A., Mühlenbock, P., Brosché, M., Van Breusegem, F. and Kangasjärvi, J. (2016) Spreading the news: subcellular and organellar reactive oxygen species production and signalling. *Journal of Experimental Botany* 67, 3831–3844. doi:10.1093/jxb/erw080.

Miller, E.W., Albers, A.E., Chang, C.J., Pralle, A. and Isacoff, E.Y. (2005) Boronate-based fluorescent probes for imaging cellular hydrogen peroxide. *Journal of American Chemical Society* 127, 16652–16659.

Miller, G. and Mittler, R. (2006) Could heat shock transcription factors function as hydrogen peroxide sensors in plants? *Annals of Botany* 98, 279–288.

Mittler, R., Vanderauwera, S., Gollery, M. and Van Breusegem, F. (2004) Reactive oxygen gene network of plants. *Trends in Plant Sciences* 9, 490–498.

Mojovic, M., Vuletic, M. and Bacic, G.G. (2005) Detection of oxygen-centered radicals using EPR spin-trap DEPMPO. The effect of oxygen. *Annals of New York Academy of Sciences* 1048, 471–475.

Møller, I.M. (2001) Plant mitochondria and oxidative stress. Electron transport, NADPH turnover and metabolism of reactive oxygen species. *Annual Reviwes of Plant Physiology and Plant Molecular Biology* 52, 561–591.

Møller, I.M., Jensen, P.E. and Hansson, A. (2007) Oxidative modifications to cellular components in plants. *Annual Reviews of Plant Biology* 58, 459–481.

Moran, J.F., Becana, M., Iturbe-Ormaetxe, I., Frechilla, S., Klucas, R.V. and Aparicio-Tejo, P. (1994) Drought induces oxidative stress in pea plants. *Planta* 194, 346–352.

Moran, J.F., Klucas, R.V., Grayer, R.J., Abian, J. and Becana, M. (1997) Complexes of iron with phenolic compounds from soybean nodules and other legume tissues: prooxidant and antioxidant properties. *Free Radicals in Biology and Medicine* 22, 861–870.

Morelli, R., Russo-Volpe, S., Bruno, N. and Lo Scalzo, R. (2003) Fenton-dependent damage to carbohydrates: free radicals scavenging activity of some simple sugars. *Journal of Agricultural and Food Chemistry* 51, 7418–7425.

Moschou, P.N., Sarris, P.F., Skandalis, N., Andriopoulou, A.H., Paschalidis, K.A., Panopoulos, N.J. and Roubelakis-Angelakis, K.A. (2009) Engineered polyamine catabolism preinduces tolerance of tobacco to bacteria and oomycetes. *Plant Physiology* 149, 1970–1981.

Munemasa, S., Hossain, M.A., Nakamura, Y., Mori, I.C. and Murata, Y. (2011) The *Arabidopsis* calcium-dependent protein kinase, CPK6, functions as a positive regulator of methyl jasmonate signaling in guard cells. *Plant Physiology* 155, 553–561.

Murata, Y., Pei, Z.-M., Mori, I.C. and Schroeder, J.I. (2001) Abscisic acid activation of plasma membrane Ca^{2+} channels in guard cells requires cytosolic NAD(P)H and is differentially disrupted upstream and downstream of reactive oxygen species production in *abi1-1* and *abi2-1* protein phosphatase 2C mutants. *The Plant Cell* 13, 2513–2523.

Myouga, F., Hosoda, C., Umezawa, T., Iizumi, H., Kuromori, T., Motohashi, R., Shono, Y., Nagata, N., Ikeuchi, M. and Shinozaki, K. (2008) A heterocomplex of iron superoxide dismutases defends chloroplast nucleoids against oxidative stress and is essential for chloroplast development in *Arabidopsis*. *The Plant Cell* 20, 3148–3162.

Nawkar, G.M., Maibam, P., Park, J.H., Sahi, V.P., Lee, S.Y. and Kang, C.H. (2013) UV-induced cell death in plants. *International Journal of Molecular Science* 14, 1608–1628.

Noctor, G., Mhamdi, A. and Foyer, C.H. (2016) Oxidative stress and antioxidative systems: recipes for successful data collection and interpretation. *Plant Cell and Environment* 39, 1140–1160.

Onda, Y. (2013) Oxidative protein-folding systems in plant cells. *International Journal of Cell Biology*, ID 585431.

Ortiz de Montellano, P.R. (1998) Heme oxygenase mechanism: evidences for an electrophilic, ferric peroxide species. *Accounts of Chemical Research* 31, 543–549.

Osakabe, Y., Yamaguchi-Shinozaki, K., Shinozaki, K. and Tran, L.S. (2013) Sensing the environment: key roles of membrane-localized kinases in plant perception and response to abiotic stress. *Journal of Experimental Botany* 64, 445–458.

Outten, F.W. and Theil, E.C. (2009) Iron-based redox switches in biology. *Antioxidants and Redox Signaling* 11, 1029–1046.

Patel, T.K. and Williamson, J.D. (2016) Mannitol in plants, fungi, and plant-fungal interactions. *Trends in Plant Science* 21, 486–497. doi: 10.1016/j.tplants.2016.01.006.

Pei, Z.M., Murata, Y., Benning, G., Thomine, S., Klusener, B., Allen, G.J., Grill, E. and Schroeder, J.I. (2000) Calcium channels activated by hydrogen peroxide mediate abscisic acid signalling in guard cells. *Nature* 406, 731–734.

Pitzschke, A. and Hirt, H. (2009) Disentangling the complexity of mitogen-activated protein kinases and reactive oxygen species signaling. *Plant Physiology* 149, 606–615.

Pitzschke, A., Forzani, C. and Hirt, H. (2006) Reactive oxygen species signaling in plants. *Antioxidants and Redox Signaling* 8, 1757–1764.

Plattner, H. and Verkhratsky, A. (2015) The ancient roots of calcium signalling evolutionary tree. *Cell Calcium* 57, 123–132.

Pospíšil, P. (2009) Production of reactive oxygen species by photosystem II. *Biochimica et Biophysica Acta – Bioenergetics* 1787, 1151–1160.

Pospíšil, P. (2014) The role of metals in production and scavenging of reactive oxygen species in photosystem II. *Plant Cell Physiology* 55, 1224–1232.

Pospíšil, P., Arato, A., Krieger-Liszkay, A. and Rutherford, A.W. (2004) Hydroxyl radical generation by photosystem II. *Biochemistry* 43, 6783–6792.

Pospíšil, P., Šnyrychová, I., Kruk, J., Strzałka, K. and Nauš, J. (2006) Evidence that cytochrome b559 is involved in superoxide production in Photosystem II: effect of synthetic short-chain plastoquinones in a cytochrome b559 tobacco mutant. *Biochemical Journal* 397, 321–327.

Pou, S., Hassett, D.J., Britigan, B.E., Cohen, M.S. and Rosen, G.M. (1989) Problems associated with spin trapping oxygen-centered free radicals in biological systems. *Analytical Biochemistry* 177, 1–6.

Pryor, W.A. and Squadrito, G.L. (1995) The chemistry of peroxynitrite: a product from the reaction of nitric oxide with superoxide. *American Journal of Physiology* 268, 699–722.

Przybyla, D., Göbel, C., Imboden, A., Hamberg, M., Feussner, I. and Apel, K. (2008) Enzymatic, but not non-enzymatic $^1O_2$-mediated peroxidation of polyunsaturated fatty acids forms part of the EXECUTER1-dependent stress response program in the flu mutant of *Arabidopsis thaliana*. *Plant Journal* 54, 236–248.

Ranty, B., Aldon, D., Cotelle, V., Galaud, J.P., Thuleau, P. and Mazars, C. (2016) Calcium sensors as key hubs in plant responses to biotic and abiotic stresses. *Frontiers in Plant Sciences* 16, 327.

Renew, S., Heyno, E., Schopfer, P. and Liszkay, A. (2005) Sensitive detection and localization of hydroxyl radical production in cucumber roots and *Arabidopsis* seedlings by spin trapping electron paramagnetic resonance spectroscopy. *Plant Journal* 44, 342–347.

Reumann, S. and Weber, A.P.M. (2006) Plant peroxisomes respire in the light: some gaps of the photorespiratory C_2 cycle have become filled – others remain. *Biochimica et Biophysica Acta* 1763, 1496–1510.

Reumann, S., Babujeea, L., Maa, C., Wienkoop, S., Siemsena, T., Antonicellia, G.E., Lüdera, R.N.F., Weckwerth, W. and Jahn, O. (2007) Proteome analysis of *Arabidopsis* leaf peroxisomes reveals novel targeting peptides, metabolic pathways, and defense mechanisms. *The Plant Cell* 19, 3170–3193.

Rhee, S.G., Chang, T.S., Jeong, W. and Kang, D. (2010) Methods for detection and measurement of hydrogen peroxide inside and outside of cells. *Molecules and Cells* 29, 539–549.

Rinalducci, S., Murgiano, L. and Zolla, L. (2008) Redox proteomics: basic principles and future perspectives for the detection of protein oxidation in plants. *Journal of Experimental Botany* 59, 3781–3801.

Robinson, N.J and Winge, D.R. (2010) Copper metallochaperones. *Annual Review of Biochemistry* 79, 537–562.

Rodrigo-Moreno, A., Andrés-Colás, N., Poschenrieder, C., Gunsé, B., Peñarrubia, L. and Shabala, S. (2013) Calcium- and potassium-permeable plasma membrane transporters are activated by copper in *Arabidopsis* root tips: linking copper transport with cytosolic hydroxyl radical production. *Plant Cell and Environment* 36, 844–855.

Rodríguez, A.A., Grunberg, K.A. and Taleisnik, E. (2002) Reactive oxygen species in the elongation zone of maize leaves are necessary for leaf extension. *Plant Physiology* 129, 1627–1632.

Rodríguez, A.A., Lascano, R., Bustos, D. and Taleisnik, E. (2007) Salinity-induced decrease in NADPH oxidase activity in the maize leaf blade elongation zone. *Journal of Plant Physiology* 164, 223–230.

Rodriguez, M.C.S, Petersen, M. and Mundy, J. (2010) Mitogen-activated protein kinase signaling in plants. *Annual Review of Plant Biology* 61, 621–649.

Romero-Puertas, M.C., Palma, J.M., Gómez, M., del Río, L.A. and Sandalio, L.M. (2002) Cadmium causes the oxidative modification of proteins in pea plants. *Plant Cell and Environment* 25, 677–686.

Ross, A.B. (1985) Reactivity of HO_2/O_2^- radicals in aqueous solution. *Journal of Physical Chemistry* 14, 1041–1091.

Rubbo, H., Radi, R., Trujillo, M., Telleri, R., Kalyanaraman, B., Barnes, S., Kirk, M. and Freeman, B.A. (1994) Nitric oxide regulation of superoxide and peroxynitrite-dependent lipid peroxidation. Formation of novel nitrogen-containing oxidized lipid derivatives. *Journal of Biological Chemistry* 269, 26066–26075.

Rubbo, H., Trostchansky, A. and O'Donnell, V.B. (2009) Peroxynitrite-mediated lipid oxidation and nitration: mechanisms and consequences. *Archives of Biochemistry and Biophysics* 484, 167–172.

Sandalio, L.M. and Romero-Puertas, M.C. (2015) Peroxisomes sense and respond to environmental cues by regulating ROS and RNS signalling networks. *Annals of Botany* 116, 475–485.

Sandalio, L.M., Rodríguez-Serrano, M., Romero-Puertas, M.C. and del Río, L.A. (2008) Imaging of reactive oxygen species and nitric oxide *in vivo* in plant tissues. *Methods of Enzymology* 440, 397–409.

Santosa, C.V.D. and Reya, P. (2006) Plant thioredoxins are key actors in the oxidative stress response. *Trends in Plant Science* 11, 329–334.

Sawyer, D.T. and Gibian, M.J. (1979) The chemistry of superoxide ion. *Tetrahedron* 35, 1471–1481.

Sawyer, D.T., Kang, C., Llobet, A. and Redman, C. (1993) Fenton reagents (1:1 Fe^{II} L_x/HOOH) react via $[L_xFe^{II}OOH(BH^+)]$ (1) as hydroxylases (RH-ROH), not as generators of free hydroxyl radicals (•OH). *Journal of American Chemical Society* 115, 5817–5818.

Sawyer, D.T., Sobkowiak, A. and Matsushita, T. (1996) Metal $[ML_x;$ M = Fe, Cu, Co, Mn]/hydroperoxide-induced activation of dioxygen for the oxygenation of hydrocarbons: oxygenated Fenton chemistry. *Accounts of Chemical Research* 29, 409–416.

Sayre, L.M., Perry, G. and Smith, M.A. (2008) Oxidative stress and neurotoxicity. *Chemical Research in Toxicology* 21, 172–188.

Schutzendubel, A. and Polle, A. (2002) Plant responses to abiotic stresses: heavy metal-induced oxidative stress and protection by mycorrhization. *Journal of Experimental Botany* 53, 1351–1365.

Schweitzer, C. and Schmidt, R. (2003) Physical mechanisms of generation and deactivation of singlet oxygen. *Chemical Reviews* 103, 1685–1757.

Seon, P. and Ja-Eun, K. (2002) Potassium efflux during apoptosis. *Journal of Biochemistry and Molecular Biology* 35, 41–46.

Šeršeň, F. and Kráľová, K. (2013) EPR spectroscopy – a valuable tool to study photosynthesizing organisms exposed to abiotic stresses. In: Dubinsky, Z. (ed.) *Agricultural and Biological Sciences. Photosynthesis*. Intech, Rijeka, Croatia, pp. 247–283.

Shabala, S., Cuin, T.A., Prismall, L. and Nemchinov, L.G. (2007) Expression of animal CED-9 anti-apoptotic gene in tobacco modifies plasma membrane ion fluxes in response to salinity and oxidative stress. *Planta* 227, 189–197.

Shacter, E. (2000) Quantification and significance of protein oxidation in biological samples. *Drug Metabolism Reviews* 32, 307–326.

Shahid, M., Pourrut, B., Dumat, C., Nadeem, M., Aslam, M. and Pinelli, E. (2014) Heavy-metal-induced reactive oxygen species: phytotoxicity and physicochemical changes in plants. *Reviews on Environmental and Contamination Toxicology* 232, 1–44.

Shen, B., Jensen, R.G. and Bohnert, H.J. (1997) Increased resistance to oxidative stress in transgenic plants by targeting mannitol biosynthesis to chloroplasts. *Plant Physiology* 113, 1177–1183.

Shi, H. and Chan, Z. (2014) Improvement of plant abiotic stress tolerance through modulation of the polyamine pathway. *Journal of Integrative Plant Biology* 56, 114–121.

Shibata, M., Oikawa, K., Yoshimoto, K., Kondo, M., Mano, S., Yamada, K., Hayashi, M., Sakamoto, W., Ohsumi, Y. and Nishimura, M. (2013) Highly oxidized peroxisomes are selectively degraded via autophagy in *Arabidopsis*. *The Plant Cell* 25, 4967–4983.

Shulaev, V. and Oliver, D.J. (2006) Metabolic and proteomic markers for oxidative stress. New tools for reactive oxygen species research. *Plant Physiology* 141, 367–372.

Shumbe, L., Chevalier, A., Legeret, B., Taconnat, L., Monnet, F. and Havaux, M. (2016) Singlet oxygen-induced cell death in *Arabidopsis* under high-light stress is controlled by OXI1 kinase. *Plant Physiology* 170, 1757–1771.

Sierla, M., Rahikainen, M., Salojärvi, J., Kangasjärvi, J. and Kangasjärvi, S. (2013) Apoplastic and chloroplastic redox signaling networks in plant stress responses. *Antioxidant Redox Signalling* 18, 2220–2239.

Sies, H. (1993) Strategies of antioxidant defense. *European Journal of Biochemistry* 215, 213–219.

Sies, H. and Cadenas, E. (1985) Oxidative stress: damage to intact cells and organs. *Philosophical Transactions of the Royal Society* 311, 617–631.

Smirnoff, N. (1993) The role of active oxygen in the response of plants to water deficit and desiccation. *New Phytologist* 125, 27–58.

Smirnoff, N. (2005) *Antioxidants and Reactive Oxygen Species in Plants*. Blackwell Publishing Ltd, Oxford, UK.

Šnyrychová, I., Pospíšil, P. and Nauš, J. (2006) Reaction pathways involved in the production of hydroxyl radicals in thylakoid membrane: EPR spin-trapping study. *Photochemical and Photobiological Sciences* 5, 472–476.

Sono, M., Roach, M.P., Coulter, E.D. and Dawson, J.H. (1996) Heme-containing oxygenases. *Chemical Reviews* 96, 2841–2887.

Sosan, A., Svistunenko, D., Straltsova, D., Tsiurkina, K., Smolich, I., Lawson, T., Subramaniam, S., Golovko, V., Anderson, D., Sokolik, A., Colbeck, I. and Demidchik, V. (2016) Engineered silver nanoparticles are sensed at the plasma membrane and dramatically modify the physiology of *Arabidopsis thaliana* plants. *Plant Journal* 85, 245–257.

Stadtman, E.R. and Levine, R.L. (2000) Protein oxidation. *Annals of New York Academy of Sciences* 899, 191–208.

Steffens, B., Steffen-Heins, A. and Sauter, M. (2013) Reactive oxygen species mediate growth and death in submerged plants. *Frontiers in Plant Sciences* 4, 179.

Sumimoto, H. (2008) Structure, regulation and evolution of Nox-family NADPH oxidases that produce reactive oxygen species. *FEBS Journal* 275, 3249–3277.

Suzuki, N., Koussevitzky, S., Mittler, R. and Miller, G (2012) ROS and redox signalling in the response of plants to abiotic stress. *Plant, Cell and Environment* 35, 259–270.

Tadolini, B. (1988) The influence of polyamine-nucleic acid complexes on Fe^{2+} autoxidation. *Molecular and Cellular Biochemistry* 83, 179–185.

Takagi, D., Takumi, S., Hashiguchi, M., Sejima, T. and Miyake, C. (2016) Superoxide and singlet oxygen produced within the thylakoid membranes both cause photosystem I photoinhibition. *Plant Physiology* 171, 1626–1634. doi: http://dx.doi.org/10.1104/pp.16.00246.

Takahashi, S. and Badger, M.R. (2011) Photoprotection in plants: a new light on photosystem II damage. *Trends in Plant Sciences* 16, 53–60.

Tang, X.D., Santarelli, L.C., Heinemann, S.H. and Hoshi, T. (2004) Metabolic regulation of potassium channels. *Annual Reviews of Physiology* 66, 131–159.

Tanou, G., Job, C., Rajjou, L., Arc, E., Belghazi, M., Diamantidis, G., Molassiotis, A. and Job, D. (2009) Proteomics reveals the overlapping roles of hydrogen peroxide and nitric oxide in the acclimation of citrus plants to salinity. *Plant Journal* 60, 795–804.

Tirone, F. and Cox, J.A. (2007) NADPH oxidase 5 (NOX5) interacts with and is regulated by calmodulin. *FEBS Letters* 581, 1202–1208.

Torres, M.A. and Dangl, J.L. (2005) Functions of the respiratory burst oxidase in biotic interactions, abiotic stress and development. *Current Opinion in Plant Biology* 8, 397–403.

Torres, M.A., Onouchi, H., Hamada, S., Machida, C., Hammond-Kosak, K.E. and Jones, J.D.G. (1998) Six *Arabidopsis thaliana* homologues of the human respiratory burst oxidase (*gp91phox*). *Plant Journal* 14, 365–370.

Torres, M.A., Dangl, J.L. and Jones, J.D.G. (2002) *Arabidopsis* gp91phox homologues, AtrbohD and AtrbohF are required for accumulation of reactive oxygen intermediates in the plant defense response. *Proceedings of the National Academy of Sciences USA* 99, 517–522.

Torres, M.A., Jones, J.D.G. and Dangl, J.L. (2005) Pathogen-induced NADPH oxidase-derived reactive oxygen intermediates suppress spread of cell death in *Arabidopsis thaliana*. *Nature Genetics* 37, 1130–1134.

Trebst, A. and Depka, B. (1997) Role of carotene in the rapid turnover and assembly of photosystem II in *Chlamydomonas reinhardtii*. *FEBS Letters* 400, 359–362.

Trouverie, J., Vidal, G., Zhang, Z., Sirichandra, C., Madiona, K., Amiar, Z., Prioul, J.L., Jeannette, E., Rona, J.P. and Brault, M. (2008) Anion channel activation and proton pumping inhibition involved in the plasma membrane depolarization induced by ABA in *Arabidopsis thaliana* suspension cells are both ROS dependent. *Plant Cell Physiology* 49, 1495–1507.

Tuteja, N., Singh, M.B., Misra, M.K., Bhalla, P.L. and Tuteja, N. (2001) Molecular mechanisms of DNA damage and repair: progress in plants. *Critical Reviews of Biochemistry and Molecular Biology* 36, 337–397.

van Breusegem, F.V., Bailey-Serres, J. and Mittler, R. (2008) Unraveling the tapestry of networks involving reactive oxygen species in plants. *Plant Physiology* 147, 978–984.

Vandelle, E. and Delledonne, M. (2011) Peroxynitrite formation and function in plants. *Plant Science* 181, 534–539.

Vanderauwera, S., Suzuki, N., Miller G., van de Cotte, B., Morsa, S., Ravanat, J.-L., Hegie, A., Triantaphylidès, C., Shulaev, V., Van Montagu, M.C.E., van Breusegem, F. and Mittler, R. (2011) Extranuclear protection of chromosomal DNA from oxidative stress. *Proceedings of the National Academy of Sciences USA* 108, 1711–1716.

Veselá, A. and Wilhelm, J. (2002) The role of carbon dioxide in free radical reactions of the organism. *Physiological Research* 51, 335–339.

Wang, S., Assmann, S.M. and Fedoroff, N.V. (2008) Characterization of the *Arabidopsis* heterotrimeric G protein. *Journal of Biological Chemistry* 283, 13913–13922.

Wang, Z., Rhee, D.B., Lu, J., Bohr, C.T., Zhou, F., Vallabhaneni, H., de Souza-Pinto, N.C. and Liu, Y. (2010) Characterization of oxidative guanine damage and repair in mammalian telomeres. *PLoS Genetics* 6, e1000951.

Williams, H.C. and Griendling, K.K. (2007) NADPH oxidase inhibitors: new antihypertensive agents? *Journal of Cardiovascular Pharmacology* 50, 9–16.

Wong, C.M., Cheema, A.K., Zhang, L. and Suzuk, Y.J. (2008) Protein carbonylation as a novel mechanism in redox signaling. *Circulation Research* 102, 310–318.

Wong, H.L., Pinontoan, R., Hayashi, K., Tabata, R., Yaeno, T., Hasegawa, K., Kojima, C., Yoshioka, H., Iba, K., Kawasaki, T. and Shimamotoa, K. (2007) Regulation of rice NADPH oxidase by binding of Rac GTPase to its N-terminal extension. *The Plant Cell* 19, 4022–4034.

Wu, H., Shabala, L., Zhou, M. and Shabala, S. (2015) Chloroplast-generated ROS dominate NaCl-induced K^+ efflux in wheat leaf mesophyll. *Plant Signalling and Behaviour* 10, e1013793.

Wu, W., Wang, Y., Lee, S.C. Lan, W. and Luan, S. (2010) Regulation of ion channels by the calcium signaling network in plant cells. In: Demidchik, V. and Maathuis, F.J.M. (eds) *Ion Channels and Plant Stress Responses*. Springer, Berlin, pp. 111–136.

Xing, T., Wang, X.J., Malik, K. and Miki, B.L. (2001) Ectopic expression of an *Arabidopsis* calmodulin-like domain protein kinase-enhanced NADPH oxidase activity and oxidative burst in tomato protoplasts. *Molecular Plant-Microbe Interactions* 14, 1261–1264.

Yamazaki, T., Kawai, C., Yamauchi, A. and Kuribayashi, F. (2011) A highly sensitive chemiluminescence assay for superoxide detection and chronic granulomatous disease diagnosis. *Tropical Medicine and Health* 39, 41–45.

Yoshiyama, K.O., Sakaguchi, K. and Kimura, S. (2013) DNA damage response in plants: conserved and variable response compared to animals. *Biology* 2, 1338–1356.

Zagorchev, L., Seal, C.E., Kranner, I. and Odjakova, M. (2013) A central role for thiols in plant tolerance to abiotic stress. *International Journal of Molecular Sciences* 14, 7405–7432.

Zepeda-Jazo, I., Velarde-Buendía, A.M., Enríquez-Figueroa, R., Bose, J., Shabala, S., Muñiz-Murguía, J. and Pottosin, I.I. (2011) Polyamines interact with hydroxyl radicals in activating Ca^{2+} and K^+ transport across the root epidermal plasma membranes. *Plant Physiology* 157, 2167–2180.

Zhang, H.X., Zhang, F.Q., Xia, Y., Wang, G.P. and Shen, Z.G. (2010) Excess copper induces production of hydrogen peroxide in the leaf of *Elsholtzia haichowensis* through apoplastic and symplastic CuZn-superoxide dismutase. *Journal of Hazardous Materials* 178, 834–843.

Zhao, H., Joseph, J., Fales, H.M., Sokolovski, E.A., Levine, R.L., Vasquez-Vivar, J. and Kalyanaraman, B. (2005) Detection and characterization of the product of hydroethidine and intracellular superoxide by HPLC and limitations of fluorescence. *Proceedings of the National Academy of Sciences USA* 102, 5727–5732

4 Plant Responses to Chilling Temperatures

Eric Ruelland*

*Institute of Ecology and Environmental Sciences of Paris,
Centre National de la Recherche Scientifique, France*

Abstract

Plants are submitted to a chilling stress when exposed to low, non-freezing temperatures. Some are able to cope with this stress and acquire chilling tolerance; in some species, the exposure to this stress will even trigger developmental responses. Other (chilling-sensitive) species will not be able to cope properly with the low temperature and will develop chilling symptoms that can lead to plant death. The acquisition of chilling tolerance is associated with huge changes in metabolite contents, such as the accumulation of soluble sugars, dehydrins, RNA chaperones and an increase in detoxification activities against reactive oxygen species (ROS). These changes in cellular components are mostly due to a transcriptome rearrangement. They mean that chilling has been perceived and transduced to the nucleus. Chilling is not perceived by a single mechanism in plants but at different sensory levels that are the very biological processes disturbed by the temperature downshift. Once perceived, chilling stress is transduced. An increase in cytosolic calcium is the major transducing event that will then regulate the activity of many signalling components, including phospholipases and protein kinases. This will end in changes in gene expression. The best-documented pathway leading to gene induction in response to cold is the C-repeat binding factor (CBF) pathway. However, other factors have recently been identified as participating in the low-temperature regulatory network.

4.1 Introduction

Being immobile, plants are daily submitted to changes in the factors that contribute to their environment. These factors can be abiotic, such as nutrient and water availability, light, temperature; or biotic, such as aggression by pathogens, insects or herbivorous animals. This chapter deals with chilling: that is, exposure to low, non-freezing temperatures. Freezing (exposure to sub-zero temperatures) is covered in detail in Chapter 11 of this volume. Exposure to chilling temperatures is not always detrimental. For instance, chilling is a key environmental cue in the process of flowering through vernalization, the exposure to such cold temperatures being necessary to accelerate the transition from vegetative growth to the reproductive phase (Zografos and Sung, 2012). Chilling of imbibed seeds (e.g. stratification) can be necessary to break seed dormancy. Chilling is also needed for the induction of bud dormancy in the autumn, for example in species such as pear and apple, and at the same time also for breaking bud dormancy at the end of winter (Heide and Prestrud, 2005). Finally, in some species chilling will also induce the tolerance to freezing temperatures in a process called cold-hardening or cold-acclimation, a phenomenon that occurs during the autumn months

* eric.ruelland@upmc.fr

© CAB International 2017. *Plant Stress Physiology*, 2nd Edition (ed. S. Shabala)

97

in temperate climates, thus preparing plants to cope with freezing in winter. Yet, for all plants, the exposure to chilling, or to a temperature downshift, will be first experienced as a stressful situation. Some plants will be able to cope with this stress and acquire chilling tolerance; in some species, as stated above, the exposure to this stress will trigger developmental responses. Other (chilling-sensitive) plants will not be able to cope properly with the chilling exposure. In this chapter, we will mainly focus on plant responses to chilling as a stress. Chilling-sensitive plants will be used to illustrate the chilling injuries; chilling-resistant plants will allow us to illustrate the responses that enable a plant to cope with chilling stress.

Chilling has been recognized for many years as a stress having a major impact on plant growth, particularly visible on crop species, many of which (e.g. maize, rice, tomato, soybean) are of tropical or subtropical origin (Boyer, 1982). The fact that the temperature range for chilling is quite large is essentially due to the fact that the threshold of chilling sensitivity is not the same for all plants. A simple glance at the distribution of plant communities (either wild or cultivated) on the surface of the planet, in parallel to that of temperatures, shows that this parameter is a major determinant in the distribution of plant species (Woodward et al., 2004). However, the level of chilling sensitivity can also vary within a species, and be dependent on ecotypes. In sensitive plants, chilling is responsible for macroscopic modifications that are visible as injuries and symptoms that can be studied at the cell level and demonstrate the big impact of low temperature on subcellular ultrastructures. This impact is a consequence of the various changes induced by chilling at the molecular level: a reduction in the rates of enzymatic reactions; changes in protein conformation and destabilization of protein complexes, including cytoskeleton; perturbations in photosynthesis and in the balance between metabolic reactions; rigidification of membranes; accumulation of reactive oxygen species (ROS); and stabilization of RNA secondary structures. Some of these effects are part of the temperature sensor that exists in plants and will be described in more detail below. Once the chilling temperature has been perceived, signalling pathways will be activated. Over the past 10 years, major components of the signalling pathway triggered by chilling have

been deciphered in *Arabidopsis thaliana* Col-0, a chilling-resistant ecotype. As discussed below, these elements are responsible for large modifications in gene expression that allow the plant to adapt to the new environmental conditions.

4.2 Chilling Sensitivity

4.2.1 Spectrum of chilling sensitivities

Chilling sensitivity has been identified, not surprisingly, in plants from tropical or subtropical climates, but also in species from temperate latitudes (Lyons, 1973). Among crop plants, typically sensitive species are cotton, cowpea, groundnut, maize and rice, as well as mung bean and tomato. Pea and spinach, on the other hand, are chilling-resistant. However, large variations in sensitivity can be observed within one plant species, between varieties or cultivars. This can be illustrated by the example of tomato. The cultivated tomato species, *Lycopersicon esculentum*, is strongly affected by a growth temperature of 10–12°C. In contrast, the wild tomato *Lycopersicon hirsutum* (originally from the west coast of South America), is present as ten ecotypes that are widely variable in their sensitivity to chilling. The most resistant ecotype grows at 3000 m in altitude, at temperatures lower than 10°C at night, while the most sensitive one is found on the tropical coast of Ecuador, where the temperature never drops below 18°C (Marcellin, 1992). As to wheat, the winter varieties can be sown in autumn whereas the spring varieties, more productive but also more sensitive to cold, are sown when the winter is over.

Another factor that controls chilling sensitivity is physiological age. In some species, such as the orchid *Phalaenopsis*, adult leaves have become insensitive to cold, whereas younger leaves are sensitive (Marcellin, 1992). Chilling sensitivity can also vary with the developmental stage. Dry seeds of sensitive species can be stored at chilling temperature with no harm; 1-day-old maize seedlings are unharmed by 4 days at 4°C, whereas 3-day-old seedlings have acquired maximum sensitivity (Lyons, 1973).

One growth phase that is particularly sensitive to cold is the reproductive phase. The effects of chilling stress on the reproductive phase (Fig. 4.1)

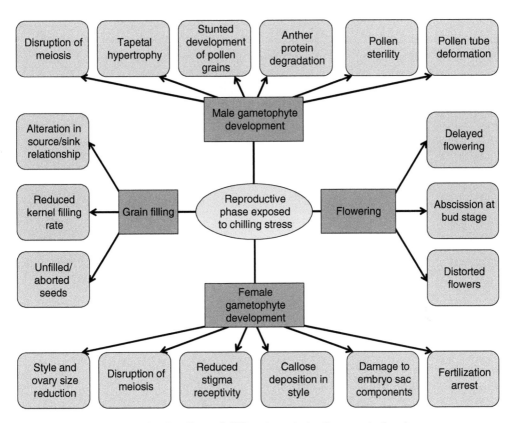

Fig. 4.1. Flow diagram showing the effects of chilling stress during the reproductive phase processes (from Thakur *et al.*, 2010).

vary according to the plant species. In spring wheat (*Triticum aestivum*) plants showed significant reductions in fertility when exposed to prolonged chilling temperatures in controlled environment experiments. In this species, the stage corresponding to meiosis within the anthers is the most sensitive one. Abnormalities such as plasmolysis and cytomixis increased in frequency between chilled and control plants. These were associated with death of developing pollen cells, and could contribute to loss of fertility (Barton *et al.*, 2014). Other explanations have been proposed to understand the basis of reproductive failure under chilling stress. Cold-induced reductions in enzyme activity, as well as a reconfiguration of the metabolome of cold-stressed cells, have been suggested as resulting in impaired supply of sugars to the tapetum cells and pollen grains because of a decrease in conversion of sucrose into glucose and fructose by the tapetum-specific invertase.

The cold-induced increase in the concentration of the plant hormone abscisic acid would also be involved by reducing the nutrient flow to the tapetum cells (Thakur *et al.*, 2010). Low temperatures also affect pollen tube growth, by altering tip-localized production of ROS and endocytosis (Gao *et al.*, 2014). Other organs that are susceptible to cold are fruits, especially those from tropical and subtropical species. For many of them, chilling stress will be more severe once the fruit has been separated from the fruit-bearing plant. The storage temperature will therefore be crucial in order to avoid chilling injuries, even if the development stage at which the fruit has been harvested, as well as the growing conditions of the fruit-bearing plant (particularly temperature), will also play a role in the length of storage life. Lower temperature limits for temperate fruits such as apple are 0–4°C; 8°C for subtropical fruits such as citrus, avocado and pineapple; and about

12°C for banana (Lyons, 1973). Fruits can indeed be classified into three categories, ranging from chilling-resistant (apricot, cherry, kiwifruit, strawberry, pear, some cultivars of apple), to moderately sensitive (avocado, lemon, some cultivars of mango, apple and orange) or very sensitive (banana, lime, passion fruit, guava, papaya, grapefruit, some cultivars of mango and melon) (Marcellin, 1992).

4.2.2 Macroscopic effects of chilling on sensitive species: stunted stature

When grown at chilling temperatures, plants exhibit a stunted stature. The reduction in growth concerns both aerial parts of the plants and roots. The mechanisms responsible for this are not well understood. It might be linked to reduction in active primary metabolism rate. Yet it might not only be caused by thermodynamical reasons (e.g. cold temperatures lowering enzymatic activities of metabolic reactions). Hormones, including salicylic acid or auxins, have been shown to be involved in the cold-triggered stunted stature of overwintering plants, as will be detailed in Section 4.7.1.

4.2.3 Macroscopic effects of chilling on sensitive species: chilling injuries

As in the case of other stresses, the severity of the symptoms observed will depend on the severity of the stress applied. In the case of germinating cotton seeds, the earlier manifestation of chilling stress is a slowing down of the germination process (Marcellin, 1992). The apex of the radicle usually degenerates, and therefore an exposure of cotton germinations to chilling will be responsible for severe decreases in yield. More generally, chilling results in a loss of water from the tissues, resulting in wilted plants that can recover once placed at warmer temperatures provided chilling has not been applied for too long. More symptoms include the appearance of necrotic areas, surface pitting and external discoloration. Banana fruits exposed to severe chilling when green will turn black. In cucumber fruits, the pitting resulting from the collapse of subsurface cells will lead to invasion

by decay organisms (Lyons, 1973). Other symptoms include rubbery texture and irregular ripening (tomato; Van Dijk et al., 2006), lignification (loquat fruit; Cai et al., 2006), and woolliness and internal breakdown (peach; Crisosto et al., 1999).

4.2.4 Effects of chilling on cellular ultrastructures

Although the response of sensitive plants to chilling is variable, at the ultrastructural level the symptoms are similar. The subcellular compartment that is mainly affected by chilling is the chloroplast, where the first manifestations of injury are a distortion and swelling of thylakoids as well as a reduction in the size and number of starch granules. In addition, chilling will induce the formation of vesicles from the envelope, forming the so-called peripheral reticulum. This phenomenon is more marked when plants are submitted to chilling in the light than in the dark, where the extent of the symptoms mentioned above will be dependent on the chilling sensitivity of the plant species. If chilling is prolonged, more severe modifications involving unstacking of grana and eventually disappearance of the chloroplast envelope have been observed. In the cold-sensitive species maize, cucumber and tobacco, no alterations of mitochondria have been observed even when chloroplasts were deeply affected by chilling. On the other hand, in the very sensitive *Episcia reptans*, exposure to 5°C for 6 h led to mitochondrial swelling and dilatation. Alterations of mitochondrial ultrastructure therefore seem to be restricted to very sensitive species. In a few other cases, other morphological changes have been observed in the nucleus (condensation of chromatin; *Saintpaulia ionantha* cv. Ritali), alterations in the appearance of the nucleolus (*T. aestivum* cv. Ducat) and in the Golgi apparatus and endoplasmic reticulum, which became dilated (*Vigna radiata*, *E. reptans*). Kratsch and Wise (2000) proposed a classification of plant species into three groups, according to their sensitivity to chilling as well as to the cold-induced ultrastructural changes (Table 4.1).

Table 4.1. Relative chilling sensitivity of selected species and observed chilling-induced symptoms. Adapted from Kratsch and Wise, 2000.

Species type	Common name	Observation
Chilling-resistant		
Arabidopsis thaliana	Arabidopsis	No chilling injury observed unless another
Brassica oleracea	Cabbage	stress factor is simultaneously imposed.
Cephaloziella exiliflora	Liverwort	Plastids and mitochondria remain intact
Hordeum vulgare	Barley	
Secale cereale	Rye	
Selaginella spp.	Spikemoss	
Spinacia oleracea	Spinach	
Triticum aestivum	Wheat	
Chilling-sensitive		
Cucumis sativum	Cucumber	Chloroplast swelling, thylakoid dilation,
Fragaria virginiana	Strawberry	randomly tilted grana stacks, formation of
Glycine max	Soybean	peripheral reticulum, serpentine-like
Lycopersicon esculentum	Tomato	thylakoids and accumulation of lipid droplets
Zea mays	Maize	in the stroma all observed. Chloroplasts
Nicotiana tabacum	Tobacco	disintegrate with prolonged chilling.
Paspalum dilatatum	Dallis grass	No injury in mitochondria
Phaseolus vulgaris	Common bean	
Pisum sativum	Pea	
Sorghum spp.	Sorghum	
Extremely chilling-sensitive		
Ephedra vulgaris	Ephedra	Chilling-induced injury also observed in
Episcia reptans	Flame violet	mitochondria. Rapid chilling injury results in
Gossypium hirsutum	Cotton	cell lysis
Saintpaulia ionantha	African violet	
Vigna radiata	Mung bean	

4.2.5 Effects of chilling on biochemical and molecular processes

Lowering temperatures will thermodynamically reduce the kinetics of metabolic reactions (Arrhenius' law). As to protein conformation, exposure to low temperatures will increase the likelihood that non-polar side chains of proteins become exposed to the aqueous medium of the cell. This will directly affect the stability and solubility of many proteins or protein complexes (Fig. 4.2A). The effects of chilling both on enzymatic rates and on protein stability will affect enzymes involved in metabolism. The resulting temperature dependency of enzyme activities is generally very different among enzymes. Therefore, the subtle equilibrium of different enzymes in metabolic cycles, cascades and redox chains is disturbed when the temperature is lowered. The activity corresponding to the limiting step of a metabolic chain may change under a shift in temperature. In consequence, pools of certain metabolic intermediates increase while others shrink. This can be illustrated by the effects of chilling on photosynthesis-related processes. During photosynthesis, CO_2 combines with ribulose-1,5-bisphosphate to produce glycerate-3-phosphate, which is reduced into triose-phosphates. Most triose-phosphate is retained in the chloroplast, with the remainder mainly being exported to the cytosol and converted into sucrose through the synthesis of phosphorylated intermediates. The final conversion steps to sucrose release inorganic phosphate (Pi) that returns to the chloroplast for ATP synthesis. Exposure to low temperatures results in a decrease in triose-phosphate utilization that leads to the accumulation of phosphorylated metabolites, which in turn leads to a decrease in Pi availability in chloroplasts. Thus, chilling stress does induce a very early depletion

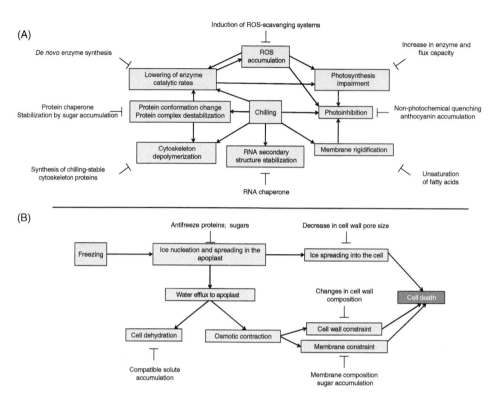

Fig. 4.2. Effects of chilling and freezing on cellular processes and the cellular responses to chilling that will lead to chilling tolerance and freezing tolerance. The effects of cold, either chilling (A) or freezing (B), are indicated in boxes. The responses that are triggered during an exposure to chilling and which will participate in alleviating the chilling stress (A) or the freezing stress (B) are indicated with a line with a bar (adapted from Ruelland et al., 2009).

in Pi. This Pi status of the plastids in turn reduces the availability of the electron acceptor glycerate-1,3-bisphosphate in the photosynthetic reduction cycle (Calvin cycle). This is the Pi-limitation of photosynthesis observed at the early stage of chilling. This limitation in photosynthesis will create an imbalance in photostasis, which is the balance between the energy input through photochemistry (i.e. the light energy harvested) and the subsequent use of this energy through metabolism and growth. The rates of the photochemical reactions of charge separation are essentially insensitive to temperature in the biologically relevant range (0–50°C) and exceed those of thermochemical processes such as the intersystem electron transport and CO_2 fixation. This results in an increase in the excitation pressure on photosystem II that reflects the more reduced state of plastoquinone A, the

primary electron acceptor in the photosystem II. This imbalance in photostasis also favours the accumulation of ROS such as singlet oxygen, hydroxyl radical or hydrogen peroxide, since the normal electron acceptor is not available. This ROS accumulation is further amplified by the fact that the activities of the scavenging enzymes are thermodynamically lowered by low temperatures; the scavenging systems will then not be able to counterbalance ROS formation. The accumulation of ROS will have deleterious effects. It will lead to peroxidation of membrane lipids; membrane integrity will be disrupted and this will result in ion leakage. Lowering the temperature will also cause membranes to be more rigid, thereby leading to a disturbance of all membrane processes (e.g. the opening of ion channels, membrane-associated electron transfer reactions, etc.). In addition, the low temperatures will favour the formation of secondary structures in RNA, which can disturb translational processes (Fig. 4.2A).

Because exposure to chilling temperatures can lead to tolerance to freezing stress, the main consequences of exposure to freezing will be detailed here. During freezing, ice is initially formed in the apoplastic space. The water potential in this compartment therefore decreases. This leads to an exit of water from the cell into the extracellular compartment, resulting in cell dehydration and osmotic contraction. The osmotic contraction can lead to apposition of membranes, such as plasma membranes and membranes of the chloroplast envelopes. This apposition can be associated with membrane rearrangement. The cell wall needs to be fluid enough to allow the osmotic contraction with no breakage of the cell wall or of the associated plasma membrane. Most of the cell death associated with freezing stress occurs because during thawing the cell wall or the membranes will not be able to resume their former shape (for instance because of the membrane rearrangements during freezing or because the cell wall is not elastic enough) (Fig. 4.2B). In addition, ice can spread into the cell, which also will lead to cell death (Ruelland *et al.*, 2009).

In conclusion, chilling, or a temperature downshift, will induce major changes in the cell at the molecular level. These changes can occur very quickly after the exposure to cold. Temperature is a physical parameter, and most of the cellular processes will be affected by temperature changes for thermodynamical reasons. Plants are not all equal when confronted with such perturbations. Chilling-sensitive plants will not be able to cope with those alterations and will develop changes in the cell ultrastructure and at whole-organ levels that can be considered as chilling injuries. Yet some plants will be able to trigger acclimating responses and will acquire chilling tolerance. In some cases, this tolerance will allow the plants to fully cope with the chilling conditions: these are the chilling-resistant plants. The responses triggered by chilling temperatures are detailed below.

4.3 The Responses to Chilling Temperatures

Chilling induces a plethora of responses in plants. Some of these will participate in the acquisition of chilling tolerance and, in some species, in freezing-tolerance. This is the molecular explanation of cold-hardening or cold-acclimation; that is, the fact that an exposure to chilling temperatures may, in some species, result in an improved freezing-tolerance. Therefore, in the description of the physiological importance of the responses to chilling that we describe below, both the impact of the responses on chilling and/or freezing-tolerance may be discussed.

4.3.1 Stress-related proteins

Dehydrins

Many of the genes encoding these proteins were first characterized as being responsive to cold, drought and/or abscissic acid (ABA). Most of these proteins have therefore been named COR (cold-responsive), LTI (low temperature-induced), RAB (responsive to abscisic acid), KIN (cold-induced) or ERD (early responsive to dehydration). They include dehydrins, which define group II of late embryogenesis abundant (LEA) proteins. Despite the large body of evidence of the importance of these proteins in the stress response, the *in vivo* protective mechanisms are not fully known. Their role is not necessarily linked to the tolerance to chilling temperatures, but rather to freezing ones. Dehydrins may indeed have a role in freezing-tolerance, possibly by preventing the membrane destabilization that occurs during the osmotic contraction associated with freezing. For instance, the small model dehydrin (*Vitis riparia* K2) is able to bind to liposomes containing phosphatidic acid and to protect the liposomes from fusing after freeze–thaw treatment (Clarke *et al.*, 2015). Dehydrins will bind to membranes as peripheral membrane proteins, since the protein sequences are highly hydrophilic and contain many charged amino acids. Because of this, dehydrins in solution are intrinsically disordered proteins, meaning they have no well-defined secondary or tertiary structure. Yet, dehydrins have been shown to gain structure when bound to ligands such as membranes (Graether and Boddington, 2014). The conserved lysine-rich segments are involved in the binding of the dehydrins to a membrane (Clarke *et al.*, 2015). In addition, some dehydrins might possess cryoprotective properties, have antifreeze activity or act as oxygen radical scavengers (Bies-Ethève *et al.*, 2008).

Chilling-induced proteins with antifreeze properties

Overwintering plants secrete antifreeze proteins (AFPs) to provide freezing-tolerance. AFPs bind irreversibly to the surface of ice and prevent ice nucleation and ice crystal coalescence. Plants have no constitutive AFP activity. This activity is induced by chilling, and plays a role in freezing-tolerance. AFPs act by directly interacting with ice *in planta* and by reducing freezing injury by slowing the growth and recrystallization of ice (Griffith *et al.*, 2005). Proteins with AFP properties have been isolated from 11 plants. While most of them are apoplastic, some, such as the AFPs from *Prunus persica* and *Forsythia suspensa*, are intracellular. Indeed, immunogold localization of *P. persica* AFP showed its presence in the chloroplast and nucleus in addition to the cytoplasm (Wisniewski *et al.*, 1999). This might indicate alternative roles for these proteins, probably in inhibition of intracellular ice nucleators. Interestingly, while most AFPs are homologous to the pathogenesis-related (PR)-proteins (such as thaumatin-like proteins, class I and class II chitinases and β-1,3-glucanase), AFPs from *P. persica* and *F. suspensa* showed sequence homology with dehydrins (Gupta and Deswal, 2014).

Cold-shock proteins and RNA-binding proteins

One of the major problems that living organisms have to deal with when exposed to low temperatures is the formation and stabilization of RNA secondary structures that may prevent efficient translation and transcription. Bacteria have RNA chaperones that destabilize RNA secondary structures, enabling efficient translation at low temperature. Such bacteria chaperones are named cold-shock proteins (CSPs) and are composed of a single nucleic acid binding domain (of about 70 amino acid residues) called the cold-shock domain (CSD). The CSD consists of a five-stranded β-barrel containing two consensus RNA-binding motifs. Plants also synthetize proteins with CSDs. All the plant CSD-bearing proteins analysed so far have the CSD at their N-terminus and they also contain a large glycine-rich region interspersed with CCHC zinc fingers at their C-terminus (Sasaki and Imai, 2012). CSPs in plants are induced by chilling temperatures.

In *A. thaliana*, the transcript level of *AtGRP2* (a CSP) increases markedly during cold stress. Root growth at 11°C of plants overexpressing GRP2 was higher than that of wild-type plants. GRP2 is capable of melting RNA secondary structures *in vivo*, and could be involved in RNA processing and/or translation ability under cold conditions (Kim *et al.*, 2007). However, the roles of such proteins are not limited to the response to chilling. For instance, transgenic rice plants (*Oryza sativa*) that express *A. thaliana* AtGRP2 displayed phenotypes similar to that of wild-type plants under high salt or cold stress conditions, but they showed much higher recovery rates and grain yields compared with the wild-type plants under drought stress conditions (Yang *et al.*, 2014).

4.3.2 Lipids

Membranes are composed of proteins and lipids. Glycerolipids are the most abundant lipids in biological membranes. Commonly, glycerolipids are built from a glycerol backbone whose two alcohol residues (*sn-1* and *sn-2*) are esterified by a fatty acid. The third hydroxyl (*sn-3*) can be linked to galactose-derived molecules, forming galactolipids, which are typical of chloroplast membranes (Fig. 4.3A). Monogalactosyldiacylglycerol and digalactosyldiacylglycerol are both galactolipids. In extra plastidial membranes, glycerolipids are mostly phosphoglycerolipids (Fig. 4.3B), that is, glycerolipids for which the third hydroxyl group of glycerol (*sn-3*) is bound through a phosphodiester link to a polar head such as choline, glycerol or ethanolamine (leading to phosphatidylcholine, phosphatidylglycerol or phosphatidylethanolamine, respectively). Other lipids that can be found in plant membranes are sterols and sphingolipids. Sphingolipids (Fig. 4.3C) are a class of lipids built from sphingoid bases. Sphingoid bases are also named long-chain bases. They result from the condensation of serine with a palmitoyl-CoA, resulting in an 18 carbon aliphatic chain. The long-chain bases can be N-acylated (on the N of the serine) by acyl-CoA to form ceramides, which therefore possess two aliphatic chains. The $-CH_2OH$ group of the serine can be linked to polar heads, leading to complex sphingolipids. Examples of complex sphingolipids are glucosylceramides or glycosylinositolphosphoceramides. With two aliphatic

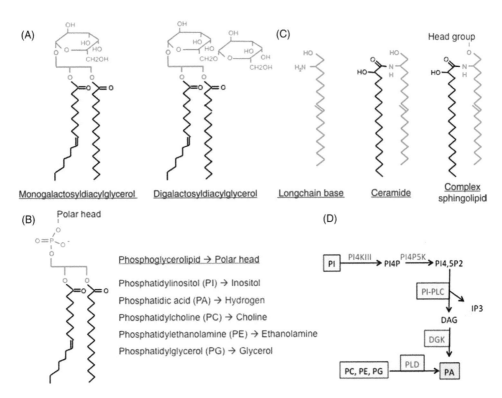

Fig. 4.3. Schematic representation of main lipids in plants. Galactolipids (A) are glycerolipids present in chloroplast membranes. The polar head of monogalactosyldiacyglycerol is one galactosyl group, while the polar head of digalactosyldiacyglycerol is a digalactosyl group. For phosphoglycerolipids (B), the polar head is bound through a phosphodiester link to the diacylglycerol backbone. The sphingolipids (C) are built from a long-chain base that results from a condensation between a serine and a palmitate. Long-chain bases can be hydroxylated and/or unsaturated. A fatty acid can be condensed to the long-chain base, leading to a ceramide that can be further linked to the polar head, thus leading to complex sphingolipids. (D) Main metabolic reactions of the phosphoglycerolipid signalling pathway. DAG, diacylglycerol; DGK, diacyglycerolkinase; IP3, inositol-triphosphate; PC, phosphatidylcholine; PE, phosphatidylethanolamine; PG, phosphatidylglycerol; PLD, phospholipase D; PI-PLC, phosphoinositide-dependent phospholipase C; PI4,5P2, phosphatidylinositol-,-bisphosphate; PI4KIII, type III-phosphatidylinositol kinase.

chains and a polar head, the sphingolipid structure can be compared to that of glycerolipids.

It is well documented that exposure to chilling induces important changes in the membrane composition of plant cells. In chloroplasts, for example, cold exposure results in a decrease in the level of monogalactosyldiacylglycerol and an increase in the level of digalactosyldiacylglycerol, both in the inner and outer envelopes. Cold exposure also leads to a decrease in plastidial phosphatidylcholine but only in the outer envelope. In *Solanum commersonii*, chilling exposure was correlated with an increase of phospholipids in the plasma membrane, primarily due to an increase in phosphatidylethanolamine.

Such an increase in the ratio phosphatidylethanolamine to phosphatidylcholine has also been observed in spring and winter rye, but not in *A. thaliana* (Uemura and Steponkus, 1997; Ruelland *et al.*, 2009).

The changes in levels of the different glycerolipid classes are not the only way chilling temperatures affect membranes. Indeed, in *A. thaliana*, the proportion of the different membrane phospholipids does not vary significantly during chilling exposure. However, there is a marked difference in their fatty acid composition. Fatty acids are commonly named through a 'number of carbons in the fatty acid chain:- number of unsaturation (i.e. double bonds)' nomenclature. For instance, the 18:1-fatty acid (oleic acid) has 18 carbons and one double bond. The proportion of di-unsaturated species such as 18:1/18:3, 18:2/18:2 and 18:2/18:3 increases in both phosphatidylcholine and phosphatidylethanolamine, and the proportion of monounsaturated species such as 18:0/18:3 and 16:0/18:3 decreases. This increase of polyunsaturated fatty acids seems to be a general effect of cold-acclimation, observed in many other plant species (Ruelland *et al.*, 2009). The level of unsaturation of lipids has a role in the response to cold, especially to chilling temperatures. The *fad2* mutant of *A. thaliana* is deficient in the activity of the reticulum-localized 18:1-desaturase, leading to plants with reduced levels of polyunsaturated fatty acids. Their growth characteristics at 22°C were very similar to that of the wild-type plants. After transfer to 6°C, rosette leaves of the mutants gradually died, before the plants themselves died (Miquel *et al.*, 1993). Similarly, the *ads2* mutant plants, mutated in ACYL-LIPID DESATURASE2 enzyme, are similar to the wild-type under standard growth conditions but display a dwarf and sterile phenotype when grown at 6°C. They also show increased sensitivity to freezing temperatures. It is suggested that ADS2 encodes a 16:0 desaturase of monogalactosyldiacylglycerol and phosphatidylglycerol; the *ads2* mutant plants at 6°C have reduced levels of 16:1, 16:2, 16:3 and 18:3 and higher levels of 16:0 and 18:0 fatty acids compared with the wild-type plants (Chen and Thelen, 2013).

Monogalactosyldiacylglycerol and phosphatidylglycerol are chloroplast lipids. The saturation level of glycerolipids is known to be important for the chilling sensitivity of the chloroplast. The *A. thaliana* fatty acid biosynthesis1 (*fab1*) mutant has increased levels of the saturated fatty acid 16:0 due to decreased activity of 3-ketoacyl-acyl carrier protein (ACP) synthase II. In *fab1* leaves, phosphatidylglycerol, the major chloroplast phospholipid, contains up to 45% molecular species with only 16:0-, 16:1- and 18:0-fatty acids, compared with less than 10% in wild-type *A. thaliana*. When exposed to low temperatures (2°–6°C) for long periods, *fab1* plants do suffer collapse of photosynthesis, degradation of chloroplasts and eventually death. A screen for suppressors of this low-temperature phenotype has identified four lines mutated in genes en-

coding enzyme pathway for glycerolipid synthesis. Two such suppressor lines are *act1* and *lpat1*, which are the chloroplast acyl-ACP: glycerol-3-phosphate acyltransferase and the chloroplast acyl-ACP:lysophosphatidic acid acyltransferase, respectively. Two other such suppressor lines are *gly1* and *fad6*, which have reduced glycerol-3-phosphate supply to the prokaryotic pathway and are deficient in the chloroplast 16:1/18:1 fatty acyl desaturase, respectively. All four of the suppressor loci result in reductions in the proportion of phosphatidylglycerol molecular species with 16:0, 16:1 and 18:0 fatty acids relative to *fab1*, reinforcing the conclusion that these molecular species of phosphatidylglycerol are responsible for chilling sensitivity of *fab1* (Gao *et al.*, 2015). It can also be seen that increased levels of unsaturated fatty acids in phosphatidylglycerol of thylakoid membranes alleviated chilling stress-induced inhibition of photosynthetic rates in transgenic rice seedlings and tomato plants. Phosphatidylglycerol is the only phosphoglycerolipid of thylakoid membranes. The role of phosphatidylglycerol in the assembly of photosystems has thus been extensively studied. For instance, the specific binding of phosphatidylglycerol to the D1 protein stabilizes the photosystem II reaction centre and is required for dimerization of photosystem II. Phosphatidylglycerol is also likely to be involved in the core complex of this photosystem. Photoinhibition of photosystem II can occur when the rate of photodamage is higher than the rate of repair (Moon *et al.*, 1995). A transgenic tobacco in which the unsaturation of fatty acids in phosphatidylglycerol has been reduced demonstrates that fatty acid unsaturation stimulates the repair but has no effect on the damage to photosynthesis of higher plants. The recovery process after photoinduced damage to the photosystem II reaction centre may involve several complicated steps, which could include proteolytic degradation of the damaged Dl protein, removal of the degraded D1 protein, synthesis of the precursor to the Dl protein, processing of the precursor protein and the reassembly of the photosystem II complex with the new D1 protein. Phosphatidylglycerol in thylakoid membranes is preferentially involved in protein-lipid interactions (Domonkos *et al.*, 2004). Thus, it is quite probable that changes in the extent of unsaturation of fatty acids in phosphatidylglycerol can modify the molecular environment of the photosystem II reaction centre complex, thereby affecting the turnover of Dl protein in the photosystem II complex. However, the specific step that is rate-determining and is accelerated by enhanced unsaturation of fatty acids in phosphatidylglycerol remains to be identified (Moon et al., 1995). It can also be noted that leaves of transgenic tobacco plants with decreased levels of fatty acid unsaturation in phosphatidylglycerol exhibited also a higher sensitivity of the photosystem I photochemistry at high light under low temperatures. Moreover, the extent of photosystem photoinhibition exhibited a very good relationship with the amount of unsaturated fatty acids. This was associated with a much higher intersystem electron pool size, suggesting over-reduction of the plastoquinone pool in the mutants (Ivanov et al., 2012).

What is the molecular link between chilling sensitivity and fatty acid saturation levels in phosphoglycerolipids? Due to a thermodynamical effect, a temperature drop will lead to a decrease in molecular movements in membranes; this corresponds to rigidification (loss of fluidity). The nature of the fatty acids of the lipids within the membranes influences membrane fluidity. The presence of double bonds in the fatty acids increases the fluidity of the membrane. Thus, one of the consequences of the increase in desaturation might be to counteract the cold-induced membrane rigidification. Conversely, saturated fatty acids might lead to the membrane becoming too rigid, which can jeopardize biological process taking place in the membrane (such as mobility of proteins). During temperature cycles, plants actively manage plasma membrane fluidity (or viscosity) to counteract the direct effects of temperature by changing the proportion of unsaturated fatty acids. This phenomenon is named *homeoviscous adaptation*: by modification of their fatty acid unsaturation, plants try to keep their membrane fluidity constant. Yet, in *Brassica napus*, despite significant changes in their lipid composition upon cold exposure, the endoplasmic reticulum membranes showed only a partial physico-chemical adaptation (as determined by measurement of membrane fluidity parameters such as local microviscosity of acyl chains and lipid lateral diffusion). This means when these

parameters – corresponding to membrane fluidity – are measured at 4°C in membranes of plants grown at 4°C, the fluidity does not equal that measured at 22°C in membranes of plants grown at 22°C. Yet, the trends towards homeoviscous adaptation is detected (Tasseva et al., 2004). Plants mutated in the fatty acid desaturase, such A. thaliana fad2 mutant, cannot achieve such regulation of membrane viscosity during warm/cold cycles (Martinière et al., 2011). Whether the alteration in membrane fluidity of mutants with altered composition in fatty acids of phosphatidylglycerol is responsible for the photoinhibition of photosystems, and more specifically in altered repair capacity of photosystem II, needs to be established. It has to be recalled that changes in fatty acid unsaturations are not the only mechanism to adjust membrane fluidity. Changes in the lipid to protein ratio (w/w) are an example of an alternative mechanism.

The changes in the lipid composition of plasma membranes and chloroplast envelopes have also been proposed as having a role in the acquisition of freezing tolerance by chilling exposure: they may prevent freeze-induced membrane damage by stabilizing the bilayer lamellar configuration (Ruelland et al., 2009). It has also been suggested that membrane fluidity is a key process for proper association of the H-subunit of Mg-chelatase, CHLCH, a key enzyme of chlorophyll biosynthesis, in the chloroplast membranes during chilling (Kindgren et al., 2015).

With respect to the sphingolipids (ceramides; glucosylceramides), the most saturated species were depleted in chilling-treated A. thaliana plants (6°C for 10 days) (Tarazona et al., 2015). Interestingly, up to 90% of sphingolipid long-chain bases in A. thaliana leaves contain a double bond at the carbon 8 of the aliphatic chain due to the activity of a sphingoid long-chain base-Δ8-desaturase (SLD). Arabidopsis plants mutated in AtSLD1 and AtSLD2 showed no detectable long-chain base-Δ8-unsaturation. This is accompanied by a 50% reduction in glucosylceramide levels and a corresponding increase in glycosylinositolphosphoceramides. The double sld1sld2 mutants lacked apparent growth phenotypes under optimal conditions, but displayed a chlorotic appearance and early senescence when grown at chilling temperatures. These results illustrate the importance of sphingolipids in the acclimation of plants to chilling (Chen et al., 2012).

Concerning the sterol lipids, chilling treatment was accompanied by an increase in acylated steryl glycosides and steryl esters containing three unsaturations. However, the role of these changes in the acclimation of plants to chilling is far from being understood (Tarazona *et al.*, 2015).

Finally, while triacylglycerols are well documented as reserve lipids in seeds, they have been shown to accumulate in stress conditions (chilling, but also drought) in leaves. This concerns long-chain fatty acid triacylglycerols. This raises new questions: do triacylglycerols have a role as an energy reserve and/or are they a form of storage for fatty acids from lipids degraded during chilling (such as monogalactosyldiacylglycerol)? Triacylglycerols could also have a role in the acquisition of freezing-tolerance during chilling treatment. Using different *A. thaliana* accessions with different freezing-tolerance capacities, it was shown that the more freezing-tolerant accessions generally accumulated more triacylglycerols during chilling treatments than the freezing-sensitive accessions, although exceptions were clearly present (Degenkolbe *et al.*, 2012).

It therefore appears that one major strategy by which plants adapt to temperature decrease is to increase the degree of unsaturation of membrane lipids. Changes in lipid unsaturation are complex and require large energy inputs, raising the question whether this strategy can be adopted by plants in ecosystems and environments with frequent alterations between high and low temperatures. Indeed, in nature, temperature change can be divided into two types: (i) frequent temperature alteration, in which the temperature rises and falls rapidly and there is a daily temperature cycle that switches between high and low temperatures and lasts for several seasons, such as those of alpine screes and deserts; and (ii) infrequent temperature alteration in which the temperature rises and falls slowly, and change between high and low temperatures is a seasonal cycle, as for temperate zones. Experimental data suggest that plants submitted to frequent important changes in temperature do not adjust their membrane lipid composition by adjusting the degree of fatty acid saturation, but by lipid turnover via the exchange of head groups, a process that is rapid and incurs a low energy cost (Zheng *et al.*, 2011).

4.3.3 Sugars

Many studies have shown that the content of leaf soluble sugars, including sucrose and raffinose, and of sugar alcohols such as galactinol, increases dramatically during chilling exposure (Kaplan and Guy, 2005). In a study with *A. thaliana* rosettes in response to a 24-h cold treatment (6°C), trehalose and *myo*-inositol were shown to increase about twofold in response to the cold treatment; sucrose increased about sixfold, whereas maximum levels of maltose, fructose, glucose and raffinose were 20–70-fold greater in the cold-treated samples than in the controls (22°C) (Sicher, 2011).

Understanding the metabolic pathways involved in sugar accumulation during treatment is the object of intensive research. It has, nevertheless, been established that carbohydrate accumulation relies in part on active carbon assimilation, i.e. active photosynthesis. Carbohydrate accumulation in *A. thaliana* leaves during a cold treatment was enhanced by increased irradiance, required long photoperiods and was inhibited by darkness or by the electron transport inhibitor, DCMU. Indeed, photosynthesis remains active during chilling stress. Sicher (2011) calculated that, in *A. thaliana*, rates of carbon assimilation increased 17% on average in response to chilling temperatures. The assimilated CO_2 contributed to sucrose, glucose and fructose accumulation; it was estimated that 42% of the carbon from assimilation was allocated towards soluble carbohydrate accumulation (Sicher, 2011). In addition, there was also evidence for a role of starch degradation in carbohydrate accumulation. Using a starchless (*pgm1*) mutant, hexose accumulation was delayed 6 h when compared to wild-type plants, which suggests that starch hydrolysis may have an important function during the initial response of *Arabidopsis* to chilling. In the starchless mutant, raffinose accumulation was abolished, suggesting that raffinose accumulation only depends on starch hydrolysis. Accordingly, maltose (a product of the hydrolysis of starch) and raffinose were the only soluble carbohydrates that accumulated in response to chilling in the dark (Sicher, 2011).

β-Amylase catalyses starch breakdown to generate maltose, which can be incorporated into sugar metabolism. This is consistent with the results obtained in *A. thaliana*. A gene encoding a β-amylase, *BMY8* (At4g17090), was induced specifically in response to 4°C treatment. *BMY8* RNAi lines with lower *BMY8* expression exhibited a starch-excess phenotype, and a dramatic decrease in maltose accumulation during a 6-h cold-shock at 4°C. The decreased maltose content was also accompanied by decreased glucose, fructose and sucrose content in the *BMY8* RNAi plants, consistent with the roles of β-amylase and maltose in transitory starch metabolism (Kaplan and Guy, 2005). Data obtained with other species also document the role of starch degradation in the accumulation of soluble carbohydrate during chilling. In *Poncirus trifoliata* the expression of *PtrBAM1*, a gene encoding a chloroplast-localizing β-amylase, was induced by cold. Overexpression of *PtrBAM1* in tobacco (*Nicotiana nudicaulis*) increased β-amylase activity, promoted starch degradation and enhanced the contents of maltose and soluble sugars. Under cold stress, higher accumulation of soluble sugars was observed in the overexpressing lines when compared with the wild-type plants. The tobacco overexpressing lines exhibited enhanced tolerance to cold at chilling and even freezing temperatures (Peng *et al.*, 2014).

The chilling treatments do not only lead to an increase in carbohydrate levels; they also lead to changes in carbohydrate allocation. In *A. thaliana* plants submitted to 4°C for 14 days, 29% of the raffinose was detected in the plastids while only 12% was detected in these compartments before cold exposure. Sucrose accumulated in plastids, starting from a ratio of about 13–14% in control plants and reaching 33–41% in the chilling exposed leaves. Both sugars are predominantly found in the cytosol of control plant cells (Knaupp *et al.*, 2011).

There seems to be a causal link between the chilling-induced modulation of sucrose metabolism and the acquisition of freezing-tolerance by chilling exposure. As already explained, during freezing, ice is initially formed in the apoplastic space, and this leads to an exit of water from the cell into the extracellular compartment (Fig. 4.2B). To avoid this cell dehydration it is necessary to decrease the water potential of the cell. The accumulation of carbohydrates within the cell contributes to that by diminishing the difference in water potentials between the apoplastic space and the solution within the cell. Second, sugars may be active cryoprotectants towards membranes.

The water associated with membranes is required to create the hydrophilic environment necessary to stabilize lipids in a bilayer. During freeze-induced dehydration, non-reducing sugars, such as sucrose or trehalose, can replace the lost water in creating this hydrophilic environment. Finally, sugars can also have a role in protecting integral membrane protein complexes. In addition, they have been proposed as having anti-ice nucleation and ROS-scavenging properties (Ruelland *et al.*, 2009). It thus seems that sugar accumulation during chilling stress is related to the acquisition of freezing-tolerance during chilling treatment, the so-called *cold-hardening* (or *cold-acclimation*). Sugar accumulation indeed appears to be a main determinant of the acquisition of freezing-tolerance during cold-acclimation. In a study with 54 *A. thaliana* accessions, contents in fructose, glucose, sucrose and especially content in raffinose were positively correlated with acquired freezing-tolerance, indicating at least an important role of the sugars in the natural variation in *A. thaliana* freezing-tolerance. Among the *Arabidopsis* galactinol synthase genes, which encode the enzyme catalysing the committing step in the raffinose biosynthesis pathway, only the expression of *GolS3* was correlated with both freezing-tolerance and raffinose content, indicating that this gene is the most important galactinol synthase one for cold-acclimation (Zuther *et al.*, 2012). The chilling-triggered increase in the contents of fructose, glucose, sucrose and raffinose is therefore an important predictor of freezing-tolerance in cold-acclimated *A. thaliana* (Korn *et al.*, 2010).

The role of the different accumulated carbohydrates may differ and each carbohydrate may have a dedicated role. Comparing a raffinose synthase mutant of *A. thaliana* with its corresponding wild-type revealed that a lack of raffinose has no effect on electrolyte leakage (Zuther *et al.*, 2004). However, electrolyte leakage specifically reports the loss of semi-permeability of the plasma membrane; this test does not necessarily detect damage to other cellular structures (Thomashow, 1999). Raffinose appears not to be involved in the protection of plasma membrane, a role achieved by other carbohydrates. It was shown that raffinose accumulated in plastids during chilling temperatures. *In situ* chlorophyll fluorescence showed that the maximum quantum yield of photosystem II photochemistry (Fv/Fm) in cold-acclimated leaves subjected to freeze–thaw cycles was significantly lower in the raffinose synthase mutant than in the wild type. Raffinose thus appears to be involved in stabilizing photosystem II of cold-acclimated leaf cells against damage during freezing. However, freezing-tolerance of photosystem II did increase significantly during cold-acclimation in raffinose synthase mutant plants (even though less than in wild-type plants). This demonstrates that raffinose accumulation is not the only mode of stabilizing the photosynthetic machinery against freezing damage. Sucrose, which also accumulates in plastids during chilling treatment, is likely to participate in protection of photosystems (Knaupp et al., 2011).

Moreover, sugar accumulation can be species-specific. The leaf contents of fructose, glucose, sucrose and raffinose were linearly correlated with the acquisition of freezing-tolerance in A. thaliana accessions. In Thellungiella accessions, a positive correlation with acclimated freezing-tolerance was only observed for sucrose, while a negative correlation was even detected for fructose. The Thellungiella accessions did not accumulate raffinose to the same extent as A. thaliana (Lee et al., 2012).

4.3.4   Compatible solutes other than sugars

In addition to soluble sugars, compatible solutes are a heterogeneous group of molecules comprising amino acids (Pro, Ala, Gly, Ser) and polyamines. Compatible solutes are low molecular weight organic molecules that are produced in response to many stresses such as desiccation, osmotic stress or low-temperature stress. It is, for instance, well documented that proline robustly accumulates under chilling stress. Polyamines are organic compounds having two or more primary amino groups. In A. thaliana submitted for 14 days to a temperature of 4°C, an increase in putrescine, ornithine and citrulline (the precursors of polyamines) is observed (Cook et al., 2004). Physiologically, compatible solutes have no adverse metabolic effects even at very high concentrations. As already mentioned for sugars, they may act by decreasing the water potential, thus preventing excessive cell dehydration,

counterbalancing the osmotic effect of ice formation in the apoplastic space. They may also stabilize proteins, assist refolding of peptides and stabilize membranes (Ruelland et al., 2009). Recent studies using different accessions of one species (e.g. A. thaliana) have tried to establish whether compatible solubles accounted for the natural variation of acquired freezing resistance in plants. In a study with 54 A. thaliana accessions, proline content could not be correlated with freezing-tolerance after chilling treatment. Moreover, plants of accessions Oy-0 and WS (accessions from Oystese, Norway and Wassilewskija, Russia, respectively) have extremely low proline accumulation but a high freezing-tolerance after acclimation (Zuther et al., 2012). Proline does not appear to be necessary for high freezing-tolerance of A. thaliana. Concerning polyamines, in a study with different accessions of A. thaliana or Thellungiella sp. the levels of free putrescine or spermidine either increased during chilling treatment or remained unaltered. However, the levels of spermine were much higher in non-acclimated Thellungiella leaves and were drastically reduced during chilling treatment; while in A. thaliana accessions the spermine level was generally lower in control plants, and decreased in only a few accessions. Interestingly, non-significant correlations could be found between freezing-tolerance after chilling treatments and spermidine or spermine levels in Thellungiella accessions, but a negative correlation was detected with the levels of free spermine in those accessions. In A. thaliana, no correlation between freezing-tolerance after chilling treatments and levels of any polyamine could be detected (Lee et al., 2012). This implies that polyamines cannot be used as a marker of chilling response leading to freezing-tolerance. Nevertheless, putrescine is an essential component of the cold-acclimation process in A. thaliana. Indeed, mutants defective in putrescine biosynthesis (adc1 and adc2 mutant plants) display reduced freezing-tolerance compared to wild-type plants. This is correlated with reduced expression of NCED3, a key gene involved in ABA biosynthesis, and downregulation of ABA-regulated genes in both adc1 and adc2 mutant plants under cold stress. Putrescine controls the levels of ABA in response to low temperature by modulating ABA biosynthesis and gene expression (Cuevas et al., 2008).

4.3.5 Photosynthesis

As explained in Section 4.2.4, cold temperatures will lead to an early imbalance between the light energy harvested by antennae and its use by biochemical processes. Two strategies can participate in re-establishing a new photostatic state: (i) photostasis can be achieved by increasing the sink capacity through increased rates of processes that consume reductants and fixed carbon, such as respiration, N-assimilation and ultimately growth; and (ii) photostasis can be achieved by decreasing the efficiency of light absorption and trapping either by increasing non-photochemical quenching through the xanthophyll cycle and/or reducing the size of the light-harvesting complexes. These two strategies are used in plants; one strategy will be favoured over the other depending on the plant ecophysiology (Hüner *et al.*, 2012). In evergreen conifers, such as *Pinus sylvestris*, cold-acclimation induces cessation of primary growth: conifers enter dormancy during winter. The cessation of growth correlates with a decrease in the sink demand for photoassimilates, i.e. with a decreased capacity for energy utilization. To attain photostasis under these conditions, overwintering conifers exhibit long-term changes in the organization of the photosynthetic apparatus that include a decrease in the number of functional photosystem II reaction centres, a loss of light-harvesting chlorophylls and the formation of a large thylakoid/protein aggregate involving the light-harvesting complex II and the photosystems I and II (Savitch *et al.*, 2002). This aggregation is associated with an increased formation of zeaxanthin, through the xanthophyll cycle. It is the cycle by which violaxanthin is rapidly and reversibly de-epoxidized into zeaxanthin via the intermediate antheraxanthin. Antheraxantin and zeaxanthin are quenching pigments, unable to pass their excitation to chlorophyll *a*. This is a protective mechanism that lowers the delivery of energy to photosystem II, and is a part of non-photochemical quenching. The formation of the large winter-induced aggregated state is fully reversible upon warming in the spring, with the result that photosynthesis is rapidly resumed without the immediate *de novo* synthesis of chlorophyll (Ivanov *et al.*, 2001).

In contrast to overwintering evergreen conifers, overwintering herbaceous plants such as winter wheat, rye, brassicas and *Arabidopsis*, continue to grow and develop during chilling exposure, thereby maintaining a high demand for photoassimilates. Yet, in the short term, these annual cold-tolerant plants show a decrease in CO_2 uptake capacity when exposed to chilling temperatures. This is the Pi-limitation of photosynthesis described in Section 4.2.4 that is due to the thermodynamic effects of cold on enzymatic reactions. This results in photostasis imbalance. Plants will tend to diminish this imbalance by converting photosystem II from a state of efficient light harvesting to a state of high thermal dissipation, through the xanthophyll cycle described above. However, this non-photochemical quenching is not necessary for long-term adaptation to chilling stress in overwintering herbaceous plants. Indeed, over the long term (days to weeks), the strategy of increasing the sink demand will be favoured. Photosynthesis recovers in cold-acclimating winter annuals and this is associated with the production of new leaves better adapted to the new thermal regime. This recovery in photosynthesis efficiency is associated with an increase in thylakoid quinone A content, with a concomitant increase in the apparent size of the intersystem electron donor pool to photosystem I, in the capacity for ROS scavenging, and in the content and activity of a number of Calvin cycle enzymes. Associated with this recovery in plastid metabolism, cold-grown plants exhibit an increase in sucrose–phosphate synthase activity and activation state and an increase in the cytosolic hexose–phosphate pool, resulting in increased sucrose biosynthesis. These changes in thylakoid membrane processes – coupled to increases in enzymatic contents and flux capacities in both the stroma and the cytosol, in combination – result in recovery of photosynthesis at low temperatures to rates equivalent to that of plants grown under permissive warm conditions (Strand *et al.*, 2003; Hüner *et al.*, 2012).

The fact that a functional chloroplast and a proper recovery of photosynthetic activity is necessary for the cold-acclimation process (and therefore for appropriate response to chilling) is also proved by mutants in the magnesium chelatase: that is, the first enzyme unique to the chlorophyll-specific branch of the porphyrin biosynthetic pathway. In *A. thaliana*, *gun5-1* is a mutant in the *CHLH* gene encoding the H-subunit of magnesium chelatase. All new leaves formed in the *gun5* mutant under low temperatures showed a striking pale appearance while in wild-type plants the new leaves showed a darker green phenotype than leaves grown at control temperatures. Besides, following exposure to low temperature, the *gun5* mutant displayed severely disrupted thylakoid membrane structures. After cold-acclimation for 3 days, wild-type plants were considerably more tolerant to freezing temperatures than the *gun5* plants. However, the role of photosynthesis does not seem to rely on the biosynthesis of sucrose. Indeed, *gun5* accumulates similar levels of soluble sugars compared to wild type following exposure to low temperatures. In contrast, the reduced cold-acclimation in the *gun5* mutant is accompanied by reduced protein translation during low temperatures. How an impaired chloroplast function inhibits *de novo* protein synthesis at low temperature is not clear. It is interesting to note that *in vitro* Mg-ProtoIX, the product of CHLH, directly binds to LOS1 protein, a translation elongation factor. The phenotype in cold of *los1-1* resembles that of the gun mutant (Kindgren *et al.*, 2015).

4.3.6 Anthocyanins

Anthocyanin accumulation can contribute to the adaptation of photosynthesis to chilling. These pigments accumulate in leaves and stems in response to low temperature and changes in light intensity. They are synthesized through the phenylpropanoid pathway that is controlled by key enzymes which include phenylalanine ammonia-lyase and chalcone synthase. Phenylalanine ammonia-lyase and chalcone synthase mRNAs accumulate in leaves of *A. thaliana* upon exposure to low temperature in a light-dependent manner. In a study using 54 *A. thaliana* accessions covering a wide range of freezing-tolerance, it was shown that most flavonols and anthocyanins accumulated upon cold exposure, but the extent of accumulation varied strongly among the accessions. Anthocyanins were only present at very low levels under control, warm growth conditions; in these conditions the flavonol composition was dominated by three abundant kaempferol derivatives. After 2 weeks at 4°C a general increase of flavonoids was observed. Among the flavonols, quercitin derivatives increased the most. The flavonol biosynthesis genes were highly cold-induced

in the majority of accessions while many anthocyanin biosynthetic genes were strongly down-regulated in the most freezing-sensitive accessions and induced in the most tolerant accessions. Gene expression data suggest a major role for post-transcriptional mechanisms in the regulation of the accumulation of flavonoids in cold-stressed *A. thaliana*. The expression of genes related to flavonoid metabolism was poorly correlated with flavonoid metabolites, indicating an important role of post-transcriptional regulation in flavonoid metabolism, especially for the flavonol pathway. Only the pool sizes of a few flavonols (three quercitin derivatives and one kæmpferol derivative) and two anthocyanins were correlated with freezing-tolerance after cold-acclimation. This illustrates the importance of flavonoid metabolism in *A. thaliana* freezing-tolerance and points to the importance of post-transcriptional mechanisms in the regulation of flavonoid metabolism in response to cold (Schulz *et al.*, 2015).

Anthocyanin accumulation has been reported to protect developing pine needles from photoinhibition during growth of seedlings at low temperatures, due to light trapping, which decreased chlorophyll excitation by blue light (Harvaux and Kloppstech, 2001). Some flavonoids that are accumulated in the cold are able to depress the freezing point of plant cells or tissues and contribute to the deep supercooling ability of xylem parenchyma cells in katsura trees (Kasuga *et al.*, 2008). Farinose flavonoids are excreted to the surfaces of aerial organs of *Primula* species, for instance, where they inhibit ice crystallization and thereby improve tolerance to mild freezing temperatures (Isshiki *et al.*, 2014).

4.3.7 Reactive oxygen species

Cold exposure induces an oxidative stress. The main source of reactive oxygen species (ROS) is the chloroplast. ROS can be produced at the acceptor side of photosystem II or at the donor side of photosystem I (Fig. 4.4). At low temperature, the enzymatic systems that normally detoxicate ROS will be less efficient. This will lead to an increase in ROS production that will need to be detoxified via the induction of ROS defence mechanisms. It appears that one major ROS detoxification system during chilling is the ascorbate/glutathione cycle

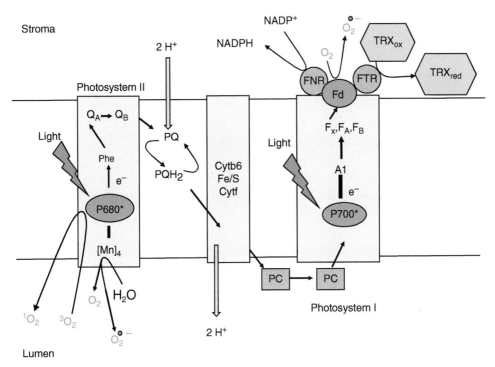

Fig. 4.4. The electron transfer chain in thylakoid membranes. P680, photosystem II primary electron donor; P700, photosystem I primary electron donor; $[Mn]_4$, $[Mn]_4$ cluster; Q_A, quinone Q_A; Q_B, quinone Q_B; PQ, plastoquinone pool; Cytb6, cytochrome b6; Cytf, cytochrome f; F_x, F_A, F_B, iron-sulfur centres of photosystem I; PC, plastocyanin; Fe/S, Fe/S centres; Fd, ferredoxin; FNR, ferredoxin NADP reductase; FTR, ferredoxin thioredoxin reductase; TRX_{ox}, oxidized thioredoxin; $TRXr_{ed}$, reduced thioredoxin. The reactive oxygen species produced at the donor side of photosystem II, the redox status of PQ and/or cytochrome b6/f, the redox status of thioredoxin and the reactive oxygen species produced at the acceptor side of photosystem I can have signalling roles, meaning they can be upstream-documented responses to chilling.

(Distelbarth *et al.*, 2013). The detoxification of hydrogen peroxide produced in the chloroplasts indeed relies exclusively on the activity of ascorbate peroxidase bound to thylakoid membranes in the vicinity of photosystem I. The dehydroascorbate thus produced can be reduced back to ascorbate using the reducing power of reduced glutathione that needs to be regenerated through glutathione reductase. A correlation can be made between chilling resistance and activation of the ascorbate/glutathione cycle. The responses of antioxidative defence systems to chilling were studied in four cultivars of rice (*Oryza sativa*). After a chilling stress of 5 days at 8°C, the two chilling-tolerant cultivars assayed presented a much lower level of electrolyte leakage and H_2O_2 content than the two chilling-sensitive cultivars. The activities

of ascorbate peroxidase and contents of antioxidants (ascorbic acid and reduced glutathione) were measured in the course of the 5 days of stress treatment. All enzymatic activities and antioxidant contents remain constant in the two chilling-tolerant cultivars while they greatly decreased in the chilling-sensitive ones in the course of the 5 days of chilling. The results indicated that tolerance to chilling in rice is associated with the enhanced capacity of the antioxidative system (Guo *et al.*, 2006). Accordingly, transgenic tobacco overexpressing glutathione reductase had improved tolerance to cold (Le Martret *et al.*, 2011). Similar results were obtained for rice overexpressing ascorbate peroxidase and having improved chilling tolerance (Sato *et al.*, 2011). In maize species, low-temperature tolerance was higher in *Zea diploperennis* having an ascorbate pool twice that of the chilling-sensitive *Zea mays* (Hull *et al.*, 1997). This indicates a major role of the ascorbate/glutathione cycle in the response to cold. It can be noted that *A. thaliana* mutants lacking the thylakoid-bound ascorbate peroxidase, stromal ascorbate peroxidase or both do not exhibit enhanced stress symptoms under low temperature; maybe the plastidial ascorbate peroxidase is important only for sudden onset of oxidative stress and can be substituted in the long term by other mechanisms, such as 2-cys-peroxiredoxin (Kangasjärvi *et al.*, 2008).

Finally, other ROS defence systems, such as the ones relying on superoxide dismutase and catalase are also likely to participate in detoxification of chilling-induced ROS increase (Guo *et al.*, 2006).

It is possible that early exposure to chilling temperatures can have positive impact on later chilling exposures because it will have induced the ROS defence system. In winter wheat plants, such cold priming is associated with more effective oxygen scavenging systems in chloroplasts and mitochondria, as exemplified by the increased activities of superoxide dismutase, ascorbate peroxidase and catalase, resulting in a better maintenance in homeostasis of ROS production (Li *et al.*, 2014).

4.3.8   Cell walls

Plants form two types of cell wall that differ in function and in composition. Primary walls surround growing and dividing plant cells. These walls provide mechanical strength but must also expand to allow the cell to grow and divide. A much thicker and stronger secondary wall is deposited once the cells have ceased to grow and is characterized by the incorporation of lignin. The plant cell walls are composed of cellulose microfibrils interlaced with non-cellulosic cross-linking polysaccharides, and embedded in a physiologically active pectin matrix and cross-linked with structural proteins. Pectins participate in the mechanical strength, porosity, adhesion and stiffness of the cell wall. Structural changes in cell wall components are mediated through the activity of cell wall-modifying enzymes that play a major role in controlling cell wall plasticity/rheology. Chilling temperatures reduce mung bean and sweet potato plant cell wall elongation.

This could partly be due to the increase in cell wall thickness and rigidity observed, for instance, in oilseed rape plants, grape stem, oak and cranberry leaves in response to low temperatures. In suspension culture of grape cells, and in apple, cold-acclimation increases cell wall strength and decreases the pore size of cell walls. In oilseed rape leaves, the cold exposure leads to an increase in pectins and pectin methylesterase activity. Transcriptomic and proteomic analyses have shown that most cell wall-related genes or proteins identified in response to cold stress correspond to pectin remodelling enzymes, including pectin methylesterase, polygalacturonase, pectin acetylesterase and pectate lyase (Baldwin *et al.*, 2014). However, it appears more and more that the cell wall responses to chilling temperatures differ according to species and to cultivars within species. For instance, exposure to chilling temperatures of a pea cultivar resulted in increased abundance of polymers, whereas in another one it resulted in a substantial decrease in pectic polymers and in strong increase in xylans and glucuronoxylans (Baldwin *et al.*, 2014). In a study aiming at characterizing lignification in the stem of two sugarcane genotypes grown under low temperatures, lignin content was significantly increased in the young cortex of one cultivar of plants at chilling temperatures but it was reduced in the mature cortex of the other.

These differences in responses to chilling were correlated with differences in freezing sensitivity. For instance, the pea cultivar with increased abundance of pectic polymers after chilling exposure was freezing tolerant while the other was freezing sensitive (Baldwin *et al.*, 2014). Concerning lignification, decrease in lignin content is correlated with increased freezing-tolerance. Tolerant to Chilling and Freezing 1 (TCF1) is a cold-induced nuclear protein. Loss of TCF1 function leads to reduced lignin content and enhanced freezing-tolerance. Plants with knocked-down BLUE-COPPER-BINDING PROTEIN (BCB) expression (amiRNA-BCB) under cold-acclimation had reduced lignin accumulation and increased freezing-tolerance. The *pal1pal2* double mutant (with a lignin content reduced by 30% compared to wild type) also showed a freezing-tolerant phenotype (Ji *et al.*, 2015).

How to link cell wall composition and freezing-tolerance? Galactan and arabinan side chains of pectins may play a role in the wall by modulating the properties of pectins, thus affecting water binding. Arabinan side chains could maintain cell wall flexibility during dehydration and rehydration events. This has been shown with resurrection plants (Moore *et al.*, 2013). Methylation of pectin would also contribute to more flexible cell walls. In contrast, lignification of cell walls makes them less flexible. As already explained, a freezing/thawing cycle corresponds to a dehydration/rehydration cycle. For cell dehydration to occur during freezing, cells should undergo cell volume reduction or cell deformation. Cell walls that are too rigid may not allow such cell deformation, resulting in negative pressures in the cells, which might compromise cell viability (cavitation, rupture of protoplasm) (Rajashekar and Lafta, 1996).

In summary, the exposure to chilling temperatures induces different responses that will lead to chilling tolerance (Fig. 4.2A) and/or freezing-tolerance (Fig. 4.2B). Among these responses are the accumulation of hydrophilic proteins, the accumulation of sugars and compatible solutes, changes in membrane lipid composition and cell wall composition, the induction of protein-chaperone and RNA-chaperone synthesis. Metabolic readjustments and the induction of ROS-scavenging systems are also necessary to cope with cold.

4.4   How Chilling Temperatures are Sensed

The fact that an exposure to chilling induces cellular responses is a clear indication that chilling has been perceived. Once perceived, chilling triggers signalling pathways that will activate transcription factors which will lead to a major rearrangement of the transcriptome, thus allowing the responses that we have listed above to occur. But first, how is chilling perceived? Temperature is a physical parameter. In contrast to hormones, for which perception is the binding of hormones to their receptors, the notion of perception in the case of temperature is difficult. It is the step when a physical parameter is converted into biochemical parameters. We define here perception as the most upstream event(s) controlling downstream signals. The following paragraphs detail the various perception steps.

4.4.1   Changes in membrane fluidity trigger responses to chilling

One of the best-documented levels of temperature sensing is the change in membrane fluidity associated with a temperature decrease. Membranes are moving mosaics of proteins and lipids. Lipids can flip-flop between monolayers, diffuse within the plane of a monolayer and rotate around their own axes, their acyl chains also rotating around C–C bonds. Each type of motion is thermodynamically driven by its own temperature dependence, i.e. its activation energy. As the temperature decreases, these movements slow down, making membranes more rigid. Membrane rigidification upon a temperature downshift has been shown to trigger several cold-associated cell responses in plants. Cold responses can be mimicked at ambient temperature or enhanced in the cold by membrane rigidification agents such as dimethylsulfoxide. Conversely, cold responses are inhibited by fluidizing membranes using chemicals such as benzylalcohol. Confirming these data, several cold-induced processes are enhanced in mutant plants characterized by membranes that are more rigid than in the wild-type (Ruelland and Zachowski, 2010).

4.4.2   Protein conformation changes with temperature

A temperature downshift, as well as a temperature increase, can lead to protein unfolding. These conformational changes can be a level of cold perception. In barley, the DNA-binding activity of a transcription factor, HvCBF2, is temperature dependent. In a cell-free system, this factor does not bind its *cis*-acting element at 30°C, but its binding activity gradually increases as the temperature of the *in vitro* assay decreases to 2°C. This increase in activity is likely to be due to a cold-induced change in conformation (Xue, 2003).

4.4.3   Chilling temperatures induce disassembly of the cytoskeleton

A drop in temperature induces depolymerization of both microtubules and microfilaments. These cytoskeletal disequilibria appear necessary for several cold responses. In rapeseed, the cold-activation of the *BN115* promoter was inhibited when microtubules and microfilaments were chemically stabilized; in *Medicago sativa* cells, calcium influx (see Section 4.5.1) and *cas30* expression at 4°C were also prevented by a microfilament stabilizer. Conversely, these responses to cold were mimicked at 25°C by microtubule or microfilament destabilizers (Janská *et al.*, 2010). It is now accepted that plant microtubules, in addition to their role in cell division and axial cell expansion, convey a sensory function that is relevant for the perception of mechanical membrane stress. Chilling temperatures, which will lead to membrane rigidification, represent such a stress. Phospholipases D (see Section 4.5.3) are proposed to be important players in the microtubule-dependent sensory hub (Nick, 2013).

4.4.4 Effects of chilling temperatures on metabolic reactions including photosynthesis

As explained in Section 4.2.4, chilling will result in an early depletion in Pi, in an increased excitation pressure on photosystem II and in ROS accumulation. Interestingly, these phenomena control downstream signalling events. Phosphate depletion is a signal and several genes induced by low temperatures are also induced by depletion in phosphate. Several traits of the response to chilling, such as development of freezing-tolerance, the accumulation of sugar or the expression of sucrose biosynthesis enzymes are exaggerated in the Arabidopsis *pho1* mutant having low concentrations of Pi in shoots, while these traits are impaired in the *pho2* mutant characterized by a higher Pi content than wild-type plants (Ruelland and Zachowski, 2010). These results are an indication of a direct role of free Pi concentration as a signal participating in the response of cells to chilling.

Concerning the photostasis imbalance, it will increase the excitation pressure on photosystem II; it reflects the relative reduction state of Q_A, the primary plastoquinone electron acceptor of photosystem II (Fig. 4.4). Indeed, cold-induced excitation pressure on photosystem II has been correlated with responses to chilling. For instance, the accumulation of cold-induced transcripts and the compact plant morphology typical of an exposure to chilling have all been found to depend on an increase in photosystem II

excitation pressure rather than growth temperature per se: they can be mimicked at ambient temperature by increasing the excitation pressure on photosystem II simply following an increase in light intensity (Ruelland and Zachowski, 2010). However, not all aspects of the cold response depend on this parameter and several genes have been identified that are inducible by cold per se, irrespectively of the photosystem II pressure (Ndong *et al.*, 2001).

The way photosystem II pressure is sensed is not clearly understood. The increased excitation pressure on photosystem II could be sensed through the concomitant increase in ROS production. Indeed, a higher photosystem II excitation pressure is likely to increase the production of ROS through the generation of singlet oxygen. In addition, energy imbalance can be sensed by other components of the chloroplast electron transfer chain. The reduced state of the plastoquinone intersystem pool and of the cytochrome $b_6 f$ components can trigger structural changes that may initiate cellular responses. Photosystem I also has a major role in the transduction of energy imbalance. H_2O_2 can be generated at the reducing site of photosystem I (Fig. 4.4). ROS have signalling roles, and some cold-dependent genes are also responsive to oxidative stress: in rice, 60% of chilling stress-induced genes are also triggered by H_2O_2 (Yun *et al.*, 2010). It has to be noted that ROS-scavenging enzymes will be induced by chilling (see Section 4.3.6), but as a mid-term response; this induction will participate in the lowering of the level of ROS that had been produced as an early effect of chilling. The early accumulation of the ROS produced by the photosynthetic machinery at photosystem I or II, under chilling temperatures, is a sensing device. Finally, one of the electron acceptors of photosystem I is ferredoxin, which will participate in the reduction of thioredoxin via a ferredoxin/thioredoxin reductase (Fig. 4.4). Thioredoxins mediate the reductive activation of many proteins, including enzymes of the primary carbon metabolism, but also of signalling enzymes. Chilling and light treatment are expected to cause a lowering of the redox poise of chloroplastic thioredoxins. This will impact their capacity to activate target proteins, by reduction of their disulfide bridges (Hutchison *et al.*, 2000). These examples illustrate that the cold-induced lowering of metabolic reactions must be considered as perception, as it is a direct consequence of the temperature downshift and is upstream of signalling processes such as Pi depletion, the increase in photosystem II excitation pressure, ROS production and changes in chloroplastic thioredoxins redox status.

In conclusion, we can see that cold is not perceived by a single mechanism in plants but at different sensory levels, that are the very biological processes disturbed by the temperature change itself. These processes are not necessarily independent one from another. For instance, the effect of cold on metabolic reactions is due to both the thermodynamical effects and the temperature effect on protein conformation. Because of the membrane/cytoskeleton continuum, changes in the membrane physical state and cytoskeleton destabilization could be interdependent. The conformation of membrane-embedded proteins could

be affected by changes of membrane fluidity. Even though the individualization of the temperature-sensing devices is somewhat artificial, it helps in visualizing what occurs when a cell is submitted to a cold stress (Fig. 4.5). Not all the temperature-sensing devices will be activated concomitantly. Some, such as membrane rigidification, will respond very quickly; while others, such as cytoskeleton destabilization, can take a few minutes or hours to achieve. Cold sensing thus implies a succession of switches that will be turned on sequentially. Besides, the temperature threshold for switching on these devices is not necessarily the same. The second aspect of the sensing system that is not well understood is the way these devices are switched off. As the temperature-sensing devices are triggered by cellular disequilibria, it is the creation of a new equilibrium that is likely to turn them off. Finally, both the temperature change rate

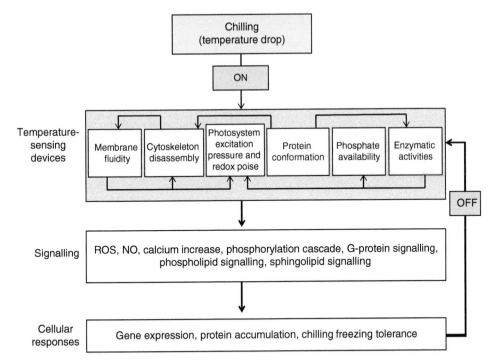

Fig. 4.5. Schematic representation of the chilling-sensing machinery in plants. Chilling-sensing devices in plants are the very cellular processes disturbed by cold: membrane fluidity; the status of cytoskeleton assembly; photosystem excitation pressure and redox poise; protein conformation; and enzymatic activities (metabolic imbalance) are affected by chilling and are upstream of the cellular responses to cold. These components can be interlinked (black arrows). The signalling pathways downstream of the sensing steps can also influence these sensing steps. Ultimately, the cellular responses activated in response to cold will participate in switching off the chilling-sensing devices (adapted from Ruelland and Zachowski, 2010). NO, nitric oxide; ROS, reactive oxygen species.

and the absolute temperature are integrated in the cellular response. The rate of temperature change might be the determining factor when the exposure to a temperature change is short, or when the response monitored is a rapid one. An example of this is calcium entry, which occurs nearly instantaneously after the temperature drops (see Section 4.5.1). In contrast, one might suppose that the longer the exposure, the more important the absolute temperature becomes as a factor. The complexity of this 'temperature-sensing machinery' will lead to an appropriate response to the chilling stress experienced by the plant (Ruelland and Zachowski, 2010).

4.5 Signal Transduction Pathways in Response to Chilling

In the section above, some of the signalling events of the response to chilling had to be described because, by definition, perception is the most upstream event controlling downstream signalling and responses. We will now detail these signalling events more closely.

4.5.1 Calcium

An immediate increase in cytosolic calcium ($[Ca^{2+}]_{cyt}$) is one of the major signalling events triggered by an exposure to chilling. In *A. thaliana* guard cells, prior to stimulation, $[Ca^{2+}]_{cyt}$ was measured at 128 nM while cold caused an immediate increase that peaked between 300 and 1100 nM (Dodd *et al.*, 2006). This increase in $[Ca^{2+}]_{cyt}$ is mainly due to calcium entry from the extracellular space. A major determinant controlling calcium entry is not the absolute temperature per se but the rate of temperature decrease. An internal component of $[Ca^{2+}]_{cyt}$ increase also exists: it has been shown that chilling induced an increase in calcium near microdomains corresponding to the cytosolic face of the vacuole. This increase could involve inositol-triphosphate ($InsP_3$) and phospholipase C (see Section 4.5.3). After the increase in $[Ca^{2+}]_{cyt}$, the subsequent return to resting levels is achieved by Ca^{2+} pumps and antiporters, such as the tonoplast Ca^{2+}/H^+ antiporters (Ruelland *et al.*, 2009).

Calcium increase during cold exposure is necessary for the accumulation of cold-responsive

transcripts. When calcium entry is inhibited, either by chelating extracellular calcium or by inhibiting plasma membrane calcium channels, the induction of transcript accumulation by cold is inhibited. Reversely, in a mutant invalidated in a tonoplast Ca^{2+}/H^+ antiporter necessary for the re-establishment of $[Ca^{2+}]_{cyt}$ to basal levels, an enhanced expression of *CBF* genes and their corresponding targets (*KIN1*, *LTI78* and *COR47*) in response to low temperature is monitored (Janská *et al.*, 2010).

Calcium acts via calcium-regulated proteins or calcium-interacting proteins. Several studies have shown that the calmodulins are involved in gene induction by low temperature. W7, an inhibitor of calmodulins and calcium-dependent protein kinases, inhibits the cold-induction of *KIN* genes. Several transcription factors also interact with calmodulins. This is the case of the so-called calmodulin-binding transcription activators (CAMTA factors), whose importance in the induction of genes to cold has been documented (see Section 4.6.2). Calcium can also be sensed by Calcineurin B-like proteins (CBL). These proteins are calcium sensors which interact with and activate a group of Ser/Thr protein kinases known as calcineurin B-like-interacting protein kinase (CIPK). In *A. thaliana*, overexpression of *CBL1* induces the expression of cold-responsive genes in conditions of no stress. Calcium can also activate calcium-dependent protein kinases (Section 4.5.2), phospholipase C and certain types of phospholipase D (Section 4.5.3) (Ruelland *et al.*, 2009).

In summary, an increase in cytosolic calcium is a major response to a cold-shock, as shown by the fact that, when it is inhibited, the downstream responses to cold (gene induction, cold-acclimation) are also inhibited. This central role is explained by the fact that calcium, by itself or through calmodulin or CBL proteins, can control many signalling effectors.

4.5.2 Protein kinases/protein phosphatases

During a cold exposure some proteins are phosphorylated whereas others are dephosphorylated, as a very early response that can occur in less than 5 min. In different plant species, calcium-dependent protein kinases (CDPKs) have been shown to be activated in response to chilling; they are involved in signalling steps leading to gene expression. In plants, the 'mitogen-activated protein kinase' (MAPK) module is also very important. MAPKs are activated by phosphorylation by the action of MAPK-kinases (MAPKK) which themselves are phosphorylated by MAPKK-kinases (MAPKKK). In *A. thaliana*, MKK2 – a MAPKK – activates MPK4 and MPK6. These activations are required for cold-induced gene expression. The involvement of the MAPK module in the response to cold has also been documented in wheat and rice (Ruelland *et al.*, 2009).

Protein phosphatases are also involved in the change in phosphoproteome triggered by a chilling stress. In *A. thaliana* the transgenic antisense plants directed against *PP2CA*, a Ser/Thr protein phosphatase, displayed an accelerated development of freezing-tolerance following a chilling treatment when compared with the wild-type plants. Furthermore, the expression of cold-induced genes such as *LTI78*, *RAB18* and *RCI2A* was detected earlier in transgenic antisense plants against *PP2CA* compared to wild-type plants (Tähtiharju and Palva, 2001), suggesting that AtPP2CA is a negative regulator of cold responses.

4.5.3   Lipid signalling

Phospholipase D (PLD, Fig. 4.3D) catalyses the hydrolysis of structural phospholipids such as phosphatidylethanolamine or phosphatidylcholine into phosphatidic acid. In *A. thaliana* suspension cells, cold activates PLD in the very first minutes of cold exposure. PLD-produced phosphatidic acid appears to be necessary for full induction of cold-responsive genes such as *LTI78*, *LTI30* and *HVA22*. Interestingly, one of these genes is *CBF3*, suggesting that the CBF pathway (see Section 4.6.1) is downstream of PLD activation (Ruelland *et al.*, 2009).

Phospholipase C (PI-PLC) catalyses the hydrolysis of phosphatidylinositol-bis-phosphate into diacylglycerol and inositol 1,4,5-triphosphate (InsP$_3$) (Fig. 4.3D). PI-PLC activation after an exposure to 0°C has been documented in *Brassica napus* and *A. thaliana* suspension cells (Ruelland *et al.*, 2009). PI-PLCs are strictly calcium-dependent and consequently cold-induced PLC activation is dependent on calcium entry. PI-PLC-generated InsP$_3$ can be subjected

to a series of phosphorylation and dephosphorylation reactions, leading to different inositol polyphosphates that could have a signalling role, most likely through calcium homeostasis. To date no InsP$_3$ receptor has been identified in plants. When the *AtIpk2beta* gene, encoding an Arabidopsis inositol polyphosphate 6-/3-kinase, was constitutively overexpressed in tobacco (*Nicotiana tabacum*), the resulting plants exhibited improved tolerance to freezing with no acclimation period at chilling temperatures (Yang *et al.*, 2008). The other product of PI-PLC activity, diacylglycerol, can be phosphorylated into phosphatidic acid by the action of a diacylglycerol kinase. While putative protein targets of phosphatidic acid are known, including protein kinases or protein phosphatases (Wang *et al.*, 2006), which of these targets are indeed involved in the response to chilling is still to be unravelled. No diacylglycerol-interacting proteins have been identified in plants.

In parallel to these pathways involving phospholipases, other signalling pathways generating lipid messengers are active. We have already mentioned the sphingolipids as structural components of membranes (Section 4.3.2). Sphingolipids are also signalling lipids and have been shown to be involved in the chilling responses. For instance, a rapid and transient formation of phosphorylated ceramide occurs in cold-stressed *A. thaliana* plantlets and cultured suspension cells. This formation is strongly impaired in a mutant mutated in ceramide kinase (*acd5* mutant) (Dutilleul *et al.*, 2015). Chilling also leads to phosphorylation of a long-chain base, phytosphingosine, into phytosphingosine-1-phosphate. This phosphorylation was abolished in the *Arabidopsis* LCB kinase *lcbk2* mutant, but not in other mutants of long-chain base kinases (*lcbk1* and *sphk1* mutants). Interestingly, the *lcbk2* mutant plant presented a constitutive AtMPK6 activation at 22°C. Therefore the phosphorylation of phytosphingosine might act on chilling stress responses by affecting the MAPK pathway (Dutilleul *et al.*, 2012).

4.5.4   Reactive oxygen species and nitric oxide

Some of the effects of ROS (O$_2^-$, H$_2$O$_2$, OH·) in the response to chilling have already been described (see Section 4.4.4). ROS can have a signalling role. As already mentioned, in rice, 60% of chilling stress-induced genes are also triggered by H_2O_2 (Yun *et al.*, 2010). In addition, Maruta *et al.* (2012) were able to silence the thylakoid membrane-bound ascorbate peroxidase (tAPX), key enzyme of the ascorbate/glutathione cycle. When the expression of tAPX was silenced in leaves, levels of oxidized protein in chloroplasts increased in the absence of stress. These silenced plants are good tools to investigate the involvement of the H_2O_2 signalling derived from chloroplasts in the regulation of the cold response. When subjected to cold stress (continuous light of 100 µmol of photons m^{-2} s^{-1}, 4°C) for 2 weeks, the APX-silenced lines exhibited brown leaves and decreased Fv/Fm under cold stress, but symptoms not seen in the control plants under cold stress. These cold-sensitive phenotypes of the APX-silenced plants were not observed under low light, because the production rate of chloroplastic H_2O_2 appears to be limited under low light. These findings demonstrated that H_2O_2 signalling derived from chloroplasts is necessary for the cold response (Maruta *et al.*, 2012).

Low temperatures have been shown to result in nitric oxide (NO) production, mainly via the action of nitrate reductase. The level of NO was positively correlated with freezing-tolerance. The formation of phytosphingosine phosphate and ceramide phosphate was negatively regulated by NO upon chilling, while that of phosphatidic acid does not appear to be dependent on it (Cantrel *et al.*, 2011). The cellular action of NO can be through nitrosylation of cysteine residues in proteins that is stimulated by a chilling stress (Puyaubert and Baudouin, 2014). Cold-induced S-nitrosylation has been shown to be responsible for about 40% inactivation of Rubisco, an important step of cold-induced photosynthetic inhibition (Cantrel *et al.*, 2011, and references within). In *Brassica juncea* it was shown that chilling temperatures increased dehydroascorbate reductase and glutathione S-transferase activity via S-nitrosylation, suggesting a central role of NO in ROS detoxification. S-nitrosylation was also detected in proteolytic enzymes (e.g. aspartyl/aspartic proteases) and cell wall-modifying and metabolic enzymes (cruciferin-like and α-l-arabinofuranosidase) (Sehrawat and Deswal, 2014).

4.5.5 Two-component system signalling

The *A. thaliana* two-component signalling system is composed of sensor histidine kinases, histidine phosphotransfer proteins and response regulators. It mediates the cytokinin response. Cold significantly induced the expression of a subset of type A ARABIDOPSIS RESPONSE REGULATORS (*ARR*) genes. ARABIDOPSIS HISTIDINE KINASE2 (*AHK2*) and *AHK3* are involved in the cold response of these type A *ARR* genes. Indeed, the *ahk2ahk3* double mutant exhibits reduced induction by chilling of the type A *ARR* genes. Concerning histidine phosphotransfer proteins, AHP2, AHP3 and AHP5 play positive roles in the cold-inducible expression of type A ARRs. Taken together, these results suggest that *ARR1* mediates cold signal via *AHP2*, *AHP3* or *AHP5* from *AHK2* and *AHK3* to express type A ARRs. Concerning the role of type A ARRs per se, *arr5* and *arr7* mutants showed statistically significant increased freezing-tolerance, after cold-acclimation. Conversely, the overexpression of the cold-inducible *ARR7* in *Arabidopsis* resulted in a hypersensitivity response to freezing temperatures under cold-acclimated conditions. These results indicate that cold-inducible type A *ARR* genes may function as a negative regulator of cold signalling (Jeon *et al.*, 2010; Jeon and Kim, 2013). In contrast, the type B *ARR1* play a positive role in cold transduction. Besides, *ARR1* positively controls the cold expression of type A *ARR* genes, the *arr1* mutant plants showing greatly reduced cold-responsive expression of type A *ARR* genes. The signal receiver domain of *ARR1* is necessary for cold-responsive expression of type A ARRs. Transcriptomic analyses further document the involvement of two components signalling in response to cold. The transcriptomic responses of *ahk2ahk3* double mutant were compared to those of wild-type plants. Many genes had their cold expression reduced in *ahk2ahk3* double mutant thus revealing a new cold-responsive gene network regulated downstream of *AHK2* and *AHK3*. *CBF3*, but not *CBF1* nor *CBF2*, was less cold-induced in the *ahk2ahk3* double mutant (Jeon and Kim, 2013).

However, the phenotypes of mutants in the two-component system are not necessarily linked to cytokinins. Indeed, cytokinin analysis of *A. thaliana* exposed to cold did not show significant changes in cytokinin levels, at least during the first 4 h of cold. The 35S:AtCKX2-2 plants overexpress cytokinin oxidase and contain less than 20% of total zeatin and less than 40% of total cytokinin compared with wild-type plants. AtCKX2-2 is the most cytokinin-deficient 35S:AtCKXtransgenic *Arabidopsis* among the 35S:AtCKX plants generated. When treated with cold for 0, 1, 2 or 4 h, these cytokinin-deficient 35S:AtCKX2-2 plants displayed no difference to wild-type plants in the expressions of *ARR5*, *ARR6*, *ARR7* and *CBF1* (Jeon *et al.*, 2010). It is interesting to note that HKs have been shown to be involved in the response to cold in *Synechocystis* where they would act to transduce membrane rigidification triggered by cold. Therefore, it is worth investigating the possibility that the two-component system is activated independently of cytokinins in higher plants, but through cold-triggered membrane rigidification.

4.5.6 G-protein signalling

G-protein-coupled receptors (GPCRs), also known as seven-transmembrane domain receptors, are good candidates to sense cold, presumably through membrane rigidification and/or cytoskeleton reorganization. They are coupled to G-proteins. G-proteins are composed of α, β and γ subunits. In the inactive state, the α subunit binds a GDP. When GPCRs are activated they then activate an associated G-protein by exchanging its bound GDP for a GTP. The G-protein α subunit, together with the bound GTP, can then dissociate from the β and γ subunits to further affect intracellular signalling proteins or target functional proteins. The regulator of G-protein signalling (RGS), acting as GTPase-accelerating protein, promotes the hydrolysis of the GTP bound to α subunit of G-proteins. In plants it is proposed that the regulation of the G-proteins activity is more dependent on RGS than on GPCRs.

G-protein signalling components have now been implicated in the responses to chilling stress. Chakraborty and colleagues (2015) used the *gcr1-5* (mutated in one *Arabidopsis* GCPR) single mutant, the *gpa1-5* mutant (mutated in *Arabidopsis* G-protein α subunit) and the *gpa1-5gcr1-5* double mutant. The seedlings were placed at 4°C overnight and their transcriptomic responses analysed. Some genes induced by cold

(e.g. *ZAT11*, *RD26*, *ERF6*) appear to be more induced in the mutant genotypes and some cold-repressed genes (e.g. *YLS9*) were more repressed in the mutant genotype. This might suggest that G-protein signalling is involved in the fine-tuning of chilling-triggered gene-responses, by attenuating them. Consistently, when wild-type plants were submitted to cold stress a slight increase in proline accumulation was detected. This increase was much more pronounced in the mutant plants. The catalase, superoxide dismutase and ascorbate peroxidase activities were higher in the mutant plants than in wild-type plants after chilling treatment. The cold treatment was also correlated with a decrease in the relative water content of the wild-type plants. The mutant plants had not such a decrease in relative water content, confirming that the G-protein signalling has a role in attenuating the adaptive response to chilling; when absent (mutant genotypes) the plants respond better and maintain plant fitness (Chakraborty *et al.*, 2015).

The importance of G-protein signalling in fine-tuning of the chilling responses is also illustrated by studies on rice. Rice is sensitive to chilling and can be grown only in certain climate zones. Human selection of *japonica* rice has extended its growth zone to regions with lower temperature. The quantitative trait locus *COLD1*, which confers chilling tolerance in *japonica* rice, was identified. Seedlings were exposed to chilling temperature (4°C) and subsequently returned to 30°C. Rice plants with chilling tolerance were defined as those that could re-differentiate new leaves or continue growing leaves when returned to normal conditions after treatment with chilling stress. Overexpression of the *COLD1* gene in *japonica* rice significantly enhances chilling tolerance, whereas rice lines with deficiency in *COLD1* are sensitive to cold. *COLD1* encodes an RGS that localizes on the plasma membrane and endoplasmic reticulum. It interacts with the G-protein α subunit to accelerate G-protein GTPase activity (Ma *et al.*, 2015).

Among G-proteins, putative target proteins are phosphatidylinositol-phospholipase C (PI-PLC) and phospholipase D (PLD), but also calcium channels (Ma *et al.*, 2015); however, the link between the cold-activation of PLDs and PI-PLCs and G-protein signalling in the response to chilling is still to be established. The way chilling stress can activate GPCRs also needs to be established.

In summary, an increase in cytosolic calcium is a major event that occurs very early after a decrease in temperature. Calcium is able to regulate the activity of many signalling components, including phospholipases and protein kinases, ultimately leading to the triggering of cold-induced induction of gene expression or repression. However, many questions remain. The proteins regulated by phospholipase-produced second messengers in the response to cold, and even the identity of some of the messengers themselves, are not yet known. It is very likely that protein kinases or protein phosphatases will be found among them. Furthermore, the relationship between the MAPK module and CDPKs has not been deciphered. Finally, an important aspect of cold signalling is that these signalling events occur at a special time and in a special place, i.e. cell type or organelle. It is the integration of this spatio-temporal web that will lead to the cellular response to chilling.

4.6 Transcriptional Cascades Activated by Chilling Temperatures

4.6.1 Identification of the CBF transcription factors

Cold exposure induces a dramatic change in gene transcription. This is due to the activation of different genetic pathways. The most detailed and documented molecular pathway leading to cold-induced gene expression is the CBF regulon. CBFs (C-repeat-Binding Factors, also named DREB1 for Drought Responsive Element-Binding factor 1) are transcription factors whose DNA-binding domains are of the APETALA 2/Ethylene-Responsive Element-Binding Protein (AP2/EREBP) type. They act through the binding to the cis-acting element rCCGAC (r stands for any purine base), designated the C-repeat (CRT) or Drought Responsive element (DRE). CBFs were first identified in *Arabidopsis*, but CBF homologues have now been found in both monocots and dicots (Ruelland *et al.*, 2009).

In *A. thaliana*, the CBF/DREB1 family is composed of CBF1 to CBF4, DDF1 and DDF2 proteins. The transcripts of *CBF1* to *CBF3* accumulate in response to cold and also, to a lesser extent, in response to ABA. In contrast, when monitored during the first 8 h of stress treatment, *CBF4* is induced by drought and ABA but not by cold; while *DDF1* and *DDF2* are induced by NaCl treatments. Therefore, in *A. thaliana*, the cold-induction of genes containing a promoter-localized CRT motif is explained mainly by the cold-induction of the *CBF1*, *CBF2* and *CBF3* genes. The CBF1–3 factors are induced at the transcriptional level very quickly, 15 min after exposure to cold; accumulation generally peaks at 3 h, whereas after 24 h transcripts are no longer detectable. Microarray analysis of transgenic *A. thaliana* plants ectopically expressing CBFs revealed a constitutive expression of downstream cold-responsive genes, including transcription factors such as RAP2.1, RAP2.7 and ZAT10. These factors will in turn regulate gene expression. The CBF regulon is therefore composed of sub-regulons activated sequentially (Medina *et al.*, 2011).

It does not seem that the three CBFs in *A. thaliana* have necessarily the same target genes. In a study with 54 *A. thaliana* accessions, the expression of *CBF1* and *CBF3* correlated with that of CBF target genes such as *COR6.6*, *COR15A* and *COR47*, while the expression of *CBF2* did not correlate with the expression of these genes. This suggests that the regulon of CBF2 differs from that of the other two CBFs (Zuther *et al.*, 2012). The gene expression was performed in non-acclimated plants, but a basal *CBF* expression exists so this correlation approach is pertinent. The conclusion that CBF2 does not necessarily control the same genes as CBF1 and CBF3 was also reached by a genetic approach invalidating the expression of one or more CBFs (Novillo *et al.*, 2007).

4.6.2 Positive regulation of the CBF expression

CBF3 expression is positively regulated by ICE

Because CBFs are mostly controlled at the transcriptional level, it is important to understand how their expression is controlled in response to cold. In *A. thaliana*, the cold-induction of *CBF3* is controlled by Inducer of Cold

Expression 1 (ICE1). The gene encoding ICE1 is constitutively expressed. It encodes a nuclear MYC basic Helix–Loop–Helix (bHLH) protein. In the *ice1* mutant, chilling and freezing tolerances are impaired. Conversely, overexpression of ICE1 in wild-type plants increases the cold-induced expression of the *CBF* regulon and improves the tolerance of the plants to freezing. However, overexpression alone does not switch on the *CBF3* regulon at ambient temperature, meaning that a cold-induced post-translational modification might be necessary for ICE1 activity as a transcriptional activator (Fig. 4.6; Medina *et al.*, 2011).

ICE1 is regulated by phosphorylation, sumoylation and ubiquitination

It has been suggested that the ability of ICE1 to activate gene transcription in response to cold may be dependent on its phosphorylation. In addition, ICE1 can be SUMOylated by SIZ1, a SUMO E3 ligase that facilitates conjugation of SUMO proteins (small ubiquitin-like modifiers) to protein substrates. It is the sumoylated form of ICE1 that must be active for induction of *CBF3* expression, because sumoylation may either activate and/or stabilize ICE1. *In vitro*, sumoylation of recombinant ICE1 reduces its polyubiquitination. ICE1 can indeed be ubiquinated by HOS1, a

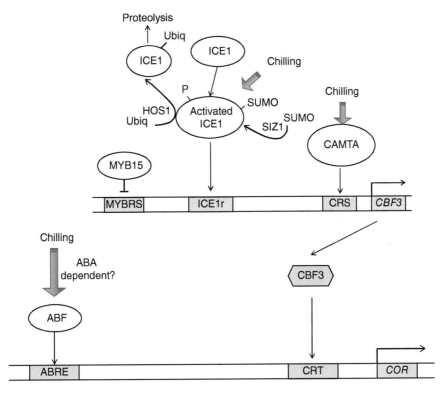

Fig. 4.6. Regulation of *CBF* and *COR* expression. This figure depicts the transcriptional cascade leading to cold-induced gene expression as it can be proposed in *Arabidopsis thaliana*. We use *CBF3* as a CBF model. The cold-induction of *CBF3* is mediated by ICE1 (inducer of CBF expression 1). ICE1 proteins can be subjected to post-translational modifications, such as phosphorylation, ubiquitination or sumoylation. *CBF1-3* induction is repressed by MYB15. *MYB15* is induced by cold and its induction is repressed by the sumoylated form of ICE1. CAMTA positively regulates the cold expression of *CBF3*. CBFs will activate the cold expression of *COR* genes whose promoters contain the CRT *cis*-element. *COR* gene induction can also be activated through ABF proteins that will bind the ABRE *cis*-element. Some ABFs are indeed induced by cold. CRS, CAMTA recognition sequence; ICEr, ICE binding region; MYBRS, MYB recognition sequence; P, phosphoryl group; S, SUMO; Ubiq, ubiquitin. The relative positions of all promoter *cis*-elements are arbitrary.

RING E3 ligase. This ubiquitination directs ICE1 for degradation in response to cold. It therefore appears that cold stress responses in *A. thaliana* are attenuated by a ubiquitination/proteasome pathway in which HOS1 mediates the degradation of the ICE1 protein, and possibly also that of the ICE1-like proteins upstream of *CBF1* and *CBF2* (Fig. 4.6; Ruelland *et al.*, 2009).

The CAMTA proteins act as inducers of CBF expression

In addition, calmodulin-binding transcriptional activators CAMTA1, CAMTA2 and CAMTA3 have been shown to function together to positively regulate the cold-induction of *CBF2, CBF1*. CAMTA could directly link the cold-induced increase in cytosolic calcium and gene expression. CAMTAs bind vCGCGb or vCGTGb DNA regulatory motifs (where v stands for A, C or G; and b stands for C, G or T). Using a camta1/2/3 triple mutant, it was evaluated that CAMTA proteins contribute to circa 15% of the genes that are cold-induced at 24 h of exposure at 4°C. Most of the genes that are cold-regulated via CAMTAs do not belong to a CBF regulon (i.e. are downstream of a CBF). CAMTAs may act not only on CBFs, but also on other transcription factors (Kim *et al.*, 2013).

Other positive regulators

Other factors acting as inducers of CBF expression have been identified. ICE2, a MYC-like bHLH factor with high homology with ICE1, was shown to activate *CBF1* expression. Constitutive *ICE2* expression led to induced freezing-tolerance (Kurbidaeva *et al.*, 2014).

4.6.3 Negative regulation of the CBF expression

Molecular analysis of a *cbf2* null mutant of *A. thaliana* suggests that CBF2 is a negative regulator of *CBF1* and *CBF3* gene expression during the cold response. In addition, the *CBF1-3* genes are negatively regulated by MYB15 by a direct interaction with the *CBF* promoter sequences (Fig. 4.6). *MYB15* is induced by cold, with a peak of transcripts reached 12 h after the beginning of the cold treatment. Expression of *MYB15*

could be regulated by ICE1, since sumoylation of ICE1 during cold-acclimation leads to a reduction in *MYB15* expression. This raises the possibility that SIZ1-mediated sumoylation of ICE1 might be required for the induction of *CBF3* and repression of *MYB15*, allowing a fine-tuning of CBF expression. In *Arabidopsis*, ZAT12, a C2H2 zinc finger transcription factor, also appears to function as a negative regulator of *CBFs*. It is induced by cold, concomitantly with *CBFs*. Transgenic overexpression of *ZAT12* decreases the expression of *CBFs* in response to a cold stress (Medina *et al.*, 2011).

4.6.4 The CBF regulon is not the only chilling-responsive regulon

Identification of 'first-wave' chilling-induced transcription factors others than CBFs

Twenty-seven genes that encoded transcription factors were induced in parallel with *CBF1*, *CBF2* and *CBF3* in *A. thaliana* submitted to chilling. They represented 'first-wave' transcription factors, whose chilling induction could be detected as soon as 30 min after chilling exposure. The overexpression of some of these factors, namely *HSFC1, ZAT12, ZAT10, ZF* or *CZF1*, at control temperature, leads to the expression of identified cold-responsive genes, a strong argument for these transcription factors having a functional importance in the chilling response. Besides, many cold-responsive genes could be assigned to the regulon of more than one of these 'first-wave' transcription factors. The 'first-wave' transcription factors also appear to be dependent on one another: ZF is a member of the CBF2 regulon, ZAT12 is a member of the HSFC1 regulon and ZAT10 is a member of the CZF1 regulon. These factors thus form a complex chilling regulatory network. It can even be questioned whether CBF factors have a central role. In an elegant experiment, the CRT *cis*-elements of promoters are occupied by a truncated version of CBF2 unable to trans-activate gene expression; the response through the CBFs is thus inhibiting. It was shown that the vast majority of cold-responsive genes assigned to the CBF2 regulon were inhibited by less than 50% when the CBF-dependent response was inhibited (Park *et al.*, 2015).

4.6.5 CBFs and cold-hardening

In the plant species able to achieve it, one of the major processes induced by chilling temperatures is cold-hardening, i.e. the induction of freezing-tolerance. Since different genetic pathways are induced by chilling, it is interesting to investigate which are necessary for the acquisition of freezing-tolerance. Several studies demonstrate the importance of the CBF regulon in the cold-hardening process. The successful downregulation of the cold-induction of both *CBF1* and *CBF3* led to a 60% reduction in the acquisition of freezing-tolerance by chilling treatment (Novillo *et al.*, 2007). Conversely, the constitutive overexpression of CBF1 or CBF3 in *Arabidopsis* plants induces an increased tolerance to freezing. The fact that overexpressing CBF leads to a constitutive freezing-tolerance has been reported in *Thlaspi arvense, Oryza sativa, Lolium perenne* and *Brassica napus* (Ruelland *et al.*, 2009; Medina *et al.*, 2011).

In *Triticum monococcum*, a cluster of 11 CBF genes has been mapped to the Frost resistance-2 locus on chromosome 5. Interestingly, the transcript levels of TmCBF16, TmCBF12 and TmCBF15 were already upregulated by a treatment at 15°C in the frost-tolerant 'G3116' cultivar, but not in the frost-sensitive 'DV92' cultivar, where a temperature of 10°C is necessary for induction of these genes. The higher threshold temperature for induction of these CBF genes in the 'G3116' cultivar could result in an earlier initiation of the cold-acclimation process and therefore in a better resistance to subsequent freezing temperatures (Knox *et al.*, 2008).

However, the expression of *CBF* genes is not necessarily what distinguishes natural variants with distinct abilities to acquire freezing-tolerance. In a study with 54 *A. thaliana* accessions, the expressions of *CBF1, CBF2* or *CBF3* (after 14 days of acclimation at 4°C) could not be correlated with freezing-tolerance, even though the expression of *CBF* target genes such as *COR6.6, COR15A, COR78* and *GolS3* was closely correlated with freezing-tolerance acquisition after cold-acclimation (Zuther *et al.*, 2012). The *CBF* genes being early responsive genes, it might not be so surprising not to detect such a correlation. After 14 days at 4°C, CBFs do not necessarily act

in the maintenance of cold-responsive gene expression, and other regulatory mechanisms are likely to be at play. Yet no clear correlation could be found between freezing-tolerance after acclimation and the expression maximum during the first 3–5 h of the *CBF* genes after transfer to low temperature (McKhann *et al.*, 2008). Even though we cannot rule out that a correlation would have been found using *CBF* expression at chilling temperature in an intermediate period (between 5 h and 14 days, a big gap), it seems that the absolute level of expression of *CBFs* is not a good criterion to determine the ability of an *A. thaliana* cultivar to develop freezing-tolerance. Similarly, it appears that the six CBFs of *Populus balsamifera* accounted for only a small amount of the variation in freezing resistance across latitude and the growing season (Menon *et al.*, 2015).

This does not mean that CBFs are not important for establishing freezing-tolerance. We can consider two *A. thaliana* ecotypes, one collected from Sweden (SW) and the other from Italy (IT). The SW ecotype is more tolerant to freezing than the IT ecotype. The genetic difference between the two ecotypes responsible for the difference in freezing-tolerance was shown to map to a region that includes the CBF locus. The cold-induction of most CBF regulon genes is lower in IT plants than in SW plants, and this is due to the IT *CBF2* gene encoding a non-functional CBF2 protein (Gehan *et al.*, 2015). It is interesting to note that the lower freezing-tolerance of the 'Cape Verde Island' accession has been associated with a deletion in the promoter of CBF2, leading to low gene expression (Alonso-Blanco *et al.*, 2005). This could indicate the importance of CBF2 in natural variation of freezing-tolerance.

In the experiment in which the CBF-dependent responses were inhibited by the overexpression of a truncated version of CBF2, this inhibition reduced the acquisition of freezing-tolerance upon chilling treatment. Yet an increase in freezing-tolerance upon chilling is still observed, confirming that pathways other than CBFs participate in cold-acclimation (Park *et al.*, 2015).

ZAT6 is a zinc finger protein that is cold-induced. It is not regulated through the CBF pathway. A strong correlation was found between the expression of ZAT6 and the acquired freezing resistance (Zuther et al., 2012).

4.6.6 MicroRNAs and the plant response to chilling

MicroRNAs (miRNAs) are a class of small, non-coding RNAs (sRNAs) that are 20–24 nucleotides in length. The RNAs participate in regulating gene expression at the post-transcriptional level. miRNAs, as part of a 'silencing complex', target protein-coding mRNAs; the complex cleaves the targeted messenger RNAs at specific positions or repress its translation. Chilling-responsive miRNAs have been detected in several plant species, such as A. thaliana, Brachypodium, Prunus persica, Populus tomentosa, Oryza sativa and Zea mays. In Solanum habrochaites, a chilling-resistant wild tomato, 192 miRNAs have been shown to increase in the response to chilling stress, while 205 decreased. The target genes of the miRNAs can be predicted. Target gene functional analysis showed that most target genes played positive roles in the chilling response, primarily by regulating the expression of anti-stress proteins, antioxidant enzyme and genes involved in cell wall formation (Cao et al., 2014).

In conclusion, cold exposure will trigger dramatic changes in transcript levels. The best-documented genetic pathway leading to gene induction in response to cold is the CBF pathway (Fig. 4.6). This pathway has been identified in both chilling-resistant and chilling-sensitive plants. In cold-resistant plants and those able to cold acclimate, the CBF proteins will activate the transcription of genes encoding proteins with major roles in cold tolerance and freezing resistance. The CBF pathway is not thought to be the only one triggered by cold exposure. Nevertheless, descriptions of CBF-independent pathways are scarce. Finally, in field or natural conditions, it is likely that transcriptomic changes are more complex since experimental conditions used to decipher the response to cold simplify the number of variables in the environment. In a last section, we will see that the response to chilling may be influenced by other environmental or internal parameters. More specifically, hormones and light influence the response to chilling. This response appears to be circadian-gated.

4.7 Crosstalk With Hormones and Light

4.7.1 Hormones

Abscisic acid (ABA)

ABA is commonly considered a 'stress hormone'. It does accumulate under cold exposure, possibly via the transcriptional induction of NCED3, a key gene in ABA biosynthesis. The levels of ABA reached after cold exposure are, nevertheless, lower that those measured under drought. However, the increase in ABA is necessary for the full response to cold to occur. In the A. thaliana frs1 mutant, deficient in ABA biosynthesis, several genes such as RAB18 or RCI2A are less cold-induced than in the wild-type plants. In the adc1 and adc2 mutants, also characterized by a lower cold-induced ABA accumulation, the expression of RD22 and RD29B in response to cold is altered. The frs1, adc1 and adc2 mutants are all characterized by lower freezing-tolerance before and after cold-acclimation (Cuevas et al., 2008). This strongly suggests that ABA might be necessary for plants to acquire full freezing-tolerance after an exposure to chilling. The effect of ABA in response to chilling might be dependent on ABA-driven gene expression. The cold-accumulated ABA might trigger the ABRE pathway already described. But ABA could also participate in the activation of the CBF regulon: ABA induces CBF4 expression, and CBF4 in turn may trans-activate CRT-containing promoters. However, the CBF pathway is considered to be mainly ABA-independent, because CBF1-3 are poorly induced by ABA compared to cold, and the induction by cold of CBF target genes is not compromised in mutants deficient in ABA. It is likely that a pathway involving the Abscisic Acid Responsive Element, PyACGTGGC (ABRE) cis-element plays a role in the response to chilling. The promoter sequences of many cold-responsive genes contain the ABRE cis-element. These elements can be recognized by basic leucine zipper (bZIP) transcription factors. In A. thaliana, ABRE-binding factors 1 and 4 (ABF1 and 4) are induced by cold and could thus participate in the cold-activated transcriptional cascades. In soybean, SGBF-1 is such a bZIP transcription factor binding to ABRE. Its binding and transcriptional activities are enhanced when it interacts with

SCOF-1, a nuclear zinc finger protein. Chilling induces expression of SCOF-1. Constitutive over-expression of SCOF-1 induced the expression of known cold-responsive genes and enhanced cold tolerance of non-acclimated transgenic *A. thaliana* and tobacco plants. These results suggest that the role of SCOF-1 as a positive regulator of gene expression in response to chilling is mediated by ABRE via a protein interaction with SGBF-1 (Ruelland *et al.*, 2009).

Salicylic acid

As already mentioned, overwintering plants such as *A. thaliana* display a stunted stature when grown at chilling temperature. Interestingly, such a phenotype seems to be controlled by salicylic acid (SA). SA and its glucoside derivative accumulate at 5°C in this plant. This chilling-induced SA biosynthesis proceeds through the isochorismate synthase (ICS) pathway, with cold-induction of ICS1 (which encodes ICS), and two genes encoding transcription factors – CBP60g and SARD1 – that positively regulate ICS1 paralleling SA accumulation (Kim *et al.*, 2013). Transgenic and mutant plants unable to accumulate SA had a biomass 2.7-fold higher than that of wild-type plants after 2 months at 5°C (Scott *et al.*, 2004). A metabolic study using wild-type plants and plants mutated in their SA accumulation could be evidence that the biomass of the plants grown at 5°C was positively correlated with fumarate but negatively correlated with malate and glutamine. The roles of these metabolites in relation to growth and their control by SA still need to be understood. Malate and glutamine are intermediates of the central metabolism; their level can be relatively depleted in conditions and/or genotypes associated with higher growth. Fumarate can be produced from malate by a cytosolic fumarase. The expression of *A. thaliana FUM2* gene correlates with fumarate level at 5°C in the plants assayed (Scott *et al.*, 2014, p. 201).

Another study reinforces the importance of SA in relation to the response to chilling. Kim *et al.* (2013) compared the changes in gene expression that occurred in wild-type plants and the sid2–1 mutant that does not synthesize SA in chilling conditions. Of the genes that were induced by cold in wild-type plants, one-quarter (27%) showed reduced cold-induction in *sid2–1*

plants. This group was highly enriched in genes associated with SA signalling, including 'defence response', 'innate immune response' and 'response to salicylic acid stimulus'. The promoters were not enriched in the CBF-binding site, suggesting that they are not part of the CBF pathway. While SA appears to contribute significantly to chilling-triggered transcriptomic changes, SA biosynthesis at low temperature does not contribute to freezing-tolerance (Kim *et al.*, 2013).

Finally, SA is known to induce enzymes of the ROS-scavenging systems, such as glutathione reductase and peroxidase. Thus the increase in SA level might help the plant to cope with the chilling-induced accumulation of ROS. Exposing seedling radicles to 0.5 mM SA 24 h before chilling at 2.5°C for 1–4 days reduced the chilling-induced increase in electrolyte leakage from maize and rice leaves and cucumber hypocotyls (Ruelland *et al.*, 2009), highlighting the protective role of SA in the response to chilling stress.

Auxins and cytokinins

Plants exhibit reduced root growth when exposed to low temperature. This is due to a reduction in meristem size and in cell number, certainly through the repression of the division potential of meristematic cells. Cold leads to a reduction in auxin accumulation in roots, and this correlates with the repressed expression of *PIN1/3/7* and auxin biosynthesis-related genes. In roots, auxin and cytokinin signalling are intermingled. Roots of *arr1-3 arr12-1* seedlings were less sensitive than wild-type roots to low temperature, in terms of changes in root length and meristem cell number. This correlated with a lesser reduction in *arr1-3 arr12-1* roots than in wild-type roots of the levels of *PIN1/3* transcripts and of the auxin level. These data suggest that low temperature inhibits root growth by reducing auxin accumulation via the *ARABIDOPSIS RESPONSE REGULATOR ARR1/12* (Zhu *et al.*, 2015).

Gibberellins (GA)

Another hint that the reduction of growth rate of plants submitted to cold is a controlled, regulated process comes from the fact that a dwarf phenotype can be obtained at normal temperatures by overexpressing CBF1 in *A. thaliana*.

The CBF1 overexpressing plants are also characterized by the accumulation of DELLA proteins, a small family of growth-restraining proteins that are part of the GA signalling pathway. In plants containing physiological levels of biologically active GAs, DELLA proteins are degraded, but when GA levels are low, DELLA proteins accumulate, restrain growth and cause a dwarf phenotype. The CBF1 overexpressing plants show reduced levels of biologically active GAs, due to an increased expression of two genes encoding enzymes converting biologically active GAs into inactive forms. This results in an increase in DELLA proteins, which in turn cause a dwarf phenotype. These data indicate that the cold-induced GA decrease is a downstream effect of CBF induction, which leads to growth inhibition via the DELLA pathway (Thomashow, 2010).

Jasmonate

The production of endogenous jasmonate was triggered by cold treatment. Cold-induction of genes acting in the CBF/DREB1 signalling pathway was upregulated by jasmonate. Several jasmonate ZIM-domain (JAZ) proteins, the repressors of jasmonate signalling, physically interact with ICE1 and ICE2 transcription factors. JAZ1 and JAZ4 repress the transcriptional function of ICE1, thereby attenuating the expression of its regulon. Consistent with this, overexpression of JAZ1 or JAZ4 represses freezing stress responses of *A. thaliana*. Taken together, there is evidence that jasmonate functions as a critical upstream signal of the ICE-CBF pathway (Hu *et al.*, 2013).

Ethylene

Ethylene accumulates in bean (*Phaseolus vulgaris*), winter rape (*Brassica napus*) and tomato (*Solanum lycopersicum*) exposed to chilling temperatures. In *A. thaliana*, when plants are exposed at 4°C, a rapid increase in ethylene is detected as soon as after 3 h of cold exposure; it is no longer detected after 1 day. This increase is dependent on ACC SYNTHASES. The cold-induced accumulation of ethylene in winter rye (*Secale cereale*) promotes the synthesis of antifreezing proteins (references in Catalá *et al.*, 2014). The *A. thaliana* ethylene overproducer mutant *eto1-3* has enhanced freezing-tolerance

(Catalá and Salinas, 2015). RCI1A, a 14-3-3 protein, is involved in the regulation of ethylene level in *A. thaliana*. RCI1A protein indeed interacts physically with different ACC SYNTHASE (ACS) isoforms (involved in the biosynthesis of ethylene) leading to decreased ACS stability. RCI1A thus restrains ethylene level, and the *rci1a* mutant is characterized by a constitutive high ethylene level, both in control and in low-temperature conditions. RCI1A is induced by cold through a CBF-independent pathway. It is therefore considered to participate in establishing adequate levels of ethylene in *Arabidopsis* under both standard and low-temperature conditions. These levels are required to promote proper cold-induced gene expression and freezing-tolerance before and after cold-acclimation (Catalá and Salinas, 2015).

The role of ethylene in the response to chilling is likely to be dependent on cultivation conditions. Indeed, while the *A. thaliana* ethylene overproducer mutant *eto1-3* was shown to have enhanced freezing-tolerance by Catalá and colleagues, Shi and colleagues (2012) found that this mutant displayed reduced tolerance to freezing temperatures. Catalá and colleagues work on plants grown in soil while Shi and colleagues work on plants grown *in vitro*. Such dependency on growing conditions may be true for all hormones.

4.7.2 Light

The acquisition of cold hardiness by herbaceous species requires light. In *A. thaliana*, a treatment at 3°C under light (100 µmol.photons.m^{-2}.s^{-1}) upregulated twice as many genes as a treatment at 3°C in the dark. Light could influence cold-hardening through chloroplast functions. In barley (*Hordeum vulgare*), the *albina* and *xantha* mutants are blocked in different steps of chloroplast development. Only about 11% of the genes regulated by cold in the wild type are regulated to a similar extent in the mutants and could be assigned as chloroplast-independent cold-regulated genes. About 67% of cold-regulated genes in the wild type were not regulated by cold in any of the mutants and were considered as chloroplast-dependent cold-regulated genes (Svensson *et al.*, 2006; Soitamo *et al.*, 2008). These chloroplast-dependent signalling pathways could be related to photosystem II excitation pressure, as already discussed, and/or could be related to ROS production in chloroplasts in conditions of low temperature and light.

Light quality has also been shown to play a role in mediating the cold response. No COR14b accumulation was found after 7 days of treatment when etiolated barley (*H. vulgare*) plants were cold hardened in the dark. However, if – prior to the cold treatment – the etiolated plants were subjected to red or blue light pulses of 5 min, COR14b accumulated during the subsequent cold treatment in the dark. Neither far-red nor green light pulses promoted COR14b accumulation. Using several copies of a cold-responsive element (CRT *cis*-element) fused to *GUS* as a reporter construct, phytochromes were subsequently shown to mediate the light signalling of cold-induction via the CRT *cis*-element (Kim *et al.*, 2002).

4.7.3 The circadian gate

Light being involved in the response to cold, the question naturally arises whether the low-temperature input into the CBF regulatory hub might be influenced by the clock.

At a warm temperature *CBF* transcript levels are low but oscillate. The peak of expression occurs 8 h after dawn both in *A. thaliana* plants grown in short-day conditions (8 h light/16 h dark) and in plants grown in long-day conditions (16 h light/8 h dark). Yet the *CBF* transcript levels at 8 h after dawn were three- to fivefold higher in short-day plants than in long-day plants. This corresponds to a downregulation of the CBF pathway under long-day conditions (Lee and Thomashow, 2012). Moreover, the cold-induction of *CBF1*, *CBF2* and *CBF3* is 'gated' by the clock. If *A. thaliana* plants that had been grown at 24°C for 14 days on a 12-h photoperiod (12 h light/12 h dark) were transferred to low temperature, the increase in *CBF1*, *CBF2* and *CBF3* transcript levels is much greater in plants exposed at low temperature 4 h after dawn, than in plants exposed to low temperature 16 h after dawn. Therefore there are three aspects of the regulation of the *CBF* expression by the circadian clock: (i) oscillation of the basal *CBF* expression, at warm temperatures; (ii) inhibition of CBF by long-day conditions; and (iii) regulation of the cold expression of the *CBFs*.

The molecular components involved in the circadian gating are starting to be unravelled. One of them, the basic Helix–Loop–Helix protein PIF7, is a transcriptional repressor. It functions redundantly with PIF4: while neither *pif4* nor *pif7* single mutants were affected in the *CBF2* expression under long- or short-day conditions, in the *pif4pif7* double mutant the repression of the circadian regulation of *CBF* in long-day conditions no longer occurred. PIF4 and PIF7 bind to G-box motifs found in the *CBF1* and *CBF2* promoter and to the E-box in the *CBF3* promoter (Lee and Thomashow, 2012). This regulation involves interactions of PIF7 with TOC1, a component of the central circadian oscillator; and with PHYB, a red light photoreceptor (Thomashow, 2010). The circadian regulation also involves direct positive action of two transcription factors that are core components of the clock: CIRCADIAN CLOCK-ASSOCIATED 1 (CCA1) and LATE ELONGATED HYPOCOTYL (LHY). In plants carrying the *cca1-11/lhy-21* double mutation, the cold-induction of *CBF1*, *CBF2* and *CBF3* was greatly impaired, and circadian regulation of the *CBF* genes was significantly reduced or even eliminated.

Finally, Pseudo Response Regulators (PRRs), key components of the core feedback loops that constitute the circadian clock, are also important for the regulation of the CBFs. When plants were grown at basal temperature, the transcript levels of *CBF1*, *CBF2* and *CBF3* remained high throughout the day in plants carrying the *prr9-11/prr7-10/prr5-10* triple mutation (Nakamichi *et al.*, 2009). Similarly, the cold-induction of *CBF1*, *CBF2* and *CBF3* in the triple-mutant plants stand at about the peak levels observed in wild-type plants regardless of the time of day at which the mutant plants were exposed to low temperature. Interestingly, in a study with 54 *A. thaliana* accessions, the expression levels of all tested *CBF* and *COR* genes in non-acclimated plants closely correlated with the expression of *Pseudo Response Regulator 5* (PRR5). After acclimation (14 days at 4°C), the expression of *PRR5* was only correlated with that of *CBF*, and not with *COR* gene expression. This illustrates the central importance of PRR (and more precisely PRR5) in the expression of CBFs (Zuther *et al.*, 2012).

In addition, it should be noted that the spike in cytosolic calcium produced in response to low temperature is also gated by the clock. In guard cells, the cold-induced $[Ca^{2+}]_{cyt}$ transients were significantly higher during the mid-photoperiod than at the beginning or end of the subjective day. The cold-induced $[Ca^{2+}]_{cyt}$ was lowest during the subjective night (Dodd *et al.*, 2006).

4.8 Conclusion

Even though many of the events involved in the response of plants to chilling have been or are being unravelled, several key questions remain. The first is the basis for sensitivity. Indeed, some plant species are resistant, while others are sensitive. Why is that so? From the studies published so far, one explanation for sensitivity is transcriptional regulation of key genes. This has been suggested in wheat, where the temperature threshold for induction of the *Frost-resistant 2* locus is too low in the sensitive cultivar to participate in frost tolerance. In *A. thaliana*, the presence of WRKY34 in pollen grains contributes to the chilling sensitivity of pollen as measured by viability and germination rate. Interestingly, WRKY34 is a transcription factor that appears to be a negative regulator of the CBF pathway (Zou *et al.*, 2010).

It has to be kept in mind that plants do not possess a temperature receptor. They are able to respond because they sense chilling and transform this perception into biochemical signals. The sensor or, more appropriately, *sensors*, are the very cellular perturbations and subsequent disequilibria that are caused by chilling. Variations in the perception steps, i.e. membrane remodelling, protein conformation changes and disassembly of the cytoskeleton – as well as metabolic reactions related to photosynthesis – could also differ between sensitive and resistant plants. However, no study has focused or determined if particular alleles associated with sensitivity could correspond, for example, to modifications in protein conformation changes leading to defects in signal transduction. The same is true for metabolic reactions related to photosynthesis. New studies focusing on sensitivity-related alleles are therefore needed.

Downstream from the sensing steps is the triggering of the signalling pathway, whose different levels are summarized in Fig. 4.7. Calcium entry clearly appears to be a master switch of the signalling pathways. Many signals are involved, many of them not specific to the response to chilling. The relationships between these different elements, as well as the basis for the specificity of the downstream response to chilling, largely remain to be deciphered. In molecular terms, the CBF transcriptional pathway is the best-documented genetic pathway leading to transcript rearrangement; but other pathways that are independent from the CBF also exist, and researchers are beginning to identify them. Barah *et al.*, (2013) studied the gene response of plants from ten *A. thaliana* ecotypes. Using the expression of plants submitted 3 h at 10°C, they could construct an *in silico* transcriptional regulatory network. The network contained 1275 nodes with 7720 connections, which included 178 transcriptions factors and 1331 target genes. This illustrates the complexity of the gene response to early chilling, and that much lies beyond the CBF molecular pathway (Barah *et al.*, 2013). Attention should be paid to the first-wave transcription factors that are induced in parallel to CBF. Many of these are zinc finger transcription factors such as ZAT10 and ZAT12. It can be noted that in a study with many *A. thaliana* accessions, *ZAT12* expression strongly correlated with leaf sugar content, specifically in chilling conditions. However, under chilling conditions, the leaf sugar content correlated even more strongly with *ZAT6* expression. *ZAT6* expression correlated with acquired freezing-tolerance, while the expression of the *CBF* genes did not (Zuther *et al.*, 2012). The importance of the transcription factors and their place in the regulatory network needs further documentation. In particular, what pathways are independent, which ones are related and what regulations exist between them? Indeed, as exemplified with the first-wave transcription factors, the chilling response network appears very integrated.

Another aspect of the chilling response that has been largely obliterated so far is the so-called *spatio-temporal web* leading to chilling resistance. If the correct succession of activations over time is clearly necessary, including at the threshold level of gene induction, one question that has largely been overlooked is whether some responses are organ-specific; and within one organ, whether all tissues, and therefore all cells, respond in a similar way. If not, signals for adaptation could be transmitted from an emitting cell to a receiving one, within a leaf blade for instance,

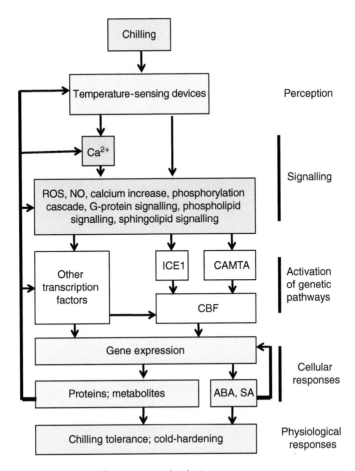

Fig. 4.7. General scheme of the chilling response in plants.

or between roots and shoots (or vice versa). For instance, the chilling of roots has an impact on the proteome of shoot tissues in rice seedlings (Neilson *et al.*, 2013). If some aspects of the response to chilling appear and are modelled as cell-autonomous, it may be necessary to envisage the existence of a systemic signal in the response to chilling at the whole-plant level.

An interesting point that is starting to be investigated is the crosstalk between the response to cold and to pathogens. Perennial or overwintering annual plants can be the targets of disease during the winter season. In particular, so-called snow mould diseases can appear. They are caused by psychrophilic fungi known as snow moulds. The major snow mould diseases for winter cereals include pink snow mould (*Microdochium nivale*), *Sclerotinia borealis*, grey snow mould (*Typhula incarnata*) and speckled snow

mould (*Typhula ishikariensis*). These fungi favour near-freezing temperatures, and dark and humid conditions for growth and plant infection. Interestingly, the chilling exposure of spring barley (*Hordeum vulgare*), meadow fescue (*Festuca pratensis*) and oilseed winter rape (*Brassica napus* var. *oleifera*) induces resistance to their specific pathogens (Plazek and Zur, 2003). It is true that some cold-induced PR proteins display both antifungal and antifreeze activities, suggesting a dual function in protecting plants from overwintering stresses (Kuwabara and Imai, 2009). New data have given hints on the cold and disease resistance crosstalk. For instance, cold stimulates the proteolytic activation of a plasma membrane-tethered NAC transcription factor, NTL6. The transcriptionally active NTL6 protein enters the nucleus, where it induces a subset of PR genes such as *PR1*, *PR2* and *PR5*. The cold-induction of

PR1 disappeared in the RNAi plants with reduced NTL6 activity. The NTL6-mediated cold-induction of the *PR* genes is independent of SA (Seo *et al.*, 2010). In triticale, the cultivars Hewo and Magnat differ in their ability to develop resistance to *M. nivale* after chilling treatment; Hewo is able to develop resistance after cold treatment while Magnat is susceptible to infection despite hardening. In cv. Hewo, cold treatment resulted in superhydrophobicity of the leaf epidermis, preventing both adhesion of *M. nivale* hyphae to the leaves and direct penetration of the epidermis, while the altered chemical composition of the cell wall protected the plants against tissue digestion by the fungus (Szechyńska-Hebda *et al.*, 2013). Finally, another aspect of cold and disease resistance crosstalk is the fact that chilling temperatures increase SA level.

An important point is that the response to chilling stress might differ between species or even accession within one species. For instance, while sucrose levels were correlated with freezing-tolerance in *A. thaliana* and *Thellungiella* sp. accessions, the raffinose level only correlated with freezing-tolerance in *A. thaliana* (Lee *et al.*, 2012). Conversely, proline levels correlated with freezing-tolerance of *Thellungiella* sp. accessions, but not in *A. thaliana* (Lee *et al.*, 2012). Similarly, a negative correlation was detected with the levels of free spermine in *Thellungiella* sp. accessions, but not in *A. thaliana* (Lee *et al.*, 2012). To illustrate that within one species the strategy for the response to chilling can differ, is the study with two chilling-resistant rice (*Oryza sativa*) varieties, Sijung and Jumli Marchi, for which transcriptomic responses are not overlapping. In the first one the initial molecular responses seem to be mainly targeted at strengthening the cell wall and plasma membrane; whereas, for the second, the protection of chloroplast translation and detoxification are the main responses (Lindlöf *et al.*, 2015).

Limited understanding of all the molecular events involved in the plant response to chilling does not prevent the search for alleles or loci improving tolerance in crops. Quantitative trait loci (QTL) analysis has been conducted in rice, soybean, tomato, wheat and maize, among others, to understand the basis of and identify molecular markers for chilling tolerance, mostly in vegetative but also in reproductive tissues. This approach may lead to the identification of new loci or relationships playing a major role in chilling resistance. The search in crop species for genes homologous to that identified in model plants such as *A. thaliana* may also allow the efforts of many laboratories and breeders to better decipher the response of various plant species to chilling, and help in reducing the consequences of this major environmental stress on plants.

Acknowledgements

I thank Dr A. Zachowski and Dr J. Rochet for careful reading of the manuscript.

References

Alonso-Blanco, C., Gomez-Mena, C., Llorente, F., Koornneef, M., Salinas, J. and Martínez-Zapater, J.M. (2005) Genetic and molecular analyses of natural variation indicate *CBF2* as a candidate gene for underlying a freezing tolerance quantitative trait locus in Arabidopsis. *Plant Physiology* 139, 1304–1312.

Baldwin, L., Domon, J.-M., Klimek, J.F., Fournet, F., Sellier, H., Gillet, F., Pelloux, J., Lejeune-Hénaut, I., Carpita, N.C. and Rayon, C. (2014) Structural alteration of cell wall pectins accompanies pea development in response to cold. *Phytochemistry* 104, 37–47.

Barah, P., Jayavelu, N., Rasmussen, S., Nielsen, H., Mundy, J. and Bones, A.M. (2013) Genome-scale cold stress response regulatory networks in ten *Arabidopsis thaliana* ecotypes. *BMC Genomics* 14, 722.

Barton, D.A., Cantrill, L.C., Law, A.M.K., Phillips, C.G., Sutton, B.G. and Overall, R.L. (2014) Chilling to zero degrees disrupts pollen formation but not meiotic microtubule arrays in *Triticum aestivum* L. *Plant, Cell & Environment* 37, 2781–2794.

Bies-Ethève, N., Gaubier-Comella, P., Debures, A., Lasserre, E., Jobet, E., Raynal, M., Cooke, R. and Delseny, M. (2008) Inventory, evolution and expression profiling diversity of the LEA (late embryogenesis abundant) protein gene family in *Arabidopsis thaliana*. *Plant Molecular Biology* 67, 107–124.

Boyer, J.S. (1982) Plant productivity and environment. *Science* 218, 443–448.

Cai, C., Xu, C., Shan, L., Li, X., Zhou, C., Zhang, W., Ferguson, I. and Chen, K. (2006) Low temperature conditioning reduces postharvest chilling injury in loquat fruit. *Postharvest Biology and Technology* 41, 252–259.

Cantrel, C., Vazquez, T., Puyaubert, J., Rezé, N., Lesch, M., Kaiser, W.M., Dutilleul, C., Guillas, I., Zachowski, A. and Baudouin, E. (2011) Nitric oxide participates in cold-responsive phosphosphingolipid formation and gene expression in *Arabidopsis thaliana*. *New Phytologist* 189, 415–427.

Cao, X., Wu, Z., Jiang, F., Zhou, R. and Yang, Z. (2014) Identification of chilling stress-responsive tomato microRNAs and their target genes by high-throughput sequencing and degradome analysis. *BMC Genomics* 15, 1130.

Catalá, R. and Salinas, J. (2015) The Arabidopsis ethylene overproducer mutant eto1-3 displays enhanced freezing tolerance. *Plant Signaling & Behavior* 10, e989768.

Catalá, R., López-Cobollo, R., Mar Castellano, M., Angosto, T., Alonso, J.M., Ecker, J.R. and Salinas, J. (2014) The Arabidopsis 14-3-3 protein RARE COLD INDUCIBLE 1A links low-temperature response and ethylene biosynthesis to regulate freezing tolerance and cold acclimation. *The Plant Cell* 26, 3326–3342.

Chakraborty, N., Singh, N., Kaur, K. and Raghuram, N. (2015) G-protein signaling components gcr1 and gpa1 mediate responses to multiple abiotic stresses in *Arabidopsis*. *Frontiers in Plant Science* 6, 1000.

Chen, M. and Thelen, J.J. (2013) ACYL-LIPID DESATURASE2 is required for chilling and freezing tolerance in Arabidopsis. *The Plant Cell* 25, 1430–1444.

Chen, M., Markham, J.E. and Cahoon, E.B. (2012) Sphingolipid Δ8 unsaturation is important for glucosylceramide biosynthesis and low-temperature performance in Arabidopsis. *The Plant Journal* 69, 769–781.

Clarke, M.W., Boddington, K.F., Warnica, J.M., Atkinson, J., McKenna, S., Madge, J., Barker, C.H. and Graether, S.P. (2015) Structural and functional insights into the cryoprotection of membranes by the intrinsically disordered dehydrins. *Journal of Biological Chemistry* 290, 26900–26913.

Cook, D., Fowler, S., Fiehn, O. and Thomashow, M.F. (2004) A prominent role for the CBF cold response pathway in configuring the low-temperature metabolome of *Arabidopsis*. *Proceedings of the National Academy of Sciences USA* 101, 15243–15248.

Crisosto, C., Mitchell, F. and Ju, Z. (1999) Susceptibility to chilling injury of peach, nectarine, and plum cultivars grown in California. *Hortscience* 34, 1116–1118.

Cuevas, J.C., López-Cobollo, R., Alcázar, R., Zarza, X., Koncz, C., Altabella, T., Salinas, J., Tiburcio, A.F. and Ferrando, A. (2008) Putrescine is involved in Arabidopsis freezing tolerance and cold acclimation by regulating abscisic acid levels in response to low temperature. *Plant Physiology* 148, 1094–1105.

Degenkolbe, T., Giavalisco, P., Zuther, E., Seiwert, B., Hincha, D.K. and Willmitzer, L. (2012) Differential remodeling of the lipidome during cold acclimation in natural accessions of *Arabidopsis thaliana*. *The Plant Journal* 72, 972–982.

Distelbarth, H., Nägele, T. and Heyer, A.G. (2013) Responses of antioxidant enzymes to cold and high light are not correlated to freezing tolerance in natural accessions of *Arabidopsis thaliana*. *Plant Biology (Stuttgart, Germany)* 15, 982–990.

Dodd, A.N., Jakobsen, M.K., Baker, A.J., Telzerow, A., Hou, S.-W., Laplaze, L., Barrot, L., Poethig, R.S., Haseloff, J. and Webb, A.A.R. (2006) Time of day modulates low-temperature Ca^{2+} signals in Arabidopsis. *The Plant Journal* 48, 962–973.

Domonkos, I., Malec, P., Sallai, A., Kovacs, L., Itoh, K., Shen, G., Ughy, B., Bogos, B., Sakurai, I., Kis, M., Strzalka, K., Wada, H., Itoh, S., Farkas, T. and Gombos Z. (2004) Phosphatidylglycerol is essential for oligomerization of photosystem I reaction center. *Plant Physiology* 134, 1471–1478.

Dutilleul, C., Benhassaine-Kesri, G., Demandre, C., Rézé, N., Launay, A., Pelletier, S., Renou, J.-P., Zachowski, A., Baudouin, E. and Guillas, I. (2012) Phytosphingosine-phosphate is a signal for AtMPK6 activation and Arabidopsis response to chilling. *New Phytologist* 194, 181–191.

Dutilleul, C., Chavarria, H., Rézé, N., Sotta, B., Baudouin, E. and Guillas, I. (2015) Evidence for ACD5 ceramide kinase activity involvement in *Arabidopsis* response to cold stress. *Plant, Cell & Environment* 38, 2688–2697.

Gao, Y.-B., Wang, C.-L., Wu, J.-Y., Zhou, H.-S., Jiang, X.-T., Wu, J. and Zhang, S.-L. (2014) Low temperature inhibits pollen tube growth by disruption of both tip-localized reactive oxygen species and endocytosis in *Pyrus bretschneideri* Rehd. *Plant Physiology and Biochemistry* 74, 255–262.

Gao, J., Wallis, J.G. and Browse, J. (2015) Mutations in the prokaryotic pathway rescue the *fatty acid biosynthesis1* mutant in the cold. *Plant Physiology* 169, 442–452.

Gehan, M.A., Park, S., Gilmour, S.J., An, C., Lee, C.-M. and Thomashow, M.F. (2015) Natural variation in the C-repeat binding factor cold response pathway correlates with local adaptation of Arabidopsis ecotypes. *The Plant Journal* 84, 682–693.

Graether, S.P. and Boddington, K.F. (2014) Disorder and function: a review of the dehydrin protein family. *Frontiers in Plant Science* 5, 576.

Griffith, M., Lumb, C., Wiseman, S.B., Wisniewski, M., Johnson, R.W. and Marangoni, A.G. (2005) Antifreeze proteins modify the freezing process in planta. *Plant Physiology* 138, 330–340.

Guo, Z., Ou, W., Lu, S. and Zhong, Q. (2006) Differential responses of antioxidative system to chilling and drought in four rice cultivars differing in sensitivity. *Plant Physiology and Biochemistry* 44, 828–836.

Gupta, R. and Deswal, R. (2014) Antifreeze proteins enable plants to survive in freezing conditions. *Journal of Biosciences* 39, 931–944.

Harvaux, M. and Kloppstech, K. (2001) The protective functions of carotenoid and flavonoid pigments against excess visible radiation at chilling temperature investigated in *Arabidopsis npq* and *tt* mutants. *Planta* 213, 953–966.

Heide, O.M. and Prestrud, A.K. (2005) Low temperature, but not photoperiod, controls growth cessation and dormancy induction and release in apple and pear. *Tree Physiology* 25, 109–114.

Hu, Y., Jiang, L., Wang, F. and Yu, D. (2013) Jasmonate regulates the INDUCER OF CBF EXPRESSION-C-REPEAT BINDING FACTOR/DRE BINDING FACTOR1 cascade and freezing tolerance in *Arabidopsis*. *The Plant Cell* 25, 2907–2924.

Hull, M.R., Long, S.P. and Jahnke, L.S. (1997) Instantaneous and developmental effects of low temperature on the catalytic properties of antioxidant enzymes in two *Zea* species. *Australian Journal of Plant Physiology* 24, 337.

Hüner, N.P.A., Bode, R., Dahal, K., Hollis, L., Rosso, D., Krol, M. and Ivanov, A.G. (2012) Chloroplast redox imbalance governs phenotypic plasticity: the "grand design of photosynthesis" revisited. *Frontiers in Plant Science* 3, 255.

Hutchison, R.S., Groom, Q. and Ort, D.R. (2000) Differential effects of chilling-induced photooxidation on the redox regulation of photosynthetic enzymes. *Biochemistry* 39, 6679–6688.

Isshiki, R., Galis, I. and Tanakamaru, S. (2014) Farinose flavonoids are associated with high freezing tolerance in fairy primrose (*Primula malacoides*) plants. *Journal of Integrative Plant Biology* 56, 181–188.

Ivanov, A.G., Sane, P.V., Zeinalov, Y., Malmberg, G., Gardeström, P., Huner, N.P. and Öquist, G. (2001) Photosynthetic electron transport adjustments in overwintering Scots pine (*Pinus sylvestris* L.). *Planta* 213, 575–585.

Ivanov, A.G., Allakhverdiev, S.I., Huner, N.P.A. and Murata, N. (2012) Genetic decrease in fatty acid unsaturation of phosphatidylglycerol increased photoinhibition of photosystem I at low temperature in tobacco leaves. *Biochimica et Biophysica Acta (BBA) - Bioenergetics* 1817, 1374–1379.

Janská, A., Marsík, P., Zelenková, S. and Ovesná, J. (2010) Cold stress and acclimation – what is important for metabolic adjustment? *Plant Biology (Stuttgart, Germany)* 12, 395–405.

Jeon, J. and Kim, J. (2013) Arabidopsis response Regulator1 and Arabidopsis histidine phosphotransfer Protein2 (AHP2), AHP3, and AHP5 function in cold signaling. *Plant Physiology* 161, 408–424.

Jeon, J., Kim, N.Y., Kim, S., Kang, N.Y., Novák, O., Ku, S.-J., Cho, C., Lee, D.J., Lee, E.-J., Strnad, M. and Kim, J. (2010) A subset of cytokinin two-component signaling system plays a role in cold temperature stress response in *Arabidopsis*. *The Journal of Biological Chemistry* 285, 23371–23386.

Ji, H., Wang, Y., Cloix, C., Li, K., Jenkins, G.I., Wang, S., Shang, Z., Shi, Y., Yang, S. and Li, X. (2015) The *Arabidopsis* RCC1 family protein TCF1 regulates freezing tolerance and cold acclimation through modulating lignin biosynthesis. *PLoS Genetics* 11, e1005471.

Kangasjärvi, S., Lepistö, A., Hännikäinen, K., Piippo, M., Luomala, E.-M., Aro, E.-M. and Rintamäki, E. (2008) Diverse roles for chloroplast stromal and thylakoid-bound ascorbate peroxidases in plant stress responses. *Biochemical Journal* 412, 275–285.

Kaplan, F. and Guy, C.L. (2005) RNA interference of Arabidopsis beta-amylase8 prevents maltose accumulation upon cold shock and increases sensitivity of PSII photochemical efficiency to freezing stress. *The Plant Journal* 44, 730–743.

Kasuga, J., Hashidoko, Y., Nishioka, A., Yoshiba, M., Arakawa, K. and Fujikawa, S. (2008) Deep supercooling xylem parenchyma cells of katsura tree (*Cercidiphyllum japonicum*) contain flavonol glycosides exhibiting high anti-ice nucleation activity. *Plant, Cell & Environment* 31, 1335–1348.

Kim, H.-J., Kim, Y.-K., Park, J.-Y. and Kim, J. (2002) Light signalling mediated by phytochrome plays an important role in cold-induced gene expression through the C-repeat/dehydration responsive element (C/DRE) in *Arabidopsis thaliana*. *Plant Journal* 29, 693–704.

Kim, J.Y., Park, S.J., Jang, B., Jung, C.-H., Ahn, S.J., Goh, C.-H., Cho, K., Han, O. and Kang, H. (2007) Functional characterization of a glycine-rich RNA-binding protein 2 in *Arabidopsis thaliana* under abiotic stress conditions. *The Plant Journal* 50, 439–451.

Kim, Y., Park, S., Gilmour, S.J. and Thomashow, M.F. (2013) Roles of CAMTA transcription factors and salicylic acid in configuring the low-temperature transcriptome and freezing tolerance of Arabidopsis. *The Plant Journal* 75, 364–376.

Kindgren, P., Dubreuil, C. and Strand, Å. (2015) The recovery of plastid function is required for optimal response to low temperatures in *Arabidopsis*. *PLoS ONE* 10, e0138010.

Knaupp, M., Mishra, K.B., Nedbal, L. and Heyer, A.G. (2011) Evidence for a role of raffinose in stabilizing photo-system II during freeze–thaw cycles. *Planta* 234, 477–486.

Knox, A.K., Li, C., Vágújfalvi, A., Galiba, G., Stockinger, E.J. and Dubcovsky, J. (2008) Identification of candidate CBF genes for the frost tolerance locus Fr-Am2 in *Triticum monococcum*. *Plant Molecular Biology* 67, 257–270.

Korn, M., Gärtner, T., Erban, A., Kopka, J., Selbig, J. and Hincha, D.K. (2010) Predicting *Arabidopsis* freezing toler-ance and heterosis in freezing tolerance from metabolite composition. *Molecular Plant* 3, 224–235.

Kratsch, H.A. and Wise, R.R. (2000) The ultrastructure of chilling stress. *Plant, Cell and Environment* 23, 337–350.

Kurbidaeva, A., Ezhova, T. and Novokreshchenova, M. (2014) *Arabidopsis thaliana* ICE2 gene: phylogeny, struc-tural evolution and functional diversification from ICE1. *Plant Science* 229, 10–22.

Kuwabara, C. and Imai, R. (2009) Molecular basis of disease resistance acquired through cold acclimation in overwintering plants. *Journal of Plant Biology* 52, 19–26.

Lee, C.-M. and Thomashow, M.F. (2012) Photoperiodic regulation of the C-repeat binding factor (CBF) cold accli-mation pathway and freezing tolerance in *Arabidopsis thaliana*. *Proceedings of the National Academy of Sciences USA* 109, 15054–15059.

Lee, Y.P., Babakov, A., de Boer, B., Zuther, E. and Hincha, D.K. (2012) Comparison of freezing tolerance, compat-ible solutes and polyamines in geographically diverse collections of *Thellungiella* sp. and *Arabidopsis thaliana* accessions. *BMC Plant Biology* 12, 131.

Le Martret, B., Poage, M., Shiel, K., Nugent, G.D. and Dix, P.J. (2011) Tobacco chloroplast transformants express-ing genes encoding dehydroascorbate reductase, glutathione reductase, and glutathione-S-transferase, ex-hibit altered anti-oxidant metabolism and improved abiotic stress tolerance. *Plant Biotechnology Journal* 9, 661–673.

Li, X., Cai, J., Liu, F., Dai, T., Cao, W. and Jiang, D. (2014) Cold priming drives the sub-cellular antioxidant systems to protect photosynthetic electron transport against subsequent low temperature stress in winter wheat. *Plant Physiology and Biochemistry* 82, 34–43.

Lindlöf, A., Chawade, A., Sikora, P. and Olsson, O. (2015) Comparative transcriptomics of Sijung and Jumli Marshi rice during early chilling stress imply multiple protective mechanisms. *PloS ONE* 10, e0125385.

Lyons, J.M. (1973) Chilling injury in plants. *Annual Review of Plant Physiology* 24, 445–466.

Ma, Y., Dai, X., Xu, Y., Luo, W., Zheng, X., Zeng, D., Pan, Y., Lin, X., Liu, H., Zhang, D., Xiao, J., Guo, X., Xu, S., Niu, Y., Jin, J., Zhang, H., Xu, X., Li, L., Wang, W., Qian, Q., Ge, S. and Chong, K. (2015) COLD1 confers chilling tol-erance in rice. *Cell* 160, 1209–1221.

Marcellin, P. (1992) Les maladies physiologiques du froid. In: Côme, D. (ed.) *Les végétaux et le froid.* Herman, Paris, pp. 53–105.

Martinière, A., Shvedunova, M., Thomson, A.J.W., Evans, N.H., Penfield, S., Runions, J. and McWatters, H.G. (2011) Homeostasis of plasma membrane viscosity in fluctuating temperatures. *New Phytologist* 192, 328–337.

Maruta, T., Noshi, M., Tanouchi, A., Tamoi, M., Yabuta, Y., Yoshimura, K., Ishikawa, T. and Shigeoka, S. (2012) H2O2-triggered retrograde signaling from chloroplasts to nucleus plays specific role in response to stress. *The Journal of Biological Chemistry* 287, 11717–11729.

McKhann, H.I., Gery, C., Bérard, A., Lévêque, S., Zuther, E., Hincha, D.K., De Mita, S., Brunel, D. and Téoulé, E. (2008) Natural variation in CBF gene sequence, gene expression and freezing tolerance in the Versailles core collection of *Arabidopsis thaliana*. *BMC Plant Biology* 8, 105.

Medina, J., Catalá, R. and Salinas, J. (2011) The CBFs: three arabidopsis transcription factors to cold acclimate. *Plant Science* 180, 3–11.

Menon, M., Barnes, W.J. and Olson, M.S. (2015) Population genetics of freeze tolerance among natural popula-tions of *Populus balsamifera* across the growing season. *New Phytologist* 207, 710–722.

Miquel, M., James, D., Dooner, H. and Browse, J. (1993) *Arabidopsis* requires polyunsaturated lipids for low-temperature survival. *Proceedings of the National Academy of Sciences USA* 90, 6208–6212.

Moon, B.Y., Higashi, S., Gombos, Z. and Murata, N. (1995) Unsaturation of the membrane lipids of chloroplasts stabilizes the photosynthetic machinery against low-temperature photoinhibition in transgenic tobacco plants. *Proceedings of the National Academy of Sciences USA* 92, 6219–6223.

Moore, J.P., Nguema-Ona, E.E., Vicré-Gibouin, M., Sørensen, I., Willats, W.G.T., Driouich, A. and Farrant, J.M. (2013) Arabinose-rich polymers as an evolutionary strategy to plasticize resurrection plant cell walls against desiccation. *Planta* 237, 739–754.

Nakamichi, N., Kusano, M., Fukushima, A., Kita, M., Ito, S., Yamashino, T., Saito, K., Sakakibara, H. and Mizuno, T. (2009) Transcript profiling of an Arabidopsis pseudo response regulator arrhythmic triple mutant reveals a role for the circadian clock in cold stress response. *Plant and Cell Physiology* 50, 447–462.

Ndong, C., Danyluk, J., Huner, N.P. and Sarhan, F. (2001) Survey of gene expression in winter rye during changes in growth temperature, irradiance or excitation pressure. *Plant Molecular Biology* 45, 691–703.

Neilson, K.A., Scafaro, A.P., Chick, J.M., George, I.S., Van Sluyter, S.C., Gygi, S.P., Atwell, B.J. and Haynes, P.A. (2013) The influence of signals from chilled roots on the proteome of shoot tissues in rice seedlings. *PROTEOMICS* 13, 1922–1933.

Nick, P. (2013) Microtubules, signalling and abiotic stress. *The Plant Journal* 75, 309–323.

Novillo, F., Medina, J. and Salinas, J. (2007) *Arabidopsis* CBF1 and CBF3 have a different function than CBF2 in cold acclimation and define different gene classes in the CBF regulon. *Proceedings of the National Academy of Sciences USA* 104, 21002–21007.

Park, S., Lee, C.-M., Doherty, C.J., Gilmour, S.J., Kim, Y. and Thomashow, M.F. (2015) Regulation of the Arabidopsis CBF regulon by a complex low-temperature regulatory network. *The Plant Journal* 82, 193–207.

Peng, T., Zhu, X., Duan, N. and Liu, J.-H. (2014) PtrBAM1, a β-amylase-coding gene of *Poncirus trifoliata*, is a CBF regulon member with function in cold tolerance by modulating soluble sugar levels. *Plant, Cell & Environment* 37, 2754–2767.

Plazek, A. and Zur, I. (2003) Cold-induced plant resistance to necrotrophic pathogens and antioxidant enzyme activities and cell membrane permeability. *Plant Science* 164, 1019–1028.

Puyaubert, J. and Baudouin, E. (2014) New clues for a cold case: nitric oxide response to low temperature: nitric oxide and cold response. *Plant, Cell & Environment* 37, 2623–2630.

Rajashekar, C.B. and Lafta, A. (1996) Cell-wall changes and cell tension in response to cold acclimation and exogenous abscisic acid in leaves and cell cultures. *Plant Physiology* 111, 605–612.

Ruelland, E. and Zachowski, A. (2010) How plants sense temperature. *Environmental and Experimental Botany* 69, 225–232.

Ruelland, E., Vaultier, M., Zachowski, A. and Hurry, V. (2009) Cold signalling and cold acclimation in plants. *Advances in Botanical Research* 49, 35–150.

Sasaki, K. and Imai, R. (2012) Pleiotropic roles of cold shock domain proteins in plants. *Frontiers in Plant Science* 2, 116.

Sato, Y., Masuta, Y., Saito, K., Murayama, S. and Ozawa, K. (2011) Enhanced chilling tolerance at the booting stage in rice by transgenic overexpression of the ascorbate peroxidase gene, OsAPXa. *Plant Cell Reports* 30, 399–406.

Savitch, L.V., Leonardos, E.D., Krol, M., Jansson, S., Grodzinski, B., Huner, N.P.A. and Öquist, G. (2002) Two different strategies for light utilization in photosynthesis in relation to growth and cold acclimation. *Plant, Cell and Environment* 25, 761–771.

Schulz, E., Tohge, T., Zuther, E., Fernie, A.R. and Hincha, D.K. (2015) Natural variation in flavonol and anthocyanin metabolism during cold acclimation in *Arabidopsis thaliana* accessions. *Plant, Cell & Environment* 38, 1658–1672.

Scott, I.M., Clarke, S.M., Wood, J.E. and Mur, L.A.J. (2004) Salicylate accumulation inhibits growth at chilling temperature in Arabidopsis. *Plant Physiology* 135, 1040–1049.

Sehrawat, A. and Deswal, R. (2014) S-Nitrosylation analysis in *Brassica juncea* apoplast highlights the importance of nitric oxide in cold-stress signaling. *Journal of Proteome Research* 13, 2599–2619.

Seo, P.J., Kim, M.J., Park, J.-Y., Kim, S.-Y., Jeon, J., Lee, Y.-H., Kim, J. and Park, C.-M. (2010) Cold activation of a plasma membrane-tethered NAC transcription factor induces a pathogen resistance response in Arabidopsis. *The Plant Journal* 61, 661–671.

Shi, Y., Tian, S., Hou, L., Huang, X., Zhang, X., Guo, H. and Yang, S. (2012) Ethylene signaling negatively regulates freezing tolerance by repressing expression of *CBF* and type-A *ARR* genes in Arabidopsis. *The Plant Cell* 24, 2578–2595.

Sicher, R. (2011) Carbon partitioning and the impact of starch deficiency on the initial response of Arabidopsis to chilling temperatures. *Plant Science* 181, 167–176.

Soitamo, A.J., Piippo, M., Allahverdiyeva, Y., Battchikova, N. and Aro, E.-M. (2008) Light has a specific role in modulating *Arabidopsis* gene expression at low temperature. *BMC Plant Biology* 8, 13.

Strand, A., Foyer, C.H., Gustafsson, P., Gardestrom, P. and Hurry, V. (2003) Altering flux through the sucrose biosynthesis pathway in transgenic *Arabidopsis thaliana* modifies photosynthetic acclimation at low temperatures and the development of freezing tolerance. *Plant, Cell and Environment* 26, 523–535.

Svensson, J.T., Crosatti, C., Campoli, C., Bassi, R., Stanca, A.M., Close, T.J. and Cattivelli, L. (2006) Transcriptome analysis of cold acclimation in barley albina and xantha mutants. *Plant Physiology* 141, 257–270.

Szechyńska-Hebda, M., Hebda, M., Mierzwiński, D., Kuczyńska, P., Mirek, M., Wędzony, M., van Lammeren, A. and Karpiński, S. (2013) Effect of cold-induced changes in physical and chemical leaf properties on the resistance of winter triticale (× *Triticosecale*) to the fungal pathogen *Microdochium nivale*. *Plant Pathology* 62, 867–878.

Tähtiharju, S. and Palva, T. (2001) Antisense inhibition of protein phosphatase 2C accelerates cold acclimation in *Arabidopsis thaliana*. *The Plant Journal* 26, 461–470.

Tarazona, P., Feussner, K. and Feussner, I. (2015) An enhanced plant lipidomics method based on multiplexed liquid chromatography-mass spectrometry reveals additional insights into cold- and drought-induced membrane remodeling. *The Plant Journal* 84, 621–633.

Tasseva, G., de Virville, J.D., Cantrel, C., Moreau, F. and Zachowski, A. (2004) Changes in the endoplasmic reticulum lipid properties in response to low temperature in *Brassica napus*. *Plant Physiology and Biochemistry* 42, 811–822.

Thakur, P., Kumar, S., Malik, J.A., Berger, J.D. and Nayyar, H. (2010) Cold stress effects on reproductive development in grain crops: an overview. *Environmental and Experimental Botany* 67, 429–443.

Thomashow, M.F. (1999) Plant cold acclimation: freezing tolerance genes and regulatory mechanisms. *Annual Review of Plant Physiology and Plant Molecular Biology* 50, 571–599.

Thomashow, M.F. (2010) Molecular basis of plant cold acclimation: insights gained from studying the CBF cold response pathway. *Plant Physiology* 154, 571–577.

Uemura, M. and Steponkus, P.L. (1997) Effect of cold acclimation on the lipid composition of the inner and outer membrane of the chloroplast envelope isolated from rye leaves. *Plant Physiology* 114, 1493–1500.

Van Dijk, C., Boeriu, C., Peter, F., Stolle-Smits, T. and Tijskens, L.M.M. (2006) The firmness of stored tomatoes (cv. Tradiro). 1. Kinetic and near infrared models to describe firmness and moisture loss. *Journal of Food Engineering* 77, 575–584.

Wang, X., Devaiah, S.P., Zhang, W. and Welti, R. (2006) Signaling functions of phosphatidic acid. *Progress in Lipid Research* 45, 250–278.

Wisniewski, M., Webb, R., Balsamo, R., Close, T.J., Yu, X.-M. and Griffith, M. (1999) Purification, immunolocalization, cryoprotective, and antifreeze activity of PCA60: a dehydrin from peach (*Prunus persica*). *Physiologia Plantarum* 105, 600–608.

Woodward, F.I., Lomas, M.R. and Kelly, C.K. (2004) Global climate and the distribution of plant biomes. *Philosophical Transactions of the Royal Society of London, Series B, Biological Sciences* 359, 1465–1476.

Xue, G.-P. (2003) The DNA-binding activity of an AP2 transcriptional activator HvCBF2 involved in regulation of low-temperature responsive genes in barley is modulated by temperature. *The Plant Journal* 33, 373–383.

Yang, L., Tang, R., Zhu, J., Liu, H., Mueller-Roeber, B., Xia, H. and Zhang, H. (2008) Enhancement of stress tolerance in transgenic tobacco plants constitutively expressing AtIpk2beta, an inositol polyphosphate 6-/3-kinase from *Arabidopsis thaliana*. *Plant Molecular Biology* 66, 329–343.

Yang, D.H., Kwak, K.J., Kim, M.K., Park, S.J., Yang, K.-Y. and Kang, H. (2014) Expression of *Arabidopsis* glycine-rich RNA-binding protein AtGRP2 or AtGRP7 improves grain yield of rice (*Oryza sativa*) under drought stress conditions. *Plant Science* 214, 106–112.

Yun, K.-Y., Park, M.R., Mohanty, B., Herath, V., Xu, F., Mauleon, R., Wijaya, E., Bajic, V.B., Bruskiewich, R. and de Los Reyes, B.G. (2010) Transcriptional regulatory network triggered by oxidative signals configures the early response mechanisms of japonica rice to chilling stress. *BMC Plant Biology* 10, 16.

Zheng, G., Tian, B., Zhang, F., Tao, F. and Li, W. (2011) Plant adaptation to frequent alterations between high and low temperatures: remodelling of membrane lipids and maintenance of unsaturation levels: lipid changes in high-low temperature alteration. *Plant, Cell & Environment* 34, 1431–1442.

Zhu, J., Zhang, K.-X., Wang, W.-S., Gong, W., Liu, W.-C., Chen, H.-G., Xu, H.-H. and Lu, Y.-T. (2015) Low temperature inhibits root growth by reducing auxin accumulation via ARR1/12. *Plant Cell Physiology* 56, 727–736.

Zografos, B.R. and Sung, S. (2012) Vernalization-mediated chromatin changes. *Journal of Experimental Botany* 63, 4343–4348.

Zou, C., Jiang, W. and Yu, D. (2010) Male gametophyte-specific WRKY34 transcription factor mediates cold sensitivity of mature pollen in *Arabidopsis*. *Journal of Experimental Botany* 61, 3901–3914.

Zuther, E., Büchel, K., Hundertmark, M., Stitt, M., Hincha, D.K. and Heyer, A.G. (2004) The role of raffinose in the cold acclimation response of *Arabidopsis thaliana*. *FEBS Letters* 576, 169–173.

Zuther, E., Schulz, E., Childs, L.H. and Hincha, D.K. (2012) Clinal variation in the non-acclimated and cold-acclimated freezing tolerance of *Arabidopsis thaliana* accessions. *Plant, Cell & Environment* 35, 1860–1878.

5 High-temperature Stress in Plants: Consequences and Strategies for Protecting Photosynthetic Machinery

Anjana Jajoo[1],* and Suleyman I. Allakhverdiev[2,3,4],*

[1]*School of Life Sciences, Devi Ahilya University, Indore India;* [2]*Controlled Photobiosynthesis Laboratory, Institute of Plant Physiology, Russian Academy of Sciences, Moscow, Russia;* [3]*Institute of Basic Biological Problems, Russian Academy of Sciences, Pushchino, Russia;* [4]*Department of Plant Physiology, Faculty of Biology, M.V. Lomonosov Moscow State University, Moscow, Russia*

Abstract

The increasing temperature of the Earth is a very significant consequence of present climatic conditions. High temperature may lead to reduced plant growth and limited crop yield. Photosynthesis is a key phenomenon that contributes substantially to the growth and development of the plant. At the same time, it is one of the most susceptible metabolic processes to any kind of environmental stress. The process of photosynthesis involves various components, such as photosynthetic pigments and photosystems, the electron transport system and CO_2 reduction pathways, and any damage at any level caused by a stress may reduce the overall photosynthetic capacity or efficiency of a plant. In this chapter we describe in detail the high-temperature-induced damage to pigments, photosystems and the components of the electron transport chain; alteration in the activities of various enzymes of carbon-reduction pathways and in the properties of thylakoid membranes; production of reactive oxygen species and heat-shock proteins; and regulation of the genes involved in the mechanism of photosynthesis, particularly in agricultural plants.

5.1 Introduction

In the last few decades, increasing anthropogenic activities have led to global warming and climate change, which are having adverse effects on plant growth. At the same time, there is more demand for food because of the growing world population. The rise in the temperature of the Earth is a very significant consequence of present climatic conditions. As a result of increased greenhouse gases, the Earth's temperature may increase between 1.1°C and 6.4°C in the next century (Yamori *et al.*, 2013). This is posing a serious threat to crop plants as it inhibits the growth and overall yield. Thus it is a big challenge to achieve higher crop yields under adverse climatic conditions, and scientists are trying to explore the ways to increase stress tolerance in plants, especially crop plants.

High temperatures may lead to reduced plant growth and limit crop yield. It has been calculated that there may be approximately a 17% decrease in yield per 1.0°C increase in average growing season temperature (Lobell and Asner, 2003). World agriculture needs a new 'green revolution' to increase crop yields to meet

* anjanajajoo@hotmail.com; suleyman.allakhverdiev@gmail.com

growing food demands (Fischer and Edmeades, 2010). These stress factors clearly merit urgent attention, and there is a need to understand thoroughly the physiological mechanisms in plants that are significantly affected by high-temperature stress.

Although plant growth is controlled by several physiological, biochemical and molecular processes, photosynthesis is an important phenomenon that contributes substantially to the growth and development of plants (Ashraf and Harris, 2013). Photosynthesis is not only one of the most important physiological processes in plants but, at the same time, it has also proved to be a good marker for stress since mostly it is affected before other processes by most of the stresses. Enhancing photosynthesis seems to be a promising approach for increasing crop yield. However, to reach this goal, it is essential to understand the processes that limit photosynthesis under a variety of growth conditions, and also to know how well the process of photosynthesis can acclimate to predicted changes in temperature.

Photosynthesis is the only process capable of converting solar energy into a usable chemical form of energy. This energy is then used by a number of metabolic processes in plants. The mechanism of photosynthesis involves several components, including photosynthetic pigments and photosystems, the electron transport chain and CO_2 reduction pathways. Any damage at any level caused by a stress may reduce the overall photosynthetic capacity or efficiency of a plant.

In nature, plants are exposed to various types of abiotic stresses, among which heat stress has an independent mode of action on the physiological and metabolic functioning of the plant cell.

Although heat stress is often aggravated by combined abiotic stresses such as high temperature, drought and salinity, it is also important to improve our understanding of the independent action of high temperature alone. The sensitivity to high temperatures in plants depends on the stage of plant development, ranging from vegetative to reproductive stages. A brief list of responses to high temperature in plants is shown in Fig. 5.1. The effects of high temperature on plants depend on species and genotype, and include many inter- and intraspecific variations (Barnabás et al., 2008; Sakata and Higashitani, 2008; Bita and Gerats, 2013).

High temperature causes, among other effects, inhibition of seed germination, reduction in plant growth, improper development, alteration in photosynthesis, alteration in phenology, decreased crop yield, poor quality and oxidative stress (Hasanuzzaman et al., 2013). Heat stress leads to a shortened life cycle and diminished plant productivity by inducing changes in the processes of respiration and photosynthesis (Barnabás et al., 2008). Thermal stress exhibits its early effects in the form of structural alterations in chloroplast protein complexes and reduced enzyme activity (Ahmad et al., 2010).

In all plant species, a direct correlation is found between the ability to carry out efficient leaf

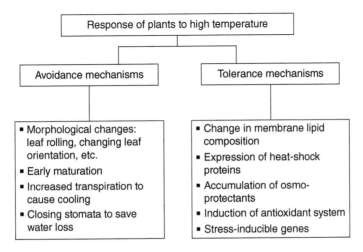

Fig. 5.1. A summary of effects of high temperature stress on plants.

gas exchange and heat stress. As mentioned above, the decrease in the amount of active ribulose-1, 5-bisphosphate carboxylase/oxygenase (Rubisco) may be responsible for the significant negative response of temperature on net photosynthesis (Salvucci and Crafts-Brandner, 2004). Harmful reactive oxygen species (ROS) are generated due to inhibition of carbon fixation and subsequently decreased oxygen evolution. This also results in inhibition of the repair mechanism of the damaged photosystem. Hence, manipulation of leaf photosynthesis and photosynthate partitioning may be a good approach to develop crops with better productivity in a high-temperature environment (Ainsworth and Ort, 2010). Generally a heat-tolerant variety is characterized by higher rate of photosynthesis, increased thermostability of thylakoid membrane and ability to avoid heat (Nagarajan et al., 2010; Scafaro et al., 2010).

This chapter describes the recent findings on responses, adaptation and tolerance to high temperature, particularly on photosynthetic machinery, and describes the various approaches plants undertake in order to tolerate high temperature (HT) stress.

5.2   Targets

The response of a plant to HT varies with the severity and duration of the temperature, as well as the plant type. Extreme HT may result in damage to cellular components within minutes, leading to complete collapse of cell organization (Hasanuzzaman et al., 2013). The stability of proteins and membranes, as well as efficiency of various enzymatic reactions in the cell, are differentially affected by HT. Among the physiological processes in plants, photosynthesis is one of those most sensitive to HT. In the photosynthetic apparatus the major targets of HT stress are the photosynthetic pigments, thylakoid membranes, photosystem II, carbon fixation reactions, etc. A summary of the targets and tolerance mechanisms in the photosynthetic machinery under high-temperature stress is given in Fig. 5.2, and we will discuss each in detail.

5.2.1   Photosynthetic pigments

Chlorophyll a is the major photosynthetic pigment in higher plants. Plants exposed to HT show reduced chlorophyll (Chl) biosynthesis (Efeoglu and Terzioglu, 2009). This may be due to impairment in the process of Chl synthesis or its acceleration in the degradation of Chl, or to both. Most of the effects are due to impairment of enzymes involved in Chl biosynthesis. The activity of the first enzyme of the pyrrole biosynthetic pathway, 5-aminolevulinate dehydratase (ALAD), decreased under HTs (Tewari and Tripathy, 1998;

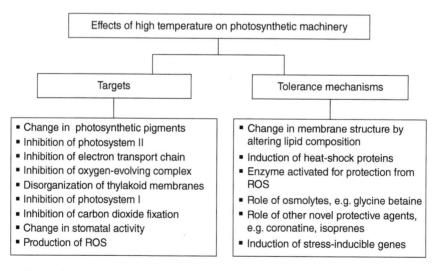

Fig. 5.2. Effects of high temperature stress on photosynthetic machinery. ROS, reactive oxygen species.

Mohanty et al., 2006). In response to high-temperature stress (38/28°C), a decrease in total Chl content (18%), Chl a content (7%), Chl a/b ratio (3%), sucrose content (9%) and an increase in reducing sugar content (47%) and leaf soluble sugars content (36%) was reported in soybean (Hasanuzzaman et al., 2013). Reduction in pigment content may ultimately lead to inhibition of electron transport and thereby reduced photosynthetic capacity in most plants (Ashraf and Harris, 2013).

5.2.2 Photosystem II

Photosystem II (PSII) is regarded as a very thermosensitive component of the photosynthetic apparatus (Srivastava et al., 1997). This is probably because of: (i) increased fluidity of thylakoid membranes at HTs leading to dislodgement of PSII light-harvesting complexes (LHC); and (ii) the dependence of PSII functional integrity on electron dynamics. The water-oxidizing complex (WOC), PSII reaction centre and the LHC are the first to be impaired by HT (Salvucci et al., 2001).

This has been shown to be mainly due to the dissociation of Ca^{2+}, Mn^{2+} and Cl^- from the PSII complex and the subsequent release of extrinsic 18, 24 and 33 kDa polypeptides from the thylakoid membranes (Allakhverdiev and Murata, 2004; Wise et al., 2004). As a consequence, inhibition or inactivation of PSII, thermal damage of D1 protein and production of ROS both in light and in dark occur (Balint et al., 2006). Among the intrinsic proteins of PSII, moderate heat exposure (40°C for 30 min) of spinach thylakoids could cleave only the D1 protein, producing a 9-kDa C-terminal and 23-kDa N-terminal fragments, while D2, CP43 or CP47 remained uncleaved (Yoshioka et al., 2006). Heat-stress effects are also associated with thylakoid membrane integrity, ion conductivity and phosphorylation activity.

HT causes reorganization of the thylakoid membranes leading to a structural change in D1 and D2 proteins, ultimately causing inhibition of Q_A–Q_B electron transfer (Cao and Govindjee, 1990). High-temperature stress leads to down-regulation of the quantum efficiency of PSII by decreasing the rate of primary charge separation, destabilization of charge separation and disconnecting some minor antenna from PSII

(Mathur et al., 2011a; Janka et al., 2013; Agrawal and Jajoo, 2015). Structural and functional variations in PS II, known as PS II heterogeneity, also change with HT (Mathur et al., 2011b).

Various Chl a fluorescence parameters have proved to be excellent indicators for high-temperature stress in PSII. It is reported that the antenna size, maximal fluorescence (Fm) and Fv/Fm decreased, while initial fluorescence (Fo) and dissipation in the form of heat increased (Mathur et al., 2011a). For most plants with leaf temperatures above 35°C, the initial decrease in photochemical efficiency is accompanied by a high stimulation of non-photochemical quenching due to increased energy dissipation as heat (Kana et al., 2008). An increase in energy dissipation at HT reduces the energy available for photochemistry under stress conditions (Mathur et al., 2011a). Heat treatment of the plants leads to an inhibition of electron transfer at various sites which include the plastoquinol (PQH_2) oxidation site at Cytochrome (Cyt) b6/f complex.

5.2.3 Electron transport chain

The primary cause for inhibition of photosynthesis under heat stress due to the heat sensitivity of PSII is exhibited in terms of damage to the photosynthetic electron transport chain (ETC) (Berry and Björkman, 1980). However, the precise site of damage of ETC under heat stress remains unclear. To reveal this issue, chlorophyll a fluorescence and modulated 820 nm reflection were simultaneously detected in sweet sorghum (Brestic and Zivcak, 2013). At 43°C, there was a significant increase in the J step in the chlorophyll a fluorescence transient, suggesting that electron transport was inhibited beyond primary quinone Q_A of PSII. When the temperature was increased to 48°C, the maximum quantum yield for primary photochemistry and the electron transport from the PSII donor side decreased remarkably, which greatly limited the electron flow to photosystem I (PSI), and inhibited PSI re-reduction. The efficiency of electron transfer from plastoquinol (PQH_2) to the PSI acceptor side increased significantly at 48°C due to greater inhibition of electron transport before PQH_2. Thus, the fragment from Q_A to PQH_2 was found to be the most heat sensitive in the ETC between PSII and PSI in sweet sorghum. One day after the stress, the leaves could not recover and died. The irreversible damage to PSII at this temperature might have resulted from heat-induced protein denaturation and lipid oxidation (Mohanty et al., 2007; Murata et al., 2007). The decrease in electron transport between PSII and PSI may play a protective role in reducing the photodamage of PSI.

A study by Zhang et al. (2014) demonstrated that ETR was not significantly changed in the light reaction; however, the expression of four genes (PETA, PETB, PETM and ATPA) for the redox chain was upregulated at 6 h of heat stress. This observation implied that, at this time point, cyclic electron transport might be induced by HT.

5.2.4 Oxygen evolving complex

The manganese (Mn) ion is an important component of WOC of photosystem II and plays a critical role in the photolysis of water. HT above 45°C caused dissociation of Mn cluster of PSII (Tiwari et al., 2008), probably because of some perturbation in the polypeptides of the OEC. At HT (47°C), release of 18 kDa protein was found to be the major cause for the release of essential Ca ions from the Mn_4Ca complex (Pospisil et al., 2003; Barra et al., 2005). In response to

high-temperature stress (45°C), the traditional OJIP curve was changed to an OJKIP curve due to an additional K band observed at 300 ms (Mathur et al., 2011a) (Fig. 5.3). The appearance of a K band is correlated with the inhibition of OEC (Lazar and Pospisil, 1999), inhibition of electron transport from pheophytin to quinine (Q_A) (Guisse et al., 1995), changes in the structure of the LHC of PSII and partial uncoupling of the OEC. The K band arises when electron flow to the acceptor side exceeds electron flow from the donor side, leading to oxidation of the RC (Strasser et al., 2004). So far, this K band has only been observed under high-temperature stress conditions.

5.2.5 Thylakoid membranes

The structure and fluidity of thylakoid membranes depends on their composition and on temperature. An increase in the fluidity of lipid membranes is observed under high-temperature conditions. This may be because of weakening of hydrogen bonds between fatty acids. Because of disruption of membrane-bound proteins, the increased membrane fluidity is also associated with increased membrane permeability. For this reason, change in membrane permeability is often used as an assay to test for extent of membrane damage due to heat stress.

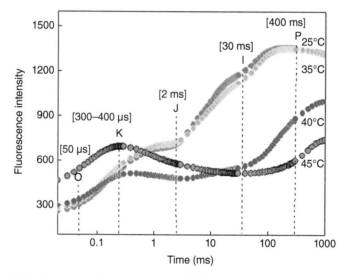

Fig. 5.3. Chlorophyll a fluorescence induction kinetics (OJIP curves) as measured in wheat plants after giving different temperature treatments (adapted from Mathur et al., 2011a).

Membrane-associated processes such as photo-synthesis and membrane transport are generally inhibited first due to HT (Xu *et al.*, 2006). As discussed earlier, sensitivity of PSII to HT is attributed partly due to its close association with the thylakoid membrane. The changes in membrane fluidity act as a signal or trigger to initiate other stress responses in the cell during heat stress, along with effects on metabolic function (Mittler *et al.*, 2012).

5.2.6   Photosystem I

Photosystem I has been shown to be much more heat stable than PSII. Brestic and Zivcak (2013) showed in sweet sorghum that at 43°C, photo-chemical capacity of PSI was not influenced even under severe heat stress at 48°C. This resulted in prolonged PSI oxidation and PSI re-reduction did not reach normal level. The inhibition of electron transport between PSII and PSI can be a protective mechanism to protect from PSI photoinhibition under heat stress.

Results of moderately HTs stimulate PSI activity *in vivo* and *in vitro* and cause increased thylakoid proton conductance and increased cyclic electron flow (CEF) around PSI (Sharkey, 2005; Agrawal *et al.*, 2015). When PSII activity is severely diminished, such as under HT, this stimulation of CEF around PSI could be an adaptive process, producing ATP. Heat stress also significantly increases the dark reduction of PQs (Bukhov and Carpenter, 2004) and enhances the trans-thylakoid proton gradient, which was interpreted through a stimulation of CEF around PSI.

Heat significantly alters the redox balance of the components of electron transport away from PSII and toward PSI. Yamane *et al.* (2000) reported a flow of electrons from the stroma to the plastoquinone pool in the dark at 36°C. PSI was more reduced by heat stress while PSII and the stroma became more oxidized at HT. This indicates that the redox balance of different electron carriers of the photosynthetic ETC can change in opposite directions.

5.2.7   CO$_2$ fixation

There is a decline in plant photosynthesis when the temperature becomes higher than the optimum range. HTs affect light reactions adversely and hence lead to a decrease in the ATP levels in chloroplasts, which consequently suppress the synthesis of ribulose 1,5-bisphosphate by the C$_3$ cycle. Carbon flow through Rubisco is impaired and thus rates of whole-leaf photosynthesis decline severely. The effects of short-term HT temperature stress on photosynthesis and grain filling varied among genotypes and the developmental stages of plants when exposed to the stress. High-temperature stress during the initiation of grain-filling stages was found to have a profound effect on CO$_2$ assimilation in rice leaves, mainly because of low stomatal conductance (Diaz and Varon, 2013). Heat-tolerant cultivars of rice were associated with high levels of photosynthesis rates in leaves.

Many of the key enzymes of the carbon fixation cycle (i.e. the Calvin–Bensen cycle) are thermolabile and are impaired or denatured at HT. The catalytic activity of Rubisco is inhibited during moderate heat stress, partly due to the thermal sensitivity of Rubisco activase (RCA). In some species, heat-stable forms of RCA are produced, which play an important role in acclimation to HT (Yamori *et al.*, 2013).

Enzyme kinetics and modelling studies assigned the thermolability of RCA as a primary cause for the inhibition of photosynthesis under moderate temperature stress (Kurek *et al.*, 2007). RCA is more sensitive to thermal stress in comparison to Rubisco (Crafts-Brandner and Salvucci, 2000). Transgenic *Arabidopsis* plants that expressed normal levels of RCA were more tolerant to moderate heat stress in comparison to plants having sub-optimal levels of RCA (Salvucci *et al.*, 2006).

Several *Arabidopsis thaliana* RCA1 (short isoform) variants exhibiting improved thermo-stability were generated by gene shuffling technology. This approach was used to test the hypothesis whether thermostable RCA can improve photosynthesis under HTs (Kurek *et al.*, 2007). Wild-type RCA1 and selected thermo-stable RCA1 variants were introduced into an *Arabidopsis* RCA deletion (Drca) line. A long-term growth test at either constant 26°C or daily 4-h 30°C exposure was carried out. It was observed that the transgenic lines with the thermostable RCA1 variants exhibited higher photosynthetic rates, better biomass, and increased seed yields in comparison to lines expressing wild-type RCA1, and a slight improvement when compared with untransformed Arabidopsis plants. These results established that RCA is a major limiting factor in plant photosynthesis under moderately elevated temperatures and it may prove to be a potential target for genetic manipulation in order to improve crop plants productivity under HT conditions. There have been many efforts to engineer less temperature-sensitive forms of RCA in order to increase the thermal range of crop species, but it still not clear if alteration of one component of the photosynthetic system may lead to an increase in overall heat tolerance (Sharkey, 2005; Allakhverdiev et al., 2008).

Three genes of the Calvin–Benson cycle involved in carboxylation were significantly repressed at HT, implying that repression of carboxylation processes might be the main reason for the decline in Rubisco activity (Zhang et al., 2014).

5.2.8   Stomatal activity

Stomatal limitation contributes up to 60% of the reduction in the rate of photosynthesis at HT. Non-stomatal limitations in photosynthesis during HT may be because of detrimental effect on ribulose bis phosphate (RuBP, a substrate for Rubisco) carboxylation and regeneration (Greer and Weedon, 2014). Investigation of the photosynthetic process during HT indicated that stomatal conductance is the main cause for the decrease in leaf stomatal conductance (Gs) and intercellular CO_2 concentration (Ci) from 3 h to 12 h of high-temperature treatment (Zhang et al., 2014).

Stomatal conductance and net photosynthesis (Pn) are inhibited by moderate heat stress in many plant species due to a decrease in the activation state of Rubisco. Temperature changes directly affect vapour pressure density (VPD) and can also result in changes in plant hydraulic conductance and water supply to the leaf surface.

5.2.9   Production of reactive oxygen species

Various physiological processes are damaged when plants are exposed to HT. High-temperature stress may inactivate enzymes and inhibit several metabolic pathways, which leads to accumulation of undesirable and harmful ROS such as singlet oxygen ($^1O_2$), superoxide radical ($O_2^{·-}$),

hydrogen peroxide (H_2O_2) and hydroxyl radical ($^·OH$), responsible for oxidative stress (Asada, 2006; Hasanuzzaman et al., 2013). Although ROS are generated in other organelles such as peroxisomes and mitochondria, the major sites of ROS generation in chloroplasts are the reaction centres of PSI and PSII (Soliman et al., 2011). When PSI and PSII do not work optimally under stress conditions, the excess electrons serve as a source of ROS. Among the ROS, singlet oxygen is formed by several photooxidation reactions, for example through the Mehler reaction in chloroplasts, during ETC in mitochondria and during photorespiration in glyoxisomes. A hydroxyl radical is formed through several reactions, such as: (i) the reaction of H_2O_2 with $O_2^{·-}$ (Haber–Weiss reaction); (ii) reactions of H_2O_2 with Fe^{2+} (Fenton reaction); and (iii) decomposition of O_3 in the apoplastic space (Moller et al., 2007; Karuppanapandian et al., 2011a). The singlet oxygen is formed during PS II electron transfer reactions and photoinhibition in the chloroplasts (Nishiyama et al., 2004; Huang and Xu, 2008; Karuppanapandian et al., 2011b). The hydroxyl radical is not considered to have a direct signalling function although the products of its reactions have been shown to elicit signalling responses (Moller et al., 2007; Karuppanapandian et al., 2011a). Hydroxyl radicals can potentially react with all types of biomolecules, including proteins, lipids, pigments and DNA (Moller et al., 2007; Karuppanapandian et al., 2011a). The singlet oxygen is capable of directly oxidizing some biomolecules such as proteins, DNA and polyunsaturated fatty acids. Oxidative stress through peroxidation of membrane components leading to instability of a membrane by protein denaturation is a major consequence of thermal stress (Camejo et al., 2006). A functional inhibition in the light reaction of photosynthesis even under moderate HTs was reported to induce oxidative stress. It happens because of ROS production caused by increased leakage of electrons from the thylakoid membrane (Hasanuzzaman et al., 2013; Schmitt et al., 2014).

5.3   Tolerance Mechanisms

Since plants cannot move, they have evolved a number of mechanisms to tolerate HTs and several other stresses (Allakhverdiev et al., 2008;

Mohanty *et al.*, 2012). Some of the mechanisms have been exploited by plant breeders to develop crops having more heat resistance. Each species exhibits a different capacity to tolerate heat stress. Those plants that are irreversibly damaged by temperatures between 30 and 40°C are known as 'heat-sensitive' plants, while those that get damaged above 40°C are known as 'heat resistant'. The ability to acclimate is a key aspect of tolerance to heat stress, and the mechanisms described below are typically upregulated in plants exposed to HT (Allakhverdiev *et al.*, 2010). A schematic illustration of heat-induced signal transduction mechanisms and processes involved in acclimation of plants to HT is presented in Fig. 5.4.

5.3.1 Membrane state, structure and composition

In order to tolerate HT, thylakoid membrane fluidity must be maintained within a biologically functional range in the plants (i.e. membrane thermostability). Membrane composition decides the extent to which the fluidity of the membrane increases with temperature.

More fluid membranes are formed from lipids having unsaturated fatty acid chains, short fatty acid chains or a low sterol content, and such membranes are less stable at HTs. The tolerance of membranes to heat stress can be improved by increasing the proportion of saturated lipids and thus altering the composition of certain lipids (Allakhverdiev *et al.*, 2009).

In cyanobacteria, at HT, there is a change in the lipid composition which includes an increase in the proportion of saturated lipids (Los and Murata, 2004). The activity of heat-shock proteins may regulate some changes in the physical properties of membranes as well (Sharkey, 2005).

By gene manipulation, heat tolerance in *Arabidopsis*, tobacco and soybean was increased by altering lipid composition (Murakami *et al.*, 2000). Transgenic tobacco plants with a reduced proportion of trienoic fatty acids in the chloroplast membranes conferred heat tolerance to the plants. In wild-type plants, HT of 45°C for 5 min reduced photosynthesis by 50% while the transgenic plants with reduced content of trienoic acids were unaffected at this temperature (Murakami *et al.*, 2000).

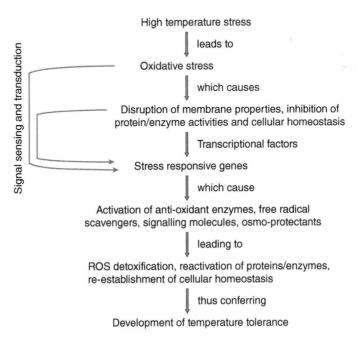

Fig. 5.4. Schematic illustration of heat-induced signal transduction mechanism and development of heat tolerance in plants (adapted from Hasanuzzaman *et al.*, 2013). ROS, reactive oxygen species.

5.3.2 Heat-shock proteins

The expression of most genes engaged in general metabolic reactions is downregulated within minutes of HT. However, a group of stress-responsive genes is also actively upregulated (Sage and Kubien, 2007; Mathur *et al.*, 2014). These genes are known as heat-shock proteins (HSP), and they occur in all organisms. In plants, expression of HSPs varies in different tissues and in cell compartments. As a direct response to heat, HSPs use a novel transcription factor.

On exposure to HT the maximum expression of HSP is achieved within 1–2 h and diminishes after 6–8 h. After this the transcription and translation of other genes resume since the cell environment is modified enough (Larkindale *et al.*, 2005a; Allakhverdiev *et al.*, 2008).

Intracellular proteins are protected by the constitutive expression of HSPs under HT. It helps to preserve their stability and function through protein folding and thus acts as a chaperone (Chang *et al.*, 2007).

The translocation of proteins across cell membranes is facilitated by the activity of chaperones (Hasanuzzaman *et al.*, 2013). The HSPs are heterogeneous in nature and new proteins are being discovered every day, as known from recent researches. The upregulation of HSP improves tolerance as well as recovery from heat stress. For example in chloroplasts isolated from tomato, addition of purified HSP significantly improved the heat tolerance by protecting electron transport through PSII (Allakhverdiev *et al.*, 2008). Five classes of HSP, with the numbers indicating the molecular weight (Larkindale *et al.*, 2005b) have been described: HSP60, HSP70, HSP100,

HSP90 and HSP40 (or small sHSP). HSP60 and HSP70 act as chaperone proteins in plants and other organisms, and also help to stabilize membranes and thus prevent loss of membrane permeability. HSP90 are less characterized in plants, and probably interact with signal transduction proteins which are partly involved in the overall heat-stress response. In conjunction with HSP60 and HSP70, HSP100 acts as a chaperone protein and probably also has other undefined roles. Plants lacking HSP 100 cannot acclimate during heat stress although they can grow normally at normal temperatures. A diverse range of small HSPS (sHSPs) having very low molecular mass of 12–40 kDa is found in plants (Morrow and Tanguay, 2012). sHSPs are quite important but are less characterized in comparison to HSP60 and HSP70. These comprise a diverse group including several gene families which are targeted to different cellular compartments such as the cytosol, chloroplast and mitochondria. The function of many of sHSPs is still obscure. Some may be functioning as chaperone protein, while others help in maintaining membrane stability. The HSP70 and HSP60 proteins have a fundamental role during HT and are among the highly conserved proteins in nature (Hong *et al.*, 2003; Kultz, 2003).

Some HSPs may increase the proteolytic activity of ubiquitin to remove damaged proteins and thus help to clean the cell.

This removal of potentially toxic protein aggregates is a key event for acclimation to heat stress. Indeed, heat stress also triggers the upregulation of ubiquitin itself (Larkindale *et al.*, 2005b). A clear demonstration of the functions of major HSPs involved in stress tolerance in plants is presented in Table 5.1.

Table 5.1. A brief account of the basic function of various heat-shock proteins (HSP) involved in heat-stress tolerance in plants (from Hasanuzzaman *et al.*, 2013, with modifications).

Major classes of heat-shock protein	Functions
HSP100	ATP-dependent dissociation and degradation of aggregate protein
HSP90	Works as ATP-dependent co-regulator of signal transduction complexes linked to heat stress and manages protein folding
HSP70, HSP40	Primary stabilization of newly formed proteins, ATP-dependent binding and release
HSP60, HSP10	ATP-dependent specialized folding machinery
HSP20 or small HSP (sHSP)	Oligomeric complexes having high molecular weight serving as cellular matrix for stabilization of unfolded proteins are formed by sHSP. HSP100, HSP70 and HSP40 are needed for its release

5.3.3 Protection from reactive oxygen species

ROS such as hydroxyl radicals, superoxide radicals and hydrogen peroxide are produced at PS II reaction centres under thermal stress. These are then scavenged by some antioxidant enzymes such as superoxide dismutase (SOD) (Bukhov and Mohanty, 1999; Sairam et al., 2000). Various antioxidant components involved in gene expression, and regulation of processes such as growth and abiotic stress responses help the plant to protect itself from the harmful effects of ROS (Abiko et al., 2005). The importance of the need for ROS detoxication for survival of the cell under stress conditions is evident from the fact that antioxidant components are found in almost all compartments of the cell (Iba, 2002; Mittler et al., 2004; Asada, 2006). Some of the important antioxidant components are ascorbic acid or glutathione, and enzymes such as SOD, ascorbate peroxidase (APX), catalase (CAT) or glutathione peroxidase (GPX) which scavenge ROS (Balla et al., 2009).

Some of the antioxidant systems are damaged at HT while some are activated. This can be regarded as a part of the HT response (Larkindale et al., 2005a; Sharkey, 2005; Allakhverdiev et al., 2008). There is an inhibition in energy generation as ROS cause damage at the ETCs in chloroplasts and mitochondria (Foyer and Noctor, 2009; Bita and Gerats, 2013).

ROSs have proved to be a direct cause of damage to cellular components at several levels, but they also play a key role as molecular signals by linking responses of the plant to pathogen infection and environmental stresses (Gechev et al., 2006). ROS/redox signalling networks in organelles such as chloroplasts and mitochondria largely regulate adaptation of the plant to abiotic stresses. These signals play a role in the complex networking between different cellular components under stress conditions by controlling critically important processes such as energy metabolism, transcription and translation (Mittler et al., 2011). Ultimately, transduction of the heat signal and expression of genes related to HSPs are carried out by ROS production (Königshofer et al., 2008).

Wheat cultivars having high activity of the enzymes glutathione S-transferase (GST), APX and CAT exhibited better tolerance to heat stress and protection against ROS production (Balla et al.,

2009). The tolerance of the wheat varieties to stress was correlated with the level of antioxidants. In five wheat genotypes, at all stages, a positive correlation was observed between the activity of antioxidant enzymes and chlorophyll content, while a negative correlation was observed between level of antioxidant enzymes and membrane injury index (Almeselmani et al., 2006).

5.3.4 Role of osmolytes

Osmotic adjustments in the cell occur by the participation of primary metabolites. An adaptive mechanism in plants exposed to HT is accumulation of osmo-protectants (Sakamoto and Murata, 2000; Bita and Gerates, 2013). Accumulation of osmo-protectants such as proline, glycine betaine and soluble sugars is essential to regulate osmotic activities. They protect cellular structures from HT by maintaining the stability of membranes and the cell water balance, and by buffering the cellular redox potential (Sakamoto and Murata, 2002; Allakhverdiev et al., 2007; Farooq et al., 2008). It was confirmed through transgenic approaches that over-production of proline has a beneficial effect on a plant under HT as it is corroborated with a more negative osmotic potential of the leaf as well as high production of protective xanthophyll pigments under HT (Dobra et al., 2010).

Glycine betaine is an important compatible solute in plants facing HT (Sakamoto and Murata, 2002). Glycine betaine production in chloroplasts has been shown to maintain the activation of Rubisco by preventing thermal inactivation of RCA (Allakhverdiev et al., 2008). Some plants, such as maize and sugarcane, exhibited high levels of glycine betaine under HT, while no glycine betaine was produced naturally in some other plant species such as rice, mustard, Arabidopsis and tobacco under stress conditions (Quan et al., 2004; Wahid and Close, 2007).

An important physiological trait associated with HT tolerance is the high availability of carbohydrates (Liu and Huang, 2000). Sucrose is the major end product of photosynthesis and is translocated through the phloem from leaves (source) to sink organs. Plant development and stress response are regulated by sucrose and its products through carbon allocation and sugar signalling (Roitsch and González, 2004).

High activity of invertases in cell wall and vacuole, along with increased import of sucrose into young tomato fruit, has been found to contribute to heat tolerance by increasing the strength of sink organs and sugar signalling activities in heat-tolerant tomato genotypes (Li et al., 2012). In the same way, they indicate a mechanism to allocate carbohydrate content in developing and mature pollen grains under HT, which is an important factor in determining pollen quality (Firon et al., 2006). Sugars have also been shown to act as antioxidants in plants (Lang-Mladek et al., 2010). At high concentrations sucrose acts as a ROS scavenger, while at low concentrations it acts as a signalling molecule (Sugio et al., 2009).

Plants protect themselves from oxidative damage under HT by enhancing the synthesis of secondary metabolites. HT induces accumulation of phenols in the plants either by stimulating their biosynthesis or by inhibiting their oxidation, as shown in watermelon. This could be a mechanism of acclimation of the plant against HT (Rivero et al., 2001). Anthocyanin pigments have a protective role under UV radiation stress where they work like a UV screen. However, under HT, accumulation of anthocyanin decreases leaf osmotic potential, which results in more uptake and transpirational loss of water. Together, all these factors enable the leaves to respond quickly and efficiently to changing environmental conditions (Wahid et al., 2007).

Carotenoid pigments also have a role in protecting plants from several kinds of stresses. For example, terpenoids such as isoprene or tocopherol and xanthophyll pigment contribute to stabilize and photo-protect the lipid phase of the thylakoid membranes (Velikova et al., 2005; Camejo et al., 2006). Several plant growth regulators, such as abscisic acid (ABA), salicylic acid (SA), ethylene (ET), cytokinins (CK) and auxins (AUX), have been shown to play an important role in conferring thermotolerance to plants (Kotak et al., 2007).

5.4   The Role of Other Novel Protective Agents

1. Coronatine (COR): *Pseudomonas syringae* produces a phytotoxin known as coronatine. It has a structure similar to jasmonates, which has several diverse roles in plant defence mechanisms (Uppalapati et al., 2005). Coronatine can manipulate plant hormone signalling to access nutrients and counteract defence responses. In addition, coronatine has been shown to affect nitrogen metabolism and chloroplast ultrastructure in plants and thus maintains photosynthetic efficiency and reduces crop yield loss under HT. Zhou et al. (2015) showed that, under HT, the photosynthetic rate of coronatine pretreated plants was 20.1% higher than in control plants. Mature grain was small and shrivelled under HT, but kernels were plump in coronatine pretreated plants, with thousand kernel weight (TKW) 8.2% higher than in the control. Twenty-two coronatine inducible proteins were found distributed in the chloroplast envelope, stroma and thylakoid membrane. Some, such as the 30S ribosomal protein S1, are involved in the biosynthetic pathway of chlorophyll. OsFTSH2 has a key role in catabolic process of PSII-associated light-harvesting complex II (LHCII). Coronatine pretreatment increased the stability of the chloroplast ultrastructure and membrane to maintain a higher photosynthesis under HT. In brief, coronatine mainly affects nitrogen metabolism by regulating protein processes and chloroplast ultrastructure to maintain photosynthetic performance in wheat under HT.

2. Isoprenes: Isoprene is a small hydrocarbon sometimes released from plants in large quantities (Sharkey, 2005). It is particularly associated with certain tree species, such as eucalyptus. The production of isoprene reduces the inhibition of photosynthesis during moderate HT, probably by associating with the thylakoid membrane to increase hydrophobic interactions and protect membrane function (Sharkey, 2005). Isoprene emitters may prove to be good candidates under HT by improving PSII performance and reduced dissipation of energy in the form of heat. Such plants will best suit higher temperatures under changing climatic conditions (Pollastri et al., 2014).

3. Irradiation: Plant scientists have been fascinated by the protective effect of natural solar irradiation in plants under HT; however, the basic mechanism of protection has yet to be unravelled. The protective role of irradiation may be because of the involvement of several factors such as the activation state of Rubisco, HSP accumulation and ROS-scavenging activity (Buchner *et al.*, 2015). It is hypothesized that, under present climatic conditions, longer periods of exposure to sub-lethal HT may lead to long-term impairment of the photosynthetic process. The presence of natural solar radiation during exposure to sub-lethal heat may help to protect photosynthetic functions; however, much research is required in this field.

5.4.1 Stress-inducible genes

Plants have the capacity of adapting to a varied range of temperatures by reprogramming their transcriptome, proteome and metabolome (Qi *et al.*, 2011; Sánchez-Rodríguez *et al.*, 2011). Plants can acclimate to HT by repairing some heat-sensitive components effectively and by preventing further heat injury, maintaining cellular homeostasis at the same time. Abundant production of HSPs is the most important characteristic of thermotolerance in plants (Vierling, 1991). Being a multigenic character, heat tolerance is regulated by several metabolic and biochemical traits such as the activity of antioxidants, unsaturation of membrane lipids, expression and translation of genes, stability of protein and accumulation of compatible solutes (Kaya *et al.*, 2001). As certain genotypes are found to be more tolerant, it is clear that plant responses to HT depend on genotypic parameters since some genotypes were found to be more tolerant in comparison to others (Prasad *et al.*, 2006; Challinor *et al.*, 2007).

Induction of HSP70 largely regulates thermotolerance in plants under HT. The enhanced expression of HSP70 was reported to assist in translocation, proteolysis, translation, folding, aggregation and refolding of denatured proteins (Gorantla *et al.*, 2007; Zhang *et al.*, 2010). In addition, HSP101 has an important role in thermotolerance (Gurley, 2000).

Elongation factor (EF-Tu) has a role in protein synthesis in chloroplasts under HT and exhibits chaperone activity (Fu *et al.*, 2008); its accumulation contributes to stress tolerance in plants under HT (Fu *et al.*, 2008; Prasad *et al.*, 2008). This was proved in the cultivars expressing a higher EF-Tu under HT stress and which were found to be more tolerant (Pressman *et al.*, 2002).

5.5 Conclusions and Future Prospects

Heat-tolerant plants exhibit acclimation of the photosynthetic process under HT stress. They do so by adjusting the composition of membranes, protein synthesis and metabolic regulation. Acclimation is also mediated by an increase in the expression of HSP. The up- or downregulation of genes involved in the photosynthetic process depends on the type, intensity and duration of the stress factor. Thus, a knowledge of the expression patterns of such genes may be helpful to develop transgenic lines of different crops with better photosynthetic performance under stress conditions. Manipulating over- or underexpression of particular genes related to photosynthesis may help to alleviate bottlenecks in photosynthesis. Thus, understanding the mechanisms of temperature acclimation of photosynthesis in different species and under different conditions is essential for faster improvements in crop production.

Acknowledgements

The authors thank Ms Margarita V. Rodionova for her valuable comments and corrections, and Ms Versha Tripathi for helping with the references. This research was supported by the Joint Indo-Russian project from the Department of Science and Technology, India (DST/RUS/RF-BR/P-173) to AJ. SIA was supported by grants from the Russian Foundation for Basic Research, and by the Molecular and Cell Biology Programs of the Russian Academy of Sciences.

References

Abiko, M., Akibayashi, K., Sakata, T., Kimura, M., Kihara, M., Itoh, K. and Higashitani, A. (2005) High-temperature induction of male sterility during barley (*Hordeum vulgare* L.) anther development is mediated by transcriptional inhibition. *Sexual Plant Reproduction* 18(2), 91–100.

Agrawal, D. and Jajoo, A. (2015) Investigating primary sites of damage in Photosystem II in response to high temperature. *Indian Journal of Plant Physiology* 20, 304–309.

Agrawal, D., Allakhverdiev, S.I. and Jajoo, A. (2015) Cyclic electron flow plays an important role in protection of spinach leaves (*Spinacia oleracia*) under high temperature stress. *Russian Journal of Plant Physiology*, 63(2), 210–215.

Ahmad, M.S.A., Ashraf, M. and Ali, Q. (2010) Soil salinity as a selection pressure is a key determinant for the evolution of salt tolerance in Blue Panicgrass (*Panicum antidotale* Retz.). *Flora - Morphology, Distribution, Functional Ecology of Plants* 205(1), 37–45.

Ainsworth, E.A. and Ort, D.R. (2010) How do we improve crop production in a warming world? *Plant Physiology* 154, 526–530.

Allakhverdiev, S.I. and Murata, N. (2004) Environmental stress inhibits the synthesis *de novo* of proteins involved in the photodamage-repair cycle of Photosystem II in *Synechocystis* sp. PCC 6803. *Biochimica et Biophysica Acta (Bioenergetics)* 1657, 23–32.

Allakhverdiev, S.I., Los, D.A., Mohanty, P., Nishiyama, Y. and Murata, N. (2007) Glycinebetaine alleviates the inhibitory effect of moderate heat stress on the repair of photosystem II during photoinhibition. *Biochimica et Biophysica Acta* 1767, 1363–1371.

Allakhverdiev, S.I., Kreslavski, V.D., Klimov, V.V., Los, D.A., Carpentier, R. and Mohanty, P. (2008) Heat stress: an overview of molecular responses in photosynthesis. *Photosynthesis Research* 98, 541–550.

Allakhverdiev, S.I., Los, D.A. and Murata, N. (2009) Regulatory roles in photosynthesis of unsaturation fatty acids in membrane lipids. In: Wada, H. and Murata, N. (eds) *Lipids in Photosynthesis: Essential and Regulatory Functions. Advances in Photosynthesis and Respiration.* Springer, Dordrecht, The Netherlands, pp. 373–388.

Allakhverdiev, S.I., Kreslavski, V.D., Fomina, I.R., Los, D.A., Klimov, V.V., Mimuro, M., Mohanty, P. and Carpentier, R. (2010) Inactivation and repair of photosynthetic machinery under heat stress. In: Guruprasad, K.N., Itoh, S. and Mohanty, P. (eds) *Photosynthesis: Overviews on Recent Progress and Future Perspective.* Narosa Publishing House, New Delhi, pp. 187–214.

Almeselmani, M., Deshmukh, P.S., Sairam, R.K., Kushwaha, S.R. and Singh, T.P. (2006) Protective role of antioxidant enzymes under high temperature stress. *Plant Science* 171(3), 382–388.

Asada, K. (2006) Production and scavenging of reactive oxygen species in chloroplasts and their functions. *Plant Physiology* 141, 391–396.

Ashraf, M. and Harris, P.J.C. (2013) Photosynthesis under stressful environments: an overview. *Photosynthetica* 51(2), 163–190.

Balint, I., Bhattacharya, J., Perelman, A., Schatz, D., Moskovitz, Y., Keren, N. and Schwarz, R. (2006) Inactivation of the extrinsic subunit of photosystem II, PsbU, in *Synechococcus* PCC 7942 results in elevated resistance to oxidative stress. *Federation of European Biochemical Societies* 580, 2117–2122.

Balla, K., Bencze, S., Janda, T. and Veisz, O. (2009) Analysis of heat stress tolerance in winter wheat. *Acta Agronomica Hungarica* 57(4), 437–444.

Barnabás, B., Jäger, K. and Fehér, A. (2008) The effect of drought and heat stress on reproductive processes in cereals. *Plant Cell Environment* 31, 11–38.

Barra, M., Haumann, M. and Dau, H. (2005) Specific loss of the extrinsic 18 KDa protein from Photosystem II upon heating to 47°C causes inactivation of oxygen evolution likely due to Ca release from the Mn complex. *Photosynthesis Research* 84, 231–237.

Berry, J.A. and Bjorkman, O. (1980) Photosynthetic response and adaptation to temperature in higher plants. *Annual Review of Plant Physiology* 31, 491–543.

Bita, C.E. and Gerats, T. (2013) Plant tolerance to high temperature in a changing environment: scientific fundamentals and production of heat stress-tolerant crops. *Frontiers in Plant Science* 4, 273.

Brestic, M. and Zivcak, M. (2013) PSII fluorescence techniques for measurement of drought and high temperature stress signal in crop plants: protocols and applications. In: Rout, G.R. and Das, A.B. (eds) *Molecular Stress Physiology of Plants.* Springer, Berlin, pp. 87–131.

Buchner, O., Stoll, M., Karadar, M., Kranner, I. and Neuner, G. (2015) Application of heat stress in situ demonstrates a protective role of irradiation on photosynthetic performance in alpine plants. *Plant & Cell Environment* 38, 812–826.

Bukhov, N. and Carpentier, R. (2004) Alternative photosystem I-driven electron transport routes: mechanisms and functions. *Photosynthesis Research* 82, 17–33.

Bukhov, N.G. and Mohanty, P. (1999) Elevated temperature stress effects on photosystems: characterization and evaluation of the nature of heat induced impairments. In: Singhal, G.S., Renger, G., Sopory, S.K., Irrgang, K.D. and Govindjee (eds) *Concepts in Photobiology.* Springer, The Netherlands, pp. 617–648.

Camejo, D., Jiménez, A., Alarcón, J.J., Torres, W., Gómez, J.M. and Sevilla, F. (2006) Changes in photosynthetic parameters and antioxidant activities following heat-shock treatment in tomato plants. *Functional Plant Biology* 33(2), 177–187.

Cao, J. and Govindjee, (1990) Chlorophyll *a* fluorescence transient as an indicator of active and inactive Photosystem II in thylakoid membranes. *Biochimica et Biophysica Acta* 1015, 180–188.

Challinor, A.J., Wheeler, T.R., Craufurd, P.Q., Ferro, C.A.T. and Stephenson, D.B. (2007) Adaptation of crops to climate change through genotypic responses to mean and extreme temperatures. *Agriculture, Ecosystems & Environment* 119(1), 190–204.

Chang, H.C., Tang, Y.C., Hayer-Hartl, M. and Hartl, F.U. (2007) Snap shot: molecular chaperones, part I. *Cell* 128(1), 212.

Crafts-Brandner, S.J. and Salvucci, M.E. (2000) Rubisco activase constrains the photosynthetic potential of leaves at high temperature and CO_2. *Proceedings of the National Academy of Sciences USA* 97,13430–13435.

Diaz, H.R. and Varon, G.G. (2013) Response of rice plants to heat stress during initiation of panicle primordia or grain-filling phases. *Journal of Stress Physiology and Biochemistry* 9(3), 318–325.

Dobra, J., Motyka, V., Dobrev, P., Malbeck, J., Prasil, I.T., Haisel, D. and Vankova, R. (2010) Comparison of hormonal responses to heat, drought and combined stress in tobacco plants with elevated proline content. *Journal of Plant Physiology* 167(16), 1360–1370.

Efeoglu, B. and Terzioglu, S. (2009) Photosynthetic responses of two wheat varieties to high temperature. *EurAsian Journal of BioScience* 3, 97–106.

Farooq, M., Basra, S.M.A., Wahid, A., Cheema, Z.A., Cheema, M.A. and Khaliq, A. (2008) Physiological role of exogenously applied glycinebetaine to improve drought tolerance in fine grain aromatic rice (*Oryza sativa* L.). *Journal of Agronomy and Crop Science* 194(5), 325–333.

Firon, N., Peet, M.M., Pharr, D.M., Zamski, E., Rosenfeld, K., Althan, L. and Pressman, E. (2006) Pollen grains of heat tolerant tomato cultivars retain higher carbohydrate concentration under heat stress conditions. *Scientia Horticulturae* 109, 212–217.

Fischer, R.A. and Edmeades, G.O. (2010) Breeding and cereal yield progress. *Crop Science* 50, 85–98.

Foyer, C.H. and Noctor, G. (2009) Redox regulation in photosynthetic organisms: signaling, acclimation, and practical implications. *Antioxidants & Redox Signaling* 11(4), 861–905.

Fu, J., Momcilovic, I., Clemente, T.E., Nersesian, N., Trick, H.N. and Ristic, Z. (2008) Heterologous expression of a plastid EF-Tu reduces protein thermal aggregation and enhances CO_2 fixation in wheat (*Triticum aestivum*) following heat stress. *Plant Molecular Biology* 68(3), 277–288.

Gechev, T.S., Van Breusegem, F., Stone, J.M., Denev, I. and Laloi, C. (2006) Reactive oxygen species as signals that modulate plant stress responses and programmed cell death. *Bioessays* 28(11), 1091–1101.

Gorantla, M., Babu, P.R., Lachagari, V.R., Reddy, A.M.M., Wusirika, R., Bennetzen, J.L. and Reddy, A.R. (2007) Identification of stress-responsive genes in an indica rice (*Oryza sativa* L.) using ESTs generated from drought-stressed seedlings. *Journal of Experimental Botany* 58(2), 253–265.

Greer, D.H. and Weedon, M.M. (2014) Does the hydrocooling of *Vitis vinifera* cv. Semillon vines protect the vegetative and reproductive growth processes and vine performance against high temperatures? *Functional Plant Biology* 41(6), 620–633.

Guisse, B., Srivastava, A. and Strasser, R.J. (1995) The polyphasic rise of the chlorophyll *a* fluorescence (OKJIP) in heat-stressed leaves. *Archives in Science Geneve* 48, 147–160.

Gurley, W.B. (2000) HSP101: a key component for the acquisition of thermotolerance in plants. *Plant Cell* 12, 457–460.

Hasanuzzaman, M., Nahar, K., Alam, M.M., Roychowdhury, R. and Fujita, M. (2013) Physiological, biochemical, and molecular mechanisms of heat stress tolerance in plants. *International Journal of Molecular Sciences* 14(5), 9643–9684.

Hong, S.W., Lee, U. and Vierling, E. (2003) Arabidopsis *hot* mutants define multiple functions required for acclimation to high temperatures. *Plant Physiology* 132(2), 757–767.

Huang, B. and Xu, C. (2008) Identification and characterization of proteins associated with plant tolerance to heat stress. *Journal of Integrative Plant Biology* 50(10), 1230–1237.

Iba, K. (2002) Acclimative response to temperature stress in higher plants: approaches of gene engineering for temperature tolerance. *Annual Review of Plant Biology* 53(1), 225–245.

Janka, E., Korner, O., Rosenqvist, E. and Ottosen, C.O. (2013) High temperature stress monitoring and detection using chlorophyll *a* fluorescence and infrared thermography in chrysanthemum (*Dendranthema grandiflora*). *Plant Physiology & Biochemistry* 67, 87–94.

Kana, B.D., Gordhan, B.G., Downing, K.J., Sung, N., Vostroktunova, G., Machowski, E.E., Tsenova, L., Young M., Kaprelyants, A., Kaplan, G. and Mizrahi, V. (2008) The resuscitation-promoting factors of *Mycobacterium tuberculosis* are required for virulence and resuscitation from dormancy but are collectively dispensable for growth *in vitro*. *Molecular Microbiology* 67(3), 672–684.

Karuppanapandian, T., Wang, H.W., Prabakaran, N., Jeyalakshmi, K., Kwon, M., Manoharan, K. and Kim, W. (2011a) 2, 4-dichlorophenoxyacetic acid-induced leaf senescence in mung bean (*Vigna radiata* L. Wilczek) and senescence inhibition by co-treatment with silver nanoparticles. *Plant Physiology and Biochemistry* 49(2), 168–177.

Karuppanapandian, T., Moon, J.C., Kim, C., Manoharan, K. and Kim, W. (2011b) Reactive oxygen species in plants: their generation, signal transduction, and scavenging mechanisms. *Australian Journal of Crop Science* 5, 709–725.

Kaya, H., Shibahara, K.I., Taoka, K.I., Iwabuchi, M., Stillman, B. and Araki, T. (2001) *FASCIATA* genes for chromatin assembly factor-1 in *Arabidopsis* maintain the cellular organization of apical meristems. *Cell* 104(1), 131–142.

Königshofer, H., Tromballa, H.W. and Loppert, H.G. (2008) Early events in signalling high-temperature stress in tobacco BY2 cells involve alterations in membrane fluidity and enhanced hydrogen peroxide production. *Plant Cell and Environment* 31(12), 1771–1780.

Kotak, S., Larkindale, J., Lee, U., von Koskull-Döring, P., Vierling, E. and Scharf, K.D. (2007) Complexity of the heat stress response in plants. *Current Opinion in Plant Biology* 10(3), 310–316.

Kultz, D. (2003) Evolution of the cellular stress proteome: from monophyletic origin to ubiquitous function. *Journal of Experimental Biology* 206(18), 3119–3124.

Kurek, I., Chang, T.K., Bertain, S.M., Madrigal, A., Liu, L., Lassner, M.W. and Zhu, G. (2007) Enhanced thermostability of *Arabidopsis* Rubisco activase improves photosynthesis and growth rates under moderate heat stress. *Plant Cell* 19, 3230–3241.

Lang-Mladek, C., Popova, O., Kiok, K., Berlinger, M., Rakic, B., Aufsatz, W. and Luschnig, C. (2010) Transgenerational inheritance and resetting of stress-induced loss of epigenetic gene silencing in *Arabidopsis*. *Molecular Plant* 3(3), 594–602.

Larkindale, J., Mishkind, M. and Vierling, E. (2005a) Plant responses to high temperature. In: Jenks, M.A. and Hasegawa, P.M. (eds) *Plant Abiotic Stress*. Blackwell Publishing, Oxford, UK, pp. 100–144.

Larkindale, J., Hall, J.D., Knight, M.R. and Vierling, E. (2005b) Heat stress phenotypes of Arabidopsis mutants implicate multiple signaling pathways in the acquisition of thermotolerance. *Plant Physiology* 138(2), 882–897.

Lazar, D. and Pospisil, P. (1999) Mathematical simulation of chlorophyll *a* fluorescence rise measured with 3-[30,40-dichlorophenyl]-1,1-dimethylurea-treated barley leaves at room and high temperature. *European Biophysics Journal* 28, 468–477.

Li, M., Ji, L., Yang, X., Meng, Q. and Guo, S. (2012) The protective mechanisms of CaHSP26 in transgenic tobacco to alleviate photoinhibition of PSII during chilling stress. *Plant Cell Reports* 31(11), 1969–1979.

Liu, X. and Huang, B. (2000) Carbohydrate accumulation in relation to heat stress tolerance in two creeping bentgrass cultivars. *Journal of the American Society for Horticultural Science* 125, 442–447.

Lobell, D.B. and Asner, G.P. (2003) Climate and management contributions to recent trends in U.S. agricultural yields. *Science* 299, 1032.

Los, D.A. and Murata, N. (2004) Membrane fluidity and its roles in the perception of environmental signals. *Biochimica et Biophysica Acta* 1666, 142–157.

Mathur, S., Jajoo, A., Mehta, P. and Bharti, S. (2011a) Analysis of elevated temperature induced inhibition of photosystem II by using chlorophyll *a* fluorescence induction kinetics in wheat leaves (*Triticum aestivum*). *Plant Biology* 13, 1–6.

Mathur, S., Allakhverdiev, S.I. and Jajoo, A. (2011b) Analysis of high temperature stress on the dynamics of antenna size and reducing side heterogeneity of Photosystem II in wheat leaves (*Triticum aestivum*). *Biochimica et Biophysica Acta* 1807, 22–29.

Mathur, S., Agrawal, D. and Jajoo, A. (2014) Photosynthesis: response to high temperature stress. *Journal of Photochemistry and Photobiology B: Biology* 137, 116–126.

Mittler, R., Vanderauwera, S., Gollery, M. and Van Breusegem, F. (2004) The reactive oxygen gene network in plants. *Trends in Plant Science* 9, 490–498.

Mittler, R., Vanderauwera, S., Suzuki, N., Miller, G., Tognetti, V.B., Vandepoele, K. and Van Breusegem, F. (2011) ROS signaling: the new wave? *Trends in Plant Science* 16(6), 300–309.

Mittler, R., Finka, A. and Goloubinoff, P. (2012) How do plants feel the heat? *Trends in Biochemical Sciences* 37(3), 118–125.

Mohanty, S., Baishna, B.G. and Tripathy, C. (2006) Light and dark modulation of chlorophyll biosynthetic genes in response to temperature. *Planta* 224, 692–699.

Mohanty, P., Allakhverdiev, S.I. and Murata, N. (2007) Application of low temperatures during photoinhibition allows characterization of individual steps in photodamage and the repair of photosystem II. *Photosynthesis Research* 94, 217–224.

Mohanty, P., Kreslavski, V.D., Klimov, V.V., Los, D.A., Mimuro, M., Carpentie, R. and Allakhverdiev, S.I. (2012) Heat stress: susceptibility, recovery and regulation. In: Eaton-Rye, J.J., Tripathy, B.C. and Sharkey, T.D. (eds) *Photosynthesis: Plastid Biology, Energy Conversion and Carbon Assimilation*. Springer, Dordrecht, The Netherlands, pp. 251–274.

Moller, I.M., Jensen, P.E. and Hansson, A. (2007) Oxidative modifications to cellular components in plants. *Annual Review of Plant Biology* 58, 459–481.

Morrow, G. and Tanguay, R.M. (2012) Small heat shock protein expression and functions during development. *The International Journal of Biochemistry and Cell Biology* 44(10), 1613–1621.

Murakami, Y., Tsuyama, M., Kobayashi, Y., Kodama, H. and Iba, K. (2000) Trienoic fatty acids and plant tolerance of high temperature. *Science* 287, 476–479.

Murata, N., Takahashi, S., Nishiyama, Y. and Allakhverdiev, S.I. (2007) Photoinhibition of photosystem II under environmental stress. *Biochimica et Biophysica Acta* 1767, 414–421.

Nagarajan, S., Jagadish, S.V.K., Hari Prasad, A.S., Tomar, A.K., Anand, A., Pal, M. and Aggarwal, P.K. (2010) Effect of night temperature and radiation on growth, yield, and grain quality of aromatic and non-aromatic rice. *Agriculture Ecosystems and Environment* 138, 274–281.

Nishiyama, Y., Allakhverdiev, S.I., Yamamoto, H., Hayashi, H. and Murata, N. (2004) Singlet oxygen inhibits the repair of Photosystem II by suppressing the translation elongation of the D1 protein in *Synechocystis* sp. PCC 6803. *Biochemistry* 43, 11321–11330.

Pollastri, S., Tsonev, T. and Loreto, F. (2014) Isoprene improves photochemical efficiency and enhances heat dissipation in plants at physiological temperatures. *Journal of Experimental Botany* 65(6), 1565–1570.

Pospisil, P., Haumann, M., Dittmer, J., Sole, V.A. and Dau, H. (2003) Stepwise titration of the tetra-manganese complex of photosystem II to a binuclear $Mn_2(\mu\text{-}O)_2$ complex in response to a temperature jump: a time-resolved structural investigation employing X-ray absorption spectroscopy. *Biophysics Journal* 84, 1370–1386.

Prasad, P.V.V., Boote, K.J., Allen, L.H., Sheehy, J.E. and Thomas, J.M.G. (2006) Species, ecotype and cultivar differences in spikelet fertility and harvest index of rice in response to high temperature stress. *Field Crops Research* 95(2), 398–411.

Prasad, P.V., Pisipati, S.R., Mutava, R.N. and Tuinstra, M.R. (2008) Sensitivity of grain sorghum to high temperature stress during reproductive development. *Crop Science* 48(5), 1911–1917.

Pressman, E., Peet, M.M. and Pharr, D.M. (2002) The effect of heat stress on tomato pollen characteristics is associated with changes in carbohydrate concentration in the developing anthers. *Annals of Botany* 90(5), 631–636.

Qi, Y., Wang, H., Zou, Y., Liu, C., Liu, Y., Wang, Y. and Zhang, W. (2011) Over-expression of mitochondrial heat shock protein 70 suppresses programmed cell death in rice. *FEBS Letters* 585(1), 231–239.

Quan, R., Shang, M., Zhang, H., Zhao, Y. and Zhang, J. (2004) Engineering of enhanced glycine betaine synthesis improves drought tolerance in maize. *Plant Biotechnology Journal* 2(6), 477–486.

Rivero, R.M., Ruiz, J.M., García, P.C., López-Lefebre, L.R., Sánchez, E. and Romero, L. (2001) Resistance to cold and heat stress: accumulation of phenolic compounds in tomato and watermelon plants. *Plant Science* 160(2), 315–321.

Roitsch, T. and González, M.C. (2004) Function and regulation of plant invertases: sweet sensations. *Trends in Plant Science* 9(12), 606–613.

Sage, R.F. and Kubien, D.S. (2007) The temperature response of C_3 and C_4 photosynthesis. *Plant Cell and Environment* 30, 1086–1106.

Sairam, R.K., Srivastava, G.C. and Saxena, D.C. (2000) Increased antioxidant activity under elevated temperatures: a mechanism of heat stress tolerance in wheat genotypes. *Biologia Plantarum* 43(2), 245–251.

Sakamoto, A. and Murata, N. (2000) Genetic engineering of glycinebetaine synthesis in plants: current status and implications for enhancement of stress tolerance. *Journal of Experimental Botany* 51(342), 81–88.

Sakamoto, A. and Murata, N. (2002) The role of glycine betaine in the protection of plants from stress: clues from transgenic plants. *Plant Cell and Environment* 25(2), 163–171.

Sakata, T. and Higashitani, A. (2008) Male sterility accompanied with abnormal anther development in plants–genes and environmental stresses with special reference to high temperature injury. *International Journal of Plant Biology* 2, 42–51.

Salvucci, M.E. and Crafts-Brandner, S.J. (2004) Inhibition of photosynthesis by heat stress: the activation state of Rubisco as a limiting factor in photosynthesis. *Physiologia Plantarum* 120, 179–186.

Salvucci, M.E., Osteryoung, K.W., Crafts-Brandner, S.J. and Vierling, E. (2001) Exceptional sensitivity of Rubisco activase to thermal denaturation *in vitro* and *in vivo*. *Plant Physiology* 127, 1053–1064.

Salvucci, M.E., DeRidder, B.P., and Portis Jr, A.R. (2006) Effect of activase level and isoform on the thermotolerance of photosynthesis in *Arabidopsis*. *Journal of Experimental Botany* 57, 3793–3799.

Sánchez-Rodríguez, E., Moreno, D.A., Ferreres, F., del Mar Rubio-Wilhelmi, M. and Ruiz, J.M. (2011) Differential responses of five cherry tomato varieties to water stress: changes on phenolic metabolites and related enzymes. *Phytochemistry* 72(8), 723–729.

Scafaro, A.P., Haynes, P.A. and Atwell, B.J. (2010) Physiological and molecular changes in *Oryza meridionalis* Ng., a heat-tolerant species of wild rice. *Journal of Experimental Botony* 61, 191–202.

Schmitt, F.J., Renger, G., Friedrich, T., Kreslavski, V.K., Zharmukhamedov, S.K., Los, D.A., Kuznetsov, V.V. and Allakhverdiev, S.I. (2014) Reactive oxygen species: re-evaluation of generation, monitoring and role in stress-signaling in phototrophic organisms. *Biochimica et Biophysica Acta* 1837, 835–848.

Sharkey, T.D. (2005) Effects of moderate heat stress on photosynthesis: importance of thylakoid reactions, rubisco deactivation, reactive oxygen species, and thermotolerance provided by isoprene. *Plant Cell and Environment* 28, 269–277.

Soliman, W.S., Fujimori, M., Tase, K. and Sugiyama, S.I. (2011) Oxidative stress and physiological damage under prolonged heat stress in C_3 grass Lolium perenne. *Grassland Science* 57, 101–106.

Srivastava, A., Guissre, B., Greppin, H. and Strasser, R.J. (1997) Regulation of antenna structure and electron transport in Photosystem II of *Pisum sativum* under elevated temperature probed by the fast polyphasic chlorophyll *a* fluorescence transient: OKJIP. *Biochimica et Biophysica Acta* 1320, 95–106.

Strasser, R.J., Tsimilli-Michael, M. and Srivastava, A. (2004) Analysis of the chlorophyll *a* fluorescence transient. In: Papageorgiou, G.C. and Govindjee (eds) *Advances in Photosynthesis and Respiration Series, Chlorophyll a Fluorescence: A Signature of Photosynthesis.* Springer, Dordrecht, The Netherlands, pp. 321–362.

Sugio, A., Dreos, R., Aparicio, F. and Maule, A.J. (2009) The cytosolic protein response as a subcomponent of the wider heat shock response in *Arabidopsis. The Plant Cell* 21(2), 642–654.

Tewari, A.K. and Tripathy, B.C. (1998) Temperature-stress-induced impairment of chlorophyll biosynthetic reactions in cucumber and wheat. *Plant Physiology* 117, 851–858.

Tiwari, A., Jajoo, A. and Bharti, S. (2008) Heat induced changes in the EPR signal of tyrosine D (YOxD): a possible role of Cytochrome b559. *Journal of Bioenergetics and Biomembrane* 40, 237–243.

Uppalapati, S.R., Ayoubi, P., Weng, H., Palmer, D.A., Mitchell, R.E., Jones, W. and Bender, C.L. (2005) The phytotoxin coronatine and methyl jasmonate impact multiple phytohormone pathways in tomato. *The Plant Journal* 42, 201–217.

Velikova, V., Pinelli, P., Pasqualini, S., Reale, L., Ferranti, F. and Loreto, F. (2005) Isoprene decreases the concentration of nitric oxide in leaves exposed to elevated ozone. *New Phytologist* 166(2), 419–426.

Vierling, E. (1991) The roles of heat shock proteins in plants. *Annual Review of Plant Biology* 42(1), 579–620.

Wahid, A. and Close, T.J. (2007) Expression of dehydrins under heat stress and their relationship with water relations of sugarcane leaves. *Biologia Plantarum* 51(1), 104–109.

Wahid, A., Gelani, S., Ashraf, M. and Foolad, M.R. (2007) Heat tolerance in plants: an overview. *Environmental and Experimental Botany* 61(3), 199–223.

Wise, R.R., Olson, A.J., Schrader, S.M. and Sharkey, T.D. (2004) Electron transport is the functional limitation of photosynthesis in field-grown Pima cotton plants at high temperature. *Plant Cell and Environment* 27, 717–724.

Xu, S., Li, J., Zhang, X., Wei, H. and Cui, L. (2006) Effects of heat acclimation pretreatment on changes of membrane lipid peroxidation, antioxidant metabolites, and ultrastructure of chloroplasts in two cool-season turf grass species under heat stress. *Environmental and Experimental Botany* 56(3), 274–285.

Yamane, Y., Shikanai, T., Kashino, Y., Koike, H. and Satoh, K. (2000) Reduction of Q_A in the dark: another cause of fluorescence F_0 increases by high temperatures in higher plants. *Photosynthesis Research* 63, 23–34.

Yamori, W., Hikosaka, K. and Way, D.A. (2013) Temperature response of photosynthesis in C_3, C_4, and CAM plants: temperature acclimation and temperature adaptation way. *Photosynthesis Research* 119, 101–117.

Yoshioka, M., Uchida, S., Mori, H., Komayama, K., Ohira, S., Morita, N., Nakanishi, T. and Yamamoto, Y. (2006) Quality control of photosystem II. Cleavage of reaction center D1 protein in spinach thylakoids by FtsH protease under moderate heat stress. *Journal of Biological Chemistry* 281(31), 21660–21669.

Zhang, J.X., Wang, C., Yang, C.Y., Wang, J.Y., Chen, L., Bao, X.M. and Liu, J. (2010) The role of arabidopsis AtFes1A in cytosolic Hsp70 stability and abiotic stress tolerance. *The Plant Journal* 62(4), 539–548.

Zhang, Y. Guanter, L., Berry, J.A., Joiner, J., Van Der Tol, C., Huete, A., Gitelson, A., Voigt, M. and Kohler, P. (2014) Estimation of vegetation photosynthetic capacity from space-based measurements of chlorophyll fluorescence for terrestrial biosphere models. *Global Change Biology* 20, 3727–3742.

Zhou, Y., Zhang, M., Li, J., Li, Z., Tian, X. and Duan, L. (2015) Phytotoxin coronatine enhances heat tolerance via maintaining photosynthetic performance in wheat based on Electrophoresis and TOF-MS analysis. *Scientific Reports* 5, 13870.

6 Flooding Stress Tolerance in Plants

Chiara Pucciariello* and Pierdomenico Perata
PlantLab, Institute of Life Sciences, Scuola Superiore Sant'Anna, Pisa, Italy

Abstract

Global warming is associated with an increase in submergence and flooding events, which makes many ecosystems worldwide vulnerable to low oxygen stress. Water submersion can severely affect crop production, since it drastically reduces the oxygen needed for plant respiration, and thus survival. Plants tolerant to flooding have evolved morphological, physiological and biochemical adaptations to oxygen deficiency. In the plant biology model species *Arabidopsis thaliana*, considerable progress has been made in terms of understanding the molecular aspects governing these responses and the sensing mechanism of an oxygen shortage has been identified. Many studies on oxygen deprivation stress have focused on rice (*Oryza sativa*), since it is one of the crops that adapts best to a flooded environment. Besides being able to germinate under submergence, rice varieties display different mechanisms for successful survival. Agronomically, the study of rice strategies to survive flooding in ecotypes that have adapted to extreme environments shows great potential in the context of climate change and the increasing global need for food.

6.1 Introduction

Many ecosystems worldwide are vulnerable to either progressive or rapid flooding, such as areas close to watercourses or exposed to monsoons. Global warming is associated with an increase in flooding events characterized by their unexpected occurrence, regimes and localization. At the same time, the increase in the global demand for food requires a large increase in crop yields, especially in marginal agricultural areas. Unusual water submersion resulting in excessively wet or flooded soils can severely affect crop production and is also coupled with a modification in plant distribution in natural ecosystems (Bailey-Serres and Voesenek, 2008).

Plants are aerobic organisms and need oxygen (O_2) to survive, thus they suffer severely from O_2 deprivation. Water submersion drastically reduces O_2 availability since it diffuses slowly in water, dropping to concentrations that restrict aerobic respiration. Water is a remarkable barrier to general gas diffusion, leading to concomitant phenomena such as restricted access to CO_2 and ethylene entrapment in submerged organs (Voesenek and Bailey-Serres, 2015). A flooded environment can also suffer from low light, thus reducing photosynthesis, and from soil modification resulting in high concentrations of toxic compounds (Voesenek and Bailey-Serres, 2015). Consequently, an underwater scarcity of O_2 alone may not always be the major cause

---

* Corresponding author: c.pucciariello@sssup.it

of injuries due to submergence; instead, a combination of different types of stress eventually lead to restrictions in ATP synthesis and carbohydrate reserves.

Crop plants are usually very sensitive to submergence, thus new genetic and agronomic strategies are needed in order to increase production in flood-prone regions. Many studies on O_2 deprivation stress have focused on flooding-tolerant genotypes of rice (*Oryza sativa*). The characterization of tolerance-related molecular traits could represent a major step towards the successful breeding of tolerant varieties, which also have high yields (Bailey-Serres and Voesenek, 2010). However, rice ecotypes vary considerably in their responses to flooding, and only a limited number of varieties can withstand a prolonged complete submergence (Xu *et al.*, 2006). Adaptive responses to specific water regimes and unexploited resources of wild rice relatives, when compatible with breeding, have the potential to mitigate the environmental adversity aggravated by climate change (Hattori *et al.*, 2009). Indeed, rice cultivation is widespread and the local adaptation of plants grown in marginal areas and selected by farmers may be crucial to isolate agronomically interesting genes.

Initial efforts were aimed at describing morphological, physiological and biochemical aspects of plant adaptation to O_2 shortage. Considerable progress has since been made in terms of understanding the molecular aspects governing these responses. In the plant biology model species *Arabidopsis thaliana*, the molecular mechanism that senses O_2 and guides the metabolism towards low O_2 adaptation has been identified (Gibbs *et al.*, 2011; Licausi *et al.*, 2011). Ethylene Responsive Factors (ERFs) of group VII are characterized by a conserved motif at the N-end protein and are a target of O_2-dependent proteolysis. On the other hand, group VII ERFs are stable under low O_2 and can activate downstream genes related to plant tolerance.

Plant adaptation to submergence has been classified into two main strategies: low oxygen quiescence syndrome (LOQS) and low oxygen escape syndrome (LOES) (Bailey-Serres and Voesenek, 2008; Colmer and Voesenek, 2009). A major feature of LOQS is the reduction in underwater shoot growth, to conserve substrate availability until the water recedes. On the other

hand, plants with LOES are characterized by fast underwater growth to quickly reach the water surface and re-establish a gas exchange (Bailey-Serres and Voesenek, 2008). This trait is successful only in genotypes capable of reaching the air before consuming all the energy available. The genetic mechanism behind LOQS and LOES have been clarified in lowland rice (Fukao *et al.*, 2006; Xu *et al.*, 2006) and deepwater rice varieties (Hattori *et al.*, 2009), respectively. In both cases, ERFs of group VII play a central role in guiding differing rice adaptations to low O_2.

This chapter focuses first on the major metabolic rearrangements consequent to flooding stress, thus low O_2 in plants, and discusses the major findings related to the various adaptation mechanisms. The low-O_2 sensing mechanism relying on the N-end dependent protein stability is detailed, together with other molecular regulation events in the cell state. The different strategies to survive O_2 deprivation in rice are then discussed in order to give an overview of the most important findings regarding the molecular aspects behind LOQS and LOES responses.

6.2 Plant Metabolism Under Low Oxygen

Oxygen is the final acceptor of electrons in the mitochondrial respiratory chain. This is the last step in plant respiration that leads to major ATP synthesis, which is necessary for all ATP-demanding processes, such as cell division. With an O_2 deficit, a tight regulation of ATP production occurs through the shift of respiration from aerobic to anaerobic. This change relies primarily on glycolysis and fermentation to generate ATP and regenerate NAD^+ to sustain glycolysis, respectively. However, because of glycolysis inefficiency in ATP production, an energy crisis ensues. Fermentation yields only 2–4 mol ATP per mol hexose compared to the 30–36 mol ATP produced by aerobic respiration (Gibbs and Greenway, 2003). Early reports described an increase in glycolytic ATP production under anoxia, named the 'Pasteur effect' (Neal and Girton, 1955; Vartapetian, 1982). However, under complete anoxia, energy production can vary between 3% and 37.5% compared to production in normoxic conditions (Licausi and Perata, 2009). This variation depends not only on the efficiency of the glycolytic flux, but also on the amount of starch or sucrose available for use as a substrate and/or the plant's capacity to catabolize them (Gibbs and Greenway, 2003).

Under anaerobic conditions, a short initial lactic and a long-lasting ethanol fermentation occur after glycolysis (Perata and Alpi, 1993). Lactic fermentation is a one-step reaction from pyruvate to lactate, catalysed by lactate dehydrogenase (LDH) with the regeneration of NAD$^+$ for re-use by glycolysis. Ethanolic fermentation is a two-step process that also regenerates NAD$^+$, in which pyruvate is first decarboxylated to acetaldehyde by pyruvate decarboxylase (PDC), and acetaldehyde is subsequently converted to ethanol by alcohol dehydrogenase (ADH) (Fig. 6.1). The accumulation of lactate as an end-product of fermentation can impair plant survival under water, since its dissociation contributes to cell cytosol acidification (Davies et al., 1974; Davies, 1980). Following the 'pH-stat' hypothesis, a reduction in the cytosolic pH value by lactate would appear to limit the lactate production itself and favour PDC, thus channelling the regeneration of NAD$^+$ towards ethanol synthesis via the ADH activity (Davies et al., 1974). High LDH activity seems to stimulate ethanolic fermentation, suggesting that lactic fermentation is either required to trigger or to favour ethanolic fermentation (Dolferus et al., 2008). Overexpression of LDH1 in Arabidopsis plants resulted in an improved root survival under low O$_2$ stress, while knockout LDH1 mutants showed a reduced survival under the same conditions (Dolferus et al., 2008). Dolferus et al. (2008) also showed that Arabidopsis plants are, surprisingly, able to release lactate into the growth medium, thus preventing the accumulation of toxic levels in the cells.

Ethanol toxicity cannot be considered as a primary cause of low O$_2$ injuries. In fact cell membranes are permeable to ethanol and it can spread to the extracellular medium (Davies et al., 1974). Acetaldehyde, which is an intermediate product of ethanol fermentation, is toxic (Perata and Alpi, 1991). Acetaldehyde can also be metabolized through its conversion to acetate by aldehyde dehydrogenase (ALDH) to enter the TCA, by reducing NAD$^+$ to NADH. However, an enhanced activity of ALDH under hypoxia would deplete glycolysis of the NAD$^+$ pool (Kürsteiner et al., 2003). The release of the fermentative products ethanol and acetaldehyde has been monitored in many plants under low O$_2$, and the results suggest that activation of ethanolic fermentation is one of the strategies for plants to survive under anoxia (Table 6.1).

An alternative hypothesis of the switch from lactic to ethanolic fermentation relies on the 'PDH/PDC stat'. This suggests an increase in pyruvate (subsequent to aerobic respiration being blocked under O$_2$ deprivation), which becomes available for the PDC reaction. In fact, PDC has a higher K$_m$ than pyruvate dehydrogenase (PDH), which catalyses its conversion to acetyl CoA (Tadege et al., 1999). Studies of the ethanol fermentation pathway in transgenic plants suggested that PDC is the metabolic control point of the alcohol fermentation pathway. While Arabidopsis plants overexpressing ADH1 did not show any increase in plant survival and ethanol concentration under low O$_2$, PDC1 and PDC2 overexpressing plants improved hypoxia tolerance and showed an increase in ethanol production (Ismond et al., 2003). However, ADH1 null mutants displayed a decreased tolerance to low O$_2$ and a reduction in ethanol production, suggesting that a normal level of ADH1 expression is critical for plant survival (Ismond et al., 2003).

Other end products of the anaerobic metabolism have also been reported. Ala is one of the molecules that accumulates under anaerobic conditions without harmful consequences (Reggiani et al., 1988). It is thought to have a role in pH balance regulation under low O$_2$ and to contribute to reducing pyruvate levels to avoid glycolysis inhibition (Rocha et al., 2010). However, the reaction for Ala formation does not consume NADH. Ala is produced by the action of Ala aminotransferase (AlaAT) which catalyses the reversible reaction of pyruvate and glutamate interconversion to Ala and 2-oxoglutarate (Streeter and Thompson, 1972; Reggiani et al., 1988; Vanlerberghe et al., 1990). In Lotus japonicus 2-oxoglutarate has been suggested to contribute to ATP generation via succinate production through a reorganization of the TCA cycle (Rocha et al., 2010). During flooding, a rapid induction of the expression of the AlaAT gene as well as an increase in the

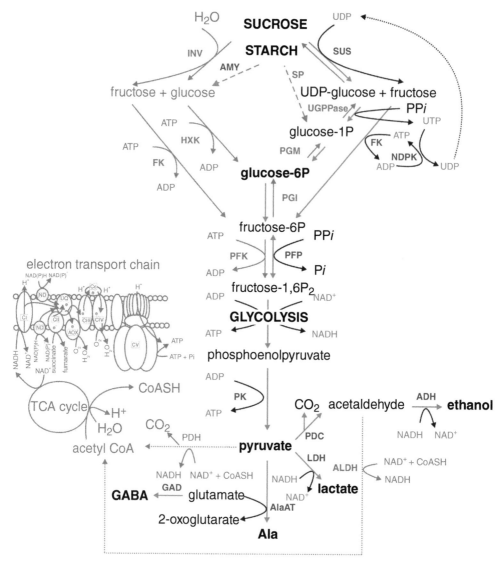

Fig. 6.1. Metabolic shift under O_2 deprivation. Red arrows indicate reactions promoted during O_2 deprivation. Abbreviations: ADH, alcohol dehydrogenase; AlaAT, alanine aminotransferase; ALDH, aldehyde dehydrogenase; AMY, amylases; AOX, alternative oxidase; C, complex (I, II, III, IV, V); Cc, cytochrome *c*; CoASH, coenzyme A; FK, fructokinase; GAD, glutamic acid decarboxylase; HXK, hexokinase; INV, invertase; LDH, lactate dehydrogenase; ND, alternate NAD(P)H dehydrogenases; NDPK, nucleoside-diphosphate kinase; PFK, ATP-dependent phosphofructokinase; PFP, PPi-dependent phosphofructokinase, PDC, pyruvate decarboxylase; PDH, pyruvate dehydrogenase; PGI, phosphoglucoisomerase; PGM, phosphoglucomutase; PK, pyruvate kinase; SP, starch phosphorylase; SUS, sucrose synthase; UGPPase, UDP-glucose pyrophosphorylase; UQ, ubiquinone; UTP, uridine triphosphate. Part of this figure has been redrawn with modification from Bailey-Serres and Voesenek, 2008.

activity of the enzyme has been observed (Good and Crosby, 1989; Good and Muench, 1992; Muench and Good, 1994). *AlaAT* belongs to the hypoxic gene core in both *Arabidopsis* and rice (Mustroph *et al.*, 2009, 2010). Another means of Ala production is by γ-aminobutyric acid transaminase (GABA-T), which uses pyruvate as a co-substrate. The two

Table 6.1. Effect of anoxia on ethanol and acetaldehyde concentration in various plants under O_2 deprivation.

Experiment	Species	Ethanol	Acetaldehyde	Reference
24 h incubation in anaerobic conditions	Woody plant leaves *Quercus alba* *Liquidambar styraciflua* *Fraxinus americana* *Fraxinus pennsylvanica* *Populus deltoides* *Pinus taeda*	175 ± 76 ng ml headspace^{-1} 241 ± 97 ng ml headspace^{-1} 227 ± 110 ng ml headspace^{-1} 188 ± 117 ng ml headspace^{-1} 187 ± 43 ng ml headspace^{-1} 226 ± 13 ng ml headspace^{-1}	35 ± 9 ng ml headspace^{-1} 53 ± 26 ng ml headspace^{-1} 26 ± 8 ng ml headspace^{-1} 32 ± 25 ng ml headspace^{-1} 130 ± 63 ng ml headspace^{-1} 10 ± 10 ng ml headspace^{-1}	Kimmerer and MacDonald, 1987
4 h anoxia	Tomato roots (*Lycopersicon esculentum* M., var UC 82b)	0 nmol mg fw^{-1} +100 mM suc: ± 0.2 nmol mg fw^{-1} +100 mM glu: ± 13 nmol mg fw^{-1}		Germain *et al.*, 1997
Anoxia	Rice coleoptiles (*Oryza sativa* L.) cv. Calrose	3 days air + 2 days anoxia: 0.21 ± 0.02 µl g fw^{-1}min^{-1} 3 days hypoxia + 3 days anoxia: $\pm0.20\pm0.01$ µl g fw^{-1}min^{-1} 5 days anoxia: 0.11 ± 0.01 µl g fw^{-1}min^{-1}		Gibbs *et al.*, 2000
	cv. IR22	3 days air + 2 days anoxia: 0.09 ± 0.01 µl g fw^{-1}min^{-1} 3 days hypoxia + 3 days anoxia: 0.07 ± 0.00 µl g fw^{-1}min^{-1} 5 days anoxia: 0.08 ± 0.01 µl g fw^{-1}min^{-1}		
63 days of anoxic incubation	*Acorus calamus* Roots Rhizome Leaves	±80 µmol g fw^{-1} ±20 µmol g fw^{-1} ±10 µmol g fw^{-1}		Schlüter and Crawford, 2001
	Iris pseudacorus Roots Rhizome Leaves	±60 µmol g fw^{-1} ±20 µmol g fw^{-1} ±20 µmol g fw^{-1}		

Continued

Table 6.1. Continued.

Experiment	Species	Ethanol	Acetaldehyde	Reference
14 h anaerobic conditions in the dark	Rice seedlings (*Oryza sativa* L.)	cv. FR13A, sub-tolerant: 40±9 µl g fw⁻¹ h⁻¹ cv. CT6241, sub-intolerant: 39±6 µl g fw⁻¹ h⁻¹	0.9±0.13 µl g FW⁻¹ h⁻¹ 1.40±0.3 µl g FW⁻¹ h⁻¹	Boamfa *et al.*, 2003, 2005
12 h anaerobic conditions in the light	Rice seedlings (*Oryza sativa* L.)	cv. FR13A, sub-tolerant: 8±0.9 µl g fw⁻¹ h⁻¹ cv. CT6241, sub-intolerant: 11±1 µl g fw⁻¹ h⁻¹	0.05±0.013 µl g FW⁻¹ h⁻¹ 0.10±0.02 µl g FW⁻¹ h⁻¹	
4 h anaerobic conditions in the light	Rice seedlings (*Oryza sativa* L. cv. Cigalon) Wheat seedlings (*Triticum aestivum* L. cv. Alcedo)	root: ±8 µl g fw⁻¹ h⁻¹ shoot: ±20 µl g fw⁻¹ h⁻¹ root: ±1 µl g fw⁻¹ h⁻¹ shoot: ±0 µl g fw⁻¹ h⁻¹	root: ±0.25 µl g fw⁻¹ h⁻¹ shoot: ±0.30 µl g fw⁻¹ h⁻¹ root: ±0.075 µl g FW⁻¹ h⁻¹ shoot: ±0.075 µl g fw⁻¹ h⁻¹	Mustroph *et al.*, 2006a
4 h anaerobic conditions in the dark	Rice seedlings (*Oryza sativa* L. cv. Cigalon) Wheat seedlings (*Triticum aestivum* L. cv. Alcedo)	root: ±0.15 µl g fw⁻¹ h⁻¹ shoot: ±2.5 µl g fw⁻¹ h⁻¹ root: ±0.12 µl g fw⁻¹ h⁻¹ shoot: ±0.5 µl g fw⁻¹ h⁻¹	root: ±3 µl g fw⁻¹ h⁻¹ shoot: ±50 µl g fw⁻¹ h⁻¹ root: ±2 µl g FW⁻¹ h⁻¹ shoot: 0 µl g fw⁻¹ h⁻¹	Mustroph *et al.*, 2006b
48 h anoxia	Rice coleoptiles (*Oryza sativa* L.) cv. Leulikelash cv. Asahimoki cv. Nipponbare cv. Yukihikari	±500 nmol coleoptile⁻¹ ±500 nmol coleoptile⁻¹ ±1200 nmol coleoptile⁻¹ ±1200 nmol coleoptile⁻¹		Kato-Noguchi and Morokuma, 2007
4 h anoxic incubation	Wheat seedlings root (*Triticum aestivum* L. cv. Alcedo)	roots in anoxia, shoots in air: 149±7 µg g fw⁻¹ plants in N atmosphere in light: 150±26 µg g fw⁻¹ plants in N atmosphere in dark 316±63 µg g fw⁻¹		Mustroph and Albrecht, 2007
24 h anoxic stress	Coleoptiles Barley (*Hordeum vulgare* L. cv. Ichibanboshi) Oat (*Avena sativa* L. cv. Victory) Rice (*Oryza sativa* L. cv. Nipponbare)	±9 µmol g FW⁻¹ ±9 µmol g FW⁻¹ ±29 µmol g FW⁻¹		Kato-Noguchi *et al.*, 2010

FW, fresh weight.

Ala production pathways may be redundant under low O_2 (Rocha *et al.*, 2010).

Besides Ala, an increase in GABA has also been observed (Reggiani, 1999). Since glutamate is a common precursor of both Ala and GABA, it is believed to play a central role in anaerobic amino acids metabolism. Glutamate was found to decrease significantly after 2 h of anoxia in *Arabidopsis* (Branco-Price *et al.*, 2008). Both Ala and GABA accumulation may help to minimize the decrease in cytosolic pH and to reduce carbon/nitrogen loss, occurring via ethanol or lactate production, which could be rapidly reused after re-oxygenation (Mustroph *et al.*, 2014). Analysis of distinct shoot and root metabolite responses to low O_2 identified the production of Ala, GABA and lactate in roots that probably suffer severely under the stress, while the shoots seem to be less affected (Mustroph *et al.*, 2014).

Cells survive hypoxia as long as carbohydrate substrate remains available. Metabolic reconfiguration under low O_2 includes sucrose degradation and, in some plant species, starch breakdown (see Section 6.6). Sucrose is driven from photosynthetic tissues towards sink organs where it is cleaved to hexoses in order to sustain metabolic processes. The enzymatic pathways used to degrade sucrose include sucrose synthases (SUS), which convert sucrose and uridine diphosphate (UDP) into UDP-glucose and fructose; and invertases (INV), which catalyse the hydrolysis of sucrose into fructose and glucose. A common idea is that the SUS route is preferred under low O_2, since it requires less energy. Indeed, many reports have indicated *SUS* genes induction under hypoxia (Springer *et al.*, 1986; Richard *et al.*, 1991; Albrecht *et al.*, 1993; Biemelt *et al.*, 1999; Baud *et al.*, 2004). Sucrose degraded via the INV route requires two ATP molecules to be converted into hexose-phosphates, since fructose and glucose are subsequently phosphorylated by hexo- and fructokinases, using ATP or UTP as energy donors (Renz *et al.*, 1993). The SUS pathway requires only one molecule of PPi since UDP-glucose is converted to glucose-1-P in a reaction catalysed by UDP-glucose pyrophosphorylase (UGPPase), which is PPi-dependent (Huber and Akazawa, 1986; Stitt, 1998). However, *Arabidopsis sus1/sus4* mutants have been shown to be as tolerant to low O_2 as the wild-type plants under anoxia, hypoxia and submergence, also displaying a similar increase

in ethanol production and reduction in ATP/ADP ratio under the treatments in comparison to normoxia (Santaniello *et al.*, 2014). Only under waterlogging did *sus1/sus4* show a higher sensitivity than the wild type to the stress, as also observed by Bieniawska *et al.* (2007). Also in this condition, the level of ethanol and the ATP/ADP ratio were similar to the wild type. However, a higher INV activity in the *sus1/sus4* mutants was observed, which is likely to have compensated for the lack of SUS activity. Thus, both INV and SUS may contribute to ethanol production under low O_2 (Santaniello *et al.*, 2014).

6.3   The Morphological Adaptation of Plants to Oxygen Shortage

LOES and LOQS are the main strategies that plants use to survive low O_2. LOES has been described in many plants, LOQS only in some lowland tolerant rice varieties belonging to the *indica* subfamily (Xu *et al.*, 2006), along with a few rice relatives (Niroula *et al.*, 2012) and the wild dicot species *Rumex acetosa* (van Veen *et al.*, 2013). The promotion of shoot elongation by submergence, which is part of the LOES, is known to occur in wetland and amphibious species over a wide taxonomic range, e.g. *Rumex palustris*, *Ranunculus sceleratus*, *Nymphoides peltata*, *Potamogeton pectinatus* and *P. distinctus* (Summers and Jackson, 1994; He *et al.*, 1999; Ishizawa *et al.*, 1999; Summers *et al.*, 2000; Sato *et al.*, 2002; Cox *et al.*, 2003; Mommer *et al.*, 2005; Jackson, 2008). Elongation is spectacular in the internodes of deepwater rice which rapidly elongate under submergence (Hattori *et al.*, 2009).

Plants display several adaptive traits to ensure survival under low O_2, often associated with LOES. These adaptive responses are genotype specific and include an altered petiole/internode elongation rate, cell ultrastructure modifications, the development of lateral and adventitious roots, and the formation of aerenchyma, together with the metabolic adaptations that were described in the previous section. These traits vary in importance depending on the water regimes and the time and depth extent of the submergence (Colmer and Voesenek, 2009).

Aerenchyma is hypothesized to be of fundamental importance under many flooding regimes

(Colmer and Voesenek, 2009). It is characterized by intercellular gas-filled spaces in plant roots and shoots that favour the longitudinal transport of O_2 from air to submerged organs (Fig. 6.2) (Drew *et al.*, 1979; Kawase, 1981). In plants, aerenchyma is formed through two different processes named schizogeny and lysigeny (Sachs, 1882). Schizogenous aerenchyma involves cell wall reorganization and cell separation, and is characteristic of *Rumex* spp. (Laan *et al.*, 1989), *Epilobium parviflorum* (Seago *et al.*, 2005), *Acorus calamus* and *Epilobium hirsutum* (Armstrong and Armstrong, 1994). Lysigenous aerenchyma is formed as a consequence of a programmed cell death (PCD) event (Campbell and Drew, 1983; Gunawardena *et al.*, 2001; Evans, 2003) and has been described in barley (Arikado and Adachi, 1955), wheat (Trought and Drew, 1980), rice (Justin and Armstrong, 1991), and maize (He *et al.*, 1994, 1996; Gunawardena *et al.*, 2001). Some species such as *Sagittaria lancifolia* can also form lysigeny and schizogeny simultaneously, although in different tissues (Schussler and Longstreth, 1996).

Although aerenchyma develops further when the soil becomes waterlogged, in some cases it is already present in well-drained conditions (Armstrong, 1971; Pradhan *et al.*, 1973; Das and Jat, 1977) and is constitutive in deepwater and lowland rice stem and leaf sheaths (Steffens *et al.*, 2011). The mechanisms responsible for aerenchyma have not yet been fully clarified. However, it is known that ethylene is involved, which accumulates in submerged organs (Kawase, 1972, 1978; Könings and Jackson, 1979; Justin and Armstrong, 1991; He *et al.*, 1996; Zhou *et al.*, 2002; Lenochova *et al.*, 2009; Geisler-Lee *et al.*, 2010). In hypoxic roots of maize, exogenous ethylene applications induce aerenchyma formation while ethylene inhibitors repress its development (Drew *et al.*, 1981; Könings, 1982; Jackson *et al.*, 1985). In addition, both 1-aminocyclopropane-1-carboxylic acid (ACC) synthase activity and ACC concentrations have been found to be high in hypoxic maize roots (Atwell *et al.*, 1988; He *et al.*, 1994; Geisler-Lee *et al.*, 2010). However, aerenchyma formation does not always require ethylene, as described for the roots of the wetland plant *Juncus effusus* (Visser and Bögemann, 2006). In rice stems, Steffens *et al.* (2011) demonstrated that aerenchyma forms in response to ethylene and hydrogen peroxide (H_2O_2), in a dose-dependent manner. In deepwater and lowland rice internodes, ethylene promotes the formation of reactive oxygen species (ROS), which drive aerenchyma formation (Steffens *et al.*, 2011). The production of lysigenous aerenchyma in *Arabidopsis* under hypoxia has also been shown to require both ethylene and H_2O_2 signalling (Mühlenbock *et al.*, 2007). The identification of aerenchyma formation-associated genes expressed in maize roots under waterlogging revealed the presence of mechanisms linked to the generation and scavenging of ROS, Ca^{2+} signalling and cell wall loosening and degradation (Rajhi *et al.*, 2011). In rice under submergence, ethylene/H_2O_2 also play a combined role in the aerenchyma formation of leaf sheaths with a dissimilar regulation depending on the plant survival strategy (Parlanti *et al.*, 2011).

The volume of aerenchyma formed in submerged plant tissues depends on the species, as well as the cultivar/accession and environmental conditions (Colmer, 2003a). Wheat submergence-tolerant cultivars have significantly higher root porosities under submergence than sensitive cultivars (Huang *et al.*, 1994a,b, 1995a,b). Justin and Armstrong (1987) studied 91 plant species from wetlands, intermediate and non-wetland habitats, and found a correlation between submergence tolerance and the extent of aerenchyma development.

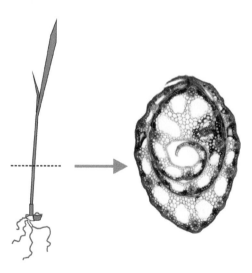

Fig. 6.2. Cross-section of rice leaf sheath showing aerenchyma formation.

The effectiveness of aerenchyma can be increased by the formation of a barrier to radial O_2 loss (ROL) at the outer cell layers of roots, which inhibits O_2 diffusion, thus loss, from the roots to the surrounding anaerobic soil (Armstrong, 1979; Visser et al., 2000; Colmer, 2003a,b). The ROL barrier also enhances the O_2 movement towards the root apex, thus promoting a deeper rooting in waterlogged soils (Colmer and Voesenek, 2009). A further role of the ROL barrier is to limit the entry of toxic compounds, generated in highly reduced wet environments (Armstrong, 1979; Armstrong et al., 1996; Armstrong and Armstrong, 2005). This barrier can be induced by other environmental constraints, such as salinity and drought (Enstone et al., 2003). In some plants, ROL is constitutive (e.g. Juncus effusus) (Visser et al., 2000), while in others it increases under stagnant conditions (e.g. Oryza sativa, Lolium multiflorum, Hordeum marinum) (McDonald et al., 2002; Colmer, 2003b; Garthwaite et al., 2003). Some crops, such as wheat and maize, are not able to form a tight ROL barrier, which results in a reduction of yield in waterlogged soil (Watanabe et al., 2013). The ROL barrier results from the deposition of suberin and lignin in the root exodermis cell wall, which forms a physical resistance to O_2 diffusion (Shiono et al., 2011). Shiono and colleagues (2014) identified specific genes upregulated in the outer part of rice roots during the formation of the ROL barrier, using laser microdissection to isolate specific cell types. The analysis revealed the expression of several genes encoding suberin biosynthesis enzymes (Shiono et al., 2014).

Many plants under water submergence show a hyponastic growth of leaves (e.g. Rumex palustris, Ranunculus repens, Caltha palustris, Arabidopsis thaliana) (Ridge, 1987; Cox et al., 2003; Millenaar et al., 2005). This leaf reorientation is driven by the unequal growth rates of adaxial and abaxial petiole sides (Cox et al., 2004). Hyponastic growth is potentially important for reaching the air surface above the water column; however, reorientation can help in capturing light and preventing ground cover (Colmer and Voesenek, 2009). Ethylene is the key component in the complex regulatory network of hyponastic growth in submerged and waterlogged Arabidopsis (Millenaar et al., 2005, 2009; van Zanten et al., 2010; Rauf et al., 2013)

and Rumex palustris plants (Heydarian et al., 2010). In Arabidopsis, ethylene maintains a tight control of the core cell cycle regulator CYCLINA2;1, which mediates cell proliferation. Such control has been identified through the study of Enhanced Hyponasty-D (EHY-D) activation-tagged plants, which show high levels of hyponasty under exogenous ethylene application (Polko et al., 2012). However, some sensitive plant species show epinastic leaf growth under water (e.g. Helianthus annuus, Nicotiana tabacum, Solanum lycopersicum). This kind of growth may reduce the dehydrating effects of the decrease in root water conductance observed under waterlogging (Jackson, 2002).

Underwater ethylene accumulation also promotes adventitious root formation in some plant species. In rice, ethylene leads to the death of epidermal cells overlying the adventitious root primordia, through the mechanical force generated by root emergence. This is likely to happen through the generation of ROS signals for cell death (Steffens et al., 2012). In tomato plants, ethylene works in combination with auxin to induce adventitious root formation (Vidoz et al., 2010).

The formation of lenticels (plant openings that allow gas exchange) on the stem, and the development of pneumatophores (specialized roots that grow out from the water surface) can also increase the amount of O_2 that reaches the underwater organs (Kozlowski, 1984). In addition to these morphological traits, the ability of the plants to form a leaf gas film termed 'plant plastron' (which is similar to the plastrons of aquatic insects), on submerged leaf surfaces, can improve submergence tolerance (Raven, 2008). This gas film enables continuous gas exchange via stomata, bypassing cuticle resistance. The film thus promotes gas exchanges, enhancing CO_2 uptake for photosynthesis under daylight and enhancing O_2 uptake for respiration during dark periods (Colmer and Pedersen, 2008; Pedersen et al., 2009).

6.4   The Oxygen-sensing Mechanism

In bacteria, fungi and animals, specific O_2 sensing regulatory systems and molecules have been described (for a review see Bailey-Serres and Chang, 2005). In mammals, the hypoxia inducible factor (HIF) 1 transcriptional complex plays a major role in low O_2 sensing. HIF1 is a heterodimeric factor made up of the hypoxia-induced HIF1α and the constitutively expressed HIF1β subunits. Under aeration, HIF1 is not active, since HIF1α is hydroxylated at two prolyl residues by the prolyl 4-hydroxylase enzyme, which requires O_2 as a co-substrate. The aerobically hydroxylated HIF1α is degraded through ubiquitin-mediated proteasomal degradation (Acker et al., 2006). Thus, reduced O_2 availability limits the rate of HIF1α degradation with the subsequent activation of the transcriptional complex, which is then translocated to the nucleus to activate downstream genes (Semenza, 2007).

In plants, no HIF1 orthologues have been found to date but a particular plant low O_2-sensing mechanism has been identified (Gibbs et al., 2011; Licausi et al., 2011) (Fig. 6.3). ERF transcription factors of group VII have been found to harbour a MetCys motif at the N-end of the protein, which is the target for proteasomal-dependent degradation. Under normoxia, the Cys secondary residue of the N-end protein is oxidated by plant cysteine oxidase (PCO1 and PCO2) enzymes that use O_2 as a co-substrate

(Weits et al., 2014) after the constitutive Met cleavage by a Met amino peptidase (MAP) enzyme. Oxidated Cys are arginylated by an arginyl T-RNA transferase (ATE) and then targeted to the proteasome through recognition by the PROTEOLYSIS 6 (PRT6) enzyme. Under low O_2, the group VII ERF is stabilized and migrates to the nucleus to activate downstream genes related to tolerance, such as ADH, PDC1 and HB1. The most powerful gene activators involved in low O_2 sensing are probably RAP2.2 and RAP2.12 (Bui et al., 2015). The RAP2.12 mechanism of action includes a state in which the protein is also present under normoxia but protected by degradation through the action of membrane-associated Acyl-CoA binding proteins (ACBP1 and ACBP2) (Licausi et al., 2011). RAP2.12 is thus active when O_2 recedes, but de novo synthesis of the TF also occurs (Kosmacz et al., 2015). A further level of regulation is provided by the antagonistic TF HYPOXIA RESPONSE ATTENU-ATOR1 (HRA1) whose interplay with RAP2.12 leads to a finely tuned response to fluctuating hypoxia (Giuntoli et al., 2014).

Together with O_2, nitric oxide (NO) is also required for the N-terminal Cys oxidation of ERFs VII, suggesting that both gases can influence gene expression through the N-end rule pathway (Gibbs et al., 2014).

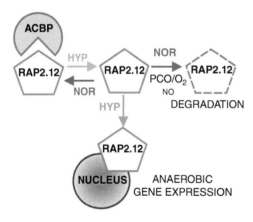

Fig. 6.3. RAP2.12 modulation by O_2 level. Under normoxia (NOR), RAP2.12 is degraded via the N-end rule pathway after oxidation of Cys by PCO1 and PCO2 enzymes, which uses O_2 as a co-substrate and requires NO. Under O_2 shortage (HYP), stable RAP2.12 can migrate from the membrane (where it is docked to ACBP1 and ACBP2 under NOR) to the nucleus where it activates the expression of anaerobic genes that are necessary for plant survival under stress.

6.5 Molecular Regulations Under Low Oxygen

Beside the direct O_2-sensing machinery, additional mechanisms of regulation have been suggested to exist in plants under O_2 shortage. The most promising cell state indicators are calcium flux, energy charge and ROS. These signalling molecules seem to be interrelated, suggesting the presence of downstream events that cross each other.

Changes in cytosolic Ca^{2+} concentration have been reported in response to various stimuli, including hormonal changes, light, biotic and abiotic stresses (for a review see Lecourieux et al., 2006). These changes may be due to a transient change in plasma membrane permeability which seems to be a common occurrence in early plant defence signalling (Atkinson et al., 1990). Calcium may also contribute to trigger low O_2 signalling in plants, since an increase in cytosolic

Ca^{2+} concentration has been observed in maize and *Arabidopsis* under hypoxia and anoxia (Subbaiah *et al.*, 1994; Sedbrook *et al.*, 1996).

Maize cells seem to require Ca^{2+} for the expression of *ADH1* (Subbaiah *et al.*, 1994). Under anoxia, the Ca^{2+} response is biphasic, consisting of a slow Ca^{2+} spike, which takes a few minutes, and a subsequent sustained Ca^{2+} elevation, which takes hours (Subbaiah *et al.*, 1994; Sedbrook *et al.*, 1996). Transient Ca^{2+} changes are decoded by an array of proteins that fall into two main classes of Ca^{2+} sensor relays and sensor responders (Lecourieux *et al.*, 2006). They give rise to downstream events such as protein phosphorylation and gene expression.

Of the Ca^{2+} sensor relays, the calcineurin B-like (CBL) proteins function through molecular interaction with CBL interacting protein kinase (CIPK) (for a review see Luan *et al.*, 2002). Rice tolerance to flooding has been shown to depend on a coordinated response to O_2 and sugar deficiency, regulated by CIPK15 (Lee *et al.*, 2009). The specific CBL that interacts with CIPK15 under O_2 deprivation has not been determined. CIPK15 regulates Snf1-related protein kinase (SnRK1) to control sugar and energy production for growth under water (Lee *et al.*, 2009).

Calcium fluxes and subsequent protein phosphorylation may also be required for the controlled generation of hydrogen peroxide (H_2O_2) (Neill *et al.*, 2002). ROS are key actors in plant responses to both biotic and abiotic stresses (for reviews, see Delledonne, 2005). Initially ROS were thought to be only involved in degenerative processes, but they have subsequently emerged as signalling actors, participating in sensing and signalling responses to different stresses. A burst of ROS has been described in *A. thaliana* plants under O_2 deprivation occurring briefly after the stress onset (Banti *et al.*, 2010; Chang *et al.*, 2012; Pucciariello *et al.*, 2012; Gonzali *et al.*, 2015).

ROS production under O_2 shortage is believed to be generated by mitochondria through the inhibition of the mETC terminal step and probably regulates the activation of mitogen-activated protein kinase MPK3, MPK4 and MPK6 (Chang *et al.*, 2012). A further hypothesis is that a ROS burst is generated by a membrane-localized NADPH-oxidase in a multiprotein complex (Gonzali *et al.*, 2015). In fact, the isoform D of the respiratory burst oxidase homologue (RBOH) proteins is transcriptionally induced under

low O_2, and *Arabidopsis* mutants impaired in the protein production (*rbohD*-) have been shown to be sensitive to anoxia (Pucciariello *et al.*, 2012).

Baxter-Burrell *et al.* (2002) showed that the activation of a RHO-like GTPase of plants (GTP-ROP) under low O_2 induces H_2O_2 accumulation. The ROP family modulates signalling cascades associated with various processes in plants (Gu *et al.*, 2004). The ROP-dependent production of H_2O_2 via an NADPH-oxidase mechanism is necessary for low O_2 tolerance (Baxter-Burrell *et al.*, 2002). Tolerance to O_2 deprivation seems to require ROP to be activated but also to be negatively regulated through feedback via a ROP GTPase activating protein (ROP-GAP), regulated by H_2O_2.

Further studies have identified the universal stress protein (USP) hypoxia responsive USP1 (HRU1) protein role, which is likely to contribute to RBOHD regulation under low O_2 (Gonzali *et al.*, 2015). Under normoxia HRU1 exists in the cytoplasm as dimers, and under anoxia it is thought to migrate to the membrane to form a complex with RBOHD and GTP-ROP (Gonzali *et al.*, 2015). A T-DNA insertional mutant that encodes a form of HRU1 which lacks a putative dimerization domain and the HRU1 null allele mutant have shown a lower tolerance to anoxia and an alteration of the H_2O_2 production shape observed instead in wild-type plants. HRU1 is a target of RAP2.12, thus highlighting a link between ROS production and the direct O_2-sensing mechanism, which needs further exploration. Heat shock transcription factors (HSF) and heat shock proteins (HSP) are highly induced under anoxia in a mechanism that overlaps with heat stress and identifies H_2O_2 as the common signalling element (Banti *et al.*, 2010). HSFs have been proposed as specific H_2O_2 sensors in plants (Miller and Mittler, 2006). *Arabidopsis* seedlings that overexpress *HsfA2* are markedly more tolerant to anoxia as well as to submergence (Banti *et al.*, 2010).

6.6   Rice Strategies to Survive Oxygen Deprivation

Rice is a semi-aquatic plant that is well adapted to surviving low O_2 environments, both when sown in paddy fields, and as an adult plant, as a consequence of natural and man-made flooding events. Rice feeds billions of people and with the increase in global food demand, a rapid increase in productivity is needed, especially in marginal lands. Ecologically, rice can be classified into three different groups: upland rice, which grows in non-irrigated fields; lowland rice, which grows in rain-fed or irrigated fields up to 50 cm deep; and deepwater rice, exposed to water exceeding 50 cm in depth (Sauter, 2000). Deepwater and lowland rice together account for 33% of global rice farmland, mainly located in India, Thailand and Bangladesh (Bailey-Serres et al., 2010).

Different water cultivation regimes have favoured the selection of local rice landraces adapted to extreme environments, thus giving rise to the enormous ecological diversity of genotypes. Around 110,000 *Oryza* genotypes, including wild species and related genera, have been collected by the International Rice Research Institute (http://irri.org) to preserve the germplasm.

In rice, the two mutually exclusive major strategies LOES and LOQS (see Section 6.3) to cope with submergence have been identified, mainly represented by the prevention of anaerobiosis through rapid growth, and tolerance to anaerobiosis through reduced growth, respectively. Each adaptation is useful under the appropriate environmental conditions in which it was developed. However, rice is an aerobic organism, thus it still suffers from O_2 deprivation. Indeed, submergence stress has been identified as the third most important constraint to rice production in Indian lowland areas (Sauter, 2000).

6.6.1 Successful rice germination under low oxygen

Rice germinates successfully under low O_2 (for a review see Magneschi and Perata, 2009). Only the coleoptile elongates, while the root fails to grow. In cereal grains, the starch stored in the seed endosperm is a major reserve. The starchy seeds are able to maintain a high energy metabolism under anaerobiosis through starch catabolism (Raymond et al., 1985). Rice harbours a complete set of starch-degrading enzymes including α and β amylases, debranching enzymes and maltases. Indeed, α-amylases have a major role in degrading native starch granules in germinating rice (Murata et al., 1968; Dunn, 1974; Sun and Henson, 1991).

α-Amylases are endo-amylolytic enzymes which catalyse the hydrolysis of α-l,4 linked glucose polymers of starch. Starch hydrolysis results in sugars that are translocated to the embryonic axis to be metabolized through glycolysis, thus generating energy and essential metabolites needed for growth. α-Amylases are not produced in anoxia-intolerant cereals such as wheat, barley, oat and rye under low O_2, which consequently suffer from sugar starvation and eventually die (Perata et al., 1992; Perata and Alpi, 1993; Guglielminetti et al., 1995; Loreti et al., 2003a).

In rice there are ten different isoforms of α-amylases, grouped into three subfamilies: Amy1, Amy2 and Amy3 (Rodriguez et al., 1992). RAMY1A is hormonally modulated by gibberellins (GA) under aerobic conditions. The GA-independency of rice anaerobic germination has been demonstrated in the GA-deficient mutant *Tan-ginbozu* (Loreti et al., 2003b). This mutant germinates under anoxia, and the expression of α-amylase genes other than the GA-induced Amy1A gene has been observed (Loreti et al., 2003b). *RAMY3D* is the main amylase that acts in anoxic rice seedlings and is anoxia induced (Perata et al., 1997; Hwang et al., 1999; Lasanthi-Kudahettige et al., 2007). *RAMY3D* is not induced by GA, since it does not possess the *cis*-acting element on the promoter region required for GA responsiveness (Morita et al., 1998; Loreti et al., 2003a). *RAMY3D*, on the other hand, is regulated by sugar starvation, which suggests a link with the reduction in soluble sugar content observed under anoxia (Guglielminetti et al., 1995; Perata et al., 1996; Loreti et al., 2003a). The expression of *RAMY3D* in rice embryos takes place 12 h after imbibition, peaks 2 days later, and then starts to decline. After 5 days of imbibition, *RAMY1A* also increases, thus suggesting that these two enzymes cooperate in anoxic starch degradation (Loreti et al., 2003a).

Crosstalk between sugar and O_2-deficiency signalling is important for rice germination under low O_2 (Lee et al., 2009). As indicated in the previous section, CIPK15 is the key regulator of carbohydrate catabolism and fermentation in rice germination under flooding. *cipk15* rice mutants were found to have extreme difficulty in germinating under water, also showing a reduced expression of *ADH* and the abolishment of *RAMY3D* mRNA accumulation (Lee et al.,

2009). The CIPK15 pathway works through Sn-RK1A, which plays a central role in the sugar signalling pathway, and the *MYBS1* transcription factor, in order to regulate *RAMY3D* expression (Lu *et al.*, 2007; Lee *et al.*, 2009). MYBS1 binds specifically to the *RAMY3D* promoter TA box element, which belongs to the sugar-responsive complex (SRC) and functions as a transcriptional activator of *RAMY3D* under sugar depletion (Lu *et al.*, 1998; Toyofuku *et al.*, 1998; Chen *et al.*, 2002). The CIPK15 pathway thus plays a key role in regulating starch catabolism under low O_2, allowing rice to germinate (Lee *et al.*, 2009).

The capacity to elongate the coleoptile is key to rice germination under low O_2. Coleoptile elongation influences the establishment of rice in submerged fields, since it enables the seedlings to make contact with the atmosphere above the water column (Huang *et al.*, 2003). Considerable variations have been observed among different rice cultivars in their capacity to elongate the coleoptile (Setter *et al.*, 1994; Magneschi *et al.*, 2009a). A screening of 165 rice cultivars to assess their ability to elongate under anoxia revealed that anoxic coleoptile length is correlated with efficient ethanol fermentation rather than being related to glycolytic and fermentative related gene expression, or with carbohydrate content (Magneschi *et al.*, 2009a).

Since coleoptile elongation under low O_2 is believed to be mostly due to cell expansion, cell wall loosening is thought to be involved. Of the enzymes involved in this process, it is likely that expansins play a role (Cosgrove, 1999; Huang *et al.*, 2000). Expansins are encoded by a multigene family composed of two subfamilies: α and β expansins. In deepwater rice, rapid internode elongation mediated by GA induces the expression of *EXPB3*, *EXPB6*, *EXPB11* and *EXPB12* (Kende *et al.*, 1998). Increases in the mRNA levels of *EXPA2* and *EXPA4* have also been observed in underwater internodes of deepwater rice (Cho and Kende, 1997). *EXPA7* and *EXPB12* have been proposed to be involved in the anoxic elongation of Nipponbare rice coleoptiles (Lasanthi-Kudahettige *et al.*, 2007). In addition, the antisense expression of *EXPA4* resulted in a reduced coleoptile elongation under submergence in the Taipei 309 cultivar (Choi *et al.*, 2003). However, the expression of 7 α and 8 β expansin genes, in two rice cultivars showing

long versus short coleoptiles when germinated under anoxia, did not show any correlation with the ability to elongate (Magneschi *et al.*, 2009b).

6.6.2 Low oxygen escape syndrome in deepwater rice

Some rice varieties exposed to submergence try to escape low O_2 stress by enhancing the elongation rate of the stem and/or leaf, in order to reach the water surface and re-establish contact with the air. As previously indicated, lowland rice varieties risk starvation and death before they make contact with the air (Bailey-Serres and Voesenek, 2008). However, deepwater rice displays an extremely fast internode elongation in response to increasing water levels (for a review see Nagai *et al.*, 2010). Deepwater rice is cultivated in rain-fed areas and tidal swamps due to its ability to elongate considerably under flooding (Catling, 1992). In these areas subjected to monsoons, paddy fields are seasonally covered with high water levels (Vergara and Mazaredo, 1975; Catling, 1992). Internodes of deepwater rice elongate with a daily increase of 20–25 cm and can reach several metres in height. Since inside the internode there is a hollow structure, air can circulate inside the plant thus reaching the underwater organs. Only a slight decrease in O_2 concentration was observed inside the culm of deepwater rice under submergence, with a concomitant rise in CO_2 (Stünzi and Kende, 1989).

Ethylene was initially suggested to play a role in internode elongation (Métraux and Kende, 1983), together with ABA and GA (Kende *et al.*, 1998). The deepwater rice response has now been partially clarified, through the analysis of the most effective major quantitative traits loci (QTL) detected in chromosome 12 (Hattori *et al.*, 2007, 2008, 2009). The two group VII ERF transcription factors SK1 and SK2, which belong to this QTL and are induced by ethylene, positively regulate internode elongation (Fig. 6.4). The overexpression of *SK1* and *SK2* in non-deepwater rice varieties promotes internode elongation, even in air (Hattori *et al.*, 2009). Gibberellins (GAs) seem to be involved in this response, since treatments of deepwater rice with GA inhibitors show an

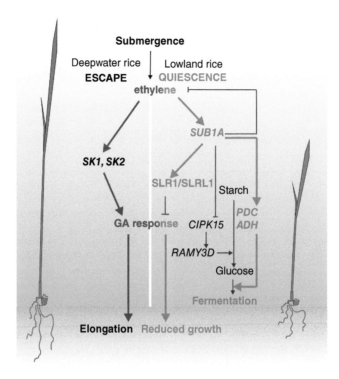

Fig. 6.4. The quiescence and escape strategies activated by rice plants under submergence (Bailey-Serres *et al.*, 2010; Nagai *et al.*, 2010). In lowland rice LOQS, submergence stimulates the production/entrapment of ethylene which activates *SUB1A-1*. *SUB1A-1* positively regulates the two GA response suppressors *SLR1-SLRL1*. Consequently, the GA signal is not activated, thus resulting in a reduced growth. *SUB1A-1* also positively regulates genes related to fermentation and negatively regulates genes related to starch breakdown. In deepwater rice LOES, *SK1* and *SK2* are activated by the ethylene produced by submergence. *SK1* and *SK2* regulate internode elongation via the GA response.

arrest in elongation (Suge, 1987; Hattori *et al.*, 2009). Ayano and colleagues (2014) identified a deepwater-dependent GA$_1$ and GA$_4$ accumulation in deepwater rice. GA feeding resulted in internode elongation under control conditions, and mutations in GA biosynthesis and signal transduction genes hampered the internode elongation, confirming a role for GA (Ayano *et al.*, 2014).

6.6.3 Quiescence strategy in tolerant lowland rice varieties

A limited number of lowland rice cultivars can tolerate more than 14 days of submergence (Mackill *et al.*, 1996). Of these, FR13A has been used to study the molecular basis of tolerance related to

the growth arrest under flooding (for a review see Bailey-Serres *et al.*, 2010). As indicated above, rice plants that use a quiescence strategy to survive flooding tend to reduce growth and metabolic activity to a minimum, thus storing energy for regrowth. QTL mapping has led to the identification of the *SUB1* locus, located near the centromere of the long arm of chromosome 9 (Xu *et al.*, 2006). *SUB1* locus comprises, together with other genes, two or three closely related *ERF* genes named *SUB1A*, *SUB1B* and *SUB1C*, which are characterized by several allelic forms (Fukao *et al.*, 2006; Xu *et al.*, 2006). *SUB1A* is indicated as the major source of flooding tolerance in rice and explains ~69% of the phenotypic variations in this trait. All *Oryza sativa* spp. harbour *SUB1B* and *SUB1C*, while *SUB1A* is present only in a few genotypes of cultivated and wild rice (Niroula *et al.*, 2012). The tolerant allele *SUB1A-1* is positively regulated by ethylene, and its mRNA level rapidly increases under submergence (Xu *et al.*, 2006). However, although SUB1A is part of group VII ERF, it is not a substrate of the N-end degradation (Gibbs *et al.*, 2011).

The presence of the *SUB1A-1* allele seems to be crucial in mediating flooding tolerance. At a molecular level, *SUB1A-1* activation reduces ethylene perception through a feedback mechanism, thus suppressing the ethylene-promoted GA-mediated induction of genes associated with carbohydrate catabolism and cell elongation (Fig. 6.4). Carbohydrate reserves are therefore conserved for plant regrowth during re-oxygenation. In addition, the expression of *ADH* and *PDC* genes is upregulated in *SUB1A-1* harbouring plants under O_2 deprivation (Fukao *et al.*, 2006; Xu *et al.*, 2006). The direct influence of *SUB1A-1* on rice submergence tolerance has been evaluated, using the introgressed line M202(*SUB1*) that contains the *SUB1A-1* gene from the FR13A cultivar. FR13A is submergence tolerant unlike the intolerant M202 cultivar (Fukao *et al.*, 2006).

The *SUB1C-1* allele may have a role in the tolerance mechanism. It is thought to act downstream of the ethylene-promoted GA-dependent signalling response, thus enhancing the shoot elongation in intolerant rice genotypes (Fukao and Bailey-Serres, 2008). The transcript abundance of *SUB1C-1* is significantly downregulated in rice accessions harbouring the *SUB1A-1* allele (Fukao *et al.*, 2006; Xu *et al.*, 2006). Tolerance via the quiescence mechanism is always associated with the presence of the *SUB1A-1/SUB1C-1* haplotype (Singh *et al.*, 2010).

SUB1A-1 increases the accumulation of the DELLA protein SLENDER RICE 1 (SLR1) and the non-DELLA protein SLR like 1 (SLRL1), both GA signalling repressors, thus limiting shoot elongation (Fukao and Bailey-Serres, 2008). Jung *et al.* (2010) compared the transcriptome of M202(SUB1) with M202 under submergence showing that *SUB1A-1* regulates multiple pathways associated with tolerance such as defence, anaerobic respiration, cytokinin-mediated senescence, ethylene-dependent gene expression and GA-mediated shoot elongation.

In addition to its role under submergence tolerance, *SUB1A-1* has been shown to improve survival under dehydration, by enhancing the

plant's ABA responsiveness and by activating stress-inducible genes (Fukao *et al.*, 2011). Submergence followed by drought often occurs in rain-fed lowlands, thus the improvement of a combined tolerance would substantially increase rice productivity in flood-prone areas. The transfer of the *SUB1* QTL by marker-assisted backcrossing into the farmer-preferred varieties led to the production of new *SUB1* megavarieties which provide submergence tolerance as well as an equally good agronomical performance. These varieties are able to produce three- to sixfold more grain than the non-*SUB1* even when a submergence event occurs (Bailey-Serres *et al.*, 2010).

6.7 Conclusions and Future Outlook

Submergence stress is one of the major constraints to plant production worldwide. Metabolic changes due to low O_2 can be crucial to plant survival, due to the low energy available for growth. Many morphological and metabolic adaptations have been described, and some traits related to tolerance have been identified. Molecular studies on *Arabidopsis* plants have

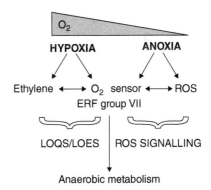

Fig. 6.5. Model of low O_2 signal mechanisms in plants. Hypoxia and anoxia are perceived by the common O_2 sensors belonging to the ERF VII group, which activate anaerobic metabolism genes. Ethylene-related signalling is mainly activated under hypoxia. This signalling regulates the ERF genes belonging to group VII which stimulate or inhibit the signal transduction paths that lead to LOES and LOQS. Anoxia activates a ROS-related pathway that includes the activation of ROS-related signalling.

revealed the existence of a specific O_2-sensing mechanism that controls genes activated under low O_2. This mechanism relates the metabolic changes directly to the O_2 level fluctuation perception (Fig. 6.5). Additional mechanisms related to changes in the cell state as a consequence of O_2 shortage have also been identified and represent further ways of regulation that probably overlap and are involved in tolerance. In rice, ethylene production/entrapment seems to govern the expression of different ERF VII which work through antithetical plant strategies to survive flooding under different water regimes/durations. From an agronomical point of view, in the context of climate change and the increasing need for food, the selection and study of ecotypes adapted to extreme environments show great potential for the future.

References

Acker, T., Fandrey, J. and Acker, H. (2006) The good, the bad and the ugly in oxygen-sensing: ROS, cytochromes and prolyl-hydroxylases. *Cardiovascular Research* 71, 195–207.

Albrecht, G., Kammerer, S., Praznik, W. and Wiedenroth, E.M. (1993) Fructan content of wheat seedlings (*Triticum aestivum* L.) under hypoxia and following re-aeration. *New Phytologist* 123, 471–476.

Arikado, H. and Adachi, Y. (1955) Anatomical and ecological responses of barley and some forage crops to the flooding treatment. *Bulletin Faculty Agriculture Mie University, Tsu City, Japan* 11, 1–29.

Armstrong, W. (1971) Radial oxygen losses from intact rice roots as affected by distance from the apex, respiration, and waterlogging. *Physiologia Plantarum* 25, 192–197.

Armstrong, W. (1979) Aeration in higher plants. *Advances in Botanical Research* 7, 225–332.

Armstrong, J. and Armstrong, W. (1994) Chlorophyll development in mature lysigenous and schizogenous root aerenchymas provides evidence of continuing cortical cell viability. *New Phytologist* 126, 493–497.

Armstrong, J. and Armstrong, W. (2005) Rice: sulphide-induced barriers to root radial oxygen loss, Fe^{2+} and water uptake, and lateral root emergence. *Annals of Botany* 96, 625–638.

Armstrong, J., Afreen-Zobayed, F. and Armstrong, W. (1996) *Phragmites* die-back: sulphide- and acetic acid-induced bud and root death, lignifications, and blockages within the aeration and vascular systems. *New Phytologist* 134, 601–614.

Atkinson, M.M., Keppler, L.D., Orlandi, E.W., Baker, C.J. and Mischke, C.F. (1990) Involvement of plasma membrane calcium influx in bacteria induction of the K^+/H^+ and hypersensitive responses in tobacco. *Plant Physiology* 92, 215–221.

Atwell, B.J., Drew, M.C. and Jackson, M.B. (1988) The influence of oxygen deficiency on ethylene synthesis, 1-aminocyclopropane-1-carboxylic acid levels and aerenchyma formation in roots of *Zea mays* L. *Physiologia Plantarum* 72, 15–22.

Ayano, M., Kani, T., Kojima, M., Sakakibara, H., Kitaoka, T. *et al.* (2014) Gibberellin biosynthesis and signal transduction is essential for internode elongation in deepwater rice. *Plant, Cell & Environment* 37, 2313–2324.

Bailey-Serres, J. and Chang, R. (2005) Sensing and signalling in response to oxygen deprivation in plants and other organisms. *Annals of Botany* 96, 507–518.

Bailey-Serres, J. and Voesenek, L.A.C.J. (2008) Flooding stress: acclimations and genetic diversity. *Annual Review of Plant Biology* 59, 313–339.

Bailey-Serres, J. and Voesenek, L.A.C.J. (2010) Life in the balance: a signaling network controlling survival of flooding. *Current Opinion in Plant Biology* 13, 489–494.

Bailey-Serres, J., Fukao, T., Ronald, P., Ismail, A., Heuer, S. *et al.* (2010) Submergence tolerant rice: SUB1's journey from landrace to modern cultivar. *Rice* 3, 138–147.

Banti, V., Mafessoni, F., Loreti, E., Alpi, A. and Perata, P. (2010) The heat-inducible transcription factor HsfA2 enhances anoxia tolerance in Arabidopsis. *Plant Physiology* 152, 1471–1483.

Baud, S., Vaultier, M.N. and Rochat, C. (2004) Structure and expression profile of the sucrose synthase multigene family in *Arabidopsis. Journal of Experimental Botany* 55, 397–409.

Baxter-Burrell, A., Yang, Z., Springer, P.S. and Bailey-Serres, J. (2002) RopGAP4-dependent Rop GTPase rheostat control of *Arabidopsis* oxygen deprivation tolerance. *Science* 296, 2026–2028.

Biemelt, S., Hajirezaei, M.R., Melzer, M., Albrecht, G. and Sonnewald, U. (1999) Sucrose synthase activity does not restrict glycolysis in roots of transgenic potato plants under hypoxic conditions. *Planta* 210, 41–49.

Bieniawska, Z., Barratt, D.H.P., Garlick, A.P., Thole, V., Kruger, N.J. *et al.* (2007) Analysis of the sucrose synthase gene family in Arabidopsis. *The Plant Journal* 49, 810–828.

Boamfa, E.I., Ram, P.C., Jackson, M.B., Reuss, J. and Harren, F.J. (2003) Dynamic aspects of alcoholic fermentation of rice seedlings in response to anaerobiosis and to complete submergence: relationship to submergence tolerance. *Annals of Botany* 91, 279–290.

Boamfa, E.I., Veres, A.H., Ram, P.C., Jackson, M.B., Reuss, J. *et al.* (2005) Kinetics of ethanol and acetaldehyde release suggest a role for acetaldehyde production in tolerance of rice seedlings to micro-aerobic conditions. *Annals of Botany* 96, 727–736.

Branco-Price, C., Kaiser, K.A., Jang, C.J.H., Larive, C.K. and Bailey-Serres, J. (2008) Selective mRNA translation coordinates energetic and metabolic adjustments to cellular oxygen deprivation and reoxygenation in *Arabidopsis thaliana*. *The Plant Journal* 56, 743–755.

Bui, L.T., Giuntoli, B., Kosmacz, M., Parlanti, S. and Licausi, F. (2015) Constitutively expressed ERF-VII transcription factors redundantly activate the core anaerobic response in *Arabidopsis thaliana*. *Plant Science* 236, 37–43.

Campbell, R. and Drew, M.C. (1983) Electron microscopy of gas space (aerenchyma) formation in adventitious roots of *Zea mays* L. subjected to oxygen shortage. *Planta* 157, 350–357.

Catling, D. (1992) *Rice in Deepwater*. The MacMillan Press Ltd, London, UK.

Chang, R., Jang, C.J., Branco-Price, C., Nghiem, P., Bailey-Serres, J. (2012) Transient MPK6 activation in response to oxygen deprivation and reoxygenation is mediated by mitochondria and aids seedling survival in *Arabidopsis*. *Plant Molecular Biology* 78, 109–122.

Chen, P.-W., Lu, C.-A., Yu, T.-S., Tseng, T.-H. *et al.* (2002) Rice α-amylase transcriptional enhancers direct multiple mode regulation of promoters in transgenic rice. *Journal of Biological Chemistry* 277, 13641–13649.

Cho, H.T. and Kende, H. (1997) Expression of expansin genes is correlated with growth in deepwater rice. *The Plant Cell* 9, 1661–1671.

Choi, D.S., Lee, Y., Cho, H.T. and Kende, H. (2003) Regulation of expansin gene expression affects growth and development in transgenic rice plants. *The Plant Cell* 15, 1386–1398.

Colmer, T.D. (2003a) Long-distance transport of gases in plants: a perspective on internal aeration and radial oxygen loss from roots. *Plant, Cell & Environment* 26, 17–36.

Colmer, T.D. (2003b) Aerenchyma and an inducible barrier to radial oxygen loss facilitate root aeration in upland, paddy and deep-water rice (*Oryza sativa* L.). *Annals of Botany* 91, 301–309.

Colmer, T.D. and Pedersen, O. (2008) Underwater photosynthesis and respiration in leaves of submerged wetland plants: gas films improve CO_2 and O_2 exchange. *New Phytologist* 177, 918–926.

Colmer, T.D. and Voesenek, L.A.C.J. (2009) Flooding tolerance: suites of plant traits in variable environments. *Functional Plant Biology* 36, 665–681.

Cosgrove, D.J. (1999) Enzymes and other agents that enhance cell wall extensibility. *Annual Review of Plant Physiology and Plant Molecular Biology* 50, 391–417.

Cox, M.C., Millenaar, F.F., Van Berkel, Y.E.M.D.J., Peeters, A.J.M. and Voesenek, L.A.C.J. (2003) Plant movement. Submergence-induced petiole elongation in *Rumex palustris* depends on hyponastic growth. *Plant Physiology* 132, 282–291.

Cox, M.C., Benschop, J.J., Vreeburg, R.A., Wagemaker, C.A., Moritz, T. *et al.* (2004) The roles of ethylene, auxin, abscisic acid, and gibberellin in the hyponastic growth of submerged *Rumex palustris* petioles. *Plant Physiology* 136, 2948–2960.

Das, D.K. and Jat, R.L. (1977) Influence of three soil-water regimes on root porosity and growth of four rice varieties. *Agronomy Journal* 69, 197–200.

Davies, D.D. (1980) Anaerobic metabolism and the production of organic acids. In: Davies, D.D. (ed.) *The Biochemistry of Plants*. Vol. 2. Academic Press, New York, pp. 581–611.

Davies, D.D., Grego, S. and Kenworthy, P. (1974) The control of the production of lactate and ethanol by higher plants. *Planta* 118, 297–310.

Delledonne, M. (2005) NO news is good news for plants. *Current Opinion in Plant Biology* 8, 390–396.

Dolferus, R., Wolansky, M., Carroll, R., Miyashita, Y., Ismond, K. *et al.* (2008) Functional analysis of lactate dehydrogenase during hypoxic stress in *Arabidopsis*. *Functional Plant Biology* 35, 131–140.

Drew, M.C., Jackson, M.B. and Giffard, S. (1979) Ethylene-promoted adventitious rooting and development of cortical air spaces (aerenchyma) in roots may be adaptive responses to flooding in *Zea mays* L. *Planta* 147, 83–88.

Drew, M.C., Jackson, M.B., Giffard, S.C. and Campbell, R. (1981) Inhibition by silver ions of gas space (aerenchyma) formation in adventitious roots of *Zea mays* L. subjected to exogenous ethylene or to oxygen deficiency. *Planta* 153, 217–224.

Dunn, G. (1974) A model for starch breakdown in higher plants. *Phytochemistry* 13, 1341–1346.

Enstone, D.E., Peterson, C.A. and Ma, F. (2003) Root endodermis and exodermis: structure, function, and responses to the environment. *Journal of Plant Growth Regulation* 21, 335–351.

Evans, D.E. (2003) Aerenchyma formation. *New Phytologist* 161, 35–49.

Fukao, T. and Bailey-Serres, J. (2008) Submergence tolerance conferred by *Sub1A* is mediated by SLR1 and SLRL1 restriction of gibberellin responses in rice. *Proceedings of the National Academy of Sciences USA* 105, 16814–16819.

Fukao, T., Xu, K., Ronald, P.C. and Bailey-Serres, J. (2006) A variable cluster of ethylene response factor- like genes regulates metabolic and developmental acclimation responses to submergence in rice. *The Plant Cell* 18, 2021–2034.

Fukao, T., Yeung, E. and Bailey-Serres, J. (2011) The submergence tolerance regulator SUB1A mediates crosstalk between submergence and drought tolerance in rice. *The Plant Cell* 23, 412–427.

Garthwaite, A.J., von Bothmer, R. and Colmer, T.D. (2003) Diversity in root aeration traits associated with water-logging tolerance in the genus *Hordeum*. *Functional Plant Biology* 30, 875–889.

Geisler-Lee, J., Caldwell, C. and Gallie, D.R. (2010) Expression of the ethylene biosynthetic machinery in maize roots is regulated in response to hypoxia. *Journal of Experimental Botany* 61, 857–871.

Germain, V., Ricard, B., Raymond, P. and Saglio, P.H. (1997) The role of sugars, hexokinase, and sucrose synthase in the determination of hypoxically induced tolerance to anoxia in tomato roots. *Plant Physiology* 114, 167–175.

Gibbs, J. and Greenway, H. (2003) Mechanisms of anoxia tolerance in plants. I. Growth, survival and anaerobic catabolism. *Functional Plant Biology* 30, 1–47.

Gibbs, J., Morrell, S., Valdez, A., Setter, T.L. and Greenway, H. (2000) Regulation of alcoholic fermentation in cole-optiles of two rice cultivars differing in tolerance to anoxia. *Journal of Experimental Botany* 51, 785–796.

Gibbs, D.J., Lee, S.C., Isa, N.M., Gramuglia, S., Fukao, T. *et al.* (2011) Homeostatic response to hypoxia is regulated by the N-end rule pathway in plants. *Nature* 479, 415–418.

Gibbs, D.J., Md Isa, N., Movahedi, M., Lozano-Juste, J., Mendiondo, G.M. *et al.* (2014) Nitric oxide sensing in plants is mediated by proteolytic control of group VII ERF transcription factors. *Molecular Cell* 53, 369–379.

Giuntoli, B., Lee, S.C., Licausi, F., Kosmacz, M., Oosumi, T. *et al.* (2014) A trihelix DNA binding protein counter-balances hypoxia-responsive transcriptional activation in Arabidopsis. *PLoS Biology* 12, e1001950.

Gonzali, E., Loreti, E., Cardarelli, F., Novi, G., Parlanti, S. *et al.* (2015) The Universal Stress Protein HRU1 mediates ROS homeostasis under anoxia. *Nature Plants* 1, 15151.

Good, A.G. and Crosby, W.L. (1989) Anaerobic induction of alanine aminotransferase in barley root tissue. *Plant Physiology* 90, 1305–1309.

Good, A.G. and Muench, D.G. (1992) Purification and characterization of an anaerobically induced alanine ami-notransferase from barley roots. *Plant Physiology* 99, 1520–1525.

Gu, Y., Wang, Z. and Yang, Z. (2004) ROP/RAC GTPase: an old new master regulator for plant signaling. *Current Opinion in Plant Biology* 7, 527–536.

Guglielminetti, L., Yamaguchi, J., Perata, P. and Alpi, A. (1995) Amylolytic activities in cereal seeds under aerobic and anaerobic conditions. *Plant Physiology* 109, 1069–1076.

Gunawardena, A.H., Pearce, D.M., Jackson, M.B., Hawes, C.R. and Evans, D.E. (2001) Characterisation of pro-grammed cell death during aerenchyma formation induced by ethylene or hypoxia in roots of maize (*Zea mays* L.). *Planta* 212, 205–214.

Hattori, Y., Miura, K., Asano, K., Yamamoto, E., Mori, H. *et al.* (2007) A major QTL confers rapid internode elong-ation in response to water rise in deepwater rice. *Breeding Science* 57, 305–314.

Hattori, Y., Nagai, K., Mori, H., Kitano, H., Matsuoka, M. *et al.* (2008) Mapping of three QTLs that regulate inter-node elongation in deepwater rice. *Breeding Science* 58, 39–46.

Hattori, Y., Nagai, K., Furukawa, S., Song, X.J., Kawano, R. *et al.* (2009) The ethylene response factors SNORKEL1 and SNORKEL2 allow rice to adapt to deep water. *Nature* 460, 1026–1030.

He, C.J., Morgan, P.W. and Drew, M.C. (1994) Induction of enzymes associated with lysigenous aerenchyma for-mation in roots of *Zea mays* during hypoxia or nitrogen starvation. *Plant Physiology* 105, 861–865.

He, C.J., Finlayson, S.A., Drew, M.C., Jordan, W.R. and Morgan, P.W. (1996) Ethylene biosynthesis during aeren-chyma formation in roots of maize subjected to mechanical impedance and hypoxia. *Plant Physiology* 112, 1679–1685.

He, J.B., Bögemann, G.M., van de Steeg, H.M., Rijnders, J.G., Voesenek, L.A. *et al.* (1999) Survival tactics of *Ranunculus* species in river floodplains. *Oecologia* 118, 1–8.

Heydarian, Z., Sasidharan, R., Cox, M.C.H., Pierik, R., Voesenek, L.A.C.J. *et al.* (2010) A kinetic analysis of hyponastic growth and petiole elongation upon ethylene exposure in *Rumex palustris*. *Annals of Botany* 106, 429–435.

Huang, B., Johnson, J.W., NeSmith, D.S. and Bridges, D.C. (1994a) Growth, physiological and anatomical re-sponses of two wheat genotypes to waterlogging and nutrient supply. *Journal of Experimental Botany* 45, 193–202.

Huang, B., Johnson, J.W., NeSmith, D.S. and Bridges, D.C. (1994b) Root and shoot growth of wheat genotypes in response to hypoxia and subsequent resumption of aeration. *Crop Science* 34, 1538–1544.

Huang, B., Johnson, J.W., NeSmith, D.S. and Bridges, D.C. (1995a) Nutrient accumulation and distribution of wheat genotypes in response to waterlogging and nutrient supply. *Plant and Soil* 173, 47–54.

Huang, B., Johnson, J.W., NeSmith, D.S. and Bridges, D.C. (1995b) Responses of squash to salinity, waterlogging and subsequent drainage: II root and shoot growth. *Journal of Plant Nutrition* 18, 141–152.

Huang, J., Takano, T. and Akita, S. (2000) Expression of α-expansin genes in young seedlings of rice (*Oryza sativa* L.). *Planta* 211, 467–473.

Huang, S.B., Greenway, H. and Colmer, T.D. (2003) Anoxia tolerance in rice seedlings: exogenous glucose improves growth of an anoxia-'intolerant', but not of a 'tolerant' genotype. *Journal of Experimental Botany* 54, 2363–2373.

Huber, S.C. and Akazawa, T. (1986) A novel sucrose synthase pathway for sucrose degradation in cultured sycamore cells. *Plant Physiology* 81, 1008–1013.

Hwang, Y.S., Thomas, B.R. and Rodriguez, R.L. (1999) Differential expression of rice α-amylase genes during seedling development under anoxia. *Plant Molecular Biology* 40, 911–920.

Ishizawa, K., Murakami, S., Kawakami, Y. and Kuramochi, H. (1999) Growth and energy status of arrowhead tubers, pondweed turions and rice seedlings under anoxic conditions. *Plant, Cell & Environment* 22, 505–514.

Ismond, K.P., Dolferus, R., De Pauw, M., Dennis, E.S. and Good, A.G. (2003) Enhanced low oxygen survival in *Arabidopsis* through increased metabolic flux in the fermentative pathway. *Plant Physiology* 132, 1292–1302.

Jackson, M.B. (2002) Long-distance signalling from roots to shoots assessed: the flooding story. *Journal of Experimental Botany* 53, 175–181.

Jackson, M.B. (2008) Ethylene-promoted elongation: an adaptation to submergence stress. *Annals of Botany* 101, 229–248.

Jackson, M.B., Fenning, T.M., Drew, M.C. and Saker, L.R. (1985) Stimulation of ethylene production and gas-space (aerenchyma) formation in adventitious roots of *Zea mays* L. by small partial pressures of oxygen. *Planta* 165, 486–492.

Jung, K.-H., Seo, Y.-S., Walia, H., Cao, P., Fukao, T., Canlas, P.E. *et al.* (2010) The submergence tolerance regulator *Sub1A* mediates stress-responsive expression of *AP2/ERF* transcription factors. *Plant Physiology* 152, 1674–1692.

Justin, S.H.F.W. and Armstrong, W. (1987) The anatomical characteristics of roots and plant response to soil flooding. *New Phytologist* 106, 465–495.

Justin, S.H.F.W. and Armstrong, W. (1991) Evidence for the involvement of ethene in aerenchyma formation in adventitious roots of rice (*Oryza sativa* L.). *New Phytologist* 118, 49–62.

Kato-Noguchi, H. and Morokuma, M. (2007) Ethanolic fermentation and anoxia tolerance in four rice cultivars. *Journal of Plant Physiology* 164, 168–173.

Kato-Noguchi, H., Yasuda, Y. and Sasaki, R. (2010) Soluble sugar availability of aerobically germinated barley, oat and rice coleoptiles in anoxia. *Journal of Plant Physiology* 167, 1571–1576.

Kawase, M. (1972) Effect of flooding on ethylene concentration in horticultural plants. *Journal of the American Society for Horticultural Science* 97, 584–588.

Kawase, M. (1978) Anaerobic elevation of ethylene concentration in waterlogged plants. *American Journal of Botany* 65, 736–740.

Kawase, M. (1981) Anatomical and morphological adaptation of plants to waterlogging. *Horticultural Science* 16, 30–34.

Kende, H., van der Knaap, E. and Cho, H.T. (1998) Deepwater rice: a model plant to study stem elongation. *Plant Physiology* 118, 1105–1110.

Kimmerer, T.W. and MacDonald, R.C. (1987) Acetaldehyde and ethanol biosynthesis in leaves of plants. *Plant Physiology* 84, 1204–1209.

Könings, H. (1982) Ethylene-promoted formation of aerenchyma in seedling roots of *Zea mays* L. under aerated and non-aerated conditions. *Physiologia Plantarum* 54, 119–124.

Könings, H. and Jackson, M.B. (1979) A relationship between rates of ethylene production by roots and the promoting or inhibiting effects of exogenous ethylene and water on root elongation. *Zeitschrift fuer Pflanzenphysiologie* 92, 385–397.

Kosmacz, M., Parlanti, S., Schwarzländer, M., Kragler, F., Licausi, F. *et al.* (2015) The stability and nuclear localization of the transcription factor RAP2.12 are dynamically regulated by oxygen concentration. *Plant, Cell & Environment* 38, 1094–1103.

Kozlowski, T.T. (1984) Responses of woody plants to flooding. In: Kozlowski, T.T. (ed.) *Flooding and Plant Growth.* Academic Press, New York, USA, pp. 129–163.

Kürsteiner, O., Dupuis, I. and Kuhlemeier, C. (2003) The pyruvate decarboxylase 1 gene of Arabidopsis is required during anoxia but not other environmental stresses. *Plant Physiology* 132, 968–978.

Laan, P., Berrevoets, M.J., Lythe, S., Armstrong, W. and Blom, C.W.P.M. (1989) Root morphology and aerenchyma formation as indicators of the flood-tolerance of *Rumex* species. *Journal of Ecology* 77, 693–703.

Lasanthi-Kudahettige, R., Magneschi, L., Loreti, E., Gonzali, S., Licausi, F. *et al.* (2007) Transcript profiling of the anoxic rice coleoptile. *Plant Physiology* 144, 218–231.

Lecourieux, D., Ranjeva, R. and Pugin, A. (2006) Calcium in plant defence-signalling pathways. *New Phytologist* 171, 249–269.

Lee, K.-W., Chen, P.-W., Lu, C.-A., Chen, S., Ho, T.-H.D. *et al.* (2009) Coordinated responses to oxygen and sugar deficiency allow rice seedlings to tolerate flooding. *Science Signaling* 2, ra61.

Lenochova, Z., Soukup, A. and Votrubova, O. (2009) Aerenchyma formation in maize roots. *Biologia Plantarum* 53, 263–270.

Licausi, F. and Perata, P. (2009) Low oxygen signaling and tolerance in plants. *Advances in Botanical Research* 50, 139–198.

Licausi, F., Kosmacz, M., Weits, D.A., Giuntoli, B., Giorgi, F.M. *et al.* (2011) Oxygen sensing in plants is mediated by an N-end rule pathway for protein destabilization. *Nature* 479, 419–422.

Loreti, E., Alpi, A. and Perata, P. (2003a) α-Amylase expression under anoxia in rice seedlings: an update. *Russian Journal of Plant Physiology* 50, 737–742.

Loreti, E., Yamaguchi, J., Alpi, A. and Perata, P. (2003b) Gibberellins are not required for rice germination under anoxia. *Plant and Soil* 253, 137–143.

Lu, C.-A., Lim, E.-K. and Yu, S.-M. (1998) Sugar response sequence in the promoter of a rice α-amylase gene serves as a transcriptional enhancer. *The Journal of Biological Chemistry* 273, 10120–10131.

Lu, C.-A., Lin, C.-C., Lee, K.-W., Chen, J.-L., Huang, L.-F. *et al.* (2007) The SnRK1A protein kinase plays a key role in sugar signaling during germination and seedling growth of rice. *Plant Cell* 19, 2484–2499.

Luan, S., Kudla, J., Rodriguez-Concepcion, M., Yalovsky, S. and Gruissem, W. (2002) Calmodulins and calcineurin B-like proteins: calcium sensors for specific signal response coupling in plants. *Plant Cell* 14, S389–S400.

Mackill, D.J., Coffman, W.R. and Garrity, D.P. (1996) *Rainfed Lowland Rice Improvement.* International Rice Research Institute, Los Banos, The Philippines.

Magneschi, L. and Perata, P. (2009) Rice germination and seedling growth in the absence of oxygen. *Annals of Botany* 103, 181–196.

Magneschi, L., Kudahettige, R.L., Alpi, A. and Perata, P. (2009a) Comparative analysis of anoxic coleoptile elongation in rice varieties: relationship between coleoptile length and carbohydrate levels, fermentative metabolism and anaerobic gene expression. *Plant Biology* 11, 561–573.

Magneschi, L., Kudahettige, R.L., Alpi, A. and Perata, P. (2009b) Expansin gene expression and anoxic coleoptile elongation in rice cultivars. *Journal of Plant Physiology* 166, 1576–1580.

McDonald, M.P., Galwey, N.W. and Colmer, T.D. (2002) Similarity and diversity in adventitious root anatomy as related to root aeration among a range of wet- and dry-land grass species. *Plant, Cell & Environment* 25, 441–451.

Métraux, J.-P. and Kende, H. (1983) The role of ethylene in the growth response of submerged deep water rice. *Plant Physiology* 72, 441–446.

Millenaar, F.F., Cox, M.C., van Berkel, Y.E., Welschen, R.A., Pierik, R. *et al.* (2005) Ethylene-induced differential growth of petioles in Arabidopsis. Analyzing natural variation, response kinetics, and regulation. *Plant Physiology* 137, 998–1008.

Millenaar, F.F., van Zanten, M., Cox, M.C., Pierik, R., Voesenek, L.A. *et al.* (2009) Differential petiole growth in *Arabidopsis thaliana*: photocontrol and hormonal regulation. *New Phytologist* 184, 141–152.

Miller, G. and Mittler, R. (2006) Could heat shock transcription factors function as hydrogen peroxide sensors in plants? *Annals of Botany* 98, 279–288.

Mommer, L., Pons, T.L., Wolters-Arts, M., Venema, J.H. and Visser, E.J.W. (2005) Submergence-induced morphological, anatomical, and biochemical responses in a terrestrial species affects gas diffusion resistance and photosynthetic performance. *Plant Physiology* 139, 497–508.

Morita, A., Umemura, T., Kuroyanagi, M., Futsuhara, Y., Perata, P. *et al.* (1998) Functional dissection of a sugar-repressed α-amylase gene (*RAmy1A*) promoter in rice embryos. *FEBS Letters* 423, 81–85.

Muench, D.G. and Good, A.G. (1994) Hypoxically inducible barley alanine aminotransferase: cDNA cloning and expression analysis. *Plant Molecular Biology* 24, 417–427.

Mühlenbock, P., Plaszczyca, M., Mellerowicz, E. and Karpinski, S. (2007) Lysigenous aerenchyma formation in *Arabidopsis* is controlled by *LESION SIMULATING DISEASE1*. *The Plant Cell* 19, 3819–3830.

Murata, T., Akazawa, T. and Fukuchi, S. (1968) Enzymic mechanism of starch breakdown in germinating rice seeds: 1. An analytical study. *Plant Physiology* 43, 1899–1905.

Mustroph, A. and Albrecht, G. (2007) Fermentation metabolism in roots of wheat seedlings after hypoxic pre-treatment in different anoxic incubation systems. *Journal of Plant Physiology* 164, 394–407.

Mustroph, A., Boamfa, E.I., Laarhoven, L.J., Harren, F.J., Albrecht, G. *et al.* (2006a) Organ-specific analysis of the anaerobic primary metabolism in rice and wheat seedlings. I: Dark ethanol production is dominated by the shoots. *Planta* 225, 103–114.

Mustroph, A., Boamfa, E.I., Laarhoven, L.J., Harren, F.J., Pörs, Y. *et al.* (2006b) Organ specific analysis of the anaerobic primary metabolism in rice and wheat seedlings II: light exposure reduces needs for fermentation and extends survival during anaerobiosis. *Planta* 225, 139–152.

Mustroph, A., Zanetti, M.E., Jang, C.J., Holtan, H.E., Repetti, P.P. *et al.* (2009) Profiling translatomes of discrete cell populations resolves altered cellular priorities during hypoxia in *Arabidopsis*. *Proceedings of the National Academy of Sciences USA* 106, 18843–18848.

Mustroph, A., Lee, S.C., Oosumi, T., Zanetti, M.E., Yang, H. *et al.* (2010) Cross-kingdom comparison of transcriptomic adjustments to low-oxygen stress highlights conserved and plant-specific responses. *Plant Physiology* 152, 1484–1500.

Mustroph, A., Barding Jr, G.A., Kaiser, K.A., Larive, C.K. and Bailey-Serres, J. (2014) Characterization of distinct root and shoot responses to low-oxygen stress in Arabidopsis with a focus on primary C- and N-metabolism. *Plant, Cell & Environment* 37, 2366–2380.

Nagai, K., Hattori, Y. and Ashikari, M. (2010) Stunt or elongate? Two opposite strategies by which rice adapts to floods. *Journal of Plant Research* 123, 303–309.

Neal, M.J. and Girton, R.E. (1955) The Pasteur effect in maize. *American Journal of Botany* 42, 733–737.

Neill, S.J., Desikan, R., Clarke, A., Hurst, R.D. and Hancock, J.T. (2002) Hydrogen peroxide and nitric oxide as signalling molecules in plants. *Journal of Experimental Botany* 53, 1237–1247.

Niroula, R.K., Pucciariello, C., Ho, V.T., Novi, G., Fukao, T. *et al.* (2012) SUB1A-dependent and -independent mechanisms are involved in the flooding tolerance of wild rice species. *The Plant Journal* 72, 282–293.

Parlanti, S., Kudahettige, N.P., Lombardi, L., Mensuali-Sodi, A., Alpi, A. *et al.* (2011) Distinct mechanisms for aerenchyma formation in leaf sheaths of rice genotypes displaying a quiescence or escape strategy for flooding tolerance. *Annals of Botany* 107, 1335–1343.

Pedersen, O., Rich, S.M. and Colmer, T.D. (2009) Surviving floods: leaf gas films improve O_2 and CO_2 exchange, root aeration, and growth of completely submerged rice. *The Plant Journal* 58, 147–156.

Perata, P. and Alpi, A. (1991) Ethanol induced injuries to carrot cell: the role of acetaldehyde. *Plant Physiology* 95, 748–752.

Perata, P. and Alpi, A. (1993) Plant responses to anaerobiosis. *Plant Science* 93, 1–17.

Perata, P., Pozuetaromero, J., Akazawa, T. and Yamaguchi, J. (1992) Effect of anoxia on starch breakdown in rice and wheat seeds. *Planta* 188, 611–618.

Perata, P., Guglielminetti, L. and Alpi, A. (1996) Anaerobic carbohydrate metabolism in wheat and barley, two anoxia-intolerant cereal seeds. *Journal of Experimental Botany* 47, 999–1006.

Perata, P., Matsukura, C., Vernieri, P. and Yamaguchi, J. (1997) Sugar repression of a gibberellin-dependent signaling pathway in barley embryos. *The Plant Cell* 9, 2197–2208.

Polko, J.K., van Zanten, M., van Rooij, J.A., Marée, A.F.M., Voesenek, L.A.C.J. *et al.* (2012) Ethylene-induced differential petiole growth in *Arabidopsis thaliana* involves local microtubule reorientation and cell expansion. *New Phytologist* 193, 339–348.

Pradhan, S.K., Varade, S.B. and Kar, S. (1973) Influence of soil water conditions on growth and root porosity of rice. *Plant Soil* 38, 501–507.

Pucciariello, C., Parlanti, S., Banti, V., Novi, G. and Perata, P. (2012) Reactive oxygen species-driven transcription in Arabidopsis under oxygen deprivation. *Plant Physiology* 159, 184–196.

Rajhi, I., Yamauchi, T., Takahashi, H., Nishiuchi, S., Shiono, K. *et al.* (2011) Identification of genes expressed in maize root cortical cells during lysigenous aerenchyma formation using laser microdissection and microarray analyses. *New Phytologist* 190, 351–368.

Rauf, M., Arif, M., Fisahn, J., Xue, G.P., Balazadeh, S. *et al.* (2013) NAC transcription factor speedy hyponastic growth regulates flooding induced leaf movement in *Arabidopsis*. *The Plant Cell* 25, 4941–4955.

Raven, J.A. (2008) Not drowning but photosynthesizing: probing plant plastrons. *New Phytologist* 177, 841–845.

Raymond, P., Al-Ani, A. and Pradet, A. (1985) ATP production by respiration and fermentation, and energy charge during aerobiosis and anaerobiosis in twelve fatty and starchy germinating seeds. *Plant Physiology* 79, 879–884.

Reggiani, R. (1999) Amino acid metabolism under oxygen deficiency. *Current Topics in Phytochemistry* 2, 171–174.

Reggiani, R., Cantu, C.A., Brambilla, I. and Bertani, A. (1988) Accumulation and interconversion of amino acids in rice roots under anoxia. *Plant Cell Physiology* 29, 981–987.

Renz, A., Merlo, L. and Stitt, M. (1993) Partial purification from potato tubers of three fructokinases and three hexokinases which show differing organ and developmental specificity. *Planta* 190, 156–165.

Richard, B., Rivoal, J., Spiteri, A. and Pradet, A. (1991) Anaerobic stress induces the transcription and translation of sucrose synthase in rice. *Plant Physiology* 95, 669–674.

Ridge, I. (1987) Ethylene and growth control in amphibious plants. In: Crawford, R.M.M. (ed.) *Plant Life in Aquatic and Amphibious Habitats*. Blackwell Science Publications, Oxford, UK, pp. 53–76.

Rocha, M., Licausi, F., Araújo, W.L., Nunes-Nesi, A., Sodek, L. *et al.* (2010) Glycolysis and the tricarboxylic acid cycle are linked by alanine aminotransferase during hypoxia induced by waterlogging of *Lotus japonicus*. *Plant Physiology* 152, 1501–1513.

Rodriguez, R.L., Huang, N., Sutliff, T.D., Ranjhan, S., Karrer, E. *et al.* (1992) Organization, structure and expression of the rice α-amylase multigene family. In: Banta, S.J. and Argosino, G.S. (eds) *Rice Genetics II: Proceedings of the Second International Rice Genetics Symposium*. IRRI, Los Baños, Laguna, Philippines, pp. 417–429.

Sachs, J.A. (1882) *A Text Book of Botany*. Oxford University Press, Oxford, UK.

Santaniello, A., Loreti, E., Gonzali, S., Novi, G. and Perata, P. (2014) A reassessment of the role of sucrose synthase in the hypoxic sucrose-ethanol transition in Arabidopsis. *Plant, Cell & Environment* 37, 2294–2302.

Sato, T., Harada, T. and Ishizawa, K. (2002) Stimulation of glycolysis in anaerobic elongation of pondweed (*Potamogeton distinctus*) turion. *Journal of Experimental Botany* 53, 1847–1856.

Sauter, M. (2000) Rice in deep water: "How to take heed against a sea of trouble". *Naturwissenschaften* 87, 289–303.

Schlüter, U. and Crawford, R.M. (2001) Long-term anoxia tolerance in leaves of *Acorus calamus* L. and *Iris pseudacorus* L. *Journal of Experimental Botany* 52, 2213–2225.

Schussler, E.E. and Longstreth, D.J. (1996) Aerenchyma develops by cell lysis in roots and cell separation in leaf petioles in *Sagittaria lancifolia* (Alismataceae). *American Journal of Botany* 83, 1266–1273.

Seago, J.L., Marsh, L.C., Stevens, K.J., Soukup, A., Votrubová, O. *et al.* (2005) A re-examination of the root cortex in wetland flowering plants with respect to aerenchyma. *Annals of Botany* 96, 565–579.

Sedbrook, J.C., Kronebush, P.J., Borisy, G.G., Trewavas, A.J. and Masson, P. (1996) Transgenic aequorin reveals organ specific cytosolic Ca²⁺ responses to anoxia in *Arabidopsis thaliana* seedlings. *Plant Physiology* 111, 243–257.

Semenza, G.L. (2007) Hypoxia-inducible factor 1 (HIF-1) pathway. *Science's STKE* 2007(407), cm8.

Setter, T.L., Ella, E.S. and Valdez, A.P. (1994) Relationship between coleoptile elongation and alcoholic fermentation in rice exposed to anoxia. II. Cultivar differences. *Annals of Botany* 74, 273–279.

Shiono, K., Ogawa, S., Yamazaki, S., Isoda, H., Fujimura, T. *et al.* (2011) Contrasting dynamics of radial O₂-loss barrier induction and aerenchyma formation in rice roots of two lengths. *Annals of Botany* 107, 89–99.

Shiono, K., Yamauchi, T., Yamazaki, S., Mohanty, B., Malik, A.I. *et al.* (2014) Microarray analysis of laser-microdissected tissues indicates the biosynthesis of suberin in the outer part of roots during formation of a barrier to radial oxygen loss in rice (*Oryza sativa*). *Journal of Experimental Botany* 65, 4795–4806.

Singh, N., Dang, T., Vergara, G., Pandey, D., Sanchez, D. *et al.* (2010) Molecular marker survey and expression analyses of the rice submergence-tolerance genes *SUB1A* and *SUB1C*. *Theoretical and Applied Genetics* 121, 1441–1453.

Springer, B., Werr, W., Starlinger, P., Bennet, C.D., Zokolica, M. *et al.* (1986) The Shrunken gene on chromosome 9 of Zea maize L. is expressed in various plant tissues and encodes an anaerobic protein. *Molecular & General Genetics* 205, 461–468.

Steffens, B., Geske, T. and Sauter, M. (2011) Aerenchyma formation in the rice stem and its promotion by H₂O₂. *New Phytologist* 190, 369–378.

Steffens, B., Kovalev, A., Gorb, S.N. and Sauter, M. (2012) Emerging roots alter epidermal cell fate through mechanical and reactive oxygen species signaling. *The Plant Cell* 24, 3296–3306.

Stitt, M. (1998) Pyrophosphate as an energy donor in the cytosol of plant cells: an enigmatic alternative to ATP. *Botanica Acta* 111, 167–175.

Streeter, J.G. and Thompson, J.F. (1972) Anaerobic accumulation of γ-aminobutyric acid and alanine in radish leaves (*Raphanus sativus* L.). *Plant Physiology* 49, 572–578.

Stünzi, J.T. and Kende, H. (1989) Gas composition in the internal air spaces of deepwater rice in relation to growth induced by submergence. *Plant and Cell Physiology* 30, 49–56.

Subbaiah, C.C., Bush, D.S. and Sachs, M.M. (1994) Elevation of cytosolic calcium precedes anoxic gene expression in maize suspension-cultured cells. *The Plant Cell* 6, 1747–1762.

Suge, H. (1987) Physiological genetics of internode elongation under submergence in floating rice. *The Japanese Journal of Genetics* 62, 69–80.

Summers, J.E. and Jackson, M.B. (1994) Anaerobic conditions strongly promote extension by stems of over-wintering tubers of *Potamogeton pectinatus* L. *Journal of Experimental Botany* 45, 1309–1318.

Summers, J.E., Ratcliffe, R.G. and Jackson, M.B. (2000) Anoxia tolerance in the aquatic monocot *Potamogeton pectinatus*: absence of oxygen stimulates elongation in association with an unusually large Pasteur effect. *Journal of Experimental Botany* 51, 1413–1422.

Sun, Z. and Henson, C.A. (1991) A quantitative assessment of the importance of barley seed α-amylase, de-branching enzyme, and α-glucosidase in starch degradation. *Archives of Biochemistry and Biophysics* 284, 298–305.

Tadege, M., Dupuis, I. and Kuhlemeier, C. (1999) Ethanolic fermentation: new functions for an old pathway. *Trends in Plant Science* 4, 320–325.

Toyofuku, K., Umemura, T. and Yamaguchi, J. (1998) Promoter elements required for sugar-repression of the *RAmy3D* gene for α-amylase in rice. *FEBS Letters* 428, 275–280.

Trought, M.C.T. and Drew, M.C. (1980) The development of waterlogging damage in wheat seedlings (*Triticum aestivum* L). II. Accumulation and redistribution of nutrients by shoots. *Plant and Soil* 56, 187–199.

Vanlerberghe, G.C., Feil, R. and Turpin, D.H. (1990) Anaerobic metabolism in the N-limited green alga *Selenastrum minutum*. I. Regulation of carbon metabolism and succinate as a fermentation product. *Plant Physiology* 94, 1116–1123.

van Veen, H., Mustroph, A., Barding, G.A., Vergeer-van Eijk, M., Welschen-Evertman, R.A.M. *et al.* (2013) Two *Rumex* species from contrasting hydrological niches regulate flooding tolerance through distinct mechanisms. *The Plant Cell* 25, 4691–4707.

van Zanten, M., Basten Snoek, L., van Eck-Stouten, E., Proveniers, M.C., Torii, K.U. *et al.* (2010) Ethylene-induced hyponastic growth in *Arabidopsis thaliana* is controlled by ERECTA. *The Plant Journal* 61, 83–95.

Vartapetian, B.B. (1982) Pasteur effect visualization by electron microscopy. *Naturwissenschaften* 69, 99.

Vergara, B.S. and Mazaredo, A. (1975) Screening for resistance to submergence under greenhouse conditions. In: *Proceedings International Seminar on Deepwater Rice*. Bangladesh Rice Research Institute, Dhaka, Bangladesh, pp. 67–70.

Vidoz, M.L., Loreti, E., Mensuali, A., Alpi, A. and Perata, P. (2010) Hormonal interplay during adventitious root formation in flooded tomato plants. *The Plant Journal* 63, 551–562.

Visser, E.J.W. and Bögemann, G.M. (2006) Aerenchyma formation in the wetland plant *Juncus effusus* is independent of ethylene. *New Phytologist* 171, 305–314.

Visser, E.J.W., Colmer, T.D., Blom, C.W.P.M. and Voesenek, L.A.C.J. (2000) Changes in growth, porosity, and radial oxygen loss from adventitious roots of selected mono- and dicotyledonous wetland species with contrasting types of aerenchyma. *Plant, Cell & Environment* 23, 1237–1245.

Voesenek, L.A. and Bailey-Serres, J. (2015) Flood adaptive traits and processes: an overview. *New Phytologist* 206, 57–73.

Watanabe, K., Nishiuchi, S., Kulichikhin, K. and Nakazono, M. (2013) Does suberin accumulation in plant roots contribute to waterlogging tolerance? *Frontiers in Plant Science* 4, 178.

Weits, D.A., Giuntoli, B., Kosmacz, M., Parlanti, S., Hubberten, H.M. *et al.* (2014) Plant cysteine oxidases control the oxygen-dependent branch of the N-end-rule pathway. *Nature Communications* 5, 3425.

Xu, K., Xu, X., Fukao, T., Canlas, P., Maghirang-Rodriguez, R. *et al.* (2006) *Sub1A* is an ethylene-response-factor-like gene that confers submergence tolerance to rice. *Nature* 442, 705–708.

Zhou, Z., de Almeida Engler, J., Rouan, D., Michiels, F., Van Montagu, M. *et al.* (2002) Tissue localization of a sub-mergence-induced 1-aminocyclopropane-1-carboxylic acid synthase in rice. *Plant Physiology* 129, 72–84.

7 Adaptations to Aluminium Toxicity

Peter R. Ryan* and Emmanuel Delhaize
CSIRO Agriculture and Food, Canberra, Australia

Abstract

Soil acidity limits agricultural production globally. Acid soils pose many stresses to plants but the major factor affecting plant growth is soluble aluminium, with Al^{3+} being the most toxic form. Al^{3+} damages root cells at sites in the apoplast and in the cytosol, and these rapidly inhibit root growth. Plants have evolved mechanisms that either avoid or minimize these damaging interactions by excluding Al^{3+} from the roots and leaves or by efficiently detoxifying any Al^{3+} that enters the cytosol. The genes conferring these resistance mechanisms have now been identified in some important crop species such as wheat, barley, sorghum and rice. Rapid progress in this field over the last decade provides exciting opportunities for increasing the Al^{3+} resistance of food crops through marker-assisted selection and genetic engineering.

7.1 Soil Acidity

Approximately 30% of the Earth's ice-free land has a pH < 5.5. Acid soils are typically weathered and low in plant-available phosphate and basic cations such as calcium, magnesium, potassium and sodium. A proportion of this land is unproductive to agriculture because of poor fertility and mineral toxicities. Large tracts of land in the humid tropics, currently under forest, could be cultivated if lime and fertilizer are applied to raise the pH and improve nutrition (von Uexkull and Mutert, 1995). Increasing demands for food and fibre in South-east Asia and sub-Saharan Africa will draw more of this land into agricultural use in the coming decades.

Acidity is determined by the mineral composition of the soil and by specific ion exchanges and hydrolysis reactions. Important inorganic components in many acid soils include layer silicates, mineral oxides and soluble acids. Layer silicates are sheet-like arrangements of silicon and aluminium which are coated with iron and aluminium oxides (Thomas and Hargrove, 1984). In acid soils, aluminium (Al^{3+}) and iron (Fe^{3+}) ions are released from these minerals along with other metal ions. The high charge/size ratio of soluble Al^{3+} and Fe^{3+} ions enables them to split the H–O bonds in water by a process called hydrolysis (Eqn 7.1):

$$Al^{3+} + 6H_2O \rightarrow Al(H_2O)_6^{3+}$$
$$\xrightarrow{H_2O} [Al(H_2O)_5OH]^{2+} + H_3O^+ \quad (7.1)$$

or, more simply:

$$Al^{3+} + H_2O \rightarrow AlOH^{2+} + H^+$$

Hydrolysis is an acidifying process which generates several soluble aluminium species.

---

* Corresponding author: Peter.Ryan@csiro.au

The mole fraction of these species varies with pH: Al^{3+}, $Al(OH)^{2+}$ and $Al(OH)_2^+$ are prevalent below pH 6.0 while $Al(OH)_4^-$ (aluminate) and insoluble $Al(OH)_3$ (gibbsite) dominate at higher pH (Fig. 7.1).

Although many soils become acidic from natural processes, certain agricultural practices can accelerate these changes. The long-term cultivation of legume-based pastures is acidifying (Williams, 1980) because nitrogen-fixing plants sometimes generate more nitrate than is consumed by plants and microorganisms in the soil. Excess nitrate leaches down the profile taking with it basic cations and leaving protons in their place. Excessive applications of ammonium-based fertilizers to crops can also accelerate acidification (Guo et al., 2010) because the conversion of ammonium to nitrate (nitrification) leads to an increased risk of nitrate leaching, and the first step in this process is itself an acidifying reaction (Eqn 7.2):

$$NH_4^+ + (3/2)O_2 \rightarrow NO_2^- + 2H^+ + H_2O \quad (7.2)$$

The rhizosphere (the zone of soil around roots altered or influenced by root metabolism) is also acidified by the uptake of ammonium because H^+ efflux from roots maintains electroneutrality. Furthermore, assimilation of ammonium acidifies the cytosol, which also stimulates H^+ efflux. The continual removal of harvestable product which is slightly alkaline will eventually lower soil pH as well.

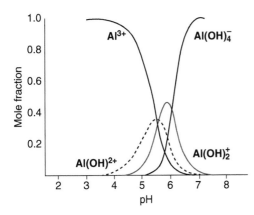

Fig. 7.1. Speciation of soluble aluminium. Distribution of soluble aluminium species showing the mole fraction of each species as a function of solution pH (adapted from Martin, 1992).

7.2 Acid Soils Limit Plant Growth and Production

Acid soils pose multiple physical and chemical stresses to plants. Plants grow poorly on these soils for two main reasons: (i) they encounter proton and metal-ion toxicities; and (ii) they run into nutrient deficiency (especially phosphorus, magnesium, calcium). Proton stress does not appear to be as damaging to plants as the soluble aluminium and manganese encountered on acid soils. Acid soils typically induce the following symptoms in plants: short and stubby roots, small dark leaves or leaves with chlorotic patches and marginal necrosis (Foy, 1984). The application of lime (calcium carbonate) and fertilizer is the most efficient way of ameliorating soil acidity. Liming works in two ways: (i) it raises the pH of the soil although it can take years and even decades to reduce sub-soil acidity; and (ii) it returns basic cations (calcium) to the soil which ameliorate aluminium stress in other ways (see later). However, farmers in developing countries with small subsistence farms rarely have the resources to apply lime. Instead they apply locally available composts and rely on crop species better adapted to acid soils (see later).

7.2.1 Al³⁺ toxicity and soil chemistry

Aluminium is the third most abundant element in the Earth's crust. Most of it occurs in minerals and complexes that are harmless to plants. At low pH the concentration of several soluble aluminium species increases in the soil solution (Fig. 7.1). Of these, the trivalent cation, Al^{3+}, is generally regarded as the major factor limiting plant growth on acid soils. Indeed, most polyvalent cations (and anions) are toxic to plants, because their high charge density promotes damaging interactions with many molecules. Several reports suggest that the aluminate anion, $Al(OH)_4^-$, may be toxic to some plant species in alkaline conditions, but this idea has not been confirmed (Eleftheriou et al., 1993; Kopittke et al., 2005; Stass et al., 2006).

Although the importance of Al^{3+} is now well established, measurements of soluble aluminium concentration, or even Al^{3+} itself, are often poor indicators of soil toxicity. The reason for this is that Al^{3+} interacts with other components in the soil that reduce its potential to injure plants. As discussed below, the presence of other cations and anions in a solution can reduce Al^{3+} toxicity by increasing ionic strength, by competing for binding sites, by changing the surface potential on charged surfaces in the apoplast or by forming harmless complexes.

Ionic strength

Dissolved ions increase the ionic strength of solutions (Eqn 7.3). This decreases the 'effective' concentration or activity of all ions present by reducing their activity coefficient γ. Activity coefficients have values between 0 and 1 and can be approximated from Eqn 7.4. The relationship between concentration and activity is provided by Eqn 7.5.

Ionic strength, I, of a solution is given by Eqn 7.3:

$$I = \tfrac{1}{2}\Sigma_j[C]_j z_j^2 \qquad (7.3)$$

where $[C]_j$ is the concentration (mol m^{-3}) of ion j with valence z for each ion in the solution.

Activity coefficient, γ, can be approximated (Eqn 7.4) from the Debye–Hükle theory from:

$$\log_{10}\gamma_\pm = \frac{0.51 z_+ z_- \sqrt{I}}{32 + \sqrt{I}} \qquad (7.4)$$

where I is ionic strength (mM), z_+ and z_- are charge numbers of the cation and anion ($z_+ z_-$ effectively becomes z^2 for the ion of interest) (Nobel, 1983).

The activity of ion X in a solution is given as (Eqn 7.5):

$$\{X\} = \gamma[X] \qquad (7.5)$$

where $\{X\}$ and $[X]$ refer to the activity and concentration of ion X, respectively, and γ is the activity coefficient.

The activities of polyvalent cations like Al^{3+} are particularly sensitive to ionic strength. For example, in a solution of 25 mol m^{-3} ionic strength the activity coefficient of a trivalent ion is 0.24. This means that for a solution containing 10 µM Al^{3+} concentration the activity of Al^{3+} is actually 2.4 µM. Therefore, for two solutions containing the same Al^{3+} concentration, $[Al^{3+}]$, the one with greater ionic strength will have a lower activity of Al^{3+}. This is important because

toxicity is more closely correlated with Al^{3+} activity in solution than Al^{3+} concentration. Estimating the activity of Al^{3+} at the surface of the root-cell membranes as described below is an even more reliable predictor of toxicity.

Cations reduce the amount of Al³⁺ accumulated near the root surface

Computer programs such as GEOCHEM (Shaff et al., 2010) can calculate the concentrations and activities of all ions and compounds in solution after accounting for ionic strength and chemical speciation. Even these accurate estimates of Al^{3+} activity in solution do not reliably predict toxicity because the activity in the bulk solution can differ substantially from the activity of Al^{3+} 'seen' by plants at the root surface. Fixed charges in the apoplast of root cells generate electrical potential differences between the surfaces carrying the charges (cell membranes and cell wall) and the bulk solution. These fixed charges are mostly negative because they derive from carboxylate groups on pectin in the cell wall, acidic amino acid residues on membrane-bound proteins and phosphate groups on membrane lipids. Note that the surface potentials are different from the electrical potential difference across the plasma membrane (between the cell cytosol and bulk solution) which is generated by Donnan equilibria and activity of the H^+-ATPase. Cations are attracted towards the negatively charged surfaces and accumulate there while soluble anions are repelled by them. The magnitude of this attraction or repulsion is exponentially related to the valency of the ions and approximated by the Nernst equation (Eqn 7.6):

$$\text{Nernst equation} \quad \frac{\{X\}_o}{\{X\}_\infty} = \exp(-zFE_o/RT)$$

$$(7.6)$$

where $\{X\}_o$ and $\{X\}_\infty$ denote the activities (mM) of ion X with valency z near the membrane surface and in the bulk solution, respectively; E_o is the electrical potential at the membrane surface (mV); R is the universal gas constant (8.3 J mol^{-1} K^{-1}); F is the Faraday constant (96.5 J mol^{-1} mV^{-1}); and T is temperature (K).

A surface potential of −19 mV on the root-cell membranes results in a tenfold greater activity of Al^{3+} near the root surface than in the bulk solution. The addition of other cations to the solution reduces the surface potentials by screening or binding with the fixed charges, and this reduces the accumulation of Al^{3+} at the root surface. These effects are more fully described by the Guoy–Chapman–Stern theory which models the behaviour of ions near charged surfaces. Therefore, solutions containing identical activities of Al^{3+} will vary in toxicity if other components of the solution affect the surface potential of the root cells. By combining a chemical speciation programme for the bulk solution (e.g. GEO-CHEM) with the Guoy–Chapman–Stern theory, the activity of Al^{3+} can be estimated at the surface of the root cells and used to successfully predict the relative toxicity of Al^{3+} in solutions of varying composition (Kinraide et al., 1992; Kinraide, 2004).

Anions chelate Al^{3+}

Some anions chelate Al^{3+} and form harmless complexes. For example, sulfate, phosphate, fluoride and various organic anions (malate, citrate and some phenolic compounds) form stable complexes with Al^{3+}. The presence of these anions in soil solution reduces the activity of soluble Al^{3+} and reduces their toxicity to plants (Kinraide, 1997).

7.2.2 Mechanisms of Al^{3+} toxicity

In simple, low pH salt solutions, micromolar concentrations of Al^{3+} can inhibit root growth within minutes or hours (Ryan et al., 1992). Longer exposures decrease root cap volume, increase root diameter and induce lesions in the epidermal and cortical tissues near the elongation zone. Short roots restrict water and nutrient uptake and reduce yields. Root apices are the most sensitive part of the root and only the terminal 2–3 mm of maize roots need be exposed to Al^{3+} for root growth to be reduced (Ryan et al., 1993). Within the apex, the 'distal transition zone' between the meristem and elongation zones appears to be the most sensitive part (Sivaguru and Horst, 1998) and callose synthesis in this region is another very early response to Al^{3+} stress (see later; Horst et al., 2010). Callose is a polysaccharide called 1,3 beta D-glucan whose synthesis is triggered by many stresses.

Al^{3+} disrupts so many different cellular functions that it is probably easier to list those that are not affected by Al^{3+} than those that are. Many studies of Al^{3+} toxicity have targeted the earliest responses with the hope of identifying a single primary lesion or interaction from which all other damage is derived. Despite considerable research it is still uncertain whether Al^{3+} initiates stress in the apoplast or whether Al^{3+} first needs to enter the cytosol to cause injury (Horst et al., 2010). The mechanism of toxicity could also differ between species and even between monocotyledons and dicotyledons due to differences in cell wall composition.

The apoplast is a prime site for Al^{3+} stress because fixed negative charges on the pectin and hemicellulose polysaccharides (e.g. xyloglucans and arabinoxylans) in the cell wall and on the membrane surfaces accumulate high concentrations of Al^{3+} and other cations. Indeed, 99.99% of the Al^{3+} absorbed by the giant-celled alga Chara corallina is retained in the cell wall (Rengel and Reid, 1997; Taylor et al., 2000) and this reflects the distribution of Al^{3+} in the roots of major crop species. High concentrations of Al^{3+} on the hemicelluloses can stiffen cell walls and inhibit cell elongation (Ma et al., 2004b). It also alters the fluidity of the root-cell membranes by binding with membrane lipids and proteins or by displacing other cations from important binding sites (Foy et al., 1978). Extracellular Al^{3+} can trigger callose production and restrict the flow of solutes and macromolecules through the apoplast (Sivaguru et al., 2006; Horst et al., 2010). It also blocks some membrane-bound transport proteins involved in nutrient uptake (e.g. Ca^{2+} and K^+; Piñeros and Tester, 1993; Gassmann and Schroeder, 1994) but nutrient deficiency per se is unlikely to explain the rapid inhibition of root growth. Al^{3+} accumulation in the wall is reduced when the fixed charges are replaced by methyl groups on pectin or o-acetyl groups on the hemicellulose polymers (J.L. Yang et al., 2011; see later).

Models for Al^{3+} toxicity often invoke some interaction with calcium (Rengel, 1992). Calcium concentrations in the apoplast and uptake by roots are rapidly reduced by Al^{3+} and these effects are greater in Al^{3+}-sensitive wheat than in resistant genotypes (Huang et al., 1992). However, low concentrations of Al^{3+} are still able to inhibit root growth without affecting apoplastic calcium or calcium uptake (Ryan *et al.*, 1994, 1997), which highlights the importance of experimental conditions. Calcium concentrations in the cytosol, $[Ca]_c$ are usually maintained below $1.0\,\mu M$ but transient increases act as signals to regulate metabolic responses and induce stress responses such as callose synthesis, which is a very sensitive indicator of Al^{3+} stress (Sivaguru *et al.*, 2000; Zhang *et al.*, 2015). Short-term effects of Al^{3+} on $[Ca]_c$ have been investigated with calcium-sensitive fluorescent compounds which are either loaded into the root cells (Zhang *et al.*, 1998) or engineered into transgenic plants (Rincon-Zachary *et al.*, 2010). With these tools, transient increases in $[Ca]_c$ have been detected in root cells after short exposures to Al^{3+} (Rincon-Zachary *et al.*, 2010). The rapidity of this response supports the idea that interactions in the apoplast are important for Al^{3+} toxicity and that $[Ca]_c$ signals could be involved in the early perception and response to Al^{3+} stress.

Although most of the aluminium absorbed by plants remains in the cell wall, a fraction crosses the plasma membrane and enters the cytosol. This was confirmed in the alga *C. corallina* where the cell wall could be physically separated from the cytosol (Rengel and Reid, 1997; Taylor *et al.*, 2000), and in the roots of land plants using other detection techniques (Lazof *et al.*, 1994). Al^{3+} may enter the cytosol by slipping through non-specific cation transporters or perhaps by endocytosis. A recent report proposes that rice has a specific transporter for Al^{3+} uptake (see later; Xia *et al.*, 2011). Once in the cytosol the combination of pH, ionic strength and ligands reduces soluble Al^{3+} to extremely low concentrations ($\sim 10^{-9}$ to 10^{-12} M). Nevertheless, even these concentrations may still damage cells because Al^{3+} has a much higher affinity for certain ligands that are normally occupied by Mg^{2+} or Fe^{2+} (Martin, 1992). Examples are the nucleoside triphosphates (ATP and GTP), which have a 10^7-fold greater affinity for Al^{3+} than for Mg^{2+} (Wang *et al.*, 1997). Al can also enter the nucleus and attach to the DNA, presumably by binding with the phosphate groups (Matsumoto, 1991; Silva *et al.*, 2000). These interactions may explain the disruption of some important cellular functions including cytoskeleton stability and cell division (Grabski and Schindler, 1995; Horst *et al.*, 1999; Sivaguru *et al.*, 2003; Rounds and Larsen, 2008).

Al^{3+} stress also triggers oxidative stress in root cells by generating harmful reactive oxygen species (ROS) such as H_2O_2 and hydroxyl radicals (Yamamoto *et al.*, 2001). These highly reactive compounds are by-products of normal metabolism but they can cause widespread damage to membranes, proteins and nucleic acids if not rapidly scavenged. ROS triggers the induction of several proteins which combat oxidative stress such as glutathione S-transferase and peroxidase. Oxidative stress also induces callose production, which in turn increases cell wall rigidity and decreases the symplastic flow of solutes and hormones between adjacent cells via the plasmadesmata (Sivaguru *et al.*, 2000; Horst *et al.*, 2010).

The close association between Al^{3+} accumulation, changes to $[Ca^{2+}]_c$ and callose production forms a central role in several hypotheses explaining the early stages of Al^{3+} injury. One of these hypotheses proposes that Al^{3+} first binds with pectin and rigidifies the cell wall and induces an oxidative burst in the apoplast. The resulting ROS triggers an increase in $[Ca^{2+}]_c$ and callose synthesis, which lead to other lesions (Jones *et al.*, 1996, 2006; Horst *et al.*, 1999, 2010). An alternative hypothesis suggests that Al^{3+} activates an anion channel in the plasma membrane that releases a glutamate-like ligand into the apoplast. This ligand binds with a receptor or calcium channel that triggers calcium influx and higher $[Ca^{2+}]_c$ (Sivaguru *et al.*, 2003). Other studies show that Al^{3+} blocks cellular signal transduction via the inositol 1,4,5-trisphosphate (Ins[1,4,5]P-3)-mediated pathway by inhibiting the cytosolic enzyme phospholipase C (Jones and Kochian, 1995). All these interactions could represent early triggers of Al^{3+} stress.

Another model for Al^{3+} toxicity emerged from the examination of *Arabidopsis* mutants. Larsen *et al.* (2005) first identified a knock-out mutation in *Arabidopsis* denoted *als3-1* that showed increased sensitivity to Al^{3+} stress. ALS3-1 is an ATP-binding cassette (ABC) transporter of unknown function that localizes to the plasma membrane in many tissues, especially around the phloem. By mutating this line a second time a suppression mutant was isolated that reversed the hypersensitivity of the *als3-1* line. This second mutation occurred in the *AtATR* gene (*ataxia telangiectasia-mutated and Rad3-related*), which is required for detecting and managing damage to DNA in higher eukaryotes (Rounds and Larsen, 2008). AtATR searches for single-stranded breaks and blocks in the replication fork. Knock-out mutations in a wild type (WT) background also increased Al^{3+} resistance compared to non-mutant controls. Therefore the hypothesis was that Al-dependent inhibition of root growth primarily arises from AtATR first detecting DNA damage and then blocking cell-cycle progression in the quiescent centre (Rounds and Larsen, 2008). Plants without normal AtATR function continue to grow during Al stress because lesions to DNA go undetected and cell division is maintained in the quiescent centre. In other words the inhibition of root growth is not a direct and inevitable consequence of damage to the root tissue or metabolism but rather an active response of the root cells to DNA damage – or possibly even to the threat of DNA damage.

In summary, Al^{3+} can damage root cells in many different ways. Some of these Al^{3+}-induced lesions occur rapidly and therefore are candidates for being the primary response to Al^{3+} stress. Evidence from *Arabidopsis* mutants suggest that the inhibition of root growth by Al^{3+} is actually controlled by the plant as a response to DNA damage and not the direct effects of Al^{3+} (see later). However, there will always be uncertainties for 'cause and effect' because so many stress responses are rapid and interconnected. The concept of a 'single' primary mechanism of toxicity may not even be a useful one because it blinkers us into considering isolated responses rather than viewing toxicity as a network of interactions, signals and responses.

7.3 Genetics of Al^{3+} Resistance

Many plants have evolved mechanisms that allow them to grow on acid soils. Species endemic to the humid tropics need some level of Al^{3+} resistance to survive the acid soils prevalent in those regions. Large variations in Al^{3+} resistance also occur among genotypes of the same species (Foy, 1988; Garvin and Carver, 2003). Genetic studies have demonstrated that Al^{3+} resistance is a multi-genic trait in many species, which means it is controlled by more than one gene (Garvin and Carver, 2003). Those studies analysed segregating populations generated with

contrasting parents or screened lines carrying chromosomal deletions or substitutions (Aniol and Gustafson, 1984; Reide and Anderson, 1996; Papernik *et al.*, 2001; Hoekenga *et al.*, 2003). The involvement of more than one gene is consistent with the multiple quantitative trait loci (QTL) for Al^{3+} resistance identified in wheat, maize, alfalfa, rice, rye and *Arabidopsis* (Kobayashi and Koyama, 2002; Xue *et al.*, 2006; Wang *et al.*, 2007b; Maron *et al.*, 2008, 2010). More recently, genome wide association mapping (GWAM) with 1055 cultivars and landraces of wheat identified markers on 15 different chromosomes that were significantly linked with Al^{3+} resistance (Raman *et al.*, 2010). Many of these loci had very small effects on the phenotype but those on chromosomes 3B, 4D and 5B were larger and highly significant. Another GWAM study in wheat identified significant markers on chromosomes 1A, 1D, 3B and 6A (Navakode *et al.*, 2014). Screens of randomly mutagenized *Arabidopsis* and rice seed have helped identify additional genes that do not explain natural genotypic variation but which are necessary for wild-type levels of Al^{3+} resistance (see later).

Despite its multigenic basis, most of the variation in Al^{3+} resistance within some species can be attributed to a single genetic locus. This is the case for major crop species like wheat, barley and sorghum. For example, many QTLs for resistance in wheat have been identified but a single locus on 4DL explains ~70% of the genetic variation among a diverse range of genotypes from all over the world (Luo and Dvořák, 1996; Reide and Anderson, 1996; Garvin and Carver, 2003; Ma *et al.*, 2004a; Raman *et al.*, 2005; Wang *et al.*, 2007b). Plant breeders can exploit this genotypic variation and with genetic markers rapidly improve the yields of important crops on acid soils (Garvin and Carver, 2003). For crops with little or no natural variation, biotechnological strategies are available for increasing their resistance, and these are discussed later.

7.4 Mechanisms of Al^{3+} Resistance

Plants have evolved two broad mechanisms for coping with Al^{3+} stress. These mechanisms either increase their ability to *exclude* Al^{3+} from their tissues or increase their *tolerance* to Al^{3+} absorbed by the roots (Taylor, 1991; Kochian *et al.*,

2004, 2015; Hiradate *et al.*, 2007). Some plants (e.g. wheat and barley) possess a single major mechanism, whereas others (e.g. buckwheat, *Fagopyrum esculentum*) appear to possess exclusion and tolerance mechanisms and these may be additive. As discussed below, numerous mechanisms of Al^{3+} tolerance and exclusion have been proposed, but the supporting evidence is more secure for some of these mechanisms than for others.

7.4.1   Mechanisms of Al^{3+} tolerance

Tolerance mechanisms enable plants to tolerate Al^{3+} once it enters the cytosol either by chelating

it to form harmless complexes, by storing it in less sensitive organelles or by repairing damage (Fig. 7.2). Tolerance mechanisms appear to be common in species endemic to regions with acid soils (e.g. the tropics) where the ability to cope with Al^{3+} stress is a prerequisite for survival. Some highly tolerant species accumulate aluminium in their leaves without adverse symptoms and their growth can even be stimulated by Al^{3+}. Notable examples include tea (*Camelia sinensis*) and buckwheat (*F. esculentum*), which accumulate aluminium to >1% dry weight in older leaves (Matsumoto *et al.*, 1976; Ma and Hiradate, 2000). Other species that accumulate Al include *Melastoma malabathricum* and *Hydrangea macrophylla*. Most of the aluminium in tea leaves

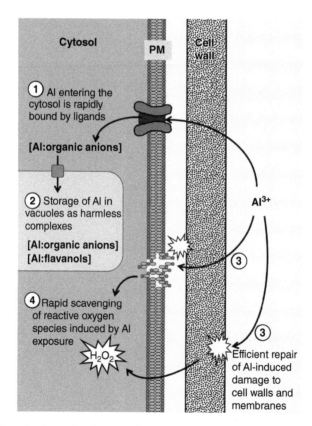

Fig. 7.2. Proposed mechanisms of resistance: Al^{3+} tolerance. The cartoon shows a plasma membrane (PM) and cell wall of cells near the root apex. Soluble Al^{3+} can damage cells by accumulating in the apoplast or by entering the cytosol. Tolerance mechanisms enable plants to reduce the potential for damaging reactions from occurring once Al^{3+} enters the cell wall or cytosol. Proposed mechanisms of Al^{3+} tolerance include: (1) binding Al^{3+} in the cytosol by organic and inorganic ligands to reduce damaging reactions from occurring; (2) storage of Al^{3+}:ligand complexes in the vacuole or other subcellular compartments; (3) rapid repair of Al^{3+}-induced damage to the cell wall and membranes; (4) efficient scavenging of reactive oxygen species induced by Al^{3+}.

resides in the apoplast (Tolra *et al.*, 2011), whereas in the leaves of *Hydrangea* and buckwheat the aluminium is bound in vacuoles by citrate and oxalate anions, respectively. Other ligands bind Al^{3+} as it moves from the roots to the leaves in the xylem stream (Ma *et al.*, 2001). The relative stability of these compounds prevents Al^{3+} from participating in other more damaging reactions.

Tolerance mechanisms can also minimize oxidative stress and repair Al^{3+}-induced damage. Al^{3+} rapidly induces the expression of genes involved in cell wall structure and function as well as glutathione S-transferase, manganese superoxide dismutase and peroxidase, all of which detoxify ROS (Ezaki *et al.*, 1995; Maron *et al.*, 2008). The induction of these genes can be viewed as Al^{3+} tolerance mechanisms even though they may not be specific to Al^{3+} stress. Indeed, transgenic plants with enhanced expression of genes encoding these proteins are slightly more resistant to Al^{3+} toxicity. For instance, the overexpression of genes encoding glutathione S-transferase, peroxidase, GDP-dissociation inhibitor, dehydroascorbate reductase, manganese superoxide dismutase or a blue copper protein in plants (*Arabidopsis*, *Brassica napus* and *Nicotiana tabacum*) increased relative root growth in media containing toxic concentrations of Al^{3+} solution by 1.5- to 2.5-fold compared to controls (Ezaki *et al.*, 2000, 2005; Basu *et al.*, 2001; Yin *et al.*, 2010). Other compounds with antioxidant activity have been implicated in the Al^{3+} tolerance of maize. Some tolerant varieties contain higher concentrations of various flavanols (caffeic acid, catechol and catechin) and accumulate more taxifolin when treated with Al^{3+} than sensitive varieties. The antioxidant activity of these compounds and their ability to bind with Al^{3+} could be protecting cells from Al^{3+} stress (Tolra *et al.*, 2009). Flavanols have also been implicated in the ability of a forage legume species *Lotus pedunculatus* to tolerate Al^{3+} uptake and accumulation because Al^{3+} is closely associated with condensed tannins in the root-cell vacuoles (Stoutjesdijk *et al.*, 2001).

7.4.2 Mechanisms of Al^{3+} exclusion

Exclusion mechanisms prevent Al^{3+} from accumulating in the symplast and minimize harmful

interactions occurring with the plasma membrane, cell wall or other targets in the apoplast. *Arabidopsis*, as well as important cereal species (wheat, rice and maize), excludes Al^{3+} from its tissues using a range of mechanisms (Fig. 7.3).

Early suggestions that the accumulation of Al^{3+} in the apoplast was correlated with pectin content of the cell walls have not been substantiated. Similarly, the differences in root-cell surface potential cannot fully explain the genotypic variation in species examined to date, although it could be a minor contributor to resistance. Al^{3+} treatment does change the hemicellulose content of cell walls and decrease the degree of *o*-acetylation. Therefore suggestions that accumulation of Al^{3+} in the cell wall, as well as sensitivity to Al^{3+} toxicity, depends on the degree to which pectin is methylated or hemicellulose is *o*-acetylated are more compelling (Zhu *et al.*, 2014). For instance, the greater accumulation of Al^{3+} by Al^{3+}-sensitive genotypes of maize, rice and common bean (*Phaseolus vulgaris*) was correlated with less methylation of the wall pectins compared to resistant genotypes (Eticha *et al.*, 2005; Stass *et al.*, 2007; Yang *et al.*, 2008). Similarly, reducing the methylation of pectin in transgenic potato by overexpressing a pectin methylesterase, resulted in greater Al^{3+} accumulation, more callose production and greater sensitivity towards Al^{3+} (Schmohl *et al.*, 2000). Similarly, knock-out mutants of the *o*-acetyltransferase gene in *Arabidopsis* reduced *o*-acetylation of xyloglucans, which led to greater Al^{3+} accumulation in the wall and increased sensitivity to Al^{3+} stress (Zhu *et al.*, 2014).

Low levels of Al^{3+} in the cell could be maintained by exporting Al^{3+} from the cytosol to the apoplast by specific transport mechanisms (see Section 7.6.1) or by vesicular trafficking and exocytosis. Indirect support for this hypothesis comes from yeast where it was demonstrated that a guanosine diphosphate (GDP) dissociation inhibitor from *Nicotiana tabacum* could complement the Al^{3+}-sensitive yeast mutant *sec19*, which is defective in vesicle transport function (Ezaki *et al.*, 2005).

A different exclusion mechanism was proposed for the highly resistant camphor tree (*Cinnamomum camphora*). Histochemical analyses identified a novel outer cell layer adjacent to the root apex which continually grows and detaches itself (Osawa *et al.*, 2011). These cells contain high concentrations of proanthocyanidin,

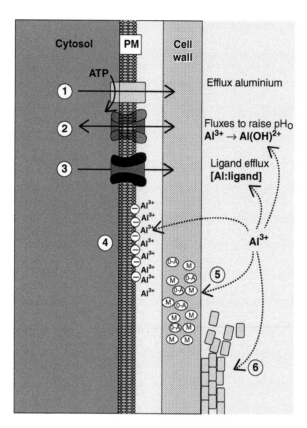

Fig. 7.3. Proposed mechanisms of resistance: Al^{3+} exclusion. The cartoon shows a plasma membrane (PM) and cell wall of cells near the root apex. Soluble Al^{3+} can damage cells by accumulating in the apoplast or by entering the cytosol. Exclusion mechanisms prevent Al^{3+} from accumulating in the symplast and avoid harmful interactions occurring with the plasma membrane, cell wall or other targets in the apoplast. Proposed mechanisms of Al^{3+} exclusion include: (1) maintenance of low Al^{3+} concentration in the cytosol by the active export of Al^{3+} from the cytosol or by vesicular transport (this might occur in the root cells or even the leaf cells); (2) transport processes across the PM that raise apoplastic pH and reduce the mole fraction of soluble Al^{3+} species; (3) efflux of ligands (especially organic anions like malate and citrate) that bind Al^{3+} in the apoplast and prevent damaging interactions occurring; (4) lower charge density on the membrane surfaces which reduce surface potentials and decrease the electrostatic accumulation of Al^{3+} at the membrane surface; (5) increased methylation (M) of pectin residues or increased o-acetylation (O-M) of the hemicellulose reduce Al^{3+} binding and accumulation in the cell wall; (6) continual growth and detachment of border cells from the root cap or other special cell layers near the root apices prevents Al^{3+} accumulation in the sensitive cells beneath.

a type of flavonoid polymer with high antioxidative properties. In other species proanthocyanidins accumulate in bark, fruit skin and seed testa where that may protect against biotic stress. The regular detachment of these proanthocyanidin-containing cells from camphor roots could shield the growing root apex from Al^{3+} and reduce its accumulation in these sensitive tissues. The excretion of other phenolics, including catechin and cyclic hydroxamates, may also contribute to Al^{3+} resistance by chelating Al^{3+} in the apoplast (Kidd et al., 2001; Poschenrieder et al., 2008). The mechanism proposed for camphor tree is similar to an earlier suggestion that border cells protect roots from Al^{3+} stress. Border cells are metabolically active cells that are released from the root cap with mucigel. Evidence in common bean (*Phaseolus vulgaris*) indicates that the border cells released from a resistant genotype are more resistant to Al^{3+} stress than are cells released from a sensitive genotype (Miyasaka and Hawes, 2001). These differences were not found in wheat (Zhu *et al.*, 2003) but the constant growth and shedding of these cells may reduce Al^{3+} accumulation in the root apices.

The Al^{3+} exclusion mechanism most thoroughly investigated in plants is the efflux of organic anions from roots. This is also the mechanism for which there is the most compelling genetic evidence linking the trait to resistance (Ma *et al.*, 2001; Ryan *et al.*, 2001; Kochian *et al.*, 2004, 2015; Delhaize *et al.*, 2007). Organic anions released from roots bind with Al^{3+} in the apoplast and reduce its accumulation in the cell wall and uptake to the cytosol. Organic anion efflux is restricted to the apical few millimetres of the root and the anions released vary between species. Malate and citrate are most common but oxalate efflux occurs from a few species (Table 7.1). This exclusion mechanism has been reported in species from the Poaceae (e.g. wheat, barley, sorghum, maize, rye), Araceae (e.g. taro), Polygonaceae (e.g. buckwheat), Brassicaceae (e.g. *Arabidopsis*) and Fabaceae (e.g. soybean, snapbean, common bean, *Cassia tora*). The evidence supporting a role for organic anion efflux in Al^{3+} exclusion is stronger for some species than others and not all studies shown in Table 7.1 have established genetic links between the anion efflux and resistance. Reports that only analysed one or two genotypes should be supported, where possible, by additional evidence such as co-segregation between efflux and Al^{3+} resistance.

The Al^{3+}-dependent efflux of malate and citrate are controlled by members of two gene families, both of which encode transport proteins. As explained below, malate efflux is encoded by members of the *ALMT* gene family and citrate efflux is encoded by members of the <u>m</u>ul<u>t</u>idrug <u>a</u>nd <u>t</u>oxic compound <u>e</u>xudation (*MATE*) gene family.

7.5 Al^{3+} Resistance Genes

7.5.1 ALMT-type genes

The *TaALMT1* gene (<u>a</u>luminium <u>a</u>ctivated <u>m</u>alate <u>t</u>ransporter) from wheat was the first Al^{3+} resistance gene to be isolated from any plant species. Wheat is not generally regarded as an ideal model species for molecular studies in view of its large and complex genome, but the thorough understanding of the genetics and physiology underlying Al^{3+} resistance in wheat proved critical in the cloning of *TaALMT1*. Near-isogenic lines (NILs) that are >99% identical to one another but differ in Al^{3+} resistance were instrumental for understanding the mechanism of resistance and for identifying the gene (Delhaize *et al.*, 1993a; Sasaki *et al.*, 2004). *TaALMT1* controls an exclusion mechanism which relies on malate efflux and it underlies the major QTL for resistance on chromosome 4DL.

TaALMT1 was identified by using subtractive hybridization of cDNAs isolated from the NILs. *TaALMT1* was found to be more highly expressed in the root apices of the resistant NIL than the sensitive NIL. Additional findings that *TaALMT1* mapped to chromosome 4DL and encoded a malate-permeable anion channel provide strong evidence that this is indeed the Al^{3+} resistance gene of wheat. Most hydropathy plots predict that TaALMT1 has six membrane-spanning domains with a long hydrophilic C-terminal region or 'tail' (Motoda *et al.*, 2007).

The function of the TaALMT1 protein was investigated in heterologous expression systems such as tobacco-suspension cells and *Xenopus laevis* oocytes (Sasaki *et al.*, 2004; Piñeros *et al.*, 2008; Zhang *et al.*, 2008). *Xenopus* oocytes are particularly convenient for analysing transport proteins because they are large cells easily impaled with electrodes. Typically, complementary RNA (cRNA) encoding the protein of interest is injected and the oocyte incubated for several days. During this period the cRNA is translated and the protein targeted to the plasma membrane. The oocyte can be further injected with possible substrates so that, to some extent, both sides of the membrane (bathing medium and cytosol) can be controlled. Electrodes inserted into the oocyte allow the experimenter to fix or 'clamp' the voltage across the membrane and at the same time to measure the current flowing across the membrane. Oocytes injected with the cRNA of interest are compared with control oocytes injected with water to ensure that any observed currents are due to the injected cRNA and not due to endogenous proteins. In the case of *TaALMT1* cRNA, negative currents (equivalent

Table 7.1. Plant species for which the efflux of organic anions from roots reportedly contributes to Al^{3+} resistance. The organic anions released and the genes that confer this trait (where these are known) are shown. These genes remain candidates only, in many cases.

Species	Organic anion released	Gene	Reference
Monocotyledons			
Barley (*Hordeum vulgare*)	Citrate	*HvAACT1*	(Zhao et al., 2003; Furukawa et al., 2007)
Yorkshire fog (*Holcus lanatus*)	Malate	*HlALMT1*	(Chen et al., 2013)
Maize (*Zea mays*)	Citrate	*ZmMATE1*	(Pellet et al., 1995; Maron et al., 2010)
Rice (*Oryza sativa*)	Citrate	*ScFRDL4*	(Yokosho et al., 2011)
Rye (*Secale cereale*)	Malate and citrate	*ScALMT* gene cluster	(Collins et al., 2008)
		ScMATE2	(Yokosho et al., 2010)
Sorghum (*Sorghum bicolor*)	Citrate	*SbMATE1*	(Magalhaes et al., 2007)
Triticale (× *Triticosecale*)	Malate and citrate		(Ma et al., 2000)
Wheat (*Triticum aestivum*)	Malate	*TaALMT1*	(Delhaize et al., 1993b; Sasaki et al., 2004)
	Citrate	*TaMATE1*	(Ryan et al., 2009)
Eudicotyledons			
Arabidopsis thaliana	Malate	*AtALMT1*	(Hoekenga et al., 2006)
	Citrate	*AtMATE1*	(Liu et al., 2009)
Aspen (*Populus tremula*)	Malate and formate		(Qin et al., 2007)
Buckwheat (*Fagopyrum esculentum*)	Oxalate		(Zheng et al., 1998)
Cassia tora	Citrate		(Ma et al., 1997)
Citrus (*Citrus sinensis*)	Citrate and malate		(L.T. Yang et al., 2011)
Common bean (*Phaseolus vulgaris*)	Citrate		(Shen et al., 2002)
Lespedeza bicolor	Malate and citrate		(Dong et al., 2008)
Oat (*Avena sativa*)	Malate and citrate		(Zheng et al., 1998)
Radish (*Raphanus sativus*)	Malate and citrate		(Zheng et al., 1998)
Rapeseed (*Brassica napus*)	Citrate	*BnALMT1* & *BnALMT2*	(Ligaba et al., 2006)
Rice bean (*Vigna umbellata*)	Citrate	*VuMATE*	(Yang et al., 2006)
Snapbean (*Phaseolus vulgaris*)	Citrate		(Miyasaka et al., 1991)
Soybean (*Glycine max*)	Citrate		(Yang et al., 2000; Silva et al., 2001)
Stylosanthes sp.	Citrate		(Li et al., 2009)
Taro (*Colocasia esculenta*)	Oxalate		(Ma and Miyasaka, 1998)
White lupin (*Lupinus albus*)	Citrate		(Wang et al., 2007a)

to anion efflux) are activated in oocytes injected with malate and incubated in solutions that contain Al^{3+}. If citrate is substituted for malate or if Al^{3+} is omitted from the bathing solution the current magnitudes reduced or disappeared. Those experiments showed that TaALMT1 is capable of transporting malate across membranes when activated by Al^{3+}, observations that fit nicely with the known physiology of the Al^{3+} resistance mechanism in wheat. Subsequent studies established that *ALMT*-like genes also

contribute to the Al^{3+} resistance of rye (Collins et al., 2008), *Arabidopsis* (Hoekenga et al., 2006; Liu et al., 2009) and *Holcus lanatus* (Chen et al., 2013).

7.5.2   MATE-type genes

The proteins facilitating citrate efflux belong to an entirely different family of proteins from those that facilitate malate efflux. Sorghum and barley show an Al^{3+}-activated efflux of citrate, and fine mapping in both species identified *MATE* genes encoding this trait. In sorghum the gene is named *SbMATE* and in barley the gene is *HvAACT1* (also known as *HvMATE1*) (Furukawa *et al.*, 2007; Magalhaes *et al.*, 2007; Wang *et al.*, 2007b). MATEs form a large family of transport proteins with a diverse range of substrates (Hvorup *et al.*, 2003). The family was originally named after certain members that confer antibiotic resistance to bacteria by exporting drugs from the cells. Heterologous expression of HvAACT1 and SbMATE in *Xenopus* oocytes shows that both proteins are capable of transporting citrate, although details of its mechanism (secondary active, etc.) are unknown. In both sorghum and barley the MATE genes are more highly expressed in Al^{3+}-resistant genotypes than in sensitive genotypes. *MATE* genes have now been implicated in the Al^{3+} resistance of *Arabidopsis* (Liu *et al.*, 2009), wheat (Ryan *et al.*, 2009), maize (Maron *et al.*, 2010) and rice (Yokosho *et al.*, 2011). In all of these cases higher expression of the *MATE* genes is associated with greater Al^{3+} resistance, indicating that transport is the rate-limiting step in these species. By contrast, in common bean (*Phaseolus vulgaris*) both Al^{3+}-resistant and Al^{3+}-sensitive genotypes show large increases in the expression of two *MATE* genes when exposed to Al^{3+} (Eticha *et al.*, 2010; Rangel *et al.*, 2010). The capacity to maintain citrate synthesis might be the important difference between Al^{3+}-sensitive and Al^{3+}-resistant genotypes of common bean, although direct evidence that the induced MATE proteins transport citrate is still lacking.

7.5.3 Gene induction and protein activation by Al^{3+}

Despite belonging to unrelated families, almost all the ALMT and MATE proteins underlying Al^{3+} resistance are activated by Al^{3+}. For example, *TaALMT1* is constitutively expressed in wheat roots so the protein is always present. However, malate efflux only begins when Al^{3+} is added to the external solution (Sasaki *et al.*, 2004). This is also true of HvACCT1, the Al^{3+} resistance MATE protein of barley (Furukawa *et al.*, 2007). By contrast, Al^{3+} induces the expression of *SbMATE* in sorghum and *AtALMT1* in

Arabidopsis, yet both proteins still appear to require Al^{3+} for their activation (Magalhaes *et al.*, 2007; Liu *et al.*, 2009). Consequently, the kinetics of organic anion efflux differ between species that express their genes constitutively and those that show gene induction. Malate efflux from wheat occurs rapidly after Al^{3+} addition, whereas citrate efflux from sorghum and rye is delayed for several hours during which transcription and translation occur (Fig. 7.4). Exactly how MATE and ALMT proteins are activated is not understood but it is proposed that Al^{3+} binds directly to specific domains on the proteins and kinase activity may also be necessary (Ligaba *et al.*, 2009). Further discussion on the structure and function of ALMTs is available elsewhere (Furuichi *et al.*, 2010; Ligaba *et al.*, 2013).

7.5.4 Evolution of organic anion efflux for Al^{3+} resistance

Species that rely on citrate or malate efflux to provide protection from Al^{3+} toxicity show higher expression of the *MATE* or *ALMT* genes in their root apices. The mechanisms driving this higher expression mostly depend on mutations to the promoter regions of these genes (Delhaize *et al.*, 2012). For instance, the promoters of *TaALMT1* in Al^{3+}-resistant wheat genotypes contain repeated blocks of sequence in different patterns, and the number of repeats is positively correlated with the level of *TaALMT1* expression (Sasaki *et al.*, 2006). These repeated sequences drive higher expression, probably because the promoter contains more copies of a transcription factor binding site. Interestingly, these same repeats could not be detected in a wide range of *Aegilops tauschii* accessions (donor of the D genome in hexaploid wheat), indicating that the mutations are likely to have occurred after the appearance of hexaploid wheat about 10,000 years ago (Ryan *et al.*, 2010). Mutations also occur in the *MATE* genes encoding citrate transporters. Examples here include *HvAACT1* in barley and *TaMATE1B* in wheat, where both promoters contain large transposable element-like insertions which drive higher expression of the genes in the root apices (Fujii *et al.*, 2012; Tovkach *et al.*, 2013). In a type of convergent evolution, random mutations have occurred in two

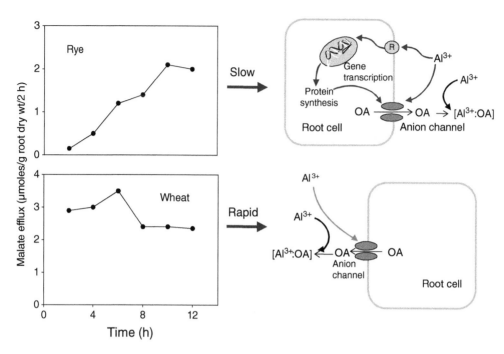

Fig. 7.4. Al^{3+}-activated efflux of organic anions from root cells. The top cartoon and graph illustrates the delayed response of organic anion efflux in species such as rye where Al^{3+} first induces the expression of the transport protein via a signal transduction pathway possibly involving a specific receptor ('R'). Once synthesized and inserted in the plasma membrane, Al^{3+} is thought to interact with the protein to activate efflux of organic anion (OA). Therefore efflux of organic anion does not start immediately Al^{3+} is added but increases through time. The lower cartoon and graph illustrate the rapid activation of organic anion efflux in species such as wheat where the anion channel is constitutively expressed. Al^{3+} is able to rapidly activate efflux by interacting directly with the pre-existing proteins. In this case the efflux of organic anion is relatively constant through time.

very different organic anion transporters to alter their function. These mutations affected the level of gene expression in the root apices so that organic anions released from those tissues protected the plants from Al^{3+} toxicity (Ryan and Delhaize, 2010). This has occurred independently in many species, suggesting the influence of a strong selection pressure. The original functions of the ALMT and MATE proteins might have included malate and citrate transport across other membranes and in other tissues. One intriguing function for ALMTs was recently proposed after it was discovered that many are inhibited by γ-aminobutyric acid (GABA) and muscimol, an angonist of GABA (Ramesh *et al.*, 2015). GABA is best known as a neurotransmitter in the animal nervous system where it binds with an anion channel called the GABA$_A$ receptor on the post-synaptic nerve. Once bound by

GABA the GABA$_A$ receptors become more permeable to Cl$^-$ ions, which hyperpolarizes the membrane and inhibits transduction of the nerve signal. Overall sequence similarity between ALMTs and GABA$_A$ receptors is very low, yet a conserved motif present in ALMTs is similar to the GABA binding site on the GABA$_A$ receptors. The intriguing suggestion that ALMT proteins function as GABA receptors in plants, and perhaps mediate signalling pathways during stress responses, highlights how little is known about this family of anion channels.

7.6   Model Species: Rice and *Arabidopsis*

The Al^{3+} resistance genes discussed above were identified because they explained a large proportion of the variation in natural populations. A similar approach is more difficult for rice because many QTL contribute to resistance. Instead, the underlying mechanisms of Al^{3+} resistance in that species are being unravelled through the identification of Al^{3+}-sensitive mutants. Mutational analysis can identify genes essential for Al^{3+} resistance but which need not be responsible for genotypic variation. Similarly, *Arabidopsis* is a model plant species that has contributed enormously towards elucidating fundamental processes in plant biology. Despite being relatively sensitive to Al^{3+}, screening of mutant populations has identified *Arabidopsis* plants that are hypersensitive to Al^{3+} stress. In addition, knock-out mutants are available for many genes in both rice and *Arabidopsis*, and specific genes can be assessed for their contribution towards Al^{3+} resistance. For example, the contribution of *ALMT* and *MATE* genes to the Al^{3+} resistance of *Arabidopsis* was established with specific knock-out of the genes homologous to those found in cereals (Hoekenga *et al.*, 2006; Liu *et al.*, 2009). The procedures for identifying mutants, isolating the underlying genes and confirming their function are similar in rice and *Arabidopsis*.

7.6.1 Model species: rice

Rice is not only an important crop in its own right but, as a monocotyledon, is a more reliable model than *Arabidopsis* for the major cereal crops such as wheat and sorghum. Rice is also among the most resistant of the small-grained cereals, a further advantage over *Arabidopsis*, which is relatively sensitive to Al^{3+} toxicity (Famoso *et al.*, 2010).

One of the first genes isolated in rice through mutant analysis was *STAR1* (sensitive to aluminium rhizotoxicity), which encodes a nucleotide binding domain of an ABC transporter (Huang *et al.*, 2009). The ABC transporters comprise a large family in plants with approximately 130 genes encoding these proteins in the rice genome alone (Schulz and Kolukisaglu, 2006). These proteins typically possess a hydrophilic nucleotide binding domain that can hydrolyse ATP as an energy source and a hydrophobic region that anchors the protein and provides the transport path across membranes. While details of the mechanism remain unclear, STAR1 appears to interact with another half-size ABC transporter in root cells and

this complex localizes to the membranes of vesicles. These vesicles reportedly accumulate UDP-glucose and release it to the apoplast, perhaps by exocytosis (Fig. 7.5). It is not certain how the proposed secretion of UDP-glucose from roots (possibly via exocytosis of vesicles) confers Al^{3+} resistance. UDP-glucose may bind with Al^{3+} or it may be the substrate for enzyme reactions that modify cell walls in order to minimize injury from Al^{3+}. It is also possible that other substrates transported by the STAR1/STAR2 complex provide protection from Al^{3+} stress.

Another rice gene isolated from mutant analysis is *ART1* (aluminium resistance transcription factor), which encodes a zinc-finger transcription factor (Yamaji *et al.*, 2009). The ART1 protein is located in the nucleus and interacts with specific regions in promoters of most genes upregulated by Al^{3+} (Tsutsui *et al.*, 2011). The *art1* mutant is very sensitive to Al^{3+} because it is unable to induce the expression of about 30 genes usually upregulated by Al^{3+} including the *STAR* genes, a *Nramp* transporter called *NRAT1*, the *MATE* transporter *FRDL4* and a magnesium transporter, *OsMGT1* (Yamaji *et al.*, 2009). NRAT1 (Nramp aluminium transporter) belongs to the NRAMP family of transporters (Xia *et al.*, 2010) and putatively transports Al^{3+} into root cells (Fig. 7.5). Curiously, mutants of *NRAT1* accumulate less Al^{3+} in roots than wild-type plants, yet are more Al^{3+}-sensitive. This contrasts with other known mechanisms that act to prevent the uptake of Al^{3+} into roots. NRAT1 is different to other NRAMP transporters characterized to date, which mostly use divalent cations (e.g. Mn^{2+}, Zn^{2+}) as substrates. Normally the uptake of Al^{3+} would be considered detrimental to cells but NRAT1 activity may be the first step in a tolerance mechanism that stores aluminium in the vacuole via ALS1, a half-sized ABC transporter located on the tonoplast. Together they reduce the concentration of Al^{3+} in the apoplast and store the cations in the vacuole out of harm's way. Further details of these genes and their functions can be found elsewhere (Kochian *et al.*, 2015). Future work needs to establish how ART1 interacts with Al^{3+} to induce gene expression because, in a heterologous expression system, ART1 can induce the expression of a reporter gene with this promoter motif in the absence of Al^{3+}.

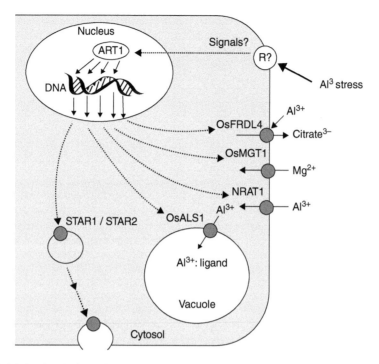

Fig. 7.5. Model showing the induction of Al^{3+} resistance genes in rice. When rice roots sense Al^{3+}, the transcription factor ART1 binds to the promoters of more than 30 genes and increases their expression. These genes contribute to the high Al^{3+} resistance of this species in different ways. Among the genes induced by ART1 are *STAR1* and *STAR2*, which encode the nucleotide binding domain of an ABC transporter, and the membrane-spanning domain of an ABC transporter, respectively. The STAR1/2 protein complex localizes to vesicular membranes where it accumulates UDP-glucose or other substrates and then releases these into the apoplast to protect against Al^{3+} toxicity (see text). *NRAT1* and *OsALS1* expression are also induced by ART1. The NRAT1 protein localizes to the plasma membrane where it transports Al^{3+} into the cell. NRAT1 forms the first step of a tolerance mechanism that requires OsALS1, another transporter located on the tonoplast. Together they reduce the concentration of Al^{3+} in the apoplast and store the Al in the vacuole. *OsFRDL4* encodes a MATE protein, which releases citrate anions from the root cells and *OsMGT1* encodes a transport protein for magnesium uptake which also provides protection from Al stress. Adapted from Delhaize *et al.* (2012).

7.6.2   Model species: *Arabidopsis*

Two genes (*AtALS1* and *AtALS3*) isolated from analysis of Al^{3+}-sensitive mutants encode half-size ABC transporters with *AtALS3* having significant sequence homology to *STAR2* from rice (see above) (Larsen *et al.*, 2005, 2007). AtALS1 is expressed in the vasculature throughout the plant, and the protein localizes to the tonoplast. A function for the protein has not yet been identified but it is possible that AtALS1 transports Al^{3+} into the vacuole. *AtALS3* is also expressed throughout the plant and its expression is induced by Al^{3+} treatment. In contrast to AtALS1, AtALS3 is found predominately at the plasma membrane where it might also transport Al^{3+}. How AtALS3 confers resistance is not understood, but it may involve redistribution of Al^{3+} within the plant away from tissues, such as root apices, that are particularly sensitive to Al^{3+}. AtALS3 encodes the transmembrane domain part of an ABC transporter and appears to interact with AtSTAR1, an *Arabidopsis* homologue of STAR1 in rice.

The *stop1* (sensitive to proton rhizotoxicity) mutant of *Arabidopsis* was originally isolated based on its hypersensitivity to low pH and was subsequently shown to be also important for

Al^{3+} resistance (Iuchi *et al.*, 2007). *STOP1* encodes a zinc-finger transcription factor that is homologous to ART1 in rice and appears to function in a similar manner except that STOP1 upregulates genes involved in combating both proton and Al^{3+} stress (Sawaki *et al.*, 2009). *AtALMT1*, *ALS3* and *AtMATE1* are other *Arabidopsis* genes controlled by STOP1 that are known to be involved in Al^{3+} resistance (see Kochian *et al.*, 2015).

An alternative to the above strategies which identified Al^{3+}-sensitive mutants is to screen for Al^{3+}-resistant mutants. In one example, mentioned above, researchers mutagenized the *als3* mutant of *Arabidopsis* (which is hypersensitive to Al^{3+}) and screened for additional mutants that suppressed the hypersensitivity to Al^{3+}. Two mutants derived from this screen were mutated in the *AtATR* gene which detects and manages damage to DNA (Rounds and Larsen, 2008). These mutants are dominant for enhanced Al^{3+} tolerance but are partial loss-of-function mutants for normal AtATR function. Al^{3+} toxicity damages DNA in cells of the root apex and it appears that it is the response mounted by the plant via AtATR that inhibits root growth (Rounds and Larsen, 2008).

7.7 Transgenic Approaches for Increasing Al^{3+} Resistance

Conventional breeding practices exploit the natural variation within a species to improve the Al^{3+} resistance of important crops. Some species lack sufficient variation to breed for Al^{3+} resistance and others can benefit from levels of resistance over and above what occurs naturally. Biotechnology has the potential to increase the Al^{3+} resistance of any species by overexpressing target genes. These genes might include those controlling the natural variation within species, those identified from mutant analysis and those that are either induced by Al^{3+} or able to confer Al^{3+} resistance to model microorganisms.

7.7.1 Enhancing the efflux of organic anions

Early attempts at increasing the efflux of organic anions from roots focused on increasing the

capacity of biosynthetic pathways to synthesize these metabolites. The genes for many of the key metabolic enzymes were readily available and could be tested in easily transformed species such as tobacco and *Arabidopsis*. Plants overexpressing citrate synthase, malate dehydrogenase and pyruvate phosphate dikinase have been assessed for their effectiveness in conferring Al^{3+} resistance. Results with these genes have been mixed and even in those cases where Al^{3+} resistance was increased, the improvement was relatively small (Ryan *et al.*, 2011). The capacity to synthesize organic anions may not be the rate-limiting step for efflux from roots to occur but rather the transport of these metabolites across the plasma membrane is important. This was demonstrated once the *TaALMT1* and *SbMATE* genes were isolated. When overexpressed in a range of plant species, *TaALMT1* enhanced the efflux of malate in an Al^{3+}-dependent manner. When expressed in barley, a crop species that is generally sensitive to acid soil, it enhanced Al^{3+} resistance to a level comparable to Al^{3+} resistant wheat (Delhaize *et al.*, 2004, 2009). When grown on acid soil the transgenic barley has improved grain yield and P-nutrition as a result of more vigorous root growth. Similarly, when *TaALMT1* is overexpressed in Al^{3+}-sensitive wheat, resistance is increased to a level comparable to that of Al^{3+}-resistant genotypes (Pereira *et al.*, 2010). Overexpression of *MATE* genes also enhances the Al^{3+} resistance of plants (Magalhaes *et al.*, 2007; Zhou *et al.*, 2014), albeit to a lower level than *TaALMT1*, and it is possible that the concomitant overexpression of two transgenes encoding different transport proteins could further enhance the Al^{3+} resistance of transgenic plants.

7.7.2 Other genes

The range of genes isolated from mutant screens of rice and *Arabidopsis* all have the potential of increasing the Al^{3+} resistance of crops and pastures. Although they have been shown to complement their respective mutants, none to date have increased the Al^{3+} resistance beyond what is found in the wild type. This may mean that the physiological processes that these genes control are not rate limiting for Al^{3+} resistance. However, it may still be possible that enhanced resistance is obtained when they are expressed in different species, and this remains to be explored.

When exposed to toxic concentrations of Al^{3+}, roots induce the expression of a wide range of genes. When a selection of these genes were overexpressed in plants, some improved in Al^{3+} resistance. Genes in this category include those associated with protecting cells from oxidative stress (e.g. superoxide dismutase and peroxidase), which reduce the level of ROS generated as a consequence of Al^{3+} toxicity. Most have only been assessed in *Arabidopsis* as a model system using agar or hydroponics, and the level of resistance achieved was relatively modest (two- to threefold increases in relative root growth). Other genes encoding a receptor kinase, auxilin-like protein, sphingolipid desaturase and a protein involved in cell death confer a relatively low level of Al^{3+} resistance when expressed in *Arabidopsis* by mechanisms that are not well understood (Ryan *et al.*, 2011).

References

Aniol, A. and Gustafson, J.P. (1984) Chromosome location of genes controlling aluminum tolerance in wheat, rye and triticale. *Canadian Journal of Genetics and Cytology* 26, 701–705.

Basu, U., Good, A.G. and Taylor, G.J. (2001) Transgenic *Brassica napus* plants overexpressing aluminium-induced mitochondrial manganese superoxide dismutase cDNA are resistant to aluminium. *Plant Cell and Environment* 24, 1269–1278.

Chen, Z.C., Yokosho, K., Kashino, M., Zhao, F.J., Yamaji, N. and Ma, J.F. (2013) Adaptation to acidic soil is achieved by increased numbers of cis-acting elements regulating *ALMT1* expression in *Holcus lanatus. The Plant Journal* 76, 10–23.

Collins, N.C., Shirley, N.J., Saeed, M., Pallotta, M. and Gustafson, J.P. (2008) An *ALMT1* gene cluster controlling aluminum tolerance at the *Alt4* locus of rye (*Secale cereale* L.). *Genetics* 179, 669–692.

Delhaize, E., Craig, S., Beaton, C.D., Bennet, R.J., Jagadish, V.C. and Randall, P.J. (1993a) Aluminum tolerance in wheat (*Triticum aestivum* L.). 1. Uptake and distribution of aluminum in root apices. *Plant Physiology* 103, 685–693.

Delhaize, E., Ryan, P.R. and Randall, P.J. (1993b) Aluminum tolerance in wheat (*Triticum aestivum* L.). 2. Aluminum-stimulated excretion of malic acid from root apices. *Plant Physiology* 103, 695–702.

Delhaize, E., Ryan, P.R., Hebb, D.M., Yamamoto, Y., Sasaki, T. and Matsumoto, H. (2004) Engineering high-level aluminum tolerance in barley with the *ALMT1* gene. *Proceedings of the National Academy of Sciences USA* 101, 15249–15254.

Delhaize, E., Gruber, B.D., Pittman, J.K., White, R.G., Leung, H., Miao, Y.S., Jiang, L.W., Ryan, P.R. and Richardson, A.E. (2007) A role for the *AtMTP11* gene of Arabidopsis in manganese transport and tolerance. *Plant Journal* 51, 198–210.

Delhaize, E., Taylor, P., Hocking, P.J., Simpson, R.J., Ryan, P.R. and Richardson, A.E. (2009) Transgenic barley (*Hordeum vulgare* L.) expressing the wheat aluminium resistance gene (*TaALMT1*) shows enhanced phosphorus nutrition and grain production when grown on an acid soil. *Plant Biotechnology Journal* 7, 391–400.

Delhaize, E., Ma, J.F. and Ryan, P.R. (2012) Transcriptional regulation of aluminium tolerance genes. *Trends in Plant Science* 17, 341–348.

Dong, X.Y., Shen, R.F., Chen, R.F., Zhu, Z.L. and Ma, J.F. (2008) Secretion of malate and citrate from roots is related to high Al-resistance in *Lespedeza bicolor. Plant and Soil* 306, 139–147.

Eleftheriou, E.P., Moustakas, M. and Fragiskos, N. (1993) Aluminate-induced changes in morphology and ultrastructure of *Thinopyrum* roots. *Journal of Experimental Botany* 44, 427–436.

Ezaki, B., Yamamoto, Y. and Matsumoto, H. (1995) Cloning and sequencing of the cDNAs induced by aluminum treatment and Pi starvation in cultured tobacco cells. *Physiologia Plantarum* 93, 11–18.

Ezaki, B., Gardner, R.C., Ezaki, Y. and Matsumoto, H. (2000) Expression of aluminum-induced genes in transgenic Arabidopsis plants can ameliorate aluminum stress and/or oxidative stress. *Plant Physiology* 122, 657–665.

Ezaki, B., Sasaki, K., Matsumoto, H. and Nakashima, S. (2005) Functions of two genes in aluminium (Al) stress resistance: repression of oxidative damage by the *AtBCB* gene and promotion of efflux of Al ions by the *NtGDI1* gene. *Journal of Experimental Botany* 56, 2661–2671.

Eticha, D., Stass, A. and Horst, W.J. (2005) Cell-wall pectin and its degree of methylation in the maize root-apex: significance for genotypic differences in aluminium resistance. *Plant Cell and Environment* 28, 1410–1420.

Eticha, D., Zahn, M., Bremer, M., Yang, Z.B., Rangel, A.F., Rao, I.M. and Horst, W.J. (2010) Transcriptomic analysis reveals differential gene expression in response to aluminium in common bean (*Phaseolus vulgaris*) genotypes. *Annals of Botany* 105, 1119–1128.

Famoso, A.N., Clark, R.T., Shaff, J.E., Craft, E., Mccouch, S.R. and Kochian, L.V. (2010) Development of a novel aluminum tolerance phenotyping platform used for comparisons of cereal Al tolerance and investigations into rice Al tolerance mechanisms. *Plant Physiology* 153, 1678–1691. DOI:10.1104/pp.110.156794.

Foy, C.D. (1984) Physiological effects of hydrogen, aluminum, and manganese toxicities in acid soil. In: Adams, F. (ed.) *Soil Acidity and Liming*. American Society of Agronomy, Crop Science Society, American Society of Soil Science, Madison, Wisconsin.

Foy, C.D. (1988) Plant adaptation to acid, aluminum-toxic soils. *Communications in Soil Science and Plant Analysis* 19, 959–987.

Foy, C.D., Chaney, R.L. and White, M.C. (1978) Physiology of metal toxicity in plants. *Annual Review of Plant Physiology and Plant Molecular Biology* 29, 511–566.

Fujii, M., Yokosho, K., Yamaji, N., Saisho, D., Yamane, M., Takahashi, H., Sato, K., Nakazono, M. and Ma, J.F. (2012) Acquisition of aluminium tolerance by modification of a single gene in barley. *Nature Communications* 3. doi: 10.1038/ncomms1726.

Furuichi, T., Sasaki, T., Tsuchiya, Y., Ryan, P.R., Delhaize, E. and Yamamoto, Y. (2010) Extracellular hydrophilic carboxy-terminal domain regulates the activity of TaALMT1, the aluminum-activated malate transport protein of wheat. *The Plant Journal* 64, 47–55.

Furukawa, J., Yamaji, N., Wang, H., Mitani, N., Murata, Y., Sato, K., Katsuhara, M., Takeda, K. and Ma, J.F. (2007) An aluminum-activated citrate transporter in barley. *Plant and Cell Physiology* 48, 1081–1091.

Garvin, D.F. and Carver, B.F. (2003) The role of the genotype in tolerance to acidity and aluminum toxicity. In: Rengel, Z. (ed.) *Handbook of Soil Acidity*. Marcel Dekker Inc., New York, pp. 387–406.

Gassmann, W. and Schroeder, J.I. (1994) Inwardly-rectifying K^+ channels in roots hairs of wheat. A mechanism for aluminum-sensitive low-affinity K^+ uptake and membrane potential control. *Plant Physiology* 105, 1399–1408.

Grabski, S. and Schindler, M. (1995) Aluminum induces rigor within the actin network of soybean cells. *Plant Physiology* 108, 897–901.

Guo, J.H., Liu, X.J., Zhang, Y., Shen, J.L., Han, W.X., Zhang, W.F., Christie, P., Goulding, K.W.T., Vitousek, P.M. and Zhang, F.S. (2010) Significant acidification in major Chinese croplands. *Science* 327, 1008–1010.

Hiradate, S., Ma, J.F. and Matsumoto, H. (2007) Strategies of plants to adapt to mineral stresses in problem soils. *Advances in Agronomy* 96, 65–132.

Hoekenga, O.A., Vision, T.J., Shaff, J.E., Monforte, A.J., Lee, G.P., Howell, S.H. and Kochian, L.V. (2003) Identification and characterization of aluminum tolerance loci in Arabidopsis (Landsberg *erecta* × Columbia) by quantitative trait locus mapping. A physiologically simple but genetically complex trait. *Plant Physiology* 132, 936–948.

Hoekenga, O.A., Maron, L.G., Piñeros, M.A., Cançado, G.M.A., Shaff, J., Kobayashi, Y., Ryan, P.R., Dong, B., Delhaize, E., Sasaki, T., Matsumoto, H., Yamamoto, Y., Koyama, H. and Kochian, L.V. (2006) AtALMT1, which encodes a malate transporter, is identified as one of several genes critical for aluminum tolerance in *Arabidopsis*. *Proceedings of the National Academy of Sciences USA* 103, 9738–9743.

Horst, W.J., Schmohl, N., Kollmeier, M., Baluska, F. and Sivaguru, M. (1999) Does aluminium affect root growth of maize through interaction with the cell wall – plasma membrane – cytoskeleton continuum? *Plant and Soil* 215, 163–174.

Horst, W.J., Wang, Y.X. and Eticha, D. (2010) The role of the root apoplast in aluminium-induced inhibition of root elongation and in aluminium resistance of plants: a review. *Annals of Botany* 106, 185–197.

Huang, C.F., Yamaji, N., Mitani, N., Yano, M., Nagamura, Y. and Ma, J.F. (2009) A bacterial-type ABC transporter is involved in aluminum tolerance in rice. *Plant Cell* 21, 655–667.

Huang, J.W.W., Shaff, J.E., Grunes, D.L. and Kochian, L.V. (1992) Aluminum effects on calcium fluxes at the root apex of aluminum-tolerant and aluminum-sensitive wheat cultivars. *Plant Physiology* 98, 230–237.

Hvorup, R.N., Winnen, B., Chang, A.B., Jiang, Y., Zhou, X.F. and Saier, M.H. (2003) The multidrug/oligosaccharidyl-lipid/polysaccharide (MOP) exporter superfamily. *European Journal of Biochemistry* 270, 799–813.

Iuchi, S., Koyama, H., Iuchi, A., Kobayashi, Y., Kitabayashi, S., Ikka, T., Hirayama, T., Shinozaki, K. and Kobayashi, M. (2007) Zinc finger protein STOP1 is critical for proton tolerance in *Arabidopsis* and coregulates a key gene in aluminum tolerance. *Proceedings of the National Academy of Sciences USA* 104, 9900–9905.

Jones, D.L. and Kochian, L.V. (1995) Aluminum inhibition of the inositol 1,4,5-trisphosphate signal-transduction pathway in wheat roots: a role in aluminum toxicity. *Plant Cell* 7, 1913–1922.

Jones, D.L., Gilroy, S. and Kochian, L.V. (1996) Effect of aluminum, oxidative, mechanical and anaerobic stress on cytoplasmic Ca^{2+} homeostatis in Arabidopsis root hairs. *Plant Physiology* 111, 78–78.

Jones, D.L., Blancaflor, E.B., Kochian, L.V. and Gilroy, S. (2006) Spatial coordination of aluminium uptake, production of reactive oxygen species, callose production and wall rigidification in maize roots. *Plant Cell and Environment* 29, 1309–1318.

Kidd, P.S., Llugany, M., Poschenrieder, C., Gunse, B. and Barcelo, J. (2001) The role of root exudates in aluminium resistance and silicon-induced amelioration of aluminium toxicity in three varieties of maize (*Zea mays* L.). *Journal of Experimental Botany* 52, 1339–1352.

Kinraide, T.B. (1997) Reconsidering the rhizotoxicity of hydroxyl, sulphate, and fluoride complexes of aluminium. *Journal of Experimental Botany* 48, 1115–1124.

Kinraide, T.B. (2004) Possible influence of cell walls upon ion concentrations at plasma membrane surfaces. Toward a comprehensive view of cell-surface electrical effects upon ion uptake, intoxication, and amelioration. *Plant Physiology* 136, 3804–3813.

Kinraide, T.B., Ryan, P.R. and Kochian, L.V. (1992) Interactive effects of Al^{3+}, H$^+$ and other cations on root elongation considered in terms of cell-surface electrical potential. *Plant Physiology* 99, 1461–1468.

Kobayashi, Y. and Koyama, H. (2002) QTL analysis of Al tolerance in recombinant inbred lines of *Arabidopsis thaliana*. *Plant and Cell Physiology* 43, 1526–1533.

Kochian, L.V., Hoekenga, O.A. and Piñeros, M.A. (2004) How do crop plants tolerate acid soils? Mechanisms of aluminum tolerance and phosphorus efficiency. *Annual Review of Plant Biology* 55, 459–493.

Kochian, L.V., Piñeros, M.A., Liu, J. and Magalhaes, J.V. (2015) Plant adaptation to acid soils: the molecular basis for crop aluminum resistance. *Annual Review of Plant Biology* 66, 571–598.

Kopittke, P.M., Menzies, N.W. and Blamey, F.P.C. (2005) Rhizotoxicity of aluminate and polycationic aluminium at high pH. *Plant and Soil* 266, 177–186.

Larsen, P.B., Geisler, M.J.B., Jones, C.A., Williams, K.M. and Cancel, J.D. (2005) *ALS3* encodes a phloem-localized ABC transporter-like protein that is required for aluminum tolerance in Arabidopsis. *Plant Journal* 41, 353–363.

Larsen, P.B., Cancel, J., Rounds, M. and Ochoa, V. (2007) Arabidopsis *ALS1* encodes a root tip and stele localized half type ABC transporter required for root growth in an aluminum toxic environment. *Planta* 225, 1447–1458.

Lazof, D.B., Goldsmith, J.G., Rufty, T.W. and Linton, R.W. (1994) Rapid uptake of aluminum into cells of intact soybean root tips (a microanalytical study using secondary-ion mass-spectrometry). *Plant Physiology* 106, 1107–1114.

Li, X.F., Zuo, F.H., Ling, G.Z., Li, Y.Y., Yu, Y.X., Yang, P.Q. and Tang, X.L. (2009) Secretion of citrate from roots in response to aluminum and low phosphorus stresses in *Stylosanthes*. *Plant and Soil* 325, 219–229.

Ligaba, A., Katsuhara, M., Ryan, P.R., Shibasaka, M. and Matsumoto, H. (2006) The *BnALMT1* and *BnALMT2* genes from rape encode aluminum-activated malate transporters that enhance the aluminum resistance of plant cells. *Plant Physiology* 142, 1294–1303.

Ligaba, A., Kochian, L. and Piñeros, M. (2009) Phosphorylation at S384 regulates the activity of the TaALMT1 malate transporter that underlies aluminum resistance in wheat. *Plant Journal* 60, 411–423.

Ligaba, A., Dreyer, I., Margaryan, A., Schneider, D.J., Kochian, L. and Piñeros, M. (2013) Functional, structural and phylogenetic analysis of domains underlying the Al sensitivity of the aluminum-activated malate/anion transporter, TaALMT1. *The Plant Journal* 76, 766–780.

Liu, J.P., Magalhaes, J.V., Shaff, J. and Kochian, L.V. (2009) Aluminum-activated citrate and malate transporters from the MATE and ALMT families function independently to confer Arabidopsis aluminum tolerance. *Plant Journal* 57, 389–399.

Luo, M.C. and Dvořák, J. (1996) Molecular mapping of an aluminum tolerance locus on chromosome 4D of Chinese Spring wheat. *Euphytica* 91, 31–35.

Ma, J.F. and Hiradate, S. (2000) Form of aluminium for uptake and translocation in buckwheat (*Fagopyrum esculentum* Moench). *Planta* 211, 355–360.

Ma, J.F., Zheng, S.J. and Matsumoto, H. (1997) Specific secretion of citric acid induced by Al stress in *Cassia tora* L. *Plant and Cell Physiology* 38, 1019–1025.

Ma, Z. and Miyasaka, S.C. (1998) Oxalate exudation by taro in response to Al. *Plant Physiology* 118, 861–865.

Ma, J.F., Taketa, S. and Yang, Z.M. (2000) Aluminum tolerance genes on the short arm of chromosome 3R are linked to organic acid release in triticale. *Plant Physiology* 122, 687–694.

Ma, J.F., Ryan, P.R. and Delhaize, E. (2001) Aluminium tolerance in plants and the complexing role of organic acids. *Trends in Plant Scence* 6, 273–278.

Ma, J.F., Nagao, S., Sato, K., Ito, H., Furukawa, J. and Takeda, K. (2004a) Molecular mapping of a gene responsible for Al-activated secretion of citrate in barley. *Journal of Experimental Botany* 55, 1335–1341.

Ma, J.F., Shen, R.F., Nagao, S. and Tanimoto, E. (2004b) Aluminum targets elongating cells by reducing cell wall extensibility in wheat roots. *Plant and Cell Physiology* 45, 583–589.

Magalhaes, J.V., Liu, J., Guimaraes, C.T., Lana, U.G.P., Alves, V.M.C., Wang, Y.H., Schaffert, R.E., Hoekenga, O.A., Piñeros, M.A., Shaff, J.E., Klein, P.E., Carneiro, N.P., Coelho, C.M., Trick, H.N. and Kochian, L.V. (2007) A gene in the multidrug and toxic compound extrusion (*MATE*) family confers aluminum tolerance in sorghum. *Nature Genetics* 39, 1156–1161.

Maron, L.G., Kirst, M., Mao, C., Milner, M.J., Menossi, M. and Kochian, L.V. (2008) Transcriptional profiling of aluminum toxicity and tolerance responses in maize roots. *New Phytologist* 179, 116–128.

Maron, L.G., Piñeros, M.A., Guimaraes, C.T., Magalhaes, J.V., Pleiman, J.K., Mao, C.Z., Shaff, J., Belicuas, S.N.J. and Kochian, L.V. (2010) Two functionally distinct members of the MATE (multi-drug and toxic compound extrusion) family of transporters potentially underlie two major aluminum tolerance QTLs in maize. *Plant Journal* 61, 728–740.

Martin, R.B. (1992) Aluminum speciation in biology. *Ciba Foundation Symposia* 169, 5–25.

Matsumoto, H. (1991) Biochemical mechanism of the toxicity of aluminum and the sequestration of aluminum in plant cells. In: Wright, R.J., Baligar, V.C. and Murrmann, R.P. (eds) *Plant–Soil Interactions at Low pH*. Kluwer Academic Publishers, Dordrecht, The Netherlands, pp. 825–838.

Matsumoto, H., Hirasawa, E., Morimura, S. and Takahashi, E. (1976) Localization of aluminum in tea leaves. *Plant and Cell Physiology* 17, 627–631.

Miyasaka, S.C. and Hawes, M.C. (2001) Possible role of root border cells in detection and avoidance of aluminum toxicity. *Plant Physiology* 125, 1978–1987.

Miyasaka, S.C., Buta, J.G., Howell, R.K. and Foy, C.D. (1991) Mechanism of aluminum tolerance in snapbeans – root exudation of citric acid. *Plant Physiology* 96, 737–743.

Motoda, H., Sasaki, T., Kano, Y., Ryan, P.R., Delhaize, E., Matsumoto, H. and Yamamoto, Y. (2007) The membrane topology of ALMT1, an aluminum-activated malate transport protein in wheat (*Triticum aestivum*). *Plant Signaling and Behavior* 2, 467–472.

Navakode, S., Neumann, K., Kobiljski, B., Lohwasser, U. and Börner, A. (2014) Genome wide association mapping to identify aluminium tolerance loci in bread wheat. *Euphytica* 198, 401–411.

Nobel, P.S. (1983) *Biophysical Plant Physiology*. W.H. Freeman and Company, San Francisco, California.

Osawa, H., Endo, I., Hara, Y., Matsushima, Y. and Tange, T. (2011) Transient proliferation of proanthocyanidin-accumulating cells on the epidermal apex contributes to highly aluminum-resistant root elongation in camphor tree. *Plant Physiology* 155, 433–446.

Papernik, L.A., Bethea, A.S., Singleton, T.E., Magalhaes, J.V., Garvin, D.F. and Kochian, L.V. (2001) Physiological basis of reduced Al tolerance in ditelosomic lines of Chinese Spring wheat. *Planta* 212, 829–834.

Pellet, D.M., Grunes, D.L. and Kochian, L.V. (1995) Organic-acid exudation as an aluminium-tolerance mechanism in maize (*Zea mays* L.). *Planta* 196, 788–795.

Pereira, J.F., Zhou, G., Delhaize, E., Richardson, T. and Ryan, P.R. (2010) Engineering greater aluminium resistance in wheat by over-expressing *TaALMT1*. *Annal of Botany* 106, 205–214.

Piñeros, M. and Tester, M. (1993) Plasma-membrane Ca^{2+} channels in roots of higher roots and their role in aluminum toxicity. *Plant and Soil* 156, 119–122.

Piñeros, M.A., Cançado, G.M.A. and Kochian, L.V. (2008) Novel properties of the wheat aluminum tolerance organic acid transporter (TaALMT1) revealed by electrophysiological characterization in *Xenopus* oocytes: Functional and structural implications. *Plant Physiology* 147, 2131–2146.

Poschenrieder, C., Gunse, B., Corrales, I. and Barcelo, J. (2008) A glance into aluminum toxicity and resistance in plants. *Science of the Total Environment* 400, 356–368.

Qin, R.J., Hirano, Y. and Brunner, I. (2007) Exudation of organic acid anions from poplar roots after exposure to Al, Cu and Zn. *Tree Physiology* 27, 313–320.

Raman, H., Zhang, K.R., Cakir, M., Appels, R., Garvin, D.F., Maron, L.G., Kochian, L.V., Moroni, J.S., Raman, R., Imtiaz, M., Drake-Brockman, F., Waters, I., Martin, P., Sasaki, T., Yamamoto, Y., Matsumoto, H., Hebb, D.M., Delhaize, E. and Ryan, P.R. (2005) Molecular characterization and mapping of *ALMT1*, the aluminium-tolerance gene of bread wheat (*Triticum aestivum* L.). *Genome* 48, 781–791.

Raman, H., Stodart, B., Ryan, P.R., Delhaize, E., Emebiri, L., Raman, R., Coombes, N. and Milgate, A. (2010) Genome-wide association analyses of common wheat (*Triticum aestivum* L.) germplasm identifies multiple loci for aluminium resistance. *Genome* 53, 957–966.

Ramesh, S., Tyerman, S.D., Xu, B., Bose, J., Kaur, S., Conn, V., Domingos, P., Ullah, S., Wege, S., Shabala, S., Feijó, J., Ryan, P.R. and Gilliham, M. (2015) GABA signalling modulates plant growth by directly regulating the activity of plant-specific anion transporters. *Nature Communications* 6, 7879, doi: 10.1038ncomms8879.

Rangel, A.F., Rao, I.M., Braun, H.P. and Horst, W.J. (2010) Aluminum resistance in common bean (*Phaseolus vulgaris*) involves induction and maintenance of citrate exudation from root apices. *Physiologia Plantarum* 138, 176–190.

Reide, C.R. and Anderson, J.A. (1996) Linkage of RFLP markers to an aluminum tolerance gene in wheat. *Crop Science* 36, 905–909.

Rengel, Z. (1992) Disturbance of cell Ca^{2+} homeostasis as a primary trigger of Al toxicity syndrome. *Plant Cell and Environment* 15, 931–938.

Rengel, Z. and Reid, R.J. (1997) Uptake of Al across the plasma membrane of plant cells. *Plant and Soil* 192, 31–35.

Rincon-Zachary, M., Teaster, N.D., Sparks, J.A., Valster, A.H., Motes, C.M. and Blancaflor, E.B. (2010) Fluorescence resonance energy transfer-sensitized emission of yellow cameleon 3.60 reveals root zone-specific calcium signatures in Arabidopsis in response to aluminum and other trivalent cations. *Plant Physiology* 152, 1442–1458.

Rounds, M.A. and Larsen, P.B. (2008) Aluminum-dependent root-growth inhibition in *Arabidopsis* results from AtATR-regulated cell-cycle arrest. *Current Biology* 18, 1495–1500.

Ryan, P.R. and Delhaize, E. (2010) The convergent evolution of aluminium resistance in plants exploits a convenient currency. *Functional Plant Biology* 37, 275–284.

Ryan, P.R., Shaff, J.E. and Kochian, L.V. (1992) Aluminum toxicity in roots: Correlation between ionic currents, ion fluxes and root elongation in Al-tolerant and Al-sensitive wheat cultivars. *Plant Physiology* 99, 1193–1200.

Ryan, P.R., Ditomaso, J.M. and Kochian, L.V. (1993) Aluminum toxicity in roots: Investigation of spatial sensitivity and the role of the root cap. *Journal of Experimental Botany* 44, 437–446.

Ryan, P.R., Kinraide, T.B. and Kochian, L.V. (1994) Al^{3+}-Ca^{2+} interaction in rhizo-toxicity. I. Inhibition of root growth is not caused by reduction in calcium uptake. *Planta* 192, 98–103.

Ryan, P.R., Reid, R.J. and Smith, F.A. (1997) Direct evaluation of the Ca^{2+}-displacement hypothesis for Al toxicity. *Plant Physiology* 113, 1351–1357.

Ryan, P.R., Delhaize, E. and Jones, D.L. (2001) Function and mechanism of organic anion exudation from plant roots. *Annual Review of Plant Physiology and Plant Molecular Biology* 52, 527–560.

Ryan, P.R., Raman, H., Gupta, S., Horst, W.J. and Delhaize, E. (2009) A second mechanism for aluminum resistance in wheat relies on the constitutive efflux of citrate from roots. *Plant Physiology* 149, 340–351.

Ryan, P.R., Raman, H., Gupta, S., Sasaki, T., Yamamoto, Y. and Delhaize, E. (2010) The multiple origins of aluminium resistance in hexaploid wheat include *Aegilops tauschii* and more recent *cis* mutations to *TaALMT1*. *The Plant Journal* 64, 446–455.

Ryan, P.R., Tyerman, S.D., Sasaki, T., Furuichi, T., Yamamoto, Y., Zhang, W.H. and Delhaize, E. (2011) The identification of aluminium-resistance genes provides opportunities for enhancing crop production on acid soils. *Journal of Experimental Botany* 62, 9–20.

Sasaki, T., Yamamoto, Y., Ezaki, B., Katsuhara, M., Ahn, S.J., Ryan, P.R., Delhaize, E. and Matsumoto, H. (2004) A wheat gene encoding an aluminum-activated malate transporter. *Plant Journal* 37, 645–653.

Sasaki, T., Ryan, P.R., Delhaize, E., Hebb, D.M., Ogihara, Y., Kawaura, K., Noda, K., Kojima, T., Toyoda, A., Matsumoto, H. and Yamamoto, Y. (2006) Sequence upstream of the wheat (*Triticum aestivum* L.) *ALMT1* gene and its relationship to aluminum resistance. *Plant and Cell Physiology* 47, 1343–1354.

Sawaki, Y., Iuchi, S., Kobayashi, Y., Ikka, T., Sakurai, N., Fujita, M., Shinozaki, K., Shibata, D., Kobayashi, M. and Koyama, H. (2009) STOP1 regulates multiple genes that protect Arabidopsis from proton and aluminum toxicities. *Plant Physiology* 150, 281–294.

Schmohl, N., Pilling, J., Fisahn, J. and Horst, W.J. (2000) Pectin methylesterase modulates aluminium sensitivity in *Zea mays* and *Solanum tuberosum*. *Physiologia Plantarum* 109, 419–427.

Schulz, B. and Kolukisaglu, H.U. (2006) Genomics of plant ABC transporters: The alphabet of photosynthetic life forms or just holes in membranes? *Febs Letters* 580, 1010–1016.

Shaff, J.E., Schultz, B.A., Craft, E.J., Clark, R.T. and Kochian, L.V. (2010) GEOCHEM-EZ: a chemical speciation program with greater power and flexibility. *Plant and Soil* 330, 207–214.

Shen, H., Yan, X.L., Wang, X.R. and Zheng, S.L. (2002) Exudation of citrate in common bean in response to aluminum stress. *Journal of Plant Nutrition* 25, 1921–1932.

Silva, I.R., Smyth, T.J., Moxley, D.F., Carter, T.E., Allen, N.S. and Rufty, T.W. (2000) Aluminum accumulation at nuclei of cells in the root tip: fluorescence detection using lumogallion and confocal laser scanning microscopy. *Plant Physiology* 123, 543–552.

Silva, I.R., Smyth, T.J., Raper, C.D., Carter, T.E. and Rufty, T.W. (2001) Differential aluminum tolerance in soybean: An evaluation of the role of organic acids. *Physiologia Plantarum* 112, 200–210.

Sivaguru, M. and Horst, W.J. (1998) The distal part of the transition zone is the most aluminum-sensitive apical root zone of maize. *Plant Physiology* 116, 155–163.

Sivaguru, M., Fujiwara, T., Samaj, J., Baluska, F., Yang, Z.M., Osawa, H., Maeda, T., Mori, T., Volkmann, D. and Matsumoto, H. (2000) Aluminum-induced 1-3-beta-D-glucan inhibits cell-to-cell trafficking of molecules through plasmodesmata. A new mechanism of aluminum toxicity in plants. *Plant Physiology* 124, 991–1005.

Sivaguru, M., Pike, S., Gassmann, W. and Baskin, T.I. (2003) Aluminum rapidly depolymerizes cortical microtubules and depolarizes the plasma membrane: evidence that these responses are mediated by a glutamate receptor. *Plant and Cell Physiology* 44, 667–675.

Sivaguru, M., Horst, W.J., Eticha, D. and Matsumoto, H. (2006) Aluminum inhibits apoplastic flow of high-molecular weight solutes in root apices of *Zea mays* L. *Journal of Plant Nutrition and Soil Science-Zeitschrift Fur Pflanzenernahrung Und Bodenkunde* 169, 679–690.

Stass, A., Wang, Y., Eticha, D. and Horst, W.J. (2006) Aluminium rhizotoxicity in maize grown in solutions with Al^{3+} or $Al(OH)_4^-$ as predominant solution Al species. *Journal of Experimental Botany* 57, 4033–4042.

Stass, A., Kotur, Z. and Horst, W.J. (2007) Effect of boron on the expression of aluminium toxicity in *Phaseolus vulgaris*. *Physiologia Plantarum* 131, 283–290.

Stoutjesdijk, P.A., Sale, P.W. and Larkin, P.J. (2001) Possible involvement of condensed tannins in aluminium tolerance of *Lotus pedunculatus*. *Australian Journal of Plant Physiology*, 28, 1063–1074.

Taylor, G.J. (1991) Current views of the aluminum stress response: the physiological basis of tolerance. *Current Topics in Plant Biochemistry and Physiology* 10, 57–93.

Taylor, G.J., Mcdonald-Stephens, J.L., Hunter, D.B., Bertsch, P.M., Elmore, D., Rengel, Z. and Reid, R.J. (2000) Direct measurement of aluminum uptake and distribution in single cells of *Chara corallina*. *Plant Physiology* 123, 987–996.

Thomas, G.W. and Hargrove, W.L. (1984) The chemistry of soil acidity. In: Adams, F. (ed.) *Soil Acidity and Liming.* American Society of Agronomy, Crop Science Society, American Society of Soil Science, Madison, Wisconsin.

Tolra, R., Barcelo, J. and Poschenrieder, C. (2009) Constitutive and aluminium-induced patterns of phenolic compounds in two maize varieties differing in aluminium tolerance. *Journal of Inorganic Biochemistry* 103, 1486–1490.

Tolra, R., Vogel-Mikus, K., Hajiboland, R., Kump, P., Pongrac, P., Kaulich, B., Gianoncelli, A., Babin, V., Barcelo, J., Regvar, M. and Poschenrieder, C. (2011) Localization of aluminium in tea (*Camellia sinensis*) leaves using low energy X-ray fluorescence spectro-microscopy. *Journal of Plant Research* 124, 165–172.

Tovkach, A., Ryan, P.R., Richardson, A.E., Lewis, D., Rathjen, T.M., Ramesh, S., Tyerman, S.D. and Delhaize, E. (2013) Transposon-mediated alteration of *TaMATE1B* expression in wheat roots confers constitutive citrate efflux from root apices. *Plant Physiology* 161, 880–892.

Tsutsui, T., Yamaji, N. and Ma, J.F. (2011) Identification of a cis-acting element of ART1, a zinc-finger transcription factor for aluminum tolerance in rice. *Plant Physiology* 156, 925–931.

von Uexkull, H.R. and Mutert, E. (1995) Global extent, development and economic impact of acid soils. *Plant and Soil* 171, 1–15.

Wang, B.L., Shen, J.B., Zhang, W.H., Zhang, F.S. and Neumann, G. (2007a) Citrate exudation from white lupin induced by phosphorus deficiency differs from that induced by aluminum. *New Phytologist* 176, 581–589.

Wang, J.P., Raman, H., Zhou, M.X., Ryan, P.R., Delhaize, E., Hebb, D.M., Coombes, N. and Mendham, N. (2007b) High-resolution mapping of the Alp locus and identification of a candidate gene HvMATE controlling aluminium tolerance in barley (*Hordeum vulgare* L.). *Theoretical and Applied Genetics* 115, 265–276.

Wang, X.G., Nelson, D.J., Trindle, C. and Martin, R.B. (1997) Experimental and theoretical analysis of Al^{3+} complexation with ADP and ATP. *Journal of Inorganic Biochemistry* 68, 7–15.

Williams, C.H. (1980) Soil acidification under clover pasture. *Australian Journal of Experimental Agriculture and Animal Husbandry* 20, 301–310.

Xia, J., Yamaji, N., Kasai, T. and Ma, J.F. (2010) Plasma membrane-localized transporter for aluminum in rice. *Proceedings of the National Academy of Sciences USA* 107, 18381–18385.

Xia, J., Yamaji, N., Kasai, T. and Ma, J.F. (2011) Plasma membrane-localized transporter for aluminum in rice. *Proceedings of the National Academy of Sciences USA* 107, 18381–18385.

Xue, Y., Wan, J.M., Jiang, L., Liu, L.L., Su, N., Zhai, H.Q. and Ma, J.F. (2006) QTL analysis of aluminum resistance in rice (*Oryza sativa* L.). *Plant and Soil* 287, 375–383.

Yamaji, N., Huang, C.F., Nagao, S., Yano, M., Sato, Y., Nagamura, Y. and Ma, J.F. (2009) A zinc finger transcription factor ART1 regulates multiple genes implicated in aluminum tolerance in rice. *Plant Cell* 21, 3339–3349.

Yamamoto, Y., Kobayashi, Y. and Matsumoto, H. (2001) Lipid peroxidation is an early symptom triggered by aluminum, but not the primary cause of elongation inhibition in pea roots. *Plant Physiology* 125, 199–208.

Yang, J.L., Zhang, L., Li, Y.Y., You, J.F., Wu, P. and Zheng, S.J. (2006) Citrate transporters play a critical role in aluminium-stimulated citrate efflux in rice bean (*Vigna umbellata*) roots. *Annals of Botany* 97, 579–584.

Yang, J.L., Li, Y.Y., Zhang, Y.J., Zhang, S.S., Wu, Y.R., Wu, P. and Zheng, S.J. (2008) Cell wall polysaccharides are specifically involved in the exclusion of aluminum from the rice root apex. *Plant Physiology* 146, 602–611.

Yang, J.L., Zhu, X.F., Peng, Y.X., Zheng, C., Li, G.X., Liu, Y., Shi, Y.Z. and Zheng, S.J. (2011) Cell wall hemicellulose contributes significantly to aluminum adsorption and root growth in Arabidopsis. *Plant Physiology* 155, 1885–1892.

Yang, L.T., Jiang, H.X., Tang, N. and Chen, L.S. (2011) Mechanisms of aluminum-tolerance in two species of citrus: Secretion of organic acid anions and immobilization of aluminum by phosphorus in roots. *Plant Science* 180, 521–530.

Yang, Z.M., Sivaguru, M., Horst, W.J. and Matsumoto, H. (2000) Aluminium tolerance is achieved by exudation of citric acid from roots of soybean (*Glycine max*). *Physiologia Plantarum* 110, 72–77.

Yin, L.N., Wang, S.W., Eltayeb, A.E., Uddin, M.I., Yamamoto, Y., Tsuji,W., Takeuchi, Y., Tanaka, K. (2010) Overexpression of dehydroascorbate reductase, but not monodehydroascorbate reductase, confers tolerance to aluminum stress in transgenic tobacco. *Planta* 231, 609–621.

Yokosho, K., Yamaji, N. and Ma, J.F. (2010) Isolation and characterisation of two *MATE* genes in rye. *Functional Plant Biology* 37, 296–303.

Yokosho, K., Yamaji, N. and Ma, J.F. (2011) An Al-inducible *MATE* gene is involved in external detoxification of Al in rice. *Plant Journal* 68, 1061–1069.

Zhang, W.H., Rengel, Z. and Kuo, J. (1998) Determination of intracellular Ca^{2+} in cells of intact wheat roots: loading of acetoxymethyl ester of Fluo-3 under low temperature. *Plant Journal* 15, 147–151.

Zhang, W.H., Ryan, P.R., Sasaki, T., Yamamoto, Y., Sullivan, W. and Tyerman, S.D. (2008) Characterization of the TaALMT1 protein as an Al^{3+}-activated anion channel in transformed tobacco (*Nicotiana tabacum* L.) cells. *Plant and Cell Physiology* 49, 1316–1330.

Zhang, H., Shi, W.L., You, J.F., Bian, M.D., Qin, X.M., Yu, H., Liu, Q., Ryan, P.R. and Yang, Z.M. (2015) Transgenic *Arabidopsis thaliana* plants expressing a β-1,3-glucanase from sweet sorghum (*Sorghum bicolor* L.) show reduced callose deposition and increased tolerance to aluminium toxicity. *Plant, Cell and Environment* 38, 1178–1188.

Zhao, Z.Q., Ma, J.F., Sato, K. and Takeda, K. (2003) Differential Al resistance and citrate secretion in barley (*Hordeum vulgare* L.). *Planta* 217, 794–800.

Zheng, S.J., Ma, J.F. and Matsumoto, H. (1998) Continuous secretion of organic acids is related to aluminium resistance during relatively long-term exposure to aluminium stress. *Physiologia Plantarum* 103, 209–214.

Zhou, G., Pereira, J.F., Delhaize, E., Zhou, M., Magalhaes, J.V. and Ryan, P.R. (2014) Enhancing the aluminium tolerance of barley by expressing the citrate transporter genes *SbMATE* and *FRD3*. *Journal of Experimental Botany* 65, 2381–2390.

Zhu, M.Y., Ahn, S. and Matsumoto, H. (2003) Inhibition of growth and development of root border cells in wheat by Al. *Physiologia Plantarum* 117, 359–367.

Zhu, X.F., Sun, Y., Zhang, B.C. and Mansoori, N. (2014) TRICHOME BIREFRINGENCE-LIKE27 affects aluminum sensitivity by modulating the O-acetylation of xyloglucan and aluminum-binding capacity in *Arabidopsis*. *Plant Physiology* 166, 181–189.

8 Plant Stress under Non-optimal Soil pH

Andre Läuchli[†] and Stephen R. Grattan*
Department of Land, Air and Water Resources,
University of California, Davis, California, USA

Abstract

Most soils cultivated for crop production fall within the pH range of pH 6–8, where nutrient availability to the plant is typically optimal. Profoundly acid soils (pH < 5.5) and alkaline soils (pH > 8), however, fall outside this optimal pH range and pose challenges for the plant such as low nutrient availability, ion toxicities and nutrient imbalances. The characteristics of acid and alkaline soils are described. Among the alkaline soils one needs to differentiate between calcareous (pH > 7.5) and sodic (exchangeable sodium percentage, ESP > 15) soils, as they present a different set of challenges. Most nutrients are not equally available to plants across the pH spectrum. Several mineral nutrients are severely affected in these non-optimal pH soils, particularly Ca, K, P and Fe. The reactions of plants to these nutrient elements under extreme soil pH conditions are discussed in detail, with emphasis on plant growth, morphological, physiological and membrane transport processes. Finally, a special case is presented of the recently discovered complex interactions between salinity, boron-toxicity and pH in plants.

8.1 Introduction

The pH of the soil solution has a profound influence on nutrient availability, nutrient uptake and ion toxicity to plants. The vast majority of soils that are cultivated for crop production around the world fall within the neutral, slightly acid and slightly basic pH range (i.e. pH 6–8). This is the general range where nutrient availability is optimal. However, there are those soils where the pH falls far from this normal range and these, if not corrected to an adequate range, can pose adverse health effects on plants. Soils that are highly acidic (pH < 5.5) or highly alkaline (pH > 8) present a spectrum of challenges for the plant, including nutrient availability, ion toxicities and nutrient imbalances influencing the ion relations and nutrition within the plant itself.

The pH of the soil solution is an indication of the activity of the hydrogen ion (H^+). The pH numerical value is expressed on a negative log scale such that a one-unit increase or decrease corresponds to a tenfold increase or decrease in the hydrogen ion activity. For example, a change of soil-solution pH from 6 to 8 corresponds to a 100-fold decrease in the H^+ activity.

Plants often differ in their response when grown on soils with extreme pH values. Native plants that have adapted on acid soils are referred to as calcifuges while those that have adapted on alkaline soils are called calcicoles (Lee, 1998). Most crop plants are neither and grow optimally in soils that are near neutral. Therefore, these extreme soils can provide challenges for the grower as the plant often suffers nutritional disorders and toxicities should the

---

* Corresponding author: srgrattan@ucdavis.edu

soil not be adjusted to a more optimal pH range. These extreme soils (5.5 > pH > 8.5) are the focus of this chapter.

8.2 Characteristics of Acid and Alkaline Soils

Although the parent rock material has a strong influence, the distribution of acid soils and alkaline soils throughout the globe are largely correlated with climate patterns and biomes. Acid soils cover about 30% of the world's ice-free land, most of which is forested soils, where only 4.5% of that is used for arable crops while even a smaller fraction is used for perennial tropical crops (von Uexküll and Mutert, 1995). Many acid soils are characterized by highly leached environments ranging from tropical rain forests to temperate climates such as those found in South America, Canada, Central Africa, northern Eurasia, southern Asia and parts of Australia (FAO, 2007). In these areas, soils weather readily and much of the basic cations (Ca^{2+}, Mg^{2+} and K^+) are leached from the profile leaving behind more stable materials rich in Fe and Al oxides and higher concentrations of acidic cations (H^+, Al^{3+}). Highly leached soils in tropical climates are usually classified as ultisols, andisols and oxisols while acid soils that dominate cold and temperate climates in the northern belt are characterized as spodisols, alfiisols, inceptisols and histosols (von Uexküll and Mutert, 1995). Because acid soils occur in both mineral and organic soils, the question has been pondered for decades whether poor plant response is attributed directly to H^+ toxicity, Al^{3+} and/or Mn^{2+} toxicities, nutrient deficiencies or a combination (Arnon and Johnson, 1942). In most cases it is likely to be a combination of factors, particularly Al^{3+} toxicity and nutrient deficiencies; however, some argue that, in highly acidic organic soils where absence of Al^{3+} allows solution pH to drop precipitously, H^+ is the proximal cause (Kidd and Proctor, 2001). Most of the acid soils that are important from a horticultural perspective are those in the tropical regions.

At the other extreme are the arid and semi-arid soils that are generally alkaline and most often calcareous in nature (FAO, 2007). These are soils that are commonly found in climates where the evaporative demand exceeds the annual precipitation. They are prevalent in the Middle East, south-western USA, parts of South America, the Mediterranean areas of Europe and large parts of Asia, Africa and Australia. Many are classified as aridisols or entisols but there are other soil orders that have alkaline pH.

Both extremes – highly acidic and alkaline soils – are very important to global food production since many of these soils produce much of the crops we consume. A diagram that describes the particular nature and properties of acid and alkaline soils is provided in Fig. 8.1. An understanding of the mechanisms of nutrient solubility, plant availability, ion toxicities and ion interactions is critical for optimizing crop production in these challenging soils.

8.2.1 Acid soils

Acid soils are characterized by having a soil-solution pH below neutrality (pH < 7); but strongly acidic soils, those with a pH < 5.5, can be detrimental to the plant either by imposing nutrient deficiencies or ion toxicities (Marschner, 1995). In highly acidic mineral soils, the activity of the H^+ ion is very high, which in itself can be toxic to plants and affect ion transport mechanisms across cell membranes such as membrane-bound ATPases. At the same time, acid soils increase the solubility of Mn^{2+} and Al^{3+} leading to plant toxicity (Fig. 8.1).

Aluminium and manganese toxicity harm the plant differently. Al^{3+} toxicity damages roots, adversely affecting their development, which in turn limits nutrient and water uptake (Rout et al., 2001). Although Al^{3+} toxicity varies among plants, its toxicity to roots results in damage to the apical meristems, similar to the effect of Ca^{2+} deficiency, giving the root system a stubby appearance (Delhaize and Ryan, 1995). As pH of the soil solution decreases, Al-hydroxide species are deceased with a corresponding increase in the free Al^{3+} species. The mechanism of Al^{3+} toxicity in plants is unclear but it may involve some interaction with Ca^{2+} since apoplastic Ca^{2+} is displaced by Al^{3+} and inhibits its uptake (Delhaize and Ryan, 1995). Al^{3+} has over a 500-fold affinity for certain phospholipids over Ca^{2+}, thereby potentially affecting membrane function (Marschner, 1995).

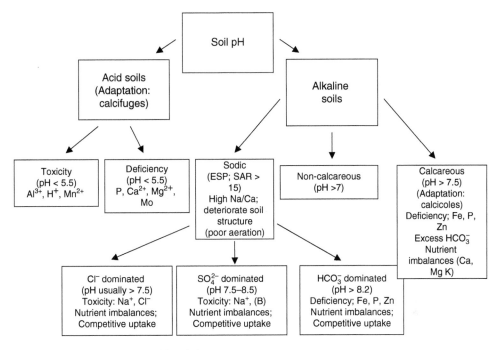

Fig. 8.1. Soils with non-optimal pH and plant responses.

While the actual mechanisms regarding Al^{3+} toxicity to plants are poorly understood, they involve exclusionary mechanisms as well as internal tolerance mechanisms. Rout *et al.* (2001) also point out that differential tolerance is likely to involve differences in the structure and function of roots and root cells. For example, transport proteins embedded in cell membranes that exude organic anions, responsible for sequestering free Al^{3+} to form non-toxic Al complexes, have been key to understanding Al-tolerance among cereal crops (Schroeder *et al.*, 2013). Therefore, genetic engineering of plants with more anion-exuding, membrane-bound transport proteins could improve Al^{3+} tolerance.

Mn^{2+} toxicity, on the other hand, is controlled by the scion's ability to control this trace metal from accumulating in leaves (Marschner, 1995). Interestingly, the presence of Mg^{2+} has been shown to increase the tolerance to high manganese in shoot tissue of wheat and increased the plant's ability to discriminate against Mn^{2+} ions (Goss and Carvalho, 1992). Therefore, both Al^{3+} and Mn^{2+} toxicity can be alleviated by increasing the soil-solution pH and perhaps by addition of calcium and magnesium, respectively.

Some plants, 'the calcifuges', have evolved in these extreme acid environments. These plants tolerate low levels of Ca^{2+} and high acidity. Clearly they have adapted a mechanism to Al^{3+} tolerance but adaptation can occur either to 'tolerance of stress' or 'avoidance of stress' (Marschner, 1995). At the same time, these tolerant species must have a mechanism allowing for highly efficient uptake and utilization of Ca^{2+}, Mg^{2+} and P, which occur in very low concentration in these soils. However, the vast majority of plants cannot tolerate these acidic conditions and would require soil amendments to increase soil pH and nutrient concentration in order to perform well.

Acid soils are also often deficient in Ca^{2+}, Mg^{2+}, P and Mo (Marschner, 1995) (Fig. 8.1). With the high activity of H^+ in the soil solution, the loading of the divalent cations into the apoplast of root cortical cells decreases. Moreover, acid soils are characteristic of highly leached soils where these bases have been naturally depleted from the soils, thereby aggravating the deficiency.

8.2.2 Alkaline soils

Alkaline soils are those having a soil-solution pH above neutrality (pH > 7). However, alkaline (or basic) soils can often be confused with those characterized as calcareous, sodic or even saline soils even though the vast majority of these soils are also alkaline (Fig. 8.1). A short description of the definitions of each follows.

Calcareous soils are those containing sufficient free $CaCO_3$ and other carbonates, such as dolomite, to effervesce visibly or audibly when treated with cold 0.1M HCl (SSSA, 2011). These soils usually contain from 10 to nearly 1000g kg^{-1} $CaCO_3$ equivalent. Therefore, the chemistry of these soils reflects calcium carbonate reactions. This does not imply that non-calcareous soils do not contain $CaCO_3$ and other carbonates. The quantities that do exist in these non-calcareous soils are just not sufficient to observe effervescence. Because of the high amounts of $CaCO_3$, soil-solution pH values of calcareous soils almost always fall in the alkaline range due to equilibrium reactions between calcium and bicarbonate in the presence of CO_2 (D. Suarez, Director of USDA–ARS Salinity Laboratory, personal communication, 2011).

There is a clear distinction between salinity and sodicity (Läuchli and Grattan, 2012). Saline soils are related to the salt concentration and sodic soils to the salt composition where soils are dominated by Na$^+$ over Ca^{2+} and Mg^{2+}. Salinity refers to the concentration of salts in the irrigation water or soil that is sufficiently high to adversely affect crop yields or crop quality. This is based strictly on the colligative property of the soil solution regardless of its ionic composition. These adverse responses are caused by high concentrations of salts that lower the osmotic potential of the soil solution (i.e. osmotic effects) or by high concentrations of specific ions such as chloride or sodium that can cause specific injury to the crop (i.e. specific-ion effects). Sodicity, on the other hand, is related to the proportion of sodium in the water, or adsorbed to the soil surface, relative to calcium and magnesium. High ratios can contribute to the deterioration of soil physical properties, which can indirectly affect plants via crusting, reduced infiltration, increased soil strength and reduced aeration resulting in anoxic or hypoxic conditions for roots.

Sodicity has been described in different ways (Jurinak and Suarez, 1990). The sodicity of soil is characterized by the percentage of the ESP occupied by sodium. The sodicity of the water, on the other hand, is a measure of the sodium adsorption ratio (SAR). SAR is defined as:

$$SAR = \frac{Na^+}{\sqrt{(Ca^{2+} + Mg^{2+})}} \qquad (8.1)$$

where concentrations of all cations are molarities. The ESP and SAR are related to one another and for most practical purposes are numerically equivalent in the range of 3–30 (Richards, 1954).

Sodic soils are often alkaline but the degree of alkalinity can vary depending upon the mineralogy and accompanying anion. Some sodic soils are dominated by chloride (such as many of those in coastal environments), others by sulfate (such as California's San Joaquin Valley), and even others by bicarbonate (such as those in parts of India). Although all of these soil types are alkaline, those dominated by bicarbonate are typically the most alkaline. The higher the alkalinity, the greater the challenge in terms of adequate mineral nutrition. The high Na/Ca ratio in sodic soils not only affects ion uptake and interactions at the membrane level but also adversely affects soil structure, which has secondary effects on plants (lower infiltration rates, poor aeration, higher soil strength, etc.). SAR values less than 5 are recommended to maintain soil physical conditions and ensure optimal water-infiltration rates (D. Suarez, Director, USDA–ARS Salinity Laboratory, personal communication, 2016).

The term 'alkali soil', a soil previously described with a pH of 8.5 or higher or with an exchangeable sodium ratio greater than 0.15, is no longer used in SSSA publications (SSSA, 2011). A 'saline–alkali soil' has a combination of harmful quantities of salts and either a high alkalinity or high content of exchangeable sodium, or both, so distributed in the profile that the growth of most crop plants is reduced. Likewise, the term 'saline–alkali soil' is no longer used in SSSA publications.

8.3 Non-optimal pH: Nutrient Availability

Nutrients are not equally available to plants across the pH spectrum but can vary dramatically as the pH increases from acidic to basic conditions. The general nutrient solubility and availability has been represented in a number of figures over the years. One of the many variations of similar figures found in the literature is represented in Fig. 8.2. In this figure the bandwidth of each of the nutrients represented in the figure indicates the relative availability for that particular nutrient in the soil solution (Lucas and Davis, 1961 as illustrated by Epstein and Bloom, 2005).

The optimal availability for most of the nutrients is found in the slightly acidic pH range, but individual nutrient availability varies with pH (Fig. 8.2). Sulfur, calcium and molybdenum are somewhat similar in that the availability is relatively less in acid environments but increases with increasing pH. Conversely, zinc is more soluble in highly acidic conditions but its availability decreases with increasing pH. Iron and potassium have optimal availability over a wide pH range but are less available under both acidic and basic conditions. Others such as boron, manganese and phosphorus are similar in that availability is minimal under slightly alkaline conditions (e.g. pH 7–8) but the availability is greater in both higher and lower pH solutions. Nitrogen and pH availability is a bit more complicated since availability depends upon the predominance of NO_3^- and NH_4^+ species in the soil solution and microbial reactions responsible for nitrification and N fixation; their activity is also pH dependent. However, availability is generally highest in slightly acid to slightly alkaline pH

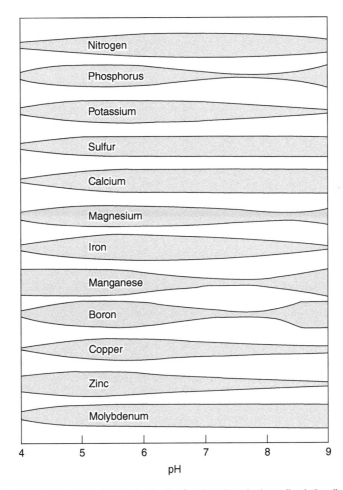

Fig. 8.2. The effect of pH on the availability to plants of various ions in the soil solution (Lucas and Davis, 1961 as illustrated by Epstein and Bloom, 2005).

ranges. Therefore, there is not one specific pH range where the availability of all nutrients is optimal, but the slightly acidic range provides the greatest availability for most of the mineral nutrient elements.

The pH of the soil solution varies not only in space, but with time as well, as do other soil characteristics. As a result, plants can acquire or be subjected to deprivation of nutrients depending upon time of year or specific location in the plant root zone. The soil chemical, physical and biological characteristics are rarely homogeneous in the root zone and therefore root growth and nutrient acquisition occurs in the profile under the most optimal conditions. The partial pressure of CO_2 in the soil air spaces can be many times that of atmospheric CO_2, especially where microbial activity and respiration is high (D. Suarez, USDA–ARS Salinity Laboratory, 2011, personal communication). This elevated CO_2 forms carbonic acid in the soil solution, lowering the pH below what it would be in equilibrium with atmospheric CO_2. At certain times of the year, soil temperature and water content will vary and these conditions affect microbial activity and nutrient availability, particularly if there are short durations of saturated soil when the Eh (reduction potential or the activity of electrons) of the soil solution changes, affecting the solubility and availability of oxygen or certain nutrients. Therefore, the plant may acquire particular nutrients from specific parts of the root zone or during certain times of the years where conditions are most favourable.

Although the pH of the soil solution has either a direct or indirect influence on nutrient availability, there are a several mineral nutrients that are severely affected, particularly in agricultural environments. Those particularly affected are Ca, K, P and Fe and these will be discussed separately below.

8.3.1 Calcium (Ca) nutrition in relation to non-optimal soil pH

Calcium is an important macronutrient element with a multitude of functions as a structural cell wall component, a second messenger in signal transduction, an enzyme activator, and mineralized as calcium carbonate and calcium oxalate

(see Reddy and Reddy, 2002; White and Broadley, 2003; Hirschi, 2004; Epstein and Bloom, 2005; Demidchik, 2012). In general, calcium is readily available to most plants over a wide range in pH. It is a component of many primary and secondary minerals in soils but its availability to plants occurs only in the soluble Ca^{2+} form. In both highly acidic and alkaline soils, the free Ca^{2+} form is often low, but for different reasons. At low pH, most soils are Ca deficient due to the leaching of basic cations from the soil. Also in acid soils, there is a reduction in loading of polyvalent cations in the apoplasm of root cortical cells, not only Ca^{2+} but also Mg^{2+}, Zn^{2+} and Mn^{2+} (Marschner, 1995). In addition, with decreasing pH into the acidic range, uptake of cations (including Ca^{2+}) by plant roots is inhibited because of the impairment of net extrusion of H^+ by the plasma membrane-bound H^+-ATPase (Marschner, 1995). At high pH, on the other hand, the total calcium content in most alkaline soils is generally very high but the free Ca^{2+} is low due to formation of insoluble Ca-based minerals such as calcium phosphates and carbonates. However, the relative availability in alkaline soil shown in Fig. 8.2 indicates that it is readily available, which may not be the case. For these different reasons, both highly acidic and highly alkaline soils can be Ca deficient. In both cases, Ca^{2+} deficiency expressed in apical meristems and young leaves can occur, leading to distortions in growing tips and stunted growth (Epstein and Bloom, 2005).

Calcium deficiency can occur in alkaline conditions where soils are also saline–sodic. Sodium ions have been shown to cause disturbances in calcium nutrition (Läuchli and Grattan, 2012). Nutritional disorders involving other elements may be linked to the effects of salinity on the transport and metabolism of calcium. Plant roots respond to external Na^+ within seconds (Knight et al., 1997). High external Ca^{2+} concentrations may mitigate the effects of saline–sodic conditions. Inadequate concentrations of Ca^{2+} may adversely affect membrane function and growth within minutes of Na^+ application (Epstein, 1961; Läuchli and Epstein, 1970; Cramer et al., 1988). Different genotypes may have widely different responses. Examples of salinity–calcium interactions can be found in a review chapter by Läuchli and Grattan (2012).

On a cellular level, Ca^{2+} is involved in a signalling network which regulates the Na^+

concentration in the cytosol in response to the external Na$^+$ level (Hirschi, 2004). In this signalling network Ca^{2+} is indirectly involved in the regulation of a Na$^+$/H$^+$-antiporter at the plasma membrane which mediates Na$^+$ extrusion from the cytoplasm to the outside of the cell. This Na$^+$/H$^+$-antiporter is likely to be influenced by the external pH in the substrate. Detailed reviews of the role of Ca^{2+} in the regulation of salinity stress have been presented by Maathuis (2007) and Läuchli and Grattan (2012).

There is also evidence that Ca^{2+} plays an important role in reactive oxygen species (ROS) signalling. Research has shown that ROS signalling is based on a Ca^{2+}-mediated mechanism allowing the plant cells to interpret a ROS-encoded message (Demidchik, 2012). However, little is currently known about the specific reactions of this mechanism.

The correction of calcium disorders can readily be made in both Ca^{2+} deficient acid and alkaline soils. This can be done by applying direct calcium suppliers or indirect calcium suppliers (Hanson et al., 2006). In acid soils adding inexpensive calcium supplements (direct calcium suppliers) such as lime or gypsum provides a readily available source of Ca^{2+}. The lime has benefits in that it will dissolve in the acid soil, raise the pH and form free Ca^{2+}. Another amendment may be calcium nitrate when both nutrients (calcium and nitrogen) are in short supply. In alkaline soils, acidifying amendments (indirect calcium suppliers) are often the preferred choice. Here these amendments form acid and solubilize calcium minerals to liberate free Ca^{2+} in the soil solution. There are a number of acidifying amendments, including strong acids such as sulfuric acid or urea sulfuric acid. Other amendments such as elemental sulfur and lime sulfur gradually acidify the soil via microbial transformations. Both the addition of lime to acid soils and acidifying amendments to alkaline soils have been effective strategies for correcting non-optimal soil pH environments to more favourable pH conditions.

8.3.2 Potassium (K) nutrition in relation to non-optimal soil pH

Potassium, an important macronutrient element, is a dominant cellular osmoticum and plays an essential role in facilitating turgor-driven solute transport processes in the plant (Marschner, 1995). It is also vital for several other processes such as enzyme activation, protein synthesis, participation in membrane transport, neutralization of anions, maintenance of osmotic potential and sustaining the overall water relations in the plant (Hsiao and Läuchli, 1986; Epstein and Bloom, 2005). With regard to water relations in the plant, the specific role of K$^+$ in the opening of stomata via guard cell turgor regulation deserves special mention (see Humble and Hsiao, 1969; Talbott and Zeiger, 1996; Talbott and Zeiger, 1998). In addition, K$^+$ is significant in plant salt tolerance as the maintenance of a high cytosolic K$^+$/Na$^+$ ratio appears to be a critical component of salt tolerance (Shabala and Cuin, 2007; Maathuis, 2014).

Potassium concentrations are diminished in soil solutions that are either highly acidic or highly alkaline (Fig. 8.2), potentially leading to K$^+$ deficiency in plants. Under acidic conditions, most of the major cations in the soil solution are in very low concentrations, including K$^+$ (Marschner, 1995). However, unlike Ca^{2+}, lower pH has less influence on loading K$^+$ into root cortical cells (i.e. less competition due to high Mn^{2+} and Al^{3+}). Therefore, at low pH, K$^+$ uptake and transport is not as reduced as that of the divalent cations Ca^{2+} and Mg^{2+}. This could potentially increase the K/(Ca + Mg) concentration ratio in shoot tissue, thereby increasing the risk of grass tetany to ruminants grazing on such forages (Marschner, 1995). At high pH, K$^+$ deficiency can also occur. K$^+$ deficiency is particularly prevalent in sodic soils where the Na$^+$/Ca^{2+} ratio in the soil solution is high. Na$^+$ is not only known to compete for K$^+$ uptake, but higher levels of K$^+$ in the tissue are often required for optimal growth (Grattan and Grieve, 1999). Therefore, K$^+$ deficiency can occur in non-optimal pH conditions, in soil solutions that are either too acidic or too alkaline.

8.3.3 Phosphorus (P) nutrition in relation to non-optimal soil pH

The availability of soil P to plants also depends on the pH of the soil. Soils with a pH between about 7 and 8 are low in available P and can induce P deficiency in plants (Fig. 8.2). Alkaline soils, both calcareous and non-calcareous, fall in this low P availability category and, in addition, acid soils with a pH < 5.5 also induce P deficiency. The main characteristics of acid versus alkaline soils have been described earlier in this chapter (see Fig. 8.1). In this section the focus is on key responses and adaptations of plants (primarily crop plants) to non-optimal pH soils, the emphasis being on P. There is extensive literature coverage of this topic, and the reader is referred, among others, to the following book chapters and reviews (Marschner, 1995; Smith, 2001; Ramírez-Rodríguez et al., 2005; Lynch and Brown, 2006).

Low soil P availability is a primary constraint to plant growth on Earth (Lynch and Brown, 2006). In general, P deficiency in plants growing in alkaline soils is caused primarily by very low total available P and low soil moisture levels, when the mobility of P is limited and root growth is restricted (Marschner, 1995). The P in the soil solution that is plant-available occurs as phosphate (P_i), and concentrations are usually less than 1 μM (Reisenauer, 1966). The overall plant response to P deficiency is reduced shoot growth rate (Lynch et al., 1991), which manifests itself by a change in the carbon partitioning, causing an increase in root/shoot ratio that provides a relatively larger root system for more effective P acquisition from the soil (Smith, 2001).

Plants express an array of phenotypic traits that improve acclimation (in an individual plant) or adaptation (over many generations) to low P availability. These traits can be expressed as morphological or metabolic/biochemical changes. Morphological changes are exhibited by increased biomass allocation to the roots and to specific root classes, and by root architectural traits that enhance topsoil foraging, such as adventitious rooting, lateral root branching, proliferation of finer roots and root hairs, mycorrhizal symbioses and others (Lynch and Brown, 2006). Root hairs have been demonstrated to be involved in P uptake and there is direct evidence for such root hair involvement (Gahoonia and Nielsen, 1998). Moreover, mutants of barley lacking root hairs were shown to have impaired P uptake (Gahoonia and Nielsen, 2003). Metabolic changes are directed towards reducing metabolic costs of soil exploration via formation of cortical aerenchyma, altered respiratory pathways and

P solubilizing root exudates such as carboxylates. All these traits contribute to root growth as a key trait for P efficiency. The latter is defined as little or no reduction in plant productivity under low P compared with high P availability (Lynch, 1995). The considerable variety of adaptive responses to low P availability for both adapted and non-adapted genotypes is summarized in Fig. 8.3. More detailed reviews cover plant responses to low P availability, emphasizing cellular and biochemical processes (Abel et al., 2002) and specific adaptations such as cluster roots (Neumann and Martinoia, 2002) and mycorrhizas (Smith et al., 2003). Cluster roots are specialized root structures common in Proteaceae and some legumes. These structures are composed of dense clusters of determinate lateral roots through which organic acids are released, thus increasing the availability of soil P in the localized regions of the cluster (Gardner et al., 1983).

The components of root respiration show dramatic changes in roots of plants that are sensitive to low P availability, as for example in common bean (Table 8.1). Total respiration can be categorized into three components: growth respiration, ion uptake respiration and maintenance respiration. The latter component supports the maintenance of existing tissue, which is crucial for plant survival under stress. In contrast, growth respiration supports growth of new tissues, and ion uptake respiration drives uptake and assimilation of ions and nutrients (Lynch and Brown, 2006). In bean grown at high P level, approximately 50% of total root respiration was in support of maintenance and about 30% was available to support growth. However, when grown for 1 month at a low P level, maintenance respiration in the root increased to about 90% of total respiration and very little respiratory energy was available to support growth and ion uptake (Table 8.1). Thus, respiratory changes are important phenotypic traits in plant response to low P availability, since they influence the metabolic costs of growth and nutrient uptake by the root. Interestingly, a P-efficient common bean genotype had a 50% lower maintenance respiration in the root than a less P-efficient genotype (Lynch and Ho, 2004). Reduced maintenance respiration under P stress can make more fixed carbon available for root growth.

As is shown in Table 8.1, ion uptake respiration in the root was very low when bean plants

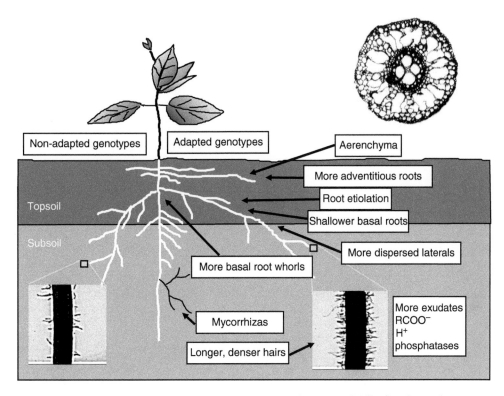

Fig. 8.3. Adaptive responses of root systems to low soil phosphorous availability (Lynch, 2011).

Table 8.1. Root respiration under low phosphorus availability in common bean (from Lynch and Brown, 2006).

		Per cent of total root respiration		
Days after planting	P level	Growth respiration	Ion uptake respiration	Maintenance respiration
14	High	29	14	57
28	High	25	11	64
14	Low	19	9	72
28	Low	6	4	89

were grown at low P level. Not surprising, net P uptake can be severely curtailed at low ambient P levels. Net P uptake consists of two components: P influx and P efflux (Elliott *et al.*, 1984; Smith, 2001). The contribution of P efflux to net uptake is variable but can be significant, even at adequate external P concentrations (Elliott *et al.*, 1984). In severely P-stressed plants, efflux of P can be similar to P influx, resulting in only negligible net P uptake (Bieleski and Läuchli, 1992).

As has been discussed above, the complex responses and adaptations of plants to non-optimal soil pH pose a major challenge to efforts for enhancing P efficiency in crop plants. Yet, breeding crops for low-fertility soils, particularly soils that have low P availability, is likely to become a priority for plant scientists in this century, as increasing malnutrition in many parts of the world makes a second green revolution based on crops tolerant of low soil fertility a high priority (Lynch, 2007).

8.3.4 Iron (Fe) nutrition in relation to non-optimal soil pH

The pH of the soil has a profound influence on the availability of Fe to plants (see Fig. 8.2). Whereas the optimum soil pH for the availability of this important micronutrient element is about pH 5.5–6.5, the high pH of calcareous soils (pH > 7.5) can cause Fe deficiency in many plant species. Under some conditions where acidic soils are saturated, Fe toxicity can also occur. However, Fe toxicity is not very common and therefore, the focus of this section is on responses and adaptations of plants to low Fe availability. For coverage of the whole spectrum from Fe deficiency to Fe toxicity and tolerance, the interested reader is referred to the reviews by Marschner (1995), Salt (2001), Fodor (2002), Lindberg and Greger (2002), Ramírez-Rodríguez et al. (2005) and Liphadzi and Kirkham (2006).

Fe deficiency is a worldwide problem in crop production, particularly in plant species grown on calcareous soils with low Fe availability (Vose, 1982). Among crop species there are genotypical differences in response to low Fe availability. These different responses are primarily related to differences in the ability of the plant to mobilize Fe in the rhizosphere. The solubility of Fe in well-aerated soils is controlled by the dissolution and precipitation of ferric oxides according to the reaction:

$$Fe(OH)_3 + 3H^+ = Fe^{3+} + 3H_2O \qquad (8.2)$$

Hence, the Fe^{3+} concentration depends on the soil pH. Increasing the pH from 4 to 8 causes the Fe^{3+} concentration to decrease from 10^{-8} to 10^{-20} M. Fe^{3+} hydrolyses readily in aqueous media to produce a series of hydrolysis products, mainly $Fe(OH)_2^+$, $Fe(OH)_3$ and $Fe(OH)_4^-$. The sum of the hydrolysis products plus Fe^{3+} gives the total soluble inorganic Fe. According to Fig. 8.4, the minimum solubility of total inorganic Fe is attained in the pH range of 7.4–8.5 and is fixed at 10^{-11} to 10^{-10} M in most well-aerated soils (Lindsay and Schwab, 1982). The actual measured Fe concentrations in soil solutions are usually in the range of 10^{-8} to 10^{-6} M and therefore higher than the calculated estimates. This higher Fe-solubility is mainly due to soluble organic complexes (chelated Fe). However, in well-aerated soils low in organic matter the Fe concentration in the soil solution is in the range of 10^{-8} to 10^{-7} M and thus is lower than the required amount for adequate plant growth (see Fig. 8.4; Römheld and Marschner, 1986a). The chelated Fe is the dominant form of transport to the root surface and is therefore important for determining Fe uptake. Enhanced Fe^{3+} reduction is one of the mechanisms required to increase Fe availability to plants grown in soils with suboptimal concentrations of chelated Fe in the soil solution. The root–soil interface (rhizosphere) plays a significant role in the actual Fe availability to the plant. Lower pH in the rhizosphere, compared to the bulk soil solution, causes a mobilization of Fe from Fe-oxides at the root surface.

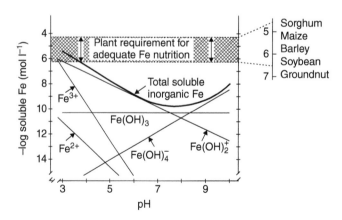

Fig. 8.4. Solubility of inorganic iron species in equilibrium with iron oxides (synthetic Fe(III) chelates) in well aerated soils in comparison with the requirement of soluble iron at the root surface of various plant species (Marschner, 1995).

The uptake of Fe from the soil is diffusion-controlled and is highly dependent on new root growth and development of root hairs (Barber, 1984).

There are several factors that can contribute to Fe deficiency (Fe-chlorosis) in plants: poor soil aeration and high concentration of HCO_3^- in the soil, inefficient Fe uptake by the plant and inhibition of root-to-shoot Fe transport. On the other hand, release of organic acids from the root contributes to Fe-mobilization in the soil and to higher Fe-solubility because of H^+ extrusion. In calcareous soils, high soil moisture and poor soil aeration will aggravate Fe-chlorosis, as gas exchange is impaired and bicarbonate accumulates according to the equation:

$$CaCO_3 + CO_2 + H_2O = Ca^{2+} + 2HCO_3^- \quad (8.3)$$

Bicarbonate impairs the response of roots to low Fe availability and contributes to chlorosis (Römheld and Marschner, 1986a; Marschner, 1995).

Plants evolved two specific mechanisms to mobilize Fe and make it available for uptake by their roots (Römheld and Marschner, 1986a,b; Rogers and Guerinot, 2002). Both mechanisms are enhanced under Fe deficiency. Most researchers in this field labelled these two mechanisms 'strategy I' and 'strategy II' but Epstein and Bloom (2005) argued in favour of 'process I' and 'process II'. Using the latter terminology, process I for Fe absorption is characteristic of dicots such as tomato, soybean and pea and non-graminaceous monocots. In process I, protons are extruded and acidify the rhizosphere. Fe^{3+} is reduced to Fe^{2+} at the plasma membrane of root cells by an inducible Fe^{3+} reductase. Fe^{2+} is then transported across the plasma membrane by an Fe^{2+}-specific transporter. In dicots, Fe is taken up preferentially as Fe^{2+} (Chaney et al., 1972). In process II of Fe absorption, which is expressed in grasses such as maize, barley and oat, phytosiderophores are synthesized and released into the rhizosphere where they complex Fe^{3+}, but no Fe^{3+} reduction to Fe^{2+} takes place. The Fe^{3+}-siderophore complex is then transported across the plasma membrane of root cells. Phytosiderophores are non-proteinogenic amino acids and chemically belong to the mucigenic acids (MA) (Ma and Nomoto, 1996; Mori, 1999). The pathway for the biosynthesis of MAs is known and described in detail by Ramírez-Rodríguez et al. (2005).

Interesting morphological changes occur in roots of plants under Fe deficiency. Well known is the formation of abundant root hairs (Römheld and Marschner, 1981) and the induction of rhizodermal transfer cells (Kramer et al., 1980; Landsberg, 1982). Transfer cells are defined as parenchyma cells with conspicuous ingrowths of secondary cell wall material that protrude into the cell lumina and increase the cell surface and plasma membrane surface area (Ramírez-Rodríguez et al., 2005). Transfer cells can be observed in plant species that respond to Fe deficiency by enhanced H^+ release and acidification of the root medium (process I-type plants). Römheld and Kramer (1983) therefore argued that transfer cells are most likely to be the sites of H^+ efflux pumps in roots of Fe-deficient plants.

In most species the transfer cell-bearing zone is located primarily in root hair regions characterized by mature xylem vessels, thus allowing an efficient long-distance transport of Fe (Landsberg, 1994). From their structural characteristics and anatomical distribution, the formation of transfer cells is now considered a mechanism for increasing the rate of solute transport at the interface of apoplast and symplast (Schmidt, 1999).

As reviewed by Offler et al. (2002), transfer cell induction is not an Fe-deficiency-specific response but has also been observed in response to P deficiency (Landsberg, 1982; Schikora and Schmidt, 2002). Transfer cells may be induced in response to an external stress in cells, presumably as a secondary response to an increased demand for solute transport. Evidence obtained using tomato and Arabidopsis mutants suggests a separate regulation of morphological (transfer cell and root hair formation) and physiological (induction of Fe-reduction activity) responses to Fe deficiency (Schikora and Schmidt, 2002). Induction of transfer cells and root hairs in tomato roots occurred in the absence of adequate Fe. P deficiency induced similar responses (Schikora and Schmidt, 2002). In a split-root study on tomato it was revealed that the frequency of transfer cells was higher in the low P-half of the root, but the density of H^+-ATPase was enhanced only in the high P-half of the split roots. This suggests that formation of transfer cells was directly controlled by the external P_i concentration, while ATPase expression was indirectly regulated by the internal nutrient status of the plant (Schikora and Schmidt, 2002). In contrast to the results for tomato seedlings, Fe- and P-deficient *Arabidopsis* seedlings developed only root hairs but not transfer cells (Schikora and Schmidt, 2001). These observations suggest that transfer cell formation is only one morphological response to low Fe availability.

The role of P in Fe-chlorosis is complex (Marschner, 1995) and not yet well understood. An example for the complexity of P–Fe interactions is the following, reported for a *Banksia* species (Handreck, 1991). In this plant species an increase in P supply suppressed the formation of proteoid roots and simultaneously induced Fe-deficiency chlorosis. Elevated P levels in chlorotic leaves are probably the result of leaf growth inhibition and thus a consequence rather than the cause of Fe-chlorosis (Mengel *et al.*, 1984; Marschner, 1995).

8.4 Boron (B) Availability, Toxicity and pH Interactions

The relative uptake of boron by plants is directly related to the pH of the soil solution, and is also directly related to the fraction of non-dissociated boric acid. For example, as the pH of the soil solution increased, the relative B uptake by barley roots decreased in a pattern similar to the reduction in the fraction of non-dissociated $B(OH)_3$ in the substrate (Oertli and Grgurevic, 1975) (Fig. 8.5). However, the relative uptake function may be dependent on the concentration of the

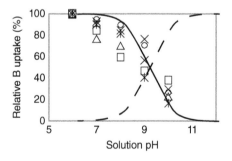

Fig. 8.5. Relative uptake of boron by barley roots in relation to external solution pH. Key for B concentrations in mg l^{-1}: □ 1.0; △ 2.5; × 5.0; ✳ 7.5; and ○ 10.0. The solid line represents the percentage of undissociated $B(OH)_3$ and the broken line represents the percentage of borate $(B(OH)_4^-)$ (modified after Oertli and Grgurevic, 1975; Marschner, 1995).

boron in the external solution. When the data are examined more closely, as the B concentration increased from 1.0 to 10.0 mg/l, the uptake function with increasing pH progressively transformed from a linear reduction to what was more closely related to the speciation of non-dissociated $B(OH)_3$. That is, the higher the boron concentration in the substrate, the closer the relative uptake followed the fractionation of the non-dissociated boric acid. Therefore, there are likely to be several mechanisms responsible for boron uptake such as passive diffusion across the plasmalemma, facilitated transport across intrinsic membrane proteins and energy-dependent transport through a high-affinity uptake system (Dannel *et al.*, 2002; Takano *et al.*, 2008).

Other abiotic stresses may also influence boron uptake. The influence of pH on the combined stresses imposed by salinity and excess boron on the fresh yields of both broccoli and cucumber are presented in Fig. 8.6 (Smith *et al.*, 2005; Smith, 2009). Both experiments were conducted in controlled sand-tank environments where the pH was maintained at slightly acidic (pH 6) or slightly alkaline (pH 8) conditions, and salinity and boron increased from low to high concentrations. There were both similarities and differences in the effects and interaction of these stresses on the head yields of broccoli and the fruit yields of cucumber.

Generally, plants performed better in slightly acidic conditions and increases in both salinity and boron reduced yields, but not in all cases. At high pH, an increase from low (2–3 dS/m) to high (8 dS/m) salinity reduced yields by about half at low boron (< 1 mg/l). However, at high boron, this increase in salinity did not reduce yields as much in cucumber and actually increased yields in broccoli. As salinity is increased further to 14 dS/m (data not shown), broccoli yields declined markedly, forming an inverted parabolic function with increased salinity. In this case, it is likely that two mechanisms are responsible for the low yield observed at both low and high salinity. In slightly acid environments, increased salinity reduced yields but not significantly for broccoli grown at high boron (21 mg/l). Similarly, an increase in boron concentration reduced yields of both crops at low salinity; but, at higher salinity (8 dS/m), increased boron only slightly reduced cucumber yields and did not affect broccoli yields. Tissue boron concentration

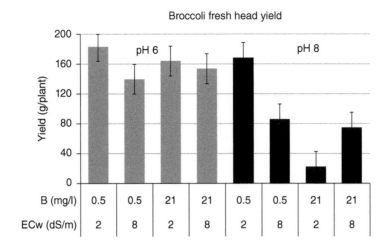

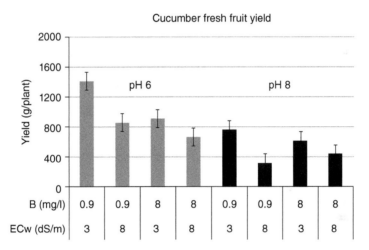

Fig. 8.6. pH–salinity–B interactions: yield responses of broccoli and cucumber (adapted from Smith *et al.*, 2005; Smith, 2009).

was a poor indicator of plant performance but investigators did not examine soluble species in plant tissues, such as those described by Wimmer *et al.* (2002). It may be that certain boron species internally could be more toxic than other species. Clearly work of that nature is warranted to better understand the complexity of the salinity–B–pH interaction.

References

Abel, S., Ticconi, C.A. and Delatorre, C.A. (2002) Phosphate sensing in higher plants. *Physiologia Plantarum* 115, 1–8.

Arnon, D.I. and Johnson, C.M. (1942) Influence of hydrogen ion concentration on the growth of higher plants under controlled conditions. *Plant Physiology* 17(4), 525–539.

Barber, S.A. (1984) *Soil Nutrient Bioavailability: A Mechanistic Approach.* Wiley, New York.

Bieleski, R.L. and Läuchli, A. (1992) Phosphate uptake, efflux and deficiency in the water fern, *Azolla. Plant, Cell and Environment* 15, 665–673.

Chaney, R.L., Brown, J.C. and Tiffin, L.O. (1972) Obligatory reduction of ferric chelates in iron uptake by soybeans. *Plant Physiology* 50, 208–213.

Cramer, G.R., Epstein, E. and Läuchli, A. (1988) Kinetics of root elongation of maize in response to short-term exposure to NaCl and elevated calcium concentration. *Journal of Experimental Botany* 39, 1513–1522.

Dannel, F., Pfeffer, J. and Romheld, V. (2002) Update on boron in higher plants – uptake, primary translocation and compartmentation. *Plant Biology* 4, 193–204.

Delhaize, E. and Ryan, P.R. (1995) Aluminum toxicity and tolerance in plants. *Plant Physiology* 107, 315–321.

Demidchik, V. (2012) Reactive oxygen species and oxidative stress in plants. In: Shabala, S. (ed.) *Plant Stress Physiology*, 1st edn. CAB International, Wallingford, UK, pp. 24–58.

Elliott, G.C., Lynch, J. and Läuchli, A. (1984) Influx and efflux of P in roots of intact maize plants: double-labeling with ^{32}P and ^{33}P. *Plant Physiology* 76, 336–341.

Epstein, E. (1961) The essential role of calcium in selective cation transport by plant cells. *Plant Physiology* 36, 437–444.

Epstein, E. and Bloom, A.J. (2005) *Mineral Nutrition of Plants: Principles and Perspectives*, 2nd edn. Sinauer Associates, Sunderland, Massachusetts.

Fodor, F. (2002) Physiological responses of vascular plants to heavy metals. In: Prasad, M.N.V. and Strzalka, K. (eds) *Physiology and Biochemistry of Metal Toxicity and Tolerance in Plants*. Kluwer Academic Publishers, Dordrecht, The Netherlands, pp. 149–177.

Food and Agricultural Organization (FAO) (2007) *Land Resources*. FAO, Rome. Available at: http://www.fao.org/nr/land/soils/en/ (accessed 14 April 2011).

Gahoonia, T.S. and Nielsen, N.E. (1998) Direct evidence on participation of root hairs in phosphorus (^{32}P) uptake from soil. *Plant and Soil* 198, 147–152.

Gahoonia, T.S. and Nielsen, N.E. (2003) Phosphorus (P) uptake and growth of a root hairless barley mutant (bald root barley, brb) and wild type in low- and high-P soils. *Plant Cell and Environment* 26, 1759–1766.

Gardner, W.K., Barber, D.A. and Parbery, D.G. (1983) The acquisition of phosphorus by *Lupinus albus* L. III. The probable mechanism by which phosphorus movement in the soil/root interface is enhanced. *Plant and Soil* 70, 107–124.

Goss, M.J. and. Carvalho, M.J.G.P.R. (1992) Manganese toxicity: the significance of magnesium for the sensitivity of wheat plants. *Plant and Soil* 139, 91–98.

Grattan, S.R. and Grieve, C.M. (1999) Salinity-mineral nutrient relations in horticultural crops. *Scientia Horticulturae* 78, 127–157.

Handreck, K.A. (1991) Interactions between iron and phosphorus in the nutrition of *Banksia ericifolia* L.f. var. *ericifolia* (Proteaceae) in soil-less potting media. *Australian Journal of Botany* 39, 373–384.

Hanson, B.R., Grattan, S.R. and Fulton, A. (2006) *Agricultural Salinity and Drainage*. Division of Agriculture and Natural Resources Publication 3375. Revised edn. University of California, Oakland, California.

Hirschi, K.D. (2004) The calcium conundrum: both versatile nutrient and specific signal. *Plant Physiology* 136, 2438–2442.

Hsiao, T.C. and Läuchli, A. (1986) Role of potassium in plant-water relations. In: Tinker, B. and Läuchli, A. (eds) *Advances in Plant Nutrition*, Vol. 2. Praeger, New York, pp. 281–312.

Humble, G.D. and Hsiao, T.C. (1969) Specific requirement of potassium for light activated opening of stomata in epidermal strips. *Plant Physiology* 44, 230–234.

Jurinak, J.J and Suarez, D.L. (1990) The chemistry of salt-affected soils and waters. In: Tanji, K.K. (ed.) *Agricultural Salinity Assessment and Management*. ASCE Manuals and Reports on Engineering Practice, No. 71. ASCE, New York, pp. 42–63.

Kidd, P.S. and Proctor, J. (2001) Why plants grow poorly on very acid soils: are ecologists missing the obvious? *Journal of Experimental Botany* 52(357), 791–799.

Knight, H., Trewavas, A.J. and Knight, M.R. (1997) Calcium signaling in *Arabidopsis thaliana* responding to drought and salinity. *Plant Journal* 12, 1067–1078.

Kramer, D., Römheld, V., Landsberg, E.C. and Marschner, H. (1980) Induction of transfer-cell formation by iron deficiency in the root epidermis of *Helianthus annuus* L. *Planta* 147, 335–339.

Landsberg, E.C. (1982) Transfer cell formation in the root epidermis: a perspective for Fe efficiency? *Journal of Plant Nutrition* 5, 415–432.

Landsberg, E.C. (1994) Transfer cell formation in sugar beet roots induced by latent iron deficiency. *Plant and Soil* 165, 197–205.

Läuchli, A. and Epstein, E. (1970) Transport of potassium and rubidium in plant roots: the significance of calcium. *Plant Physiology* 45, 639–641.

Läuchli, A. and Grattan, S.R. (2012) Plant responses to saline and sodic conditions. In: Wallender, W.W. and Tanji, K.K. (eds) *Agricultural Salinity Assessment and Management*. ASCE Manuals and Reports on Engineering Practice, No. 71, 2nd edn. American Society of Civil Engineers, New York, pp. 169–205.

Lee, J.A. (1998) The calcicole – calcifuge problem revisited. *Advances in Botanical Research* 29, 1–30.

Lindberg, S. and Greger, M. (2002) Plant genotypic differences under metal deficient and enriched conditions. In: Prasad, M.N.V. and Strzalka, K. (eds) *Physiology and Biochemistry of Metal Toxicity and Tolerance in Plants*. Kluwer Academic Publishers, Dordrecht, The Netherlands, pp. 357–393.

Lindsay, W.L. and Schwab, A.P. (1982) The chemistry of iron in soils and its availability to plants. *Journal of Plant Nutrition* 5, 821–840.

Liphadzi, M.S. and Kirkham, M.B. (2006) Physiological effects of heavy metals on plant growth and function. In: Huang, B. (ed.) *Plant–Environment Interactions*, 3rd edn. Taylor & Francis, Boca Raton, Florida, pp. 243–269.

Lucas, R.E. and Davis, J.F. (1961) Relationship between pH values of organic soils and availabilities of 12 plant nutrients. *Soil Science* 92, 177–182.

Lynch, J.P. (1995) Root architecture and plant productivity. *Plant Physiology* 109, 7–13.

Lynch, J.P. (2007) Roots of the second green revolution. Turner Review. *Australian Journal of Botany* 55, 493–512.

Lynch, J.P. (2011) Root phenes for enhanced soil exploration and phosphorus acquisition: tools for future crops. *Plant Physiology* 156, 1041–1049.

Lynch, J.P. and Brown, K.M. (2006) Whole plant adaptations to low phosphorus availability. In: Huang, B. (ed.) *Plant–Environment Interactions*, 3rd edn. Taylor & Francis, Boca Raton, Florida, pp. 209–242.

Lynch, J. and Ho, M. (2004) Rhizoeconomics: carbon costs of phosphorus acquisition. *Plant and Soil* 269, 45–56.

Lynch, J., Läuchli, A. and Epstein, E. (1991) Vegetative growth of the common bean in response to phosphorus nutrition. *Crop Science* 31, 380–387.

Ma, J.F. and Nomoto, K. (1996) Effective regulation of iron acquisition in graminaceous plants. The role of mugineic acids as phytosiderophores. *Physiologia Plantarum* 97, 609–617.

Maathuis, F.J.M. (2007) Root signaling in response to drought sand salinity. In: Jenks, M.A., Hasegawa, P.A. and Jain, S.M. (eds) *Advances in Molecular Breeding towards Salinity and Drought Tolerance*. Springer, Dordrecht, The Netherlands, pp. 317–331.

Maathuis, F.J.M. (2014) Sodium in plants: perception, signaling, and regulation of sodium fluxes. *Journal of Experimental Botany* 65, 849–858.

Marschner, H. (1995) *Mineral Nutrition of Higher Plants*, 2nd edn. Academic Press, New York.

Mengel, K., Bubl, W. and Scherer, H.W. (1984) Iron distribution in vine leaves with HCO_3 induced chlorosis. *Journal of Plant Nutrition* 7, 715–724.

Mori, S. (1999) Iron acquisition by plants. *Current Opinion in Plant Biology* 2, 250–253.

Neumann, G. and Martinoia, E. (2002) Cluster roots – an underground adaptation for survival in extreme environments. *Trends in Plant Science* 7, 162–167.

Oertli, J.J. and Grgurevic, E. (1975) The effect of pH on the absorption of boron by excised barley roots. *Agronomy Journal* 67, 278–280.

Offler, C.E., McCurdy, D.W., Patrick, J.W. and Talbot, M.J. (2002) Transfer cells: cells specialized for a special purpose. *Annual Review of Plant Biology* 54, 431–445.

Ramírez-Rodríguez, V., López-Bucio, J. and Herrera-Estrella, L. (2005) Adaptive responses in plants to non-optimal soil pH. In: Jenks, M.A. and Hasegawa, P.M. (eds) *Plant Abiotic Stress*. Blackwell Publishing, Oxford, UK, pp. 145–170.

Reddy, A.J. and Reddy, V.S. (2002) Calcium as a second messenger in stress signal transduction. In: Pessarakli, M. (ed.) *Handbook of Plant and Crop Physiology*, 2nd edn. Dekker, New York, pp. 697–733.

Reisenauer, H.M. (1966) Mineral nutrients in soil solution. In: Altman, P.L., Dittmer, D.S. (eds) *Environmental Biology*. Federation of American Societies for Experimental Biology, Bethesda, Maryland, pp. 507–508.

Richards, L.A. (ed.) (1954) *Diagnosis and Improvement of Saline and Alkali Soils*. Agriculture Handbook No. 60. United States Department of Agriculture, Washington, DC.

Rogers, E.E. and Guerinot, M.L (2002) Iron acquisition in plants. In: Templeton, D.M. (ed.) *Molecular and Cellular Iron Transport*. Dekker, New York, pp. 359–393.

Römheld, V. and Kramer, D (1983) Relationship between proton efflux and rhizodermal transfer cells induced by iron deficiency. *Zeitschrift für Pflanzenphysiologie* 113, 73–83.

Römheld, V. and Marschner, H. (1981) Iron deficiency stress induced morphological and physiological changes in the root tips of sunflower. *Physiologia Plantarum* 53, 354–360.

Römheld, V. and Marschner, H. (1986a) Mobilization of iron in the rhizosphere of different plant species. In: Tinker, B. and Läuchli, A. (eds) *Advances in Plant Nutrition*, Vol. 2. Praeger, New York, pp. 155–204.

Römheld, V. and Marschner, H. (1986b) Evidence for a specific uptake system for iron phytosiderophores in roots of grasses. *Plant Physiology* 80, 175–180.

Rout, G.R., Samantaray, S. and Das, P. (2001) Aluminium toxicity in plants: a review. *Agronomie* 21, 3–21.

Salt, D. (2001) Responses and adaptations of plants to metal stress. In: Hawkesford, M.J. and Buchner, P. (eds) *Molecular Analysis of Plant Adaptation to the Environment*. Kluwer, Dordrecht, The Netherlands, pp. 159–179.

Schikora, A. and Schmidt, W. (2001) Acclimative changes in root epidermal cell fate in response to Fe and P deficiency. A specific role for auxin? *Protoplasma* 218, 67–75.

Schikora, A and Schmidt, W. (2002) Formation of transfer cells and H$^+$-ATPase expression in tomato roots under P and Fe deficiency. *Planta* 215, 304–311.

Schmidt, W. (1999) Mechanisms and regulation of reduction-based uptake in plants. *New Phytologist* 141, 1–26.

Schroeder, J.I., Delhaize, E., Frommer, W.B., Guerinot, M.L., Harrison, M.J., Herrera-Estrella, L., Horie, T., Kochian, L.V., Munns, R., Nishizawa, N.K., Tsay, Y.-F. and Sanders, D. (2013) Using membrane transporters to improve crops for sustainable food production. *Nature* 497, 60–66.

Shabala, S. and Cuin, T.A. (2007) Potassium transport and plant salt tolerance. *Physiologia Plantarum* 133, 651–669.

Smith, F.W. (2001) Plant responses to nutritional stresses. In: Hawkesford, M.J. and Buchner, P. (eds) *Molecular Analysis of Plant Adaptation to the Environment*. Kluwer, Dordrecht, The Netherlands, pp. 249–269.

Smith, S.E., Smith, F.A. and Jakobsen, I. (2003) Mycorrhizal fungi can dominate phosphate supply to plants irrespective of growth responses. *Plant Physiology* 133, 16–20.

Smith, T.E. (2009) Salinity, boron and pH interactions in broccoli: impacts on yield, biomass, water use, boron uptake and tissue ion relations. PhD thesis, University of California, Davis, California.

Smith, T.E., Grieve, C.M., Grattan, S.R., Poss, J.A., Läuchli, A.E. and Suarez, D.L. (2005) Alkalinity enhances B-tolerance in cucumber. *International Salinity Forum: Managing Saline Soils and Water: Science, Technology and Social Issues*. Riverside, California, April 2005, pp. 25–27.

SSSA (Soil Science Society of America) (2011) *Glossary of Soil Science Terms*. SSSA, Madison, Wisconsin. Available at: https://www.soils.org/publications/soils-glossary (accessed 15 December 2015).

Takano, J., Miwa, K. and Fujiwara, T. (2008) Boron transport mechanism: collaboration of channels and transporters. *Trends in Plant Science* 13, 451–457.

Talbott, L.D. and Zeiger, E. (1996) Central roles for potassium and sucrose in guard-cell osmoregulation. *Plant Physiology* 111, 1051–1057.

Talbott, L.D. and Zeiger, E. (1998) The role of sucrose in guard cell osmoregulation. *Journal of Experimental Botany* 49, 329–337.

von Uexküll, H.R and Mutert, E. (1995) Global extent, development and economic impact of acid soils. *Plant and Soil* 171, 1–15.

Vose, P.B. (1982) Iron nutrition in plants: a world overview. *Journal of Plant Nutrition* 5, 233–249.

White, P.J. and Broadley, M.R. (2003) Calcium in plants. *Annals of Botany* 92, 487–511.

Wimmer, M.A., Mühling, K.H., Läuchli, A. Brown, P.H. and Goldbach, H.E. (2002) Boron toxicity: the importance of soluble boron. In: Goldbach, H., Rerkasem, B., Wimmer, M.A., Brown, P.H., Thellier, M. and Bell, R.W. (eds) *Boron in Plant and Animal Nutrition*. Kluwer Academic/Plenum Publishers, New York, pp. 241–253.

9 Desiccation Tolerance

**Jill M. Farrant*, Keren Cooper, Halford J.W. Dace,
Joanne Bentley and Amelia Hilgart**
*Department of Molecular and Cell Biology,
University of Cape Town, Rondebosch, South Africa*

Abstract

Desiccation tolerance is the ability to survive loss of 90% of cellular water or dehydration to tissue water concentrations of ≤ 0.1 g $H_2O.g^{-1}$ dry mass. It is relatively common in reproductive structures such as seeds (termed orthodox), but is rare in vegetative tissues, occurring in some 135 angiosperm species (termed resurrection plants). In this chapter we present an overview of the stresses associated with desiccation and review the current mechanisms proposed to explain how orthodox seeds and resurrection plants tolerate such water loss. Physiological, biochemical and molecular processes involved in protection from mechanical stress, oxidative damage and metabolic disruptions are discussed and similarities between seeds and resurrection plants are drawn. Protective mechanisms unique to vegetative tissues are presented and differences among species are discussed. We review the biogeographical distribution and evolution of angiosperm resurrection plants and propose that the developmentally regulated programme of acquisition of desiccation tolerance in seeds is utilized in the acquisition of tolerance in vegetative tissues of resurrection plants, possibly in response to environmentally regulated rather than developmental cues.

9.1 Introduction

The now commonly held definition of desiccation tolerant (DT) is the ability of an organ or organism to survive loss of more than 90% of its cellular water (corresponding to a tissue water concentration of or below 0.1 g H_2O/g dry mass and a water potential of ≤ -100 MPa) for extended periods and to recover full metabolic competence upon rehydration (Bewley, 1979; Vertucci and Farrant, 1995; Walters et al., 2002). It differs from what is often termed 'drought tolerance' which is essentially the ability to survive periods of limited water availability (and thus lower water potentials of ≤ 1 MPa) either by maintenance of high internal water concentrations

(technically drought resistant) or tolerance of the loss of some water (between 10 and 70%, depending on the species) for short periods only (Höfler et al., 1941; Iljin, 1957; Walters et al., 2002; Moore et al., 2009). The distinction between drought and DT lies in the nature and extent of protection mechanisms implemented by the plant. In the former, these include mechanisms that enable sequestration and maintenance of internal water supplies and/or brief periods of survival in the face of some loss of water. In the latter, these include mechanisms that enable organisms to survive without water, i.e. in an anhydrobiotic state, for prolonged periods.

In the plant kingdom DT is common in reproductive tissues such as spores, pollen and

---

* Corresponding author: jill.farrant@uct.ac.za

seeds (being present in 90% of angiosperm seeds) (reviewed in Berjak *et al.*, 2007; Leprince and Buitink, 2010; Gaff and Oliver, 2013) and there is a relatively high occurrence thereof in vegetative tissues of non-tracheophytes, such as algae (an estimated 35 species from 25 genera), bryophytes (an estimated 25% of species) and lichens (all species) (Kappen and Valadares, 1999;

Oliver *et al.*, 2000, 2005; Alpert, 2006; Proctor *et al.*, 2007; Kranner *et al.*, 2009; Holzinger and Karsten, 2013; Stark *et al.*, 2013; Yobi *et al.*, 2013). However, DT in vegetative tissue is rare in sporophytes of pteridophytes (64 species, or 1% of ferns) and in angiosperms (135 species or 0.04% of flowering plants; Fig. 9.1) and is completely absent in extant gymnosperms (Gaff, 1989;

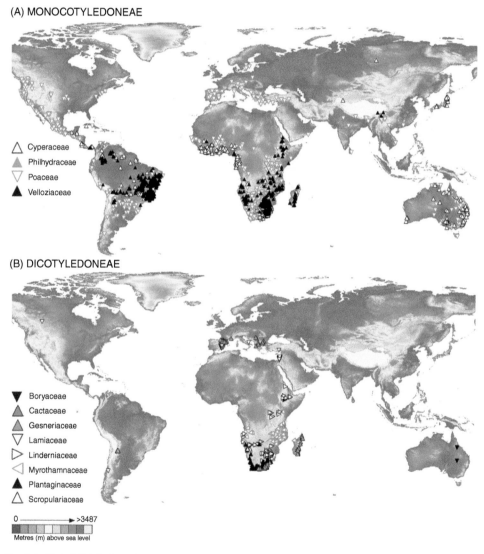

Fig. 9.1. Global distribution of monocotyledonous (A) and dicotyledonous (B) resurrection plants. Georeferenced locality points for most desiccation-tolerant (DT) species within genera, where available, representing the most well-studied DT families, were obtained from the Global Biodiversity Information Facility (www.gbif.org). Using Quantum (Q) GIS (www.qgis.org), the locality points were plotted onto a 2.5 arc-second (corresponding to a resolution of 4 km²) digital elevation model (DEM) obtained from www.worldclim.org.

Alpert and Oliver, 2002; Proctor and Pence, 2002; Farrant, 2007; Gaff and Oliver, 2013). The mechanisms of vegetative DT differ between the lower and higher orders. In the former, desiccation occurs very rapidly and protection prior to drying is minimal and constitutive. Survival is thought to be based largely on rehydration-induced repair processes (Oliver et al., 1998, 2005; Alpert and Oliver, 2002; Yobi et al., 2013). In angiosperms, while some repair is probably inevitable, considerable and complex protection mechanisms are laid down during drying (Gaff, 1989; Farrant, 2007; Oliver, 2007; Moore et al., 2009; Blomstedt et al., 2010; Dinakar and Bartels, 2013; Mitra et al., 2013; Rakić et al., 2014). Because of the relative importance of angiosperms to our agricultural practices, considerable research has been conducted on mechanisms of DT in seeds (termed orthodox) and more recently on vegetative tissues of the few species (commonly called resurrection plants), that exhibit such characteristics. The ability of orthodox seeds to survive desiccation has been exploited by humans to store them for prolonged periods for economic and conservation purposes. Understanding how vegetative tissues of angiosperms survive considerable water loss could be used for biotechnological application in the production of drought-tolerant crops. Indeed, it is becoming increasingly evident that many of the protection systems instituted in vegetative tissues of resurrection plants are similar to those described for orthodox seeds (Illing et al., 2005; Leprince and Buitink, 2010; Suarez Rodriguez et al., 2010; Farrant and Moore, 2011). Thus, in this chapter we will review the stresses associated with water loss; and the physiological, biochemical and – where relevant – molecular responses instituted in resurrection plants and orthodox seeds that are purported to serve as protection against such loss. We include aspects of biogeographical distribution of angiosperm resurrection plants and conclude with an evolutionary perspective of the phenomenon of DT in angiosperms.

9.2 Aspects of Ecology, Biogeography and Evolution of Angiosperm Resurrection Plants

The majority (approximately 90%; Porembski and Barthlott, 2000) of resurrection plants are chasmophytic ('rock-dwelling') inselberg specialists.

The term 'inselberg' describes the solitary, and usually monolithic, mountains or rocky platforms that rise abruptly from the surrounding plains, often disjunct across the landscape (Bornhardt, 1900). The majority of these inselbergs are stable and ancient granitic and gneissic crystalline continental shields, dating back to the Cambrian and Pre-Cambrian era (Porembski et al., 1998).

The inselberg environment is harsh but hosts a variety of habitats and microhabitats for plants that are able to withstand the extreme conditions. These habitats include growth on extremely thin edaphic soils, seasonally water-filled rock pools, shallow depressions and bowls, monocotyledonous mats, ephemeral and wet flush vegetation (Barthlott et al., 1996), as well as rocky crevices on steep slopes or exposed cliffs (Porembski and Barthlott, 2000; Behnke et al., 2013). Inhabitants of inselbergs might endure multiple abiotic stresses, including: a high evaporation rate, intense solar radiation, and frequent and/or extended periods of severe water deficit, as well as a high degree of geographic isolation. All of these factors contribute to their extreme microclimate specialization to niches inhospitable for most vascular plants, resulting in vegetation that is strikingly different to the immediate surrounding plains.

Resurrection plants occupying these habitats might experience a few or all of these stresses, and the selective pressures within the microhabitats influence the particular adaptive characteristics. For example, Chamaegigas intrepidus (Linderniaceae) is a dicotyledonous resurrection plant that grows in ephemeral rock pools in highly seasonally arid Namibia. Its habitat is characterized by extreme dehydration during the dry season lasting up to 11 consecutive months; scorching temperatures and intensive solar radiation; extreme nutrient deficiencies, especially with regard to nitrogen; high diurnal oscillations in the pH of the pool water; and multiple cycles of flooding and drying (Heilmeier and Hartung, 2014). In comparison, the only known example of a succulent resurrection plant, Blossfeldia liliputana (Cactaceae), grows in nearly soilless shaded rock crevices on steep inaccessible slopes along the Andean mountain chain, avoiding direct solar radiation and experiencing frequent nocturnal condensation (Barthlott and Porembski, 1996). Conversely, Vellozia crinita (Velloziaceae) grows on low water-holding capacity milky quartz gravel and sand, known to be poor in carbon and other nutrients, in the Brazilian *campos rupestres* where it exhibits a mound-forming growth strategy and resists high wind speeds (Alves and Kolbek, 2010).

Interestingly, the inselberg is often the one remaining undisturbed ecosystem in an increasingly fragmented landscape disturbed by deforestation and agriculture. Due to this inherently disjunct and isolated nature of inselberg habitats, there is often a high degree of morphological variation within the same species across its range. One example is the mat-forming *Afrotrilepis pilosa* (Cyperaceae), which is present on most inselbergs between Gabon and Senegal across a distribution of more than 3000 km, and is remarkably morphologically divergent across its range (Porembski and Barthlott, 2000).

There are some 13 families of angiosperm resurrection plants, the largest number of species being found in the cosmopolitan families Poaceae and Cyperaceae (Fig. 9.1). Angiosperm resurrection plants are most abundant towards the tropics in south-eastern Africa and Madagascar, eastern South America and western Australia (Gaff, 1977, 1987; Fig. 9.1). Although there is an abundance of suitable inselberg habitats present in Australia, there are notably fewer resurrection species there than there are present in southern Africa. One explanation for this has been that, from a geological perspective, the flora of Australia has been subjected to arid and pluvial cycles only since the Tertiary, while southern Africa has been exposed to these for a much longer period, since the Cretaceous (Lazarides, 1992). Resurrection flora is comparatively poorly represented in North America, and consists mainly of pteridophytes (Kappen and Valladares, 2007), which share their ability to synthesize rehydrins with the more primitive bryophytes (Oliver *et al.*, 2000).

There are also few documented examples of resurrection plants from the East. One example, however, is the Chinese endemic *Acanthochlamys bracteata* (Velloziaceae). Mello-Silva *et al.* (2011) suggest an origin of Velloziaceae in Africa, with the vicariant single-species representative *A. bracteata* in China being the product of the splitting of the Indian plate from Gondwanaland around 115 million years ago (mya). The later splitting of the African and South American plates, around 100 mya, initiated the vicariant distribution of the African-centred *Xerophyta* with South American Velloziaceae (Mello-Silva *et al.*, 2011).

Phylogenetic analyses of the major groups of land plants indicate that vegetative DT is likely to have evolved initially within the primitive bryophytes, being lost later in the tracheophytes and then evolving independently across multiple unrelated lineages, including a 'back-evolution' into ferns and *Selaginella* (Selaginellaceae; Oliver *et al.*, 2000). Oliver *et al.* (2000) suggest that the evolution of land plants from their aquatic environments was crucially dependent on the evolution of vegetative DT, but that this strategy was costly as metabolic rates in DT plants are low compared to desiccation-sensitive (DS) plants. Evolution would have then favoured the advantages offered by increased growth rates (Oliver *et al.*, 2000). Furthermore, they suggest that the trait of DT in seeds may have then evolved secondarily as a product of the more primitive forms of DT.

Within the angiosperms, Oliver *et al.* (2000) estimate that DT evolved eight times, though a later estimate of 13 times is made by Gaff and Oliver (2013) based on the observation that the trait is present in 13 unrelated angiosperm families. They hypothesize that harsh and arid environmental conditions were likely to have triggered the convergent evolution of DT across multiple unrelated families and, in each independent appearance of the trait, the particular adaptive response to dehydration has been modified.

9.3 Challenges Associated With the Study of DT

In seeds, DT is acquired during the mid- to late stages of maturation, being initiated concurrently with the accumulation of complex reserves and being completed just prior to the onset of maturation drying proper. DT is lost during germination, usually coincident with radical elongation. Understanding of metabolic processes associated with DT is thus confounded by metabolism associated with development and dormancy acquisition. Furthermore, the timing of onset and loss of DT can differ among seed tissues, which are rarely studied separately. Experiments in which developmental aspects were separated from DT acquisition in seeds of the genomic model organisms *Medicago truncatula* and *Arabidopsis thaliana*, by allowing germination to the point of radicle protrusion and re-establishing DT by subjecting them to a mild osmotic stress (c.−1.5 MPa) and/or abscisic acid (ABA) treatments, have added considerable 'omic' data on what genes and proteins might be important in the attainment of developmental DT (Buitink *et al.*, 2003, 2006; Boudet *et al.*, 2006; Maia *et al.*, 2011, 2014).

Although such overlapping metabolic processes are not a hindrance in the study of resurrection plants, other drawbacks have limited our attainment as yet of a clear and comprehensive understanding of DT. Due to constraints associated with collection of plants in and transport from remote areas, as well as the difficulty experienced in the propagation of some species (by transplanting, tissue culture or availability of seed), there are frequently limits to the number of biological replicates for research studies. In addition, insufficient attention is often paid to light intensities, day length and temperatures under which plants are subject to dehydration and rehydration events, and time of day in which samples are taken (overlooking possibly the effect of circadian rhythms), making comparisons among species and even within a species difficult. Disparities in drying rate among plants and in similar tissues of the same plants (including old versus young leaves) can cause further challenges in sampling, resulting in large variations in water content and metabolic parameters tested. Furthermore, once water loss is initiated it can occur very rapidly (Farrant, 2007), usually over several hours, making sampling at regular water content intervals at the same time of day (to account for daily metabolic cycles) problematic. In some cases, in order to obtain more even drying rates, researchers excise leaves or leaf discs from a hydrated plant and dry these. While in species such as *Boea hygroscopica* and *Myrothamnus flabellifolia* such tissues can survive at least one cycle of dehydration without the presence of roots (Jiang *et al.*, 2007; Mitra *et al.*, 2013; Ma *et al.*, 2015) this is not a common phenomenon and begs the question: what is the role of roots in establishment and recovery from desiccation in such species? Indeed, in many species (particularly monocots) leaves do not acquire DT unless some drying has occurred while still attached to the roots (Gaff and Loveys, 1992; Whittaker *et al.*, 2004). There is currently a paucity of information on the attainment of DT in roots and their contribution to tolerance of the whole plant, due to difficulties associated with root studies. Sampling of roots

compromises the plant for subsequent measurements in a drying course and adhering soil particles with associated microbes cannot be easily removed since washing of roots changes their water content. Furthermore, roots of many resurrection plants (typified in the *Xerophyta* species) appear to be recalcitrant to extraction procedures conventionally used for plant molecular studies (Kamies *et al.*, 2010; Kamies, 2011; Waters, 2015). A further factor, which can impede comparisons among species worked on in different laboratories, is the choice of method used to report tissue water content. In the seed literature, it is common to calculate absolute water contents gravimetrically, which is then expressed as g H_2O/g dry mass (essentially being a measure of water concentration). In literature pertaining to resurrection plants the relative water content (RWC) is more commonly used, in which absolute water content of the sampled tissue is expressed relative to the maximal water content it can hold at full turgidity. This measure is favoured because it allows comparison among tissue types and species with respect to metabolic events occurring upon loss of a given proportion of total water content. However, it is here that some discrepancies in calculation of RWC can arise. In principle, this is calculated as: RWC (%) = [(W–DW) / (TW–DW)] × 100, where W is the sample fresh weight; TW is the sample turgid weight and DW is the sample dry weight. TW is determined independently at each water content determination. Many authors, ourselves included, calculate TW only on fully turgid tissues immediately prior to the start of any drying down experiment and thus this value reflects RWC normalized to that of the population of plants under study, under the environmental conditions in which those plants are maintained for the duration of the study. In making comparisons among studies, it is thus important to ascertain how the water content measurements were undertaken. A final caution to bear in mind when conducting research on resurrection plants is that of 'cellular memory'. We have observed that, if the time between dry down and recovery experiments on any one plant or population of plants is short, responses to the subsequent dehydration cycle occur more rapidly due possibly to a cellular memory of previous dehydration events. At the opposite extreme, in some species (usually small dicots such as the *Craterostigma* spp. and *Boea hygroscopica*), long-term maintenance of hydrated plants under stress-free conditions can result in the inability of plants to tolerate extreme water loss unless some priming (ABA and/or osmoticum) is first applied (Furini *et al.*, 1997; Phillips *et al.*, 2002; Zhu *et al.*, 2015). Interestingly, such treatments induce similar subcellular events to those observed in the early induction of DT in germinating seedlings of the monocot resurrection plant *Xerophyta viscosa* (Lyall *et al.*, 2014) as well as the re-induction of DT in germinating seedlings of *M. truncatula* and *A. thaliana* (Boudet *et al.*, 2006; Maia *et al.*, 2011, 2014).

DT is a complex multigenic and multifactorial phenotype (Moore *et al.*, 2009; Oliver *et al.*, 2010a; Gechev *et al.*, 2013; Xiao *et al.*, 2015), and gaining a full understanding of this phenomenon requires, at best, an integrative systems biology approach to fully appreciate the adaptive responses to extreme water deficit. This takes into consideration changes in the transcriptome, proteome, metabolome and lipidome, contextualized by input from biochemical, biophysical and physiological studies. To date, this has not been fully achieved for any one species, although considerable information is available for species such as *Boea hygrometrica* (reviewed in Mitra *et al.*, 2013; Xiao *et al.*, 2015; Zhu *et al.*, 2015), *Craterostigma plantagineum* (reviewed in Suarez Rodrigues *et al.*, 2010; Dinakar and Bartels, 2013), *Haberlea rhodopensis* (Moyankova *et al.*, 2014; Djilianov *et al.*, 2015); *Myrothamnus flabellifolia* (Moore *et al.*, 2006, 2007a,b, 2009); *Sporobolus stapfianus* (reviewed in Le *et al.*, 2007; Gaff *et al.*, 2009; Oliver *et al.*, 2011a,b), *Ramonda serbica*, *Ramonda nathaliae* (reviewed in Rakić *et al.*, 2014), *Xerophyta humilis* and *X. viscosa* (reviewed in Collett *et al.*, 2004; Farrant, 2007; Farrant *et al.*, 2015). Most resurrection plants are polyploid (hexaploid or octaploid) with large genomes and sequencing, as well as transformation thereof for functional analysis, has not yet been readily achieved. It is thus not surprising that the only genome reported to date is that of the tetraploid dicotyledonous resurrection plant *B. hygrometrica* (Xiao *et al.*, 2015). While this work will go some way to filling gaps in our current understanding of some of molecular processes underlying DT in this species, it is imperative that draft genomes of other species, monocots in particular, be made available. To this end, we are currently

sequencing the genomes of the octoaploid monocot *X. viscosa*.

It is important to be aware of the above limitations in data interpretation. That having been said, we do have a fair understanding of what enables DT in seeds and vegetative tissues of angiosperm resurrection plants. In this review, we have attempted to better understand the similarities in the desiccation response in seeds and vegetative tissues.

9.4 Stresses Associated with Water Loss

In understanding DT, it is important to recognize the effect of loss of water on physical and metabolic processes. In the light of that we can assess how seeds and resurrection plants are able to prevent, slow down and/or repair the deleterious reactions induced by the removal of water. Figure 9.2 shows the sequence of stresses associated with water deficit as well as the cellular and metabolic responses in orthodox seeds and angiosperm resurrection plants during attainment of DT. The various hydration levels calculated for seeds (Vertucci and Farrant, 1995; Walters *et al.*, 2005; Berjak *et al.*, 2007) and for leaf tissues of a number of resurrection plants (Gaff and Oliver, 2013) have been associated with five hydration levels (V–I), which have been proposed as a tool to analyse the processes involved in DT. The boundaries, however, are not completely distinct. Nevertheless, orthodox seeds and resurrection plants survive the loss of all but some level I water, the degree of tolerance of loss of the latter being species-dependent and can be confounded by conditions under which desiccation occurs and tissues are maintained (Walters *et al.*, 2005; Gaff and Oliver, 2013).

Since water has multiple and various roles in supporting life, it is not surprising that there are numerous stresses associated with its loss. It has a structural role providing mechanical stabilization at the cellular level by filling intracellular spaces resulting in turgor pressure (Iljin, 1957; Levitt, 1980). At the molecular level water provides hydrophobic and hydrophilic associations and controls intermolecular distances that determine the conformation of macromolecules and their partitioning within organelles.

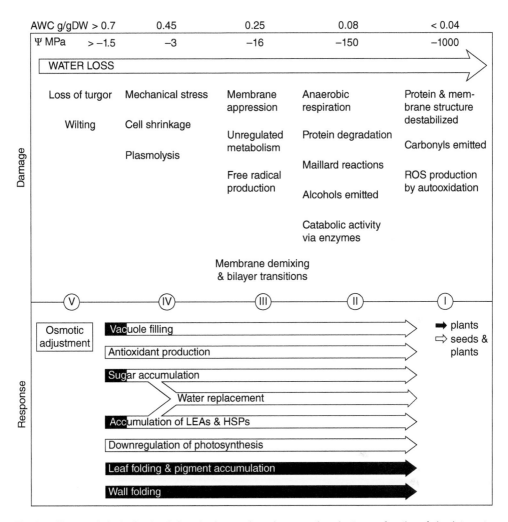

Fig. 9.2. Changes in hydration levels in orthodox seeds and resurrection plants as a function of absolute water content (g.H_2O.g DW^{-1}) and water potentials of tissues (modified from Walters *et al.*, 2005; Berjak *et al.*, 2007; and Gaff and Oliver, 2013). Associated metabolic damage to water loss is shown. Responses of seeds and plants to these stresses are shown, with those common to seeds and plants indicated in white and those specific to resurrection plants in black. HSPs, heat shock proteins; LEAs, late embryogenesis abundant proteins.

Tissues are fully hydrated at the start of hydration and loss of this water results in loss of turgor (Levitt, 1980). Further loss of water within level IV results in cell shrinkage and mechanical stress in which tension is placed on the plasmalemma as it shrinks from plasmodesmatal attachments to the cell wall. The ultimate rupture of the plasmalemma allows entry of extracellular hydrolases and cell death. In many species wall collapse occurs, which is equally lethal (Walters *et al.*, 2002). As water deficit gets progressively worse, loss of the aqueous medium results in membrane appression, demixing and lipid bilayer transitions, protein degradation and destabilization of macromolecules and membrane structures (Vertucci and Farrant, 1995; Walters *et al.*, 2002, 2005).

Water also plays a role in controlling metabolism as it is a reactant and product of many reactions. It provides the fluid matrix that allows diffusion of substances to reactive sites. Its loss profoundly affects the nature of biochemical reactions and thus metabolism. This happens progressively, with free radical production and unregulated metabolism being initiated, followed by anaerobic respiration, catabolic activity, alcohol emission and Maillard reactions, with free radical production via auto-oxidation and emission of carbonyls occurring at low water contents (Vertucci and Farrant, 1995; Walters *et al.*, 2002, 2005). The properties of water also change with its progressive loss from tissues (Vertucci, 1990; Walters *et al.*, 2002, 2005). Level V water behaves as it would in a dilute solution but the aqueous matrix becomes more viscous, having the properties of a syrup (level IV), rubber (level III) and glass (levels II and I) (Walters *et al.*, 2002, and references therein). The stability of the latter may be critical to survival of desiccation and the length of time an organism can remain in the desiccated state (Walters *et al.*, 2005).

9.5 Protection Against Mechanical Stress

9.5.1 Vacuole filling and water replacement

Although osmotic adjustment (to prevent departure of water from the cells/vacuoles) during the initial phase of water loss occurs in most plants (including those that are DS) this is insufficient to prevent the strain related to severe water loss and desiccation.

In seed tissues, progressive storage reserve accumulation within vacuoles and cytoplasm fills the cells with dry matter, minimizing cell shrinkage and plasmalemma withdrawal (Fig. 9.3i, A and B). For most orthodox seeds the timing of maximum dry matter accumulation and acquisition of maximum DT are coincident and, conversely, the loss of DT upon germination is associated with mobilization of these reserves and increased vacuolation within seedling tissues (reviewed in Vertucci and Farrant, 1995; Berjak *et al.*, 2007; Leprince and Buitink, 2010). Re-establishment of DT in germinating radicles of *M. truncatula* and *A. thaliana* resulted in the upregulation of genes involved in sucrose and seed storage reserve production (Buitink *et al.*, 2006; Maia *et al.*, 2011), indicating the importance of such metabolism in DT, most likely in minimizing mechanical stress associated with severe water loss. Seeds that do not acquire DT

during maturation or even post shedding are termed recalcitrant. Most of these do accumulate storage reserves in a similar manner to orthodox seeds, these being required as an energy source for germination, and this presumably can minimize mechanical stresses associated with desiccation. It has been noted that there is a range in degree of reserve accumulation in seeds of such species and that, particularly the extent of vacuolation, is correlated with the amount of water loss tolerated before viability is lost (Berjak *et al.*, 1989; Farrant *et al.*, 1989; Kermode, 1990; Vertucci and Farrant, 1995).

The mechanical stabilization of cells by means of reserve accumulation in seeds shares similarities to the strategy of angiosperm resurrection plants, where water is replaced in cells with compatible solutes (Fig. 9.3i C–F). Replacement of water in vacuoles within dry tissues of resurrection plants was first suggested based on ultrastructural observations that vacuoles continued to take up a large proportion of the cytoplasmic space despite there being no bulk water available in tissues, the remaining water being purely structurally associated (Farrant, 2000; van der Willigen *et al.*, 2001; Farrant *et al.*, 2007; Moore *et al.*, 2007b). Given the accumulation of proteins and solutes observed in numerous studies in resurrection plant tissues, it seemed likely that these vacuoles were being filled with non-aqueous media. The content of vacuoles from desiccated leaves of *Eragrostis nindensis* was analysed after non-aqueous extraction and was shown to contain proline, sucrose and protein in equal proportions (van der Willigen *et al.*, 2004). Vacuoles from both hydrated and dry leaves of *M. flabellifolia* contain 3, 4, 5 tri-o-galloylquinic acid, but this chemical increases to entirely fill the vacuoles in dry leaves (Moore *et al.*, 2005, 2007b; Fig. 9.3D).

9.5.2 Cell wall folding

For some angiosperm resurrection plants, mechanical stabilization is further enhanced by considerable and reversible wall folding (Fig. 9.3ii). The extent of wall folding varies among species however. In *Craterostigma wilmsii* (Fig. 9.3F), where this is almost exclusively used as a form of mechanical stabilization, the mechanism of folding appears to involve both structural and

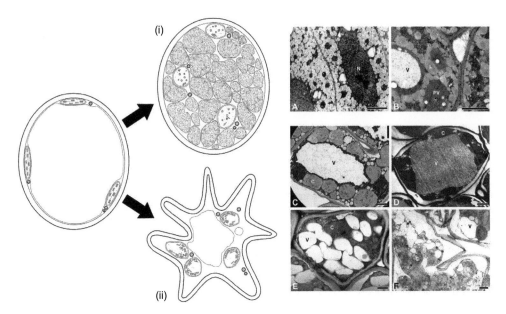

Fig. 9.3. Mechanical stabilization in plant tissues by means of (i) water replacement in vacuoles (vacuole filling); and/or (ii) regulated wall folding. A–F, transmission electron micrographs of embryonic tissues of *Podocarpus falcatus* (A) and *Craterostigma wilmsii* (B) and mesophyll cells from dry leaf tissues of *Mohria caffrorum* (C), *Myrothamnus flabellifolia* (D), *Xerophyta humilis* (E) and *Craterostigma wilmsii* (F). C, chloroplast; N, nucleus; V, vacuole; L, lipid body; PV, protein vacuole. Scale bar equivalent to 2µm.

biochemical changes that are reversed on rehydration (Vicré *et al.*, 1999, 2004). On drying there is a reduction in glucose and an increase in galactose substitutions to the xyloglucan chains and it has been proposed that cleavage, or partial cleavage of the long-chained xyloglucan units into shorter, more flexible ones, allows for wall folding. Furthermore, increased abundance of cell wall expansins in *Craterostigma plantagineum* during initial drying has been proposed to contribute towards wall flexibility (Louise and Simon, 2004). Further water loss results in an increase in wall-associated Ca^{2+} and since this ion plays an important role in cross-linking wall polymers, such as acid pectins, it has been proposed that this serves to stabilize walls in the dry state and, more importantly, prevent mechanical stress of rehydration. *Craterostigma* spp. rehydrate rapidly due to their small size, and initial movement of water is mainly apoplastic (Sherwin and Farrant, 1996). If walls hydrate and unfold before cell volume is regained, plasmalemma tearing and further subcellular damage could occur (reviewed in Vicré *et al.*, 2004). In species such as *E. nindensis*, *M. flabellifolia* and

several of the *Xerophyta* spp., in which some wall folding occurs (vacuole filling being extensive and presumably the predominant means of mechanical stabilization), there appear to be no notable biochemical wall changes upon drying. The leaf cell walls of these plants have constitutively high proportions of arabinose, present as arabino galactan proteins (AGPs) and in association with pectins in dicots and xyloglucans in monocots (Moore *et al.*, 2006, 2012; Plancot *et al.*, 2014). Interestingly, the DS species *Eragrostis tef*, while having similar chemical wall constituents as its DT relative *E. nindensis*, has significantly lower levels of xyloglucan-associated arabinose. In the resurrection fern *Mohria caffrorum* which displays seasonal DT, being tolerant in the dry season but sensitive in the wet season (Farrant *et al.*, 2009), the walls of the tolerant fronds have a higher proportion of AGPs than the sensitive form, and these increase significantly during drying (Moore *et al.*, 2012). Since arabinose polymers are highly mobile, allow wall flexibility (Foster *et al.*, 1996; Renard and Jarvis, 1999) and have a high water-absorbing capacity (Goldberg *et al.*, 1989; Belton, 1997), which would be important for rehydration, we have proposed that such constitutively high levels allow constant preparedness for dehydration–rehydration in these resurrection plants (Moore *et al.*, 2009, 2012).

9.6   Protection Against Metabolic Stress

9.6.1   Formation of reactive oxygen species and toxic compounds, and prevention of oxidative stress

Various metabolic stresses are also initiated with loss of water, and at very low water contents become lethal for DS tissues. These include formation of reactive oxygen species (ROS) and other cytotoxic compounds such as hydrogen peroxide (H_2O_2) and methylglyoxal (MG), which form as a natural consequence of metabolic processes in chloroplasts, mitochondria and peroxisomes (Fig. 9.4; Halliwell and Gutteridge, 1999; Apel and Hirt, 2004; Bailly, 2004; Hussain *et al.*, 2011; review by Das *et al.*, 2015). Interestingly, formation of both MG and H_2O_2 can result in ROS formation. For example, research by Saito *et al.* (2011), which focused on MG as a source of cellular injury during various abiotic stresses, has shown that MG produced in the chloroplast leads to the production of ROS, specifically superoxide.

Peroxisomes are sites of H_2O_2 formation during normal cellular functioning, as a product of the oxidation of glycolate during photorespiration (reviewed by Noctor *et al.*, 2014). In the presence of high ROS concentrations, the formation of nitric oxide (itself an oxidant) in peroxisomes can also lead to the production of reactive nitrogen species (RNS) such as peroxynitrite (del Río, 2015). The action of various oxidases in this organelle can also lead to the formation of ROS (Das *et al.*, 2015), but metabolism involving electron transport in the mitochondria and chloroplasts make these major sites of ROS production (Fig. 9.4). Photosynthesis, in particular, is very sensitive to water deficit. Electron leakage during photosynthetic electron transport and the formation of singlet oxygen are significantly increased when cells of photosynthetic tissues suffer water loss, and this has frequently been

cited as a primary cause of damage and resultant plant death in most species (Seel *et al.*, 1992; Smirnoff, 1993; Kranner and Birtić, 2005).

Many seeds contain photoheterotrophic plastids, the photosynthetic activity of which contributes oxygen and metabolic re-assimilation of CO_2 released by reserve biosynthesis (Rolletschek *et al.*, 2004, 2005; Ruuska *et al.*, 2004) as well as energy metabolism (Borisjuk *et al.*, 2005). This activity is switched off at the onset of maturation drying (Fait *et al.*, 2006; Bewley *et al.*, 2013), probably during the loss of types V and IV water. Whether this is the consequence of drying, or a prerequisite for DT (by minimizing the amount of photosynthetically produced ROS) is not clear. Angiosperm resurrection plants downregulate photosynthesis at water contents of between 80 and 65% RWC depending on the species, minimizing photosynthetically produced ROS (Sherwin and Farrant, 1998; Tuba *et al.*, 1998; Farrant, 2000; van der Willigen *et al.*, 2001; Farrant *et al.*, 2003; Illing *et al.*, 2005; Rapparini *et al.*, 2015). Downregulation of photosynthesis is achieved by one of two mechanisms termed poikilochlorophylly and homoiochlorophylly (Gaff, 1989; Smirnoff, 1993; Sherwin and Farrant, 1998; Tuba *et al.*, 1998; Farrant, 2000).

Poikilochlorophyllous types, many of which are monocots, break down chlorophyll and dismantle thylakoid membranes during dehydration (Tuba *et al.*, 1993a,b, 1998; Farrant, 2000). Breakdown of photosystem II (PSII), which is responsible for the water-splitting, oxygen-evolving and thus oxidizing reactions of photosynthesis, is a highly effective strategy to minimize damaging levels of ROS formation. Indeed, it has been shown that poikilochlorophyllous species are able to retain viability in the dry state for far longer than homoiochlorophyllous ones (Tuba *et al.*, 1998; Proctor and Tuba, 2002). The drawback of this strategy is that reassembly of the photosynthetic apparatus on rehydration requires coordinated transcription and *de novo* translation (Dace *et al.*, 1998; Collett *et al.*, 2003; Ingle *et al.*, 2008) and poikilochlorophyllous plants require longer periods after rehydration to resume normal growth and development. Homoiochlorophyllous species retain most of their chlorophyll (the amount retained depending on the light levels under which the plants are dried) and thylakoid membranes remain intact in the dry state, although it has been observed in certain

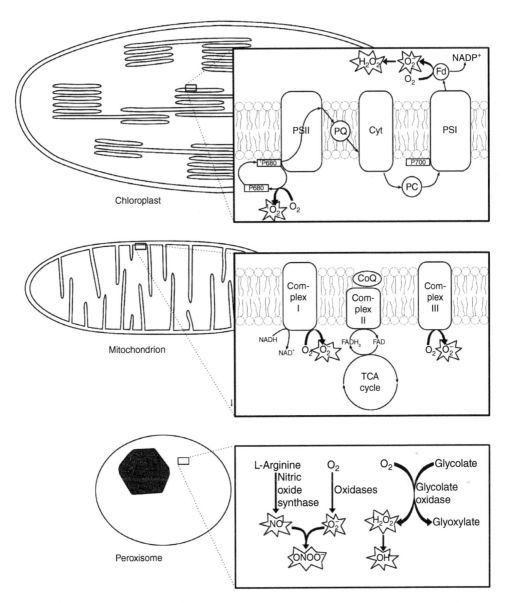

Fig. 9.4. Schematic diagram of reactive oxygen species (ROS) formation as by-products of aerobic metabolism within the chloroplast, mitochondrion and peroxisome. Ground state oxygen is converted to more reactive ROS forms either by energy transfer giving rise to singlet oxygen ($^1O_2$), by electron transfer reactions resulting in the sequential reduction to superoxide (O_2^-) and hydrogen peroxide (H_2O_2) or by the action of oxidases. In the peroxisome, normal metabolism (conversion of glycolate to glyoxylate and production of nitric oxide) leads to formation of H_2O_2 and reactive nitrogen species (RNS).

species that there is partial dismantling or re-organization of the photosynthetic complexes that facilitate reversible photosynthetic shut-down (Sárvári et al., 2014; Charuvi et al., 2015). Despite this, due to the presence of chlorophyll,

ROS production from photoactivated chlorophyll uncoupled from metabolic dissipation mechan-isms is inevitable and so homoiochlorophyllous plants utilize various mechanisms to prevent light–chlorophyll interaction (Sherwin and

Farrant, 1998; Farrant, 2000; Farrant et al., 2003, 2009). This is achieved by leaf folding and shading of inner leaves (e.g. the *Craterostigma* spp.) or adaxial surfaces (e.g. *Myrothamnus flabellifolia, Mohria caffrorum*) from light. Surfaces that remain exposed to light may have reflective hairs and/or waxes and there is an accumulation of antho-cyanin, xanthophyll pigments and polyphenols, all of which act as 'suncreens' reflecting back photosynthetically active light, masking chloro-phyll and acting as antioxidants or mechanisms for dissipation of excess excitation energy (Smirnoff, 1993; Sherwin and Farrant, 1998; Farrant, 2000; Farrant et al., 2003, 2009; Georgieva et al., 2007, 2009; Moore et al., 2007a,b; Jahns and Holz-warth, 2012).

9.6.2 Antioxidant protection against reactive oxygen species and toxic compounds

While these mechanisms minimize ROS produc-tion associated with photosynthesis, ROS pro-duction associated with other metabolic processes is exacerbated with acute water loss. This is ini-tially accompanied by increased production and/or upregulation of activities of what are termed 'classical' (Kranner and Birtić, 2005) or 'housekeeping' antioxidants (Illing et al., 2005), so called because they are present in all plants and are crucial to maintenance of cellular homeostasis under day-to-day conditions and in protection against a myriad of abiotic and biotic stresses (for an overview, see Elstner and Oss-wald, 1994). The difference between desiccation-tolerant and desiccation-sensitive tissues is that the former are able to maintain their antioxidant potential in the dry state such that these same antioxidants can be utilized during the early stages of rehydration, thus protecting against the ROS associated with reconstitution of full metabolism. In DS tissues such antioxidants are compromised by drying below critical water contents (Illing et al., 2005; Kranner et al., 2006; Farrant et al., 2007). Furthermore, desiccation-tolerant tissues utilize antioxidants not reported in DS tissues. Such extensive protection is probably required to effectively cope with ROS production via Maillard and auto-oxidation reactions.

Classical/housekeeping antioxidants in-clude the water-soluble glutathione (g-glutamyl-cysteinylglycine; GSH) and ascorbic acid (Asc) (Noctor and Foyer, 1998); the lipid-soluble to-cochromanols and β-carotene (Munne-Bosch and Alegre, 2002; review by Long et al., 2015); and enzymes such as superoxide dismutase (SOD), catalase (CAT), ascorbate peroxidase (AP) and other peroxidases, mono- and dehydroascorbate reductases and glutathione reductase (GR) (Elst-ner and Osswald, 1994; Bailly, 2004; Kranner and Birtić, 2005). Typically, studies investigat-ing their role in DT follow concentrations of par-ticular antioxidant pools and/or activities of the antioxidant enzymes. Although the antioxidant responses vary greatly among species, research has found either no change in antioxidant levels or increased concentrations and/or activity (Cruz de Carvalho, 2013; Noctor et al., 2014).

Maintenance of antioxidant potential in-volves regeneration and/or *de novo* synthesis of antioxidants, and activities of enzymes in-volved in these processes are also monitored. There is interaction between the various anti-oxidant systems (Foyer and Halliwell, 1976; Bailly, 2004; Kranner and Birtić, 2005) and thus the amount or activity of any one antioxi-dant can be influenced by others. Furthermore, the changes in activity of various antioxidants can fluctuate in space (between tissues and within sub-compartments within cells) and time, as water is lost and DT is attained. Due to this complexity, changes in antioxidant status should be monitored at regular intervals dur-ing drying in order to identify trends and fluctuations in concentrations and enzyme ac-tivities that might be important to the attain-ment of DT. Such rigorous studies are rarely undertaken and consequently there is consid-erable variation in reports from the literature among desiccation-tolerant seeds and resur-rection plants with respect to the nature and extent of upregulation of the various house-keeping antioxidants, and the water contents during a dehydration/rehydration time course at which the observed changes occur (reviewed for example in Farrant, 2000; Farrant et al., 2003; Illing et al., 2005). However, what does appear to be consistent in desiccation-tolerant tissues is that the antioxidant enzymes AP, GR, CAT and SOD retain the ability (in *in vitro* as-says) to detoxify ROS even at RWC of < 10%, suggesting that there is some protection of these proteins that prevents their denaturation and maintains the native state in dry conditions. This was not the case in DS species and it has been proposed that it is this ability in DT systems that is a unique mechanism of DT (Illing *et al.*, 2005; Farrant, 2007). Kranner *et al.* (2006) have proposed that glutathione is key to survival of desiccation in a variety of DT systems. It was shown that the half-cell redox potential ($E_{GSSG/2GSH}$) can be used as a marker for plant stress and values of ≥ 160 mV correlate with loss of viability. Since ROS production continues in dry tissues (albeit at a reduced rate due to restricted mobility in glassy states) the authors propose that ultimate failure of this antioxidant system triggers programmed cell death in dry stored seeds, resurrection plants and other desiccation-tolerant systems. In support of this hypothesis, the maintenance of DT in *Zea mays* seeds treated with PEG showed a correlation with the levels of GR (Huang and Song, 2013). Kranner and Birtić (2005) demonstrated that the critical $E_{GSSG/2GSH}$ in the resurrection plant *M. flabellifolia* occurred after 8 months of dry storage. In studies in which the longevity of this and other species (*C. wilmsii* and *X. humilis*) in the dry state was measured, loss of viability did correlate with loss of activity of antioxidant enzymes (Farrant, 2007), illustrating the importance of maintenance of antioxidant potential to attainment and maintenance of DT.

Although the role of Asc in DT has not been as clearly demonstrated (and indeed questioned with respect to DT in seeds (Bailly, 2004)), other than for its involvement in the ascorbate–glutathione scavenging of H_2O_2, from recent data we believe that it might play an important role in DT of angiosperm resurrection plants. Transcription of *VTC2*, the gene coding for the first committed step to Asc synthesis (Linster *et al.*, 2007; Linster and Clarke, 2008) is upregulated in the resurrection plant *X. viscosa* when the plants are dried below 60% RWC, and mRNA levels remain high in the desiccated plant and during early stages of rehydration (Bresler, 2010). Ascorbate levels in roots and leaves of this plant follow the same trend (Kamies *et al.*, 2010) and we propose that elevated ascorbate levels are maintained during drying and early rehydration by a combination of *de novo* synthesis and regeneration of ascorbate by AP, which itself retains the ability to remain active (Farrant *et al.*, 2007;

Kamies *et al.*, 2010). This compares well with data by Suarez Rodriguez *et al.* (2010), who show the AP transcript to be more abundant in the desiccated leaves of *C. plantagineum* with increasing levels in rehydrated leaves.

In addition to the protection afforded by housekeeping antioxidants, resurrection plants have the ability to induce, *de novo*, antioxidants such as 1- and 2-cys-peroxiredoxins, glyoxylase I family proteins, zinc metallothionine, metallothionine-like antioxidants, oxidoreductases and several members of the aldehyde dehydrogenases in response to desiccation (Velasco *et al.*, 1994; Blomstedt *et al.*, 1998; Kirch *et al.*, 2001; Chen *et al.*, 2002; Mowla *et al.*, 2002; Collett *et al.*, 2004; Illing *et al.*, 2005; Ingle *et al.*, 2007; Mulako *et al.*, 2008; Walford, 2008). Some of these are reported to occur in orthodox seeds (Aalen, 1999; Stacy *et al.*, 1999; Boudet *et al.*, 2006; Buitink *et al.*, 2006; Mulako *et al.*, 2008; Walford, 2008; Shin *et al.*, 2009) and transcripts of 1-cys peroxiredoxin, aldehyde dehydrogenase and aldehyde reductase were again accumulated during the re-induction of DT in *M. truncatula* seeds (Buitink *et al.*, 2006; Leprince and Buitink, 2010). During reserve accumulation and maturation drying, an increase in glyoxylase has been observed in seeds of rice (Xu *et al.*, 2008; Lee and Koh, 2011), maize (Wang *et al.*, 2014) and *M. truncatula* (Gallardo *et al.*, 2007). This enzyme is required for the detoxification of methylglyoxal (Thornalley, 2003). These compounds do not appear to be upregulated in the vegetative tissues of DS tissues (Aalen, 1999; Mulako *et al.*, 2008) and we believe they play an important role in attainment of DT. Transcription of 1 cys-perioxiredoxin and metallothione-like antioxidants occurs between 50 and 60% RWC and translation at water contents below this. Both transcripts and proteins disappear rapidly during rehydration either prior to or upon reaching full turgor (Mowla *et al.*, 2002; Collett *et al.*, 2004; Mulako *et al.*, 2008). We propose that such antioxidants serve to scavenge ROS and neutralize toxic metabolic intermediates that occur as a consequence of unregulated metabolism associated with severe water deficit.

Increasingly, the classical view of antioxidant systems is being reformed as it becomes clear that many primary and secondary metabolites are involved in free radical scavenging (Keunen *et al.*, 2013; Yobi *et al.*, 2013; Noctor *et al.*, 2014).

Disaccharides, raffinose family oligosaccharides (RFOs), fructans, sugar alcohols (e.g. galactinol), polyols and certain amino acids have been identified as being active in ROS-scavenging and particularly hydroxyl-scavenging (Nishizawa *et al.*, 2008; Stoyanova *et al.*, 2011; Hernandez-Marin and Martínez, 2012; Tatsimo *et al.*, 2012; Peshev *et al.*, 2013; Yobi *et al.*, 2013), although they cannot be recycled like GSH and Asc.

Many polyphenols and flavonoids/ols are believed to have antioxidant properties (Smirnoff, 1993; Wang *et al.*, 1996; Kahkonen *et al.*, 1999) and, indeed, recent work has supported this (Veljovic-Jovanovic *et al.*, 2008; Mihaylova *et al.*, 2013). In seeds there have been few studies reported on the role of polyphenols as antioxidants, possibly because they have been bred out of species that are most studied. In a comprehensive study on polyphenols in *M. flabellifolia* it was demonstrated that dry leaves contain high levels (up to 50% of the leaf dry weight) of 3, 4, 5 tri-*O*-galloylquinic acid, which acts as a potent antioxidant (Moore *et al.*, 2005). Although this polyphenol is predominantly located in the vacuole it has been proposed to act as an antioxidant reservoir linked to the cytoplasmic antioxidants and functioning as a redox buffer (Moore *et al.*, 2007b). A survey conducted on seven other resurrection plants showed that total polyphenol content was lower than in *M. flabellifolia*, and varied among the species (Farrant *et al.*, 2007). However the relative antioxidant potential as determined by ferric reducing/antioxidant power (FRAP) and 2,2-diphenyl-1-picrylhydrazyl (DPPH) antioxidant activity assays was higher in the resurrection plants than in related DS species. In *R. serbica*, polyphenols and the activity of polyphenol oxidase increases during desiccation (Sgherri *et al.*, 2004; Veljovic-Jovanovic *et al.*, 2008). A proteome study of *B. hygrometrica* showed that proteins relating to phenolic metabolism are upregulated on drying in leaves of this species (Jiang *et al.*, 2007). Recently, phenolic extracts from the resurrection plant *H. rhodopensis* showed significant reducing power (Mihaylova *et al.*, 2013). The scale of the contribution of such antioxidants to DT in resurrection plants is unknown, but they are necessary, in combination with other antioxidants, to best facilitate protection of photosynthetic vegetative tissues from ROS stress associated with desiccation.

9.6.3   Water replacement and glass formation

Progressive loss of water (level IV and below) results in metabolic stress related to cytoplasmic crowding. The cytoplasm becomes increasingly viscous, proteins begin to denature and membrane fusion occurs (Vertucci and Farrant, 1995). It has been proposed that desiccation-tolerant organisms counteract cytoplasmic crowding by replacing water with compatible solutes capable of substituting for the hydrogen bonds lost due to dehydration (Fig. 9.5). This water replacement hypothesis presupposes that these molecules are able to stabilize macromolecules in their native configuration during desiccation (Crowe *et al.*, 1986, 1989). Additional stabilization of the subcellular milieu is believed to be achieved via cytosolic vitrification (Vertucci and Farrant, 1995; Hoekstra *et al.*, 2001; Walters *et al.*, 2002). Solutes believed responsible for replacement and stabilization include: (i) sucrose and oligosaccharides (reviewed in Berjak *et al.*, 2007; Farrant, 2007); and (ii) proteins, particularly late embryogenesis abundant (LEA) proteins (reviewed in Hoekstra *et al.*, 2001; Illing *et al.*, 2005; Mtwisha *et al.*, 2006; Berjak *et al.*, 2007) and small heat shock proteins (Almoguera and Jordano, 1992; Alamillo *et al.*, 1995; Wehmeyer *et al.*, 1996; Mtwisha *et al.*, 2006).

A recent complementary hypothesis is based on the observation that certain sugars and organic acids readily form ionic liquid solvents in the absence of water, and that these natural deep eutectic solvents (NADES) stably dissolve amphipathic molecules including many peptides and proteins (Choi *et al.*, 2011). Demonstrating the presence of NADES *in vivo* is challenging and thus the roles, if any, of such ionic liquids in plant DT have not yet been convincingly demonstrated. However, *in vitro* studies do show that NADES formed from molar ratios of suc-fru-glc may play a role in maintaining the functions of, for example, antioxidant enzymes at low water contents (Fan, 2014). It is also highly possible that they play a role in facilitating the biophysical changes observed in tissues of DT organisms during desiccation.

9.6.4   Sucrose and oligosaccharides

The accumulation of non-reducing sugars and certain oligosaccharides in both seed development

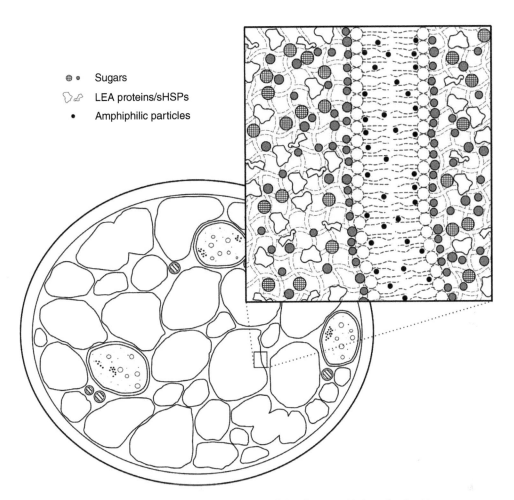

Sugars

LEA proteins/sHSPs

Amphiphilic particles

Fig. 9.5. Schematic 2-dimensional diagram of an intracellular glassy matrix hypothesized to occur through the process of vitrification at low water contents due to accumulation of sugars, LEA and small heat shock proteins (among others), stabilizing the subcellular milieu in the dry state as well as amphiphilic compounds (Hoekstra *et al.*, 2001).

(reviewed in Vertucci and Farrant, 1995; Berjak *et al.*, 2007) and during dehydration of resurrection plant tissues (Illing *et al.*, 2005; reviewed in Farrant, 2007) has been correlated with the acquisition of DT.

Sucrose increases in response to desiccation in all angiosperm resurrection plants studied to date, with a large proportion accumulating after the cessation of photosynthesis, below a leaf RWC of 60% (Farrant, 2007). It has therefore been proposed that this results from an alteration in carbon partitioning and is not a product of carbon assimilation (Illing *et al.*, 2005). The source of carbon has been investigated in some resurrection species and has been proposed to partially originate from conversion reserve carbohydrates,

for example octulose (Norwood *et al.*, 2000), and starch (Whittaker *et al.*, 2007). Although to a lesser extent than sucrose, oligosaccharides – particularly raffinose and to some degree stachyose – also accumulate in resurrection plants during drying. However, due to the variation in the quantity accumulated, it is thought that oligosaccharides, together with other compatible solutes, function interactively in protection against desiccation (reviewed in Farrant, 2007).

Seeds show a universal increase in sucrose during development and this has again been linked with DT. However, as mentioned previously, it is difficult to separate accumulation due to normal developmental metabolism from that associated with DT, particularly as it is required as a substrate for germination in all seeds, including DS (recalcitrant) seeds (Berjak *et al.*, 2007). Although it has been widely held that the primary source of sucrose in developing seeds is transport from the parent plant, recent studies on model systems that attempted to separate development from DT acquisition indicated that sucrose accumulation results from metabolic switches within the seed itself. Transcriptome and metabolite profiling studies have indicated that starch and lipids are the main reserves mobilized in *M. truncatula* (Buitink *et al.*, 2006) and metabolite profiling in *A. thaliana* suggests lipid is mobilized (Fait *et al.*, 2006) for sucrose production. Bogdana and Zagdańska (2009) also recorded significant changes in enzyme activity involved in sucrose synthesis and hydrolysis in response to dehydration in wheat seedlings, confirming the involvement of sugar metabolism in the regulation of dehydration tolerance. Total sugar content and sucrose levels were found to be highly variable with no simple correlation to DT in a study on seed embryos in a variety of species, representing three seed storage categories: orthodox, intermediate and recalcitrant

(Steadman *et al.*, 1996). Similarly, raffinose contents show contradicting results (Still *et al.*, 1994; Bochicchio *et al.*, 1997; Black *et al.*, 1999) and its role in glass formation (discussed below) is debated (Buitink *et al.*, 2000). Instead, sugar composition in seeds, specifically the ratio of sucrose to raffinose, has been implicated as the critical aspect in the acquisition of DT (Steadman *et al.*, 1996). This hypothesis has been corroborated by several studies. Figure 9.6 shows a summary of the data from these studies in addition to the ratios for hydrated and dry tissue in a variety of resurrection plants. In seeds, despite some variability, a value of sucrose-to-raffinose family oligosaccharide (RFO) was proposed in the order of 7:1 for several species of orthodox seeds. Recalcitrant seeds were recorded as having a 12:1 ratio (Horbowicz and Obendorf, 1994; Lin and Huang, 1994; Steadman *et al.*, 1996). A decreased ratio in orthodox seeds compared to recalcitrant seeds can clearly be seen in Fig. 9.6. Although the clustering is not as evident for hydrated tissue of desiccation-tolerant plants, there is a definite trend towards a lower ratio in dry tissue. If the change in ratio for individual

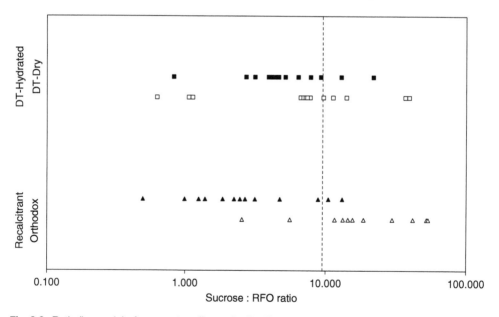

Fig. 9.6. Ratio (log scale) of sucrose to raffinose family oligosaccharides (raffinose and stachyose) in embryo or embryonic axis of mature recalcitrant (Δ) and orthodox (▲) seeds (taken from Lin and Huang, 1994, Horbowicz and Obendorf, 1994 and Steadman *et al.*, 1996) and hydrated (□) and dry (■) desiccation-tolerant leaf tissue (taken from Ghasempour *et al.*, 1998, Peters *et al.*, 2007 and Farrant, unpublished data). Vertical divider (dotted line) represents the midpoint between critical values separating tolerance from non-tolerance in seeds as suggested by Steadman *et al.* (1996).

desiccation-tolerant plants is examined (Fig. 9.7), however, it is apparent that though sucrose, raffinose and stachyose increase, most show a decrease in sucrose-to-RFO from hydrated to dry tissue.

Despite the acknowledged importance in protection against DT and the evident benefit of the presence of a reserve carbon source for rehydration and germination, the exact role(s) of sugars in DT has yet to be fully elucidated. We have already discussed the osmoprotectant and stabilization functions above. Allied to this is the hypothesis that sugars can facilitate the formation of a glass (vitrification), which acts to further stabilize the subcellular milieu (Vertucci and Farrant, 1995). Although sucrose is thought to be the main glass-forming component, raffinose has been suggested as important in preventing sucrose crystallization (Leopold and Vertucci, 1986). It has also been proposed that the formation of non-reducing sugars indirectly acts to remove monosaccharides which are involved in the ROS-forming Maillard-type reactions (reviewed

in Farrant, 2007). This is supported by observations of declining monosaccharide levels during maturation and drying in seeds (Vertucci and Farrant, 1995; Steadman et al., 1996) and resurrection plants, respectively (Farrant, 2007). This decline in reducing sugar levels is not universal, however, with DT grasses tending to retain high concentrations of glucose and fructose in the dry state (Ghasempour et al., 1998). More recently, studies have suggested additional roles for sugars. Sucrose and sucrosyl oligosaccharides (including RFOs) have been linked to the protection against oxidative stress, acting as antioxidants in a ROS-scavenging capacity (Van den Ende and Valluru, 2009). Also, by means of sugar-sensing, soluble sugars are thought to contribute to the regulation of gene expression involved in plant growth and metabolism. These sugars act as signals in pathways associated with stress tolerance to moderate cellular metabolism, growth and plant development responses under oxidative stress (reviewed in Rosa et al., 2009).

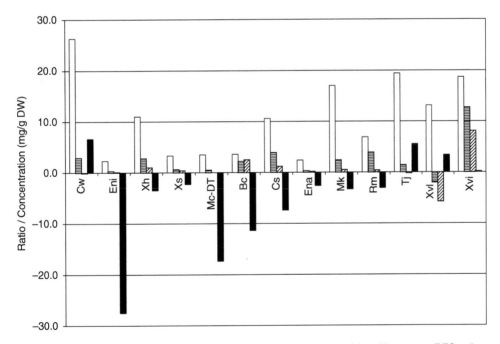

Fig. 9.7. Changes in concentration of sucrose (□), raffinose (■), stachyose (▨) and in sucrose:RFO ratio (■) (calculated from Ghasempour et al., 1998; Peters et al., 2007 and Farrant, unpublished data) from hydrated to dry tissue of the following resurrection plants. Cw: *Craterostigma wilmsii*; Eni: *Eragrostis nindensis*; Xh: *Xerophyta humilis*; Xs: *Xerophyta schlechteri*; Mc-DT: *Mohria caffrorum*, desiccation tolerant form; Bc: *Borya constricta*; Cs: *Coleochloa setifera*; Ena: *Eragrostiella nardoides*; Mk: *Microchloa kunthii*; Rm: *Ramonda myconi*; Tj: *Tripogon jacquemontii*; Xvl: *Xerophyta villosa*; Xvi: *Xerophyta viscosa*. DW, dry weight.

9.6.5 Lipids and other small molecular species

The loss of cellular water during desiccation affects the fluidity and stability of lipid bilayer membranes. DT requires adaptation to these physical changes, and resurrection plants have been shown to make significant changes to membrane composition of various cellular compartments in response to water loss. Gasulla *et al.* (2013) showed that in the resurrection plants *C. plantagineum* and *Lindernia brevidens*, monogalactosyldiacylglycerols (MGDG) in the thylakoid membrane are broken down, and used to synthesize phospholipids and oligogalactolipids. In addition, both *C. plantagineum* and *L. brevidens*, but not DS control species, accumulated phosphatidylinositol (PI) lipids. Their abundant accumulation suggested these may contribute to the stability of the anhydrobiotic state. The enrichment of PI and oligogalactolipids provides a layer of hydroxyl-rich moieties on the membrane surface that may contribute hydrogen-bonding capacity, reduce the impact of rising salt concentrations and interact with the vitrifying cytomatrix in such a way as to stabilize the membrane bilayer.

In addition, lipids may be involved in cellular signalling and control of gene expression. Gasulla *et al.* (2013) suggest that phosphatidic acid (PA) and PI fulfil roles both as metabolic intermediates, and as stress signals, with the signalling pathway mediated by phospholipase D. This pathway has been shown by Munnik and Vermeer (2010) and others reviewed by Okazaki and Saito (2014) to be involved in plant water stress responses in a variety of species.

Maintenance of a stable anhydrobiotic state in plant tissues is supported by the accumulation of other small molecular species, in addition to carbohydrates and lipids. For example, in *M. fla-bellifolia*, the polyphenol 3,4,5,tri-o-galloylquinic acid is observed to accumulate to levels of up to 74% by mass in dehydrated leaf tissue, where it accumulates predominantly in vacuoles and has been shown to stabilize tonoplast membranes (Moore *et al.*, 2005). As a general principle (discussed above), all resurrection plants must achieve vacuolar stabilization by filling vacuoles with some suitable material during dehydration. Similar quinic acid derivatives, including caffeoylquinic acids, have been observed in other DT plants, notably the resurrection plant *Barbacenia purpurea*

(Suguiyama *et al.*, 2014). Although *in vivo* localization of these metabolites was not demonstrated, their role as antioxidants and potential contributors to stability of cytoplasmic glasses has been proposed.

Numerous other small metabolites have been observed to accumulate in resurrection plants, including the γ-glutamyl amino acids (Yobi *et al.*, 2013), polyols including sorbitol and xylitol, and aromatic amino acids (Yobi *et al.*, 2012). Many of these compounds are hydroxyl-rich, and could contribute to water replacement and the stabilization of the anhydrobiotic cytomatrix. Others, including the aromatic amino acids, have amphipathic characters that are likely to help maintain the solubility of macromolecules of intermediate polarity within the sugar-rich vitreous matrix. Furthermore, γ-glutamyl amino acids and some polyols are effective ROS scavengers and/or participate in pathways involving antioxidant metabolism (Smirnoff, 1998; Yobi *et al.*, 2103). It is thus possible that many of these metabolites have dual roles in the broader context of DT.

9.6.6 Late embryogenesis abundant proteins

As the name suggests, LEA proteins were first identified due to their abundant accumulation (4% of total cellular protein, Roberts *et al.*, 1993) during the late stages of seed development coincident with the onset of DT (e.g. Galau *et al.*, 1986; Blackman *et al.*, 1992, 1995; Baker *et al.*, 1995; Russouw *et al.*, 1995; Manfre *et al.*, 2006). They have been found subsequently in many bacteria (Garay-Arroyo *et al.*, 2000; Cytryn *et al.*, 2007), nematodes and tardigrades (Goyal *et al.*, 2005; Tunnacliffe and Wise, 2007) and are widely present in the plant kingdom, both in seeds and vegetative tissues (reviewed *inter alia* in Cuming, 1999; Berjak *et al.*, 2007; Moore *et al.*, 2009; Leprince and Buitink, 2010). While their expression appears to be correlated with abiotic stresses such as water deficit, osmotic and temperature stresses (all of which do affect subcellular water status) (Wise and Tunnacliffe, 2004; Illing *et al.*, 2005), there is still little understanding of the precise roles of these proteins *in vivo*, and few investigations study the proteins themselves. This is a consequence of the fact that they are intrinsically disordered, non-catalytic proteins which are unfolded in aqueous solutions, making it experimentally difficult to assign a structure and determine function. Various groups have attempted a system of classification using a variety of strategies including molecular weight, expression patterns and bioinformatics (reviewed in Amara *et al.*, 2014). While there appear to be some commonalities between the groupings, LEAs show a great deal of variation, and inconsistent labelling or no mention of the classification method used makes discussion and comparative analysis of these proteins extraordinarily challenging.

Predicted functions, based on RNA sequence information indicating Gly-rich, charged and polar amino acid content include: (i) water replacement molecules and/or hydration buffers; (ii) ion sequesterers; (iii) chaperonins and/or heat shields; (iv) protein/membrane anti-aggregants; and (v) promoters of vitrification (Bray, 1997; Hoekstra *et al.*, 2001; Wise and Tunnacliffe, 2004; Bartels, 2005; Goyal *et al.*, 2005; Mtwisha *et al.*, 2006; Berjak *et al.*, 2007; Chakrabortee *et al.*, 2007; Tunnacliffe and Wise, 2007; Oliver *et al.*, 2009). The following functions have been shown experimentally through either *in vivo* or *in vitro* experimentation: binding to phospholipids (Koag *et al.*, 2009; Eriksson *et al.*, 2011; Hundertmark *et al.*, 2011; Clarke *et al.*, 2015); radical scavenging (Hara *et al.*, 2004); binding to water and ions (Bokor *et al.*, 2005; Tompa *et al.*, 2006); phosphorylation (Alsheikh *et al.*, 2003, 2005; Röhrig *et al.*, 2006); binding to calcium (Heyen *et al.*, 2002; Kruger *et al.*, 2002); protection of enzymes (Kovacs *et al.*, 2008; Liu *et al.*, 2011); binding to cytoskeletons (Rahman *et al.*, 2011, 2013); binding to nucleic acids (Hara *et al.*, 2009); and sugar glass stabilization (Wolkers *et al.*, 2001; Shih *et al.*, 2004).

Functional characterization has been carried out in piecemeal fashion with only a few functions analysed for a small number of LEAs. Of all of the LEA proteins, dehydrins, or Group 2 LEAs, are the best characterized and to date have exclusively been found in plants (Röhrig *et al.*, 2006, 2008), where they are reported to have a variety of functions (Hara, 2010; Fig. 9.8). Like the dehydrins, many other LEAs have been proposed to have multiple functions, something that is thought to be due to their intrinsic disorder which allows for binding across a multitude of substrates under various physiochemical

environments, such as would occur in a dehydrating cell as well as upon rehydration (Tunnacliffe and Wise, 2007).

LEA genes are among the most differentially expressed and highly upregulated, as shown in recent transcriptomics studies on orthodox seeds and resurrection plants, and this is supported by proteomic studies where these have been performed (Table 9.1).

Publicly available *Arabidopsis* genome and microarray data have shown that 74% of LEAs are expressed during seed development. Fifteen of the 35 LEAs analysed were seed specific and a further six LEAs were expressed at their maximum during seed development when DT is acquired, but were also expressed in response to other abiotic stresses (Illing *et al.*, 2005; Table 9.1). Both transcriptome and proteome studies on *M. truncatula* have shown a number of LEA-like proteins in mature seeds of which expression is upregulated when desiccation is re-induced during germination (Boudet *et al.*, 2006; Buitink *et al.*, 2006; Maia *et al.*, 2011; Table 9.1). At least two, a group 1 LEA (Em6; PF00477) and a group 5 LEA (PM25; PF04927), were confirmed to be associated with DT, the remainder being proposed to be associated with osmotic stress and drought tolerance, since their induction in this and other species appears to occur at higher hydration levels of more than 0.8 gH_2O/g FW (Black *et al.*, 1999; Reyes *et al.*, 2005; Boudet *et al.*, 2006). It is likely that these are indeed required for DT but are needed at earlier stages of water loss. In the resurrection plant *X. humilis*, 16 of 55 genes shown to be upregulated during desiccation were annotated as LEAs (Collett *et al.*, 2004). This microarray study tested only 424 cDNAs randomly selected from an 11K normalized library made from leaf and root tissues of the plant and this, together with the observation that the transcripts only became evident once the water content fell below 50% RWC (0.44 gH_2O/g FW) with transcripts disappearing early during the rehydration process (Illing *et al.*, 2005), indicates the potential importance of these in acquisition of DT in this species. Further microarray profiling in which genes from desiccated roots, leaves and seeds of *X. humilis* were compared showed considerable enrichment of LEA family proteins in all of these tissues, of which three, a group 1 LEA (At2g40170), a group 6 LEA (At3g22490) and a LEA10 (At1g04560), were among the *Arabidopsis* seed-specific genes; and

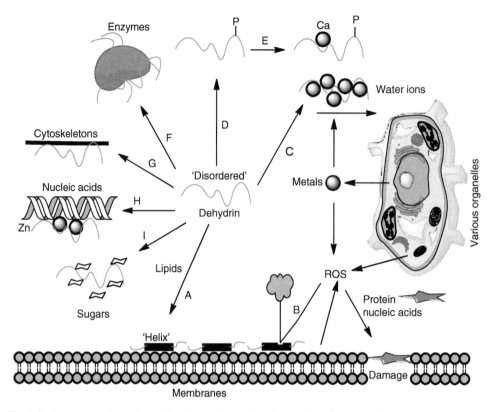

Fig. 9.8. Experimentally confirmed functions of dehydrins from various *in vivo* and *in vitro* studies. Functions are indicated by A, membrane stabilization; B, radical scavenging; C, binding to water and ions; D, phosphorylation; E, binding to calcium; F, protection of enzymes; G, binding to cytoskeletons; H, binding to nucleic acids; and I, sugar glass stabilization. This figure was adapted from Hara, 2010.

Table 9.1. Studies of late embryogenesis abundant (LEA) proteins from model seed species and desiccation-tolerant angiosperms.

Tissue	Species	Type of study	Citation
Seed	*Arabidopsis thaliana*	G, T	Illing *et al.*, 2005; Hundertmark and Hincha, 2008; Walford, 2008; Costa *et al.*, 2015
	Medicago truncatula	P, T	Boudet *et al.*, 2006; Buitink *et al.*, 2006; Dias *et al.*, 2011; Bai *et al.*, 2012
Plant	*Boea hygrometrica*	G, G, T	Zhang *et al.*, 2012; Xiao *et al.*, 2015; Zhu *et al.*, 2015
	Bryum argenteum	T	Gao *et al.*, 2015
	Craterostigma plantagineum	T	Piatkowski *et al.*, 1990; Röhrig *et al.*, 2006; Suarez Rodriguez *et al.*, 2010; Giarola *et al.*, 2015
	Haberlea rhodopensis	T	Gechev *et al.*, 2013
	Myrothamnus flabellifolia	P, T	Koonjul *et al.*, 1999; Ma *et al.*, 2015
	Physcomitrella patens	P	Wang *et al.*, 2009
	Selaginella tamariscina	P	Wang *et al.*, 2010
	Sporobolus stapfianus	T	Blomstedt *et al.*, 1998; Neale *et al.*, 2000; Le *et al.*, 2007
	Xerophyta humilis	P, T	Collett *et al.*, 2004; Illing *et al.*, 2005: Walford, 2008; Liu *et al.*, 2009; Shen, 2014; Waters, 2015
	Xerophyta viscosa	P, T	Mundree and Farrant, 2000; Ndima *et al.*, 2001; Abdalla and Rafudeen, 2012

Study type is indicated as G, genome; P, protein/proteome; and T, transcript/transcriptome.

one, a LEA3 (At1g52690), was highly upregulated during DT in seeds of this species (Walford, 2008). A further four group 3 LEAs (At2g42560, At3g53040, At4g15910, At4g21020), a group 4 LEA (At1g72100), a group 7 LEA (At3g53770) and a group 8 LEA (At1g01470) were found to be upregulated in desiccated vegetative and seed tissues of *X. humilis*, although these did not fall into the *Arabidopsis* seed-specific LEAs identified by Illing *et al.* (2005). Transcriptome and proteome studies on other resurrection plants have all reported the presence of LEA genes associated with desiccation (see Table 9.1). These studies have ranged in the numbers of genes tested, and so there is variability in the final numbers of LEAs and the groups to which they belong. However, in all studies they form a high percentage of the genes tested, with LEAs from groups 2 and 3 predominating. The early analyses described were the foundation for the exploration of LEA expression in current studies.

As global -omics analyses are implemented on a wider scale with greater application of sequencing techniques such as RNAseq and genome sequencing of resurrection species, comparative analyses of unique sequences are being conducted to understand how individual LEA profiles compare across DT species as opposed to merely modelling species such as *A. thaliana*. Ma *et al.* (2015) showed a number of sequence homologies between the LEA genes of the resurrection plant *M. flabellifolia* with those of *B. hygrometrica* and *H. rhodopensis* (Suarez Rodriguez *et al.*, 2010; Gechev *et al.*, 2013). Global transcriptome and proteome studies additionally allow for targeted expression studies such as that by Giarola *et al.* (2015) and Waters (2015) as well as structural and functional analysis of interesting LEAs (Ginbot, 2011; Waters, 2015). Also, these can be conducted at a higher variety of RWCs which contributes greatly to our understanding of the role LEAs play in DT.

Functional studies of LEA transgenic expression across a variety of organisms have been conducted to determine how individual LEAs contribute to DT (reviewed in Leprince and Buitink, 2010). In general, transformation studies of such LEAs from DT species have reported a general increase in water deficit tolerance (Swire-Clark and Marcotte, 1999; Brini *et al.*, 2007; Amara *et al.*, 2014). Additionally, in studies where LEA genes are removed, seeds and

plants have reduced longevity and DT (Hundertmark *et al.*, 2011). While LEA expression has been cited in many cases to be upregulated during periods of desiccation in both orthodox seeds and resurrection tissues, they are also expressed in many DS species (Shao *et al.*, 2005; Battaglia *et al.*, 2008; Hundertmark *et al.*, 2011; Bai *et al.*, 2012). This further confirms the evolutionary lineage of LEA proteins, but also suggests that LEAs alone cannot make the vegetative tissues of a plant DT.

9.6.7 Heat shock proteins

There are a number of heat shock proteins (HSPs) spanning various molecular weights including HSP100, HSP90, HSP70, HSP60 and HSP20, although they show significant conservation among them. As the name suggests, they are induced by heat. Regulation of HSPs occurs through heat stress transcription factors (HSFs) (Scharf *et al.*, 2012). Experiments in various species have shown that the HSFs upregulated during desiccation are in turn regulated by dehydration-response element binding proteins (DREBs) in *A. thaliana* (Sakuma *et al.*, 2006) and, interestingly, in sunflower seeds but not in vegetative tissues (Almoguera *et al.*, 2009).

The small HSPs (sHSPs, 15-42 kDa) are the most prominent HSPs in plants and can mostly be classed from consensus sequences into 11 classes, distinguishing their subcellular localizations from their transit/target/signal sequences located at the N-terminal ends, and include locations within the cytosol/nucleus, peroxisome, endoplasmic reticulum and the mitochondria (Waters, 2013). Their N-terminal domains are highly variable and, as with LEAs, their intrinsically disordered nature makes them difficult to characterize (McDonald *et al.*, 2012; Basha *et al.*, 2013). However, it is proposed that the N-terminal is the functional region of these proteins and is thus important for future investigations. Unlike other HSP families, sHSPs are ATP-independent (Mymrikov *et al.*, 2011; Bondino *et al.*, 2012), which makes them more energetically viable in the intermediate water contents of dehydrating and rehydrating cells where energy metabolism is at a minimum.

As with LEAs, HSPs are intensively synthesized during seed development prior to desiccation, and persist in the dry state, indicating a role in the acquisition of DT (Vierling, 1991; Coca et al., 1994; Wehmeyer et al., 1996; Kermode and Finch-Savage, 2002; Kalemba and Pukacka, 2007). Seeds of M. truncatula are reported to have two HSP genes (HSP 83 and HSP 18.2), transcripts of which accumulate during the acquisition of DT, both during maturation and after re-establishment of DT in germinating radicles (Buitink et al., 2006). Proteomic analysis has revealed upregulation in various DT plants (Alamillo et al., 1995; Ingle et al., 2007; Röhrig et al., 2008; Oliver et al., 2010b; Abdalla and Rafudeen, 2012; Gechev et al., 2013; Zhang et al., 2013; Chen et al., 2014; Ma et al., 2015) as well as in several mosses (Oliver et al., 2009; Wang et al., 2009; Cruz de Carvalho et al., 2014). In C. plantagineum, it has been shown that sHSPs are constitutively present in the vegetative tissues but levels increase in response to water stress and heat shock. Additionally, exogenous application of the stress hormone ABA induces the expression of sHSPs and the acquisition of DT in previously DS C. plantagineum callus (Alamillo et al., 1995). In a microarray analysis in which mRNA transcripts from hydrated leaves and roots were compared with those in desiccated leaves, roots and seeds of X. humilis, we identified two sHSPs (At1g535340, At3g57340) that were upregulated in all the dry tissues. In a study of the heat stable proteome of leaves of this species, two of the 33 proteins that appeared de novo on desiccation were HSPs, one a sHSP (HSP26.7b, a chloroplast low molecular weight heat shock protein), while the other was annotated as HSP70 (Liu et al., 2009). Characterization of the nuclear proteome of X. viscosa showed that there was considerable upregulation of a 17.6 kDa sHSP upon dehydration (Abdalla et al., 2010) and a member of the HSP90 family (Grp94) was found to be induced by desiccation and heat stress in leaves of this species (Walford et al., 2004). The most extensive study of sHSPs in a resurrection plant was conducted in B. hygrometrica by Zhang et al. (2013), where the mRNA expression levels of ten sHSPs previously found to be upregulated during desiccation were monitored across a series of different treatments. Of the ten sHSPs, six were classified as cytosolic/nuclear, one as mitochondrial, one as chloroplastic and two as orphan in this classification schema. Their desiccation treatment revealed differential expression of sHSPs across the dehydration treatment where two sHSPs were most highly upregulated during early-stage dehydration, and six sHSPs were most highly upregulated during late-stage dehydration. In both cases the upregulated sHSPs were classified as cytosolic/nuclear or orphan. ABA treatment increased the expression of two of the sHSPs, suggesting that not all sHSPs are ABA regulated in B. hygrometrica. While the chloroplastic and mitochondrial sHSPs expression levels did not appear to be upregulated by any particular treatment, they were constitutively expressed in control tissues, suggesting that they may play a role as maintenance chaperones in these tissues. To date there is little experimental evidence that points to a specific role for HSPs in DT, but they are thought to offer a general protective role in the dry state based on their chaperone-like activity. In this regard they may act to minimize inappropriate interactions among molecules, and so enable maintenance of protein structure in the dry state, and may also facilitate appropriate refolding upon rehydration (Alpert and Oliver, 2002; Mtwisha et al., 2006). Nakamoto and Vigh (2007) also showed that sHSPs can bind an almost equal mass of substrate protein. This would be beneficial in maintaining the structure of DS globular proteins during desiccation, as it could prevent aggregation. Recent studies suggest that sHSPs can assist insoluble protein aggregates to solubilize again and that they have a high affinity for binding denatured substrates (Eyles and Gierasch, 2010; Tyedmers et al., 2010; Garrido et al., 2012). While there are few studies focused on the water contents at which these proteins are upregulated and/or induced, it is likely that they protect in all phases of water loss from type IV and below, and that some at least offer constitutive protection.

9.6.8 What we know about cell signalling in DT

As with many responses to abiotic stress, signal transduction in DT plant cells appears to be initiated mostly by protein kinases and phosphatases, which in turn initiate the activity of a number of regulatory genes. These then regulate various signalling pathways including those mediated by various phytohormones, lipids, sugars and other molecules. While there are exceptions to this model, current DT research generally follows the schema outlined in Bartels *et al.* (2007).

Protein kinases were shown to be important to the DT strategy in the proteomic study of *Physcomitrella patens* (Wang *et al.*, 2009) and *Sporobolus stapfianus* (Oliver *et al.*, 2010b). Transcriptome studies have also shown the importance of protein kinases such as that of Ma *et al.* (2015), which showed increased transcript levels of 484 protein kinases expressed in *M. flabellifolia* during dehydration. This led them to suggest that protein kinases may play a key regulatory role in drought tolerance adaptation in this species. Suarez Rodriguez *et al.* (2010) showed that of the top five upregulated transcripts during dehydration, a protein phosphatase was highly upregulated in *C. plantagineum*. Ultimately, understanding of the pathways initiated by these mechanisms will occur through analysis of the phosphoproteome, such as was done in *C. plantagineum* (Röhrig *et al.*, 2006, 2008) and *Selaginella moellendorffii* (Chen *et al.*, 2014).

Hussain *et al.* (2011) have explored the various functional transgenic studies conducted on a number of regulatory genes, including transcription factors, found to increase drought tolerance in various crop plants and *A. thaliana*. Many of these studies focus on ABA responsive element binding proteins (AREBs) and DREBs, among others, which ultimately showed increased tolerance to drought in various crop and model organisms.

ABA is the best described and characterized phytohormone involved in DT. Its biosynthesis and regulation serve as some of the best conserved stress responses across all plants, as extensively characterized in studies of the model moss *Physcomitrella patens* and the model angiosperm *A. thaliana* (Cuming and Stevenson, 2015). Studies of the leaves of several resurrection plants have shown manyfold upregulation of ABA during desiccation in leaf tissue (*C. plantagineum*, Bartels *et al.*, 1990; *M. flabellifolia*, Gaff and Loveys, 1992; *Selaginella tamariscina*, Wang *et al.*, 2010) as well as the onset of DT with exogenous exposure to ABA (*C. plantagineum*, Bartels *et al.*, 1990; *Polypodium virginianum*, Reynolds and Bewley, 1993; *S. stapfianus*, Ghasempour *et al.*, 2001). As was suggested by Ramanjulu and Bartels (2002) and shown experimentally by Ma *et al.* (2015) in *M. flabellifolia*, much of the signalling during DT

can be split into two categories, ABA-dependent and ABA-independent. Various studies indicate that signalling through DREBs as well as various other hormones appears to regulate an important number of the ABA-independent pathways in orthodox seeds (Ramanjulu and Bartels, 2002; Sakuma *et al.*, 2006; Dias *et al.*, 2011; Terrasson *et al.*, 2013; Verdier *et al.*, 2013) and DT plants (Djilianov *et al.*, 2013; Ma *et al.*, 2015).

The most comprehensive study of phytohormone expression in a resurrection plant was conducted by Djilianov *et al.* (2013) in *H. rhodopensis*. Their study considered the expression of ABA, jasmonic acid (JA), salicylic acid (SA), dihyrdrophaseic acid (DPA), phaseic acid, indole-3-acetic acid (IAA) and IAA-aspartate (IAA-Asp), as well a range of cytokines. The changes in phytohormone concentrations of most of these across dehydration as well as rehydration stages in leaf and root tissues suggests a probable signalling role for these and other molecules. While some work has been done in this area, gene regulation remains a critically important area for future research in DT and particularly for effective potential DT crops.

9.7 Conclusion

This review has shown that there are considerable similarities in the mechanisms of DT in orthodox seeds and resurrection plants. These include, *inter alia*, vacuole filling, antioxidant production, synthesis of sucrose and production by certain RFOs of protective proteins such as LEAs and sHSPs, many of which are similarly regulated by ABA-dependent and ABA-independent pathways. However, some of the mechanisms are also unique to plants, such as wall folding, leaf folding, pigment production and photosynthetic downregulation. We propose that the developmentally regulated programme of acquisition of DT in seeds is utilized in the acquisition of tolerance in vegetative tissues of resurrection plants, possibly in response to environmentally regulated rather than developmental cues. Additional plant-specific mechanisms, such as those associated with desiccation-induced photosynthetic stresses, have been acquired in these plants. The differences among them is likely to be due to independent evolution among different lineages, DT in angiosperms having been reported to have appeared in at least 13 independent phylogenetic lineages (Gaff and Oliver, 2013).

In order to gain deeper insight into the mechanism of DT, and the similarities among seeds and resurrection plants, it is important that future studies should, *inter alia*:

- be conducted under reproducible conditions of light, relative humidity and temperature and be standardized for each resurrection plant species at conditions as close as possible to those occurring in the natural environment;
- carry out changes in water content and associated physiological/biochemical/molecular

measurements at more frequent intervals to capture as best as possible all the changes occurring during a drying and rehydration time course;

- study the individual tissues in seeds, or just the embryonic axis; measurements should include monitoring of parameters at the start of or just prior to the onset of DT and at its completion; and
- encompass more systems studies on individual species (seeds and vegetative tissues) that enable an integrated understanding of changes in response to water deficit stress from the molecular to whole-plant ecophysiological levels.

References

Aalen, R.B. (1999) Peroxiredoxin antioxidants in seed physiology. *Seed Science Research* 9, 285–295.
Abdalla, K.O. and Rafudeen, M.S. (2012) Analysis of the nuclear proteome of the resurrection plant *Xerophyta viscosa* in response to dehydration stress using iTRAQ with 2DLC and tandem mass spectrometry. *Journal of Proteomics* 75, 2361–2374.
Abdalla, K.O., Baker, B. and Rafudeen, M.S. (2010) Proteomic analysis of nuclear proteins during dehydration of the resurrection plant *Xerophyta viscosa*. *Plant Growth Regulation* 62, 279–292.
Alamillo, J., Almoguera, C., Bartels, D. and Jordano, J. (1995) Constitutive expression of small heat shock proteins in vegetative tissues of the resurrection plant *Craterostigma plantagineum*. *Plant Molecular Biology* 29, 1093–1099.
Almoguera, C. and Jordano, J. (1992) Developmental and environmental concurrent expression of sunflower dry-seed stored low-molecular-weight heat-shock protein and LEA mRNAs. *Plant Molecular Biology* 19, 781–792.
Almoguera, C., Prieto-Dapena, P., Díaz-Martín, J., Espinosa, J.M., Carranco, R. and Jordano, J. (2009) The HaD-REB2 transcription factor enhances basal thermotolerance and longevity of seeds through functional interaction with HaHSFA9. *BMC Plant Biology* 9, 75.
Alpert, P. (2006) Constraints of tolerance: why are desiccation-tolerant organisms so small or rare? *Journal of Experimental Biology* 209, 1575–1584.
Alpert, P. and Oliver, M.J. (2002) Drying without dying. In: *Black, M.* and Pritchard, H.W. (eds) *Desiccation and Survival in Plants – Drying without Dying*. CAB International, Wallingford, UK, pp. 3–43.
Alsheikh, M.K., Heyen, B.J. and Randall, S.K. (2003) Ion binding properties of the dehydrin ERD14 are dependent upon phosphorylation. *Journal of Biological Chemistry* 278, 40882–40889.
Alsheikh, M.K., Svensson, J.T. and Randall, S.K. (2005) Phosphorylation regulated ion-binding is a property shared by the acidic subclass dehydrins. *Plant, Cell and Environment* 28, 1114–1122.
Alves, R.J.V. and Kolbek, J. (2010) Vegetation strategy of *Vellozia crinita* (Velloziaceae). *Biologia* 65, 254–264.
Amara, I., Zaidi, I., Masmoudi, K., Ludevid, M.D., Pagès, M., Goday, A. and Brini, F. (2014) Insights into late embryogenesis abundant (LEA) proteins in plants: from structure to the functions. *American Journal of Plant Sciences* 5, 3440–3455.
Apel, K. and Hirt, H. (2004) Reactive oxygen species: metabolism, oxidative stress, and signal transduction. *Annual Review of Plant Biology* 55, 373–399.
Bai, Y., Yang, Q., Kang, J., Sun, Y., Gruber, M. and Chao, Y. (2012) Isolation and functional characterization of a *Medicago sativa* L. gene, MsLEA3-1. *Molecular Biology Reports* 39, 2883–2892.
Bailly, C. (2004) Active oxygen species and antioxidants in seed biology. *Seed Science Research* 14, 93–107.
Baker, E.H., Bradford, K.J., Bryant, J.A. and Rost, J.L. (1995) A comparison of desiccation-related proteins (dehydrin and QP47) in peas (*Pisum sativum*). *Seed Science Research* 5, 185–193.
Bartels, D. (2005) Desiccation tolerance studied in the resurrection plant *Craterostigma plantagineum*. *Integrative and Comparative Biology* 45, 696–701.
Bartels, D., Schneider, K., Terstappen, G., Piatkowski, D. and Salamini, F. (1990) Molecular cloning of abscisic acid-modulated genes which are induced during desiccation of the resurrection plant *Craterostigma plantagineum*. *Planta* 181, 27–34.

Bartels, D., Phillips, J. and Chandler, J. (2007) Desiccation tolerance: gene expression, pathways, and regulation of gene expression. In: Jenks, M.A. and Wood, A.J. (eds) *Plant Desiccation Tolerance*. Blackwell Publishing Ltd, Oxford, UK, pp. 115–148.

Barthlott, W. and Porembski, S.T. (1996) Ecology and morphology of *Blossfeldia liliputana* (Cactaceae): a poikilohydric and almost astomate succulent. *Botanica Acta* 109, 161–166.

Barthlott, W., Porembski, S., Szarzynski, J. and Mund, J.P. (1996) Phytogeography and vegetation of tropical inselbergs. In: Guillaumet, J.-L., Belin, M. and Puig, H. (eds) *Phytogéographie Tropicale: Réalités et Perspectives*. ORSTOM, Paris, France, pp. 15–24.

Basha, E., Jones, C., Blackwell, A.E., Cheng, G., Waters, E.R., Samsel, K.A., Siddique, M., Pett, V., Wysocki, V. and Vierling, E. (2013) An unusual dimeric small heat shock protein provides insight into the mechanism of this class of chaperones. *Journal of Molecular Biology* 425, 1683–1696.

Battaglia, M., Olvera-Carrillo, Y., Garciarrubio, A., Campos, F. and Covarrubias, A.A. (2008) The enigmatic LEA proteins and other hydrophilins. *Plant Physiology* 148, 6–24.

Behnke, H., Hummel, E., Hillmer, S., Sauer-Gürth, H., Gonzalez, J. and Wink, M. (2013) A revision of African Velloziaceae based on leaf anatomy characters and rbcL nucleotide sequences. *Botanical Journal of the Linnean Society* 172, 22–94.

Belton, P.S. (1997) NMR and the mobility of water in polysaccharide gels. *International Journal of Biological Macromolecules* 21, 81–88.

Berjak, P., Farrant, J.M. and Pammenter, N.W. (1989) The basis of recalcitrant seed behaviour. In: Taylorson, R.B. (ed.) *Recent Advances in the Development and Germination of Seeds*. Plenum Press, New York.

Berjak, P., Farrant, J.M. and Pammenter, N.W. (2007) Seed desiccation-tolerance mechanisms. In: Jenks, M.A. and Wood, A.J. (eds) *Plant Desiccation Tolerance*. Blackwell Publishing Ltd, Oxford, UK, pp. 151–192.

Bewley, J.D. (1979) Physiological aspects of desiccation tolerance. *Annual Review of Plant Physiology* 30, 195–238.

Bewley, J.D., Bradford, K.J., Hilhorst, H.W.M. and Nonogaki, H. (2013) *Seeds: Physiology of Development, Germination and Dormancy*. Springer, New York.

Black, M., Corbineau, F., Gee, H. and Côme, D. (1999) Water content, raffinose, and dehydrins in the induction of desiccation tolerance in immature wheat embryos. *Plant Physiology* 120, 463–472.

Blackman, S.A., Obendorf, R.L. and Leopold, A.C. (1992) Maturation proteins and sugars in desiccation tolerance of developing soybean seeds. *Plant Physiology* 100, 225–230.

Blackman, S.A., Obendorf, R.L. and Leopold, A.C. (1995) Desiccation tolerance in developing soybean seeds: the role of stress proteins. *Physiologia Plantarum* 93, 630–638.

Blomstedt, C.K., Gianello, R.D., Hamill, J.D., Neale, A.D. and Gaff, D.F. (1998) Drought-stimulated genes correlated with desiccation tolerance of the resurrection grass *Sporobolus stapfianus*. *Plant Growth Regulation* 24, 153–161.

Blomstedt, C., Griffiths, C., Fredericks, D., Hamill, J., Gaff, D. and Neale, A. (2010) The resurrection plant *Sporobolus stapfianus*: an unlikely model for engineering enhanced plant biomass? *Plant Growth Regulation* 62, 217–232.

Bochicchio, A., Vernieri, P., Puliga, S., Murelli, C. and Vazzana, C. (1997) Desiccation tolerance in immature embryos of maize: sucrose, raffinose and the ABA-sucrose relation. In: Ellis, R.H., Black, M., Murdoch, A.J. and Hong, T.D. (eds) *Basic and Applied Aspects of Seed Biology*. Kluwer Academic Publishers, Dordrecht, The Netherlands, pp. 13–22.

Bogdana, J. and Zagdańska, B. (2009) Alterations in sugar metabolism coincide with a transition of wheat seedlings to dehydration intolerance. *Environmental and Experimental Botany* 66, 186–194.

Bokor, M., Csizmók, V., Kovács, D., Bánki, P., Friedrich, P., Tompa, P. and Tompa, K. (2005) NMR relaxation studies on the hydrate layer of intrinsically unstructured proteins. *Biophysical Journal* 88, 2030–2037.

Bondino, H.G. and Valle, E.M. (2012) Evolution and functional diversification of the small heat shock protein/α-crystallin family in higher plants. *Planta* 235, 1299–1313.

Borisjuk, L., Nguyen, T.H., Neuberger, T., Rutten, T., Tschiersch, H., Claus, B., Feussner, I., Webb, A.G., Jakob, P., Weber, H., Wobus, U. and Rolletschek, H. (2005) Gradients of lipid storage, photosynthesis and plastid differentiation in developing soybean seeds. *New Phytologist* 167, 761–776.

Bornhardt, W. (1900) *Zur oberflächengestaltung und geologie Deutsch Ostafrikas*. Reimer, Berlin, Germany.

Boudet, J., Buitink, J., Hoekstra, F.A., Rogniaux, H., Larré, C., Satour, P. and Leprince, O. (2006) Comparative analysis of the heat stable proteome of radicles of *Medicago truncatula* seeds during germination identifies late embryogenesis abundant proteins associated with desiccation tolerance. *Plant Physiology* 140, 1418–1436.

Bray, E.A. (1997) Plant responses to water deficit. *Trends in Plant Science* 2, 48–54.

Bresler, A.P. (2010) Molecular characterisation of XvVTC2, a gene coding for a GDP-L-galactose phosphorylase from *Xerophyta viscosa*. MSc thesis, University of Cape Town, Cape Town, South Africa.

Brini, F., Hanin, M., Lumbreras, V., Amara, I., Khoudi, H., Hassairi, A., Pages, M. and Masmoudi, K. (2007) Overexpression of wheat dehydrin DHN-5 enhances tolerance to salt and osmotic stress in *Arabidopsis thaliana*. *Plant Cell Reports* 26, 2017–2026.

Buitink, J., Hemming, M.A. and Hoekstra, F.A. (2000) Is there a role for oligosaccharides in seed longevity? An assessment of intracellular glass stability. *Plant Physiology* 122, 1217–1224.

Buitink, J., Ly Vu, B., Satour, P. and Leprince, O. (2003) The re-establishment of desiccation tolerance in germinated radicles of *Medicago truncatula* Gaertn. seeds. *Seed Science Research* 13, 273–286.

Buitink, J., Leger, J.J., Guisle, I., Ly Vu, B., Wuillème, S., Lamirault, G., Le Bars, A., Le Meur, N., Becker, A., Küster, H. and Leprince, O. (2006) Transcriptome profiling uncovers metabolic and regulatory processes occurring during the transition from desiccation-sensitive to desiccation-tolerant stages in *Medicago truncatula* seeds. *The Plant Journal* 47, 735–750.

Chakrabortee, S., Boschetti, C., Walton, L.J., Sarkar, S., Rubinsztein, D.C. and Tunnacliffe, A. (2007) Hydrophilic protein associated with desiccation tolerance exhibits broad protein stabilization function. *Proceedings of the National Academy of Sciences USA* 104, 18073–18078.

Charuvi, D., Nevo, R., Shimoni, E., Navey, L., Zia, A., Farrant, J.M., Kirchoff, H. and Reich, Z. (2015) Photoprotection conferred by changes in photosynthetic protein levels and organization during dehydration of a homoiochlorophyllous resurrection plant. *Plant Physiology* 167, 1554–1465.

Chen, X., Zeng, Q. and Wood, A. (2002) The stress-responsive *Tortula ruralis* gene *ALDH21A1* describes a novel eukaryotic aldehyde dehydrogenase protein family. *Journal of Plant Physiology* 159, 667–684.

Chen, X., Chan, W.L., Zhu, F.Y. and Lo, C. (2014) Phosphoproteomic analysis of the non-seed vascular plant model *Selaginella moellendorffii*. *Proteome Science* 12, 16–29.

Choi, Y.H., van Spronsen, J., Dai, Y., Verberne, M., Hollmann, F., Arends, I.W.C.E., Witkamp, G. and Verpoorte, R. (2011) Are natural deep eutectic solvents the missing link in understanding cellular metabolism and physiology? *Plant Physiology* 156, 1701–1705.

Clarke, M.W., Boddington, K.F., Warnica, J.M., Atkinson, J., McKenna, S., Madge, J., Barker, C.H. and Graether, S.P. (2015) Structural and functional insights into the cryoprotection of membranes by the intrinsically disordered dehydrins. *Journal of Biological Chemistry* 290, 26900–26913.

Coca, M.A., Almoguera, C. and Jordano, J. (1994) Expression of sunflower low-molecular-weight heat-shock proteins during embryogenesis and persistence after germination: localization and possible functional implications. *Plant Molecular Biology* 25, 479–492.

Collett, H., Butowt, R., Smith, J., Farrant, J. and Illing, N. (2003) Photosynthetic genes are differentially transcribed during the dehydration-rehydration cycle in the resurrection plant, *Xerophyta humilis*. *Journal of Experimental Botany* 54, 2593–2595.

Collett, H., Shen, A., Gardner, M., Farrant, J.M., Denby, K.J. and Illing, N. (2004) Towards transcript profiling of desiccation tolerance in *Xerophyta humilis*: Construction of a normalized 11 k *X. humilis* cDNA set and microarray expression analysis of 424 cDNAs in response to dehydration. *Physiologia Plantarum* 122, 39–53.

Costa, M.C.D., Righetti, K., Nijveen, H., Yazdanpanah, F., Ligterink, W., Buitink, J. and Hilhorst, H.W.M. (2015) A gene co-expression network predicts functional genes controlling the re-establishment of desiccation tolerance in germinated *Arabidopsis thaliana* seeds. *Planta* 242, 435–449.

Crowe, L.M., Womersley, C., Crowe, J.H., Reid, D., Appel, L. and Rudolph, A. (1986) Prevention of fusion and leakage in freeze-dried liposomes by carbohydrates. *Biochimica et Biophysica Acta* 861, 131–140.

Crowe, J.H., Hoekstra, F.A., Crowe, L.M., Anchordoguy, T.J. and Drobnis, E. (1989) Lipid phase transitions measured in intact cells with Fourier transform infrared spectroscopy. *Cryobiology* 26, 76–84.

Cruz de Carvalho (2013) Drought stress and reactive oxygen species. Production, scavenging and signalling. *Plant Signalling Behaviour* 3, 156–165.

Cruz de Carvalho, R., Bernardes da Silva, A., Soares, R., Almeida, A.M., Coelho, A.V., Marques da Silva, J. and Branquinho, C. (2014) Differential proteomics of dehydration and rehydration in bryophytes: evidence towards a common desiccation tolerance mechanism. *Plant, Cell and Environment* 37, 1499–1515.

Cuming, A.C. (1999) LEA proteins. In: Shewry, P.R. and Casey, R. (eds) *Seed Proteins*. Kluwer Academic Publishers, Dordrecht,The Netherlands, pp. 753–779.

Cuming, A.C. and Stevenson, S.R. (2015) From pond slime to rain forest: the evolution of ABA signalling and the acquisition of dehydration tolerance. *New Phytologist* 206, 5–7.

Cytryn, E.J., Sangurdekar, D.P., Streeter, J.G., Franck, W.L., Chang, W., Stace, G., Emerich, D.W., Joshi, T., Xu, D. and Sadowsky, M.J. (2007) Transcriptional and physiological responses of *Bradyrhizobium japonicum* to desiccation-induced stress. *Journal of Bacteriology* 189, 6751–6762.

Dace, H., Sherwin, H.W., Illing, N. and Farrant, J.M. (1998) Use of metabolic inhibitors to elucidate mechanisms of recovery from desiccation stress in the resurrection plant *Xerophyta humilis*. *Plant Growth Regulation* 24, 171–177.

Das, P., Nutan, K.K., Singla-Pareek, S.L. and Pareek, A. (2015) Oxidative environment and redox homeostasis in plants: dissecting out significant contribution of major cellular organelles. *Frontiers in Environmental Science* 2, 1–11.

del Río, L.A. (2015) ROS and RNS in plant physiology: an overview. *Journal of Experimental Botany* 66, 2827–2837.

Dias, P.M.B., Brunel-Muguet, S., Dürr, C., Huguet, T., Demilly, D., Wagner, M-H. and Teulat-Merah, B. (2011) QTL analysis of seed germination and pre-emergence growth at extreme temperatures in *Medicago truncatula*. *Theoretical and Applied Genetics* 122, 429–444.

Dinakar, C. and Bartels, D. (2013) Desiccation tolerance in resurrection plants: new insights from transcriptome, proteome and metabolome analysis. *Frontiers in Plant Science* 4, 482.

Djilianov, D.L., Dobrev, P.I., Moyankova, D.P., Vankova, R., Georgieva, D.T., Gajdošová, S. and Motyka, V. (2013) Dynamics of endogenous phytohormones during desiccation and recovery of the resurrection plant species *Haberlea rhodopensis*. *Journal of Plant Growth Regulation* 32, 564–574.

Djilianov, D., Ivanov, S., Georgieva, T., Moyankova, S., Berjov, S., Petrova, G., Mladenov, P., Chrisov, N., Hristozova, N., Peshev, D., Tchorbadjieva, M., Alexieva, V., Tosheva, A., Nikolova, M., Ionkova, I. and van den Ende, W. (2015) A holistic approach to resurrection plants. *Haberlea rhodopensis* – a case study. *Biotechnology and Biotechnology* 23, 1414–1416.

Elstner, E.F. and Osswald, W. (1994) Mechanisms of oxygen activation during plant stress. *Proceedings of the Royal Society Edinburgh* 102, 131–154.

Eriksson, S.K., Kutzer, M., Procek, J., Grobner, G. and Harryson, P. (2011) Tunable membrane binding of the intrinsically disordered dehydrin Lti30, a cold-induced plant stress protein. *The Plant Cell* 23, 2391–2404.

Eyles, S.J. and Gierasch, L.M. (2010) Nature's molecular sponges: small heat shock proteins grow into their chaperone roles. *Proceedings of the National Academy of Sciences USA* 107, 2727–2728.

Fait, A., Angelovici, R., Less, H., Ohad, I., Urbanczyk-Wochniak, E., Ferni, A.R. and Galili, G. (2006) *Arabidopsis* seed development and germination is associated with temporally distinct metabolic switches. *Plant Physiology* 14, 839–854.

Fan, C. (2014) Investigation into the formation and protective functioning of natural deep eutectic solvents (NADES) *in vitro*. BSc thesis, University of Cape Town, Cape Town, South Africa.

Farrant, J.M. (2000) Comparison of mechanisms of desiccation tolerance among three angiosperm resurrection plants. *Plant Ecology* 151, 29–39.

Farrant, J.M. (2007) Mechanisms of desiccation tolerance in angiosperm resurrection plants. In: Jenks, M.A. and Wood, A.J. (eds) *Plant Desiccation Tolerance*. Blackwell Publishing, Oxford, UK, pp. 51–90.

Farrant, J.M. and Moore, J.P. (2011) Programming desiccation-tolerance: from plants to seeds to resurrection plants. *Current Opinions in Plant Biology* 14, 340–345.

Farrant, J.M., Pammenter, N.W. and Berjak, P. (1989) Germination-associated events and the desiccation sensitivity of recalcitrant seeds – a study on three unrelated species. *Planta* 178, 189–198.

Farrant, J.M., Bartsch, S., Loffell, D., Van der Willigen, C. and Whittaker, A. (2003) An investigation into the effects of light on the desiccation of three resurrection plants species. *Plant Cell and Environment* 26, 1275–1286.

Farrant, J.M., Brandt, W. and Lindsey, G.G. (2007) An overview of mechanisms of desiccation tolerance in selected angiosperm resurrection plants. *Plant Stress Journal* 1, 72–84.

Farrant, J.M., Lehner, A., Cooper, K. and Wiswedel, S. (2009) Desiccation tolerance in the vegetative tissues of the fern *Mohria caffrorum* is seasonally regulated. *The Plant Journal* 57, 65–79.

Farrant, J.M., Dace, H.J.W., Cooper, K., Hilgart, A., Peton, N., Mundree, S.G, Rafudeen, M.S. and Thomson, J.A. (2015) A molecular physiological review of vegetative desiccation tolerance in the resurrection plant *Xerophyta viscosa* (Baker) with reference to biotechnological application for the production of drought tolerant cereals. *Planta* 242, 407–426.

Foster, T.J., Ablett, S., McCann, M.C. and Gidley, M.J. (1996) Mobility resolved ^{13}C-NMR spectroscopy of primary plant cell walls. *Biopolymers* 39, 51–66.

Foyer, C.H. and Halliwell, B. (1976) The presence of glutathione and glutathione reductase in chloroplasts: a proposed role in ascorbic acid metabolism. *Planta* 133, 21–25.

Furini, A., Koncz, C., Salamini, F. and Bartels, D. (1997) High level transcription of a member of a repeated gene family confers dehydration tolerance to callus tissue of *Craterostigma plantagineum*. *EMBO Journal* 16, 3599–3608.

Gaff, D.F. (1977) Desiccation tolerant vascular plants of Southern Africa. *Oecologia* 31, 95–109.

Gaff, D.F. (1987) Desiccation tolerant plants in South America. *Oecologia* 74, 133–136.

Gaff, D.F. (1989) Responses of desiccation tolerant 'resurrection' plants to water deficit. In: Kreeb, K.H., Richter, H. and Hinckley, T.M. (eds) *Adaptation of Plants to Water and High Temperature Stress*. Academic Publishing, The Hague, The Netherlands, pp. 207–230.

Gaff, D.F. and Loveys, B.R. (1992) Abscisic acid levels in drying plants of a resurrection grass. *Transactions of Malaysian Society Plant Physiology* 3, 286–287.

Gaff, D.F. and Oliver, M.J. (2013) The evolution of desiccation tolerance in angiosperm plants: a rare yet common phenomenon. *Functional Plant Biology* 40, 315–328.

Gaff, D.F., Blomstedt, C.K., Neale, A.D, Le, T.N., Hamill, J.D. and Ghasempour, H.R. (2009) *Sporobolus stapfianus*, a model desiccation tolerant grass. *Functional Plant Biology* 36, 589–599.

Galau, G.A., Hughes, D.W. and Dure III, L. (1986) Abscisic acid induction of cloned cotton late embryogenesis-abundant (*Lea*) mRNAs. *Plant Molecular Biology* 7, 155–170.

Gallardo, K., Firnhaber, C., Zuber, H., Hericher, D., Belghazi, M., Henry, C., Kuster, H. and Thompson, R. (2007) A combined proteome and transcriptome analysis of developing *Medicago truncatula* seeds: evidence for metabolic specialization of maternal and filial tissues. *Molecular and Cellular Proteomics* 6, 2165–2179.

Gao, B., Zhang, D., Li, X., Yang, H., Zhang, Y. and Wood, A. (2015) *De novo* transcriptome characterization and gene expression profiling of the desiccation tolerant moss *Bryum argenteum* following rehydration. *BMC Genomics* 16, 416.

Garay-Arroyo, A., Colmenero-Flores, J.M., Garciarrubio, A. and Covarrubias, A.A. (2000) Highly hydrophilic proteins in prokaryotes and eukaryotes are common during conditions of water deficit. *Journal of Biological Chemistry* 275, 5668–5674.

Garrido, C., Paul, C., Seigneuric, R. and Kampinga, H.H. (2012) The small heat shock proteins family: the long forgotten chaperones. *International Journal of Biochemistry and Cell Biology* 44, 1588–1592.

Gasulla, F., vom Dorp, K., Dombrink, I., Zähringer, U., Gisch, N., Dörmann, P. and Bartels, D. (2013) The role of lipid metabolism in the acquisition of desiccation tolerance in *Craterostigma plantagineum*: a comparative approach. *The Plant Journal* 75, 726–741.

Gechev, T.S., Benina, M., Obata, T., Tohge, T., Sujeeth, N., Minkov, I., Hille, J., Temanni, M.R., Marriott, A.S., Berg-ström, E., Thomas-Oates, J., Antonio, C., Mueller-Roeber, B., Schippers, J.H.M., Fernie, A.R. and Toneva, V. (2013) Molecular mechanisms of desiccation tolerance in the resurrection glacial relic *Haberlea rhodopensis*. *Cellular and Molecular Life Sciences* 70, 689–709.

Georgieva, K., Szigeti, Z., Savarti, E., Gaspar, L., Maslenkova, L., Peeva, V, Peli, E. and Tuba, Z. (2007) Photosyn-thetic activity of homoiochlorophyllous desiccation tolerant plant *Haberlea rhodopensis* during dehydration and rehydration. *Planta* 225, 955–964.

Georgieva, K., Röding, A. and Büchel, C. (2009) Changes in some thylakoid membrane proteins and pigments upon desiccation of the resurrection plant *Haberlea rhodopensis*. *Journal of Plant Physiology* 166, 1520–1528.

Ghasempour, H.R., Gaff, D.F., Williams, R.D. and Gianello, R.D. (1998) Contents of sugars in leaves of drying des-iccation tolerant flowering plants, particularly grasses. *Plant Growth Regulation* 24, 185–191.

Ghasempour, H.R., Anderson, E.M. and Gaff, D.F. (2001) Effects of growth substances on the protoplasmic drought tolerance of leaf cells of the resurrection grass, *Sporobolus stapfianus*. *Australian Journal of Plant Physiology* 28, 1115–1120.

Giarola, V., Challabathula, D. and Bartels, D. (2015) Quantification of expression of dehydrin isoforms in the des-iccation tolerant plant *Craterostigma plantagineum* using specifically designed reference genes. *Plant Science* 236, 103–115.

Ginbot, Z.G. (2011) Characterization of two, desiccation-linked, group 1 LEA proteins from the resurrection plant *Xerophyta humilis*. PhD thesis, University of Cape Town, Cape Town, South Africa.

Goldberg, R., Morvan, C., Hervé du Penhoat, C. and Michen, V. (1989) Structure and properties of acidic polysac-charides of mung bean hypocotyls. *Plant Cell Physiology* 30, 163–173.

Goyal, K., Pinelli, C., Maslen, S.L., Rastogi, R.K., Stephens, E. and Tunnacliffe, A. (2005) Dehydration-regulated processing of late embryogenesis abundant protein in a desiccation-tolerant nematode. *FEBS Letters* 579, 4093–4098.

Halliwell, B. and Gutteridge, J.M.C. (1999) *Free Radicals in Biology and Medicine*, 3rd edn. Oxford University Press, Oxford, UK.

Hara, M. (2010) The multifunctionality of dehydrins. *Plant Signaling Behavior* 5, 503–508.

Hara, M., Fujinaga, M. and Kuboi, T. (2004) Radical scavenging activity and oxidative modification of citrus dehydrin. *Plant Physiology and Biochemistry* 42, 657–662.

Hara, M., Shinoda, Y., Tanaka, Y. and Kuboi, T. (2009) DNA binding of citrus dehydrin promoted by zinc ion. *Plant Cell And Environment* 32, 532–541.

Heilmeier, H. and Hartung, W. (2014) The aquatic resurrection plant *Chamaegigas intrepidus* – adaptation to mul-tiple abiotic stresses and habitat isolation. *Botanica Serbica* 38, 69–80.

Hernandez-Marin, E. and Martínez, A. (2012) Carbohydrates and their free radical scavenging capability: a the-oretical study. *The Journal of Physical Chemistry* 116, 9668–9675.

Heyen, B.J., Alsheikh, M.K., Smith, E.A., Torvik, C.F., Seals, D.F. and Randall, S.K. (2002) The calcium-binding activity of a vacuole-associated, dehydrin-like protein is regulated by phosphorylation. *Plant Physiology* 130, 675–687.

Hoekstra, F.A., Golovian, E.A. and Buitink, J. (2001) Mechanisms of plant desiccation tolerance. *Trends in Plant Science* 6, 431–438.

Höfler, K. (1941) Uber die austrocknungsfahigkeit des protoplasmas. *Berichte der deutschen botanischen Geselschaft* 60, 94–106.

Holzinger, A. and Karsten, U. (2013) Desiccation stress and tolerance in green algae: consequences for ultrastructure, physiological and molecular mechanisms. *Frontiers in Plant Science* 4, 327.

Horbowicz, M. and Obendorf, R. (1994) Seed desiccation tolerance and storability: dependence on flatulence-producing oligosaccharides and cyclitols – review and survey. *Seed Science Research* 4, 385–405.

Huang, H. and Song, S. (2013) Change in desiccation tolerance of maize embryos during development and germination at different water potential PEG-6000 in relation to oxidative process. *Plant Physiology and Biochemistry* 68, 61–70.

Hundertmark, M. and Hincha, D.K. (2008) LEA (late embryogenesis abundant) proteins and their encoding genes in *Arabidopsis thaliana*. *BMC Genomics* 9, 118.

Hundertmark, M., Dimova, R., Lengefeld, J., Seckler, R. and Hincha, D.K. (2011) The intrinsically disordered late embryogenesis abundant protein LEA18 from *Arabidopsis thaliana* modulates membrane stability through binding and folding. *Biochimica et Biophysica Acta* 1808, 446–453.

Hussain, S.S., Kayani, M.A. and Amjad, M. (2011) Transcription factors as tools to engineer enhanced drought stress tolerance in plants. *Biotechnology Progress* 27, 297–306.

Iljin, W.S. (1957) Drought resistance in plants and physiological processes. *Annual Review of Plant Physiology* 3, 341–363.

Illing, N., Denby, K., Collett, H., Shen, A. and Farrant, J.M. (2005) The signature of seeds in resurrection plants: a molecular and physiological comparison of desiccation tolerance in seeds and vegetative tissues. *Integrative and Comparative Biology* 45, 771–787.

Ingle, R.A., Schmidt, U., Farrant, J.M., Mundree, S.G. and Thompson, J.A. (2007) Proteomic analysis of leaf proteins during dehydration of the resurrection plant *Xerophyta viscosa*. *Plant Cell and Environment* 30, 435–446.

Ingle, R.A., Collett, H., Cooper, K., Takahashi, Y., Farrant, J.M. and Illing, N. (2008) Chloroplast biogenesis during rehydration of the resurrection plant *Xerophyta humilis*: parallels to the etioplast-chloroplast transition. *Plant, Cell and Environment* 31, 1813–1824.

Jahns, P. and Holzwarth, A.R. (2012) The role of xanthophyll cycle and of lutein in photoprotection of photosystem II. *Biochimica et Biophysica Acta* 1817, 182–193.

Jiang, G., Wang, Z., Shang, H., Yang, W., Hu, Z., Phillips, J. and Deng, X. (2007) Proteome analysis of leaves from the resurrection plant *Boea hygrometrica* in response to dehydration and rehydration. *Planta* 225, 1405–1420.

Kahkonen, M.P., Hopia, A.I., Vuorela, H.J., Ruaha, J.P., Pihlaja, K.K. and Heinonen, T.S. (1999) Antioxidant activity of plant extracts containing phenolic compounds. *Journal of Agricultural and Food Chemistry* 47, 3562–3954.

Kalemba, E.M. and Pukacka, S. (2007) Possible roles of LEA proteins and sHSPs in seed protection: a short review. *Biological Letters* 44, 3–16.

Kamies, R. (2011) A methodological investigation of the proteins present in the roots of the resurrection plant, *Xerophyta viscosa* for further proteomic analyses. MSc thesis, University of Cape Town, Cape Town, South Africa.

Kamies, R., Rafudeen, S.R. and Farrant, J.M. (2010) The use of aeroponics to investigate antioxidant activity in the roots of the resurrection plant, *Xerophyta viscosa*. *Plant Growth Regulation* 62, 203–211.

Kappen, L. and Valladares, F. (1999) Opportunistic growth and desiccation tolerance: the ecological success of poikilohydrous autotrophs. In: Pugnaire, F.I. and Valladares, F. (eds) *Handbook of Functional Plant Ecology*. Marcel Dekker, New York, pp. 10–80.

Kappen, L. and Valladares, F. (2007) Opportunistic growth and desiccation tolerance: the ecological success of poikilohydrous autotrophs. In: Pugnaire, F.I. and Valladares, F. (eds) *Functional Plant Ecology*. CRC Press, New York, pp. 7–65.

Kermode, A.R. (1990) Regulatory mechanisms involved in the transition from seed development to germination. *Plant Science* 9, 155–195.

Kermode, A.R. and Finch-Savage, W.E. (2002) Desiccation sensitivity in orthodox and recalcitrant seeds in relation to development. In: Black, M. and Pritchard, H.W. (eds) *Desiccation and Survival in Plants – Drying without Dying*. CAB International, Wallingford, UK, pp. 149–184.

Keunen, E., Peshev, D., van Gronsveld, J., van den Ende, W. and Cuypers, A. (2013) Plant sugars are crucial players in the oxidative challenge during abiotic stress: extending the traditional concept. *Plant, Cell and Environment* 36, 1242–1255.

Kirch, H.H., Nair, D. and Bartels, D. (2001) Novel ABA- and dehydration-inducible aldehyde dehydrogenase genes isolated from the resurrection plant *Craterostigma plantagineum* and *Arabidopsis thaliana*. *Plant Journal* 28, 555–567.

Koag, M.-C., Wilkens, S., Fenton, R.D., Resnik, J., Vo, E. and Close, T.J. (2009) The K-segment of maize DHN1 mediates binding to anionic phospholipid vesicles and concomitant structural changes. *Plant Physiology* 150, 1503–1514.

Koonjul, P.K., Brandt, W.W., Farrant, J.M. and Lindsey, G.G. (1999) Inclusion of polyvinylpyrrolidone in the polymerase chain reaction reverses the inhibitory effects of polyphenolic contamination of RNA. *Nucleic Acid Research* 27, 915–916.

Kovacs, D., Kalmar, E., Torok, Z. and Tompa, P. (2008) Chaperone activity of ERD10 and ERD14, two disordered stress-related plant proteins. *Plant Physiology* 147, 381–390.

Kranner, I. and Birtić, S. (2005) A modulating role for antioxidants in desiccation tolerance. *Integrative and Comparative Biology* 45, 734–740.

Kranner, I., Birtić, S., Anderson, K.M. and Pritchard, H.W. (2006) Glutathione halfcell reduction potential: a universal stress marker and modulator of programmed cell death? *Free Radical Biology and Medicine* 40, 2155–2165.

Kranner, I., Beckett, R., Hochman, A. and Nash, T.H. (2009) Desiccation-tolerance in lichens: a review. *The Bryologist* 111, 576–593.

Kruger, C., Berkowitz, O., Stephan, U.W. and Hell, R. (2002) A metal-binding member of the late embryogenesis abundant protein family transports iron in the phloem of *Ricinus communis* L. *Journal of Biological Chemistry* 277, 25062–25069.

Lazarides, M. (1992) Resurrection grasses (Poaceae) in Australia. In: Chapman, G.P. (ed.) *Desertified Grasslands: Their Biology and Management*. Academic Press, London, UK, pp. 213–234.

Le, T.N., Blomstedt, C.K., Kuang, J., Tenlen, J., Gaff, D.F., Hamill, J.D. and Neale, A.D. (2007) Desiccation-tolerance specific gene expression in leaf tissue of the resurrection plant *Sporobolus stapfianus*. *Functional Biology* 34, 589–600.

Lee, J. and Koh, H.J. (2011) A label-free quantitative shotgun proteomics analysis of rice grain development. *Proteome Science* 9, 61.

Leopold, A.C. and Vertucci, C.W. (1986) *Membranes, Metabolism and Dry Organisms*. Cornell University Press, Ithaca, New York.

Leprince, O. and Buitink, J. (2010) Desiccation tolerance: from genomics to the field. *Plant Science* 179, 554–564.

Levitt, J. (1980) *Responses of Plants to Environmental Stresses*, Vol 2. Academic Press, New York.

Lin, T. and Huang, N. (1994) The relationship between carbohydrate composition of some tree seeds and their longevity. *Journal of Experimental Botany* 45, 1289–1294.

Linster, C.L. and Clarke, S.G. (2008) L-Ascorbate biosynthesis in higher plants: the role of VTC2. *Trends in Plant Science* 13, 567–573.

Linster, C.L., Gomez, T.A., Christensen, K.C., Adler, L.N., Young, B.D., Brenner, C. and Clarke, S.G. (2007) *Arabidopsis* VTC2 encodes a GDP-L-galactose phosphorylase, the last unknown enzyme in the Smirnoff-Wheeler Pathway to ascorbic acid in plants. *The Journal of Biological Chemistry* 282, 18879–18885.

Liu, Y., Rafudeen, M.S. and Farrant, J.M. (2009) Expression and function of heat stable proteins during dehydration in *Xerophyta humilis* leaves. *South African Journal of Botany* 75, 435.

Liu, Y., Chakrabortee, S., Li, R., Zheng, Y. and Tunnacliffe, A. (2011) Both plant and animal LEA proteins act as kinetic stabilisers of polyglutamine-dependent protein aggregation. *FEBS Letters* 585, 630–634.

Long, R.L., Gorecki, M.J., Renton, M., Scott, J.K., Colville, L., Goggin, D.E., Commander, L.E., Westcott, D.A., Cherry, H. and Finch-Savage, W.E. (2015) The ecophysiology of seed persistence: a mechanistic view of the journey to germination or demise. *Biological Reviews* 90, 31–59.

Louise, J. and Simon, M.M. (2004) A role for expansins in dehydration and rehydration of the resurrection plant *Craterostigma plantagineum*. *FEBS Letters* 559, 61–65.

Lyall, R., Ingle, R.A. and Illing, N. (2014) The window of desiccation tolerance shown by early-stage germinating seedlings remains open in the resurrection plant, *Xerophyta viscosa*. *PLoS ONE* 9, e93093. doi: 10.1371/journal.pone.0093093.

Ma, C., Wang, H., Macnish, A.J., Estrada-melo, A.C., Lin, J., Chang, Y. and Reid, M.S. (2015) Transcriptomic analysis reveals numerous diverse protein kinases and transcription factors involved in desiccation tolerance in the resurrection plant *Myrothamnus flabellifolia*. *Horticulture Research* 2, 1–12.

Maia, J., Dekkers, B.J.W., Provart, N.J., Ligterink, W. and Hilhorst, H. (2011) The re-establishment of desiccation tolerance in germinated *Arabidopsis thaliana* seeds and its associated transcriptome. *PLoS ONE* 6, e29123. doi: 0.1371/journal.pone.0029123.

Maia, J., Dekkers, B.J.W., Dolle, M., Ligterink, W. and Hilhorst, H.W.M. (2014) Abscisic acid (ABA) sensitivity regulates desiccation tolerance in germinated *Arabidopsis* seeds. *New Phytologist* 203, 81–93.

Manfre, A.J., Lanni, L.M. and Marcotte Jr, W.R. (2006) The *Arabidopsis* group 1 late embryogenesis abundant protein ATEM6 is required for normal seed development. *Plant Physiology* 140, 140–149.

McDonald, E.T., Bortolus, M., Koteiche, H.A. and McHaourab, H.S. (2012) Sequence, structure, and dynamic determinants of Hsp27 (HspB1) equilibrium dissociation are encoded by the N-terminal domain. *Biochemistry* 51, 1257–1268.

Mello-Silva, R., Yara, D., Santos, D.Y.A., Salatino, M.L.F., Motta, L.B., Cattai, M.B., Sasaki, D., Lovo, J., Pita, P.B., Rocini, C., Rodrigues, C.D.N., Zarrei, M. and Chase, M.W. (2011) Five vicarious genera from Gondwana: the Velloziaceae as shown by molecules and morphology. *Annals of Botany* 108, 87–102.

Mihaylova, D., Bahchevanska, S. and Toneva, V. (2013) Examination of the antioxidant activity of *Haberlea rhodopensis* leaf extracts and their phenolic constituents. *Journal of Food Biochemistry* 37, 255–261.

Mitra, J., Xu, G., Wang, B., Meijing, L. and Deng, X. (2013) Understanding desiccation tolerance using the resurrection plant *Boea hygrometrica* as a model system. *Frontiers in Plant Science* 4, 446–455.

Moore, J., Westall, K.L., Ravenscroft, N., Farrant, J.M., Lindsey, G.G. and Brandt, W.F. (2005) The predominant polyphenol in the leaves of the resurrection plant *Myrothamnus flabellifolia*, 3,4,5 tri-O-galloylquinic acid, protects membranes against desiccation and free radical-induced oxidation. *Biochemical Journal* 385, 301–308.

Moore, J.P., Nguema-Ona, E., Chevalier, L.M., Lindsey, G.G., Brandt, W., Lerouge, P., Farrant, J.M. and Driouich, A. (2006) The response of the leaf cell wall to desiccation in the resurrection plant *Myrothamnus flabellifolia*. *Plant Physiology* 141, 651–662.

Moore, J., Lindsey, G.G., Farrant, J.M. and Brandt, W.F. (2007a) An overview of the biology of the desiccation-tolerant plant *Myrothamnus flabellifolia*. *Annals of Botany* 99, 211–217.

Moore, J.P., Hearshaw, M., Ravenscroft, N., Lindsey, G.G., Farrant, J.M. and Brandt, W.F. (2007b) Desiccation-induced ultrastructural and biochemical changes in the leaves of the resurrection plants *Myrothamnus flabellifolia*. *Australian Journal of Botany* 55, 482–491.

Moore, J.P., Le, N.T., Brandt, W.F., Driouich, A. and Farrant, J.M. (2009) Towards a systems-based understanding of plant desiccation tolerance. *Trends in Plant Science* 14, 110–117.

Moore, J.P., Nguema-Ona, E.E., Vicre-Gibouin, M., Sørensen, I., Willats, W.G.T, Driouich, A. and Farrant, J.M. (2012) Arabinose-rich polymers as an evolutionary strategy to plasticize resurrection plant cell walls against desiccation. *Planta* 237, 1–16.

Mowla, S.B., Thomson, J.A., Farrant, J.M. and Mundree, S.G. (2002) A novel stress inducible antioxidant enzyme identified from the resurrection plant *Xerophyta viscosa*. *Planta* 215, 716–726.

Moyankova, D., Mladenov, P., Berkov, S., Peshev, D., Georgieva, D. and Djilianov, D. (2014) Metabolic profiling of the resurrection plant *Haberlea rhodopensis* during desiccation and recovery. *Physiologia Plantarum* 152, 675–687.

Mtwisha, L., Farrant, J., Brandt, W. and Lindsey, G.G. (2006) Protection mechanisms against water deficit stress: desiccation tolerance in seeds as a study case. In: Ribaut, J. (ed.) *Drought Adaptation in Cereals*. Haworth Press, New York, pp. 531–549.

Mulako, I., Farrant, J.M., Collett, H. and Illing, N. (2008) HC205, a novel member of a metalloenzyme superfamily, the Vicinal Oxygen Chelate (VOC) is upregulated during desiccation in *Xerophyta humilis* (Bak) Dur and Schinz. *Journal Experimental Botany* 59, 3885–3901.

Mundree, S.G. and Farrant, J.M. (2000) Some physiological and molecular insights into the mechanisms of desiccation tolerance in the resurrection plant *Xerophyta viscosa* Baker. In: Cherry, J. (ed.) *Plant Tolerance to Abiotic Stresses in Agriculture: Role of Genetic Engineering*. Kluwer Academic Publishers, Dordrecht, The Netherlands, pp. 201–222.

Munne-Bosch, S. and Alegre, L. (2002) The function of tocopherols and tocotrienols in plants. *Critical Reviews in Plant Science* 21, 31–57.

Munnik, T. and Vermeer, J. (2010) Osmotic stress-induced phosphoinositide and inositol phosphate signalling in plants. *Plant Cell and Environment* 33, 655–669.

Mymrikov, E.V., Seit-Nebi, A.S. and Gusev, N.B. (2011) Large potentials of small heat shock proteins. *Physiological Reviews* 91, 1123–1159.

Nakamoto, H. and Vígh, L. (2007) The small heat shock proteins and their clients. *Cellular and Molecular Life Sciences* 64, 294–306.

Ndima, T.B., Farrant, J.M., Thomson, J.A. and Mundree, S.G. (2001) Molecular characterization of *XVT8*, a stress-responsive gene from the resurrection plant *Xerophyta viscosa* Baker. *Plant Growth Regulation* 35, 137–145.

Neale, A.D., Blomstedt, C.K., Bronson, P., Le, T.-J., Guthridge, K., Evans, J., Gaff, D.F. and Hammill, J.D. (2000) The isolation of genes from the resurrection grass *Sporobolus stapfianus* which are induced during severe drought stress. *Plant, Cell and Environment* 23, 265–277.

Nishizawa, A., Yabuta, Y. and Shigeoka, S. (2008) Galactinol and raffinose constitute a novel function to protect plants from oxidative damage. *Plant Physiology* 147, 1251–1263.

Noctor, G. and Foyer, C.H. (1998) Ascorbate and glutathione: keeping active oxygen under control. *Annual Review of Plant Physiology and Plant Molecular Biology* 49, 249–279.

Noctor, G., Mhamdi, A. and Foyer, C.H. (2014) The roles of reactive oxygen metabolism in drought: Not so cut and dried. *Plant Physiology* 164, 1636–1648.

Norwood, M., Truesdale, M.R., Richter, A. and Scott, P. (2000) Photosynthetic carbohydrate metabolism in the resurrection plant *Craterostigma plantagineum*. *Journal of Experimental Botany* 51, 159–165.

Okazaki, Y. and Saito, K. (2014) Roles of lipids as signaling molecules and mitigators during stress response in plants. *The Plant Journal* 79, 584–596.

Oliver, M.J. (2007) Lessons on dehydration tolerance from desiccation tolerant plants. In: Jenks, M. and Wood, A. (eds) *Plant Desiccation Tolerance*. Blackwell Publishing, Oxford, UK, pp. 11–50.

Oliver, M.J., Wood, A.J. and O'Mahony, P. (1998) "To dryness and beyond" – preparation for the dried state and rehydration in vegetative desiccation-tolerant plants. *Plant Growth Regulation* 24, 193–201.

Oliver, M.J., Tuba, Z. and Mishler, B.D. (2000) The evolution of vegetative desiccation tolerance in land plants. *Plant Ecology* 151, 85–100.

Oliver, M.J., Velten, J. and Mishler, B.D. (2005) Desiccation tolerance in bryophytes: a reflection of the primitive strategy for plant survival in dehydrating habitats? *Integrative Computational Biology* 45, 788–799.

Oliver, M.J., Hudgeons, J., Dowd, S.E. and Payton, P.R. (2009) A combined subtractive suppression hybridization and expression profiling strategy to identify novel desiccation response transcripts from *Tortula ruralis* gametophytes. *Physiologia Plantarum* 136, 437–460.

Oliver, M.J., Cushman, J.C. and Koster, K.L. (2010a) Dehydration tolerance in plants. *Methods in Molecular Biology* 639, 3–24.

Oliver, M.J., Jain, R., Balbuena, T.S., Agrawal, G., Gasulla, F. and Thelen, J.J. (2010b) Proteome analysis of leaves of the desiccation-tolerant grass, *Sporobolus stapfianus*, in response to dehydration. *Phytochemistry* 72, 1273–1284.

Oliver, M.J., Balbuena, J.R., Agrawal, G., Gasulla, F. and Thelen, J.J. (2011a) Proteome analysis of leaves of the desiccation-tolerant grass, *Sporobolus stapfianus*, in response to dehydration. *Phytochemistry* 72, 1273–1984.

Oliver, M.J., Guo, L., Ryals, A.D.C., Wone, B.W. and Cushman, J.C. (2011b) A sister group contrast using untargeted global metabolomics analysis delineates the biochemical regulation underlying desiccation tolerance in *Sporobolus stapfianus*. *Plant Cell* 23, 1231–1248.

Peshev, D., Vergauwen, R., Moglia, A., Hideg, E. and Van den Ende, W. (2013) Towards understanding vacuolar antioxidant mechanisms: a role for fructans? *Journal of Experimental Botany* 64, 1025–1038.

Peters, S., Mundree, S.G., Thomson, J.A., Farrant, J.M. and Keller, F. (2007) Protection mechanisms in the resurrection plant *Xerophyta viscosa* (Baker): both sucrose and raffinose family oligosaccharides (RFOs) accumulate in leaves in response to water deficit. *Journal of Experimental Botany* 58, 1947–1956.

Phillips, J.R., Oliver, M.J. and Bartels, D. (2002) Molecular genetics of desiccation-tolerant systems. In: Black, M. and Pritchard, H.W. (eds) *Desiccation and Survival in Plants: Drying without Dying*. CAB International, Wallingford, UK, pp. 319–341.

Piatkowski, D., Schneider, K., Salamini, F. and Bartels, D. (1990) Characterisation of five abscisic acid-responsive cDNA clones isolated from the desiccation-tolerant plant *Craterostigma plantagineum* and their relationship to other water-stress genes. *Plant Physiology* 94, 1682–1688.

Plancot, B., Vanier, G., Maire, F., Bardor, M., Lerouge, P., Farrant, J.M., Moore, J.P., Driouich, A., Vicre-Gibouin, M., Afonso, C. and Loutelier-Bourhis, C. (2014) Structural characterization of arabinoxylans from two African plant species *Eragrostis nindensis* and *Eragrostis tef* by MALDI-MS, ESI-MS, IM-MS and GC-MS. *Rapid Communications in Mass Spectrometry* 28, 908–916.

Porembski, S. and Barthlott, W. (2000) Granitic and gneissic outcrops (inselbergs) as centers of diversity for desiccation-tolerant vascular plants. *Plant Ecology* 151, 19–28.

Porembski, S., Martinelli, G., Ohlemüller, R. and Barthlott, W. (1998) Diversity and ecology of saxicolous vegetation mats on inselbergs in the Brazilian Atlantic rainforest. *Diversity and Distributions* 4, 107–119.

Proctor, M.C.F. and Pence, V. (2002) Vegetative tissues: bryophytes, vascular resurrection plants and vegetative propagules. In: Black, M. and Pritchard, H.W. (eds) *Desiccation and Survival in Plants: Drying without Dying*. CAB International, Wallingford, UK, pp. 207–238.

Proctor, M.C.F. and Tuba, Z. (2002) Poikilohydry and homoihydry: antithesis or spectrum of possibilities? *New Phytologist* 156, 327–349.

Proctor, M.C.F., Ligrone, R. and Duckett, J.G. (2007) Desiccation tolerance in the moss *Polytrichum formosum*: physiological and fine-structural changes during desiccation and recovery. *Annals of Botany* 99, 75–93.

Rahman, L.N., Smith, G.S.T., Bamm, V.V., Voyer-Grant, J.M., Moffatt, B.A., Dutcher, J.R. and Harauz, G. (2011) Phosphorylation of *Thellungiella salsuginea* dehydrins TsDHN-1 and TsDHN-2 facilitates cation-induced conformational changes and actin assembly. *Biochemistry* 50, 9587–9604.

Rahman, L.N., McKay, F., Giuliani, M., Quirk, A., Moffatt, B.A., Harauz, G. and Dutcher, J.R. (2013) Interactions of *Thellungiella salsuginea* dehydrins TsDHN-1 and TsDHN-2 with membranes at cold and ambient temperatures – surface morphology and single-molecule force measurements show phase separation, and reveal tertiary and quaternary associations. *Biochimica et Biophysica Acta* 1828, 967–980.

Rakić, T., Lazarević, M., Jovanović, Z.S., Radović, S., Silak-Yakovlev, S., Stevanović, B. and Stevanović, V. (2014) Resurrection plants of the genus *Ramonda*: prospective survival strategies – unlock further capacity of adaptation or embark on the path of evolution? *Frontiers in Plant Science* 4, 550–558.

Ramanjulu, S. and Bartels, D. (2002) Drought- and desiccation-induced modulation of gene expression in plants. *Plant, Cell and Environment* 25, 141–151.

Rapparini, F., Neri, L., Mihailova, G., Petkova, S. and Georgieva, K. (2015) Growth irradiance affects the photoprotective mechanisms of the resurrection angiosperm *Haberlea rhodopensis* Friv. in response to desiccation and rehydration at morphological, physiological and biochemical levels. *Environmental and Experimental Botany* 113, 67–79.

Renard, G.M.G.C. and Jarvis, M.C. (1999) A cross polarization magic angle spinning ^{13}C nuclear magnetic resonance study of polysaccharides in sugar beet cell walls. *Plant Physiology* 119, 1315–1322.

Reyes, J.L., Rodrigo, M.-J., Colmenero-Flores, J.M., Gil, J.-V., Garay-Arroyo, A., Campos, F., Salamini, F., Bartels, D. and Covarrubias, A.A. (2005) Hydrophilins from distant organisms can protect enzymatic activities from water limitation effects *in vitro*. *Plant, Cell and Environment* 28, 709–718.

Reynolds, T.L. and Bewley, J.D. (1993) Characterization of protein synthetic changes in a desiccation-tolerant fern, *Polypodium virginianum*. Comparison of the effects of drying, rehydration and abscisic acid. *Journal of Experimental Botany* 44, 921–928.

Roberts, J.K., DeSimone, N.A., Lingle, W.L. and Dure III, L. (1993) Cellular concentrations and uniformity of cell-type accumulation of two LEA proteins in cotton embryos. *Plant Cell* 5, 769–780.

Röhrig, H., Schmidt, J., Colby, T., Bräutigam, A., Hufnagel, P. and Bartels, D. (2006) Desiccation of the resurrection plant *Craterostigma plantagineum* induces dynamic changes in protein phosphorylation. *Plant, Cell and Environment* 29, 1606–1617.

Röhrig, H., Colby, T., Schmidt, J., Harzen, A., Facchinelli, F. and Bartels, D. (2008) Analysis of desiccation-induced candidate phosphoproteins from *Craterostigma plantagineum* isolated with a modified metal oxide affinity chromatography procedure. *Proteomics* 8, 3548–3560.

Rolletschek, H., Borisjuk, L., Radchuk, R., Miranda, M., Heim, U., Wobus, U. and Weber, H. (2004) Seed-specific expression of a bacterial phosphoenolpyruvate carboxylase in *Vicia narbonensis* increases protein content and improves carbon economy. *Plant Biotechnology Journal* 5, 211–219.

Rolletschek, H., Koch, K., Wobus, U. and Borisjuk, L. (2005) Positional cues for the starch/lipid balance in maize kernels and resource partitioning to the embryo. *The Plant Journal* 42, 69–83.

Rosa, M., Prado, C., Podazza, G., Interdonato, R., González, J.A., Hilal, M. and Prado, F.E. (2009) Soluble sugars—metabolism, sensing and abiotic stress: a complex network in the life of plants. *Plant Signalling Behaviour* 4, 388–393.

Russouw, P.S., Farrant, J., Brandt, W., Maeder, D. and Lindsey, G.G. (1995) Isolation and characterization of a heat-soluble protein from pea (*Pisum sativum*) embryos. *Seed Science Research* 5, 137–144.

Ruuska, S.A., Schwender, J. and Ohlrogge, J.B. (2004) The capacity of green oilseeds to utilize photosynthesis to drive biosynthetic processes. *Plant Physiology* 136, 2700–2709.

Saito, R., Yamamoto, H., Makino, A., Sugimoto, T. and Miyake, C. (2011) Methylglyoxal functions as hill oxidant and stimulates the photoreduction of O_2 at photosystem I: a symptom of plant diabetes. *Plant, Cell and Environment* 34, 1454–1464.

Sakuma, Y., Maruyama, K., Qin, F., Osakabe, Y., Shinozaki, K. and Yamaguchi-Shinozaki, K. (2006) Dual function of an *Arabidopsis* transcription factor DREB2A in water-stress-responsive and heat-stress-responsive gene expression. *PNAS* 103, 18822–18827.

Sárvári, É., Mihailova, G., Solti, A., Keresztes, A., Velitchkova, M. and Georgieva, K. (2014) Comparison of thylakoid structure and organization in sun and shade *Haberlea rhodopensis* populations under desiccation and rehydration. *Journal of Plant Physiology* 171, 1591–1600.

Scharf, K.-D., Berberich, T., Ebersberger, I. and Nover, L. (2012) The plant heat stress transcription factor (Hsf) family: structure, function and evolution. *Biochimica et Biophysica Acta* 1819, 104–119.

Seel, W., Hendry, G.A.F. and Lee, J.A. (1992) Effects of desiccation on some activated oxygen processing enzymes and anti-oxidants in mosses. *Journal of Experimental Botany* 43, 1031–1037.

Sgherri, C., Stevanovic, B. and Navari-Izzo, F. (2004) Role of phenolics in the antioxidative status of the resurrection plant *Ramonda serbica* during dehydration and rehydration. *Physiologia Plantarum* 22, 478–485.

Shao, H-B., Liang, Z-S. and Shao, M.-A.(2005) LEA proteins in higher plants: structure, function, gene expression and regulation. *Colloids and Surfaces B: Biointerfaces* 45, 131–135.

Shen, A. (2014) The transcriptome response of leaves of the resurrection plant *Xerophyta humilis* to desiccation. PhD thesis, University of Cape Town, Cape Town, South Africa.

Sherwin, H.W. and Farrant, J.M. (1996) Differences in rehydration of three different desiccation-tolerant species. *Annals of Botany* 78, 703–710.

Sherwin, H.W. and Farrant, J.M. (1998) Protection mechanism against excess light in the resurrection plants *Craterostigma wilmsii* and *Xerophyta viscosa*. *Plant Growth Regulation* 24, 203–210.

Shih, M.D., Lin, S.D., Hsieh, J.S., Tsou, C.H., Chow, T.Y., Lin, T.P. and Hsing, Y.I.C. (2004) Gene cloning and characterization of a soybean (*Glycine max* L.) LEA protein, GmPM16. *Plant Molecular Biology* 56, 689–703.

Shin, J., Kim, S. and An, G. (2009) Rice aldehyde dehydrogenase7 is needed for seed maturation and viability. *Plant Physiology* 149, 905–915.

Smirnoff, N. (1993) The role of active oxygen in the response of plants to water deficit and desiccation. *New Phytologist* 125, 214–237.

Smirnoff, N. (1998) Plant resistance to environmental stress. *Current Opinion in Biotechnology* 9, 214–219.

Stacy, R.A.P., Nordeng, T.W., Culiáñez-Macià, F.A. and Aalen, R.B. (1999) The dormancy-related peroxiredoxin anti-oxidant, PER1, is localized to the nucleus of barley embryo and aleurone cells. *The Plant Journal* 19, 1–8.

Stark, L.R., Greenwood, J.L., Brinda, J.C. and Oliver, M.J. (2013) The desert moss *Pterygoneurum lamellatum* (Pottiaceae) exhibits an inducible ecological strategy of desiccation tolerance: effects of rate of drying on shoot damage and regeneration. *American Journal of Botany* 100, 1522–1531.

Steadman, K.J., Pritchard, H.W. and Dey, P.M. (1996) Tissue-specific soluble sugars in seeds as indicators of storage category. *Annals of Botany* 77, 667–674.

Still, D.W., Kovach, D.A. and Bradford, K.J. (1994) Development of desiccation tolerance during embryogenesis in rice (*Oryza sativa*) and wild rice (*Zizania palustris*) (dehydrin expression, abscisic acid content and sucrose accumulation). *Plant Physiology* 104, 431–438.

Stoyanova, S., Geuns, J., Hideg, E. and Van den Ende, W. (2011) The food additives inulin and stevioside counteract oxidative stress. *International Journal of Food Sciences and Nutrition* 62, 207–214.

Suarez Rodriguez, M.C., Edsga, D., Hussain, S., Alquezar, D., Rasmussen, M., Gilbert, T., Nielsen, B.H., Bartels, D. and Mundy, J. (2010) Transcriptomes of the desiccation-tolerant resurrection plant *Craterostigma plantagineum*. *The Plant Journal* 63, 212–228.

Suguiyama, V.F., Silver, E.A., Meirelles, S.T., Centeno, D.C. and Braga, M.R. (2014) Leaf metabolite profile of the Brazilian resurrection plant *Barbacenia purpurea* Hook. (Velloziaceae) shows two time-dependent responses during desiccation and recovering. *Frontiers in Plant Science* 5, 193.

Swire-Clark, G.A. and Marcotte Jr, W.R. (1999) The wheat LEA protein Em functions as an osmoprotective molecule in *Saccharomyces cerevisiae*. *Plant Molecular Biology* 39, 117–128.

Tatsimo, S.J.N., de Dieu Tamokou, J., Havyarimana, L., Csupor, D., Forgo, P., Hohmann, J., Kuiate, J-R. and Tane, P. (2012) Antimicrobial and antioxidant activity of kaempferol rhamnoside derivatives from *Bryophyllum pinnatum*. *BMC Research Notes* 5, 158.

Terrasson, E., Buitink, J., Righetti, K., Ly Vu, B., Pelletier, S., Zinsmeister, J., Lalanne, D. and Leprince, O. (2013) An emerging picture of the seed desiccome: confirmed regulators and newcomers identified using transcriptome comparison. *Frontiers in Plant Science* 4, 1–16.

Thornalley, P.J. (2003) Glyoxalase I – structure, function and a critical role in the enzymatic defence against glycation. *Biochemical Society Transactions* 31, 1343–1348.

Tompa, P., Bánki, P., Bokor, M., Kamasa, P., Kovács, D., Lasanda, G. and Tompa, K. (2006) Protein-water and protein-buffer interactions in the aqueous solution of an intrinsically unstructured plant dehydrin: NMR intensity and DSC aspects. *Biophysical Journal* 91, 2243–2249.

Tuba, Z., Lichtenthaler, H.K., Csintalan, Z. and Pocs, T. (1993a) Regreening of the desiccated leaves of the poikilochlorophyllous *Xerophyta scabrida* upon rehydration. *Journal of Plant Physiology* 142, 103–108.

Tuba, Z., Lichtenthaler, H.K., Maroti, I. and Csintalan, Z. (1993b) Resynthesis of thylakoids and functional chloroplasts in the desiccated leaves of the poikilochlorophyllous *Xerophyta scabrida* upon rehydration. *Journal of Plant Physiology* 142, 742–748.

Tuba, Z., Proctor, M. and Csintalan, Z. (1998) Ecophysiological responses of homoiochlorophyllous and poikilochlorophyllous desiccation tolerant plants: a comparison and an ecological perspective. *Plant Growth Regulation* 24, 211–217.

Tunnacliffe, A. and Wise, M.J. (2007) The continuing conundrum of the LEA proteins. *Naturwissenschaften* 94, 791–812.

Tyedmers, J., Mogk, A. and Bukau, B. (2010) Cellular strategies for controlling protein aggregation. *Nature* 11, 777–788.

Van den Ende, W. and Valluru, R. (2009) Sucrose, sucrosyloligosaccharides and oxidative stress: scavenging and salvaging? *Journal of Experimental Botany* 60, 9–18.

van der Willigen, C., Pammenter, N.W., Mundree, S.G. and Farrant, J.M. (2001) Some physiological comparisons between the resurrection grass, *Eragrostis nindensis*, and the related desiccation-sensitive species, *Eragrostis curvula*. *Plant Growth Regulation* 35, 121–129.

van der Willigen, C., Mundree, S.G., Pammenter, N.W. and Farrant, J.M. (2004) Mechanical stabilisation in desiccated vegetative tissues of the resurrection grass *Eragrostis nindensis*: does an alpha TIP and/or sub-cellular compartmentalization play a role? *Journal of Experimental Botany* 55, 651–661.

Velasco, R., Salamini, F. and Bartels, D. (1994) Dehydration and ABA increase mRNA levels and enzyme activity of cytosolic GADPH in the resurrection plant *Craterostigma plantagineum*. *Plant Molecular Biology* 26, 541–546.

Veljovic-Jovanovic, S., Kukavica, B. and Navari-Izzo, F. (2008) Characterization of polyphenol oxidase changes induced by desiccation of *Ramonda serbica* leaves. *Physiologia Plantarum* 132, 407–416.

Verdier, J., Lalanne, D., Pelletier, S., Torres-Jerez, I., Righetti, K., Bandyopadhyay, K., Leprince, O., Chatelain, E., Vu, B.L., Gouzy, J., Gamas, P., Udvardi, M.K. and Buitink, J. (2013) A regulatory network-based approach dissects late maturation processes related to the acquisition of desiccation tolerance and longevity of *Medicago truncatula* seeds. *Plant Physiology* 163, 757–774.

Vertucci, C.W. (1990) Calorimetric studies of the state of water in seed tissues. *Biophysical Journal* 58, 1463–1471.

Vertucci, C.W. and Farrant, J.M. (1995) Acquisition and loss of desiccation tolerance. In: Kigel, J. and Galili, G. (eds) *Seed Development and Germination*. Marcel Dekker Press Inc., New York, pp. 237–271.

Vicré, M., Sherwin, H.W., Driouich, A., Jaffer, M., Jauneau, A. and Farrant, J.M. (1999) Cell wall properties of hydrated and dry leaves of the resurrection plant *Craterostigma wilmsii*. *Journal of Plant Physiology* 155, 719–726.

Vicré, M., Lerouxel, O., Farrant, J.M., Lerouge, P. and Driouich, A. (2004) Composition and desiccation induced alterations of the cell wall in the resurrection plant *Craterostigma wilmsii*. *Physiologia Plantarum* 120, 229–239.

Vierling, E. (1991) The roles of heat shock proteins in plants. *Annual Review of Plant Physiology and Plant Molecular Biology* 42, 579–620.

Walford, S.A. (2008) Activation of seed-specific genes in leaves and roots of the desiccation tolerant plant, *Xerophyta humilis*. PhD thesis, University of Cape Town, Cape Town, South Africa.

Walford, S.A., Thomson, J.A., Farrant, J.M. and Mundree, S.G. (2004) Isolation and characterisation of a novel dehydration-induced Grp94 homologue from the resurrection plant *Xerophyta viscosa*. *South African Journal of Botany* 70, 741–750.

Walters, C., Farrant, J.M., Pammenter, N.W. and Berjak, P. (2002) Desiccation and damage. In: Black, M. and Pritchard, H.W. (eds) *Desiccation and Survival in Plants – Drying without Dying*. CAB International, Wallingford, UK, pp. 263–291.

Walters, C., Hill, L.M. and Wheeler, L.M. (2005) Dying while dry: kinetics and mechanisms of deterioration in desiccated organisms. *Integrative and Comparative Biology* 45, 751–758.

Wang, H., Cao, G.H. and Prior, R.L. (1996) Total antioxidant capacity of fruits. *Journal of Agricultural Food Chemistry* 44, 701–705.

Wang, W.Q., Ye, J.Q., Rogowska-Wrzesinska, A., Wojdyla, K.I., Jensen, O.N., Møller, I.M. and Song, S.Q. (2014) Proteomic comparison between maturation drying and prematurely imposed drying of *Zea mays* seeds reveals a potential role of maturation drying in preparing proteins for seed germination, seedling vigor, and pathogen resistance. *Journal of Proteome Research* 13, 606–626.

Wang, X., Chen, S., Zhang, H., Shi, L., Cao, F., Guo, L., Xie, Y., Wang, T., Yan, X. and Dai, S. (2010) Desiccation tolerance mechanism in resurrection fern-ally *Selaginella tamariscina* revealed by physiological and proteomic analysis. *Journal of Proteome Research* 9, 6561–6577.

Wang, X.Q., Yang, P.F., Liu, Z., Liu, W.Z., Hu, Y., Chen, H., Kuang, T.Y., Pei, Z.M., Shen, S.H. and He, Y.K. (2009) Exploring the mechanism of *Physcomitrella patens* desiccation tolerance through a proteomic strategy. *Plant Physiology* 149, 1739–1750.

Waters, E.R. (2013) The evolution, function, structure, and expression of the plant sHSPs. *Journal of Experimental Botany* 64, 391–403.

Waters, R. (2015) Gene expression and structural analysis of the "LEAome" in the resurrection plant *Xerophyta humilis* (Baker). MSc thesis, University of Cape Town, Cape Town, South Africa.

Wehmeyer, N., Hernandez, L.D., Finkelstein, R.R. and Vierling, E. (1996) Synthesis of small heat-shock proteins is part of the developmental program of late seed maturation. *Plant Physiology* 112, 747–757.

Whittaker, A., Martinelli, T., Bochicchio, A., Vazzana, C. and Farrant, J. (2004) Comparison of sucrose metabolism during the rehydration of desiccation-tolerant and desiccation-sensitive leaf material of *Sporobolus stapfianus*. *Physiologia Plantarum* 122, 11–20.

Whittaker, A., Martinelli, T., Farrant, J.M., Bochicchio, A. and Vazzana, C. (2007) Sucrose phosphate synthase activity and the co-ordination of carbon partitioning during sucrose and amino acid accumulation in desiccation-tolerant leaf material of the C_4 resurrection plant *Sporobolus stapfianus* during dehydration. *Journal of Experimental Botany* 58, 3775–87.

Wise, M.J. and Tunnacliffe, A. (2004) POPP the question: what do LEA proteins do? *Trends in Plant Science* 9, 13–17.

Wolkers, W.F., McCready, S., Brandt, W.F., Lindsey, G.G. and Hoekstra, F.A. (2001) Isolation and characterization of a D-7 LEA protein from pollen that stabilizes glasses *in vitro*. *Biochimica et Biophysica Acta* 1544, 196–206.

Xiao, L., Yang, G., Zhang, L., Yang, X., Zhao, S., Ji, Z., Zhou, Q., Hu, M., Wang, Y., Chen, M., Xu, Y., Jin, H., Xiao, X., Hu, G., Bao, F., Hu, Y., Wan, P., Li, L., Deng, X., Kuang, T., Xiang, C., Zhu, J., Oliver, M.J. and He, Y. (2015) The resurrection genome of *Boea hygrometrica* : a blueprint for survival of dehydration. *Proceedings of the National Academy of Sciences USA* 112, 5833–5837.

Xu, S.B., Li, T., Deng, Z.Y., Chong, K., Xue, Y.B. and Wang, T. (2008) Dynamic proteomic analysis reveals a switch between central carbon metabolism and alcoholic fermentation in rice filling grains. *Plant Physiology* 148, 908–925.

Yobi, A., Wone, B.W.M., Xu, W., Alexander, D.C., Guo, L., Ryals, J.A., Oliver, M.J. and Cushman, J.C. (2012) Comparative metabolic profiling between desiccation-sensitive and desiccation-tolerant species of *Selaginella* reveals insights into the resurrection trait. *The Plant Journal* 72, 983–999.

Yobi, A., Wone, B.W.M., Xu, W., Alexander, D.C., Guo, L., Ryals, J.A., Oliver, M.J. and Cushman, J.C (2013) Metabolic profiling in *Selaginella lepidophylla* at various hydration states provides new insights into the mechanistic basis of desiccation tolerance. *Molecular Plant* 6, 369–385.

Zhang, T., Fang, Y., Wang, X., Deng, X., Zhang, X., Hu, S. and Yu, J. (2012) The complete chloroplast and mitochondrial genome sequences of *Boea hygrometrica*: insights into the evolution of plant organellar genomes. *PLoS ONE* 7, 30531.

Zhang, Z., Wang, B., Sun, D. and Deng, X. (2013) Molecular cloning and differential expression of sHSP gene family members from the resurrection plant *Boea hygrometrica* in response to abiotic stresses. *Biologia* 68, 651–661.

Zhu, Y., Wang, B., Phillips, J., Zhange, Z.-N., Du, H., Xu, T., Huang, L.-C., Zhange, X.-F., Xu, G.-H., Li, W.-L., Wang, Z., Wang, L., Liu, Y.-X. and Deng, X. (2015) Global transcriptome analysis reveals acclimation-primed processes involved in the acquisition of desiccation tolerance in *Boea hygrometrica*. *Plant Cell Physiology* 56, 1429–1441.

10 Ultraviolet-B Radiation: Stressor and Regulatory Signal

Marcel A.K. Jansen*

School of Biological, Earth and Environmental Sciences,
University College Cork, Ireland

Abstract

Following the discovery of ozone layer depletion in the late 1980s, large numbers of studies investigated the effects of ambient and/or enhanced levels of ultraviolet-B (UV-B) radiation on plants, animals, humans and microorganisms. Initial studies reported severe, inhibitory UV effects on plant growth and development, and these were associated with damage to genetic material and the photosynthetic machinery. This led to a strong perception that UV radiation is harmful for plants. Since that time, a conceptual U-turn has taken place in the way that UV-B effects are perceived. Under realistic UV-B exposure conditions, accumulation of UV-mediated damage is a relatively rare event. Instead, it is now recognized that UV-B is predominantly an environmental regulator that controls cellular, metabolic, developmental and stress-protection processes in plants through a dedicated UV-B photoreceptor. UV-B regulated signalling pathways control, among others, expression of hundreds of genes, the biochemical make-up and the morphology of plants and this, in turn, can alter the nutritional value, pest and disease tolerance, sexual reproduction, and hardiness of plants and plant tissues. As a consequence, UV-B radiation can impact on trophic relationships and ecosystem function, but is also a potentially valuable tool for sustainable agriculture.

10.1 Ultraviolet-B in the Biosphere

Solar radiation is the main source of energy that enables life on Earth. Plant leaves, with their high surface area to volume ratio and their characteristic optical properties, have evolved to maximize capture of incoming solar radiation for photosynthesis. The placement of leaves at right angles with respect to incoming light, in some species in conjunction with leaf heliotropism, further increases capture of photosynthetic active radiation (PAR). An unavoidable consequence of the photoautotrophic lifestyle is, however, that plants are also exposed to solar ultraviolet (UV) radiation.

10.1.1 Ultraviolet radiation and the stratospheric ozone layer

The sun emits a mixture of UV, visible and infra-red radiation. The UV wavelengths in the solar spectrum can be categorized in the UV-C (100–280 nm), UV-B (280–315 nm) and UV-A (315–400 nm) bands. UV-C, and much of the solar UV-B wavelengths, are attenuated by the stratospheric ozone layer and therefore do not reach the Earth's surface (Fig. 10.1). Already in 1970 it was shown that nitrous oxides, emitted as a result of natural processes as well as human activities, may cause ozone layer depletion (Crutzen, 1970). However, it was the realization

---

* Corresponding author e-mail: m.jansen@ucc.ie

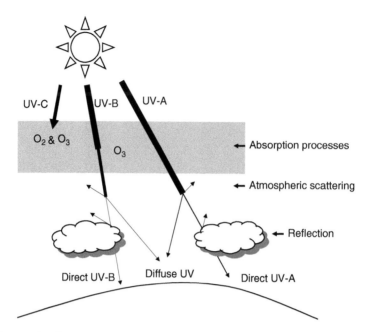

Fig. 10.1. Diagram showing attenuation of the UV-C, UV-B and UV-A wavelength bands in the Earth's atmosphere. UV wavelengths shorter than 242 nm are effectively absorbed by O_2, while wavelengths shorter than 290 nm are absorbed by both O_2 and O_3. Because of absorption processes, no solar radiation with wavelengths less than 290 nm is present in the biosphere. Scattering and reflection further alter the solar UV spectrum, and as a result a considerable proportion of the UV photons reach the Earth's surface as diffuse radiation.

that anthropogenically produced organo-halogens catalyse the breakdown of ozone at low stratospheric temperatures (Molina and Rowland, 1974), followed by the discovery of a springtime stratospheric ozone hole over the Antarctic (cf. McKenzie *et al.*, 2007), that raised profound concerns about increasing penetration of UV-B radiation in the biosphere. Concerns about the rise in UV levels were based on the increased threat of skin cancers in a broad range of species, as well as potential decreases in primary productivity. These concerns led to the ratification of the Montreal Protocol in 1987 (Montreal Protocol on Substances that Deplete the Ozone Layer), which restricted emissions of ozone-depleting substances such as chlorofluorocarbons. The state of the stratospheric ozone layer has since been regularly reviewed by the World Meteorological Organization (WMO) and the United Nations Environment Programme (UNEP). These reviews show that surface UV-B radiation levels have increased significantly in the period from 1979 to 2008 at all

latitudes, except near the equator (Ballaré *et al.*, 2011). Yet, as a direct result of the successful implementation of the Montreal Protocol, levels of ozone-depleting chemicals in the atmosphere are gradually returning to pre-1980 values. Current models predict that stratospheric ozone levels will recover, potentially reaching higher levels than prior to the onset of ozone layer depletion in the 1980s (McKenzie *et al.*, 2011). However, large uncertainties remain with respect to future UV levels in the biosphere, due – among others – to the confounding effects of global climate change on cloudiness, surface reflectivity and tropospheric aerosol loading (McKenzie *et al.*, 2011). Moreover, recent studies indicate that UV-B radiation can contribute to climate change by stimulating release of carbon and/or volatile organic compounds from plants, plant litter and soils (cf. Bornman *et al.*, 2015), while ozone depletion has been reported to directly change atmospheric and oceanic circulation, and therefore weather patterns (Robinson and Erickson,

2015). Thus, intricate feedbacks between ozone depletion and climate change have implications for plant UV responses.

10.1.2 Ultraviolet-B radiation: a dynamic environmental parameter

In order to assess the biological consequences of UV-B exposure, an appreciation of the dynamic character of this environmental parameter is necessary. UV-B fluence rates are highly variable (Fig. 10.2) and depend on geographical, temporal, meteorological and other environmental parameters (Aphalo *et al.*, 2012). Thus, time of day, season, weather conditions, atmospheric pollution, altitude, latitude and surrounding vegetation all affect the UV-B exposure levels of plants (McKenzie *et al.*, 2007). UV-B fluence rates increase during the morning, peak at solar noon and decrease throughout the afternoon (Fig. 10.2). Interestingly, it has been shown that some plant UV-B-protection responses similarly increase from dawn, peak at solar noon, while decreasing towards the evening (Barnes *et al.*, 2008). Thus, plant UV responses are closely aligned with exposure dynamics and knowledge

of the natural dynamics of UV-B exposure is essential for understanding plant UV responses. The shape of the daily UV-B fluence rate profile is relatively narrow compared to the PAR profile (i.e. UV-B is restricted to the hours around solar noon), resulting in a variable UV-B:PAR ratio that peaks at solar noon (Aphalo *et al.*, 2012). This clearly presents a technical difficulty in designing a realistic UV-B exposure experiment, but is nevertheless an important consideration as the UV-B:PAR ratio has been shown to be an important determinant of plant UV-B responses. Many UV effects are aggravated when PAR levels are low (Björn, 2015). For example, PAR and UV-B have been shown to have interactive effects on plant morphology and essential oil yield in peppermint (*Mentha* × *piperita*), such that it is concluded that production of top-quality oils requires exposure to high levels of natural, unscreened sunlight (Behn *et al.*, 2010). UV-B fluence rates also vary with geography. Highest UV-B fluence rates can be measured near the equator due to the steep angle of incoming solar radiation and because the stratospheric ozone layer is generally thinner in the tropics. Daily UV doses decrease gradually towards the poles and as a result polar regions receive approximately six times less UV compared to the equatorial zone (McKenzie *et al.*, 2007). High UV-B levels also occur at high-altitude sites, particularly when snow is present. More complex is the effect of cloud cover, which may either diminish or enhance surface UV-B levels depending on the degree of cover (McKenzie *et al.*, 2007). Thus, a picture has emerged of a highly dynamic environmental parameter, with some of the variation in fluence rates being predictable (e.g. variation linked to time of day, latitude or altitude) while other variation is much less predictable (e.g. variation linked to plant shading, atmospheric pollution or meteorological conditions) (Aphalo *et al.*, 2012).

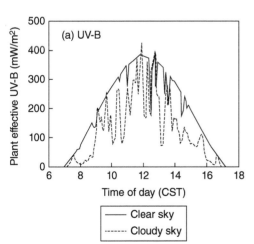

Fig. 10.2. Diurnal changes in solar UV-B radiation over two consecutive summer days in New Orleans, Louisiana, USA, which varied in degree of cloud cover. Solar UV-B was weighted according to the generalized higher plant action spectrum (reproduced from Barnes *et al.*, 2016, with permission of John Wiley and Sons).

10.1.3 Ultraviolet-B exposure studies

The dynamic character of UV-B radiation complicates the development of environmentally relevant UV-exposure experiments. This is a real concern as there is a strong awareness that the UV-B exposure strategy used determines both the environmental relevance of a study as well as, potentially, the biological outcome (Aphalo *et al.*, 2012). In the last decade, advanced UV-exposure methodologies have been developed, including appropriate UV sources, dimmers, filters, weighing factors and UV-measuring equipment. Spectral weighing factors have been developed to facilitate comparisons between studies using different light sources, as well as with natural solar UV-B (Aphalo *et al.*, 2012; Björn, 2015). These spectral weighing factors score narrow wavelength bands according to their impact on a particular biological process such as plant growth, and the total score is known as a biologically effective dose (Xu and Sullivan, 2010). In general, the damaging effects of UV radiation increase with decreasing wavelength such that shorter UV wavelengths contribute more to the total biologically effective dose. Clearly, the implication is that accurate spectral information about used UV sources is crucial when comparing studies, or when experimental data are extrapolated to field conditions.

Two distinct UV-B exposure strategies are commonly used. Laboratory UV-B exposure experiments enable the accurate study of UV-induced processes in a highly controlled manner. Such experiments have been instrumental in identifying transcription factors and other molecular and physiological elements of UV-B response pathways (Jenkins, 2014). Laboratory studies have also been useful in comparing UV tolerance of ecotypes, cultivars and/or species (Jansen *et al.*, 2010). However, in general, the results of laboratory studies cannot be automatically extrapolated to a more realistic environment due to the use of relatively high UV-B fluence rates, sometimes lack of accompanying UV-A and often low levels of PAR (Xu and Sullivan, 2010). Thus, differences in UV-B protection observed under laboratory conditions are not necessarily relevant under natural radiation conditions. Field studies whereby UV-B levels are either increased using UV-lamps, or decreased using cut-off filters, are potentially much more realistic from an environmental perspective, although care is required with respect to introduced microclimatic changes (Albert *et al.*, 2008). UV supplementation experiments typically focus on elucidating the impact of increased fluence rates of UV-B (for example mimicking thinning of the stratospheric ozone

layer), while exclusion experiments analyse the impact of current UV-B levels. In supplementation experiments, fluorescent UV-tubes are suspended above plants. In an ideal scenario, the output of these tubes is linked to ambient, solar UV-B levels generating a fixed percentage increase in incident UV-B radiation throughout the day. In exclusion experiments, plants are covered by screens with specific transmission properties, enabling removal of the UV-B or UV-B and UV-A wavelength bands from the solar spectrum (Newsham and Robinson, 2009). The term 'realistic' UV-exposure conditions is commonly used, but ill-defined, to mean those conditions that closely resemble a natural light environment. Field-based experiments are, obviously, more likely to be realistic but will suffer from irreproducible environmental variations, and typically will need to be pursued for relatively long periods in order to separate UV-B induced effects from other environmentally induced processes. The biological outcomes of several long-term UV-B studies under outdoor conditions have been reviewed, and this includes sites located in high Arctic Zackenberg (Albert *et al.*, 2008), high Arctic Svalbard (Rozema *et al.*, 2006) and subarctic Abisko (Phoenix *et al.*, 2001). These long-term experiments were specifically located in the Arctic and subarctic zones as these areas experience relatively large, seasonal increases in UV-B doses as a result of stratospheric ozone layer depletion. Of course, these experiments do not 'shed any new light' on the biological impacts of the much higher UV-B fluence rates at lower latitudes which are, however, much less affected by ozone layer depletion.

10.1.4   The importance of realistic ultraviolet-B exposure conditions

Increased awareness of realistic UV-B exposure conditions has been instrumental in obtaining realistic assessments of the biological effects of UV-B radiation on plant growth. The improved quality (i.e. realistic nature) of UV-B exposure studies during the last 15 years is an important outcome in environmental UV-B research. It is also an outcome that has led to a very substantial change in the perception of UV-B radiation effects on plants, a change in emphasis from stress towards regulation. Thus, use of advanced UV-B exposure protocols has led to a recognition that the severe stress effects reported in many older UV-B studies do not necessarily occur in the natural environment (Jansen and Bornman, 2012). There is one other important lesson to be learned from all this. Currently, much research is dedicated to the characterization of biological impacts of another type of global climate change, i.e. global warming. The lesson from three decades of UV-B research is that realistic exposure conditions should be a prime consideration for this rapidly expanding research field.

10.2 Ultraviolet-B Radiation: a Putative Environmental Stressor

Of all the solar wavelengths in the biosphere, UV-B radiation has the highest energy content per photon. As a result, UV-B photons can directly drive the photomodification of, among others, DNA, proteins and lipids. When plants are exposed to low levels of UV-B radiation, low levels of reactive oxygen species (ROS) are likely to be generated and these may contribute to the activation of UV-acclimation responses. In contrast, exposure to high UV-B doses can cause disruptions of cellular metabolism which are associated with the production of high levels of ROS (Fig. 10.3). The accumulation of ROS in tissues exposed to high doses of UV-B has been demonstrated using selective ROS probes and/or electron paramagnetic resonance (EPR) spin trapping reporters (Hideg *et al.*, 2013). Indirectly, ROS formation and/or oxidative stress has also been inferred from changes in the redox state of the ascorbate–dehydroascorbate and reduced-to-oxidized-glutathione redox pairs in UV-B exposed plants. ROS formation has been visualized as the accumulation of lipid peroxidation products such as malondialdehyde (MDA). These negative UV-B effects have been extensively reported, and in that sense these deleterious UV-B effects are 'real'. However, UV-B mediated oxidative damage is considered to be rare in healthy plants under environmentally relevant conditions (Hideg *et al.*, 2013). Thus, UV-B is a 'potential' stressor.

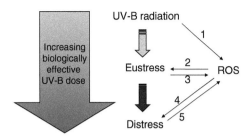

Fig. 10.3. UV-B radiation induces eustress, leading to UV-acclimation. UV-B-induced distress appears to be a relatively rare phenomenon under natural light conditions. There is some evidence that UV-B can directly induce reactive oxygen species (ROS) formation (1), although it is not clear to what extent this happens under realistic UV-B conditions. ROS that are formed may contribute to eustress and the UV acclimation response (2) or cause oxidative damage (4). Conversely, further ROS can be produced as part of the UV response of plants, either as part of UV-acclimation (3) or as a result of metabolic disruption (5) (reproduced from Hideg *et al.*, 2013, with permission of Elsevier).

10.2.1 Ultraviolet-B can cause DNA damage

Most studies of UV-B mediated damage in plants have focused on the negative impact of UV-B radiation on DNA integrity, photosynthetic function and/or biomass production. DNA directly absorbs in the UV-B part of the spectrum, giving rise to the formation of pre-mutagenic cyclobutane pyrimidine dimers (CPD) (Fig. 10.4) and, to a lesser extent, pyrimidine (6-4) pyrimidinone dimers (also known as 6-4 photoproducts, or 6-4PP) (Britt, 2004). These photoproducts block the progress of DNA and RNA polymerases along the DNA strand, thus inhibiting replication and transcription, respectively. Consequently, dimer persistence is cytotoxic and has been associated with growth inhibition. For example, under greenhouse conditions, increased UV-mediated CPD accumulation paralleled inhibition of leaf expansion (Giordano *et al.*, 2004). However, DNA damage is normally efficiently repaired by photolyases, which restore the native DNA structure through a process known as photorepair or photoreactivation (Britt, 2004). Photolyase mutants are known to be hypersensitive to

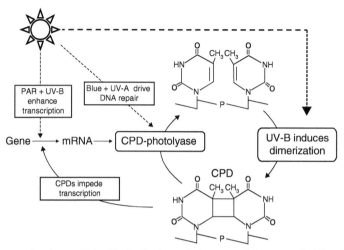

Fig. 10.4. Diagram showing multiple effects of solar radiation on cyclobutane pyrimidine dimer (CPD) formation and photorepair. Transcription of the CPD photolyase gene is minimal in the dark, but induced by PAR and UV-B wavelengths. Activity of the CPD photolyase requires blue and UV-A radiation, while CPD formation is mainly driven by UV-B. If CPD dimerization is not reversed, gene transcription will be impeded (adapted from Jansen *et al.*, 1998, with permission of Elsevier).

UV-B, indicating their important role in enabling life under solar UV-B. Photorepair is light dependent, involving the absorption of blue or UV-A photons by the chromophore of the photolyase (Fig. 10.4). The light energy is used to catalyse electron transfer, and to open the covalent bond of the DNA dimer. Expression of the CPD photolyase is also light dependent, transcription being enhanced by visible and UV-B wavelengths (Fig. 10.4) (Britt, 2004). Thus, plants appear to exploit the fact that damaging UV-B photons are always accompanied by a large surfeit of UV-A and visible wavelengths under natural light conditions. In contrast to the CPD photolyase, a 6-4pp photolyase is constitutively expressed (Britt, 2004).

An alternative process of DNA repair, nucleotide excision repair, can operate in the dark (Britt, 2004). This process involves excision of damaged bases and restoration of the DNA using the undamaged strand as a template. The relevance of this process for the repair of DNA-dimers in plants has, however, been questioned. In field experiments, ambient UV-B exposed *Gunnera magellanica* plants accumulated CPDs, but no repair was observed when plants were transferred to the dark, indicating that photorepair is the principal mechanism underlying the opening of CPDs and the restoration of the original

DNA structure in this species (Giordano *et al.*, 2003). In contrast, it was found that some of the most UV-B sensitive *Arabidopsis thaliana* mutants are those that are defective in either CPD or 6-4pp specific photolyase activity (*uvr2* and *uvr3*, respectively) or in nucleotide excision repair (*uvr1*, *uvr5*, *uvr7* and *uvh1*) (Britt and Fiscus, 2003), indicating that either process can play a role in the repair of photodamaged DNA. Interestingly, an *Arabidopsis* line overexpressing CPD photolyase accumulated significantly fewer CPDs compared to the corresponding wild type, when exposed to UV radiation, while growth was also more UV resistant (Kaiser *et al.*, 2009). This triggers the question whether photorepair activity can be limiting under natural light conditions. Photorepair is normally fast and with high fidelity, and many studies indicate that little damaged DNA does accumulate in plants kept under realistic UV-B conditions. For example, in bryophytes UV-B-induced DNA damage is largely restored within a relatively short time after UV-B exposure (Rozema *et al.*, 2005). However, this appears not always to be the case; in a UV-exclusion study it was shown that the herbaceous plant *G. magellanica* accumulates considerable levels of CPDs under near-ambient UV conditions in southernmost Patagonia. This part of Patagonia is exposed to the Antarctic hole in the ozone layer, and therefore results may reflect lack of adaptation of the local flora to increased UV-B levels (Giordano et al., 2003). Similarly, a recent meta-analysis of CPD accumulation data revealed considerable accumulation of such dimers in plants from both Arctic and Antarctic regions (Newsham and Robinson, 2009). These data indicate that photorepair can be limiting under natural conditions. One possible factor that may contribute to accumulation of CPDs is the slowdown of photolyase activity under low temperatures. In G. magellanica no CPD repair could be measured at 8°C, a finding that has considerable relevance for plants in (sub-)polar regions (Giordano et al., 2003).

There appears to be a continuous cycle of DNA damage and damage repair in plants exposed to UV-B. Realistic levels of UV-B induce DNA-dimerization, which is, normally, rapidly repaired through photolyase-mediated photorepair (Rozema et al., 2005). The metabolic cost of damage and repair cycles has not been documented, but is likely to be small as many UV studies show that realistic levels of UV-B radiation only have a minor effect on biomass accumulation. It has been estimated that the effect of increased UV-B on plant growth due to ozone layer thinning is 'unlikely to be greater than 6% yield loss, and probably much lower owing to plant acclimation/adaptation' (Ballaré et al., 2011).

10.2.2 The photosynthetic machinery as an ultraviolet-B target

A second cellular target of UV-B radiation is photosynthesis. In the late 1980s and early 1990s, many groups reported UV-B-mediated inactivation of components of the photosynthetic machinery, including the light reactions of photosynthesis, ribulose-1,5-bisphosphate carboxylase activity and chloroplast ATPase activity. Particularly widespread were reports on the UV-B-mediated inactivation of photosystem II (PS II) (Björn, 2015). Based on their absorption in the UV range, several PS II targets have been proposed including tyrosine electron donors, the Mn_4Ca-cluster of the water splitting complex and quinone electron acceptors (Vass, 2012). Notwithstanding uncertainties about the precise UV-B target, large numbers of researchers exploit the relative UV susceptibility of PSII as a tool to assess the relative level of plant UV protection. In particular, chlorophyll a fluorescence measurements have proved themselves to be convenient and non-destructive tools to assess UV protection in large numbers of samples (for example Jansen et al., 2010). However, extreme care is required when extrapolating the outcomes of such studies to natural sunlight conditions. Much of the evidence supporting the notion that UV-B radiation can impede photosynthetic activity is based on research performed under non-realistic conditions, such as excessively high levels of UV-B and/or low levels of accompanying PAR (Fiscus and Booker, 1995). Following a reappraisal of exposure protocols used, Fiscus and Booker (1995) concluded that there is actually little evidence for an overall impediment of photosynthetic oxygen evolution or CO_2 uptake under realistic UV conditions. This conclusion was subsequently widely accepted, and has dominated the UV-B field for the last decade; yet, in the last few years, new and realistic field-based studies have shown that ambient UV-B levels can negatively impact on photosynthetic performance. For example, in a UV-B exclusion field study, it was shown that ambient levels of UV-B radiation decrease photosynthetic activity in Arctic Salix arctica. The measured effect comprised decreased PS II performance and decreased net CO_2 fixation (Albert et al., 2011). Evidence of UV-B-mediated inhibition of photosynthesis is not limited to Arctic plants. A study of near-isogenic maize lines showed that maximal PS II efficiency and net CO_2 fixation were substantially decreased by ambient UV-B in plants raised under low nutrient levels, although only minor UV effects were observed in well-fertilized plants (Lau et al., 2006). Similarly, a study of the interactions between UV-B radiation and precipitation showed significantly increased UV-B stress in some samples of photosynthetic soil organisms (cyanobacteria, lichens and mosses) when the precipitation frequency, but not total amount of precipitation, was increased (Belnap et al., 2008). This finding is particularly relevant for the ecology of cryptogamic vegetation in semi-arid and arid habitats. Thus, a picture has emerged whereby inactivation of photosynthetic activity by UV-B is a rare event that can, however, occur when plants are already challenged due to the simultaneous exposure to another unfavourable environmental condition. The ecological consequences of UV-mediated inactivation of photosynthesis may be substantial for plants that are already living at the limit of their distribution and/or in an extreme environment. In this regard it is worth remembering that highest UV-B fluence rates occur near the equator, and that exposure to high UV-B levels coincides in many climates with exposure to heat and drought.

10.2.3   Exposure to multiple stressors: aggravated stress or cross-tolerance

Under otherwise favourable environmental conditions, exposure to realistic UV-B doses does not normally result in accumulation of damaged DNA and/or inactivated photosynthetic machinery. The discrepancy between potentially damaging UV-B effects on key cellular targets, vis á vis the observed modest yield losses under field conditions, can be largely attributed to the effectiveness of protection and damage repair responses. Thus, UV-B radiation stress is a 'potential' stressor that only becomes significant under specific exposure conditions such as very high UV doses, high UV-B:PAR ratio, use of non-acclimated plants or use of plants already challenged by other environmental conditions. Examples of the latter include plants exposed to extreme climatic conditions of the polar zones, nutrient deficiencies or drought (Lau et al., 2006; Belnap et al., 2008; Albert et al., 2011). Bandurska and Cieślak (2013) showed that lipid peroxidation, measured as MDA production, increased strongly when barley leaves (Hordeum vulgare) were simultaneously exposed to both UV-B and drought. Neither drought nor UV-B alone substantially affected MDA production. One might speculate that the 'double whammy' of simultaneous exposure to multiple stressors would generate numerous reports on severely deleterious UV-B effects. However, a survey of the literature yields relatively few examples of such aggravated UV stress. On the contrary, a range of multiple stress exposure studies generated evidence for induced cross-tolerance rather than enhanced damage. For example, in field experiments UV-B exposure increased cold tolerance in Rhododendron 'English

Roseum', and this involved interactions with other environmental cues such as day length and night temperatures (Chalker-Scott and Scott, 2004). UV exposure also alleviated drought symptoms such as needle loss and decreases in photosynthetic capacity in a Mediterranean pine species (Pinus pinea) under field conditions (Manetas et al., 1997), and this was related to UV-induced morphological changes (cuticle thickness) in the needles. Similarly, UV-B-induced drought tolerance has been observed in a range of herbaceous species, as well as in birch (Betula pendula) trees (Robson et al., 2015a). In Arabidopsis, UV-B-induced drought tolerance was associated with improved water retention, proline content and decreased stomatal conductance (Poulson et al., 2006), while increased levels of dehydrins have also been reported (Schmidt et al., 2000). Relatively few studies have addressed the issue of the seemingly contradictory outcomes of aggravated stress versus cross-tolerance (Bandurska and Cieślak, 2013). Li et al. (2012) showed that simultaneous exposure of soybean seedlings (Glycine max) to UV-B and cadmium led to inhibition of net photosynthesis; however, if plants had been pre-treated with either of these two stressors, negative effects were minimal. Thus, cross-tolerance is associated with acclimation responses such as the UV-induced enhanced scavenging of ROS, and other morphological, physiological and biochemical adjustments. To understand how plants can not only tolerate exposure to UV-B radiation but actually increase their tolerance to other stressors as well, it is necessary to study the processes of UV perception, signalling and acclimation.

10.3   Ultraviolet-B Perception and Signalling and its Effects on Gene Expression

Plant UV-B protection is largely based on an array of UV-B induced protection and repair responses. An important prerequisite of induced (as opposed to constitutively expressed) protection is the ability to detect and/or anticipate the presence of the stressor, as well as 'knowledge' and/or 'memory' of the relevant stress defence strategy (Badyaev, 2005). Plants can detect and respond to UV-B radiation, resulting in changes in gene expression that are distinct from those induced by other stressors (Kilian *et al.*, 2007). Plants 'measure' UV-B radiation in order to regulate UV-B defence and repair processes, but there is evidence that they also exploit UV-B radiation as a source of information about their environment. This information is used to control expression of a broad range of genes important for abiotic and/or biotic stress tolerance. It is in the area of UV-B sensing, and subsequent signal transduction, that major advances in understanding have been achieved in the past few years.

10.3.1 Ultraviolet-B-specific signal transduction pathways

Exposure to low levels of UV-B can have a major impact on plant physiology, inducing changes in gene expression, plant metabolite accumulation, morphology and defence against abiotic and biotic factors. A UV-B photoreceptor, linked to a signal transduction pathway, was postulated many years ago (cf. Wellmann, 1983) but its identity was only conclusively elucidated a few years ago (Rizzini *et al.*, 2011; Heijde and Ulm, 2012; Jenkins, 2014). The discovery of the UV-B photoreceptor, UV RESISTANCE LOCUS8 (UVR8), is undoubtedly one of the most important breakthroughs in plant UV-biology in the last decade. As a result we can now categorically state that UV-B is a specific regulator of plant growth and development. UV-B radiation shares this ability to regulate plant growth and development with other wavelength bands, including UV-A and blue light which are sensed by cryptochromes, phototropins and members of the Zeitlupe family, and red and far-red light which are perceived by phytochromes.

UVR8 is a UV-B-specific photoreceptor that can specifically perceive low fluences of such radiation (Jenkins, 2009; Heijde and Ulm, 2012). Identification of UVR8 started with the discovery of a highly UV-sensitive *Arabidopsis thaliana* mutant, which was later identified as a *uvr8* mutant (Favory *et al.*, 2009). *UVR8* encodes a seven-bladed β-propeller protein (Jenkins, 2014). Tryptophans, which are an intrinsic part of the sequence of UVR8, act as UV-B absorbing chromophores. Exciting these tryptophans with UV radiation will lead to the conversion of UVR8 from a homodimer to a monomeric form (Fig. 10.5),

rapid accumulation of the UVR8 protein in the nucleus and binding to the positive regulator of UV-B responses CONSTITUTIVE PHOTOMORPHO-GENESIS1 (COP1) (Favory *et al.*, 2009; Heijde and Ulm, 2012; Jenkins, 2014) (Fig. 10.5). The UVR8-COP1 complex associates with the chromatin regions of several UV-B-activated genes (including that of ELONGATED HYPOCOTYL5 [HY5]). HY5 expression is promoted by COP1 in association with UVR8, just like HY5 promotes *COP1* expression, thus creating a positive feedback loop (Jenkins, 2014). HY5 controls the expression of a number of UV-B induced genes under low levels of UV-B (Fig. 10.5). Regulation of expression of UV-B induced genes by HY5 can potentially happen independently of other photoreceptors such as phytochromes A and B, cryptochromes 1 and 2, and phototropins 1 and 2 (Ulm *et al.*, 2004). Nevertheless, HY5, HY5 HOMOLOG (HYH) and COP1 are all involved in multiple photomorphogenic pathways, including responses to red light, and light–dark transitions, thus integrating multiple light response pathways. A negative regulator of UVR8-induced HY5 signalling is the B-BOX ZINC FINGER PROTEIN24/SALT TOLERANCE protein (BBX24/STO). In turn, *BBX24/STO* expression can be decreased by RADICAL INDUCED CELL DEATH 1 (RCD1) (Jiang *et al.*, 2012). Direct interactions of REPRESSOR OF UV-B PHOTOMORPHOGENESIS1 and 2 (RUP1 and RUP2) with UVR8 can also impede UV-B-induced gene expression (Grüber *et al.*, 2010; Heijde and Ulm, 2012) (Fig. 10.5). Transcription of *RUP1* and *RUP2* genes is promoted by UV-B radiation, mediated through UVR8, COP1 and HY5 (Grüber *et al.*, 2010). Thus, multiple feedback loops create considerable scope for the fine regulation of gene expression and this further modulates gene expression levels.

UVR8 regulates expression of hundreds of genes including key genes involved in UV-B protection, such as enzymes of the phenylpropanoid pathway, a CPD photolyase and various enzymes involved in oxidative stress tolerance (Favory *et al.*, 2009; Jenkins 2014). Consequently, UVR8 controls a broad range of UV-mediated physiological responses, including UV-B defence, disease tolerance, flavonoid biosynthesis, suppression of hypocotyl elongation, control of cell expansion in leaves, control of stomatal development, entrainment of the circadian clock and control of photosynthetic capacity (Jenkins,

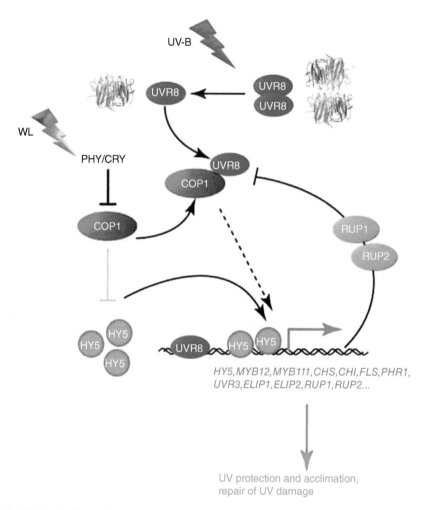

Fig. 10.5. Model of UVR8-mediated signalling. Under visible light (WL, white light), UVR8 is present mainly as a homodimer. Under these conditions, COP1 represses photomorphogenesis by promoting degradation of HY5 although this is, in turn, under the negative control of light-activated phytochromes and cryptochromes. In the presence of UV-B radiation, UVR8 monomerizes and interacts with COP1. HY5 is stabilized and UV-B responsive genes are activated. These include genes encoding chalcone isomerase (CHI) and chalcone synthase (CHS), early light-inducible protein 1 and 2 (ELIP1 and ELIP2), flavonol synthase (FLS), MYB domain proteins 12 and 111 (MYB12 and MYB111), photolyase 1 (PHR1), UV repair defective 3 (UVR3) and repressor of UV-B photomorphogenesis 1 and 2 (RUP1 and RUP2). The latter two gene products constitute negative feedback on UVR8 activity (reproduced from Heijde and Ulm, 2012, with permission of Elsevier).

2014). Given such a broad role for the UV-B photoreceptor, a pertinent question is whether the UVR8-COP1-HY5 pathway is the only UV-B signalling pathway. At the moment it is not clear how many UV-B-specific signalling pathways operate in plants. Distinct gene expression profiles have been reported following exposure to different fluence rates (Brosché and Strid, 2003) and/or different wavelength regions (Ulm et al.,

2004), and this may possibly reflect the existence of multiple UV signalling pathways. Alternatively, integration of UVR8 signalling with other photoreceptor pathways (Jenkins, 2014) may create scope for fine regulation of environmental responses. Consistently, Morales et al. (2013) showed that changes in the pattern of UVR8-mediated gene expression depended on whether plants were grown in a growth room, or under natural sunlight outdoors. Thus, the UV-B sensing system is particularly responsive, and even small changes in UV-B dose and/or spectrum can induce specific regulatory cascades (Jenkins, 2009).

Despite strong evidence supporting the regulatory aspects of UV-B radiation, it should not be overlooked that UV-B is also a potential environmental stressor, as attested in the previous paragraphs. High UV-B fluence rates can cause damage to macromolecules, including DNA, proteins and phenolics, resulting in disruptions of cellular metabolism. Cellular and/or DNA damage may, in turn, trigger transcriptional responses involving signalling molecules such as ROS and wound/defence-related factors including jasmonic acid, salicylic acid and ethylene, which are all known to control expression of specific genes (Brosché and Strid, 2003; Jenkins, 2009). Due to the involvement of wound- and defence-related factors in UV-B signalling under high fluence rates, high UV-B dose responses are not particularly stressor specific (Brosché and Strid, 2003; Jenkins, 2009), and include many genes that are also expressed in defence, wound or general stress responses. Indeed, there are similarities in gene expression pattern between plants exposed to artificially generated ROS and plants exposed to high levels of UV-B (Mackerness et al., 2001). Conversely, low UV-B doses regulate many of the genes linked to antioxidative protection via the UVR8 pathway (Hideg et al., 2013), priming antoxidative defences prior to potential exposure to high UV-B.

Thus, the tale about UV-B perception and signalling is a two-faced story. Low UV-B doses control gene expression in a specific manner, through UVR8, and often in the absence of cellular damage. In contrast, high UV-B doses cause cellular damage, which in turn leads to more generic, stress-induced changes in gene expression. In practice, these two scenarios may co-occur, depending on the UV-fluence rate, and also on the duration of the exposure, the plant acclimation state and other physiological and environmental parameters.

10.3.2 Ultraviolet-B mediated gene expression

UV-B induced changes in gene transcription have been analysed at the whole-genome level by several groups (Ulm et al., 2004; Casati et al., 2006; Hectors et al., 2007). Specific UV-B-induced changes in gene expression pattern depend, among others, on fluence rate, spectral composition, exposure kinetics (chronic or acute) and plant tissue. The complexity of UV-mediated gene regulation is demonstrated in a study in which Arabidopsis was exposed to chronic, low fluence rates of UV-B. Under these conditions, expression of chlorophyll biosynthesis genes was upregulated, but that of chlorophyll-binding proteins downregulated; several phenylpropanoid pathway genes were downregulated although polyphenol content of plants had increased; several auxin-biosynthesis and -response genes were downregulated, although some gibberellin response genes were upregulated; and many genes related to general stress responses were also downregulated (Hectors et al., 2007). Some of these UV-B-induced changes in gene expression are mediated through UVR8 (Favory et al., 2009), while others are not. Adding to the complexity of this response is the fact that some UV-induced genes are only transiently expressed, while others remain activated for long periods (Kilian et al., 2007). At present there is a gap in our understanding of the relationship between UV-induced gene transcription patterns versus induction of physiological responses. For many UV-regulated genes, it is not clear if (or to what extent) altered gene expression results in altered enzyme levels and physiological activities. In other cases it is not clear if (or how) the dynamics of gene expression affect the dynamics of physiological acclimation responses. For example, several phenylpropanoid biosynthesis genes are downregulated under chronic UV-B conditions, while flavonoids and other phenolics accumulate (Hectors et al., 2007). This may reflect a negative feedback mechanism whereby expression is downregulated once enough transcript and/or protein has accumulated, but this remains speculative.

10.4 Ultraviolet-B-mediated Physiological Responses

UV-B acclimation refers to the UV-B-induced processes that confer increased UV-B tolerance by preventing and/or repairing UV-B-mediated damage. UV-B acclimation may include increases in the capacity for UV screening, ROS scavenging and DNA-damage repair. There is good evidence that the capacity to recycle damaged proteins is also enhanced in UV-B-acclimated plants. Removal of damaged, dysfunctional proteins is vital for cellular metabolism. Increased levels of transcripts encoding ubiquitin, ubiquitin-binding proteins, proteasome proteins, proteinases and some chaperonins have been reported in maize (cf. Casati and Walbot, 2004). Perhaps the most intriguing outcome of UV-acclimation experiments is that it is not clear whether all UV-B responses can actually be classified as UV acclimation, i.e. as contributing to UV tolerance. At the metabolite level, UV-induced responses such as accumulation of specific alkaloids, polyamines and cyanogenic glycosides (cf. Jansen et al., 2008) have not been conclusively linked to UV protection. There is also little consensus about the functional significance of UV-B-induced changes in leaf and plant architecture. The UV-induced phenotype has been associated with drought (Manetas et al., 1997; Schmidt et al., 2000; Poulson et al., 2006), cold (Chalker-Scott and Scott, 2004) and heat tolerance (Teklemariam and Blake, 2003). Indeed, there is now an increasing speculation that plants use UV-B as a signal not just for UV-B protection, but also to adjust growth and development to environmental conditions other than UV-B radiation. Thus, complex UV-B-induced changes in gene expressions should not simply be perceived in the context of plant UV-B protection, but rather as part of a larger UV-B-mediated adjustment to environmental conditions. To complicate matters even further, some well-characterized UV-protection responses, such as the accumulation of UV-absorbing and ROS-scavenging phenolics, have secondary consequences that are not directly linked to UV protection.

10.4.1 Plant secondary metabolites

Large numbers of studies have shown that exposure to UV-B radiation affects (aspects of) the biochemical make-up of plants. Some of these induced changes in the concentrations of secondary plant metabolites have an adaptive role in adjusting plant metabolism to UV-B exposure. However, the ecological importance of such UV-B-induced changes goes well beyond the plants themselves. Altered concentrations of secondary metabolites may alter plant litter quality and litter decomposition, the release of volatile organic compounds that potentially contribute to climate change, the susceptibility to pests and diseases, interactions with nitrogen-fixing microorganisms and nutritional quality for human consumers (Schreiner et al., 2012; Bornman et al., 2015).

An important group of plant metabolites are the phenolics. These constitute a diverse group of aromatic molecules that are derived from phenylalanine and malonyl-coenzyme A. This group includes hydroxycinnamic acids, chalcones, flavones, flavonols, flavandiols, anthocyanins and condensed tannins. The increased accumulation of phenolic compounds is one of the best-documented UV-B-acclimation responses in plants (Jansen et al., 2008). It has long been known that some phenolics (especially flavonoids) contribute to UV protection, and this was related to their absorption in the UV part of the spectrum. However, many phenolics are also effective ROS scavengers and it has been speculated that such antioxidant activity contributes substantially to overall UV protection (Agati and Tattini, 2010). UV-induced phenolics accumulate in high concentrations in the vacuoles of epidermal cells of leaves and stems where they specifically absorb UV photons, without affecting penetration of PAR into the underlying mesophyll (Ålenius et al., 1995). Phenolics also accumulate in chloroplasts, the nucleus and in cell walls, thus decreasing UV penetration via anticlinal cell walls (Agati and Tattini, 2010). The heterogeneous distribution of phenolics across cell types and subcellular compartments was visualized for olive (Olea europaea) leaves. Leaves accumulate phenolics in lamina and trichomes, and at either site phenolics are present in the soluble fraction, or associated with either the waxy cuticle or cell walls (Liakopoulos et al., 2006). Different phenolics accumulate in different compartments, thus while quercetin was present in high concentrations in the trichomes, the leaf lamina itself contained a more complex phenolic mixture that was especially enriched in luteolin derivatives (Liakopoulos et al., 2006). The accumulation of different phenolics at different sites is likely to reflect the particular (sub-) cellular environment and also the specific physiological roles of the accumulated phenolics (Agati and Tattini, 2010).

Variations in stereochemistry, position and nature of the substitutions, degree of polymerization and linkages between the basic phenol units result in an overwhelming diversity of phenolic compounds in plants (Pourcel *et al.*, 2007). The importance of this diversity is still poorly understood, but is likely to be related to the multiple functional roles of phenolics (Agati and Tattini, 2010). Some phenolic aglycones are more upregulated during UV-B acclimation than others. Specifically, quercetin levels increase strongly during UV acclimation while those of structurally related kaempferols remain relatively constant. It has been argued that the quercetin-to-kaempferol shift mirrors the increased ROS-scavenging capacity of the former compound (Agati and Tattini, 2010). Notwithstanding the fact that UV-B induces specific phenolic derivatives, it appears that there is a degree of functional redundancy between different phenolic compounds with respect to UV protection. For example, the *Arabidopsis tt4* mutant is unable to produce flavonoids but is only moderately UV sensitive as it can accumulate UV-absorbing sinapate esters. In contrast, the *Arabidopsis tt5* and *tt6* mutants accumulate neither flavonoids nor sinapates and are highly UV sensitive (Li *et al.*, 1993). Thus, a degree of functional redundancy, possibly linked to regulatory interactions between branches of the phenylpropanoid pathway, and mediated by

MYB transcriptional activators, may result in different classes of phenolic compounds contributing to UV-B protection in different conditions. Moreover, it should be recognized that UV-B is not the only environmental parameter regulating accumulation of phenolic compounds. As a consequence, effects of UV-B exposure on phenolic metabolites are not always significant, particularly under field conditions where plants may also have been exposed to cold, drought and pathogens, all of which can induce phenolic accumulation. The roles of phenolics in, among others, biotic defence responses, abiotic stress tolerance, interactions between plants and rhizobia, attraction of pollinators, modulation of auxin transport and regulation of morphogenesis have been extensively documented (Agati and Tattini, 2010). Consequently, UV-mediated changes in both the amount and composition of the phenolic pool have important secondary consequences for plant performance in a natural environment (Paul and Gwynn-Jones, 2003).

Phenolics are not the only compounds that accumulate in UV-B exposed plant tissues. In fact, the concentrations of a broad range of antioxidants and secondary metabolites are affected by UV-B (Fig. 10.6). This includes polyamines, tocopherols, glucosinolates, cyanogenic glycosides, alkaloids and terpenoids such as xanthophylls, carotenes and mono- and diterpenes

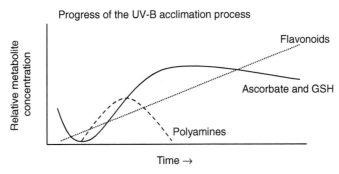

Fig. 10.6. Schematic model that shows the dynamic changes in metabolite levels that can occur during UV-B acclimation. Levels of ascorbate and glutathione (GSH) decrease during the first few days of UV-B exposure to acute UV-B stress, with the extent of this decrease depending on the severity of the imposed oxidative stress. After the initial decrease, antioxidant levels increase during acclimation to chronic UV-B conditions. Polyamines transiently accumulate during the early stages of the acclimation process (3–6 days after start of experiment), while levels of soluble flavonoids gradually increase, often reaching a steady state after c.10 days of UV-B exposure (reproduced from Jansen *et al.*, 2008, with permission of Elsevier).

(Jansen *et al.*, 2008; Schreiner *et al.*, 2012). Many upregulated secondary plant metabolites have valuable pharmaceutical or health-promoting benefits for human consumers, thus emphasizing the importance of understanding regulation of their accumulation. For example, production of the antimalarial drug artemisinin is strongly upregulated in *Artemisia annua* seedlings exposed to UV-B radiation (Pan *et al.*, 2014), and accumulation of the anticancer drug taxol is similarly upregulated in UV-B-exposed *Taxus chinensis* (Zu *et al.*, 2010). However, levels of secondary metabolites do not necessarily increase in UV-B-exposed plants. A review of the literature shows that levels of some metabolites increase following UV-B exposure, while those of others decrease, change transiently or are differently affected by low and high UV-B doses. These metabolic changes in UV-B-exposed plants are only partially characterized, but initial studies suggest a complex, highly dynamic, UV-dose-responsive system (Fig. 10.6) (Jansen *et al.*, 2008).

A substantial number of the UV-B-induced metabolites act as antioxidants, and contribute to the scavenging of ROS (Hideg *et al.*, 2013). UV-B is also known to induce both enzymatic ROS scavengers (e.g. superoxide dismutase, ascorbate peroxidase, catalase and glutathione reductase) (Gao and Zhang, 2008; Berli *et al.*, 2010) and non-enzymatic antioxidants such as ascorbate, glutathione and carotenoids, phenolics and tocopherols (Fig. 10.6) (Jansen *et al.*, 2008; Hideg *et al.*, 2013). Ascorbate is the central antioxidant in the Haliwell–Asada pathway, and its importance is underlined by the relative UV-B sensitivity of the ascorbate-deficient *Arabidopsis* mutant (*vtc1*; vitamin C1) (Gao and Zhang, 2008). Moreover, the UV-B-mediated upregulation of antioxidant defences can, at least partially, explain reported cases of cross-tolerance between UV-B and other stressors (Section 10.2.3). Thus, UV-B mediated changes in various plant metabolites have important consequences for the plants in their complex environment.

10.4.2 Ultraviolet-induced morphological alterations

UV-B-mediated changes in morphology been widely reported. Plants exposed to UV-B display a more dwarfed phenotype, with smaller but thicker leaves, shorter petioles, increased axillary branching, a shorter inflorescence and a shift in the root/shoot ratio (cf. Jansen *et al.*, 1998; Robson *et al.*, 2015b). UV-B-induced decreases in leaf expansion have been studied by different groups in different species. Some studies have reported an inhibition of cell division, while others reported decreased cell elongation without a change in cell numbers (Wargent *et al.*, 2009a; Hectors *et al.*, 2010; Robson *et al.*, 2015b). In *Arabidopsis*, leaf growth is a complex, dynamic process with cell division and both asynchronous and synchronous elongation phases along the proximo-distal axis of the leaf. Contradictory reports on the effects of UV-B radiation on cell division and elongation are at least partially due to single time point analyses of a kinetic process (Hectors *et al.*, 2010); however, it is likely that differences in experimental UV-B doses and spectra also contribute to discrepancies in obtained results. Indeed, in one of the first studies of UV-induced morphogenesis, *A. thaliana* plants exposed to three levels of UV radiation displayed a bell-shaped dose response (Fig. 10.7) (Brodführer, 1955), with stem elongation increasing when plants were exposed to low levels of UV-B, but decreasing following exposure to ambient UV levels. This may relate to UV-B-induced morphological changes being underpinned by different mechanisms at high and low UV-B doses.

UV-induced morphological alterations represent a redistribution of growth in response to an environmental signal, rather than simply a cessation of growth. It is likely that the UV-induced morphological process is to some extent systemic, as it involves coordination between tissues and organs. Several mechanisms have been proposed to explain the UV-mediated morphological response. Mutant analysis showed that the UV-B-mediated morphology is, at least partially, controlled through a UV-B-specific signalling and response pathway that centres on the UV-B photoreceptor UVR8 and the positive regulator of UV-B responses COP1 (Jenkins, 2009; Wargent *et al.*, 2009b). Consistently, neither *uvr8* nor *cop1* mutants display UV-B-induced changes in plant morphology (Favory *et al.*, 2009). The mechanism underlying UVR8-mediated morphological changes has not been elucidated, but might be related to UV-B antagonizing auxin and gibberellin activity in a UVR8-dependent manner (Hayes *et al.*, 2014). Such UVR8-mediated

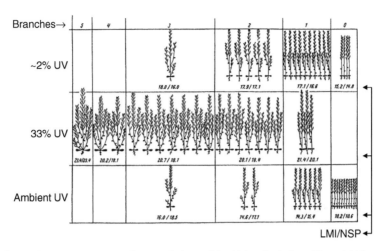

Fig. 10.7. Effects of UV-B exposure on the morphology of *Arabidopsis thaliana* 'Catania'. Plants were grown either under ambient UV conditions or under decreased UV levels, in the summer, in Davos, Switzerland. Shown are the UV effects on the number of inflorescence branches, on the length of the main inflorescence (LMI) and on the number of seed pods (NSP) produced. Increasing levels of UV initially increase both branching and inflorescence elongation, but plants raised under ambient UV are considerably shorter compared to those raised under the lowest level of UV (reproduced from Brodführer, 1955, with permission of Springer).

morphogenesis will operate under low UV doses. In contrast, it is likely that stress-mediated morphological responses will occur under higher UV doses. Stress-induced morphogenic responses (SIMRs) have been reported following exposure of plants to a broad variety of abiotic stressors. SIMRs represent a redistribution of growth that is thought to be mediated by changes in ROS and phytohormones, especially auxin (Fig. 10.8) (Potters *et al.*, 2007). Another scenario, which is specific to UV-B, centres on the correlation of CPD accumulation and decreased leaf expansion. CPD loads in UV-B-exposed *G. magellanica* were found to correlate with decreased leaf expansion (Giordano *et al.*, 2004). It has been found that the atypical E2F transcription factor DEL1 coupled CPD-photolyase expression with the onset of endoreduplication (Radziejwoski *et al.*, 2011). Thus, levels of DEL1 decreased in UV-B exposed tissues, resulting in increased CPD photolyase expression, and a switch in the cell cycle from mitosis and cytokinesis towards increased endoreduplication (Fig. 10.8). The latter switch can result in reductions in cell numbers, although these can be potentially compensated for by increased ploidy-dependent cell growth (Radziejwoski *et al.*, 2011). Other studies on

UV-B-induced morphology have focused on the physiological cross-talk between UV-B induced phenolics and the phytohormone auxin (Fig. 10.8). Flavonoids such as quercetin, kaempferol and apigenin inhibit polar auxin transport and alter local auxin accumulation *in planta*. Flavonoids also affect auxin catabolism (Agati and Tattini, 2010). Thus, it might be speculated that the UV-mediated morphology is modulated by UV-B-mediated accumulation of phenolics. At present it remains unclear which mechanisms affect morphology under environmentally relevant conditions. The existence of multiple potential mechanisms may well indicate that the UV-B-mediated morphology is subject to regulation at multiple organizational levels, perhaps reflecting the ecophysiological importance of this process. Indeed, it has been argued that UV-B-induced morphological alterations may diminish the exposure of cells or leaves to UV-B radiation. For example, leaves of short, bushy plants are more likely to be shaded and less exposed to direct UV-B within a canopy (Barnes *et al.*, 1996). UV-B levels under a closed canopy of leaves can be decreased by as much as 98–99% (Flint and Caldwell, 1998). Thus, intuitively it could be argued that UV-B-induced morphology helps reduce

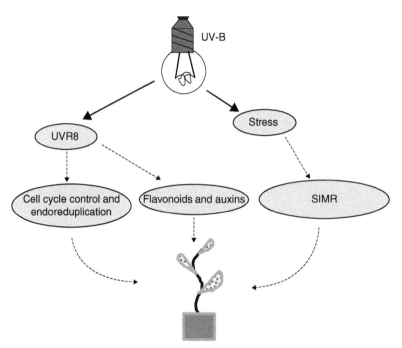

UV-acclimated plant with stocky phenotype

Fig. 10.8. Summary of morphological processes in UV-B-exposed plants. The relative stocky phenotype of the UV-B-acclimated plant can be directly mediated through the UVR8 pathway (Favory *et al.*, 2009), possibly involving UV-B-mediated changes in auxin metabolism (Hayes *et al.*, 2014), flavonoid accumulation (Hectors *et al.*, 2007) or endoreduplication (Radziejwoski *et al.*, 2011), or alternatively be controlled through generic stress-induced morphogenic responses (SIMR) (Potters *et al.*, 2007).

UV-B exposure of cells and tissues. However, this assumption is not backed up by experimental evidence. Moreover, the distinct UV responses observed at low and high UV-B doses (e.g. Brodführer, 1955), and the transient character of some of the responses (Robson and Aphalo, 2012), would argue against a simple, single functional role for UV-B-induced morphology (Robson *et al.*, 2015b).

needles (cf. Day and Neale, 2002). However, the resulting UV tolerance of conifers is thought to reflect adaptation of conifers to dry and/or cold conditions, rather than to UV radiation. Thus, the presence of effective UV-B-protection mechanisms in a particular plant species cannot be interpreted, a priori, as evidence that local UV-B levels exerted a selective pressure that contributed to the evolution of UV-B tolerance.

10.5 Ultraviolet-B Adaptation, Acclimation and Fitness Cost

Some plants are inherently more UV-B tolerant than others, due to cellular and/or morphological adaptations. For example, epidermal UV transmission varies substantially between species, and as a result penetration of incident UV-B radiation into mesophyll tissue ranges from 41% for some herbaceous dicots to virtually zero in conifer

10.5.1 Ultraviolet-B and adaptive evolution

The presence of the UV-B photoreceptor UVR8 has been confirmed in angiosperms, mosses, lycopods and green algae (Jenkins, 2014). Thus, it appears that UVR8 is highly conserved and that all plants possess a similar capability to sense UV-B radiation in the environment. Nevertheless, different plant species grow in different

UV environments, and are exposed to different UV-B doses depending on, among others, altitude and latitude. At present, little is known about the adaptive variations in UV sensing and signalling. However, a range of different studies have elucidated adaptive differences in UV-B protection. Plants growing at high altitudes are typically more UV-B tolerant than those at lower altitudes, and this has been associated with increased levels of UV-screening pigments and increased leaf thickness (Rozema *et al.*, 1997). However, enhanced UV-B tolerance measured *in situ* may simply reflect acclimation to local high UV-B levels (for example at high altitudes), and does not necessarily indicate genetic adaptation. *Ex situ* studies, whereby all plants are raised under the same UV level, can help distinguish between acclimation and adaptation. Using such an approach, it was shown that *Rumex acetosella* and *Plantago lanceolata* ecotypes as well as *Lupinus* and *Taraxacum* species collected from equatorial alpine sites were generally effectively UV protected compared with congeneric lowland and/or higher latitude ecotypes or species (Barnes *et al.*, 1987). UV-B tolerance was assessed as the relative lack of UV-B effects on leaf-gas exchange under *ex situ*, standardized environmental conditions. Similarly, a comparison of tartary buckwheat (*Fagopyrum tataricum*) ecotypes, raised under controlled field conditions, showed that UV-B protection could be statistically linked to UV-B levels at the site of origin (Yao *et al.*, 2007). Strikingly, some buckwheat ecotypes were negatively affected by supplemental UV-B, while others were positively affected for the same trait (Yao *et al.*, 2007). Positive UV-B effects were also noted in a study comparing *Silene vulgaris* ecotypes from alpine and lowland origins. UV-B exposure resulted in decreased flower and seed production in a lowland ecotype, but up to 2.5-fold more seeds per plant in an alpine ecotype (Van de Staaij *et al.*, 1997). Thus, it appears that some plants actually benefit from UV-B exposure. Evidence for increased UV protection is largely based on studies of plants from more extreme alpine and/or equatorial origins. In contrast, a comparison of 224 *A. thaliana* ecotypes originating at latitudes between 16°N and 67°N and altitudes between −4 m to +1650 m revealed significant differences in UV protection of the photosynthetic machinery, but such differences were not linked to geographic origin

(Jansen *et al.*, 2010). Remarkably, significant differences in UV tolerance were present among ecotypes originating in the same geographic area. This implies that differences in UV tolerance may have evolved in response to selective pressures other than UV-B. The notion that UV-B tolerance can be a secondary consequence of adaptation to another environmental parameter is supported by a range of studies that show that UV-B-absorbing pigments can be induced by a broad range of unfavourable environmental conditions such as wounding, herbivory, drought, low temperatures and mineral deficiencies (Treutter, 2006). For example, in *Vicia faba* plants the adaxial epidermal UV-B transmission was significantly decreased when the growth temperature was reduced by just 3°C, and this was related to the accumulation of soluble flavonoids (Bilger *et al.*, 2007). Interestingly, no such increase in epidermal absorption, with decreasing temperature, was measured in the Arctic–alpine species *Oxyria digyna* (Bilger *et al.*, 2007). The environmentally induced accumulation of flavonoids is particularly relevant in the case of elevational gradients, along which numerous environmental parameters change in parallel. Thus, despite the fact that in a small number of studies statistical relationships have been established between UV-B protection and altitude, it remains to be proved that UV-B levels were a key environmental parameter that contributed to the evolution of UV-B adaptation.

10.5.2 Constitutive and inducible ultraviolet-B protection

A survey of the literature reveals that the overwhelming majority of reports on UV-B protection refer to acclimation, i.e. the induction of specific cellular and morphological defence responses. Few studies refer to the constitutive expression of UV protection. One of the few studies that did so revealed evidence for constitutively elevated expression of UV-screening pigments in high-altitude populations of *Oenothera stricta* and *Plantago lanceolata* from Hawaii (Ziska *et al.*, 1992). Low-altitude plants from the same species could be induced to accumulate UV-screening pigments, and as a result no differences in UV-B screening between high- and low-altitude plants were observed when they were raised in the presence of UV-B radiation. In another study, a high-altitude ecotype of the duckweed *Landoltia punctata* was found to be constitutively UV protected, and this was associated with relatively thick fronds (Jansen *et al.*, 2001). A low-altitude ecotype of the same duckweed species was found to be sensitive when exposed to short-term UV-B, but this ecotype could be acclimated under low levels of UV-B such that its overall protection was similar to that of the high-altitude species (Jansen *et al.*, 2001). At the genomic level, a comparison of maize genotypes from different altitudes showed that genotypes from high altitudes display constitutive alterations in expression of UV-regulated chromatin remodelling proteins, and this may facilitate gene expression under high UV-B (Casati *et al.*, 2006). Chromatin remodelling is also thought to be involved in the upregulation and carry-over of UV-induced protection responses across multiple generations, a phenomenon that was demonstrated using *Dimorphotheca sinuata*, a desert annual from southern Africa (Musil *et al.*, 1999). The idea of an epigenetic memory of UV-B exposure has been explored (i.e. Müller-Xing *et al.*, 2014), but at present there is no conclusive evidence that such a memory exists. How plants balance constitutive and inducible stress-protection responses is poorly understood. Analysis of plant–herbivore relationships indicates that inducible responses are a cost-saving measure in comparison with the constitutive expression of defences. If herbivory varies over time, constant defence is unnecessary and selection will favour an induced defence. In contrast, selection will favour a constitutive defence when the pressure of herbivory is high. This argument is likely to be stressor specific, and will need to take into consideration the relative metabolic cost of the defence response. Therefore, the question arises: what is the cost of UV acclimation? Detailed analysis of UV-B acclimation in *Impatiens capensis* showed that plants with increased levels of phenolics were favoured under ambient UV conditions, but that selection favoured a decrease in phenolics in the absence of UV-B (Weinig *et al.*, 2004). This may reflect the metabolic cost of phenolic biosynthesis, but could also relate to secondary consequences of phenolic accumulation (see Section 10.4.1). Similarly, reductions in stem length were found to be advantageous under ambient UV-B levels, whereas increased stem elongation was favoured in the absence of UV-B exposure. Decreased stem elongation constitutes a fitness cost in the absence of UV-B exposure, presumably due to the increased likelihood of shading. Indeed, using mixtures of wheat (*Triticum aestivum*) and wild oat (*Avena fatua*), it was shown that UV-B-induced decreases in leaf area and plant height affected the relative capture of PAR and as a consequence altered the competitive balance between neighbouring plants (Barnes *et al.*, 1996) (Fig. 10.9). The importance of small UV-induced changes in morphology, and especially the relative positioning of leaf area, was neatly demonstrated in a field study where UV-B exposure did not affect plant productivity in monocultures, but did change the competitive balance in mixed cultures of wheat and wild oat (Barnes *et al.*, 1995). Similarly, above-ambient levels of UV-B were found to alter interspecific competition between broccoli (*Brassica oleracea*) and *Chenopodium album* (Furness *et al.*, 2005). These data on UV-induced morphogenesis and competitiveness indicate that the UV-B phenotype may be 'adaptive' in some environments but 'maladaptive' in other environments, thus emphasizing an advantage of inducible, carefully regulated UV-B responses.

10.6 Plant Ultraviolet-B Responses and Ecological Interactions

The awareness that UV-B rarely causes plant stress is complemented by an understanding that UV-B is an important regulator that controls, among others, interactions between plants and various other organisms. Several studies have shown that ambient levels of UV-B radiation can reduce infections of plants by microbial pathogens (Ballaré, 2014). However, major uncertainties remain concerning the underlying mechanism and especially the range of pathogens whose growth is impeded by UV-B radiation. Knowledge about the effects of UV-B radiation on plant–pest relationships is, comparatively, more extensive (Ballaré, 2014), and there is a consensus that UV-B can decrease insect herbivory. Thus, there is scope to exploit UV-B radiation responses in the context of a more sustainable horticulture.

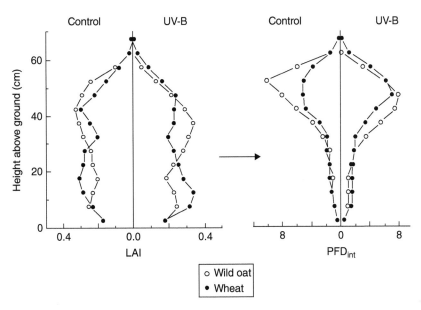

Fig. 10.9. Effects of UV-B on plant architecture and light capture. In wild oat (*Avena fatua*) raised under UV-B radiation leaves are placed slightly lower along the stem and as a result of this morphological change light capture is decreased, especially in comparison with wheat (*Triticum aestivum*), which is less affected by UV-B exposure. Such a change in relative light capture may affect the competitive balance between neighbouring plants. Shown are the leaf area index (LAI) and the canopy intercepted photon flux density (PFD$_{int}$) (reproduced from Barnes *et al.*, 1996, with permission of Elsevier).

10.6.1 Ultraviolet-B acclimation affects plant–herbivore relationships

Generally, plants raised under ambient UV-B are less susceptible to insect herbivory than plants under low or no UV-B (Paul and Gwynn-Jones, 2003; Bassman, 2004; Ballaré *et al.*, 2011; Ballaré, 2014); however, these effects are both plant-species and herbivore specific. An example of the impact of UV-B on insect herbivory comes from a study whereby broccoli (*B. oleracea*) was grown under either low or near-ambient UV-B, in the presence of specialist cabbage aphids (*Brevicoryne brassicae*) or generalist green peach aphids (*Myzus persicae*). UV-B exposure resulted in a reduction of the numbers of specialist cabbage aphids, while numbers of generalist green peach aphids were not affected (Kuhlmann and Müller, 2010). Similarly, under field conditions soybean (*Glycine max*) was less affected by stink bugs (*Nezara viridula* and *Piezodorus guildinii*) when grown under UV-transmitting filters, as compared to UV-attenuating filters, and this was associated with better seed formation (Zavala

et al., 2015). Potentially, UV-B radiation can impact on the plant, the insect, and/or on the interactions between them (Paul and Gwynn-Jones, 2003). Direct UV-B effects on insects have been reported, and these include impeded growth and increased mortality. UV-B can also act as a source of information, altering insect behaviour. For example, thrips (*Caliothrips phaseoli*) can detect and respond to UV-B, perhaps using such radiation as a source of positional information within the canopy (Mazza *et al.*, 2010). Thus, avoidance of UV-B by insects is thought to partly account for the negative association between UV-B and herbivory (Bornman *et al.*, 2015). However, in other cases it is clear that changes in the metabolic make-up of UV-B-acclimated plants alter plant–insect relationships (Zavala *et al.*, 2015). In *Nicotiana attenuata* plants, the UV-B-induced tolerance to insect herbivory was absent in transgenics in which the jasmonate signalling pathway was silenced. Jasmonate has been implicated in both herbivory and UV-B-induced signalling, and jasmonate-silenced plants contain altered levels of specific phenolics such as the caffeoyl-conjugated polyamines (Demkura et al., 2010). However, it is unlikely that UV-mediated decreases in herbivory are due solely to altered accumulation of phenolics. UV-B-induced changes have been reported in the profiles of phenolic and other secondary metabolites, nitrogen–carbon ratio and levels of defence peptides such as PR-proteins and proteinases (Paul and Gwynn-Jones 2003), and these metabolic changes are complemented by changes in cell, leaf and organismal morphology. Many UV-B-induced phytochemicals – phenolics or otherwise – are pharmacologically active, nutritionally important and/or affect colour, flavour or digestibility of plant tissues (Jansen et al., 2008). Terpenes include many general feeding deterrents and toxins, although these compounds may also act as feeding stimulants, or even as cues for host plant identification by specialized insect herbivores. Alkaloids can act as feeding stimulants for insects, but are generally deterrents for mammalian herbivores (cf. Thines et al., 2008); soluble phenolics such as rutin have been associated with insect resistance (Misra et al., 2010); while tannins have anti-feeding effects due to their protein-binding properties. The UV-B-induced phytochemicals can have opposite effects on, for example, generalist herbivores compared to specialists that co-evolved with the phytochemical-producing plant. In this regard, tritrophic interactions are particularly interesting. *Populus trichocarpa* leaves that were raised under enhanced UV-B are more susceptible to herbivory by chrysomeline beetles; in turn the beetles were more protected from predation by secondary consumers due to the accumulation of UV-induced plant salicylates (Bassman, 2004). Thus, the effects of UV-B on plant–herbivore interactions are highly complex, reflecting the species specificity of UV-induced changes in plant metabolite profile in conjunction with the highly evolved relationships between individual plant and insect species. Moreover, there are indications that plant UV-B exposure does not affect feed intake and digestion by mammalian herbivores (Thines et al., 2008). As a consequence, rather than a general effect on herbivores, it is surmised that UV-B radiation can specifically alter the species composition and diversity of communities of canopy consumers (Ballaré et al., 2011).

10.6.2 Ultraviolet-B acclimation affects plant sexual reproduction

The co-evolution of plants and pollinating insects has resulted in a highly evolved relationship between these two groups of organisms. UV-B alters this relationship in a complex manner by affecting plants, pollinating insects and the interaction between these two groups (Fig. 10.10). UV-B can enhance flower coloration, the size and number of flowers, increase nectar production and alter flowering phenology (Llorens et al., 2015). Delayed flowering was observed in some UV-B-exposed annual species, although it remains to be seen whether this is a general plant response. Given that the timing of flowering can strongly influence the reproductive success, UV effects on phenology, as well as on flower development, need further analysis. UV effects on pollinators are primarily related to the UV vision of such insects, including the detection of UV-reflectant patterns and nectar guides on the flowers (Fig. 10.10) (Llorens et al., 2015). Some studies have indicated that UV-B can decrease the number of floral visits by insects, although the actual duration of the insect visit to each single flower may increase. Plant sexual reproduction is vital from an ecological, nutritional and/or economical perspective. Therefore, it is important to elucidate how this process is moderated by solar UV-B.

10.7 What Next in Ultraviolet-B Research?

Although UV-B radiation is potentially damaging to plants, sophisticated UV-exposure experiments have shown that under realistic exposure conditions, UV-B-mediated damage is mostly minor or absent. Rather, the identification of the UV-B photoreceptor emphasizes that plant responses to UV-B are carefully regulated. Knowledge of the UVR8 signal transduction chain, and understanding of the role of UVR8 in controlling gene expression, have increased rapidly in the last few years, although major questions remain with respect to the ecological role of UVR8 under natural sunlight conditions. An intriguing question is whether the UV-B sensitivity of UVR8 varies with the latitude or

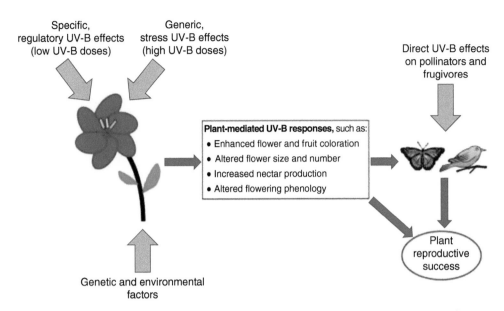

Fig. 10.10. Schematic diagram depicting possible direct and indirect effects of UV-B radiation on plant reproductive success. UV-B radiation can directly modulate plant reproductive traits and this involves either regulatory (UVR8-mediated) or stress-mediated signalling pathways. Pollinators and frugivores can also be directly affected by UV-B or, alternatively, can be indirectly affected via plant-mediated UV-responses (reproduced from Llorens *et al.*, 2015, with permission of Elsevier).

altitude where plants originate, i.e. with the prevailing UV-B dose. Another important question concerns the relative importance of UVR8 vis-à-vis other regulatory environmental influences, such as red/far-red ratios in light, or low temperatures. Despite these questions, it is clear that the regulatory UV-B responses offer novel opportunities for sustainable manipulation of plant architecture, metabolic make-up, and pest and disease susceptibility.

10.7.1 Ultraviolet-B radiation as a tool in sustainable horticulture

Our knowledge of UV-induced plant metabolites has increased rapidly (see Section 10.4.1). Much of the focus has been on generating understanding of the mechanisms that control accumulation of plant metabolites. However, there is also an emerging awareness that UV radiation can be used as a tool to manipulate plant metabolic profiles in order to improve pest and disease tolerance; architecture; and colour, flavour and/or nutritional value of crops (Schreiner *et al.*, 2012; Huché-Thélier *et al.*, 2016). For example,

Carbonell-Bejerano *et al.* (2014) showed how UV-B impacts upon grapevine berry skin phenolic composition, increasing accumulation of a range of glycosylated flavonols, some of which are associated with wine quality. Topcu *et al.* (2015) showed how broccoli (*Brassica oleracea*) plants growing under supplementary UV-B accumulated enhanced levels of the dietary glucosinolates sinigrin and glucotropaeolin. There is already a long history of exploiting regulatory plant responses in horticulture. For example, horticulturists routinely manipulate temperature to induce germination or flowering, and alter day length to promote or delay flowering. Observations of UV-B-induced abiotic stress tolerance trigger the question whether UV-B exposures can be exploited as part of a more sustainable approach to horticulture. The development of more stress-tolerant, 'stocky' plants is a major aim for a horticultural industry that wishes to improve sustainability. So far, the UV-induced phenotype has been associated with drought (Manetas *et al.*, 1997; Schmidt *et al.*, 2000; Poulson *et al.*, 2006), cold (Chalker-Scott and Scott, 2004) and/or heat tolerance (Teklemariam and Blake, 2003). The stocky phenotype is also likely to increase mechanical stress tolerance, an important parameter for an industry that transports ever greater numbers of plants across the globe. UV-B treatments can be used in protected cropping, or to pre-acclimatize glasshouse-raised plants to outdoor conditions, perhaps taking advantage of recently developed wavelength-selective cladding materials or novel LED-lighting systems, which enable the commercial manipulation of the radiation environment (Paul *et al.*, 2005). The development of protected cropping conditions where plants are more pest and/or disease tolerant due to UV-B-induced defences is similarly attractive in view of costs associated with the use of pesticides, and an aversion towards the use of crop-protection chemicals. Development of UV-B applications is, however, complex, as UV wavelengths impact on both the plant and its pathogen or pest (cf. Paul and Gwynn-Jones, 2003; Paul *et al.*, 2005) and these impacts are species specific. Yet, notwithstanding our limited understanding of what has been termed the 'photoecology of ultraviolet radiation' (Paul and Gwynn-Jones, 2003), there is a strong sense that a conceptual switch has taken place. Instead of an undesirable stressor, UV-B radiation is increasingly recognized as an interesting regulator of plant growth and development, while the focus of UV-B research has switched from damage limitation to potential exploitation.

Acknowledgements

MAKJ acknowledges funding by Science Foundation Ireland (SFI) and the support of WoB.

References

Agati, G. and Tattini, M. (2010) Multiple functional roles of flavonoids in photoprotection. *New Phytologist* 186, 786–793.

Albert, K.R., Rinnan, R., Ro-Poulsen, H., Mikkelsen, T.N., Håkansson, K.B., Arndal, M.F. and Michelsen, A. (2008) Solar ultraviolet-B radiation at Zackenberg: the impact on higher plants and soil microbial communities. *Advances in Ecological Research* 40, 421–440.

Albert, K.R., Mikkelsen, T.N., Ro-Poulsen, H., Arndal, M.F. and Michelsen, A. (2011) Ambient UV-B radiation reduces PSII performance and net photosynthesis in high Arctic *Salix arctica. Environmental and Experimental Botany* 72, 439–447.

Ålenius, C.M., Vogelmann, T.C. and Bornman, J.F. (1995) A three-dimensional representation of the relationship between penetration of u.v.-B radiation and u.v.-screening pigments in leaves of *Brassica napus. New Phytologist* 131, 297–302.

Aphalo, P.J., Albert, A., Björn, L.O., McLeod, A., Robson, T.M. and Rosenqvist, E. (eds) (2012) *Beyond the Visible: A Handbook of Best Practice in Plant UV Photobiology.* University of Helsinki, Helsinki, Finland.

Badyaev, A.V. (2005) Stress-induced variation in evolution: from behavioural plasticity to genetic assimilation. *Proceedings of the Royal Society B* 272, 877–886.

Ballaré, C.L. (2014) Light regulation of plant defense. *Annual Review of Plant Biology* 65, 335–363.

Ballaré, C.L., Caldwell, M.M., Flint, S.D., Robinson, S.A. and Bornman, J.F. (2011) Effects of solar UV radiation on terrestrial ecosystems. Patterns, mechanisms, and interactions with climate change. *Photochemical & Photobiological Sciences* 10, 226–241.

Bandurska, H. and Cieślak, M. (2013) The interactive effect of water deficit and UV-B radiation on salicylic acid accumulation in barley roots and leaves. *Environmental and Experimental Botany* 94, 9–18.

Barnes, P.W., Flint, S.D. and Caldwell, M.M. (1987) Photosynthesis damage and protective pigments in plants from a latitudinal arctic/alpine gradient exposed to supplemental UV-B radiation in the field. *Arctic and Alpine Research* 19, 21–27.

Barnes, P.W., Flint, S.D. and Caldwell, M.M. (1995) Early-season effects of supplemented solar UV-B radiation on seedling emergence, canopy structure, simulated stand photosynthesis and competition for light. *Global Change Biology* 1, 43–53.

Barnes, P.W., Ballaré, C.L. and Caldwell, M.M. (1996) Photomorphic effects of UV-B radiation on plants: consequences for light competition. *Journal of Plant Physiology* 148, 15–20.

Barnes, P.W., Flint, S.D., Slusser, J.R., Gao, W. and Ryel, R.J. (2008) Diurnal changes in epidermal UV transmittance of plants in naturally high UV environments. *Physiologia Plantarum* 133, 363–372.

Barnes, P.W., Tobler, M.A., Keefover-Ring, K., Flint, S.D., Barkley, A.E., Ryel, R.J. and Lindroth, R.L. (2016) Rapid modulation of ultraviolet shielding in plants is influenced by solar ultraviolet radiation and linked to alterations in flavonoids. *Plant, Cell & Environment* 39, 222–230.

Bassman, J.H. (2004) Ecosystem consequences of enhanced solar ultraviolet radiation: secondary plant metabolites as mediators of multiple trophic interactions in terrestrial plant communities. *Photochemistry and Photobiology* 79, 382–398.

Behn, H., Albert, A., Marx, F., Noga, G. and Ulbrich, A. (2010) Ultraviolet-B and photosynthetically active radiation interactively affect yield and pattern of monoterpenes in leaves of peppermint (*Mentha* × *piperita* L.). *Journal of Agricultural and Food Chemistry* 58, 7361–7367.

Belnap, J., Phillips, S.L., Flint, S., Money, J. and Caldwell, M. (2008) Global change and biological soil crusts: effects of ultraviolet augmentation under altered precipitation regimes and nitrogen additions. *Global Change Biology* 14, 670–686.

Berli, F.J., Moreno, D., Piccolo, P., Hespanhol-Viana, L., Silva, M.F., Bressan-Smith, R., Cavagnaro, J.B. and Bottini, R. (2010) Abscisic acid is involved in the response of grape (*Vitis vinifera* L.) cv. Malbec leaf tissues to ultraviolet-B radiation by enhancing ultraviolet-absorbing compounds, antioxidant enzymes and membrane sterols. *Plant, Cell and Environment* 33, 1–10.

Bilger, W., Rolland, M. and Nybakken, L. (2007) UV screening in higher plants induced by low temperature in the absence of UV-B radiation. *Photochemical & Photobiological Sciences* 6, 190–195.

Björn, L.O. (2015) On the history of phyto-photo UV science (not to be left in skoto toto and silence). *Plant Physiology and Biochemistry* 93, 3–8.

Bornman, J.F., Barnes, P.W., Robinson, S.A., Ballaré, C.L., Flint, S.D. and Caldwell, M.M. (2015) Solar ultraviolet radiation and ozone depletion-driven climate change: effects on terrestrial ecosystems. *Photochemical & Photobiological Sciences* 14, 88–107.

Britt, A. and Fiscus, E.L. (2003) Growth responses of *Arabidopsis* DNA repair mutants to solar irradiation. *Physiologia Plantarum* 118, 183–192.

Britt, A.B. (2004) Repair of DNA damage induced by solar UV. *Photosynthesis Research* 81, 105–112.

Brodführer, U. (1955) Der Einfluss einer abgestuften Dosierung von ultravioletter Sonnenstrahlung auf das Wachstum der Pflanzen. *Planta* 45, 1–56.

Brosché, M. and Strid, Å. (2003) Molecular events following perception of ultraviolet-B radiation by plants. *Physiologia Plantarum* 117, 1–10.

Carbonell-Bejerano, P., Diago, M.P., Martínez-Abaigar, J., Martínez-Zapater, J.M., Tardáguila, J. and Núñez-Olivera, E. (2014) Solar ultraviolet radiation is necessary to enhance grapevine fruit ripening transcriptional and phenolic responses. *BMC Plant Biology* 14, 183.

Casati, P. and Walbot, V. (2004) Rapid transcriptome responses of maize (*Zea mays*) to UV-B in irradiated and shielded tissues. *Genome Biology* 5, R16.1–16.19.

Casati, P., Stapleton, A.E., Blum, J.E. and Walbot, V. (2006) Genome-wide analysis of high-altitude maize and gene knockdown stocks implicates chromatin remodeling proteins in response to UV-B. *The Plant Journal* 46, 613–627.

Chalker-Scott, L. and Scott, J. (2004) Elevated ultraviolet-B radiation induces cross-protection to cold in leaves of *Rhododendron* under field conditions. *Photochemistry and Photobiology* 79, 199–204.

Crutzen, P.J. (1970) The influence of nitrogen oxides on the atmospheric ozone content. *Quarterly Journal of the Royal Meteorological Society* (96)408, 320–325.

Day, T.A. and Neale, P.J. (2002) Effects of UV-B radiation on terrestrial and aquatic primary producers. *Annual Review of Ecological Systems* 33, 371–396.

Demkura, P.V., Abdala, G., Baldwin, I.T. and Ballaré, C.L. (2010) Jasmonate-dependent and -independent pathways mediate specific effects of solar ultraviolet B radiation on leaf phenolics and antiherbivore defense. *Plant Physiology* 152, 1084–1095.

Favory, J.-J., Stec, A., Gruber, H., Rizzini, L., Oravecz, A., Funk, M., Albert, A., Cloix, C., Jenkins, J.I., Oakeley, E.J., Seidlitz, H.K., Nagy, F. and Ulm, R. (2009) Interaction of COP1 and UVR8 regulates UV-B-induced photomorphogenesis and stress acclimation in *Arabidopsis*. *The EMBO Journal* 28, 591–601.

Fiscus, E.L. and Booker, F.L. (1995) Is increased UV-B a threat to crop photosynthesis and productivity? *Photosynthesis Research* 43, 81–92.

Flint, S.D. and Caldwell, M.M. (1998) Solar UV-B and visible radiation in tropical forest gaps: measurements partitioning direct and diffuse radiation. *Global Change Biology* 4, 863–870.

Furness, N.H., Jolliffe, P.A. and Upadhyaya, M.K. (2005) Competitive interactions in mixtures of broccoli and *Chenopodium album* grown at two UV-B radiation levels under glasshouse conditions. *Weed Research* 45, 449–459.

Gao, Q. and Zhang, L. (2008) Ultraviolet-B-induced oxidative stress and antioxidant defense system responses in ascorbate-deficient *vtc1* mutants of *Arabidopsis thaliana*. *Journal of Plant Physiology* 165, 138–148.

Giordano, C.V., Mori, T., Sala, O.E., Scopel, A.L., Caldwell, M.M. and Ballaré, C.L. (2003) Functional acclimation to solar UV-B radiation in *Gunnera magellanica*, a native plant species of southernmost Patagonia. *Plant, Cell and Environment* 26, 2027–2036.

Giordano, C.V., Galatro, A., Puntarulo, S. and Ballaré, C.L. (2004) The inhibitory effects of UV-B radiation (280–315 nm) on *Gunnera magellanica* growth correlate with increased DNA damage but not with oxidative damage to lipids. *Plant, Cell and Environment* 27, 1415–1428.

Grüber, H., Heijde, M., Heller, W., Albert, A., Seidlitz, H.K. and Ulm, R. (2010) Negative feedback regulation of UV-B-induced photomorphogenesis and stress acclimation in *Arabidopsis*. *Proceedings of the National Academy of Sciences USA* 107, 20132–20137.

Hayes, S., Velanis, C.N., Jenkins, G.I. and Franklin, K.A. (2014) UV-B detected by the UVR8 photoreceptor antagonizes auxin signaling and plant shade avoidance. *Proceedings of the National Academy of Sciences USA* 111, 11894–11899.

Hectors, K., Prinsen, E., De Coen, W., Jansen, M.A.K. and Guisez, Y. (2007) *Arabidopsis thaliana* plants acclimated to low dose rates of ultraviolet B radiation show specific changes in morphology and gene expression in the absence of stress symptoms. *New Phytologist* 175, 255–270.

Hectors, K., Jacques, E., Prinsen, E., Guisez, Y., Verbelen, J.-P., Jansen, M.A.K. and Vissenberg, K. (2010) UV radiation reduces epidermal cell expansion in leaves of *Arabidopsis thaliana*. *Journal of Experimental Botany* 61, 4339–4349.

Heijde, M. and Ulm, R. (2012) UV-B photoreceptor-mediated signalling in plants. *Trends in Plant Science* 17, 230–237.

Hideg, É., Jansen, M.A.K. and Strid, Å. (2013) UV-B exposure, ROS, and stress: inseparable companions or loosely linked associates? *Trends in Plant Science* 18, 107–115.

Huché-Thélier, L., Crespel, L., Le Gourrierec, J., Morel, P., Sakr, S. and Leduc, N. (2016) Light signaling and plant responses to blue and UV radiations—perspectives for applications in horticulture. *Environmental and Experimental Botany* 121, 22–38.

Jansen, M.A.K. and Bornman, J.F. (2012) UV-B radiation: from generic stressor to specific regulator. *Physiologia Plantarum* 145, 501–504.

Jansen, M.A.K., Gaba, V. and Greenberg, B.M. (1998) Higher plants and UV-B radiation: balancing damage, repair and acclimation. *Trends in Plant Science* 3, 131–135.

Jansen, M.A.K., Van den Noort, R.E., Tan, M.Y.A., Prinsen, E., Lagrimini, L.M. and Thorneley, R.N.F. (2001) Phenol-oxidizing peroxidases contribute to the protection of plants from ultraviolet radiation stress. *Plant Physiology* 126, 1012–1023.

Jansen, M.A.K., Hectors, K., O'Brien, N., Guisez, Y. and Potters, G. (2008) Plant stress and human health: do human consumers benefit from UV-B acclimated crops? *Plant Science* 175, 449–458.

Jansen, M.A.K., Le Martret, B. and Koornneef, M. (2010) Variations in constitutive and inducible UV-B tolerance: dissecting photosystem II protection in *Arabidopsis thaliana* accessions. *Physiologia Plantarum* 138, 22–34.

Jenkins, G.I. (2009) Signal transduction in responses to UV-B radiation. *Annual Review of Plant Biology* 60, 407–431.

Jenkins, G.I. (2014) The UV-B photoreceptor UVR8: from structure to physiology. *The Plant Cell* 26, 21–37.

Jiang, L., Wang, Y., Li, Q.F., Björn, L.O., He, J.X. and Li, S.S. (2012) *Arabidopsis* STO/BBX24 negatively regulates UV-B signaling by interacting with COP1 and repressing HY5 transcriptional activity. *Cell Research* 22, 1046–1057.

Kaiser, G., Kleiner, O., Beisswenger, C. and Batschauer, A. (2009) Increased DNA repair in *Arabidopsis* plants overexpressing CPD photolyase. *Planta* 230, 505–515.

Kilian, J., Whitehead, D., Horak, J., Wanke, D., Weinl, S., Batistic, O., D'Angelo, C., Bornberg-Bauer, E., Kudla, J. and Harter, K. (2007) The AtGenExpress global stress expression data set: protocols, evaluation and model data analysis of UV-B light, drought and cold responses. *The Plant Journal* 50, 347–363.

Kuhlmann, F. and Müller, C. (2010) UV-B impact on aphid performance mediated by plant quality and plant changes induced by aphids. *Plant Biology* 12, 676–684.

Lau, T.S.L., Eno, E., Goldsein, G., Smith, C. and Christopher, D.A. (2006) Ambient levels of UV-B in Hawaii combined with nutrient deficiency decrease photosynthesis in near-isogenic maize lines varying in leaf flavonoids: flavonoids decrease photoinhibition in plants exposed to UV-B. *Photosynthetica* 44, 394–403.

Li, J., Ou-Lee, T.-M., Raba, R., Amundson, R.G. and Last, R.L. (1993) Arabidopsis flavonoid mutants are hypersensitive to UV-B irradiation. *The Plant Cell* 5, 171–179.

Li, X., Zhang, L., Li, Y., Ma, L., Bu, N. and Ma, C. (2012) Changes in photosynthesis, antioxidant enzymes and lipid peroxidation in soybean seedlings exposed to UV-B radiation and/or Cd. *Plant and Soil* 352, 377–387.

Liakopoulos, G., Stavrianakou, S. and Karabourniotis, G. (2006) Trichome layers versus dehaired lamina of *Olea europaea* leaves: differences in flavonoid distribution, UV-absorbing capacity, and wax yield. *Environmental and Experimental Botany* 55, 294–304.

Llorens, L., Badenes-Pérez, F.R., Julkunen-Tiitto, R., Zidorn, C., Fereres, A. and Jansen, M.A.K. (2015) The role of UV-B radiation in plant sexual reproduction. *Perspectives in Plant Ecology, Evolution and Systematics* 17, 243–254.

Mackerness, S.H.A., John, C.F., Jordan, B. and Thomas, B. (2001) Early signalling components in ultraviolet-B responses: distinct roles for different reactive oxygen species and nitric oxide. *FEBS Letters* 489, 237–242.

Manetas, Y., Petropoulou, Y., Stamatakis, K., Nikolopoulos, D., Levizou, E., Psaras, G. and Karabourniotis, G. (1997) Beneficial effects of enhanced UV-B radiation under field conditions: improvement of needle water relations and survival capacity of *Pinus pinea* L. seedlings during the dry Mediterranean summer. *Plant Ecology* 128, 101–108.

Mazza, C.A., Izaguirre, M.M., Curiale, J. and Ballaré, C.L. (2010) A look into the invisible: ultraviolet-B sensitivity in an insect (*Caliothrips phaseoli*) revealed through a behavioural action spectrum. *Proceedings of the Royal Society B* 277, 367–373.

McKenzie, R.L., Aucamp, P.J., Bais, A.F., Björn, L.O. and Ilyas, M. (2007) Changes in biologically-active ultraviolet radiation reaching the Earth's surface. *Photochemical & Photobiological Sciences* 6, 218–231.

McKenzie, R.L., Aucamp, P.J., Bais, A.F., Björn, L.O., Ilyas, M. and Madronich, S. (2011) Ozone depletion and climate change: impacts on UV radiation. *Photochemical & Photobiological Sciences* 10, 182–198.

Misra, P., Pandey, A., Tiwari, M., Chandrashekar, K., Sidhu, O.P., Asif, M.H., Chakrabarty, D., Singh, P.K., Trivedi, P.K., Nath, P. and Tuli, R. (2010) Modulation of transcriptome and metabolome of tobacco by Arabidopsis transcription factor, *AtMYB12*, leads to insect resistance. *Plant Physiology* 152, 2258–2268.

Molina, M.J. and Rowland, F.S. (1974) Stratospheric sink for chlorofluoromethanes – chlorine atomic-catalysed destruction of ozone. *Nature* 249, 810–812.

Morales, L.O., Brosché, M., Vainonen, J., Jenkins, G.I., Wargent, J.J., Sipari, N., Strid, V., Lindfors, A.V., Tegelberg, R. and Aphalo, P.J. (2013) Multiple roles for UV RESISTANCE LOCUS8 in regulating gene expression and metabolite accumulation in Arabidopsis under solar ultraviolet radiation. *Plant Physiology* 161, 744–759.

Müller-Xing, R., Xing, Q. and Goodrich, J. (2014) Footprints of the sun: memory of UV and light stress in plants. *Frontiers in Plant Science* 5, 474.

Musil, C.F., Midgley, G.F. and Wand, S.J.E. (1999) Carry-over of enhanced ultraviolet-B exposure effects to successive generations of a desert annual: interaction with atmospheric CO_2 and nutrient supply. *Global Change Biology* 5, 311–329.

Newsham, K.K. and Robinson, S.A. (2009) Responses of plants in polar regions to UVB exposure: a meta-analysis. *Global Change Biology* 15, 2574–2589.

Pan, W.S., Zheng, L.P., Tian, H., Li, W.Y. and Wang, J.W. (2014) Transcriptome responses involved in artemisinin production in *Artemisia annua* L. under UV-B radiation. *Journal of Photochemistry and Photobiology B: Biology* 140, 292–300.

Paul, N.D. and Gwynn-Jones, D. (2003) Ecological roles of solar UV radiation: towards an integrated approach. *Trends in Ecology and Evolution* 18, 48–55.

Paul, N.D., Jacobson, R.J., Taylor, A., Wargent, J.J. and Moore, J.P. (2005) The use of wavelength-selective plastic cladding materials in horticulture: understanding of crop and fungal responses through the assessment of biological spectral weighting functions. *Photochemistry and Photobiology* 81, 1052–1060.

Phoenix, G.K., Gwynn-Jones, D., Callaghan, T.V., Sleep, D. and Lee, J.A. (2001) Effects of global change on a sub-Arctic heath: effects of enhanced UV-B radiation and increased summer precipitation. *Journal of Ecology* 89, 256–267.

Potters, G., Pasternak, T.P., Guisez, Y., Palme, K.J. and Jansen, M.A.K. (2007) Stress-induced morphogenic responses: growing out of trouble? *TRENDS in Plant Science* 12, 98–105.

Poulson, M.E., Torres Boeger, M.R. and Donahue, R.A. (2006) Response of photosynthesis to high light and drought for *Arabidopsis thaliana* grown under a UV-B enhanced light regime. *Photosynthesis Research* 90, 79–90.

Pourcel, L., Routaboul, J.-M., Cheynier, V., Lepiniec, L. and Debeaujon, I. (2007) Flavonoid oxidation in plants: from biochemical properties to physiological functions. *TRENDS in Plant Sciences* 12, 29–36.

Radziejwoski, A., Vlieghe, K., Lammens, T., Berckmans, B., Maes, S., Jansen, M.A.K., Knappe, C., Albert, A., Seidlitz, H.K., Bahnweg, G., Inzé, D. and De Veylder, L. (2011) Atypical E2F activity coordinates PHR1 photolyase gene transcription with endoreduplication onset. *The EMBO Journal* 30, 355–363. doi: 10.1038/emboj.2010.313.

Rizzini, L., Favory, J.J., Cloix, C., Faggionato, D., O'Hara, A., Kaiserli, E., Baumeister, R., Schäfer, E., Nagy, F., Jenkins, G.I. and Ulm, R. (2011) Perception of UV-B by the *Arabidopsis* UVR8 protein. *Science* 332, 103–106.

Robinson, S.A. and Erickson, D.J. (2015) Not just about sunburn–the ozone hole's profound effect on climate has significant implications for Southern Hemisphere ecosystems. *Global Change Biology* 21, 515–527.

Robson, T.M. and Aphalo, P.J. (2012) Species-specific effect of UV-B radiation on the temporal pattern of leaf growth. *Physiologia Plantarum* 144, 146–160.

Robson, T., Hartikainen, S.M. and Aphalo, P.J. (2015a) How does solar ultraviolet-B radiation improve drought tolerance of silver birch (*Betula pendula* Roth.) seedlings? *Plant, Cell & Environment* 38, 953–967.

Robson, T., Klem, K., Urban, O. and Jansen, M.A.K. (2015b) Re-interpreting plant morphological responses to UV-B radiation. *Plant, Cell & Environment* 38, 856–866.

Rozema, J., Chardonnens, A., Tosserams, M., Hafkenscheid, R. and Bruijnzeel, S. (1997) Leaf thickness and UV-B absorbing pigments of plants in relation to an elevational gradient along the Blue Mountains, Jamaica. *Plant Ecology* 128, 151–159.

Rozema, J., Boelen, P. and Blokker, P. (2005) Depletion of stratospheric ozone over the Antarctic and Arctic: responses of plants of polar terrestrial ecosystems to enhanced UV-B, an overview. *Environmental Pollution* 137, 428–442.

Rozema, J., Boelen, P., Solheim, B., Zielke, M., Buskens, A., Doorenbosch, M., Fijn, R., Herder, J., Callaghan, T., Björn, L.O., Gwynn-Jones, D., Broekman, R., Blokker, P. and de Poll, W.V. (2006) Stratospheric ozone depletion: high arctic tundra plant growth on Svalbard is not affected by enhanced UV-B after 7 years of UV-B supplementation in the field. *Plant Ecology* 182, 121–136.

Schmidt, A.M., Ormrod, D.P., Livingston, N.J. and Misra, S. (2000) The interaction of ultraviolet-B radiation and water deficit in two *Arabidopsis thaliana* genotypes. *Annals of Botany* 85, 571–575.

Schreiner, M., Mewis, I., Huyskens-Keil, S., Jansen, M.A.K., Zrenner, R., Winkler, J.B., O'Brien, N. and Krumbein, A. (2012) UV-B-induced secondary plant metabolites-potential benefits for plant and human health. *Critical Reviews in Plant Sciences* 31, 229–240.

Teklemariam, T. and Blake, T.J. (2003) Effects of UVB preconditioning on heat tolerance of cucumber (*Cucumis sativus* L.). *Environmental and Experimental Botany* 50, 169–182.

Thines, N.J., Shipley, L.A., Bassman, J.H., Slusser, J.R. and Gao, W. (2008) UV-B effects on the nutritional chemistry of plants and the responses of a mammalian herbivore. *Oecologia* 156, 125–135.

Topcu, Y., Dogan, A., Kasimoglu, Z., Sahin-Nadeem, H., Polat, E. and Erkan, M. (2015) The effects of UV radiation during the vegetative period on antioxidant compounds and postharvest quality of broccoli (*Brassica oleracea* L.). *Plant Physiology and Biochemistry* 93, 56–65.

Treutter, D. (2006) Significance of flavonoids in plant resistance: a review. *Environmental Chemistry Letters* 4, 147–157.

Ulm, R., Baumann, A., Oravecz, A., Máté, Z., Ádám, É., Oakeley, E.J., Schäfer, E. and Nagy, F. (2004) Genome-wide analysis of gene expression reveals function of the bZIP transcription factor HY5 in the UV-B response of *Arabidopsis*. *Proceedings of the National Academy of Sciences USA* 101, 1397–1402.

Van de Staaij, J.W.M., Bolink, E., Rozema, J. and Ernst, W.H.O. (1997) The impact of elevated UV-B (280–320 nm) radiation levels on the reproduction biology of a highland and a lowland population of *Silene vulgaris*. *Plant Ecology* 128, 173–179.

Vass, I. (2012) Molecular mechanisms of photodamage in the Photosystem II complex. *Biochimica et Biophysica Acta (Bioenergetics)* 1817, 209–217.

Wargent, J.J., Gegas, V.C., Jenkins, G.I., Doonan, J.H. and Paul, N.D. (2009a) UVR8 in *Arabidopsis thaliana* regulates multiple aspects of cellular differentiation during leaf development in response to ultraviolet B radiation. *New Phytologist* 183, 315–326.

Wargent, J.J., Moore, J.P., Ennos, A.R. and Paul, N.D. (2009b) Ultraviolet radiation as a limiting factor in leaf expansion and development. *Photochemistry and Photobiology* 85, 279–286.

Weinig, C., Gravuer, K.A., Kane, N.C. and Schmitt, J. (2004) Testing adaptive plasticity to UV: costs and benefits of stem elongation and light-induced phenolics. *Evolution* 58(12), 2645–2656.

Wellmann, E. (1983) UV radiation in photomorphogenesis. In: Lange, O.L., Nobel, P.S., Osmond, C.B. and Ziegler, H. (eds) *Encyclopedia of Plant Physiology, New Series*. 16B. Springer, New York, pp. 745–756.

Xu, C. and Sullivan, J.H. (2010) Reviewing the technical designs for experiments with ultraviolet-B radiation and impact on photosynthesis, DNA and secondary metabolism. *Journal of Integrative Plant Biology* 52, 377–387.

Yao, Y., Xuan, Z., He, Y., Lutts, S., Korpelainen, H. and Li, C. (2007) Principal component analysis of intraspecific responses of tartary buckwheat to UV-B radiation under field conditions. *Environmental and Experimental Botany* 61, 237–245.

Zavala, J.A., Mazza, C.A., Dillon, F.M., Chludil, H.D. and Ballaré, C.L. (2015) Soybean resistance to stink bugs (*Nezara viridula* and *Piezodorus guildinii*) increases with exposure to solar UV-B radiation and correlates with isoflavonoid content in pods under field conditions. *Plant, Cell & Environment* 38, 920–928.

Ziska, L.H., Teramura, A.H. and Sullivan, J.H. (1992) Physiological sensitivity of plants along an elevational gradient to UV-B radiation. *American Journal of Botany* 79, 863–871.

Zu, Y.G., Pang, H.H., Yu, J.H., Li, D.W., Wei, X.X., Gao, Y.X. and Tong, L. (2010) Responses in the morphology, physiology and biochemistry of *Taxus chinensis* var. *mairei* grown under supplementary UV-B radiation. *Journal of Photochemistry and Photobiology B: Biology* 98, 152–158.

11 Freeze Tolerance and Avoidance in Plants

Michael Wisniewski[1],*, Ian R. Willick[2] and Lawrence V. Gusta[2]
*[1]United States Department of Agriculture Agricultural Research Service,
Kearneysville, West Virginia, USA; [2]Department of Plant Sciences,
University of Saskatchewan, Saskatoon, Saskatchewan, Canada*

Abstract

Understanding and improving the cold hardiness of plants has been an endeavour that has been pursued since the onset of studying plant biology. Cold acclimation is a multigenic, quantitative trait that involves biochemical and structural changes that affect the physiology of a plant. The type and form of freezing injury experienced by plants varies with species and their degree of freezing tolerance and/or ability to avoid freezing. Advances in biotechnology have allowed us to move beyond structure and physiology to identifying and understanding the role of specific genes and proteins. The present chapter reviews our current understanding of freezing tolerance and avoidance and emphasizes that essential to the beneficial use of modern biotechnology is a thorough understanding of plant biology in relation to cold hardiness. The use of plant biotechnology grounded in an understanding of plant biology has great potential for increasing plant productivity in a rapidly changing climate.

11.1 Introduction

'It was the best of times, it was the worst of times, it was the age of wisdom, it was the age of foolishness, it was the epoch of belief, it was the epoch of incredulity' (Charles Dickens, *A Tale of Two Cities*). This quotation can be applied to the recent history of cold-hardiness research, and perhaps the history of research in general. Newspapers carry stories of a marvellous breakthrough that will result in the development of plants more resistant to abiotic stress which will help to address the need to produce more food for a growing world population on less arable land and under more stressful environmental conditions. In a more realistic picture, however, increasing cold hardiness without impacting other favourable attributes, such as biomass and yield, has proved to be more complex and elusive.

However, the good news is that we now have the ability, through 'omic' technologies, to elucidate how plants respond to the environment on many different levels. Using these new tools, one can obtain a global picture of the genetic structure of a plant, what genes are being expressed at a particular point in time, as well as what proteins and secondary metabolites are present. Through genomics, transcriptomics, proteomics, metabolomics, hormonics and bioinformatics one can obtain incredible amounts of information about how individual tissues, organs or whole plants respond to an environmental stress. This multilevel examination and integration of data is generally referred to as systems biology. Plants can vary greatly in their ability to cold acclimate and withstand freezing temperatures (Table 11.1). The objective of cold-hardiness research is to identify the function of genes, proteins, metabolites and

---

* Corresponding author: michael.wisniewski@ars.usda.gov

Table 11.1. Variation in cold hardiness among plant types.

Hardiness range (°C)	Plant type
−3 to −9	Summer annuals, *Arabidopsis*
−10 to −40	Winter annuals, perennials (e.g. grasses)
−20 to −45	Fruit trees, native and ornamental woody plants
−45 to −196	Extreme northern and southern latitude plants (e.g. poplar, spruce, red-osier dogwood)

hormones as they relate to the ability of a plant to cold acclimate and survive the stresses associated with low temperature, dehydration and the presence of ice within their tissues (Gusta and Wisniewski, 2012). In the present review, the reader is provided with a general background to the subject of freezing tolerance and avoidance in plants and the complexities involved in defining strategies to improve cold hardiness.

11.2 Genetic Regulation of Freezing Tolerance

Reductions in yield due to freeze damage, whether it occurs in late autumn, midwinter or early spring, can be quite significant and vary from year to year. In 2007, widespread freezes during the spring months in the Eastern and Midwestern portions of the USA resulted in over US$2 billion in crop damage (Gu *et al.*, 2008). Fruit crops, field and vegetable crops, as well as ornamentals, were all affected by these frost episodes and in some cases entire plantings of perennial plants such as grapes (*Vitis vinifera*), blueberries (*Vaccinium* sp.) and raspberries (*Rubus* sp.) were completely lost. An increase in devastating spring frosts is expected as a direct result of global climate change despite overall increases in mean temperatures (Gu *et al.*, 2008). Ball and Hill (2009) have also reported that elevated atmospheric CO_2 concentrations have a negative impact on cold acclimation and enhance vulnerability to frost damage.

It has long been recognized that cold acclimation is a multigenic, quantitative trait that involves biochemical and structural changes which affect the physiology of a plant (Weiser, 1970; Levitt, 1972). There is no consensus on the number and identity of genes causally related to cold acclimation. Various reports have estimated that from < 100 to 1000s of genes are upregulated and a similar number are downregulated (Fowler and Thomashow, 2002; Hannah *et al.*, 2005; Bassett and Wisniewski, 2009). Cold acclimation is an inducible process involving a short photoperiod, low temperature (< 10°C) and moderate to high light (generally above 400 µmol/m²/s) to achieve the species' full hardiness potential. The conditions needed to achieve maximum freezing tolerance are species specific.

A family of transcription factors known either as the C-repeat binding factor (CBF) or as the dehydration-responsive binding element (DREB) have been established as one of the main genes responsible for responding to low temperatures and initiating cold acclimation (Stockinger *et al.*, 1997; Gilmour *et al.*, 2004; Maruyama *et al.*, 2004). At 20–22°C, transgenic *Arabidopsis* (*A. thaliana*) overexpressing a native CBF gene induces about 12% of the cold-induced transcriptional changes that occur in wild-type plants during the normal process of cold acclimation (Fowler and Thomashow, 2002). The number of CBF/DREB genes varies with species. For example both *Arabidopsis* and grape are reported to have four (Fowler and Thomashow, 2002; Xiao *et al.*, 2006) whereas wheat has 17 (Dhillon *et al.*, 2010). A complete description of CBF genes in several fruit crops (peach, apple and grape) and poplar was provided by Wisniewski *et al.* (2014). CBF/DREB expression in *Arabidopsis* is induced within 15 min of cold induction and attains maximum expression within 1.5–4.5 h depending on the specific CBF gene (Zarka *et al.*, 2003; Benedict *et al.*, 2006). The expression pattern of any specific CBF gene in a woody plant can vary in annual (leaves) versus perennial (bark) tissues (Benedict *et al.*, 2006) and the timing of CBF expression can be different prior to and after the onset of dormancy (Welling and Palva, 2008). In a series of papers, Wisniewski and colleagues reported on the impact of a peach CBF (*PpCBF1*) gene on the freezing tolerance, growth, and onset and release of dormancy in apple (Wisniewski *et al.*, 2011, 2015; Artlip *et al.*, 2014). These studies clearly indicated how the regulation of cold hardiness in woody plants is integrated with the regulation of dormancy and growth (Fig. 11.1). Overexpression of *PpCBF1* in apple modified the expression of *Dormancy-Associated Mad Box (DAM)* genes that are believed to regulate dormancy in woody plants, *RGL (DELLA)* genes that inhibit GA-mediated growth processes, and a homologue of the *Early Budbreak (EBB1)* gene that was originally demonstrated to regulate the timing of budbreak in poplar (Yordanov *et al.*, 2014).

CBF is undoubtedly involved in the regulation of freezing tolerance, as transgenic plants overexpressing CBF are several degrees more freezing tolerant compared to wild-type plants. The induction of any specific CBF, however, may be a function of the experimental conditions. For example, when canola (*Brassica napus* L.) plants growing in 15 cm pots were transferred directly from 23 to 4°C, they experienced a cold shock compared to plants cooled 2°C/h from 23 to 4°C (cold stress) (R.G. Trischuk and L.V. Gusta, unpublished results). CBF increased dramatically in the cold-shocked plants compared to the cold-stressed plants. After 24 h, leaf water potential was –0.3 MPa in the cold-stressed plants versus –0.7 MPa in the cold-shocked plants. A proteomic study of these plants revealed that there were at least 29 proteins that were present in the cold-shocked plants that were not observed in the cold-stressed plants. These results suggest that a portion of the CBF gene induction may reflect a cold-shock response rather than a programmed process of cold acclimation.

11.3 Role of Light in Freezing Resistance

It has long been recognized that light plays an important role in cold acclimation and freezing

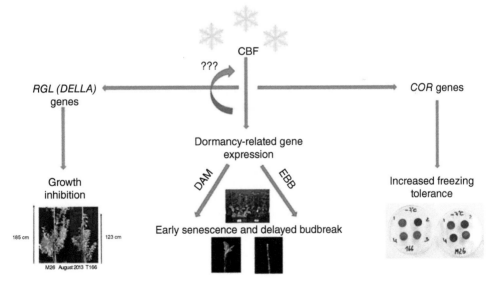

Fig. 11.1. Diagram summarizing the results obtained when a peach *CBF* (*PpCBF1*) gene was overexpressed in an apple (*Malus* × *domestica*) rootstock variety, M.26. Overexpression of the *CBF* gene (*PpCBF1*) results in altered expression of apple dormancy-related genes resulting in early senescence in the autumn and delayed budbreak in the spring. Overexpression of the *CBF* gene also leads to altered expression of *RGL* genes which results in growth inhibition. *CBF* overexpression also induces the expression of *COR* genes which results in increased freezing tolerance. The regulation of the various genes and other *CBF* genes may be due to the presence of C-repeat motifs present in the promoter region of the studied genes. CB, cold-binding factor; COR, cold-regulated gene; RGL, repressor of GA1-3. Taken from Wisniewski *et al.* (2015).

resistance (Vasilyev, 1961; Levitt, 1972; Hüner *et al.*, 1998; Öquist and Hüner, 2003). Due to the breadth of information on photobiology and energy sensing, all of its intricacies, however, will not be discussed. More comprehensive reviews of this topic have been provided by Hüner *et al.* (1998), Öquist and Hüner (2003) and Hüner *et al.* (2013).

Photoperiod duration (Weiser, 1970), light intensity and spectral distribution (Steponkus and Lanphear, 1968; Roberts, 1984; Gray *et al.*, 1997) all play a role in freezing tolerance (Fig. 11.2). Depending upon the species, low light may be inadequate for the full development of freezing tolerance. For example, Nezami *et al.* (2012) reported that in chickpeas (*Cicer arietinum*), exposure to 400 µmol/m²/s combined with low temperature increased freezing tolerance, whereas exposure to 100 µmol/m²/s was ineffective. On the other hand, when plants are unable to acclimate to high light conditions, due either to an inability to dissipate excess energy as heat via the xanthophyll cycle or to reduce the size of the light harvesting antenna, photosynthetic efficiency is reduced, resulting in dynamic photoinhibition (Hüner *et al.*, 1993; Maxwell *et al.*, 1995; Machalek *et al.*, 1996). When the rate of photo-oxidative damage to PSII reaction centres exceeds the rate of repair, then chronic photoinhibition can occur, resulting in permanent damage to the photosynthetic capacity of the plant. Some species, such as winter rye (*Secale cereale* L.), have the capacity to develop resistance to cold-induced photoinhibition (Hüner *et al.*, 1993) while less freezing-tolerant plants,

such as winter barley (*Hordeum vulgare* L.), are less effective and undergo photobleaching under conditions of high irradiance (L.V. Gusta, unpublished results). This may be one of the limiting factors that prevents the development of a high level of freezing tolerance in winter barley and cultivars of winter wheat.

Plants must maintain a balance (photostasis) between photochemical reactions (temperature insensitive) and metabolic/developmental (temperature sensitive) processes in order to avoid photoinhibiton. Photostasis can be maintained by non-photochemical quenching mechanisms, which are used by the plant to dissipate excess energy, not used in photosynthesis, as heat. This protects the PSII reaction centre from damage that can occur upon exposure to cooler temperatures through increased energy usage via photorespiration, nitrogen assimilation, and growth (Öquist and Hüner, 2003).

Plants exhibit a wide range in the maximum freezing tolerance that they can attain. This ability is dependent upon the species, cultivar, variety and whether the plant maintains active photosynthesis as a cold-tolerant species or reduces photosynthesis as a cold-sensitive species (Öquist *et al.*, 1993; Öquist and Hüner, 2003). For example, winter wheat (*Triticum aestivum* L.), rye (*Secale cereale* L.) and winter barley (*Hordeum vulgare* L.), when cold acclimated, exhibit greater carbon assimilation at low temperatures (< 10°C) (Hurry and Hüner, 1991; Hurry *et al.*, 1995; Öquist and Hüner, 2003) than cold-sensitive spring wheat (Hurry and Hüner, 1991) and spring rape (*Brassica napus* L.) (Hurry

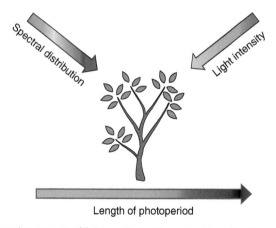

Length of photoperiod

Fig. 11.2. Diagram illustrating aspects of light quality that impact cold acclimation.

et al., 1995). Greater carbon assimilation at low, cold-acclimating temperatures is associated with increased sink capacity (Gray *et al.*, 1996; Strand *et al.*, 1999).

Enhanced carbon assimilation at low, cold-acclimating temperatures is a multi-component process requiring photosynthesis for the production of sugars. These act not only as a source of energy but also as a cryo-protectant, and as a signalling molecule triggering the induction of specific genes, proteins and metabolites, all of which contribute to increased freezing tolerance (Guy, 1990; Koch, 1996; Gray *et al.*, 1997; Smeekens, 1998; Thomashow, 1999; Rekarte-Cowie *et al.*, 2008). Cold acclimation induces the rapid accumulation of soluble carbohydrates in plants. These carbohydrates protect macromolecules and membranes in cells from degradation and injury caused by freeze-induced dehydration. Carbohydrates also serve as a source of nutrition, providing the energy source needed to drive cold acclimation, as well as recovery processes after a freezing episode (see reviews by Levitt, 1972; Gusta *et al.*, 2009).

The chloroplast also acts as a sensor of both environmental and temperature stress (Hüner *et al.*, 1998; Kosmala *et al.*, 2009; Murchie *et al.*, 2009). Wanner and Junttila (1999), studying the effect of light and temperature on cold acclimation of *Arabidopsis*, concluded that the accumulation of sucrose was essential for improving freezing tolerance. Many cold-regulated (COR) mRNAs accumulated at low temperature in the absence of light; however, there was no increase in cold hardiness. Thus, the accumulation of these cold-induced mRNAs and their translational products (proteins) alone were insufficient in establishing freezing tolerance. Successful cold acclimation required the accumulation of sucrose. It is also possible that additional levels of regulation (such as epigenetic control of genes) are regulated by photoperiod and/or light intensity.

Kosmala *et al.* (2009), studying the development of cold acclimation in perennial grass (*Festuca pratensis*) plants that varied in their level of freezing tolerance, found that photosynthesis was initially inhibited due to the degradation of photosynthetic proteins at low temperatures. They reported that plants with low freezing-tolerance potential were affected more

than plants with high freezing tolerance. These initial stages of cold acclimation, where photosynthesis plays an important role, may be crucial to the full development of frost tolerance. This relationship is extremely relevant to studies conducted in a controlled environment where plants are transferred directly from non-acclimating temperatures to low, acclimating temperatures (e.g. 23 to 4°C).

Exposure to low temperatures results in a myriad of structural, biochemical and physiological changes in plants. The observed changes include a reduction in tissue water content; and enhanced levels of sugars, antioxidants, proline, chlorophyll metabolism, and late-embryogenesis-abundant (LEA) proteins (including dehydrins) and growth-inhibiting compounds such as ABA. Aquaporins, chaperonins, water binding proteins, lipid transfer proteins, energy production, regulators of cell vesicular traffic, proteins required for PSII functionality and thylakoid integrity, and universal stress proteins can all be induced and regulated by low temperature (Guy, 1990; Thomashow, 1999; Manning *et al.*, 2007; Shinozaki and Yamaguchi-Shinozaki, 2007). Kosmala *et al.* (2009) analysed 800 protein profiles in cold-acclimating plants of *F. pratensis* that exhibit either high or low cold-hardiness potential and found that most of the detected proteins did not exhibit significant differences in their level of accumulation. Among the 41 proteins that exhibited significant differences in plants with high versus low cold-hardiness potential, half were associated with the photosynthetic apparatus. These findings highlight the important role of photosynthesis in the process of cold acclimation.

The study of Kosmala *et al.* (2009) is somewhat unique in that they characterized protein changes (presence and abundance) associated with the cold acclimation of genotypes that varied in cold-hardiness potential. Many of the findings of 'omic' studies have been based only on a single genotype or a comparison of spring and winter genotypes. When comparing winter with spring genotypes, several hundred significant differences in gene expression or protein abundance would be expected, due to the myriad of physiological and developmental differences associated with these contrasting genotypes, such as vernalization differences and low-temperature growth adjustments. Therefore, comparing winter with spring genotypes may not be the most efficient approach to identifying genes associated specifically with freezing tolerance. More relevant information could be obtained by comparing hardy winter genotypes with other winter genotypes or with closely related genotypes that have significant differences in their ability to cold acclimate. For example, winter rye can tolerate −33°C when fully acclimated; however, winter wheat can only be hardened to −24°C under ideal conditions. It would not be too surprising to find that there are not many differences in gene or protein expression between hardy lines of winter wheat and winter rye, as was reported for the genotypes of the perennial grass F. pratensis, by Kosmala et al. (2009).

Gusta et al. (2009) observed only subtle differences in the metabolites present in fully cold-acclimated 'Norstar' winter wheat (LT$_{50}$ −24°C) versus 'Puma' winter rye (LT$_{50}$ −30°C). The most apparent differences were in sugars and flavonoids. In contrast, there were large differences in the metabolite profiles of the spring wheat cultivar 'Elza' (LT$_{50}$ −9°C) and 'Norstar' winter wheat (LT$_{50}$ −24°C). Gusta et al. (1997) reported that spring wheat could tolerate −9°C when cooled at 2°C h^{-1} but that it was killed at −3°C after 48 h. In contrast, 'Norstar' winter wheat can tolerate −24°C when cooled at 2°C/h but crowns can only survive −12°C for 15 days. This suggests that winter cereals have evolved different strategies to survive long periods of sub-zero winter temperatures. In the hunt to discover the underlying mechanisms of freezing tolerance and their genetic regulation, it is essential that a full and complete understanding of the physiology of the plant is appreciated. It is also essential that we compare apples with apples and not apples with oranges.

11.4   Cold-sensing in Plants

Regulating when a plant will flower is essential to perpetuating a species. In many plant species, the ability to initiate floral development requires that a plant be exposed to low temperatures for a period of time, a process referred to as vernalization. In other plant species, this process is regulated by photoperiod, i.e. the length of day versus night. In winter cereals, the more

cold-hardy winter rye genotypes undergo vernalization at a warmer temperature than less cold-hardy winter wheat and barley varieties (Wilen et al., 1998). Winter rye also initiates cold acclimation at a warmer temperature than winter wheat, which initiates cold acclimation at a warmer temperature than winter barley. Plants that perceive a low-temperature signal at warmer temperatures have a longer period of time to cold acclimate prior to winter, and are also less sensitive to mixed temperature (cold and warm) signals during the period of cold acclimation in the autumn. Such plants can acclimate faster at warmer temperatures, assimilate more sugars, overcome photoinhibition, etc.

The accumulation of cryoprotectants and thermal hysteresis proteins (antifreeze proteins) are not always tied to vernalization. When winter wheat and winter canola were cold acclimated, de-acclimated and then re-acclimated, Trischuk et al. (2014) found that the accumulation of sucrose was tied to the vernalization response, whereas the production of dehydrins (a cryoprotective protein) was associated with both the initial cold acclimation and the subsequent re-acclimation response. The winter wheat cultivar 'Norstar' can cold acclimate to −25°C; however, if vernalization saturation is met and the plants are dehardened to −3°C, they can only reharden to −9°C. This is the same level of cold acclimation as spring wheat cultivars, which have no vernalization requirement. This suggests that vernalization controls a switch that regulates cold acclimation and growth. Once growth initiates, sugars are then utilized for growth rather than freezing tolerance. In some herbaceous species such as cauliflower (Brassica oleracea var. botrytis) vernalization is only associated with flowering and not with increases in freezing tolerance (Wurr et al., 2004).

The requirement for prolonged cold to become flowering competent has been studied extensively in winter-habit Arabidopsis (Fig. 11.3) (Chandler and Dean, 1994; Baurle and Dean, 2006; DeLucia et al., 2008; Crevillen and Dean, 2010; Liu et al., 2010). Following germination, winter-habit Arabidopsis plants develop an inhibition to flowering that is partially regulated by the expression of the FRIGIDA gene (Johanson et al., 2000), which can slowly be switched off by exposure to cold temperatures. The mechanism of FRIGIDA's repression is currently an active

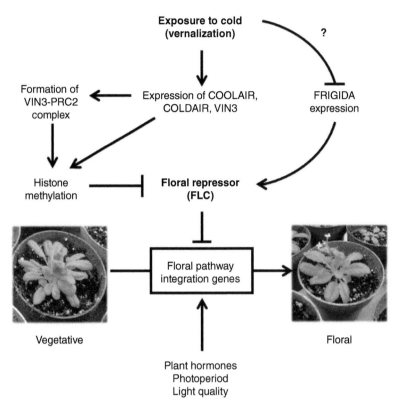

Fig. 11.3. Simplified pathway depicting the epigenetic control of vernalization in *Arabidopsis*. VIN3, vernalization insensitive 3; PRC2, polycomb reverse complex 2; FLC, flowering locus C. Drawn using information from Johanson *et al.*, 2000; Baurle and Dean, 2006; Swiezweski *et al.*, 2009; Heo and Sung, 2011.

area of research. In these *Arabidopsis* plants, vernalization represents a quantitative reduction in the expression of a floral repressor known as flowering locus C (FLC) (Baurle and Dean, 2006; Crevillen and Dean, 2010). This 'switching off' mechanism is a quantitative epigenetic event (Baurle and Dean, 2006; Crevillen and Dean, 2010; Liu *et al.*, 2010). Epigenetic memory of cold can be established by the modification of DNA through: (i) transmittable DNA methylation; (ii) chromatin modification; or (iii) non-coding RNA-based mechanisms (Mirouze and Paszkowski, 2011; King, 2015). Upon exposure to cold, winter-habit *Arabidopsis* begins to accumulate non-coding RNA, designated COOLAIR, prior to the epigenetic silencing of FLC (De Lucia *et al.*, 2008; Swiezewski *et al.*, 2009; Heo and Sung, 2011). A second non-coding RNA, designated COLDAIR, begins to accumulate halfway

through the vernalization process. COLDAIR is believed to play a role in the process of protein binding at the site on the chromatin strand required for initiating methylation of FLC (Heo and Sung, 2011). Once the plant has reached its point of maximum vernalization saturation, epigenetic silencing of FLC via histone (H3K27me3) methylation is partially controlled by increased concentrations of the vernalization-insensitive 3 (VIN3) protein, which forms a protein complex with a conserved polycomb repressive complex 2 (PRC2) that functions in chromatin silencing (Swiezewski *et al.*, 2009; Angel *et al.*, 2011). During embryogenesis, FLC-silencing by DNA methylation resets (Sheldon *et al.*, 2009). This mechanism of epigenetic gene silencing is relevant to the propagation of horticultural plants. Cuttings taken from cold-acclimated plants would remember their cold exposure. Thus, one would be able propagate plants with a pre-established resistance to cold temperatures.

Epigenetic memory of cold may also explain the clinal variation in winter hardiness and phenology observed within genetically homogeneous populations of woody plants (Dormling and Johnsen, 1992; Kvaalen and Johnsen, 2008; Yakovlev *et al.*, 2010, 2012, 2014). For example, the timing of budset and budbreak in Norway spruce (*Picea abies*) (Kvaalen and Johnsen, 2008; Yakovlev *et al.*, 2010, 2012, 2014), white spruce (*Picea glauca*) (Stoehr *et al.*, 1998) and Scots pine (*Pinus sylvestris*) (Dormling and Johnsen, 1992) have been attributed to epigenetic memory of cold. Understanding how plants perceive their environment and the role of epigenetics is still an actively evolving topic of research and will be the subject of many studies for years to come.

11.5 Ice Nucleation and Deep Supercooling

The freezing point of pure water, referred to as the homogeneous nucleation temperature, is approximately −40°C. Of course, in nature freezing of water occurs at much warmer temperatures, typically just below 0°C. This is because it is rare for water to exist in a pure state; rather, it exists as an ionic or colloidal solution. In such solutions, heterogenous ice nucleation is initiated on the surface of objects or on suspended particles. Heterogeneous nucleators are very effective at inducing ice formation and are very abundant. The temperature at which a heterogenous nucleator induces ice formation is specific to the ice-nucleating properties of that specific compound (Ashworth, 1992).

The role of heterogenous ice nucleators in inducing ice formation in plants is important because, if methods can be developed for regulating ice nucleation, significant advances could be made in limiting frost injury to both freezing-sensitive and cold-adapted plants. While all plants can supercool (i.e. have tissue temperatures that are below 0°C but are unfrozen) to some extent, the degree of supercooling varies between plant species and is influenced by the presence of ice-nucleating agents which may be of intrinsic (plant) or extrinsic (fungal or bacterial)

origin (Ashworth and Kieft, 1995; Wisniewski *et al.*, 2002a).

Starting in the late 1970s, research focused on ice-nucleating agents and their role in inducing plants to freeze at warm, sub-zero temperatures. The identification of extrinsic – especially ice-nucleation-active (INA) bacteria – and intrinsic nucleation agents and their role in the freezing process received a great deal of attention. While published research in this area diminished greatly in the 1990s, new insights were gained when high-resolution infrared thermography was employed to study the freezing process (Wisniewski *et al.*, 2014). Additionally, published reports on antifreeze proteins (which exhibit hysteresis, bind to ice crystals and affect their morphology, and have the ability to inhibit nucleating compounds) and anti-nucleators (compounds that inhibit the activity of nucleating agents but do not exhibit hysteresis) have added to the complexity of our understanding of what induces a plant to freeze (Duman and Wisniewski, 2014). Wisniewski *et al.* (2009) provided a historical view and detailed analysis of this topic.

Among the factors that influence when a plant freezes are: (i) moisture and extrinsic ice-nucleating agents; (ii) hydrophobic barriers; (iii) plant structure; (iv) ice nucleators, antifreeze proteins, sugar and anti-nucleators; and (v) tissue water content. A brief summary of the role of these factors follows.

11.5.1 Moisture and extrinsic ice-nucleating agents

Dry plants will supercool to a lower temperature than wet plants. The presence of ice-nucleating agents that are active at warm, sub-zero temperatures will influence the temperature at which a plant freezes but the nucleators must be in an aqueous solution. In order for dry plants to supercool, they must be free of intrinsic ice-nucleating agents that are active at warm, sub-zero temperatures (Wisniewski *et al.*, 2002a).

11.5.2 Hydrophobic barriers

Assuming that a plant is free of intrinsic ice-nucleating agents that are active at warm sub-zero temperatures, ice crystals must physically grow through a crack in the cuticle, a broken epidermal hair or a stoma to induce ice nucleation within a plant (Wisniewski and Fuller, 1999). Therefore, hydrophobic barriers, such as a thick, waxy cuticle or a synthetic substance applied to the plant surface, can prevent frost on the surface of a plant from initiating an internal ice-nucleating event. Figure 11.4 (top panel) illustrates the application of a synthetic hydrophobic kaolin film to the surface of a tomato plant that was then sprayed with water containing INA bacteria (*Pseudomonas syringae*, strain Cit7). The plants were then cooled at 2°C/h to −6°C (lower panel). The plants with the hydrophobic coating supercooled, did not freeze and thus survived. In contrast, uncoated plants were killed when the plants froze at −2°C since they had no freezing tolerance. As illustrated in Fig. 11.5, similar results using the hydrophobic film were obtained with potato (*Solanum tuberosum*), as well as grape (*Vitis vinifera*) and citrus (*Citrus limon*) (Wisniewski *et al.*, 2008). Aryal and Neuner (2010) reported that, in general, plants growing at high altitudes and in open places where low night-time temperatures and dew formation are more frequent, have reduced leaf wettability due to their thick, waxy cuticles. The ability to avoid ice formation plays an important role in the reproductive ecology of alpine plants (Hacker *et al.*, 2011) and in alpine ecology in general (Neuner, 2014).

11.5.3 Plant structure

While the factors that determine where ice forms and accumulates within a plant are not well understood, it is clear that ice formation within

Fig. 11.4. Tomato (*Lycopersicon esculentum*) uncoated and coated with a hydrophobic kaolin (white powder), sprayed with water containing ice-nucleating-active (INA) bacteria and subjected to freezing temperatures. The hydrophobic barrier prevented the extrinsic ice crystals from penetrating the plant tissue and inducing freezing; thus, the coated plant survived exposure to −6°C whereas the uncoated plant froze and was killed.

Fig. 11.5. Frost protection of potato (top), grape (middle) and citrus (bottom) plants coated with a hydrophobic particle film. All plants were subject to −3 to −4°C. Coated plants are on the left in the top and middle frames and on the right in the bottom frame.

a plant is not uniform but rather occurs in select sites. Where ice forms, how it propagates and how it is accommodated are all important factors that affect the ability of a plant to survive freezing, and may be as important as the ability of a cell to tolerate the dehydrative stress associated with extracellular ice formation (Wisniewski *et al.*, 1997; Workmaster *et al.*, 1999). Recent research by Kishimoto *et al.* (2014) and Ishikawa *et al.* (2015) have clearly demonstrated that different tissues in woody plants exhibit different levels of ice-nucleation activity and that this level of activity changes on a seasonal basis. In particular, bud scales surrounding and protecting shoot apical meristems have high ice-nucleation activity, ensuring that ice will form first in these tissues. Water will then migrate to the bud scales and freeze. This protects the shoot apical meristem, responsible for spring shoot growth, from being injured by the formation of large ice crystals (Ishikawa *et al.*, 2015).

11.5.4 Ice nucleators, antifreeze proteins, sugars and anti-nucleators

The compounds responsible for ice-nucleating activity in INA bacteria have been identified as proteins, and genes coding for these proteins have been identified (Warren, 1995). In contrast, compounds responsible for intrinsic ice-nucleating activity observed in many woody plants have not been identified. Their synthesis, distribution, turnover and adaptive role remain largely uncharacterized.

Antifreeze proteins (AFPs), also known as thermal hysteresis proteins (THPs), inhibit ice crystal growth in a non-colligative manner, lowering the freezing point below the melting point, thus producing thermal hysteresis. These proteins have been identified in fish, insects and plants, but do not exhibit any significant similarities in composition or structure (Duman and Olsen, 1993). AFPs are found in a wide range of overwintering plants where they inhibit the growth and recrystallization of ice that forms in intercellular spaces (Griffith and Yaish, 2004). In winter rye (*Secale cereale*), AFPs are homologous to pathogenesis-related proteins and accumulate in response to cold, short day-length, dehydration and ethylene. In rye, AFPs function by

directly interacting with ice *in planta* and reduce freezing injury by slowing the growth and recrystallization of ice (Griffith *et al.*, 2005).

Anti-nucleators (compounds that inhibit the activity of ice nucleators but do not exhibit hysteresis) have been reported from microorganisms, insects, plants and synthetic polymers (Holt, 2003). However, the practical application of these compounds in promoting supercooling has only recently been investigated (Fujikawa *et al.*, 2009). Gusta *et al.* (2004) demonstrated that extracellular fluids obtained from cold-acclimated and non-acclimated plants contain sugars, proteins and other substances that can affect the rate of ice propagation (Fig. 11.6). The specific roles of these compounds in overall freezing tolerance remains to be validated. More recently, Kasuga *et al.* (2010) demonstrated the anti-nucleating properties of flavonol glycosides, and Kuwabara *et al.* (2013) reported the anti-nucleating properties of phenols. In both cases, the compounds were hypothesized to facilitate deep supercooling in woody plants.

11.5.5 Deep supercooling

Deep supercooling is a unique characteristic of many temperate species of woody plants. It can be defined as the ability to maintain water in a liquid phase to temperatures as low as −50°C despite the presence of extracellular ice. Deep supercooling has been observed in dormant buds and xylem tissues, and requires that a barrier exists that prevents the growth of ice crystals into the cell and also prevents the rapid loss of water to extracellular ice. Reviews of deep supercooling of buds and of stem tissues can be found in Quamme (1995) and Wisniewski (1995), respectively. Early work indicated that deep supercooling of xylem tissues predominated in northern hardwood species, especially those having a ring porous xylem. Thus, due to the inherent limits of deep supercooling associated with the homogeneous nucleation temperature of water, it was suggested that species exhibiting deep supercooling would be limited to below the −40°C isotherm. However, Gusta *et al.* (1983) reported that there are some species of hardwoods that exhibit deep supercooling and do grow north of the −40°C isotherm. Despite

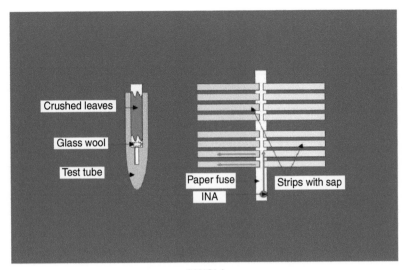

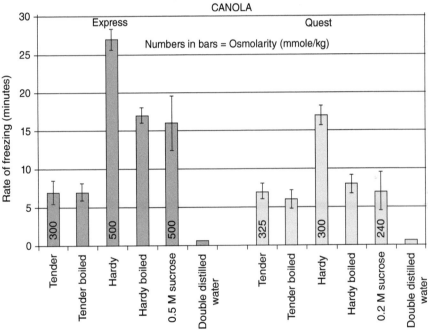

Fig. 11.6. Rates of freezing of plant extracts obtained from canola as assessed using infrared video thermography. Top panel: extracts were centrifuged from leaf tissues and applied to strips of filter paper. Freezing was initiated in the centre strip and rates of freezing were determined as the ice in the centre strip initiated freezing in each of the side strips containing the plant extracts. Lower panel: extracts obtained from cold-acclimated plants (hardy) always froze at much slower rates than extracts obtained from non-acclimated (tender) plants. Freezing rates were influenced by both sugars and proteins present in the extracts. Extracts from plants always froze at a slower rate than strips soaked in distilled water.

supercooling being so integral to the cold hardiness of many woody plants, especially fruit trees, this phenomenon is poorly understood. The majority of research in this area has been conducted by Fujikawa and colleagues (Fujikawa *et al.*, 2009). It should also be noted that freeze avoidance and freeze tolerance as overwintering adaptive strategies are not mutually exclusive.

Freeze-tolerance and freeze-avoidance mechanisms can and do exist in the same plant and often occur in different tissues that are spatially separated from each other by only a few cells to a few millimetres. Freeze-tolerance mechanisms of bark tissues and deep supercooling properties of xylem tissues in woody plants are examples of the two different strategies functioning in parallel with little to no spatial separation.

11.6   Site of Freezing Injury

It is generally accepted that cellular membranes in some species of plants are very susceptible to freezing injury. Maximov (1912) and Chandler (1913) were among the first to suggest that the plasma membrane was the primary site of freezing injury. This was based primarily on the visual symptoms of frost-sensitive plants which have a flaccid, water-soaked appearance. Injury was thought to be due to the mechanical force of expanding ice crystals. Steponkus (1984) suggested that freezing injury in frost-sensitive tissue was due to expansion-induced lysis. During a freezing event the cell undergoes plasmolysis due to the loss of water from the symplast to the apoplast. This results in a loss of membrane surface area. During thawing, the cell is unable to add membrane material back quickly enough into the plasma membrane as the cell begins to take up water and expand. Thus, the plasma membrane ruptures. Heber and Ernst (1967) provided evidence that chloroplast injury occurred when freeze-induced desiccation resulted in the uncoupling of phosphorylation from electron transport. Rajashekar *et al.* (1979) showed by nuclear magnetic resonance (NMR) imaging that the plasma membranes of cells in herbaceous plants were no longer semipermeable to Mn^{2+} when tissue was cooled below a critical temperature. These authors suggested that freezing injury occurred during cooling rather than during warming. In contrast, Palta and Li (1978), using onion cells, observed that the semipermeable properties of the plasma membrane were not altered during freezing, but rather that the active transport properties of the cells were impacted. The differences observed in these various studies may reflect differences that exist between plant species

or may be due to differences in the level of freezing tolerance in the plants that were studied.

The type of injury manifested by freezing is dependent on the freezing tolerance of the species and the specific tissue, the water content of the cells, freezing rate, duration of freezing, degree of supercooling, the rate of ice growth, plus many other factors (Gusta *et al.*, 2009). For example, tender species of potato or cucumber appear flaccid and water soaked following an episodic frost. In contrast, up to 3 weeks of regrowth are required to determine if winter cereals are alive or dead (Gusta and Fowler, 1977). Death in this case can be attributed to a loss in the ability to generate new roots (Chen *et al.*, 1983). When leaves are fully exposed to sunlight photobleaching can occur; however, if leaves are shaded by the surrounding leaf canopy there is less evidence of photobleaching (Fig. 11.7). On a sunny day, it is not uncommon for frozen plants to be to exposed to irradiances from 1500 to 2000 μ mol/m^2/s that result in photobleaching, again highlighting how difficult it can be to model real environmental conditions when conducting acclimation and freezing studies in controlled environment chambers.

11.7   Assessment of Frost and Winter Injury

Due to the complex factors that impact freezing tolerance and freezing injury it can be very difficult to develop a simple, efficient, high-throughput method to assess killing temperatures in plants. Due to this complexity, one should always bear in mind that any test for freezing tolerance is most likely relative rather than absolute. In nature, injury following an episodic frost or winter survival is best evaluated by regrowth. Unfortunately this may require weeks or months to assess. Injury to roots, which are often the most freeze-sensitive tissue, may not become apparent until the plant suffers from a lack or loss of water and nutrient uptake. Frost injury at flowering may not be immediately evident if the pollen is killed and sterility results. Secondary xylem tissues are often 10–30°C more frost sensitive than bark tissues and, even if not killed directly, are subject to cavitation resulting in a partial or complete loss of water transport. Therefore,

Fig. 11.7. Photobleaching in leaves of maize due to frost event followed by exposure to high irradiance.

winter injury in trees may not become apparent for several years (Levitt, 1972; Sakai and Larcher, 1987) as the heartwood becomes subject to fungal decay. Freezing injury to cambial tissues in woody plants can be detected by making a longitudinal cut along the stem to assess enzymatic browning (Burke *et al.*, 1976); however, initial browning may overestimate injury because it does not take into account recovery over time. Root injury, especially for winter annuals, can be evaluated by removing the plants from the soil and observing the regrowth of new white roots (Gusta and Fowler, 1977). Artefacts produced in controlled environmental chambers can and have been accepted as representative rather than relative, leading to a misrepresentation of the actual level of freezing tolerance.

   In artificial freeze tests, whole plants growing in pots, or excised plant parts, can be evaluated for frost tolerance. Plants in pots are placed in a large controlled environment room set at a temperature close to 0°C. The pots are generally placed in styrofoam boxes and covered with some form of insulation to prevent the soil and roots from freezing. When episodic frosts occur in watering, the soil does not freeze due to the large thermal mass of the soil. The temperature of the chamber is lowered to between −2°C and −3°C and the

plants are sprayed with a fine mist of water alone or water containing some form of ice nucleator. Often, ice-nucleating bacteria such as *Pseudomonas syringae* (Gusta *et al.*, 2004) or the product Snowmax® (York Snow, Victor, New York, USA), in which the active nucleator is a protein from *P. syringae* strain 31a (Chen *et al.*, 2009; Skirvin *et al.*, 2010), are sprayed on the leaves to ensure ice nucleation. The use of water alone does not guarantee that plants will freeze at −2°C to −3°C. Plants grown in controlled-environment chambers may have either no or low populations of ice-nucleating bacteria. Also, both extrinsic and intrinsic ice nucleators must be wet for ice formation. For example, excised dry mulberry leaves supercooled to −7°C, whereas ice formation occurred at −2°C in wet leaves (Kitavra, 1967). Using infrared video thermography, Gusta *et al.* (2004) observed dry canola leaves supercooled to as low as −14°C whereas wet leaves supercooled to −6°C. Just because water freezing on the leaf surface induces a leaf to freeze, this does not necessarily result in the whole plant freezing because of the existence of physical barriers that inhibit ice propagation. Wisniewski *et al.* (2002b) found that a hydrophobic particle film acted as a barrier to extrinsic ice nucleation in plants.

Once plants have been exposed to a range of freezing temperatures and allowed to slowly thaw at 4°C overnight, the extent of injury is most commonly assessed using electrolyte leakage (Fig. 11.8). The details of this protocol can be found in most reports where freezing injury has been assessed, and are extrapolated from the method reported by Zhang and Willison (1987).

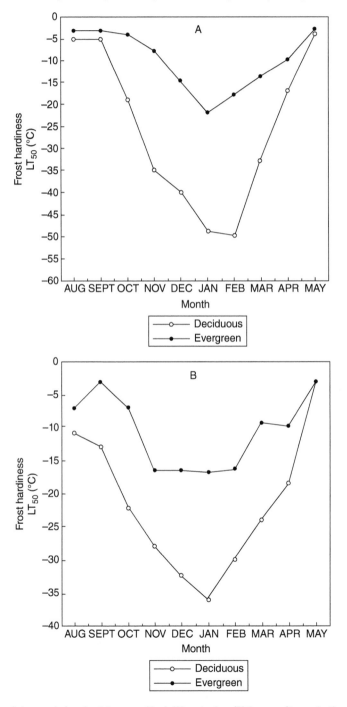

Fig. 11.8. Seasonal changes in freezing tolerance of bark (A) and xylem (B) tissues of two selections (deciduous vs evergreen) of peach (*Prunus persica*). Each data point represents the temperature at which 50% of electrolyte leakage (LT_{50}) occurred. Electrolyte leakage was determined using the method of Zhang and Willison (1987).

In nature, frost or winter injury is best evaluated by regrowth. Artificial tests to evaluate freezing injury (e.g. electrolyte leakage) should be validated by some form of regrowth test. In the case of large plants, such as trees, excised plant parts such as branches can be held in a mist bed or large test tubes containing a small amount of water and assessed for the presence of healthy or decayed tissues after 1–2 weeks.

In contrast to herbaceous and woody plants that exhibit freezing tolerance, buds and xylem tissues of woody plants that exhibit deep supercooling are often assessed using differential thermal analysis (DTA) that relies on being able to detect the exothermic event produced when water within the plant freezes and undergoes a phase transition (liquid to solid). A representative DTA analysis is presented in Fig. 11.9. The use of DTA as a common tool was developed by Quamme (1991) and a review of deep supercooling of xylem tissues can be found in Wisniewski (1995).

11.8 Summary

Cold acclimation is a multigenic, quantitative trait that involves biochemical and structural changes that affect the physiology of a plant (Weiser, 1970; Levitt, 1972). Mechanisms associated with freeze tolerance or freeze avoidance develop and are lost on an annual cycle (Fig. 11.10). When conducting studies to determine the underlying mechanisms underlying cold hardiness, it is essential to take into account that the type of freezing injury that occurs in a plant varies with species and their degree of freezing tolerance and/or ability to avoid freezing (Fig. 11.8). Different forms of freezing injury can be manifested in a plant depending on tissue water content, plant structure and architecture, degree of supercooling, the ability to resist photoinhibition, the ability to generate new roots, etc. A comprehensive understanding of the biology of the species being investigated can greatly assist in identifying the type of injury that must be addressed. The type of injury, as well as the factors limiting freezing tolerance, can change from autumn through midwinter to late spring and this also needs to be taken into account. Finally, any protocols to assess freezing injury should be designed to take into account the biology of the species in its native habitat (natural or agricultural) in order to ensure that the information derived from the study provides accurate information on the problem that is being investigated.

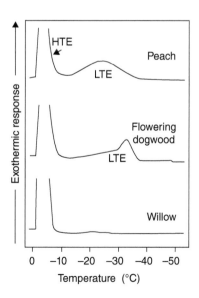

Fig. 11.9. Differential thermal analysis (DTA) profiles of three species of woody plants. Peach (*Prunus persica*) and flowering dogwood (*Cornus florida*) exhibit deep supercooling of xylem tissues while willow (*Salix babylonica*) does not. The high-temperature exotherm (HTE) represents the freezing of water in xylem-conducting elements and extracellular spaces, while the low-temperature exotherm (LTE) represents the freezing of deep supercooled xylem parenchyma cells.

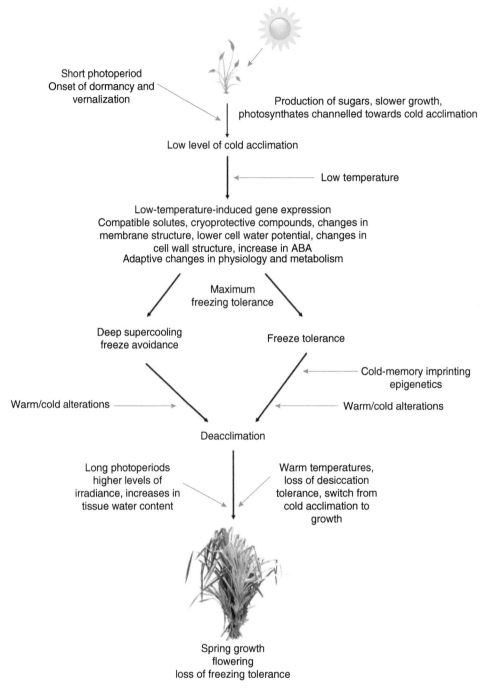

Fig. 11.10. Diagrammatic figure illustrating the major forces that drive cold acclimation in the autumn, the acquisition of maximum freezing tolerance in midwinter and the loss of freezing tolerance (deacclimation) and growth in the following spring.

References

Angel, A., Song, J., Dean, C. and Howard, M. (2011) A polycomb-based switch underlying quantitative epigenetic memory. *Nature* 476, 105–108.

Artlip, T.S., Wisniewski, M.E. and Norelli, J.L. (2014) Field evaluation of apple overexpressing a peach CBF gene confirms its effect on cold hardiness, dormancy, and growth. *Environmental and Experimental Botany* 106, 79–86.

Aryal, B. and Neuner, G. (2010) Leaf wettability decreases along an extreme altitudinal gradient. *Oecologia* 162, 1–9.

Ashworth, E.N. (1992) Formation and spread of ice in plant tissues. *Horticultural Reviews* 13, 215–255.

Ashworth, E.N. and Kieft, T.L. (1995) Ice nucleation activity associated with plants and fungi. In: Lee Jr, R., Warren, G. and Gusta, L.V. (eds) *Biological Ice Nucleation and Its Applications.* APS Press, St Paul, Minnesota, pp. 137–162.

Ball, M.C. and Hill, M.J. (2009) Elevated atmospheric CO_2 concentrations enhance vulnerability to frost damage in a warming world. In: Gusta, L., Wisniewski, M. and Tanino, K. (eds) *Plant Cold Hardiness: From the Laboratory to the Field.* CAB International, Wallingford, UK, pp. 183–189.

Bassett, C.L. and Wisniewski, M. (2009) Global expression of cold-responsive genes in fruit trees. In: Gusta, L., Wisniewski, M. and Tanino, K. (eds) *Plant Cold Hardiness: From the Laboratory to the Field.* CAB International, Wallingford, UK pp. 72–79.

Baurle, I. and Dean, C. (2006) The timing of developmental transitions in plants. *Cell* 125, 655–664.

Benedict, C., Skinner, J.S., Meng, R., Chang, Y., Bhalerao, R., Hüner, N.P.A., Finn, C.E., Chen, T.H.H. and Hurry, V. (2006) The CBF1-dependent low temperature signalling pathway, regulon and increase in freeze tolerance are conserved in *Populus* spp. *Plant Cell and Environment* 29, 1259–1272.

Burke, M.J., Gusta, L.V., Quamme, H.A., Weiser, C.J. and Li, P.H. (1976) Freezing and injury in plants. *Annual Review of Plant Physiology* 27, 507–528.

Chandler, J. and Dean, C. (1994) Factors influencing the vernalization response and flowering time of late flowering mutants of *Arabidopsis thaliana. Journal of Experimental Botany* 45, 1279–1288.

Chandler, W.H. (1913) The killing of plant tissue by low temperature. *Missouri Agricultural Experiment Station Research Bulletin* 8, pp. 171.

Chen, A., Gusta, L.V., Brûlé-Babel, A., Leach, R., Baumann, U., Fincher, G.B. and Collins, N.C. (2009) Varietal and chromosome 2H locus-specific frost tolerance in reproductive tissues of Barley (*Hordeum vulgare* L.) detected using a frost simulation chamber. *Theoretical and Applied Genetics* 119, 685–694.

Chen, T.H., Gusta, L.V. and Fowler, D.B. (1983) Freezing injury and root development in winter cereals. *Plant Physiology* 73, 773–777.

Crevillen, P. and Dean, C. (2010) Regulation of the floral repressor gene FLC: the complexity of transcription in a chromatin context. *Current Opinion in Plant Biology* 14, 1–7.

De Lucia, F., Crevillen, P., Jones, A.M., Greb, T. and Dean, C. (2008) A PHD-Polycomb Repressive Complex 2 triggers the epigenetic silencing of FLC during vernalization. *Proceedings of the National Academy of Sciences USA* 105, 16831–16836.

Dhillon, T., Pearce, S.P., Stockinger, E.J., Distelfeld, A., Li, C., Knox, A.K., Vashegyi, J., Galiba, G. and Dubcovsky, J. (2010) Regulation of freezing tolerance and flowering in temperate cereals: the VRN-1 connection. *Plant Physiology* 153, 1846–1858.

Dickens, C. (1859) *A Tale of Two Cities.* Puffin Books, London, 2009, p.3.

Dormling, I. and Johnsen, Ø. (1992) Effects of the parental environment on full-sib families of *Pinus sylvestris. Canadian Journal of Forestry Restoration* 22, 88–100.

Duman, J.G. and Olsen, T.M. (1993) Thermal hysteresis protein activity in bacteria, fungi, and phylogenetically diverse plants. *Cryobiology* 30, 322–328.

Duman, J.G. and Wisniewski, M.J. (2014) The use of antifreeze proteins for frost protection in sensitive crop plants. *Environmental and Experimental Botany* 106, 60–69.

Fowler, S. and Thomashow, M.F. (2002) Arabidopsis transcriptome profiling indicates that multiple regulatory pathways are activated during cold acclimation in addition to the CBF cold responsive pathway. *The Plant Cell* 14, 1675–1690.

Fujikawa, S., Kasuga, J., Takata, N. and Arakawa, K. (2009) Factors related to change of deep supercooling capability in xylem parenchyma cells of trees. In: Gusta, L.V., Wisniewski, M.E. and Tanino K.K. (eds) *Plant Cold Hardiness: From the Laboratory to the Field.* CAB International, Wallingford, UK, pp. 29–42.

Gilmour, S.J., Fowler, S.G. and Thomashow, M.F. (2004) Arabidopsis transcription activators CBF1, CBF2 and CBF3 have matching functional activities. *Plant Molecular Biology* 54, 767–781.

Gray, G.R., Savitch, L.V., Ivanov, A.G. and Hüner, N.P.A. (1996) Photosystem II excitation pressure and development of resistance to photoinhibition: II. Adjustment of photosynthetic capacity in winter wheat and winter rye. *Plant Physiology* 110, 61–71.

Gray, G.R., Chauvin, L.P., Sarhan, F. and Hüner, N.P.A. (1997) Cold acclimation and freezing tolerance. A complex interaction of light and temperature. *Plant Physiology* 114, 467–474.

Griffith, M. and Yaish, M.W.F. (2004) Antifreeze proteins in overwintering plants: a tale of two activities. *Trends in Plant Science* 9, 399–405.

Griffith, M., Lumb, C., Wiseman, S.B., Wisniewski, M., Johnson, R.W. and Marangoni, A.G. (2005) Antifreeze proteins modify the freezing process in planta. *Plant Physiology* 138, 330–340.

Gu, L., Hanson, P.J., Post, W.M., Kaiser, D.P., Yang, B., Nemani, R., Pallardy, S.G. and Meyers, T. (2008) The 2007 eastern US spring freeze: increased cold damage in a warming world? *BioScience* 58, 253–262.

Gusta, L.V. and Fowler, D.B. (1977) Factors affecting the cold survival of winter cereals. *Canadian Journal of Plant Science* 57, 213–219.

Gusta, L.V. and Wisniewski, M. (2012) Understanding plant cold hardiness: an opinion. *Physiologia Plantarum* 147, 4–14.

Gusta, L.V., Tyler, M.J. and Chen, T.H. (1983) Deep undercooling in woody taxa growing north of the −40°C isotherm. *Plant Physiology* 72, 122–128.

Gusta, L.V., O'Connor, B.J. and MacHutcheon, M.G. (1997) The selection of superior winter hardy genotypes using a prolonged freeze test. *Canadian Journal of Plant Science* 77, 15–21.

Gusta, L.V., Wisniewski, M., Nesbitt, N.T. and Gusta, M.L. (2004) The effect of water, sugars and proteins on the pattern of ice nucleation and propagation in acclimated and non-acclimated canola leaves. *Plant Physiology* 135, 1642–1653.

Gusta, L.V., Wisniewski, M.E. and Trischuk, R.G. (2009) Patterns of freezing in plants: the influence of species, environment and experiential procedures. In: Gusta L.V., Wisniewski, M.E. and Tanino, K.K. (eds) *Plant Cold Hardiness from the Laboratory to the Field*. CAB International, Wallingford, UK, pp. 214–225.

Guy, C.L. (1990) Cold acclimation and freezing stress tolerance: role of protein metabolism. *Annual Review of Plant Physiology and Plant Molecular Biology* 41, 187–223.

Hacker, J., Ladinig, U., Wagner, J. and Neuner, G. (2011) Inflorescences of alpine cushion plants freeze autonomously and may survive subzero temperatures by supercooling. *Plant Science* 180, 149–156.

Hannah, M.A., Heyer, A.G. and Hincha, D.K. (2005) A global survey of gene regulation during cold acclimation in *Arabidopsis thaliana*. *PLoS Genetics* 1, e26.

Heber, U. and Ernst, R. (1967) A biochemical approach to the problem of frost injury and frost hardiness. In: Asahina, E. (ed.) *Cellular Injury and Resistance in Freezing Organisms*. The Institute of Low Temperature Science, Hokkaido University, Sapporo, Japan, pp. 63–77.

Heo, J.B. and Sung, S. (2011) Encoding memory of winter by noncoding RNAs. *Epigenetics* 6, 544–547.

Holt, C.B. (2003) Substances which inhibit ice nucleation: a review. *CryoLetters* 24, 269–274.

Hüner, N.P.A., Öquist, G., Hurry, V.M., Krol, M., Falk, S. and Griffith, M. (1993) Photosynthesis, photoinhibition and low temperature acclimation in cold tolerant plants. *Photosynthesis Research* 37, 19–39.

Hüner, N.P.A., Öquist, G. and Sarhan, F. (1998) Energy balance and acclimation to light and cold. *Trends Plant Science* 3, 224–230.

Hüner, N.P.A., Bode, R., Dahal, K., Busch, F.A., Possmayer, M., Szyszka, B., Rosso, D., Ensminger, I., Krol, M., Ivanov, A.G. and Maxwell, D.P. (2013) Shedding some light on cold acclimation, cold adaptation and phenotypic plasticity. *Botany* 91, 127–136.

Hurry, V. and Hüner, N. (1991) Low growth temperature affects a differential inhibition of photosynthesis in spring and winter wheat. *Plant Physiology* 96, 491–497.

Hurry, V., Strand, A., Tobiaeson, M., Gardenstrom, P. and Öquist, G. (1995) Cold hardening of spring and winter wheat and rape results in differential effects on growth, carbon metabolism and carbohydrate content. *Plant Physiology* 109, 697–706.

Ishikawa, M., Toyomasu, T., Aoki, T. and Price, W.S. (2015) Ice nucleation activity in various tissues of *Rhododendron* flower buds: their relevance to extraorgan freezing. *Frontiers in Plant Science* 6, 149.

Johanson, U., West, J., Lister, C., Michaels, S., Amasino, R. and Dean, C. (2000) Molecular analysis of *FRIGIDA*, a major determinant of natural variation in *Arabidopsis* flowering time. *Science* 290, 344–347.

Kasuga, J., Fukushi, Y., Kuwabara, C., Wang, D., Nishioka, A., Fujikawa, E., Arakawa, K. and Fujikawa, S. (2010) Analysis of supercooling-facilitating (anti-ice nucleation) activity of flavonol glycosides. *Cryobiology* 60, 240–243.

King, G.J. (2015) Crop epigenetics and the molecular hardware of genotype × environment interactions. *Frontiers in Plant Science* 6, 968.

Kishimoto, T., Sekozawa, Y., Yamazaki, H., Murakawa, H., Kuchitsu, K. and Ishikawa, M. (2014) Seasonal changes in ice nucleation activity in blueberry stems and effects of cold treatments *in vitro*. *Environmental Experimental Botany* 106, 13–23.

Kitavra, K. (1967) Supercooling and ice formation in mulberry trees. In: Asahina, E. (ed.) *Cellular Injury and Freeze Resistance in Organisms*. Institute of Low Temperature Science, Hokkaido University, Sapporo, Japan, pp. 143–156.

Koch, K.E. (1996) Carbohydrate-modulated gene expression in plants. *Annual Review of Plant Physiology and Plant Molecular Biology* 47, 509–540.

Kosmala, A., Bocian, A., Rapacz, M., Jurczyk, B. and Zwierzkowski, Z. (2009) Identification of leaf proteins differentially accumulated during cold acclimation between *Festuca pratensis* plants with distinct levels of frost tolerance. *Journal of Experimental Botany* 60, 3595–3609.

Kuwabara, C., Wang, D., Endoh, K., Fukushi, Y., Arakawa, K. and Fujikawa, S. (2013) Analysis of supercooling activity of tannin-related polyphenols. *Cryobiology* 67, 40–49.

Kvaalen, H. and Johnsen, O. (2008) Timing of bud set in *Picea abies* is regulated by a memory of temperature during zygotic and somatic embryogenesis. *New Phytologist* 177, 49–59.

Levitt, J. (1972) *Responses of Plants to Environmental Stresses*. Academic Press, New York.

Liu, F., Marquardt, S., Lister, C., Swiezewski, S. and Dean, C. (2010) Targeted 3' processing of antisense transcripts triggers *Arabidopsis FLC* chromatin silencing. *Science* 327, 94–97.

Machalek, K.M., Davison, I.R. and Falkowski, P.C. (1996) Thermal acclimation and photo-acclimation of photosynthesis in the brown alga *Laminaria saccharina*. *Plant Cell and Environment* 19, 1005–1016.

Manning, V.A., Hardison, L.K. and Ciuffetti, L.M. (2007) *Ptr Tox A* interacts with a chloroplast-localized protein. *Molecular Plant-Microbe Interactions* 20, 168–177.

Maruyama, K., Sauma, Y., Kasuga, M., Ito, Y., Seki, M., Goda, H., Shimada, Y., Yoshida, S., Shinozaki, K. and Yamaguchi-Shinozaki, K. (2004) Identification of cold-inducible downstream genes of the *Arabidopsis* DREB1A/CB3 transcriptional factor using two microarray systems. *The Plant Journal* 38, 982–999.

Maximov, N.A. (1912) Chemische Schutzmittel der Pflanzen gegan Erfrieren. *Berliner Deutsche Botanische Geselschaft* 30, 52–65, 293, 305, 504–516.

Maxwell, D.P., Falk, S. and Hüner N.P.A. (1995) Photosystem II excitation pressure and development of resistance to photoinhibition. Light harvesting couples II abundance and zeathin content in *Chlorella vulgaris*. *Plant Physiology* 107, 687–694.

Mirouze, M. and Paszkowski, J. (2011) Epigenetic contribution to stress adaptation in plants. *Current Opinion in Plant Biology* 14, 267–274.

Murchie, E.H., Pinto, M. and Horton, P. (2009) Agriculture and the new challenges for photosynthetic research. *New Phytologist* 181, 532–552.

Neuner, G. (2014) Frost resistance in alpine woody plants. *Frontiers in Plant Science* 5, 654.

Nezami, A., Bandara, M.S. and Gusta, L.V. (2012) An evaluation of freezing tolerance of winter chickpea (*Cicer arietinum* L.) using controlled freeze tests. *Canadian Journal of Plant Science* 92, 155–161.

Öquist, G. and Huner, N.P.A. (2003) Photosynthesis of overwintering evergreen plants. *Annual Review of Plant Biology* 54, 329–355.

Öquist, G., Hurry, V.M. and Hüner, N.P.A. (1993) Low temperature effects on photosynthesis and correlation with freezing tolerance in spring and winter cultivars of wheat and rye. *Physilogia Plantarum* 101, 245–250.

Palta, J.P. and Li, P.H. (1978) Cell membrane properties in relation to freezing injury. In: Li, P.H. and Sakai, A. (eds) *Plant Cold Hardiness and Freezing Stress*. Academic Press, New York, pp. 93–115.

Quamme, H.A. (1991) Use of thermal analyses to measure the freezing resistance of grape buds. *Canadian Journal of Plant Science* 66, 945–952.

Quamme, H.A. (1995) Deep supercooling in buds of woody plants. In: Lee, R.E., Warren Jr, G.J. and Gusta, L.V. (eds) *Biological Ice Nucleation and Its Applications*. APS Press, St Paul, Minnesota, pp. 183–200.

Rajashekar, C., Gusta, L.V. and Burke, M.J. (1979) Membrane structure transition: probable relation to frost damage in hardy herbaceous species. In: Lyson, J.M., Graham, D. and Raison, J.K. (eds) *Low Temperature Stress in Crop Plants: The Role of the Membrane*. Academic Press, London, New York, pp. 255–274.

Rekarte-Cowie, I., Ebshish, O.S., Mohammed, K. and Pearce, R.S. (2008) Sucrose helps regulate cold acclimation of *Arabidopsis thaliana*. *Journal of Experimental Botany* 59, 4205–4217.

Roberts, D.W.A. (1984) The effect of light on the rosette growth habit of winter wheat. *Canadian Journal of Botany* 62, 818–822.

Sakai, A. and Larcher, W. (1987) *Frost Survival of Plants Responses and Adaptation to Freezing Stress*. Springer, Berlin.

Sheldon, C.C., Finnegan, E.J., Peacock, W.J. and Dennis, E.S. (2009) Mechanisms of gene expression by vernalization in Arabidopsis. *Plant Journal* 59, 488–498.

Shinozaki, K. and Yamaguchi-Shinozaki, K. (2007) Gene networks involved in drought stress response and tolerance. *Journal of Experimental Botany* 58, 221–227.

Skirvin, R.M., Kohler, E., Steiner, H., Ayers, D., Laughnan, A., Norton, M.A. and Warmund, M.A. (2000) The use of genetically engineered bacteria to control frost on strawberries and potatoes. Whatever happened to all of that research? *Scientia Horticulturae* 84, 179–189.

Smeekens, S. (1998) Sugar regulation of gene expression in plants. *Current Opinions in Plant Biology* 1, 230–234.

Steponkus, P.L. (1984) Role of the plasma membrane in freezing injury and cold acclimation. *Annual Review of Plant Physiology* 35, 543–584.

Steponkus, P.L. and Lanphear, F.O. (1968) The role of light in cold acclimation of *Hedra helix* L. var *Thorndale*. *Plant Physiology* 43, 151–156.

Stockinger, E.J., Gilmour, S.J. and Thomashow, M.I. (1997) *Arabidopsis thaliana* CBF1 encodes are AP2 domain-containing transcriptional factor that binds to the C-repeat/DRE, a cis-acting DNA regulatory element that stimulates transcription in response to low temperature and water deficits. *Proceedings of the National Academy of Sciences USA* 94, 1035–1040.

Stoehr, M.U., L'Hirondelle, S.J., Binder, W.D. and Webber, J.E. (1998) Parental environment after-effects on germination, growth, and adaptive traits in selected spruce families. *Canadian Journal of Forest Restoration* 28, 418–426.

Strand, A., Hurry, V., Henkes, S., Hüner, N., Gustafsson, P., Gardestrom, P. and Stitt, M. (1999) Acclimation of Arabidopsis leaves developing at low temperatures. Increasing cytoplasmic volume accompanies increased activities of enzymes in the Calvin cycle and in the sucrose-biosynthesis pathway. *Plant Physiology* 119, 1387–1397.

Swiezewski, S., Liu, F., Magusin, A. and Dean, C. (2009) Cold-induced silencing mediated by long antisense RNA from a Polycomb target. *Nature* 462, 799–802.

Thomashow, M.F. (1999) Plant cold acclimation: freezing tolerance genes and regulatory mechanisms. *Annual Review of Plant Physiology and Plant Molecular Biology* 5, 571–599.

Trischuk, R.G., Schilling, B.S., Low, N.H., Gray, G.R. and Gusta, L.V. (2014) Cold acclimation, de-acclimation and re-acclimation of spring canola, winter canola and winter wheat: the role of carbohydrates, cold-induced stress proteins and vernalization. *Environmental and Experimental Botany* 106, 156–163.

Vasilyev, J.M. (1961) *Wintering of Plants.* American Institute of Biological Sciences, Washington, DC.

Wanner, L.A. and Juntilla, O. (1999) Cold-induced freezing tolerance in Arabidopsis. *Plant Physiology* 120, 391–399.

Warren, G. (1995) Identification of *ina* genes and proteins. In: Lee Jr, R., Warren, G. and Gusta, L.V. (eds) *Biological Ice Nucleation and Its Applications.* APS Press, St Paul, Minnesota, pp. 85–100.

Weiser, C.J. (1970) Cold resistance and injury in woody plants. *Science* 169, 1269–1278.

Welling, A. and Palva, E.T. (2008) Involvement of *CBF* transcription factors in winter hardiness of birch. *Plant Physiology* 147, 1199–1211.

Wilen, R.W., Fu, P., Robertson, A.J. and Gusta, L.V. (1998) A comparison of the cold hardiness potential of spring cereals and vernalized and non-vernalized winter cereals. In: Li, P. and Chen, T. (eds) *Plant Cold Hardiness.* Plenum Press, New York, pp. 191–201.

Wisniewski, M. (1995) Deep supercooling in woody plants and the role of plant structure. In: Lee Jr, R.E., Warren, G.J. and Gusta, L.V. (eds) *Biological Ice Nucleation and Its Applications.* APS Press, St Paul, Minnesota, pp. 163–181.

Wisniewski, M. and Fuller, M.P. (1999) Ice nucleation and deep supercooling in plants: new insights using infrared thermography. In: Margesin, R. and Schinner, F. (eds) *Cold-Adapted Organisms: Ecology, Physiology, Enzymology, and Molecular Biology.* Springer, Berlin, pp. 104–118.

Wisniewski, M., Lindow, S. and Ashworth, E. (1997) Observations of ice nucleation and propagation in plants using infrared thermography. *Plant Physiology* 113, 327–334.

Wisniewski, M., Fuller, M., Glenn, D.M., Gusta, L., Duman, J. and Griffith, M. (2002a) Extrinsic ice nucleation in plants: what are the factors involved and can they be manipulated. In: Li, P.H. and Palva, E.T. (eds) *Plant Cold Hardiness: Gene Regulation and Genetic Engineering.* Kluwer Academic/Plenum Publishers, New York, pp. 211–221.

Wisniewski, M., Glenn, D.M. and Fuller, M.P. (2002b) Use of a hydrophobic particle film as a barrier to extrinsic ice nucleation in tomato plants. *Journal of the American Society of Horticultural Science* 127, 358–364.

Wisniewski, M., Glenn, D.M., Gusta, L. and Fuller, M.P. (2008) Using infrared thermography to study freezing in plants. *HortScience* 43, 1648–1651.

Wisniewski, M.E., Gusta, L.V., Fuller, M.P. and Karlson, D. (2009) Ice nucleation, propagation and deep supercooling: the lost tribes of freezing studies. In: Gusta, L.V., Wisniewski, M.E. and Tanino, K.K. (eds) *Plant Cold Hardiness: From the Laboratory to the Field.* CAB International, Wallingford, UK, pp. 1–11.

Wisniewski, M., Norelli, J., Bassett, C., Artlip, T. and Macarisin, D. (2011) Ectopic expression of a novel peach (*Prunus persica*) CBF transcription factor in apple (*Malus* × *domestica*) results in short-day induced dormancy and increased cold hardiness. *Planta* 233, 971–983.

Wisniewski, M., Nassuth, A., Teulieres, C., Marque, C., Rowland, J., Cao, P.-B. and Brown, A. (2014) Genomics of cold hardiness in woody plants. *Critical Reviews in Plant Science* 33, 92–124.

Wisniewski, M., Norelli, J. and Artlip, T. (2015) Overexpression of a peach CBF gene in apple: a model for understanding the integration of growth, dormancy, and cold hardiness in woody plants. *Frontiers in Plant Science* 6, 85.

Workmaster, B.A., Palta, J.P. and Wisniewski, M. (1999) Ice nucleation and propagation in cranberry uprights and fruit using infrared thermography. *Journal of the American Society of Horticultural Science* 124, 619–625.

Wurr, D.C.E., Fellows, J.R. and Fuller, M.P. (2004) Simulated effects of climate change on the production pattern of winter cauliflower in the UK. *Scientifica Horticulturae* 101, 359–372.

Xiao, H., Siddiqua, M., Braybrook, G. and Nassuth, A. (2006) Three grape CBF/DREB1 genes are regulated by low temperature, drought and abscisic acid. *Plant Cell & Environment* 29, 1410–1421.

Yakovlev, I.A., Fossdal, C.G. and Johnsen, Ø. (2010) MicroRNAs, the epigenetic memory and climatic adaptation in Norway spruce. *New Phytologist* 187, 1154–1169.

Yakovlev, I., Fossdal, C.G., Skroppa, T., Olsen, J.E., Jahren, A.H. and Johnsen, Ø. (2012) An adaptive epigenetic memory in conifers with important implications for seed production. *Seed Science Restoration* 22, 63–76.

Yakovlev, I.A., Lee, Y., Rotter, B., Olsen, J.E., Skrøppa, T., Johnsen, Ø. and Fossdal, C.G. (2014) Temperature-dependent differential transcriptomes during formation of an epigenetic memory in Norway spruce embryogenesis. *Tree Genetics and Genomes* 10, 355–366.

Yordanov, Y.S., Ma, C., Strauss, S.H. and Buscov, V.B. (2014) *EARLY BUD-BREAK 1 (EBB1)* is a regulator of release from seasonal dormancy in poplar trees. *Proceedings of the National Academy of Sciences USA* 111, 10001–10006.

Zarka, D.G., Vogel, J.T., Cook, D. and Thomashow, M.F. (2003) Cold induction of Arabidopsis *CBF* gene involves multiple ICE (inducer of *CBF* expression) promoter elements and a cold-regulatory circuit that is desensitized by low temperature. *Plant Physiology* 133, 910–918.

Zhang, M.I.N. and Willison, J.H.M. (1987) An improved method for the measurement of frost hardiness. *Canadian Journal of Botany* 65, 710–718.

12 Heavy-metal Toxicity in Plants

Philip J. White* and Paula Pongrac

The James Hutton Institute, Invergowrie, UK

Abstract

Heavy metals include the transition-metal elements essential to plant nutrition: iron (Fe), zinc (Zn), manganese (Mn), copper (Cu), nickel (Ni) and molybdenum (Mo), cobalt (Co) (which is required for nitrogen fixation in legumes); and the non-essential elements chromium (Cr), cadmium (Cd), mercury (Hg) and lead (Pb). All these elements are toxic to crop plants at high tissue concentrations. In agriculture, deficiencies of essential heavy-metal elements are more common than their toxicities. Nevertheless, Mn toxicity can reduce crop yields on acidic soils, and Mn and Fe toxicities occur on waterlogged or flooded soils. Toxicities can also arise in soils enriched in specific heavy metals by the weathering of the underlying rocks or anthropogenic activities. The molecular biology of heavy-metal uptake and transport within plants is well understood, and the regulatory cascades enabling heavy-metal homeostasis in plant cells and tissues are being elucidated. Cellular responses to excess heavy metals are also known. Many of these responses proceed through the generation of reactive oxygen species and involve the synthesis of antioxidant compounds and enzymes. Tolerance of high concentrations of heavy metals in the environment is brought about by restricting the entry of heavy metals to the root and their movement to the xylem, and by chelating heavy metals entering the cytoplasm and sequestering them in non-vital compartments, such as the apoplast and vacuole. The mechanisms by which certain plant species are able to hyperaccumulate heavy metals are also providing insight into the ability of plants to exclude and tolerate heavy metals in their tissues.

12.1 Heavy Metals in Plants: Essential Functions

The term 'heavy metal' is rather inexact (Duffus, 2002). However, the naturally abundant transition-metal elements, chromium (Cr), manganese (Mn), iron (Fe), cobalt (Co), nickel (Ni), copper (Cu), zinc (Zn), molybdenum (Mo), cadmium (Cd) and mercury (Hg), together with the metallic element lead (Pb), are often designated 'heavy metals'. This article will focus solely on these elements. Several of these elements (Fe, Zn, Mn, Cu, Ni, Mo) are essential to the mineral nutrition of plants, but all are toxic to crop plants at high tissue concentrations (Table 12.1).

An essential mineral nutrient is defined as an element required for plant growth and/or reproduction that cannot be replaced by another element because of a unique physiological or biochemical role (Kirkby, 2012). Adequate tissue concentrations vary greatly between mineral nutrients (Table 12.1). Shoot concentrations greater than 50–150 µg Fe/g dry matter (DM), 15–30 µg Zn/g DM, 10–20 µg Mn/g DM, 1–5 µg Cu/g DM, 0.1 to 1 µg Mo/g DM and 2–10 ng Ni/g DM are generally sufficient for

* Corresponding author: Philip.White@hutton.ac.uk

Table 12.1. The physiological functions of transition-metal elements, the chemical form in which they are acquired by plant roots, the critical leaf concentrations for their sufficiency and toxicity in typical non-tolerant crop plants, and the threshold leaf concentrations defining their 'hyperaccumulation'. The critical concentration for sufficiency is defined as the concentration in a diagnostic tissue that allows a crop to achieve 90% of its maximum yield. The critical concentration for toxicity is defined as the concentration in a diagnostic tissue above which yield is decreased by more than 10%. It should be recognized that critical tissue concentrations depend upon the tissue sampled and the exact solute composition of the soil solution, and can differ greatly both between and within plant species. The latter differences reflect both ancestral habitats and ecological strategies. Data compiled from Baker et al. (2000), Reeves and Baker (2000), Mengel et al. (2001), Shanker et al. (2005), White and Brown (2010), Fageria et al. (2011), Van der Ent et al. (2013) and White and Greenwood (2013).

Element	Form acquired	Physiological functions	Critical leaf concentrations (mg/g DM)		Hyperaccumulation (mg/g DM)
			Sufficiency	Toxicity	
Iron (Fe)	Fe^{2+}, Fe^{3+}-chelates	Photosynthesis; mitochondrial respiration; C and N metabolism; production and scavenging of reactive oxygen species; regulation of transcription and translation; hormone biosynthesis	$50\text{--}150 \times 10^{-3}$	> 0.5	–
Manganese (Mn)	Mn^{2+}, Mn-chelates	Photosystem II; enzyme activation in photosynthesis; C and N metabolism, RNA polymerase	$10\text{--}20 \times 10^{-3}$	$0.2\text{--}5.3$	10
Copper (Cu)	Cu^+, Cu^{2+}, Cu-chelates	Photosynthesis; mitochondrial respiration; C and N metabolism; protection against oxidative stress	$1\text{--}5 \times 10^{-3}$	$15\text{--}30 \times 10^{-3}$	0.3
Zinc (Zn)	Zn^{2+}, Zn-chelates	Structural stability of proteins; regulation of transcription and translation; oxidoreductases and hydrolytic enzymes	$15\text{--}30 \times 10^{-3}$	$100\text{--}700 \times 10^{-3}$	3
Nickel (Ni)	Ni^{2+}, Ni-chelates	Constituent of urease	0.01×10^{-3}	$20\text{--}30 \times 10^{-3}$	1
Molybdenum (Mo)	MoO_4^{2-}	Catalytic site of nitrate reductase, aldehyde oxidase, xanthine dehydrogenase and sulfite oxidase	$0.1\text{--}1.0 \times 10^{-3}$	1	–
Cobalt (Co)	Co^{2+}	Nitrogen fixation	Beneficial	$10\text{--}20 \times 10^{-3}$	0.3
Lead (Pb)	Pb^{2+}	–	–	$10\text{--}20 \times 10^{-3}$	1
Cadmium (Cd)	Cd^{2+}, Cd-chelates	–	–	$5\text{--}10 \times 10^{-3}$	0.1
Mercury (Hg)	–	–	–	$2\text{--}5 \times 10^{-3}$	–
Chromium (Cr)	$Cr^{3+}(Cr^{2+})$, CrO_4^{2-}	–	–	$1\text{--}2 \times 10^{-3}$	0.3

DM, dry matter.

plant growth and fecundity (White and Brown, 2010). Some of these elements (e.g. Fe, Mn, Cu, Ni) are redox-active, which allows them to perform catalytic functions, and most serve to activate or stabilize enzymes through cation–protein interactions (Hänsch and Mendel, 2009; Pilon *et al.*, 2009; Tejada-Jiménez *et al.*, 2009; White and Greenwood, 2013). Excess of these elements, however, leads to the formation of destructive reactive oxygen species (ROS) and to the ectopic binding of their cations to proteins (Hänsch and Mendel, 2009; Pilon *et al.*, 2009). For this reason, free transition-metal cations rarely exist in the cytoplasm and are chelated avidly by specific proteins (chaperones) and other organic metabolites (Krämer and Clemens, 2005; Broadley *et al.*, 2007; Hänsch and Mendel, 2009; Pilon *et al.*, 2009; Tejada-Jiménez *et al.*, 2009).

Both Fe and Cu are redox-active metals that play both exclusive and interchangeable roles in photosynthesis, mitochondrial respiration, carbon (C) and nitrogen (N) metabolism, and protection against oxidative stress (Table 12.1). Iron is an integral component of many proteins and enzymes as either: (i) a haem group, which is present in cytochromes, peroxidases and catalases; (ii) a non-haem Fe bound to sulfur atoms, which occurs in ferredoxin, nitrite reductase, sulphite reductase, glutamate synthase and iron-storage proteins; or (iii) a non-haem non-sulfur bound Fe, which is present in the plastidic Fe-storage protein ferritin (Hänsch and Mendel, 2009; Palmer and Guerinot, 2009; Pilon *et al.*, 2009; Guerinot, 2010). Up to 90% of leaf Fe is present in chloroplasts, where it is distributed equally between stroma and thylakoids (Hänsch and Mendel, 2009; Guerinot, 2010). Iron is also intimately involved in the regulation of transcription and translation, and in the biosynthesis of ethylene, gibberellic acid and jasmonic acid (Hänsch and Mendel, 2009). In *Arabidopsis thaliana*, Cu is associated with at least 100 different proteins (Krämer and Clemens, 2005). These include Cu-binding chaperones and numerous enzymes, particularly oxidases (Krämer and Clemens, 2005; Burkhead *et al.*, 2009; Hänsch and Mendel, 2009). Several of these enzymes are important in mounting plant defence responses to pests and pathogens (Burkhead *et al.*, 2009; Fageria *et al.*, 2011). About 50% of leaf Cu is

present in chloroplasts, where it participates in photosynthetic reactions (Hänsch and Mendel, 2009). Copper is also part of the ethylene receptor and is involved in Mo-cofactor biosynthesis (Hänsch and Mendel, 2009; Bittner and Mendel, 2010). In legumes, Cu deficiency reduces nodulation and N_2 fixation.

Manganese is also a redox-active metal and serves this role in several enzymes, including the water-splitting complex of photosystem II, superoxide dismutase (SOD) and oxalate oxidase (Hänsch and Mendel, 2009). It also regulates the activity of several enzymes catalysing reactions in intermediary metabolism (malic enzyme, isocitrate dehydrogenase, PEP carboxykinase), fatty acid biosynthesis, N metabolism (glutamine synthetase, arginase), the shikimic acid pathway and subsequent pathways leading to the formation of aromatic amino acids, lignins, flavonoids and the phytohormone indole acetic acid, gibberellic acid biosynthesis and RNA polymerase (Hänsch and Mendel, 2009; Williams and Pittman, 2010).

Zinc plays a major role in the regulation of gene expression and enzyme activities. In *A. thaliana*, Zn is associated with over 2300 proteins in 181 gene families, including all six classes of enzymes (Krämer and Clemens, 2005; Broadley *et al.*, 2007). These proteins include a large number of transcription factors, zinc-finger proteins, oxidoreductases and hydrolytic enzymes such as protein kinases, metalloproteases and phosphatases (Broadley *et al.*, 2007; Hänsch and Mendel, 2009; Clemens, 2010).

Nickel is required by plants for the activation of urease, which catalyses the hydrolysis of urea to carbon dioxide and ammonia, and Mo is required for the pterin-cofactor (Moco), which constitutes the catalytic site of nitrate reductase, aldehyde oxidase, xanthine dehydrogenase and sulfite oxidase (Hänsch and Mendel, 2009; Tejada-Jiménez *et al.*, 2009; Bittner and Mendel, 2010). The nitrogenase enzyme of N_2-fixing bacteria also contains Mo.

In addition to these essential transition-metal elements, Co is recognized as a beneficial element for plant growth. It is essential for N_2 fixation in legumes and can enhance drought resistance and defence against herbivores under specific circumstances (Mengel *et al.*, 2001; Pilon-Smits *et al.*, 2009).

12.2   Heavy Metals in the Environment: Conditions of Excess

High phytoavailability of heavy metals in the soil can inhibit plant growth and reduce crop yields (MacNicol and Beckett, 1985; Kabata-Pendias, 2000; Mengel et al., 2001; White and Brown, 2010; Fageria et al., 2011; Hodson and Donner, 2013; White and Greenwood, 2013). Heavy metals can be present at high concentrations in the rhizosphere solution as a result of weathering of the underlying rock (e.g. Ni, Co, Pb, Cd), soil acidity (e.g. Mn, Zn), flooding (Mn, Fe) or human activities (e.g. Cr, Co, Ni, Cu, Zn, Mo, Pb, Cd). Transition-metal elements are present in the soil solution as ionic species, uncharged salts and organic complexes (Mengel et al., 2001; Fageria et al., 2011). Transport proteins catalysing the uptake of specific ionic species and transition-metal chelates are present in the plasma membrane of root cells. Most of the Fe, and significant amounts of the Zn (75–85%), Mn (50–90%) and Cu (80–90%), in the soil solution are present as organic complexes. Some of the organic ligands complexing these transition-metal elements are secreted by plant roots (Mengel et al., 2001; Fageria, 2009). The dominant ionic forms of these transition-metal elements in solutions from aerated soils are Fe^{3+}, Zn^{2+}, Mn^{2+} and Cu^{2+} (Table 12.1; Lindsay, 1979; Mengel et al., 2001; Fageria, 2009). Molybdenum generally occurs in soil solutions as MoO_4^{2-} (Lindsay, 1979; Mengel et al., 2001; Fageria, 2009). Cadmium and Co occur in the soil solution predominantly as Cd^{2+} and Co^{2+}, but also as cation-chelates, Pb is mostly present in organic complexes or colloidal material, Cr is present as Cr^{3+} and CrO_4^{2-} or $Cr_2O_7^{2-}$, and Hg is present as organic complexes (Tyler and Olsson, 2001; Tudoreanu and Phillips, 2004; Shanker et al., 2005; Sharma and Dubey, 2006; Micó et al., 2008). The abundance of these chemical forms in the soil solution, and therefore the phytoavailability of heavy metals to plants, is often determined by the absolute concentrations of the elements in the soil, the pH and redox environment of the soil, the amount of soil organic matter and the active secretion of protons and organic compounds by plant roots into the rhizosphere (Fageria, 2009; Hodson and Donner, 2013; White and Greenwood, 2013).

The concentrations of most cations (e.g. Fe^{3+}, Zn^{2+}, Mn^{2+}, Cu^{2+}, Co^{2+}) in the soil solution generally decrease with increasing pH due to the increase in pH-dependent charge on oxides of Fe, aluminium (Al) and Mn, chelation by organic matter and precipitation as metal hydroxides (Fig. 12.1; Lindsay, 1979; Mengel et al., 2001; Fageria, 2009; Fageria et al., 2011). The concentration of Fe^{3+} decreases by about 1000-fold for each unit increase in pH from pH 4 to pH 9, compared to about 100-fold decreases in the concentrations of Zn^{2+}, Mn^{2+} and Cu^{2+} (Lindsay, 1979; Mengel et al., 2001; Fageria, 2009). This is a consequence of the precipitation of iron hydroxide, $Fe(OH)_3$, which is ultimately the most significant soil source of Fe for plant nutrition. By contrast, Mo concentration in the soil solution is lowest between pH 2 and pH 6, but often increases sharply as pH increases from pH 6 to pH 8 (Lindsay, 1979; Mengel et al., 2001; Tyler and Olsson, 2001; Fageria, 2009; Fageria et al., 2011). This occurs because the major chemical form of Mo in the soil solution, MoO_4^{2-}, is adsorbed strongly to Fe-oxides and Fe-hydroxides in acid soils. However, Mo toxicity is rarely observed in crop plants (Fageria, 2009). Small increases in the concentrations of Cr and Hg in the soil solution are also commonly observed as the soil pH is increased (Tyler and Olsson, 2001). The increase in Cr concentration in the soil solution has been attributed to the oxidation of Cr^{3+} to CrO_4^{2-}, and the increase in Hg concentration

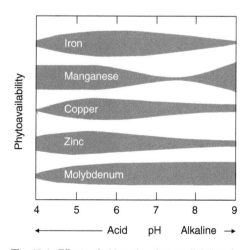

Fig. 12.1. Effects of pH on the phytoavailability of essential heavy-metal elements present in the soil solution. Adapted from Taiz and Zeigler (2006).

has been attributed to increased solubility of organic Hg-chelates (Tyler and Olsson, 2001). The concentrations of Ni, Cd and Pb in the soil solution can show complex relationships with soil pH (Tyler and Olsson, 2001). The concentrations of these elements are often lowest between about pH 6 and pH 7, but increase as the soil becomes more acidic or alkaline.

The Food and Agriculture Organization of the United Nations estimates that soil acidity restricts crop production on over 40% of the world's arable land (FAO, 2010). Crop production on acidic soils is restricted primarily by Al and Mn toxicities (Mengel et al., 2001; Kochian et al., 2004; He et al., 2005; Fageria, 2009; Fageria et al., 2011; White and Greenwood, 2013). Plants accumulate toxic Mn concentrations on acid soils largely because of the increased phytoavailability of Mn^{2+} in the soil solution (Mortvedt, 2000; Fageria et al., 2011). Tissue concentrations of 0.2 mg Mn/g DM can be toxic to some plants, although the absolute ability of a plant to tolerate Mn depends upon its genotype, growth rate and the concentration of silicic acid in the soil solution (Horst, 1988; Mortvedt, 2000). Iron, calcium (Ca) and magnesium (Mg) deficiencies are common symptoms of Mn toxicity and are thought to result from the displacement of Fe^{2+}, Ca^{2+}, Mg^{2+} and Zn^{2+} in their essential biochemical functions (Table 12.2; Bould et al., 1983; White and Brown, 2010; Fageria et al., 2011). In addition, Mn toxicity is associated with increased oxidative damage to plant tissues and the appearance of blackish-brown or red necrotic spots (Kabata-Pendias, 2000; Mengel et al., 2001). In some plant species, black precipitates of manganese dioxide (MnO_2) can be observed (Bould et al., 1983; Kabata-Pendias, 2000). Zinc toxicity can also occur in crops growing on acidic soils enriched by anthropogenic Zn inputs, such as agricultural soils treated with sewage sludge, soils of urban and peri-urban areas, and land contaminated by mining or smelting activities (Chaney, 1993; Broadley et al., 2007; White and Greenwood, 2013). Toxicity symptoms usually become visible at leaf Zn concentrations greater than about 300 µg Zn/g DM, although some crops show toxicity symptoms at concentrations lower than 100 µg Zn/g DM (Table 12.1). The symptoms of Zn toxicity include stunted growth, induced Fe-deficiency chlorosis through decreased chlorophyll synthesis

and increased chloroplast degradation, short roots and aberrant phosphorus (P), Mg, Mn and Cu uptake (Table 12.2; Foy et al., 1978; Bould et al., 1983; Chaney, 1993; Kabata-Pendias, 2000; Broadley et al., 2007, 2010). Plants lacking P often accumulate Zn, and this can lead to Zn toxicity (Broadley et al., 2007, 2010). Similarly, Cu toxicity has been observed in P-deficient plants (Wallace, 1984). Plants lacking sufficient Fe can also accumulate excess Zn (Kanai et al., 2009), and plants lacking sufficient Zn can accumulate excess Mn or Cu, which suggests physiological interactions between these elements (Brady and Weil, 2002). Genotypic differences exist both between and within crop species in their tolerance to soil acidity, rhizosphere Mn^{2+} and Zn^{2+} concentrations, and tissue Mn^{2+} and Zn^{2+} concentrations (Chaney, 1993; Broadley et al., 2007; Khabaz-Saberi et al., 2010; Fageria et al., 2011; White and Broadley, 2011; Yun et al., 2015). This provides the option to select appropriate crops, or to breed preferred crops, for cultivation on acidic soils, and chromosomal loci (QTL) affecting the sensitivity of rice to Zn and Mn toxicities (Wang et al., 2002; Dong et al., 2006) and soybean to Mn toxicity (Kassem et al., 2004) have already been identified.

Toxicities of Mn and Fe can occur on waterlogged or flooded soils as the redox potential becomes more negative (White and Brown, 2010; Fageria et al., 2011; White and Greenwood, 2013). Mn toxicity occurs in these environments because they favour the formation of Mn^{2+}, while Fe toxicity occurs because Fe^{3+} is reduced to Fe^{2+}, which increases Fe^{2+} availability and Fe uptake by plants (Mengel et al., 2001; Fageria, 2009; Fageria et al., 2011; White and Greenwood, 2013). Microbial reduction of Fe and, especially, Mn also occurs in flooded soils (Fageria, 2009). Leaf Fe concentrations in excess of 500 µg Fe/g DM are toxic to most crop plants (Table 12.1). Iron toxicity has been attributed largely to Fe-induced deficiencies of potassium (K), P, Zn, Ca and Mg (Fageria, 2009). Symptoms of Fe toxicity include dark green foliage, the appearance of orange-brown necrotic spots on older leaves, stunted growth, and the browning and blackening of roots (Table 12.2; Kabata-Pendias, 2000; Fageria, 2009). Paddy rice is particularly sensitive to Mn and Fe toxicities, although significant genetic variation exists between cultivars (Wang et al., 2002; Fageria, 2009).

Table 12.2. General symptoms of heavy-metal toxicities in non-tolerant crop plants. Data compiled from Bould *et al.* (1983), Kabata-Pendias (2000) and Fageria (2009).

Element	Symptoms	Sensitive crops
Cd (cadmium)	Brown leaf margins, reddish veins and petioles, total chlorosis, curled leaves and brown roots. Deterioration of xylem tissues. Severe plant stunting, inhibited tillering and reduced root growth	Legumes, spinach, radish, carrots, oats
Co (cobalt)	Pale green leaves, interveinal chlorosis in new leaves followed by induced Fe-deficiency interveinal necrosis, white leaf margins and tips, stunted plants with short brown roots and damaged root tips	–
Cr (chromium)	Yellow leaves, interveinal chlorosis of new leaves, necrotic spots, reduced plant height and purpling of tissues; wilting, reduced root growth	–
Cu (copper)	Dark green or bluish leaves followed by induced Fe chlorosis, young leaves chlorotic with dark-brown interveinal necrosis, stunted plants with short roots	Cereals, legumes, spinach, citrus
Fe (iron)	Dark green foliage, orange-brown necrotic spots (bronzing) of older leaves, stunted growth, browning and blackening of roots	Rice, tobacco
Hg (mercury)	Yellow leaves, leaf chlorosis and browning of leaf points, red stems, severe stunting and reduced root growth	Sugar beet, maize
Mn (manganese)	Interveinal chlorosis and necrotic lesions on old leaves, blackish-brown or red necrotic spots, accumulation of black MnO_2 particles in epidermal cells, drying leaf tips, stunted plants and reduced root growth	Cereals (e.g. barley), legumes (e.g. lucerne, beans), brassica (e.g. cauliflower, cabbage, kale, swede), lettuce, sugar beet, pineapple, potatoes, tomatoes
Mo (molybdenum)	Yellowing or browning of leaves, appearance of blue-purple or gold leaf pigments, reduced tillering and root growth	Cereals
Ni (nickel)	Grey-green leaves, induced Fe-deficiency yellow or white interveinal chlorosis and necrosis in new leaves, stunted plants with short brown roots	Cereals (e.g. oats), sugar beet and spinach beet
Pb (lead)	Dark green leaves, wilting of older leaves, stunted plants and short blackened roots	–
Zn (zinc)	Yellow leaves, chlorotic and necrotic leaf tips, interveinal leaf chlorosis, stunted plants with short roots	Cereals, leafy vegetables (e.g. spinach, beet), citrus

Specific geological formations have resulted in high soil concentrations of transition-metal elements at particular locations. For example, Ni, Co and Cr toxicities occur worldwide on soils derived from serpentine or other ultrabasic rocks (Mengel *et al.*, 2001; He *et al.*, 2005; Fageria, 2009). Similarly, soils can be enriched in Cd, Cu and Pb through the weathering of mineral outcrops containing these elements (Hodson and Donner, 2013; White and Greenwood, 2013). In sensitive plants, Ni becomes toxic at leaf concentrations above about 20 μg Ni/g DM (Table 12.1). Acute Ni toxicity induces Fe-deficiency chlorosis in plants, reduces plant height and causes root browning (Table 12.2). In cereals, characteristic yellow or white interveinal chlorosis occurs along the length of the leaves and necrosis occurs at the leaf margins. Cobalt toxicity, Cr toxicity (when caused by Cr^{3+}), Cu toxicity and Cd toxicity also exhibit the symptoms of induced Fe deficiency, which include interveinal chlorosis, then necrosis, in new leaves; reduced plant height; and root browning (Table 12.2; Bould et al., 1983; Kabata-Pendias, 2000). In general, these symptoms are more severe when plants are exposed to Cd^{2+} or Co^{2+} than when they are exposed to Cu^{2+} or Cr^{3+} (Bould et al., 1983). The thylakoid membrane of the chloroplast, and photosystem II in particular, is the primary target of Cu toxicity (Burkhead et al., 2009). The symptoms of Pb toxicity include darkening of foliage, the wilting of older leaves, a reduction in plant height and root blackening (Table 12.2).

Anthropogenic activities have increased soil concentrations of heavy metals through the mining and refining of metal ores, from the application of Cu-rich, Ni-rich or Cd-rich fertilizers, biosolids or municipal composts to agricultural soils; from the repeated use of Cu-containing fungicides, pesticides or herbicides; or from the incineration of Pb-rich coal and automobile fuel (Colburn and Thornton, 1978; Welch et al., 1991; Bacon et al., 1996; Mengel et al., 2001; He et al., 2005; Sharma and Dietz, 2006; Kirkby and Johnson, 2008; Fageria, 2009; White and Greenwood, 2013). High soil Cr concentrations have resulted from the disposal of industrial effluents to agricultural land (Brar et al., 2000; Shanker et al., 2005).

Traditional agronomic methods to reduce the uptake of heavy metals by crop plants include liming with calcite lime ($CaCO_3$), dolomite lime ($MgCO_3$) or dolomitic lime ($CaMg(CO_3)_2$); application of gypsum ($CaSO_4$) or phosphogypsum; application of organic matter; deeper tillage; and the alleviation of anaerobiosis on waterlogged or flooded soils (Fageria, 2009; Fageria et al., 2011; White and Greenwood, 2013). Increasing soil pH by liming not only decreases the concentrations of heavy-metal cations in the soil solution but also antagonizes their uptake by plant roots (Fageria, 2009; Fageria et al., 2011). The addition of organic material buffers the concentrations of cations in the soil solution due to its high cation exchange capacity, and its incorporation into soils can reduce concentrations of heavy metals in the soil solution, thereby preventing their uptake by roots and their leaching into groundwater (Fageria et al., 2011). Deeper tillage allows plant roots to explore lower soil horizons and, thereby, to avoid high concentrations of heavy metals associated with organic matter in the topsoil (White and Greenwood, 2013). However, these traditional agronomic countermeasures are often expensive and only partially or temporarily successful. Plant breeders are therefore developing crop genotypes that take up less of the toxic heavy metals and can produce a commercial yield on contaminated soils (Uraguchi and Fujiwara, 2013). The ability of food crops to exclude heavy metals, or to prevent their accumulation in edible tissues, is important not only for crop yields on contaminated soils but also for the transfer of these elements through the food chain. Excesses of heavy metals in the diets of humans and other animals can be detrimental to their health (White and Broadley, 2005; White and Brown, 2010; White et al., 2012).

12.3 The Uptake and Accumulation of Heavy Metals by Plants

12.3.1 Mobilization and uptake of heavy metals from the soils

Heavy-metal toxicities are rarely observed in agriculture. By contrast, deficiencies of essential transition-metal elements are frequently observed in plants in both natural and agricultural environments (Broadley et al., 2007; Fageria, 2009; White and Broadley, 2009; White and Brown, 2010; Fageria et al., 2011; White et al., 2012, 2013; White and Greenwood, 2013). This is often a consequence of limited phytoavailability, rather than a lack of these elements in the soil (Graham et al., 1999; Broadley et al., 2007; Fageria, 2009; White and Broadley, 2009; White and Brown, 2010; Fageria et al., 2011; White et al., 2012, 2013; White and Greenwood, 2013; Rengel, 2015). The low concentrations of heavy metals in the soil solution, together with the small diffusion coefficients of their cationic species, means that the delivery of heavy metals to the root surface is generally determined by transpiration-driven mass flow of the soil solution, rather than by local diffusion

(Barber, 1995; Sterckeman *et al.*, 2004; Fageria, 2009; White and Greenwood, 2013; White *et al.*, 2013). Plants have adapted to soils with low phytoavailability of essential transition-metal elements by evolving mechanisms to increase their concentrations in the soil solution and to accelerate their interception and uptake by roots. These mechanisms include modification of rhizosphere chemistry and biology by root exudates, the production of an extensive root system and the induction of high-affinity transport systems for the uptake of essential transition-metal elements as ions or cation-chelates (Fig. 12.2; White and Broadley, 2009; White and Greenwood, 2013; White *et al.*, 2013). These mechanisms can also increase the uptake of non-essential heavy metals, such as Co and Cd (Lux *et al.*, 2011).

When non-graminaceous plant species (Strategy I plants) lack sufficient Fe, their roots acidify the rhizosphere and release organic acids

and phenolic compounds to increase Fe^{3+} concentrations in the soil solution (Fig. 12.2; White, 2012a; Schmidt *et al.*, 2014). These compounds chelate Fe^{3+}, which is subsequently reduced to Fe^{2+} by ferric reductases in the plasma membrane of root epidermal cells and this Fe^{2+} is taken up by Fe^{2+}-transporters encoded by members of the *zinc-regulated transporter (ZRT)-, iron-regulated transporter (IRT)-like protein (ZIP)* gene family and members of the *natural resistance associated macrophage protein (NRAMP)* gene family (Palmer and Guerinot, 2009; Puig and Peñarrubia, 2009; White and Broadley, 2009; Cailliatte *et al.*, 2010; Guerinot, 2010; White and Greenwood, 2013; Pinto and Ferreira, 2015). Cereals and grasses (Strategy II plants) can take up Fe^{2+} through Fe^{2+}-transporters, but also release phytosiderophores (structural derivatives of mugineic acid) into the rhizosphere to chelate Fe^{3+} and take up Fe^{3+}-phytosiderophore complexes through homologues of the

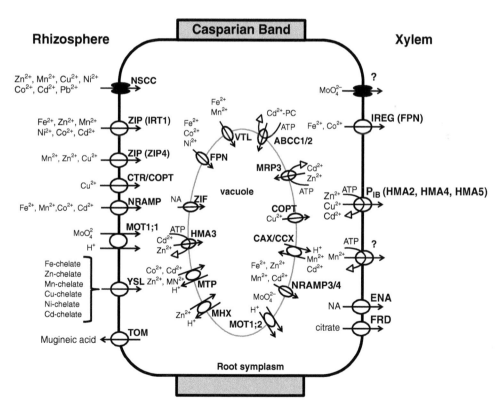

Fig. 12.2. Transport proteins in the plasma membranes and vacuoles of root cells implicated in the movement of heavy metals from the rhizosphere to the xylem through the symplasm. NA, nicotianamine.

maize yellow stripe 1 protein (YSL) present in the plasma membrane of root cells (Fig. 12.2; Palmer and Guerinot, 2009; Puig and Peñarrubia, 2009; White and Broadley, 2009; Guerinot, 2010; White and Greenwood, 2013; Pinto and Ferreira, 2015). The efflux of mugineic acids is facilitated by homologues of the rice transporter of mugineic acid, OsTOM1 (Nozoye et al., 2011). A lack of Zn, Mn or Cu also induces rhizosphere acidification, and the exudation of organic acids, phytosiderophores and enzymes capable of degrading organic compounds in the soil (Broadley et al., 2007; Palmer and Guerinot, 2009; Pilon et al., 2009; Puig and Peñarrubia, 2009; White and Broadley, 2009; Clemens, 2010; Guerinot, 2010; White et al., 2013). These elements can be taken up in cationic forms by roots of all plants, and also as phytosiderophore complexes by roots of cereals and grasses (Fig. 12.2; Grotz and Guerinot, 2006; Broadley et al., 2007; Burkhead et al., 2009; Palmer and Guerinot, 2009; Pilon et al., 2009; Pilon-Smits et al., 2009; Puig and Peñarrubia, 2009; Tejada-Jiménez et al., 2009; White and Broadley, 2009; Clemens, 2010; Krämer, 2010; Williams and Pittman, 2010; White and Greenwood, 2013). The uptake of Zn^{2+}, Mn^{2+}, Cu^{2+} and Ni^{2+} is facilitated by members of the ZIP gene family, the uptake of Mn^{2+} by orthologues of AtNRAMP1 (Socha and Guerinot, 2014) and the uptake of Cu^+ by members of the copper transporter (CTR/COPT) gene family (Fig. 12.2). Some ZIPs, such as AtIRT1 and AtNRAMP1, also transport Co^{2+} and Cd^{2+} into root cells (Clemens, 2006; Verbruggen et al., 2009b; Cailliatte et al., 2010; Williams and Pittman, 2010; Lux et al., 2011). In cereals and grasses, Zn, Mn, Cu, Ni and Cd are also taken up as phytosiderophore complexes, catalysed by YSL proteins. The activity of transport proteins catalysing the uptake of Fe, Zn, Cu and Mn, and the expression of the genes that encode these transporters, is regulated in response to plant nutritional status to ensure appropriate tissue concentrations of essential mineral elements. It is likely that divalent cations also enter root cells through non-selective cation channels (NSCC; White and Broadley, 2003; DalCorso et al., 2008; Verbruggen et al., 2009b). These channels are thought to be the principal route of Pb entry to root cells.

A proton-coupled symporter (MOT1;1), encoded by a member of the molybdate transporter (MOT1) gene family, is thought to be responsible for the uptake of molybdate by root cells (Tejada-Jiménez et al., 2009, 2013). Increasing sulfate concentration in the rhizosphere and, thereby, plant sulfur status, reduces Mo uptake and improves plant growth on soils with high Mo phytoavailability (Pasricha et al., 1977; Chatterjee et al., 1992). Chromate is also thought to be taken up by root cells through anion transporters, although their identity is unknown (Shanker et al., 2005). In brewer's yeast (Saccharomyces cerevisiae), CrO_4^{2-} is taken up through the Sul1p and Sul2p transporters (Pereira et al., 2008) and Cr toxicity is linked to Cr-induced S-starvation, which results in increased mRNA mistranslation (Holland et al., 2010).

12.3.2 Movement of heavy metals to the xylem

Heavy metals entering roots from the soil solution can reach the xylem through either intracellular (symplastic) or extracellular (apoplastic) pathways. Most essential transition-metal elements are thought to enter the root symplast through transport proteins located in the plasma membrane of epidermal and/or cortical cells and then transported across the root through the cytoplasm of root cells connected by plasmodesmata (Broadley et al., 2007; Pilon et al., 2009; Puig and Peñarrubia, 2009; Tejada-Jiménez et al., 2009; White and Broadley, 2009; Clemens, 2010; Guerinot, 2010; Williams and Pittman, 2010; Sinclair and Krämer, 2012; White and Greenwood, 2013). To protect essential cytoplasmic functions, however, transition-metal cations are generally bound by organic ligands upon entry to the symplasm.

Iron entering plant cells is bound in organic complexes. Zinc is chelated by a multitude of diverse proteins in the cytosol, including many that modulate enzymic activities or gene transcription, by glutathione (GSH), by phytochelatins and by nicotianamine (NA), and the cytosolic Zn^{2+} concentration is extremely low (Krämer and Clemens, 2005; Broadley et al., 2007; Hänsch and Mendel, 2009; Clemens, 2010; Clemens et al., 2013). Most Cu entering a plant cell is immediately bound to metallothioneins and phytochelatins, but some is chelated by chaperone proteins that deliver it to specific apo-proteins to form biologically active Cu-proteins (Hänsch and Mendel, 2009). The abundance of metallotheioneins and (occasionally) phytochelatins is increased in plants upon exposure to Cu, and this is thought to improve their Cu tolerance (Hall, 2002; Guo et al., 2008; White and Broadley, 2009). Cadmium is similarly bound to metallotheioneins, phytochelatins and GSH within the cell and the abundance of both metallotheioneins and phytochelatins, which increases when plants are exposed to Cd, appears to play a significant role in Cd detoxification (Clemens, 2006; DalCorso et al., 2008; Ernst et al., 2008; Shim et al., 2009; Lux et al., 2011; Andresen et al., 2013). Phytochelatin production also appears to contribute to Pb tolerance (Fischer et al., 2014), and when plants are exposed to Pb their phytochelatin concentrations increase (Sharma and Dubey, 2006).

The symplastic movement of heavy metals to the xylem is also restricted by their sequestration in the vacuoles of root cells (Peng and Gong, 2014). In A. thaliana, both vacuolar iron transporter 1 (AtVIT1) and AtVIT1-like (AtVTL) proteins and the ferroportin metal efflux protein AtFPN2 (AtIREG2) have been implicated in Fe sequestration in the vacuole, while AtNRAMP3 and AtNRAMP4 catalyse Fe^{2+} efflux from the vacuole (Fig. 12.2; Kim et al., 2006; Grotz and Guerinot, 2006; Puig et al., 2007; Palmer and Guerinot, 2009; Morrissey et al., 2009; White and Broadley, 2009; Gollhofer et al., 2014). Members of the cation diffusion facilitator (CDF) family, such as the metal tolerance proteins AtMTP1 and AtMTP3, and the Mg^{2+}/H^+ antiporter AtMHX, catalyse the movement of Zn^{2+} into the vacuole, while the zinc-induced facilitator 1 (AtZIF1) protein catalyses the movement of nicotianamine or Zn^{2+}-chelates into the vacuole (Haydon and Cobbett, 2007; Palmer and Guerinot, 2009; Puig and Peñarrubia, 2009; White and Broadley, 2009; Clemens, 2010; Sinclair and Krämer, 2012; Clemens et al., 2013). In A. thaliana, Mn^{2+} influx to the vacuole is catalysed by Mn^{2+}/H^+-antiporters, such as AtCAX2, AtCAX4, AtCAX5 and AtCCX3, and by AtVIT1; and Mn transport into the trans-Golgi network and pre-vacuolar compartments is catalysed by AtMTP11 and the P_{2A}-ATPase AtECA3 (Puig and Peñarrubia, 2009; Williams and Pittman, 2010;

Connorton et al., 2012; Emery et al., 2012). Carboxylic acids, such as malate or citrate, form Mn^{2+} complexes in the vacuole (Fernando et al., 2013). The influx of Cd^{2+} to the vacuole is catalysed by Cd^{2+}/H^+-antiporters, such as AtCAX2 and AtCAX4 (Clemens, 2006; Korenkov et al., 2007, 2009), in addition to heavy-metal P_{1B}-ATPases, such as AtHMA3 (Morel et al., 2009). Cadmium chelates are sequestered in the vacuole by ABC transporters, such as AtMRP3 (Tommasini et al., 1998; Clemens, 2006; DalCorso et al., 2008; Verbruggen et al., 2009b) and AtABCC1/2 (Mendoza-Cózatl et al., 2011). In many plant species, Cd tolerance is related to Cd accumulation in the vacuole (e.g. Korenkov et al., 2007; Seregin and Kozhevnikova, 2008). Zinc, Mn^{2+} and Cd^{2+} appear to be released from the vacuole through NRAMPs, including AtNRAMP3 and AtNRAMP4 (Thomine et al., 2003; Clemens, 2006; Verbruggen et al., 2009b; Williams and Pittman, 2010). The proteins catalysing Cu influx to the vacuole are unknown, but Cu-binding vegetative storage proteins appear to have a role in Cu homeostasis and Cu detoxification in the vacuole (Mira et al., 2002; Kung et al., 2006) and Cu is thought to be released from internal stores by members of the CPT/COPT family, such as AtCOPT3 and AtCOPT5 (Burkhead et al., 2009; Martinoia et al., 2012). Molybdate release from the vacuole is thought to be catalysed by proton-coupled molybdate transporters, such as AtMOT1;2 (Tejada-Jiménez et al., 2013). Chromate is thought to be reduced to Cr^{3+} in root cells and sequestered in the vacuoles of root cortical cells (Shanker et al., 2005).

Heavy-metal cations are loaded into the xylem by transport proteins located in the plasma membrane of stelar parenchyma cells (Fig. 12.2). Members of the heavy-metal P_{1B}-ATPase family, such as AtHMA2, AtHMA4 and AtHMA5, load Zn, Cu and Cd into the xylem (Mills et al., 2005; DalCorso et al., 2008; Wong and Cobbett, 2009; Verbruggen et al., 2009b; White and Broadley, 2009; Lux et al., 2011; Sinclair and Krämer, 2012; White and Greenwood, 2013; Pinto and Ferreira, 2015), and the ferroportin metal efflux protein AtFPN1 (AtIREG1) has been implicated in loading Fe and Co into the xylem (Morrissey et al., 2009). Manganese is also thought to be loaded into the xylem by an ATPase, while Mo is probably loaded into the xylem by anion transporters (Shanker *et al.*, 2005).

Most transition-metal elements are transported as cation-chelates in the xylem sap (Álvarez-Fernández *et al.*, 2014). Manganese, Zn and Co appear to be transported as cations or cation-complexes with organic acids (Welch, 1995; White and Broadley, 2009; Collins *et al.*, 2010). Iron is transported mainly as Fe^{3+}-citrate, and the loading of citrate into the xylem (catalysed by orthologues of the *A. thaliana* AtFRD3 transporter) appears to be essential for the transport of Fe to the shoot (Welch, 1995; von Wirén *et al.*, 1999; Durrett *et al.*, 2007; Puig *et al.*, 2007; Guerinot, 2010). Zinc, Cu and Ni can be transported as complexes with asparagine and histidine; and Zn, Cu, Mn and Ni can be transported as nicotianamine complexes (Welch, 1995; von Wirén *et al.*, 1999; Broadley *et al.*, 2007; Curie *et al.*, 2009). The loading of nicotianamine into the xylem is thought to be catalysed by an efflux transporter of nicotianamine (ENA; Kobayashi and Nishizawa, 2012). Chromium is thought to be transported as CrO_4^{2-} or as organic Cr^{3+}-complexes (Cary *et al.*, 1977).

Entry to the root symplasm and, thereby, selective transport of solutes to the xylem, is enforced by the formation of an endodermal Casparian band (stage I endodermis) at an early stage of root development (White, 2001). This suberized structure joins the plasma membrane of each endodermal cell and restricts the apoplastic movement of solutes to the xylem. In basal regions of the root, suberin lamellae cover the entire surface of endodermal cells (stage II endodermis), which prevents these cells taking up solutes from the apoplast (Moore *et al.*, 2002). Thick cellulose secondary walls are subsequently deposited over the suberin lamellae, which can be lignified (stage III endodermis). In addition, roots of most angiosperms form another apoplastic barrier (the exodermis) in parallel with the endodermis, which restricts the entry of solutes from the rhizosphere to the root cortex (Ma and Peterson, 2003). Several heavy metals, including Ni, Cu, Co, Cr, Cd and Pb are often (but not always) retained in roots, suggesting that their delivery to the xylem is restricted in most plants (Cary *et al.*, 1977; Polle and Schützendübel, 2003; Page and Feller, 2005; Shanker *et al.*, 2005; Clemens, 2006; Sharma and Dubey, 2006; Reeves, 2006;

Fageria, 2009; Conn and Gilliham, 2010; Lux *et al.*, 2011).

Among the heavy metals, significant amounts of Fe and Zn are thought to reach the xylem through a solely apoplastic pathway in particular plant species and/or when these elements occur at high concentrations in the soil solution (White, 2001; White *et al.*, 2002; Taiz and Zeigler, 2006; Broadley *et al.*, 2007). Since the apoplastic pathway is relatively non-selective, non-essential heavy metals such as Cd and Pb can also reach the xylem by this route (Lux *et al.*, 2004, 2011; Sharma and Dubey, 2006). Movement to the xylem solely via the apoplast is generally restricted to regions lacking a functional Casparian band, such as the extreme root tip, regions where lateral roots are being initiated (White, 2001) and the basal region of contractile roots (Lux *et al.*, 2015). The accelerated development of the root exodermis, endodermis and other extracellular barriers by high rhizosphere Cd or Pb concentrations is thought to be an adaptive response to restrict apoplastic movement of Cd and Pb to the xylem (Sharma and Dubey, 2006; Lux *et al.*, 2011).

It is noteworthy that the movement of heavy metals to the xylem can also be restricted by efflux from root cells to the rhizosphere. For example, in *A. thaliana*, the expression of *AtHMA5* is induced by high Cu and this transporter is thought to remove excess Cu from the cytosol to the apoplast (Andrés-Colás *et al.*, 2006). Similarly, the ABC transporter AtPDR8 has been implicated in Cd efflux across the plasma membrane of root epidermal cells and reduction of Cd accumulation by plants (Kim *et al.*, 2007).

12.3.3 Transport of heavy metals within the plant

Water from the xylem stream flows preferentially through the major and minor veins of the leaves to the sites of rapid transpiration (Karley *et al.*, 2000). If the concentration of a heavy metal in the xylem sap is excessive, it can accumulate at the leaf margins and cause necrosis (Table 12.2). Heavy metals are selectively retrieved from leaf veins and apoplast by competent leaf cells. Thus, heavy metals are accumulated in the vacuoles of specific leaf cells, whose identity can vary dependent upon the heavy metal in question and the plant species (Broadley *et al.*, 2007; Conn and Gilliham, 2010). As in root cells, members of the ZIP family catalyse the uptake of Zn^{2+}, Fe^{2+}, Cu^{2+} and Mn^{2+} (White and Broadley, 2009), members of the COPT family catalyse Cu uptake (Cohu and Pilon, 2010) and members of the YSL family catalyse the uptake of metal chelates by leaf cells (Guerinot, 2010). Molybdate probably enters leaf cells through phosphate or sulfate transporters. As in root cells, heavy metals entering the cytosol of shoot cells are rapidly chelated and delivered to specific apoproteins to form biologically active proteins, or sequestered in the vacuole as metal chelates (Martinoia *et al.*, 2012). In leaves the major Fe-binding protein is ferritin, which accumulates in plastids and appears to serve a protective role in plants accumulating high Fe concentrations (Ravet *et al.*, 2008; Palmer and Guerinot, 2009). Vacuoles provide both temporary storage and permanent sequestration of essential and toxic heavy metals (Martinoia *et al.*, 2012; Peng and Gong, 2014).

Heavy metals can be redistributed from stem and leaf cells through both the xylem and phloem. Selective movement of heavy metals in the phloem allows the delivery of essential elements to developing tissues, tubers, fruit and seed, while toxic elements are retained in older leaves. Iron, Zn, Cu, Ni and Mo have high to moderate mobility in the phloem, whereas Mn, Co, Cd, Pb and Cr have low mobility in the phloem (Page and Feller, 2005; Lux *et al.*, 2011; White, 2012b). The remobilization of Fe, Zn and Cu, but not Mn, is often closely related to leaf senescence (Waters *et al.*, 2009; White, 2012b). Members of the ZIP family are thought to load Zn into the phloem (Ishimaru *et al.*, 2005), while members of the YSL family, or related members of the oligopeptide transporter (OPT) family, probably load Fe, Mn, Zn and Cu into the phloem (Curie *et al.*, 2009; White and Broadley, 2009; Williams and Pittman, 2010; Zhai *et al.*, 2014). In rice, OsLCA1 has been implicated in loading Cd into the phloem (Uraguchi and Fujiwara, 2013). In general, transition-metal elements are transported to sink tissues as metal–NA complexes or in association with amino acids (glutamate, cysteine, histidine) or small proteins, such as the iron transport protein (White and Broadley, 2009; Curie *et al.*, 2009; Guerinot, 2010; Harris *et al.*, 2012; Ando *et al.*, 2013; Álvarez-Fernández *et al.*, 2014).

Phloem composition often serves as a systemic signal to indicate the nutritional status of the plant. It is possible that increased sucrose transport from the shoot to the root, which indicates a nutritional imbalance in the shoot, can increase the uptake of essential transition-metal elements both directly (through altered gene expression), and indirectly (by increasing the size of the root system) (Hammond and White, 2008). In addition, the complement of microRNAs (miRNAs) is altered by the accumulation of heavy metals in the shoot. Many of these miRNAs are transported in the phloem and regulate transcriptional cascades controlling the uptake and detoxification of heavy metals in the root and other tissues (Section 12.4; Ding and Zhu, 2009; Pilon *et al.*, 2009; Buhtz *et al.*, 2010; Kong and Yang, 2010; Cuypers *et al.*, 2011). The expression of genes encoding proteins responsible for the mobilization of transition-metal elements from soil, their uptake by plant roots and their redistribution within the plant are often upregulated during their deficiency. For example, when Fe is in short supply, genes encoding FROs, ZIPs, NRAMPs and YSLs, as well as enzymes involved in the biosynthesis of phytosiderophores and NA, are all upregulated (White and Broadley, 2009). Similarly, *ZIPs*, *HMAs*, *YSLs*, *MTPs*, *ZIF1*, *FRD3* and enzymes involved in the biosynthesis of phytosiderophores and NA are upregulated during Zn deficiency (White and Broadley, 2009).

12.4 Cellular Signals and Responses to Heavy Metal Exposure

Heavy-metal toxicity is primarily a consequence of: (i) ectopic binding of heavy-metal cations to proteins, due to their affinities for thioyl-, histidyl- and carboxyl-groups; (ii) displacement of essential cations from specific binding sites, which disrupts structural integrity, impairs ionic homeostasis and causes metabolic imbalance; and (iii) the generation of ROS, which causes oxidative stress (Section 12.1; Hall, 2002; Polle and Schützendübel, 2003; Maksymiec, 2007; Sharma and Dietz, 2009).

There are clear relationships between exposure to heavy metals in the environment, antioxidant capacity and redox homeostasis in plants, and the degree of tolerance to acute exposure to heavy metals appears to be directly related to their antioxidant capacity (Sharma and Dietz, 2009). Principal antioxidants include GSH, ascorbate, α-tocopherol and proline, and the enzymes SOD, catalase (CAT), ascorbate peroxidase (APX) and peroxiredoxin (PRX), while the enzymes glutathione peroxidase (GPX) and glutathione-S-transferase (GST), glutathione reductase (GR), monodehydroascorbate reductase (MDAR) and dehydroascorbate reductase (DHAR) contribute to the redox poise of a plant cell (Polle and Schützendübel, 2003; Mittler *et al.*, 2004; Sharma and Dietz, 2006, 2009). In *A. thaliana*, redox homeostasis is controlled by an intricate genetic network consisting of over 150 genes (Mittler *et al.*, 2004). The synthesis of these antioxidants and enzymes generally increases upon prolonged exposure to heavy metals and appears to be constitutively greater in plant species that hyperaccumulate heavy metals (Sharma and Dietz, 2006, 2009; Cuypers *et al.*, 2011; Rascio and Navari-Izzo, 2011).

Acute exposure to heavy metals, including Fe, Cu, Mn, Ni, Cd, Cr, Co, Pb and Hg, causes the rapid production of ROS and depletion of tissue GSH and homoglutathione (hGSH) causing redox imbalance (Polle and Schützendübel, 2003; Shanker *et al.*, 2005; Clemens, 2006; Sharma and Dubey, 2006; Maksymiec, 2007; Sharma and Dietz, 2009; Remans *et al.*, 2010; Cuypers *et al.*, 2011). Redox-active heavy-metal cations, such as Fe, Cu, Cr and Co, produce O_2^-, H_2O_2 and $\cdot OH$ via Haber–Weiss and Fenton reactions (Demidchik, 2010). Glutathione binds heavy metals, which can cause oxidative stress indirectly, and cellular GSH concentrations can be reduced by increased phytochelatin biosynthesis, for which GSH is a precursor. Changes in cellular redox state, and in the concentration of GSH in particular, are thought to be the primary signal eliciting responses of cellular biochemistry and gene expression to the oxidative stresses caused by heavy metals (May *et al.*, 1998; Sharma and Dietz, 2009). The predominant sources of ROS are the electron transfer activities of chloroplasts and mitochondria and oxidative metabolism in the peroxisomes. Exposure to heavy metals generally stimulates ROS production primarily in peroxisomes (Sharma and Dietz, 2009). In addition, exposure to heavy metals, such as Ni, Cd and Pb, stimulates the generation of ROS by plasma-membrane-bound NADPH oxidases, which is amplified through a cytosolic Ca^{2+} signalling cascade (Sharma and Dietz, 2009; Demidchik, 2010). The generation of ROS can also initiate cytosolic Ca^{2+} signals by activating outward-rectifying K^+ channels, HACCs and VICCs in the plasma membrane, and heavy metals can interact directly with components of these Ca^{2+} signalling cascades (White and Broadley, 2003; Maksymiec, 2007; Demidchik, 2010).

In rice roots, mitochondrial ROS production and NADPH oxidase activity are associated with the activation of mitogen-activated protein kinase (MAPK) signalling cascades by both Cu^{2+} and Cd^{2+} (Yeh *et al.*, 2007). It is thought that the activation of MAPK signalling cascades contributes to Cd tolerance in rice because Cd-tolerant cultivars possess higher MAPK activities. Heavy metals such as Cu^{2+}, Cd^{2+}, Fe^{2+}, Zn^{2+} and Pb^{2+} also initiate MAPK signalling cascades in other plant tissues and species (Jonak *et al.*, 2004; Nakagami *et al.*, 2005; Maksymiec, 2007; Sharma and Dietz, 2009; X.-M. Liu *et al.*, 2010). The MAPK signalling cascades are thought to integrate biochemical and physiological responses to diverse biotic and abiotic challenges, and specific elements are activated directly by ROS (Nakagami *et al.*, 2005; Zhang *et al.*, 2006; Sharma and Dietz, 2009). The production of ROS following exposure to excess heavy metals can also lead to the initiation of programmed cell death (Polle and Schützendübel, 2003).

Exposure to excess heavy metals often, but not always, stimulates the production of nitric oxide (NO) by plants (Xiong *et al.*, 2010; Saxena and Shekhawat, 2013). Nitric oxide is a key intermediate in signalling cascades regulating processes such as plant growth and development, fruit ripening, stomatal closure, the expression of defence-related genes and programmed cell death, probably acting in concert with classical plant growth regulators (Neill *et al.*, 2003), and application of exogenous NO appears to induce heavy-metal tolerance in plants (Xiong *et al.*, 2010). It has been proposed that NO effects this by: (i) acting as an antioxidant to scavenge ROS; (ii) modifying cell wall chemistry to increase the apoplastic accumulation of heavy metals; and (iii) initiating cellular signalling cascades that regulate the expression of genes conferring metal tolerance, possibly via cytosolic Ca^{2+} signals (Neill et al., 2003; Xiong et al., 2010).

Acclimation to both acute and prolonged exposure to heavy metals is achieved initially by the regulation of enzyme activities and subsequently by changes in gene expression. Enzyme activities are regulated by redox reactions, protein phosphorylation, changes in cytosolic Ca^{2+} concentration or pH and interactions between regulatory components (Romero-Puertas et al., 2007; Demidchik, 2010). These are thought to be initiated by oxidative stress and the production of ROS. Gene expression is regulated by signalling cascades that alter the abundance of transcription factors and miRNAs (Polle and Schützendübel, 2003; Ding and Zhu, 2009; Cuypers et al., 2011; Yang and Chen, 2013; Gupta et al., 2014; Zeng et al., 2014; Liu et al., 2015). Many recent studies have indicated important roles for miRNAs, encoded by miR genes, in plant acclimation to heavy-metal excess, particularly in response to Cu and Cd stress (Ding and Zhu, 2009; Pilon et al., 2009; Cuypers et al., 2011; Gupta et al., 2014; Zeng et al., 2014). For example, the abundance of miR398 is reduced by Cu excess. This miRNA reduces the abundance of mRNAs encoding Cu, Zn-SODs (CSD1 and CSD2) and a Cu-chaperone for these proteins (CCS1) at the post-transcriptional level, and appears to be a key element in the regulation of Cu-dependent gene expression (Ding and Zhu, 2009; Beauclair et al., 2010; Cuypers et al., 2011). A reduction in miR398 increases the abundance of these antioxidant enzymes and, since CSDs are Cu-binding proteins, also enables the sequestration of excess Cu. Another target of miRNA398 is cytochrome c oxidase, which is also a major Cu-containing protein. Similarly, the abundance of several miRNAs responds to Cd stress (Ding and Zhu, 2009). These include miRNA398, whose abundance decreases in response to Cd exposure. Decreased abundance of miR398 will increase the abundance of CSDs and alleviate oxidative stress. The abundance of this miRNA is also reduced by exposure to Hg (Yang and Chen, 2013). The targets of many of the other miRNAs whose transcription is affected by Cd exposure are transcription factors or components of protein signalling cascades implicated in acclimation to heavy-metal stresses (Ding and Zhu, 2009; Yang and Chen, 2013;

Gupta et al., 2014). In addition to miR398, the abundance of two other miRNAs (miR171 and miR373) has been observed to respond to Cd exposure in Medicago truncatula, Brassica napus and rice (Ding and Zhu, 2009). The targets of miR171 are transcripts of genes encoding scarecrow-like transcription factors, which regulate floral development. The targets of miR373 are transcripts of four F-box genes. One of these is TIR1 (transport inhibitor response 1), which is a key regulatory component of auxin signalling cascades. Plant responses to the availability of Fe also appear to be effected by the abundance of specific miRNA (Buhtz et al., 2010; Kong and Yang, 2010; Gupta et al., 2014).

The generation of ROS causes non-specific oxidative damage to proteins, lipids, nucleic acids and other cellular compounds. Damage to cell wall components is thought to be responsible for the cell wall stiffening and, in part, for the impaired cell elongation observed in plants exposed to excess heavy metals (Polle and Schützendübel, 2003). Exposure to heavy metals also induces lipid peroxidation by increasing the abundance and activity of lipoxygenases, which are potential precursors for signalling molecules such as jasmonic acid, oxilipins and volatile derivatives (Maksymiec, 2007; Sharma and Dietz, 2009; Remans et al., 2010; Cuypers et al., 2011). The concentrations of the plant growth regulators (PGR) jasmonic acid, salicylic acid, abscisic acid and ethylene increase upon exposure to heavy metals, although the role of these PGR in plant heavy-metal tolerance is still a matter for debate (Polle and Schützendübel, 2003; Maksymiec, 2007; DalCorso et al., 2008; Sharma and Dietz, 2009; Verbruggen et al., 2009a). It is possible that these PGR regulate gene expression and initiate acclimatory responses to prolonged heavy-metal exposure systemically throughout the plant.

In most plants, signalling cascades induce acclimation to excess heavy metals in the environment through: (i) the avoidance of oxidative stress, by increasing concentrations of antioxidant compounds and the activity of antioxidant enzymes; (ii) preventing the uptake of heavy metals by plant cells, by reducing the abundance or activity of transport proteins and developing barriers to solute movement in the apoplast; (iii) preventing the movement of heavy metals within the plant, from the root to the shoot in the xylem or from older leaves to phloem-fed tissues; and (iv) the detoxification of heavy-metal cations within the cell by their chelation or sequestration in extra-cytoplasmic compartments.

12.5 Preventing Heavy-metal Toxicity in Plants

12.5.1 Mechanisms of avoidance and tolerance

As observed in Section 12.3, heavy-metal toxicities are observed less frequently than deficiencies of essential transition-metal elements in crop plants, and studies of the mineral nutrition of crop plants have generally focused on improving the acquisition of these essential elements from the soil (White and Broadley, 2009; Fageria *et al.*, 2011). Nevertheless, heavy-metal toxicities can result in significant crop losses, particularly on acid or waterlogged soils (Fageria *et al.*, 2011). Traditional agronomic methods such as liming, applications of gypsum and organic matter, deeper tillage and the alleviation of anaerobiosis on waterlogged or flooded soils can be used to reduce the phytoavailability of heavy metals, but these countermeasures are generally expensive and only partially or temporarily successful (Section 12.2). Thus, farmers often cultivate traditional crops, and plant breeders are developing genotypes of preferred crops, with greater tolerance of soils with high heavy-metal phytoavailability.

Plants have evolved various mechanisms to allow them to grow on soils with high heavy-metal phytoavailability (Whiting *et al.*, 2005; Broadley *et al.*, 2007; Fageria *et al.*, 2011; Hodson and Donner, 2013). These mechanisms can be grouped into two, non-exclusive strategies: avoidance and tolerance (Fig. 12.3; Baker, 1981). Avoidance mechanisms prevent the entry of heavy metals to the plant, whereas tolerance mechanisms detoxify heavy metals within the plant. Although the evolution of both avoidance and tolerance mechanisms can be observed within decades of contamination of soils with high concentrations of transition-metal elements, most mechanisms have evolved over hundreds or millions of years (Whiting *et al.*, 2005; Reeves, 2006).

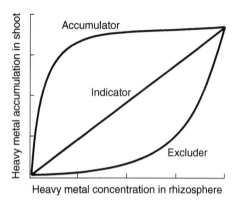

Fig. 12.3. Hypothetical relationships between heavy-metal uptake and the phytoavailiability of the heavy metal in the soil solution (Baker, 1981). Three relationships are postulated: (i) Accumulator plants show high uptake of heavy metals at low phytoavailability in the soil solution and uptake approaches a constant value with increasing heavy-metal phytoavailability; (ii) Indicator plants show a linear increase in the uptake of a heavy metal with increasing heavy-metal phytoavailability in the soil solution; and (iii) Excluder plants take up little heavy metal until a critical threshold phytoavailability is exceeded, whereupon there is a steep increase in the uptake of a heavy metal as the heavy-metal phytoavailability in the soil solution increases.

Plants have evolved various mechanisms to restrict the entry of heavy metals to the root symplasm and their movement to the xylem (van Hoof *et al.*, 2001b; Hall, 2002; Polle and Schützendübel, 2003; Shanker *et al.*, 2005; Sharma and Dubey, 2006; Broadley *et al.*, 2007; Verbruggen *et al.*, 2009a; Conn and Gilliham, 2010; Fageria *et al.*, 2011; Lux *et al.*, 2011; Rascio and Navari-Izzo, 2011). These include:

1. Avoiding patches of soil containing high concentrations of heavy metals.
2. Reducing the mobilization of heavy metals from the soil by root exudates.
3. Producing mucilage with a high cation exchange capacity or precipitating heavy metals as insoluble compounds in the rhizosphere or root apoplast.
4. Fostering beneficial associations with mycorrhizal fungi or soil microbes that sequester or immobilize heavy metals.
5. Developing physical barriers to restrict the apoplastic movement of heavy metals from the

rhizosphere to the cortex and from the cortex to the xylem.

6. Reducing the abundance or activity of transport proteins catalysing the uptake of heavy metals.

7. Pumping heavy metals entering root cells back into the apoplast or rhizosphere.

8. Restricting the symplastic delivery of heavy metals to the stele through the production of metal chelates in the cytoplasm of root cells and the sequestration of heavy metals and their chelates in the vacuoles of root cells.

9. Reducing the abundance or activity of transport proteins involved in xylem loading.

These strategies are expressed to different extents by different plant species and there is considerable intraspecific variation in these traits. This enables the selection of appropriate crops, and the breeding of genotypes of preferred crops, with greater tolerance of soils with high heavy-metal phytoavailability.

Although the preferred strategy of most plant species is to avoid the accumulation of toxic heavy metals, this is compromised by their absolute requirement for essential transition-metal elements. Thus, exclusion mechanisms must be complemented by tolerance mechanisms when plants grow in soils with high heavy-metal phytoavailability. Plants have evolved a variety of methods for tolerating high tissue concentrations of heavy metals (Kabata-Pendias, 2000; van Hoof *et al.*, 2001a; Verbruggen *et al.*, 2009a,b; Memon and Schroder, 2009; Puig and Peñarrubia, 2009; Zhao and McGrath, 2009; Fageria *et al.*, 2011; Rascio and Navari-Izzo, 2011; Hodson and Donner, 2013). These include:

1. Selective uptake into plant cells.

2. Immobilizing heavy metals in the apoplast by precipitation or fixation by ligands.

3. Accumulating heavy metals in specific cell types.

4. Producing metal chelates in the cytoplasm and sequestering heavy metals and their chelates in vacuoles.

5. Enhancing heavy-metal excretion, for example by leaf shedding, guttation or volatilization.

6. Altering enzyme complement and biochemical pathways to maintain metabolism.

The mechanisms of exclusion and tolerance of heavy metals are tightly controlled. The transcription of genes encoding: (i) enzymes involved in the biosynthesis and efflux of proteins and organic compounds that increase the phytoavailability of heavy metals in the soil; (ii) transport proteins in the plasma membrane of root cells responsible for the uptake and delivery of heavy metals to the xylem; and (iii) when plants have sufficient of these elements there is downregulation of the transcription of genes encoding enzymes involved in the synthesis of chelates and transport proteins facilitating the movement of heavy metals in the phloem (Broadley *et al.*, 2007; Pilon *et al.*, 2009; White and Broadley, 2009). By contrast, exposure to excess heavy metals upregulates the transcription of genes encoding proteins involved in the intracellular chelation and compartmentalization of heavy metals. For example, the synthesis of metallotheioneins and phytochelatins is induced by elevated tissue concentrations of Cu or Cd (Cobbett and Goldsbrough, 2002; Clemens, 2006; Persson *et al.*, 2006; Ernst *et al.*, 2008; Seregin and Kozhevnikova, 2008), and abundance of ferritin increases in response to excess Fe, which sequesters Fe and protects plant tissues from oxidative damage (Ravet *et al.*, 2008).

There is genetic variation both between and within plant species in the ability to accumulate heavy metals and also in their ability to tolerate mineral toxicities arising in acidic, alkaline, saline, sodic and calcareous soils (Fageria, 2009; White and Broadley, 2011; Tang *et al.*, 2012; White *et al.*, 2012). Marked phylogenetic effects on the accumulation of transition-metal elements by angiosperms have been observed (Broadley *et al.*, 2001, 2007; Watanabe *et al.*, 2007). Meta-analyses using data from a limited number of species suggested that a large proportion of the variation in shoot Ni (52.4%, 79 species), Zn (48.6%, 70 species), Cu (46.9%, 64 species), Cd (36.0%, 108 species), Cr (36.9%, 69 species) and Pb (31.9%, 81 species) concentrations among angiosperms occurs at the level of the family or above (Broadley *et al.*, 2001). A more recent meta-analysis using data from 365 angiosperm species indicated that 22.1% of the variation in shoot Zn concentrations among angiosperms occurred at the level of the family or above (Broadley *et al.*, 2007). A large survey of plants in their natural habitats also suggested that a substantial proportion of the variation in shoot Cu (36.0%), Zn (34.8%), Mn (29.1%) and Ni (27.3%), Co (20.7%), Fe (15.7%) and Cd (8.3%) among angiosperm species occurred at the level of family or above (Watanabe *et al.*, 2007). Angiosperm species accumulating high shoot concentrations of Ni occurred in the Brassicales, of Zn in the Caryophyllales and Brassicales, and of Cu in the Malvales and Malphigiales (Broadley *et al.*, 2001, 2007). There is also considerable variation between genotypes of a particular species to accumulate heavy metals in their above-ground parts (e.g. Wang *et al.*, 2002; Fageria, 2009; White and Broadley, 2009; Qiu *et al.*, 2010; Williams and Pittman, 2010; Lux *et al.*, 2011).

Plants that tolerate large concentrations of heavy metals in the soil can be used for the revegetation or remediation of contaminated land (Whiting *et al.*, 2005; Schwitzguébel *et al.*, 2009; Tang *et al.*, 2012). Species used commercially for the revegetation of contaminated land include ecotypes of the temperate grasses *Agrostis capillaries* and *Festuca rubra* that can be grown on mine tailings containing Pb, Zn and Cu (Smith and Bradshaw, 1979; Whiting *et al.*, 2005). Plants that have the additional ability to accumulate large concentrations of heavy metals in their tissues can be used commercially for phytomining or phytoremediation of contaminated soils (Whiting *et al.*, 2005; Schwitzguébel *et al.*, 2009; Sheoran *et al.*, 2009; Tang *et al.*, 2012).

12.5.2 Metallophytes and hyperaccumulator plants

Metallophytes are defined as species growing on soils contaminated with heavy metals (Whiting *et al.*, 2005; Roosens *et al.*, 2008; Pollard *et al.*, 2014). They are classified as being absolute (eumetallophytes) or facultative (pseudometallophytes), depending upon whether they occur only on contaminated soils or occur on both contaminated and non-contaminated soils. Metallophyte microbes commonly grow in association with the metallophyte flora (Mengoni *et al.*, 2010; Hörger *et al.*, 2013). Most metallophyte angiosperms show strong exclusion of heavy metals, but some also exhibit metal hypertolerance, which enables the hyperaccumulation of metals in their tissues.

Hyperaccumulator plants can accumulate heavy metals in their tissues to concentrations several orders of magnitude greater than that in the surrounding vegetation, without any detrimental effect on their growth or fecundity (Baker *et al.*, 2000; Reeves and Baker, 2000; Broadley *et al.*, 2001, 2007; Whiting *et al.*, 2005; Reeves, 2006; Verbruggen *et al.*, 2009a; Krämer, 2010; Lux *et al.*, 2011; Van der Ent *et al.*, 2013; Pollard *et al.*, 2014). Hyperaccumulation is defined as a shoot concentration in excess of 3 mg Zn/g DM, 10 mg Mn/g DM, 0.3 mg Cu/g DM, 1 mg Ni/g DM, 0.3 mg Co/g DM, 1 mg Pb/g DM, 0.1 mg Cd/g DM or 0.3 mg Cr/g DM in a plant growing in its natural environment (Table 12.1). These plants are of interest not only as botanical curiosities, but also for commercialization for the extraction of heavy metals from the soil for phytomining or phytoremediation of contaminated land (Whiting *et al.*, 2005; Schwitzguébel *et al.*, 2009; Sheoran *et al.*, 2009; Rascio and Navari-Izzo, 2011; Tang *et al.*, 2012; Hodson and Donner, 2013). Heavy-metal hyperaccumulation has been interpreted as an adaptive trait that: (i) serves as a deterrent to a variety of herbivores and pathogens depending on their feeding or infection strategies (Rascio and Navari-Izzo, 2011; Boyd, 2012; Fones and Preston, 2012, 2013; Hörger *et al.*, 2013; Van der Ent *et al.*, 2013, 2015; Cappa and Pilon-Smits, 2014; Pollard *et al.*, 2014); and (ii) restricts the growth of competing plant species through allelopathy (Van der Ent *et al.*, 2013, 2015; Cappa and Pilon-Smits, 2014; Pollard *et al.*, 2014).

The trait of hyperaccumulation is relatively rare among angiosperms (Table 12.3), although the number of plant species reported to hyperaccumulate heavy metals is likely to increase in the future as more specimens from metalliferous environments are analysed. It is evident that Ni hyperaccumulation (400–450 species in about 42 families; Reeves and Baker, 2000; Borhidi, 2001; Broadley *et al.*, 2001; Reeves, 2006; Verbruggen *et al.*, 2009a; Cecchi *et al.*, 2010; Krämer, 2010; Jaffré *et al.*, 2013; Van der Ent *et al.*, 2013, 2015; Cappa and Pilon-Smits, 2014) is far more common than Cu hyperaccumulation (30–35 species in about 15 families; Reeves and Baker, 2000; Broadley *et al.*, 2001; Reeves, 2006; Verbruggen *et al.*, 2009a; Krämer, 2010; Van der Ent and Reeves, 2015; Van der Ent *et al.*, 2013), Co hyperaccumulation (25–30

Table 12.3. Estimates of the number of angiosperm species hyperaccumulating heavy metals and the number of families to which they belong. Key references listing species that hyperaccumulate heavy metals are: Hossner *et al.* (1998), Broadley *et al.* (2001), Reeves (2006), Fernando *et al.* (2009, 2013), Krämer (2010), Lux *et al.* (2011), Jaffré *et al.* (2013), Van der Ent *et al.* (2013, 2015).

Heavy metal	Number of species	Number of families
Ni	400–450	42
Cu	31–35	15
Co	30	11
Mn	12–26	6–9
Zn	12–20	6
Pb	14	7
Cr	11	7
Cd	2–7	< 5

species in about 11 families; Reeves and Baker, 2000; Reeves, 2006; Verbruggen *et al.*, 2009a; Krämer, 2010; Van der Ent *et al.*, 2013), Mn hyperaccumulation (12–26 species in 6–9 families; Reeves, 2006; Fernando *et al.*, 2009, 2013; Van der Ent *et al.*, 2013) or Zn hyperaccumulation (12–20 species in 4–6 families; Broadley *et al.*, 2007; Verbruggen *et al.*, 2009a; Krämer, 2010; Van der Ent *et al.*, 2013), which are more common than Pb hyperaccumulation (14 species in seven families; Broadley *et al.*, 2001; Reeves, 2006; Verbruggen *et al.*, 2009a; Krämer, 2010; Van der Ent *et al.*, 2013), Cr hyperaccumulation (11 species in seven families; Hossner *et al.*, 1998) or Cd hyperaccumulation (2–7 species in < 5 families; Verbruggen *et al.*, 2009a; Krämer, 2010; Lux *et al.*, 2011; Van der Ent *et al.*, 2013) among angiosperms. Furthermore, it has been suggested that the number of hyperaccumulators of Cu, Co, Pb and Cr have been overestimated due to contamination of samples with soil or dust (Reeves, 2006; Faucon *et al.*, 2007; Verbruggen *et al.*, 2009a; Van der Ent *et al.*, 2013).

Hyperaccumulation of heavy metals is more prevalent in particular angiosperm families (Fig. 12.4). For example, of the 361 species of Ni hyperaccumulators surveyed by Reeves (2006), most occur in the Brassicaceae (89 taxa), Euphorbiaceae (87), Asteraceae (44), Flacourtiaceae (19), Buxaceae (17) and Rubiaceae

(15). The remaining 90 angiosperm species belong to more than 30 different families, distributed widely within the angiosperm phylogeny, suggesting that this trait has evolved independently many times (Reeves, 2006). The trait of heavy-metal hyperaccumulation is especially prevalent in the Brassicaeae, and species that hyperaccumulate Ni, Zn and Cd are present in this family (Fig. 12.4; Broadley *et al.*, 2001; Roosens *et al.*, 2008; Verbruggen *et al.*, 2009a; Cecchi *et al.*, 2010; Krämer, 2010; Koch and German, 2013; Cappa and Pilon-Smits, 2014). This has enabled comparative studies on the physiology and evolution of heavy-metal hyperaccumulation. Nickel hyperaccumulation appears to have evolved several times within the Brassicaceae, most likely through independent evolutionary events in populations colonizing ultramafic soils (Cecchi *et al.*, 2010; Krämer, 2010; Koch and German, 2013). The trait of Zn hyperaccumulation appears to have evolved at least twice: at the base of the *Noccaea/Raparia* clade and in the predecessors of *Arabidopsis halleri* (Macnair, 2003; Broadley *et al.*, 2007; Roosens *et al.*, 2008; Krämer, 2010). Since Ni hyperaccumulation also evolved at the base of the *Noccaea/Raparia* clade it has been speculated that Zn hyperaccumulation might be a modification of the Ni hyperaccumulation trait, but species that hyperaccumulate Zn do not always hyperaccumulate Ni (Krämer, 2010; Koch and German, 2013). Intriguingly, Cd hyperaccumulation also appears to have evolved several times within the Brassicaeae, and occurs in species that also exhibit Zn hyperaccumulation, such as *Noccaea* spp. and *A. halleri*, though not in all populations of these plant species (Verbruggen *et al.*, 2009a; Krämer, 2010). A few genetic loci appear to account for Zn and Cd hyperaccumulation in these species (Assunção *et al.*, 2001; Deniau *et al.*, 2006; Broadley *et al.*, 2007; Verbruggen *et al.*, 2009a; Willems *et al.*, 2010; Craciun *et al.*, 2012) and it has been proposed that the trait of heavy-metal hyperaccumulation can be gained or lost rapidly through altered expression of a few genes that are not species specific (Verbruggen *et al.*, 2009a).

The trait of hyperaccumulation is associated with greater uptake, efficient translocation from root to shoot and greater tissue tolerance of heavy metals (Broadley *et al.*, 2007; Palmer and Guerinot, 2009; Verbruggen *et al.*, 2009a;

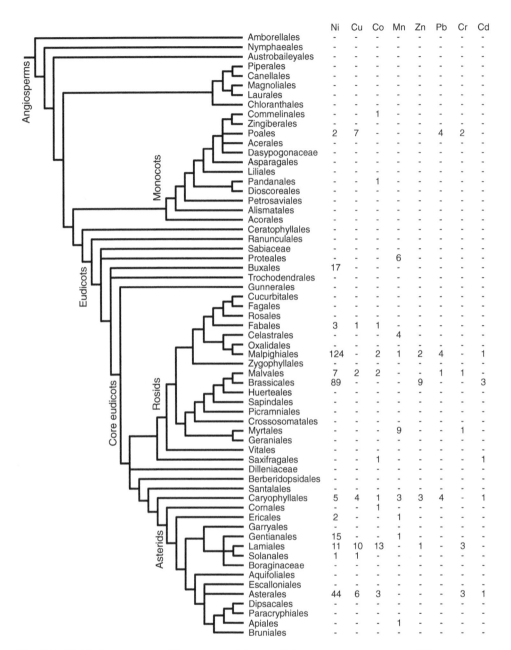

Fig. 12.4. Distribution of proposed hyperaccumulator taxa between angiosperm orders. Data comprise: 320 species of Ni hyperaccumulators abstracted from Broadley *et al.* (2001) and Reeves (2006), 31 species of Cu hyperaccumulators considered by Broadley *et al.* (2001), 26 Co hyperaccumulators listed by Reeves (2006) and Faucon *et al.* (2007), 26 Mn hyperaccumulators listed by Fernando *et al.* (2009), 15 Zn hyperaccumulators listed by Broadley *et al.* (2007), 13 Pb hyperaccumulators considered by Broadley *et al.* (2001), 10 Cr hyperaccumulators listed by Hossner *et al.* (1998) and 7 Cd hyperaccumulators listed by Lux *et al.* (2011). The phylogenetic tree of the angiosperms is reproduced from the Angiosperm Phylogeny Group III (2009).

Krämer, 2010; Lux *et al.*, 2011; Rascio and Navari-Izzo, 2011; Cappa and Pilon-Smits, 2014). Heavy-metal uptake by hyperaccumulator plants is enhanced by: (i) the proliferation of roots in patches of soil enriched in heavy metals (Schwartz *et al.*, 1999; Whiting *et al.*, 2000; Haines, 2002; Dechamps *et al.*, 2008; F. Liu *et al.*, 2010); (ii) constitutive expression of genes encoding plasma membrane transporters catalysing the uptake of heavy metals (Broadley *et al.*, 2007; Roosens *et al.*, 2008; Verbruggen *et al.*, 2009a; Krämer, 2010; Rascio and Navari-Izzo, 2011; Hanikenne and Nouet, 2011); and (iii) in some instances, increased mobilization of heavy metals from the soil by exudates from plant roots (Li *et al.*, 2011). Efficient translocation of heavy metals from the root to the shoot is facilitated by: (i) reduced sequestration in root vacuoles and constitutive expression of transport proteins, such as orthologues of AtHMA4 and YSL transporters, which catalyse loading of heavy metals into the xylem; (ii) multiplication of genes encoding transporters catalysing the transport of heavy metals into the xylem (Hanikenne *et al.*, 2008; Ó Lochlainn *et al.*, 2011; Craciun *et al.*, 2012); and (iii) constitutive expression of enzymes synthesizing compounds enabling xylem transport, such as organic acids in plants that hyperaccumulate Zn, nicotianamine in plants that hyperaccumulate Zn and Ni and histidine in plants that hyperaccumulate Ni or Co, and transporters loading these compounds into the xylem (Krämer *et al.*, 1996; Sharma and Dietz, 2006; Broadley *et al.*, 2007; Roosens *et al.*, 2008; Verbruggen *et al.*, 2009a; Rascio and Navari-Izzo, 2011; Hanikenne and Nouet, 2011; Lu *et al.*, 2013). Greater tolerance of heavy metals in shoot tissues of hyperaccumulator plants is effected by constitutive production of ligands that complex heavy metals, such as organic acids, glutathione, histidine and nicotinamine (Wycisk *et al.*, 2004; Broadley *et al.*, 2007; Roosens *et al.*, 2008; Verbruggen *et al.*, 2009a; Jaffré *et al.*, 2013), constitutive expression of transporters effecting vacuolar sequestration of heavy metals (Broadley *et al.*, 2007; Roosens *et al.*, 2008; Verbruggen *et al.*, 2009a; Hanikenne and Nouet, 2011; Rascio and Navari-Izzo, 2011) and accumulation of heavy metals in specific cell types, such as epidermal cells or trichomes (Section 12.5.3; Broadley *et al.*, 2007; Verbruggen *et al.*, 2009a). Protection

from ROS is effected by constitutive synthesis of antioxidant compounds and enzymes (Section 12.4; Sharma and Dietz, 2006, 2009; Cuypers *et al.*, 2011; Rascio and Navari-Izzo, 2011).

12.5.3 Distribution of heavy metals within shoots of hyperaccumulator plants

In a plant that hyperaccumulates a particular heavy metal, its concentration is generally greater in the above-ground biomass than in roots. Whether the heavy metal is accumulated in older or younger leaves is related to the mobility of the heavy metal in the phloem. The concentrations of heavy metals can be greater in stems and petioles than in leaves (e.g. Tian *et al.*, 2010; Zhang *et al.*, 2011; Villafort Carvahlo *et al.*, 2013), although heavy metals often accumulate at leaf margins, presumably as a consequence of transpirational water flows from the xylem (Mizuno *et al.*, 2003; Cosio *et al.*, 2005; Tappero *et al.*, 2007; Kachenko *et al.*, 2008; Huguet *et al.*, 2012). Guttation can also be observed in some hyperaccumulator plants (Mizuno *et al.*, 2003).

Significant progress has been made in resolving the distribution of heavy metals at the cellular level in hyperaccumulator plants, mainly due to greater access to techniques enabling multi-element analysis with appropriate sensitivity (<10 mg kg^{-1}) and spatial resolution (<1 μm), such as X-ray fluorescence. The application of X-ray fluorescence techniques to hyperaccumulator plants has revealed that the distribution of heavy metals within the leaf can be unique for the heavy metal in question and plant species studied (Fig. 12.5; Mesjasz-Przybyłowicz and Przybyłowicz, 2011; Sarret *et al.*, 2013; Vogel-Mikuš *et al.*, 2014).

The largest concentrations of Ni are often found in epidermal tissues and in the vasculature of Ni hyperaccumulator plants (Fig. 12.5; Bathia *et al.*, 2004; Broadhurst *et al.*, 2004; Kachenko *et al.*, 2008; Wang *et al.*, 2013; Mesjasz-Przybyłowicz *et al.*, 2016), although there are reports of Ni accumulation in leaf mesophyll tissues of *Berkheya coddii* (Mesjasz-Przybyłowicz and Przybyłowicz, 2011). Manganese has been shown to be accumulated in the upper layer of palisade mesophyll cells or in sub-epidermal

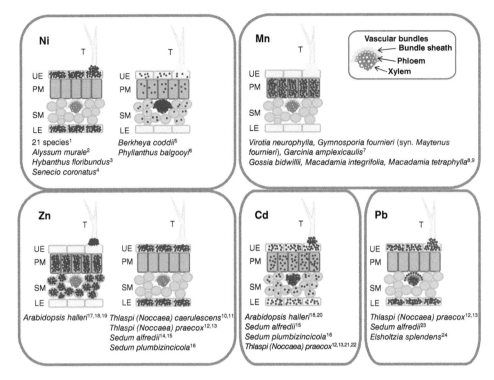

Fig. 12.5. Distribution of heavy metals within the shoots of hyperaccumulator plants. A schematic cross-section of a leaf comprising upper epidermis (UE), palisade mesophyll (PM), spongy mesophyll (SM), lower epidermis (LE), trichome (T) and vascular bundles (see inlay in Mn panel). Blue circles represent heavy metal content. [1]Bathia *et al.*, 2004; [2]Broadhurst *et al.*, 2004; [3]Kachenko *et al.*, 2008; [4]Wang *et al.*, 2013; [5]Mesjasz-Przybyłowicz and Przybyłowicz, 2011; [6]Mesjasz-Przybyłowicz *et al.*, 2016; [7]Fernando *et al.*, 2012; [8]Fernando *et al.*, 2006a; [9]Fernando *et al.*, 2006b; [10]Küpper *et al.*, 1999; [11]Frey *et al.*, 2000; [12]Vogel-Mikuš *et al.*, 2008a; [13]Vogel-Mikuš *et al.*, 2008b; [14]Tian *et al.*, 2009; [15]Tian *et al.*, 2011; [16]Hu *et al.*, 2015; [17]Zhao *et al.*, 2000; [18]Küpper *et al.*, 2000; [19]Sarret *et al.*, 2009; [20]Isaure *et al.*, 2015; [21]Pongrac *et al.*, 2010; [22]Koren *et al.*, 2013; [23]Tian *et al.*, 2010; [24]Zhang *et al.*, 2011.

large vacuolated cells (Fernando *et al.*, 2006a,b, 2012, 2013) in leaves of several Mn hyperaccumulator plants. Two patterns for the distribution of Zn within leaves of Zn hyperaccumulator plants have been described. In some Zn hyperaccumulator plants, such as *Arabidopsis halleri*, Zn is accumulated in the mesophyll tissues and at the base of trichomes (Küpper *et al.*, 2000; Zhao *et al.*, 2000; Sarret *et al.*, 2009), whereas in *Thlaspi (Noccaea) caerulescens*, *Thlaspi (Noccaea) praecox*, *Sedum alfredii* and *Sedum plumbizincicola* Zn is accumulated in vacuolated epidermal cells (Küpper *et al.*, 1999; Frey *et al.*, 2000; Vogel-Mikuš *et al.*, 2008a,b; Tian *et al.*, 2009, 2011; Hu *et al.*, 2015).

A homogeneous Cd distribution has been reported within leaves of *A. halleri* (Isaure *et al.*, 2015), although the largest Cd concentrations occur at the base of trichomes (Küpper *et al.*, 2000; Isaure *et al.*, 2015). Similarly, although Cd is present throughout leaves of *T. (N.) praecox*, the largest Cd concentrations are present in epidermal cells (Vogel-Mikuš *et al.*, 2008a,b; Pongrac *et al.*, 2010; Koren *et al.*, 2013). By contrast, Cd is mainly observed in the mesophyll of *S. alfredii* (Tian *et al.*, 2011) and *S. plumbizincicola* (Hu *et al.*, 2015) leaves. A novel strategy for Cd tolerance (and accumulation) has been reported for *Gomphrena claussenii*, in which Cd colocalizes with Ca-oxalate crystals (Villafort Carvalho *et al.*, 2015). Lead is present in the epidermis and vascular bundles of *T. (N.) praecox* collected from the field (Vogel-Mikuš *et al.*, 2008a,b) and in hydroponically grown *S. alfredii* (Tian *et al.*,

2010) and *Elsholtzia splendens* (Zhang *et al.*, 2011). In *T. (N.) praecox* (Vogel-Mikuš *et al.*, 2008a,b), *A. halleri* (S. Clemens, S. Höreth, P. Pongrac *et al.*, unpublished observations) and *Viola fargesii* (syn. *Viola principis*; Lei *et al.*, 2008) the largest Pb concentrations were observed in the bundle sheath cells in the middle vein. In *V. fargesii* Pb was also detected in the mesophyll cells of the leaf lamina (Lei *et al.*, 2008).

The distributions of Cu and Co in leaves of hyperaccumulator plants have not been studied in detail. However, Cu is apparently located in the lower epidermis in *Haumaniastrum robertii* (Siegele *et al.*, 2008) and Co and Ca appear to be colocalized at the edge of the leaf (Morrison *et al.*, 1981). The distribution of Cr in leaves of hyperaccumulator plants has not been studied.

References

Álvarez-Fernández, A., Díaz-Benito, P., Abadia, A., López-Millán, A.-F. and Abadia, J. (2014) Metal species involved in long distance metal transport in plants. *Frontiers in Plant Science* 5, 105. doi: 10.3389/fpls.2014.00105.

Ando, Y., Nagata, S., Yanagisawa, S. and Yoneyama, T. (2013) Copper in xylem and phloem saps from rice (*Oryza sativa*): the effect of moderate copper concentrations in the growth medium on the accumulation of five essential metals and a speciation analysis of copper-containing compounds. *Functional Plant Biology* 40, 89–100.

Andrés-Colás, N., Sancenón, V., Rodríguez-Navarro, S., Mayo, S., Thiele, D.J., Ecker, J.R. *et al.* (2006) The Arabidopsis heavy metal P-type ATPase HMA5 interacts with metallochaperones and functions in copper detoxification of roots. *Plant Journal* 45, 225–236.

Andresen, E., Mattusch, J., Wellenreuther, G., Thomas, G., Arroyo Abad, U. *et al.* (2013) Different strategies of cadmium detoxification in the submerged macrophyte *Ceratophyllum demersum* L. *Metallomics* 5, 1377–1386.

Angiosperm Phylogeny Group (2009) An update of the Angiosperm Phylogeny Group classification for the orders and families of flowering plants: APG III. *Botanical Journal of the Linnean Society* 161, 105–121.

Assunção, A.G.L., Da Costa Martins, P., De Folter, S., Vooijs, R., Schat, H. *et al.* (2001) Elevated expression of metal transporter genes in three accessions of the metal hyperaccumulator *Thlaspi caerulescens*. *Plant, Cell and Environment* 24, 217–226.

Bacon, J.R., Jones, K.C., McGrath, S.P. and Johnston, A.E. (1996) Isotopic character of lead deposited from the atmosphere at a grassland site in the United Kingdom since 1860. *Environmental Science and Technology* 30, 2511–2518.

Baker, A.J.M. (1981) Accumulators and excluders – strategies in the response of plants to heavy metals. *Journal of Plant Nutrition* 3, 643–654.

Baker, A.J.M., McGrath, S.P., Reeves, R.D. and Smith, J.A.C. (2000) Metal hyperaccumulator plants: a review of the ecology and physiology of a biological resource for phytoremediation of metal polluted soils. In: Terry, N. and Banuelos, G.S. (eds) *Phytoremediation of Contaminated Soil and Water*. CRC Press, Boca Raton, Florida, pp. 85–107.

Barber, S.A. (1995) *Soil Nutrient Bioavailability*, 2nd edn. Wiley, New York.

Bathia, N.P., Walsh, K.B., Orlic, I., Ashwath, N. and Baker, A.J.M. (2004) Studies on spatial distribution of nickel in leaves and stems of the metal hyperaccumulator *Stackhousia tryonii* using nuclear microprobe (micro-PIXE) and EDXS techniques. *Functional Plant Biology* 31, 1061–1074.

Beauclair, L., Yu, A. and Bouche, N. (2010) MicroRNA-directed cleavage and translational repression of the copper chaperone for superoxide dismutase mRNA in Arabidopsis. *Plant Journal* 62, 454–462.

Bittner, F. and Mendel, R.-R. (2010) Cell biology of molybdenum. In: Hell, R. and Mendel, R.-R. (eds) *Plant Cell Monographs 17, Cell Biology of Metals and Nutrients*. Springer, Berlin, pp. 119–143.

Borhidi, A. (2001) Phylogenetic trends in Ni-accumulating plants. *South African Journal of Science* 97, 544–547.

Bould, C., Hewitt, E.J. and Needham, P. (1983) *Diagnosis of Mineral Disorders in Plants. Volume 1: Principles*. HMSO, London.

Boyd, R.S. (2012) Plant defense using toxic inorganic ions: conceptual models of the defensive enhancement and joint effects hypotheses. *Plant Science* 195, 88–95.

Brady, N.C. and Weil, R.R. (2002) *The Nature and Properties of Soil*, 13th edn. Longman, Prentice-Hall, London.

Brar, M.S., Malhi, S.S., Singh, A.P., Arora, C.L. and Gill, K.S. (2000) Sewage water irrigation effects on some potentially toxic trace elements in soil and potato plants in northwestern India. *Canadian Journal of Soil Science* 80, 465–471.

Broadhurst, C.L., Chaney, R.L., Angle, J.S., Erbe, E.F. and Maugel, T.K. (2004) Nickel localization and response to increasing Ni soil levels in leaves of the Ni hyperaccumulator *Alyssum murale*. *Plant and Soil* 265, 225–242.

Broadley, M.R., Willey, N.J., Wilkins, J.C., Baker, A.J.M., Mead, A. *et al.* (2001) Phylogenetic variation in heavy metal accumulation in angiosperms. *New Phytologist* 152, 9–27.

Broadley, M.R., White, P.J., Hammond, J.P., Zelko, I. and Lux, A. (2007) Zinc in plants. *New Phytologist* 173, 677–702.

Broadley, M.R., Ó Lochlainn, S., Hammond, J.P., Bowen, H.C., Cakmak, I. *et al.* (2010) Shoot zinc (Zn) concentration varies widely within *Brassica oleracea* L. and is affected by soil Zn and phosphorus (P) levels. *Journal of Horticultural Science and Biotechnology* 85, 375–380.

Buhtz, A., Pieritz, J., Springer, F. and Kehr, J. (2010) Phloem small RNAs, nutrient stress responses, and systemic mobility. *BMC Plant Biology* 10, 64. doi: 10.1186/1471-2229-10-64.

Burkhead, J.L., Gogolin Reynolds, K.A., Abdel-Ghany, S.E., Cohu, C.M. and Pilon, M. (2009) Copper homeostasis. *New Phytologist* 182, 799–816.

Cailliatte, R., Schikora, A., Briat, J.-F., Mari, S. and Curie, C. (2010) High-affinity manganese uptake by the metal transporter NRAMP1 is essential for *Arabidopsis* growth in low manganese conditions. *Plant Cell* 22, 904–917.

Cappa, J.J. and Pilon-Smits, E.A.H. (2014) Evolutionary aspects of elemental hyperaccumulation. *Planta* 239, 267–275.

Cary, E.E., Allaway, W.H. and Olsen, O.E. (1977) Control of chromium concentrations in food plants. 1. Absorption and translocation of chromium by plants. *Journal of Agriculture and Food Chemistry* 25, 300–304.

Cecchi, L., Gabbrielli, R., Arnetoli, M., Gonnelli, C., Hasko, A. *et al.* (2010) Evolutionary lineages of nickel hyperaccumulation and systematics in European Alysseae (Brassicaceae): evidence from nrDNA sequence data. *Annals of Botany* 106, 751–767.

Chaney, R.L. (1993) Zinc phytotoxicity. In: Robson, A.D. (ed.) *Zinc in Soil and Plants*. Kluwer Academic Publishers, Dordrecht, The Netherlands, pp. 135–150.

Chatterjee, C., Nautiyal, N. and Agarwala, S.C. (1992) Excess sulfur partially alleviates copper deficiency effects in mustard. *Soil Science and Plant Nutrition* 38, 57–64.

Clemens, S. (2006) Toxic metal accumulation, responses to exposure and mechanisms of tolerance in plants. *Biochimie* 88, 1707–1719.

Clemens, S. (2010) Zn – a versatile player in plant cell biology. In: Hell, R. and Mendel, R.-R. (eds) *Plant Cell Monographs 17, Cell Biology of Metals and Nutrients*. Springer, Berlin, pp. 281–298.

Clemens, S., Deinlein, U., Ahmadi, H., Höreth, S. and Uraguchi, S. (2013) Nicotianamine is a major player in plant Zn homeostasis. *Biometals* 26, 623–632.

Cobbett, C. and Goldsbrough, P. (2002) Phytochelatins and metallothioneins: roles in heavy metal detoxification and homeostasis. *Annual Review of Plant Biology* 53, 159–182.

Cohu, C.M. and Pilon, M. (2010) Cell biology of copper. In: Hell, R. and Mendel, R.-R. (eds) *Plant Cell Monographs 17, Cell Biology of Metals and Nutrients*. Springer, Berlin, pp. 55–74.

Colburn, P. and Thornton, I. (1978) Lead pollution in agricultural soils. *European Journal of Soil Science* 29, 513–526.

Collins, R.N., Bakkaus, E., Carrière, M., Khodja, H., Proux, O. *et al.* (2010) Uptake, localization, and speciation of cobalt in *Triticum aestivum* L. (wheat) and *Lycopersicon esculentum* M. (tomato). *Environment Science & Technology* 44, 2904–2910.

Conn, S. and Gilliham, M. (2010) Comparative physiology of elemental distributions in plants. *Annals of Botany* 105, 1081–1102.

Connorton, J.M., Webster, R.E., Cheng, N. and Pittman, J.K. (2012) Knockout of multiple *Arabidopsis* cation/H⁺ exchangers suggests isoform-specific roles in metal stress response, germination and seed mineral nutrition. *PLoS ONE* 7(10), e47455. doi: 10.1371/journal.pone.0047455.

Cosio, C., DeSantis, L., Frey, B., Diallo, S. and Keller, C. (2005) Distribution of cadmium in leaves of *Thlaspi caerulescens*. *Journal of Experimental Botany* 56, 765–775.

Craciun, A.R., Meyer, C.-L., Chen, J., Roosens, N., De Groodt, R. *et al.* (2012) Variation in *HMA4* copy number and expression among *Noccaea caerulescens* populations presenting different levels of Cd tolerance and accumulation. *Journal of Experimental Botany* 63, 4179–4189.

Curie, C., Cassin, G., Couch, D., Divol, F., Higuchi, K. *et al.* (2009) Metal movement within the plant: contribution of nicotianamine and yellow stripe 1-like transporters. *Annals of Botany* 103, 1–11.

Cuypers, A., Smeets, K., Ruytinx, J., Opdenakker, K., Keunen, E. *et al.* (2011) The cellular redox state as a modulator in cadmium and copper responses in *Arabidopsis thaliana* seedlings. *Journal of Plant Physiology* 168, 309–316.

DalCorso, G., Farinati, S., Maistri, S. and Furini, A. (2008) How plants cope with cadmium: staking all on metabolism and gene expression. *Journal of Integrative Plant Biology* 10, 1268–1280.

Dechamps, C., Noret, N., Mozek, R., Draye, X. and Meerts, P. (2008) Root allocation in metal-rich patch by *Thlaspi caerulescens* from normal and metalliferous soil – new insights into the rhizobox approach. *Plant and Soil* 310, 211–224.

Demidchik, V. (2010) Reactive oxygen species, oxidative stress and plant ion channels. In: Demidchik, V. and Maathuis, F. (eds) *Ion Channels and Plant Stress Responses*. Springer, Berlin, pp. 207–232.

Deniau, A.X., Pieper, B., Ten Bookum, W.M., Lindhout, P., Aarts, M.G.M. *et al.* (2006) QTL analysis of cadmium and zinc accumulation in the heavy metal hyperaccumulator *Thlaspi caerulescens*. *Theoretical and Applied Genetics* 113, 907–920.

Ding, Y.-F. and Zhu, C. (2009) The role of microRNAs in copper and cadmium homeostasis. *Biochemical and Biophysical Research Communications* 386, 6–10.

Dong, Y., Ogawa, T., Lin, D., Koh, H.-J., Kamiunten, H. *et al.* (2006) Molecular mapping of quantitative trait loci for zinc toxicity in rice seedling (*Oryza sativa* L.). *Field Crops Research* 95, 420–425.

Duffus, J.H. (2002) "Heavy Metals" – a meaningless term? *Pure and Applied Chemistry* 74, 793–807.

Durrett, T.P., Gassmann, W. and Rogers, E.E. (2007) The FRD3-nediated efflux of citrate into the root vasculature is necessary for efficient iron translocation. *Plant Physiology* 144, 197–205.

Emery, L., Whelan, S., Hirschi, K.D. and Pittman, J.K. (2012) Protein phylogenetic analysis of Ca^{2+}/cation antiporters and insights into their evolution in plants. *Frontiers in Plant Science* 3(1). doi: 10.3389/fpls.2012.00001.

Ernst, W.H.O., Krauss, G.-J., Verkleij, J.A.C. and Wesenberg, D. (2008) Interaction of heavy metals with the sulphur metabolism in angiosperms from an ecological point of view. *Plant, Cell and Environment* 31, 123–143.

Fageria, N.K. (2009) *The Use of Nutrients in Crop Plants*. CRC Press, Boca Raton, Florida.

Fageria, N.K., Baligar, V.C. and Jones, C.A. (2011) *Growth and Mineral Nutrition of Field Crops*, 3rd edn. CRC Press, Boca Raton, Florida.

FAO (Food and Agriculture Organization of the United Nations) (2010) *Land Resources*. Available at: http://www.fao.org/nr/land/databasesinformation-systems/en (accessed 12 February 2011).

Faucon, M.-P., Ngoy Shutcha, M. and Meerts, P. (2007) Revisiting copper and cobalt concentrations in supposed hyperaccumulators from SC Africa: influence of washing and metal concentrations in soil. *Plant and Soil* 301, 29–36.

Fernando, D.R., Bakkaus, E.J., Perrier, N., Baker, A.J.M., Woodrow, I.E. *et al.* (2006a) Manganese accumulation in the leaf mesophyll of four tree species: a PIXE/EDAX localization study. *New Phytologist* 171, 751–758.

Fernando, D.R., Batianoff, G.N., Baker, A.J. and Woodrow, I.E. (2006b) *In vivo* localisation of manganese in the hyperaccumulator *Gossia bidwillii* (Benth.) N. Snow & Guymer (Myrtaceae) by cryo-SEM/EDAX. *Plant, Cell and Environment* 29, 1012–1020.

Fernando, D.R., Guymer, G., Reeves, R.D., Woodrow, I.E., Baker, A.J.M. *et al.* (2009) Foliar Mn accumulation in eastern Australian herbarium specimens: prospecting for 'new' Mn hyperaccumulators and its potential application in taxonomy. *Annals of Botany* 103, 931–939.

Fernando, D.R., Woodrow, I.E., Baker, A.J.M. and Marshall, A.T. (2012) Plant homeostasis of foliar manganese sinks: specific variation in hyperaccumulators. *Planta* 236, 1459–1470.

Fernando, D.R., Marshall, A., Baker, A.J.M. and Mizuno, T. (2013) Microbeam methodologies as powerful tools in manganese hyperaccumulation research: present status and future directions. *Frontiers in Plant Science* 4, 319. doi: 10.3389/fpls.2013.00319.

Fischer, S., Kühnlenz, T., Thieme, M., Schmidt, H. and Clemens, S. (2014) Analysis of plant Pb tolerance at realistic submicromolar concentrations demonstrates the role of phytochelatin synthesis for Pb detoxification. *Environment Science & Technology* 48, 7552–7559.

Fones, H. and Preston, G.M. (2012) The impact of transition metals on bacterial plant disease. *FEMS Microbiology Reviews* 37, 495–519.

Fones, H.N. and Preston, G.M. (2013) Trade-offs between metal hyperaccumulation and induced disease resistance in metal hyperaccumulator plants. *Plant Pathology* 62, 63–71.

Foy, C.D., Chaney, R.L. and White, M.C. (1978) The physiology of metal toxicity in plants. *Annual Review of Plant Physiology* 29, 511–566.

Frey, B., Keller, C., Zierold, K. and Schulin, R. (2000) Distribution of Zn in functionally different leaf epidermal cells of the hyperaccumulator *Thlaspi caerulescens*. *Plant, Cell and Environment* 23, 675–687.

Gollhofer, J., Timofeev, R., Lan, P., Schmidt, W. and Buckhout, T.J. (2014) Vacuolar-iron-transporter1-like proteins mediate iron homeostasis in Arabidopsis. *PLoS ONE* 9(10), e110468. doi: 10.1371/journal.pone.0110468.

Graham, R., Senadhira, D., Beebe, S., Iglesias, C. and Monasterio, I. (1999) Breeding for micronutrient density in edible portions of staple food crops: conventional approaches. *Field Crops Research* 60, 57–80.

Grotz, N. and Guerinot, M.L. (2006) Molecular aspects of Cu, Fe and Zn homeostasis in plants. *Biochimica et Biophysica Acta* 1763, 595–608.

Guerinot, M.L. (2010) Iron. In: Hell, R. and Mendel, R.-R. (eds) *Plant Cell Monographs 17, Cell Biology of Metals and Nutrients*. Springer, Berlin, pp. 75–94.

Guo, W.-J., Meetam, M. and Goldsbrough, P.B. (2008) Examining the specific contributions of individual Arabidopsis metallothioneins to copper distribution and metal tolerance. *Plant Physiology* 146, 1697–1706.

Gupta, O.P., Sharma, P., Gupta, R.K. and Sharma, I. (2014) MicroRNA mediated regulation of metal toxicity in plants: present status and future perspectives. *Plant Molecular Biology* 84, 1–18.

Haines, B.J. (2002) Zincophilic root foraging in *Thlaspi caerulescens*. *New Phytologist* 155, 363–372.

Hall, J.L. (2002) Cellular mechanisms for heavy metal detoxification and tolerance. *Journal of Experimental Botany* 53, 1–11.

Hammond, J.P. and White, P.J. (2008) Sucrose transport in the phloem: integrating root responses to phosphorus starvation. *Journal of Experimental Botany* 59, 93–109.

Hanikenne, M. and Nouet, C. (2011) Metal hyperaccumulation and hypertolerance: a model for plant evolutionary genomics. *Current Opinion in Plant Biology* 14, 252–259.

Hanikenne, M., Talke, I.N., Haydon, M.J., Lanz, C., Nolte, A. *et al.* (2008) Evolution of metal hyperaccumulation required *cis*-regulatory changes and triplication of *HMA4*. *Nature* 453, 391–395.

Hänsch, R. and Mendel, R.R. (2009) Physiological functions of mineral micronutrients (Cu, Zn, Mn, Fe, Ni, Mo, B, Cl). *Current Opinion in Plant Biology* 12, 259–266.

Harris, W.R., Sammons, R.D. and Grabiak, R.C. (2012) A speciation model of essential trace metal ions in phloem. *Journal of Inorganic Biochemistry* 116, 140–150.

Haydon, M.J. and Cobbett, C.S. (2007) A novel major facilitator superfamily protein at the tonoplast influences zinc tolerance and accumulation in Arabidopsis. *Plant Physiology* 143, 1705–1719.

He, Z.L., Yang, X.E. and Stoffella, P.J. (2005) Trace elements in agroecosystems and impacts on the environment. *Journal of Trace Elements in Medicine and Biology* 19, 125–140.

Hodson, M. and Donner, E. (2013) Managing adverse soil chemical environments. In: Gregory, P.J. and Nortcliffe, S. (eds) *Soil Conditions and Plant Growth*, 12th edn. Blackwell, Oxford, UK, pp. 195–237.

Holland, S.L., Ghosh, E. and Avery, S.V. (2010) Chromate-induced sulfur starvation and mRNA mistranslation in yeast are linked in a common mechanism of Cr toxicity. *Toxicology in Vitro* 24, 1764–1767.

Hörger, A.C., Fones, H.N. and Preston, G.M. (2013) The current status of the elemental defense hypothesis in relation to pathogens. *Frontiers in Plant Science* 4, 395. doi: 10.3389/fpls.2013.00395.

Horst, W.J. (1988) The physiology of manganese toxicity. In: Graham, R.D., Hannam, R.J. and Uren, N.C. (eds) *Manganese in Soils and Plants*. Kluwer Academic Publishers, Dordrecht, The Netherlands, pp. 75–88.

Hossner, L.R., Loeppert, R.H., Newton, R.J. and Szaniszlo, P.J. (1998) *Literature Review: Phytoaccumulation of Chromium, Uranium, and Plutonium in Plant Systems*. Amarillo National Resource Centre for Plutonium, Springfield, Virginia.

Hu, P., Wang, Y., Przybyłowicz, W.J., Li, Z., Barnabas, A., Wu, L., Luo, Y. and Mesjasz-Przybyłowicz, J. (2015) Elemental distribution by cryo-micro-PIXE in the zinc and cadmium hyperaccumulator *Sedum plumbizincicola* grown naturally. *Plant and Soil* 388, 267–282.

Huguet, S., Bert, V., Laboudigue, A., Barthes, V., Isaure, M-P. *et al.* (2012) Cd speciation and localization in the hyperaccumulator *Arabidopsis halleri*. *Environmental and Experimental Botany* 82, 54–65.

Isaure, M-P., Huguet, S., Meyer, C.-L., Castillo-Michel, H., Testemale, D. *et al.* (2015) Evidence of various mechanisms of Cd sequestration in the hyperaccumulator *Arabidopsis halleri*, the non-accumulator *Arabidopsis lyrata*, and their progenies by combined synchrotron-based techniques. *Journal of Experimental Botany* 66, 3201–3214.

Ishimaru, Y., Suzuki, M., Kobayashi, T., Takahashi, M., Nakanishi, H. *et al.* (2005) OsZIP4, a novel zinc-regulated zinc transporter in rice. *Journal of Experimental Botany* 56, 3207–3214.

Jaffré, T., Pillon, Y., Thomine, S. and Merlot, S. (2013) The metal hyperaccumulators from New Caledonia can broaden our understanding of nickel accumulation in plants. *Frontiers in Plant Science* 4, 279. doi: 10.3389/fpls.2013.00279.

Jonak, C., Nakagami, H. and Hirt, H. (2004) Heavy metal stress. Activation of distinct mitogen-activated protein kinase pathways by copper and cadmium. *Plant Physiology* 136, 3276–3283.

Kabata-Pendias, A. (2000) *Trace Elements in Soils and Plants*. CRC Press, Boca-Raton, Florida.

Kachenko, A., Siegele, R., Bathia, N.P., Singh, B. and Ionescu, M. (2008) Evaluation of specimen preparation techniques for micro-PIXE localisation of elements in hyperaccumulating plants. *Nuclear Instruments and Methods in Physics Research B* 66, 1598–1604.

Kanai, M., Hirai, M., Yoshiba, M., Tadano, T. and Higuchi, K. (2009) Iron deficiency causes zinc excess in *Zea mays*. *Soil Science and Plant Nutrition* 55, 271–276.

Karley, A.J., Leigh, R.A. and Sanders, D. (2000) Where do all the ions go? The cellular basis of differential ion accumulation in leaf cells. *Trends in Plant Science* 5, 465–470.

Kassem, M.A., Meksem, K., Kang, C.H., Njiti, V.N., Kilo, V. *et al.* (2004) Loci underlying resistance to manganese toxicity mapped in a soybean recombinant inbred line population of 'Essex' × 'Forrest'. *Plant and Soil* 260, 197–204.

Khabaz-Saberi, H., Rengel, Z., Wilson, R. and Setter, T.L. (2010) Variation of tolerance to manganese toxicity in Australian hexaploid wheat. *Journal of Plant Nutrition and Soil Science* 173, 103–112.

Kim, D.J., Kim, H., Kim, M. and Lee, J. (2007) Analysis of oxalic acid of various vegetables consumed in Korea. *Food Science and Biotechnology* 16, 650–654.

Kim, S.A., Punshon, T., Lanzirotti, A., Li, L., Alonso, J.M. *et al.* (2006) Localization of iron in *Arabidopsis* seed requires the vacuolar membrane transporter VIT1. *Science* 314, 1295–1298.

Kirkby, E. (2012) Introduction, definition and classification of nutrients. In: Marschner, P. (ed.) *Marschner's Mineral Nutrition of Higher Plants*, 3rd edn. Academic Press, London, pp. 3–5.

Kirkby, E.A. and Johnson, A.E. (2008) Soil and fertilizer phosphorus in relation to crop nutrition. In: White, P.J. and Hammond, J.P. (eds) *The Ecophysiology of Plant-Phosphorus Interactions*. Springer, Dordrecht, The Netherlands, pp. 177–223.

Kobayashi, T. and Nishizawa, N.K. (2012) Iron uptake, translocation, and regulation in higher plants. *Annual Review of Plant Biology* 63, 131–152.

Koch, M.A. and German, D.A. (2013) Taxonomy and systematics are key to biological information: *Arabidopsis, Eutrema (Thellungiella), Noccaea* and *Schrenkiella* (Brassicaceae) as examples. *Frontiers in Plant Science* 4, 267. doi: 10.3389/fpls.2013.00267.

Kochian, L., Hoekenga, O.A. and Piñeros, M.A. (2004) How do crop plants tolerate acid soils? Mechanisms of aluminum tolerance and phosphorous efficiency. *Annual Review of Plant Biology* 55, 459–493.

Kong, W.W. and Yang, Z.M. (2010) Identification of iron-deficiency responsive microRNA genes and *cis*-elements in *Arabidopsis*. *Plant Physiology and Biochemistry* 48, 153–159.

Koren, Š., Arčon, I., Kump, P., Nečemer, M. and Vogel-Mikuš, K. (2013) Influence of $CdCl_2$ and $CdSO_4$ supplementation on Cd distribution and ligand environment in leaves of the Cd hyperaccumulator *Noccaea* (*Thlaspi*) *praecox*. *Plant and Soil* 370, 125–148.

Korenkov, V., Hirschi, K., Crutchfield, J.D. and Wagner, G.J. (2007) Enhancing tonoplast Cd/H antiport activity increases Cd, Zn, and Mn tolerance, and impacts root/shoot Cd partitioning in *Nicotiana tabacum* L. *Planta* 226, 1379–1387.

Korenkov, V., King, B., Hirschi, K. and Wagner, G.J. (2009) Root-selective expression of *AtCAX4* and *AtCAX2* results in reduced lamina cadmium in field-grown *Nicotiana tabacum* L. *Plant Biotechnology Journal* 7, 219–226.

Krämer, U. (2010) Metal hyperaccumulation in plants. *Annual Review of Plant Biology* 61, 517–534.

Krämer, U. and Clemens, S. (2005) Function and homeostasis of zinc, copper, and nickel in plants. *Topics in Current Genetics* 14, 215–271.

Krämer, U., Cotter-Howells, J.D., Charnock, J.M., Baker, A.J.M. and Smith, J.A.C. (1996) Free histidine as a metal chelator in plants that accumulate nickel. *Nature* 379, 635–638.

Kung, C.C.S., Huang, W.N., Huang, Y.C. and Yeh, K.C. (2006) Proteomic survey of copper-binding proteins in *Arabidopsis* roots by immobilized metal affinity chromatography and mass spectrometry. *Proteomics* 6, 2740–2758.

Küpper, H., Zhao, F.-J. and McGrath, S.P. (1999) Cellular compartmentation of zinc in leaves of the hyperaccumulator *Thlaspi caerulescens*. *Plant Physiology* 119, 305–312.

Küpper, H., Lombi, E., Zhao, F.-J. and McGrath, S.P. (2000) Cellular compartmentation of cadmium and zinc in relation to other elements in the hyperaccumulator *Arabidopsis halleri*. *Planta* 212, 75–84.

Lei, M., Chen, T.-B., Huang, Z.-C., Wang, Y.-D. and Huang, Y.-Y. (2008) Simulateneous compartmentalization of lead and arsenic in co-hyperaccumulator *Viola principis* H. de Boiss.: an application of SRXRF probe. *Chemosphere* 72, 1491–1496.

Li, T., Di, Z., Islam, E., Jiang, H. and Yang, X. (2011) Rhizosphere characteristics of zinc hyperaccumulator *Sedum alfredii* involved in zinc accumulation. *Journal of Hazardous Materials* 185, 818–823.

Lindsay, W.L. (1979) *Chemical Equilibria in Soils*. Wiley, New York.

Liu, F., Tang, Y., Du, R., Yang, H., Wu, Q. *et al.* (2010) Root foraging for zinc and cadmium requirement in the Zn/Cd hyperaccumulator plant *Sedum alfredii*. *Plant and Soil* 327, 365–375.

Liu, W., Xu, L., Wang, Y., Shen, H., Zhu, H. *et al.* (2015) Transcriptome-wide analysis of chromium-stress responsive microRNAs to explore miRNA-mediated regulatory networks in radish (*Raphanus sativus* L.). *Scientific Reports* 5, 14024. doi: 10.1038/srep14024.

Liu, X.-M., Kim, K.E., Kim, K.-C., Nguyen, X.C., Han, H.J. *et al.* (2010) Cadmium activates *Arabidopsis* MPK3 and MPK6 via accumulation of reactive oxygen species. *Phytochemistry* 71, 614–618.

Lu, L., Tian S., Zhang, J., Yang, X., Labavitch, J.M. *et al.* (2013) Efficient xylem transport and phloem remobilization of Zn in the hyperaccumulator plant species *Sedum alfredii*. *New Phytologist* 198, 721–731.

Lux, A., Sŏttníková, A., Opatrná, J. and Greger, M. (2004) Differences in structure of adventitious roots in *Salix* clones with contrasting characteristics of cadmium accumulation and sensitivity. *Physiologia Plantarum* 120, 537–545.

Lux, A., Martinka, M., Vaculík, M. and White, P.J. (2011) Root responses to cadmium in the rhizosphere – a review. *Journal of Experimental Botany* 62, 21–37.

Lux, A., Lackovič, A., van Staden, J., Lišková, S., Kohanová, J. *et al.* (2015) Cadmium translocation by contractile roots differs from that in regular, non-contractile roots. *Annals of Botany* 115, 1149–1154.

Ma, F. and Peterson, C.A. (2003) Recent insights into the development, structure and chemistry of the endodermis and exodermis. *Canadian Journal of Botany* 81, 405–421.

Macnair, M.R. (2003) The hyperaccumulation of metals by plants. *Advances in Botanical Research* 40, 63–105.

MacNicol, R.D. and Beckett, P.H.T. (1985) Critical tissue concentrations of potentially toxic elements. *Plant and Soil* 85, 107–129.

Maksymiec, W. (2007) Signaling responses in plants to heavy metal stress. *Acta Physiologiae Plantarum* 29, 177–187.

Martinoia, E., Meyer, S., DeAngeli, A. and Nagy, R. (2012) Vacuolar transporters in their physiological context. *Annual Review of Plant Biology* 63, 183–213.

May, M.J., Vernoux, T., Leaver, C., van Montagu, M. and Inzé, D. (1998) Glutathione homeostasis in plants: implications for environmental sensing and plant development. *Journal of Experimental Botany* 49, 649–667.

Memon, A.R. and Schroder, P. (2009) Implications of metal accumulation mechanisms to phytoremediation. *Environmental Science and Pollution Research* 16, 162–175.

Mendoza-Cózatl, D., Jobe, T.O., Hauser, F. and Schroeder, J.I. (2011) Long-distance transport, vacuolar sequestration and transcriptional responses induced by cadmium and arsenic. *Current Opinion in Plant Biology* 14, 554–562.

Mengel, K., Kirkby, E.A., Kosegarten, H. and Appel, T. (2001) *Principles of Plant Nutrition*. Kluwer Academic, Dordrecht, The Netherlands.

Mengoni, A., Schat, H. and Vangronsveld, J. (2010) Plants as extreme environments? Ni-resistant bacteria and Ni-hyperaccumulators of serpentine flora. *Plant and Soil* 331, 5–16.

Mesjasz-Przybyłowicz, J. and Przybyłowicz, W.J. (2011) PIXE in metal hyperaccumulation: from soil to plants and insects. *X-Ray Spectrometry* 40, 181–185.

Mesjasz-Przybyłowicz, J., Przybyłowicz, W., Barnabas, A. and van der Ent, A. (2016) Extreme nickel hyperaccumulation in the vascular tracts of the tree *Phyllanthus balgooyi* from Borneo. *New Phytologist* 209, 1513–1526.

Micó, C., Li, H.F., Zhao, F.J. and McGrath, S.P. (2008) Use of Co speciation and soil properties to explain variation in Co toxicity to root growth of barley (*Hordeum vulgare* L.) in different soils. *Environmental Pollution* 156, 883–890.

Mills, R.F., Francini, A., da Rocha, P.S.C.F., Baccarini, P.J., Aylett, M. *et al.* (2005) The plant P_{1B}-type ATPase AtHMA4 transports Zn and Cd and plays a role in detoxification of transition metals supplied at elevated levels. *FEBS Letters* 579, 783–791.

Mira, H., Martínez, N. and Peñarrubia, L. (2002) Expression of a vegetative-storage-protein gene from *Arabidopsis* is regulated by copper, senescence and ozone. *Planta* 214, 939–946.

Mittler, R., Vanderauwera, S., Gollery, M. and Van Breusegem, F. (2004) Reactive oxygen gene network of plants. *Trends in Plant Science* 9, 490–498.

Mizuno, N., Nosaka, S., Mizuno, T., Hornie, K. and Obata, H. (2003) Distribution of Ni and Zn in the leaves of *Thlaspi japonicum* growing on ultramafic soil. *Soil Science and Plant Nutrition* 49, 93–97.

Moore, C.A., Bowen, H.C., Scrase-Field, S., Knight, M.R. and White, P.J. (2002) The deposition of suberin lamellae determines the magnitude of cytosolic Ca^{2+} elevations in root endodermal cells subjected to cooling. *Plant Journal* 30, 457–466.

Morel, M., Crouzet, J., Gravot, A., Auroy, P., Leonhardt, N. *et al.* (2009) AtHMA3, a P_{1B}-ATPase allowing Cd/Zn/Co/Pb vacuolar storage in Arabidopsis. *Plant Physiology* 149, 894–904.

Morrison, R.S., Brooks, R.R., Reebes, R.D., Malaisse, F., Horowitz, P. *et al.* (1981) The diverse chemical forms of heavy metals in tissue extracts of some metallophytes from Shaba Province, Zaïre. *Phytochemistry* 20, 455–458.

Morrissey, J., Baxter, I.R., Lee, J., Li, L., Lahner, B. *et al.* (2009) The ferroportin metal efflux proteins function in iron and cobalt homeostasis in *Arabidopsis*. *Plant Cell* 21, 3326–3338.

Mortvedt, J.J. (2000) Bioavailability of micronutrients. In: Sumner, M.E. (ed.) *Handbook of Soil Science*. CRC Press, Boca Raton, Florida, pp. D71–D88.

Nakagami, H., Pitzschke, A. and Hirt, H. (2005) Emerging MAP kinase pathways in plant stress signalling. *Trends in Plant Science* 10, 339–346.

Neill, S.J., Desikan, R. and Hancock, J.T. (2003) Nitric oxide signalling in plants. *New Phytologist* 159, 11–35.

Nozoye, T., Nagasaka, S., Kobayashi, T., Takahashi, M., Sato, Y. *et al.* (2011) Phytosiderophore efflux transporters are crucial for iron acquisition in graminaceous plants. *Journal of Biological Chemistry* 286, 5446–5454.

Ó Lochlainn, S., Bowen, H.C., Fray, R.G., Hammond, J.P., King, G.J. *et al.* (2011) Tandem quadruplication of *HMA4* in the zinc (Zn) and cadmium (Cd) hyperaccumulator *Noccaea caerulescens*. *PLoS ONE* 6(3), e17814. doi: 10.1371/journal.pone.0017814.

Page, V. and Feller, U. (2005) Selective transport of zinc, manganese, nickel, cobalt and cadmium in the root system and transfer to the leaves in young wheat plants. *Annals of Botany* 96, 425–434.

Palmer, C.M. and Guerinot, M.L. (2009) Facing the challenges of Cu, Fe and Zn homeostasis in plants. *Nature Chemical Biology* 5, 333–340.

Pasricha, N.S., Nayyar, V.K., Randhawa, N.S. and Sinha, M.K. (1977) Molybdenum accumulation in forage crops. 2. Influence of sulfur fertilization on suppression of molybdenum uptake by berseem (*Trifolium alexandrinum* L.) and oats (*Avena sativa* L.) grown on a molybdenum-toxic soil. *Plant and Soil* 46, 245–250.

Peng, J.-S. and Gong, J.-M. (2014) Vacuolar sequestration capacity and long-distance metal transport in plants. *Frontiers in Plant Science* 5, 19. doi:10.3389/fpls.2014.00019.

Pereira, Y., Lagniel, G., Godat, E., Baudouin-Cornu, P., Junot, C. *et al.* (2008) Chromate causes sulfur starvation in yeast. *Toxicological Sciences* 106, 400–412.

Persson, D.P., Hansen, T.H., Holm, P.E., Schjoerring, J.K., Hansen, H.C.B. *et al.* (2006) Multi-elemental speciation analysis of barley genotypes differing in tolerance to cadmium toxicity using SEC-ICP-MS and ESI-TOF-MS. *Journal of Analytical Atomic Spectrometry* 21, 996–1005.

Pilon, M., Cohu, C.M., Ravet, K., Abdel-Ghany, S.E. and Gaymard, F. (2009) Essential transition metal homeostasis in plants. *Current Opinion in Plant Biology* 12, 347–357.

Pilon-Smits, E.A.H., Quinn, C.F., Tapken, W. Malagoli, M. and Schiavon, M. (2009) Physiological functions of beneficial elements. *Current Opinion in Plant Biology* 12, 267–274.

Pinto, E. and Ferreira, I.M.P.L.V.O. (2015) Cation transporters / channels in plants: tools for nutrient biofortification. *Journal of Plant Physiology* 179, 64–82.

Pollard, A.J., Reeves, R.D. and Baker, A.J.M. (2014) Facultative hyperaccumulation of heavy metals and metalloids. *Plant Science* 217/218, 8–17.

Polle, A. and Schützendübel, A. (2003) Heavy metal signalling in plants: linking cellular and organismic responses. In: Hirt, H. and Shinozaki, K. (eds) *Topics in Current Genetics 4: Plant Responses to Abiotic Stress*. Springer, Berlin. pp. 187–215.

Pongrac, P., Vogel-Mikuš, K., Vavpetič, P., Tratnik, J., Regvar, M. *et al.* (2010) Cd induced redistribution of elements within leaves of the Cd/Zn hyperaccumulator *Thlaspi praecox* as revealed by micro-PIXE. *Nuclear Instruments and Methods in Physics Research B* 268, 2205–2210.

Puig, S. and Peñarrubia, L. (2009) Placing metal micronutrients in context: transport and distribution in plants. *Current Opinion in Plant Biology* 12, 299–306.

Puig, S., Andrés-Colás, N., García-Molina, A. and Peñarrubia, L. (2007) Copper and iron homeostasis in *Arabidopsis*: responses to metal deficiencies, interactions and biotechnological applications. *Plant, Cell and Environment* 30, 271–290.

Qiu, B., Zhou, W., Xue, D., Zeng, F., Ali, S. *et al.* (2010) Identification of Cr-tolerant lines in a rice (*Oryza sativa*) DH population. *Euphytica* 174, 199–207.

Rascio, N. and Navari-Izzo, F. (2011) Heavy metal hyperaccumulating plants: how and why do they do it? And what makes them so interesting? *Plant Science* 180, 169–181.

Ravet, K., Touraine, B., Boucherez, J., Briat, J.F., Gaymard, F. *et al.* (2008) Ferritins control interaction between iron homeostasis and oxidative stress in *Arabidopsis*. *Plant Journal* 57, 400–412.

Reeves, R.D. (2006) Hyperaccumulation of trace elements by plants. In: Morel, J.-L., Echevarria, G. and Goncharova, N. (eds) *Phytoremediation of Metal-Contaminated Soils*. Springer, Berlin, pp. 25–52.

Reeves, R.D. and Baker, A.J.M. (2000) Metal-accumulating plants. In: Raskin, I. and Ensley, B.D. (eds) *Phytoremediation of Toxic Metals: Using Plants to Clean Up the Environment*. Wiley, New York, pp. 193–229.

Remans, T., Opdenakker, K., Smeets, K., Mathysen, D., Vangronsveld, J. *et al.* (2010) Metal-specific and NADPH oxidase dependent changes in lipoxygenase and NADPH oxidase gene expression in *Arabidopsis thaliana* exposed to cadmium or excess copper. *Functional Plant Biology* 37, 532–544.

Rengel, Z. (2015) Availability of Mn, Zn and Fe in the rhizosphere. *Journal of Soil Science and Plant Nutrition* 15, 397–409.

Romero-Puertas, M.C., Corpas, F.J., Rodríguez-Serrano, M., Gómez, M., del Río, L.A. *et al.* (2007) Differential expression and regulation of antioxidative enzymes by cadmium in pea plants. *Journal of Plant Physiology* 164, 1346–1357.

Roosens, N.H.C.J., Willems, G. and Saumitou-Laprade, P. (2008) Using *Arabidopsis* to explore zinc tolerance and hyperaccumulation. *Trends in Plant Science* 13, 208–215.

Sarret, G., Willems, G., Isaure, M-P., Marcus, M.A., Fakra, S.C. *et al.* (2009) Zinc distribution and speciation in *Arabidopsis halleri* × *Arabidopsis lyrata* progenies presenting various zinc accumulation capacities. *New Phytologist* 184, 581–595.

Sarret, G., Pilon Smits, E.A.H., Castillo Michel, H., Isaure, M.P., Zhao, F.J. *et al.* (2013) Use of Synchrotron-based techniques to elucidate metal uptake and metabolism in plants. *Advances in Agronomy* 199, 1–82.

Saxena, I. and Shekhawat, G.S. (2013) Nitric oxide (NO) in alleviation of heavy metal induced phytotoxicity and its role in protein nitration. *Nitric Oxide* 32, 13–20.

Schmidt, H., Günther, C., Weber, M., Spörlein, C., Loscher, S. *et al.* (2014) Metabolome analysis of *Arabidopsis thaliana* roots identifies a key metabolic pathway for iron acquisition. *PLoS ONE* 9(7), e102444. doi: 10.1371/journal.pone.0102444.

Schwartz, C., Morel, J.L., Saumier, S., Whiting, S.N. and Baker, A.J.M. (1999) Root development of the zinc-hyperaccumulator plant *Thlaspi caerulescens* as affected by metal origin, content and localization in soil. *Plant and Soil* 208, 103–115.

Schwitzguébel, J.P., Kumpiene, J., Comino, E. and Vanek, T. (2009) From green to clean: a promising and sustainable approach towards environmental remediation and human health for the 21st century. *Agrochimica* 53, 209–237.

Seregin, I.V. and Kozhevnikova, A.D. (2008) Roles of root and shoot tissues in transport and accumulation of cadmium, lead, nickel, and strontium. *Russian Journal of Plant Physiology* 55, 1–22.

Shanker, A.K., Cervantes, C., Loza-Tavera, H. and Avudainayagam, S. (2005) Chromium toxicity in plants. *Environment International* 31, 739–753.

Sharma, P. and Dubey, R.S. (2006) Lead toxicity in plants. *Brazilian Journal of Plant Physiology* 17, 35–52.

Sharma, S.S. and Dietz, K.-J. (2006) The significance of amino acids and amino acid-derived molecules in plant responses and adaptation to heavy metal stress. *Journal of Experimental Botany* 57, 711–726.

Sharma, S.S. and Dietz, K.-J. (2009) The relationship between metal toxicity and cellular redox imbalance. *Trends in Plant Science* 14, 43–50.

Sheoran, V., Sheoran, A.S. and Poonia, P. (2009) Phytomining: a review. *Minerals Engineering* 22, 1007–1019.

Shim, D., Hwang, J.-U., Lee, J., Lee, S., Choi, Y. *et al.* (2009) Orthologues of the class A4 heat shock transcription factor HsfA4a confer cadmium tolerance in wheat and rice. *Plant Cell* 21, 4031–4043.

Siegele, R., Kachenko, A.G., Bathia, N.P., Wang, Y.D., Ionescu, M. *et al.* (2008) Localisation of trace metals in metal-accumulating plants using μ-PIXE. *X-Ray Spectrometry* 37, 133–136.

Sinclair, S.A. and Krämer, U. (2012) The zinc homeostasis network of land plants. *Biochimica et Biophysica Acta* 1823, 1553–1567.

Smith, R.A.H. and Bradshaw, A.D. (1979) The use of metal tolerant plant populations for the reclamation of metalliferous wastes. *Journal of Applied Ecology* 16, 595–612.

Socha, A.L. and Guerinot, M.L. (2014) Mn-euvering manganese: the role of transporter gene family members in manganese uptake and mobilization in plants. *Frontiers in Plant Science* 5,106. doi:10.3389/fpls.2014.00106.

Sterckeman, T., Perriguey, J., Caël, M., Schwartz, C. and Morel, J.L. (2004) Applying a mechanistic model to cadmium uptake by *Zea mays* and *Thlaspi caerulescens*: consequences for the assessment of the soil quantity and quality factors. *Plant and Soil* 262, 289–302.

Taiz, L. and Zeigler, E. (2006) *Plant Physiology*, 4th edn. Sinauer Associates, Sunderland, Massachusetts.

Tang, Y.-T., Deng, T.-H.-B., Wu, Q.-H., Wang, S.-Z., Qiu, R.-L. *et al.* (2012) Designing cropping systems for metal contaminated sites: a review. *Pedosphere* 22, 470–488.

Tappero, R., Peltier, E., Grafe, M., Heidel, K., Ginder-Vogel, M. *et al.* (2007) Hyperaccumulator *Alyssum murale* relies on a different metal storage mechanism for cobalt than for nickel. *New Phytologist* 175, 641–654.

Tejada-Jiménez, M., Galván, A., Fernández, E. and Llamas, Á. (2009) Homeostasis of the micronutrients Ni, Mo and Cl with specific biochemical functions. *Current Opinion in Plant Biology* 12, 358–363.

Tejada-Jiménez, M., Chamizo-Ampudia, A., Galván, A., Fernández, E. and Llamas, Á. (2013) Molybdenum metabolism in plants. *Metallomics* 5, 1191–1203.

Thomine, S., Lelièvre, F., Debarbieux, E., Schroeder, J.I. and Barbier-Brygoo, H. (2003) AtNRAMP3, a multispecific vacuolar metal transporter involved in plant responses to iron deficiency. *Plant Journal* 34, 685–695.

Tian, S.K., Lu, L.L., Yang, X.E., Labavitch, J.M., Huang, Y.Y. *et al.* (2009) Stem and leaf sequestration of zinc at the cellular level in the hyperaccumulator *Sedum alfredii*. *New Phytologist* 182, 116–126.

Tian, S., Lu, L., Yang, X., Webb, S.M., Du, Y. *et al.* (2010) Spatial imaging and speciation of lead in the accumulator plant *Sedum alfredii* by microscopically focused synchrotron X-ray investigation. *Environmental Science & Technology* 44, 5920–5926.

Tian, S., Lu, L., Labavitch, J., Yang, X., He, Z. *et al.* (2011) Cellular sequestration of cadmium in the hyperaccumulator plant species *Sedum alfredii*. *Plant Physiology* 157, 1914–1925.

Tommasini, R., Vogt, E., Fromenteau, M., Hörtensteiner, S., Matile, P. *et al.* (1998) An ABC-transporter of *Arabidopsis thaliana* has both glutathione-conjugate and chlorophyll catabolite transport activity. *Plant Journal* 13, 773–780.

Tudoreanu, L. and Phillips, C.J.C. (2004) Empirical models of cadmium accumulation in maize, rye grass and soya bean plants. *Journal of the Science of Food and Agriculture* 84, 845–852.

Tyler, G. and Olsson, T. (2001) Concentrations of 60 elements in the soil solution as related to the soil acidity. *European Journal of Soil Science* 52, 151–165.

Uraguchi, S. and Fujiwara, T. (2013) Rice breaks ground for cadmium-free cereals. *Current Opinion in Plant Biology* 16, 328–334.

Van der Ent, A. and Reeves, R.D. (2015) Foliar metal accumulation in plants from copper-rich ultramafic outcrops: case studies from Malaysia and Brazil. *Plant and Soil* 389, 401–418.

Van der Ent, A., Baker, A.J.M., Reeves, R.D., Pollard, A.J. and Schat, H. (2013) Hyperaccumulators of metal and metalloid trace elements: facts and fiction. *Plant and Soil* 362, 319–334.

Van der Ent, A., Erskine, P. and Sumail, S. (2015) Ecology of nickel hyperaccumulator plants from ultramafic soils in Sabah (Malaysia). *Chemoecology* 25, 243–259.

van Hoof, N.A.L.M., Hassinen, V.H., Hakvoort, H.W.J., Ballintijn, K.F., Schat, H. *et al.* (2001a) Enhanced copper tolerance in *Silene vulgaris* (Moench) Garcke populations from copper mines is associated with increased transcript levels of a 2b-type metallothionein gene. *Plant Physiology* 126, 1519–1526.

van Hoof, N.A.L.M., Koevoets, P.L.M., Hakvoort, H.W.J., Ten Bookum, W.M., Schat, H. *et al.* (2001b) Enhanced ATP-dependent copper efflux across the root cell plasma membrane in copper-tolerant *Silene vulgaris*. *Physiologia Plantarum* 113, 225–232.

Verbruggen, N., Hermans, C. and Schat, H. (2009a) Molecular mechanisms of metal hyperaccumulation in plants. *New Phytologist* 181, 759–776.

Verbruggen, N., Hermans, C. and Schat, H. (2009b) Mechanisms to cope with arsenic or cadmium excess in plants. *Current Opinion in Plant Biology* 12, 364–372.

Villafort Carvalho, M.T., Amaral, D.C., Guilherme, L.R.G. and Aarts, M.G.M. (2013) *Gomphrena claussenii*, the first South American metallophyte species with indicator-like Zn and Cd accumulation and extreme metal tolerance. *Frontiers in Plant Science* 4, 180. doi: 10.3389/fpls.2013.00180.

Villafort Carvalho, M.T., Pongrac, P., Mumm, R., van Arkel, J., van Aelst, A. *et al.* (2015) *Gomphrena claussenii*, a novel metal-hypertolerant bioindicator species, sequesters cadmium, but not zinc, in vacuolar oxalate crystals. *New Phytologist* 208, 763–775.

Vogel-Mikuš, K., Regvar, M., Mesjasz-Przybyłowicz, J., Przybyłowicz, W.J., Simčič, J. *et al.* (2008a) Spatial distribution of cadmium in leaves of metal hyperaccumulating *Thlaspi praecox* using micro-PIXE. *New Phytologist* 179, 712–721.

Vogel-Mikuš, K., Simčič, J., Pelicon, P., Budnar, M., Kump, P. *et al.* (2008b) Comparison of essential and non-essential element distribution in leaves of the Cd/Zn hyperaccumulator *Thlaspi praecox* as revealed by micro-PIXE. *Plant, Cell and Environment* 31, 1484–1496.

Vogel-Mikuš, K., Pongrac, P. and Pelicon, P. (2014) Micro-PIXE elemental mapping for ionome studies of crop plants. *International Journal of PIXE* 24, 217–233.

von Wirén, N., Klair, S., Bansal, S., Briat, J.-F., Khodr, H. *et al.* (1999) Nicotianamine chelates both FeIII and FeII. Implications for metal transport in plants. *Plant Physiology* 119, 1107–1114.

Wallace, A. (1984) Effect of phosphorus deficiency and copper excess on vegetative growth of bush bean plants in solution culture at two different solution pH. *Journal of Plant Nutrition* 7, 603–608.

Wang, Y.D., Mesjasz-Przybyłowicz, J., Tylko, G., Barnabas, A.D. and Przybyłowicz, W.J. (2013) Micro-PIXE analyses of frozen hydrated semi-thick biological sections. *Nuclear Instruments and Methods in Physics Research B* 306, 134–139.

Wang, Y.X., Wu, P., Wu, Y.R. and Yan, X.L. (2002) Molecular marker analysis of manganese toxicity tolerance in rice under greenhouse conditions. *Plant and Soil* 238, 227–233.

Watanabe, T., Broadley, M.R., Jansen, S., White, P.J., Takada, J. *et al.* (2007) Evolutionary control of leaf element composition in plants. *New Phytologist* 174, 516–523.

Waters, B.M., Uauy, C., Dubcovsky, J. and Grusak, M.A. (2009) Wheat (*Triticum aestivum*) NAM proteins regulate the translocation of iron, zinc, and nitrogen compounds from vegetative tissues to grain. *Journal of Experimental Botany* 60, 4263–4274.

Welch, R.M. (1995) Micronutrient nutrition of plants. *Critical Reviews in Plant Science* 14, 49–82.

Welch, R.M., Allaway, W.H., House, W.A. and Kubota, J. (1991) Geographic distribution of trace element problems. In: Mortvedt, J.J., Cox, F.R., Shuman, L.M. and Welch, R.M. (eds) *Micronutrients in Agriculture*, 2nd edn. Soil Science Society of America, Madison, Wisconsin, pp. 31–57.

White, P.J. (2001) The pathways of calcium movement to the xylem. *Journal of Experimental Botany* 52, 891–899.

White, P.J. (2012a) Ion uptake mechanisms of individual cells and roots: short distance transport. In: Marschner, P. (ed) *Marschner's Mineral Nutrition of Higher Plants*, 3rd edn. Academic Press, London, pp. 7–47.

White, P.J. (2012b) Long-distance transport in the xylem and phloem. In: Marschner, P. (ed.) *Marschner's Mineral Nutrition of Higher Plants*, 3rd edn. Academic Press, London, pp. 49–70.

White, P.J. and Broadley, M.R. (2003) Calcium in plants. *Annals of Botany* 92, 487–511.

White, P.J. and Broadley, M.R. (2005) Biofortifying crops with essential mineral elements. *Trends in Plant Science* 10, 586–593.

White, P.J. and Broadley, M.R. (2009) Biofortification of crops with seven mineral elements often lacking in human diets – iron, zinc, copper, calcium, magnesium, selenium and iodine. *New Phytologist* 182, 49–84.

White, P.J. and Broadley, M.R. (2011) Physiological limits to zinc biofortification of edible crops. *Frontiers in Plant Science* 2, 80. doi: 10.3389/fpls.2011.00080.

White, P.J. and Brown, P.H. (2010) Plant nutrition for sustainable development and global health. *Annals of Botany* 105, 1073–1080.

White, P.J. and Greenwood, D.J. (2013) Properties and management of cationic elements for crop growth. In: Gregory, P.J. and Nortcliffe, S. (eds) *Soil Conditions and Plant Growth*, 12th edn. Blackwell Publishing, Oxford, UK, pp. 160–194.

White, P.J., Whiting, S.N., Baker, A.J.M. and Broadley, M.R. (2002) Does zinc move apoplastically to the xylem in roots of *Thlaspi caerulescens*? *New Phytologist* 153, 201–207.

White, P.J., Broadley, M.R. and Gregory, P.J. (2012) Managing the nutrition of plants and people. *Applied and Environmental Soil Science*, article 104826. doi:10.1155/2012/104826.

White, P.J., George, T.S., Dupuy, L.X., Karley, A.J., Valentine, T.A. *et al.* (2013) Root traits for infertile soils. *Frontiers in Plant Science* 4, 193. doi: 10.3389/fpls.2013.00193.

Whiting, S.N., Leake, J.R., McGrath, S.P. and Baker, A.J.M. (2000) Positive responses to Zn and Cd by roots of the Zn and Cd hyperaccumulator *Thlaspi caerulescens*. *New Phytologist* 145, 199–210.

Whiting, S.N., Reeves, R.D., Richards, D.G., Johnson, M.S., Cooke, J.A. *et al.* (2005) Use of plants to manage sites contaminated with metals. In: Broadley, M.R. and White, P.J. (eds) *Plant Nutritional Genomics*. Blackwell, Oxford, UK, pp. 287–315.

Willems, G., Frérot, H., Gennen, J., Salis, P., Saumitou-Laprade, P. *et al.* (2010) Quantitative trait loci analysis of mineral element concentrations in an *Arabidopsis halleri* × *Arabidopsis lyrata petraea* F$_2$ progeny grown on cadmium contaminated soil. *New Phytologist* 187, 368–379.

Williams, L.E. and Pittman, J.K. (2010) Dissecting pathways involved in manganese homeostasis and stress in higher plants. In: Hell, R. and Mendel, R.-R. (eds) *Plant Cell Monographs 17, Cell Biology of Metals and Nutrients*. Springer, Berlin, pp. 95–117.

Wong, C.K.E. and Cobbett, C.S. (2009) HMA P-type ATPases are the major mechanism for root-to-shoot Cd translocation in *Arabidopsis thaliana*. *New Phytologist* 181, 71–78.

Wycisk, K., Kimb, E.J., Schroeder, J.I. and Krämer, U. (2004) Enhancing the first enzymatic step in the histidine biosynthesis pathway increases the free histidine pool and nickel tolerance in *Arabidopsis thaliana*. *FEBS Letters* 578, 128–134.

Xiong, J., Fu, G., Tao, L. and Zhu, C. (2010) Roles of nitric oxide in alleviating heavy metal toxicity in plants. *Archives of Biochemistry and Biophysics* 497, 13–20.

Yang, Z.M. and Chen, J. (2013) A potential role of microRNAs in plant response to metal toxicity. *Metallomics* 5, 1184–1190.

Yeh, C.M., Chien, P.S. and Huang, H.J. (2007) Distinct signalling pathways for induction of MAP kinase activities by cadmium and copper in rice roots. *Journal of Experimental Botany* 58, 659–671.

Yun, L., Larson, S.R., Jensen, K.B., Staub, J.E. and Grossl, P.R. (2015) Quantitative trait loci (QTL) and candidate genes associated with trace element concentrations in perennial grasses grown on phytotoxic soil contaminated with heavy metals. *Plant and Soil* 396, 277–296.

Zeng, H., Wang, G., Hu, X., Wang, H., Du, L. *et al.* (2014) Role of microRNAs in plant responses to nutrient stress. *Plant and Soil* 374, 1005–1021.

Zhai, Z., Gayomba, S.R., Jung, H., Vimalakumari, N.K., Piñeros, M. *et al.* (2014) OPT3 is a phloem-specific iron transporter that is essential for systemic iron signaling and redistribution of iron and cadmium in *Arabidopsis*. *The Plant Cell* 26, 2249–2264.

Zhang, J., Tian, S., Lu, L., Shohag, M.J.I., Liao, H. and Yamg, X. (2011) Lead tolerance and cellular distribution in *Elsholtzia splendens* using synchrotron radiation micro-X-ray fluorescence. *Journal of Hazardous Materials* 197, 264–271.

Zhang, T., Liu, Y., Yang, T., Zhang, L., Xu, S. *et al.* (2006) Diverse signals converge at MAPK cascades in plant. *Plant Physiology and Biochemistry* 44, 274–283.

Zhao, F.-J. and McGrath S.J. (2009) Biofortification and phytoremediation. *Current Opinion in Plant Biology* 12, 373–380.

Zhao, F.J., Lombi, E. and Breedon, T.M.S.P. (2000) Zinc hyperaccumulation and cellular distribution in *Arabidopsis halleri*. *Plant Cell and Environment* 23, 507–514.

13 Biotic Stress Signalling: Ca²⁺-mediated Pathogen Defence Programmes

Yi Ma and Gerald A. Berkowitz*

Department of Plant Science and Landscape Architecture,
University of Connecticut, Storrs, Connecticut, USA

Abstract

Plant cells sensing pathogenic microorganisms evoke defence systems that can confer resistance to infection. This immune reaction can include triggering of basal defence responses as well as programmed cell death, or hypersensitive response (HR). In both cases (basal defence and HR), pathogen perception is translated into elevated cytosolic Ca²⁺ (mediated by plasma membrane and intracellular channels) as an early step in a signalling cascade. Cyclic nucleotide-gated channels contribute to this influx of Ca²⁺ into the cell. The identification of specific steps in the signalling pathway leading from pathogen perception to generation of defence molecules in the cytosol, transcriptional reprogramming and other aspects of the plant immune response is not completely delineated at present and is an active area of current research. This chapter will present current information about this defence network.

13.1 Introduction

During their life cycles, plants are exposed to invading microorganisms including bacteria, true fungi, filamentous protist oomycetes and viruses, which are potentially pathogenic in that impairment of host plant growth, development, survival and/or reproduction could result from successful infection. Land plants are typically sessile organisms and must therefore combat pathogen invasion without physically separating themselves from the source of infection. They also lack mobile sentry cells and have to respond to such potentially injurious microbe invasion using defence strategies that preclude chemotaxis-facilitated phagocytosis (a cornerstone of innate immunity in animals). In addition to the absence of a mobile self/non-self-surveillance system,

plants lack the powerful recombinatorial- (i.e. antibody-) based adaptive immunity system of jawed vertebrates. Nonetheless, plants have a complex multi-layered system of host survival-enhancing immune responses to pathogens that involves increased transcription of pathogen defence-related genes, and can include cellular-level apoptotic programmed cell death (PCD). Immune responses are initiated upon recognition of a pathogen as 'non-self' by individual plant cells; this perception of an invading pathogen must then be translated through a signal transduction cascade into defence responses in order to maintain system fitness.

Plants share with animals the use of Ca²⁺ as a cytosolic secondary messenger molecule involved in numerous cell signalling cascades responding to abiotic and biotic stimuli. Perturbations in

---

* Corresponding author: gerald.berkowitz@uconn.edu

cytosolic Ca^{2+} homeostasis are known to be an essential early step in pathogen perception and subsequent immune response of plant cells. This review will focus on our understanding of the role Ca^{2+} plays in the signal transduction pathway leading to defence responses and immune-associated host cell apoptosis in plants. Recent advancements in, as well as outstanding questions related to, the: (i) cellular-level molecular events linking pathogen perception/recognition to altered cytosolic Ca^{2+}; (ii) cell and intracellular Ca^{2+} transporters potentially involved in microbe-associated perturbations in Ca^{2+} homeostasis; and (iii) downstream events (in the signalling and pathogen response systems) affected by altered cytosolic Ca^{2+} levels will be highlighted. A fair amount of work in these areas has been published recently and has not been included in recent reviews of pathogen defence responses. Hence the focus here on this topic should provide new perspective to the understanding of mechanisms underlying plant defence responses to the biotic stress of pathogen infection.

13.2 Biotic Stresses

Plants are exposed to biotic assault constantly from bacteria, viruses, fungi, insect herbivores and parasitic plants. Plant–microbe interactions that underlie pathogenesis, as well as defence programmes, have been studied fundamentally from both the plant and microbial perspectives. Many genes involved in either pathogenesis and plant defence have been identified and signalling pathways involved in the interaction have been delineated. As this chapter focuses primarily on the role of Ca^{2+} in plant–bacterial pathogen interaction, a brief summary of Ca^{2+} involvement in herbivore attack and plant defence will be presented.

It has been reported that, after insect feeding, there is significant cytosolic Ca^{2+} influx in the cells around the wounded zone (Maffei et al., 2004). Both the reactive oxygen species (ROS) and NO can be induced upon herbivory. These plant defence molecules may be generated downstream from Ca^{2+} signals (and concomitant membrane depolarization) during plant defence responses to herbivory. Like some pathogenic compounds released by microbes or fungi,

herbivores also introduce oral secretions into the wounded region of the plant. Orally secreted compounds from insects are able to induce Ca^{2+} elevations and membrane depolarization in plant cells (Mithöfer and Boland, 2008; Bricchi et al., 2010).

As herbivore damage causes Ca^{2+} signals in plants, the interplay of Ca^{2+} and Ca^{2+} sensor proteins after herbivore attack has been investigated. In Arabidopsis, CPK3 and CPK13 are involved in the transcriptional activation of plant defence gene PDF1.2 by phosphorylation of a heat shock transcription factor (Hsf2a) in herbivore-infested plants (Kanchiswamy et al., 2010). Studies showed that CPK3 facilitates the phosphorylation of a series of downstream transcription factors involved in herbivore responses, such as the jasmonic acid/ethylene-inducible APE/ERF domain transcription factor 1 (ERF1), the wound-inducible CZF1/ZFAR1 transcription factor and ATL2, a member of a multigene family of highly related RING-H2 zinc finger proteins that functions as E3 ubiquitin ligases as well as a potential PDF1.2 transcription regulator (see review by Arimura and Maffei, 2010). These results suggest that ubiquitination may be important for facilitating and modulating signal transduction leading to transcriptional reprogramming related to herbivore defence in plants. However, more evidence is needed to more fully elucidate the roles of Ca^{2+} in the herbivore-induced signalling network, and to link Ca^{2+} signalling to ubiquitination and other downstream responses. Some new evidence indicates that Ca^{2+}-mediated herbivore responses may involve systemic long-range signalling as well as defences locally at the site of predation (Kiep et al., 2015).

Recent studies in maize and Arabidopsis have showed that expression of genes encoding plant elicitor peptides (Peps), ZmPROPEP3 and AtPROPEP3, are rapidly induced upon treatment of the plant with oral secretions of the herbivore Spodoptera littoralis larvae (Huffaker et al., 2013; Klauser et al., 2015). ZmPep3 induced a range of herbivore defence responses, such as jasmonic acid and ethylene production, defence gene upregulation resulting in inhibition of larval growth (Huffaker et al., 2013). The Arabidopsis Pep receptor double mutant pepr1/pepr2 demonstrated impaired defence responses to herbivory (Klauser et al., 2015). It should be noted, however, that no study has yet linked the Pep/PEPR system to cytosolic Ca^{2+} elevations in response to herbivore feeding (see below for information about the Pep/PEPR system and Ca^{2+} signalling in response to pathogens).

13.3 Why Calcium?

As mentioned above, transient elevations in cytosolic Ca^{2+} are a ubiquitous component of numerous cell signalling cascades, including plant cell responses to biotic stress (pathogen invasion). It is hard to fathom (and we currently know little about) how the plant cell can make sense of so many signalling cues involving Ca^{2+}; i.e. how the cell orchestrates the correct linkage of a specific output of cell responses to the correct specific external cue that initiated the Ca^{2+} signal. With perturbations in Ca^{2+} homeostasis involved in, and at the crossroads of so many signalling cascades, how can the cell parse one cue from another? And, not incidentally, why is calcium at the crossroads? In an incisive and elegant review entitled 'The language of calcium signaling', Dodd et al. (2010) touched on the evolution of Ca^{2+} signalling, pointing out that the maintenance of sub-micromolar cytosolic Ca^{2+} is a requirement for normal cell function. From birth of the first cell (likely to have been in the seawater milieu of the oceans), to the ubiquity of Ca^{2+} signalling across all kingdoms of life, it is not a coincidence that cellular-level Ca^{2+} signalling is a paramount feature of life. Of the four cations (Na^+, Mg^{2+}, K^+ and Ca^{2+}) and the anion (Cl^-) abundant in seawater, normal cell functions most critically depend on control over cellular Ca^{2+}. As pointed out by Dodd et al. (2010), Ca^{2+} must be maintained at low concentrations to prevent the formation of insoluble Ca^{2+} salts of inorganic PO_4^{3-} and organic phosphates. Availability of soluble inorganic and organic phosphates is required for energy (adenosine phosphate) metabolism, nucleic acid synthesis and membrane (phospholipid) formation; high cytosolic Ca^{2+} would also cause cytotoxic aggregation of proteins and nucleic acids and affect membrane integrity (Case et al., 2007). Thus, as postulated by Case et al. (2007), 'the first cell' must have evolved (as evidenced by the presence even in the most primitive bacteria)

with plasma membrane Ca^{2+} pumps and Ca^{2+} exchangers to facilitate Ca^{2+} extrusion from the cytosol.

The maintenance of cytosolic Ca^{2+} homeostasis at ~100 nM is a hallmark of all life forms. At the same time, Ca^{2+}, more than the other ionic constituents of seawater, interacts with biological molecules due to specific properties such as its high affinity for carboxylate oxygen (found in amino acids), its rapid binding kinetics and its flexible bond coordination chemistry, and is an ideal molecule to act as a signal (Jaiswal, 2001). Thus it 'makes sense', from an evolutionary standpoint, that perturbations from homeostasis would then be used in the evolution of more complex cell functions as a secondary message to impart signalling cues to Ca^{2+}-sensing machinery. This machinery includes the suite of Ca^{2+} readout proteins present in cells such as proteins that change conformation upon Ca^{2+} binding (such as calmodulin and other calmodulin-like proteins) as well as enzymes with catalytic activity affected by Ca^{2+} binding (such as Ca^{2+}-dependent kinases). Hence, in this context, it makes quite a bit of sense that most cells maintain Ca^{2+} concentration gradients of 10,000–20,000 between the cytosol and plasma membrane and endomembrane-delimited compartments and, at the same time, use transient cytosolic Ca^{2+} elevations as a ubiquitous cytosolic secondary messenger.

13.4 Overview of Pathogen Defence Programmes: A Multi-layered System

Immunity, or pathogen defence responses in plants, involves two interrelated and interconnected defence mechanisms: (i) a basal reaction to microbe invasion that is a general non-self-recognition phenomenon; and (ii) a disease resistance (R) gene, or gene-for-gene mediated reaction to potential pathogens (Jones and Dangl, 2006; He et al., 2007; Hofius et al., 2007; Boller and Felix, 2009; Boller and He, 2009). In addition to immune responses displayed by cells directly challenged by pathogen attack and cells neighbouring the site of infection, plant response to pathogens includes systemic acquired resistance, which involves long-distance transport of plant-derived signalling molecules and potentiates plant defences (localized to the site of infection as well as in distal tissues) against current and subsequent pathogen attack (Grant and Lamb, 2006).

Systemic acquired resistance does not involve PCD as a plant protective response to pathogen attack, while plant immunity can be thought of as an integrated system that can lead to apoptosis in the cells at the site of infection. This PCD at the infection site (also referred to as the hypersensitive response (HR)) limits the spread of pathogens within the plant; it is one of many plant defensive strategies that limit and/or ameliorate disease. As is the case with animal PCD, in addition to Ca^{2+} as a cytosolic signalling molecule, ROS including H_2O_2, superoxide (O_2^-), hydroxyl radicals (OH^-) and nitric oxide (NO) are involved in the induction, signal cascade and execution of PCD during plant HR to pathogens.

Excellent reviews have been published recently on various aspects of plant immunity and Ca^{2+} transport and signalling (Hetherington and Brownlee, 2004; Garcia-Brugger et al., 2006; Grün et al., 2006; Jones and Dangl, 2006; Lecourieux et al., 2006; He et al., 2007; Hofius et al., 2007; Newman et al., 2007; Boller and He, 2009; Boller and Felix, 2009; Kim et al., 2009; McAinsh and Pittman, 2009; Nicaise et al., 2009; Zipfel, 2009; Dodd et al., 2010; Galon et al., 2010; Kudla et al., 2010; Verret et al., 2010; Seybold et al., 2014; Bigeard et al., 2015). Here, the focus is on the specific interface of Ca^{2+} involvement in plant defence responses. Most of the studies cited in this review were done with *Arabidopsis thaliana* plants/cells; important exceptions are noted, as discussed below.

13.4.1 Multiple pathogen-related signals evoke plant defences

Current models of plant immune defence against potentially pathogenic microbes (see reviews cited above) conceptualize the two protective systems, basal responses and *R* gene-mediated resistance, as different 'layers' of a single plant defence network reacting to (in the case of basal resistance) non-self antigens and (with *R* gene resistance) pathogen-associated virulence

proteins that typically inactivate some component of the basal defence system. Plant basal defences are activated by perception of evolutionarily conserved microbial components essential to the proper functioning of the potential pathogen: microbial (or pathogen)-associated molecular patterns (MAMPs or PAMPs, subsequently referred to here as PAMPs). PAMPs can be essential components of non-pathogenic microbes, hence the dual name. Examples of PAMPs are the bacterial flagellin and elongation factor (EF-Tu) protein, as well as a lipid constituent of Gram-negative bacteria, lipopolysaccharide (LPS). Fungal PAMPs include chitin, a component of the fungal cell wall, ergosterol and xylanase. Oomycete-derived PAMPs include β-glucans and transglutaminase.

In addition to PAMPs, activators of basal resistance include compounds released from plant cells (Boller and Felix, 2009) or from the cell wall of the invading microbe by hydrolytic enzymes of the plant or pathogen, i.e. breakdown products of an initial host–pathogen assault/defence 'skirmish'. Breakdown products derived from plants that act as signals to evoke defence responses have been termed host-associated molecular patterns (HAMPs) (Galletti et al., 2009). An example of a HAMP is oligogalacturonide, a plant cell wall fragment generated by the action of invading microbe enzymes.

In addition to plant cell breakdown products from pathogenic attack, plant cells also generate specific signalling molecules upon perception of pathogen invasion; these have been termed 'damage-associated molecular patterns', or DAMPs (Boller and Felix, 2009). An example of a plant DAMP is the AtPep family of peptides generated in *Arabidopsis* upon perception of invading pathogens. Finally, microbial (fungal and bacterial) toxins from a number of species have been found to evoke defence signalling responses in plants (Garcia-Brugger et al., 2006; Lecourieux et al., 2006; Qutob et al., 2006; Munemasa et al., 2007; Errakhi et al., 2008; Hu et al., 2009). Here, we refer to these toxins as 'elicitors' of plant immunity. Examples of such elicitors are the peptides thaxtomin A, a peptide produced by all plant-pathogenic *Streptomyces* species of bacteria (Errakhi et al., 2008), and the protein cryptogein secreted by the fungus *Phytophthora cryptogea* (Garcia-Brugger et al., 2006).

13.4.2   PAMP-triggered immunity and effector-triggered immunity

PAMP (as well as DAMP and HAMP) perception occurs at the plant cell plasma membrane (PM), facilitated by a multiplicity of pattern recognition receptor (PRR) proteins that transduce PAMP/pathogen recognition into convergent signalling within the host cell under attack (Zipfel et al., 2004; Jones and Dangl, 2006; He et al., 2007). This basal response includes metabolic and transcriptional changes in the host cell leading to PAMP-triggered immunity (PTI). Protein effectors produced by (specific races of) invading microbes can interfere with host resistance afforded by PTI. These specific gene products can enhance the virulence of the invading microbe, thus conveying effector-triggered susceptibility (ETS) to the invaded host cells under attack. 'Layered' on top of host cell basal resistance (PTI) is a secondary network of responses to the pathogen, in this case species- and/or cultivar-specific. This secondary response network is induced by interaction of a translation product from a plant-encoded (cultivar and/or species-specific) R gene either directly or indirectly with the specific protein effector produced by the pathogen. If the plant under attack has an R gene that encodes a protein that specifically interacts with the microbial effector, this gene-for-gene interaction can provide a signal that induces another network of immune responses in the plant: effector-triggered immunity (ETI). ETI conveys disease resistance to the host plant and typically includes, among the network of plant defence responses providing disease resistance, an HR to the specific race of microbe (Jones and Dangl, 2006). If ETI imparts effective resistance to the host plant through this gene-for-gene microbe:plant interaction, the microbial gene encoding the effector protein (which inactivates the PTI basal defence network of the plant) triggering this plant R gene-based immunity ironically ends up as an avirulence (avr) gene, and the pathogen–plant host interaction is termed 'incompatible'. The overview of PTI and ETI signalling pathways is shown in Fig. 13.1.

PTI and ETI have been viewed as perhaps overlapping but mechanistically distinct resistance systems, with HR classically thought of as solely an ETI response to pathogen avr gene products (Hofius et al., 2007; Boller and He,

2009). However, more recent models of plant immune responses view these response pathways as a continuum (Thomma et al., 2011). This evolving picture of PTI and ETI as more intimately related notes that they share many similar signal transduction components as well as downstream targets and presents plant immune response to pathogens harbouring avr genes as invoking an 'accelerated' immune response (e.g. Mészáros et al., 2006). Also consistent with this model are studies (Naito et al., 2008) that show application of flagellin alone can induce HR-like cell death in Arabidopsis leaves. Consistent with this finding, flagellin has been shown to activate key regulators of R gene-mediated signalling pathways as well as basal defence networks, thus suggesting an intimate overlap and linkage of the two pathogen defence responses of plants (Robatzek et al., 2006). Providing further texture to the characterization of plant immunity is the generation by plants of specific DAMPs upon pathogen recognition that elicit many of the ETI immune responses evoked by microbial avr gene products (Huffaker et al., 2006).

13.4.3   Identified pattern recognition receptor proteins in plants

Research in the area of plant biotic stress signalling and physiology has recently focused on characterizing several PM proteins that act as PAMP receptors (i.e. PRR) and initiate the cytosolic signalling that occurs in response to pathogen perception (reviewed in Boller and Felix, 2009; Nicaise et al., 2009). The signal generated upon binding of a PAMP to a PRR initiates the PTI component of plant immune signalling. Several of these PRRs have been well characterized in Arabidopsis; the FLS2 receptor binds to flagellin and the EFR receptor binds the bacterial PAMP EF-Tu. Both of these PRR proteins are leucine-rich-repeat receptor-like kinases (LRR-RLKs). The extracellular LRR domains of these proteins act as ligand (PAMP) binding sites. In the case of FLS2, it is known that PAMP binding causes a rapid association of this receptor with an auxiliary protein BAK1 (BAK1 is considered a coreceptor) that is also an LRR-RLK. Ligand-dependent phosphorylation occurs in the cytosolic kinase domains of FLS2 and BAK1. It is

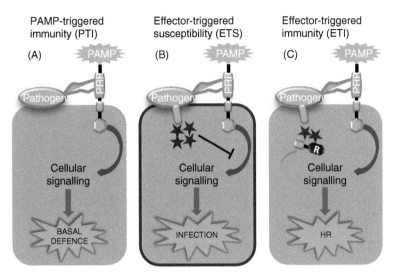

Fig. 13.1. A simple schematic diagram of pathogen:plant interactions that trigger immune responses. (A) Upon microbial attack, the plant recognizes PAMPs through their interaction with plasma membrane localized pattern-recognition receptors (PRRs). This initiates downstream signalling cascades that evoke basal defence: PAMP-triggered immunity (PTI). (B) Virulent pathogens have acquired effector molecules (red stars) that are transported into the host cell to suppress PTI, resulting in effector-triggered susceptibility (ETS) and successful infection. (C) In turn, plants have evolved the ability to generate resistance (R) proteins that recognize these attacker-specific effectors, resulting in a secondary immune response called effector-triggered immunity (ETI), which can lead to programmed cell death and HR, which prevents the spread of infection.

currently speculated that the transmission of the perception of non-self (i.e. the binding of the PAMP to the external domain of the PRR) is transmitted to the cell interior and a signalling cascade is initiated in some manner through the action of the kinase domain of the FLS2 LRR-RLK. The PAMP receptor EFR has also recently been shown to also associate with BAK1 upon binding of EF-Tu to this LRR-RLK (Schulze *et al.*, 2010) and therefore EFR signalling may follow a similar pathway to that mediated by FLS2. It is known that Ca^{2+} elevation in the cytosol is a critical and necessary component of the PTI defence responses initiated by flagellin perception upon binding of the PAMP to FLS2. The steps linking FLS2 and BAK1 phosphorylation to the initiation of Ca^{2+} signalling are still unclear. Recently, *Botrytis*-induced kinase (BIK1) was shown to associate with the FLS2 and BAK1 receptor complex to initiate plant innate immunity (Lu *et al.*, 2010). The tyrosine phosphorylation of BIK1 by BAK1 is required for the defence responses to pathogens (Lin *et al.*, 2014). In addition, BIK1 was shown to phosphorylate the

NADPH oxidase RBOHD in a Ca^{2+}-independent manner (Kadota *et al.*, 2014), suggesting that this process (also ROS generation) either is upstream of cytosolic Ca^{2+} elevation or does not require Ca^{2+}. Not all currently identified PAMP receptors are LRR-RLKs (Boller and Felix, 2009). Interestingly, work from this laboratory has characterized a possible model for how the DAMP receptor (also a LRR-RLK) AtPEPR1 is linked to a critical and necessary step of cytosolic Ca^{2+} elevation (Qi *et al.*, 2010). Other recent work (Postel *et al.*, 2010; Schulze *et al.*, 2010) has demonstrated that this DAMP receptor (AtPEPR1) also binds to the auxiliary LRR-RLK BAK1. Research from our laboratory (Ma *et al.*, 2012) suggests a functional interdependency between FLS2 and AtPEPR1 signalling; optimal immune responses to either the PAMP flagellin or the DAMP AtPep peptide require the functional presence of both PRRs (FLS2 and AtPEPR1) in the plant cell PM as well as the generation of a Ca^{2+} signal in the cytosol. We have also shown that FLS2 and AtPEPR1 signalling pathways share some common downstream components, such as CPKs, NO and ROS generation, and up-regulation of defence gene expression (Ma *et al.*, 2013). In addition, BIK1 is required for Pep1-mediated immunity to *Botrytis cinerea* (*B. cinerea*) and BIK1 is phosphorylated by PEPR1 in response to both Pep1 and ethylene (Liu *et al.*, 2013). However, whether the phosphorylation requires Ca^{2+} or is required for the resistance to *B. cinerea* is unclear.

13.5 Cytosolic Ca^{2+} Elevation is an Early and Critical Component of Pathogen Defence Systems in the Plant Cell

As mentioned above, the plant cell cytosol Ca^{2+} level is tightly regulated and maintained under homeostasis at ~100 nM (Dodd *et al.*, 2010). Ca^{2+} rise in the cytosol is an early signalling event in plant immune responses to pathogens. As is the case with other signal transduction cascades involving a rise in cytosolic Ca^{2+}, the elevation associated with biotic stress signalling is transitory and returns (within minutes or less) to homeostatic levels after initial pathogen perception (Grant *et al.*, 2000; Lecourieux *et al.*, 2006; Ma *et al.*, 2010).

The cytosolic Ca^{2+} elevation that is a critical step in plant immunity is mediated by an increase in Ca^{2+} influx rather than a decrease in Ca^{2+} efflux from the cytosol (see references in Takabatake *et al.*, 2007). Presumably, as is the case with many other Ca^{2+} signalling pathways, continued function of the PM Ca^{2+} ATPase, the vacuolar H$^+$/Ca^{2+} antiporter and endomembrane (tonoplast and endoplasmic reticulum) Ca^{2+}-ATPases provide transport systems and pathways for Ca^{2+} movement out of the cytosol. The operation of these Ca^{2+} efflux (from the cytosol) systems would return the cytosolic Ca^{2+} to homeostatic levels after pathogen-associated elevations. Recent work supports the importance of these efflux systems for biotic stress signalling and this model of multiple systems involved in shaping the transitory pathogen-initiated cytosolic Ca^{2+} elevation. Zhu *et al.* (2010) found that translational arrest of an endomembrane-localized Ca^{2+}-ATPase resulted in an increased cytosolic Ca^{2+} signal and faster onset of HR in response to a pathogen elicitor in tobacco. Translational

arrest of two tonoplast-localized Ca^{2+}-ATPases resulted in spontaneous PCD and other pathogen response-related phenotypes (Boursiac *et al.*, 2010), suggesting a similar mechanism that blocking Ca^{2+} efflux increases the strength of the signal leading to pathogen defence responses.

Plant cells are surrounded by a pectin-containing wall that provides a reservoir of Ca^{2+} in the apoplast outside the plant cell PM that can be in the order of ~1–10 mM (White and Broadley, 2003). The cell wall could act as a source of chelated Ca^{2+} juxtaposed next to the PM *in planta*, and could supply the Ca^{2+} that enters the cell during pathogen-related cytosolic Ca^{2+} elevations. The plant cell vacuole and endoplasmic reticulum could represent 90% of the symplastic volume delimited by the plant cell PM and may also contain millimolar amounts of Ca^{2+} (Lecourieux *et al.*, 2006). Thus these endomembrane systems represent another source of Ca^{2+} contributing to cytosolic rise upon perception of pathogens by plant cells.

Due to the steep Ca^{2+} concentration gradient (the inside concentration is lower than the extracellular Ca^{2+}) between the apoplast (i.e. the reservoir of Ca^{2+} associated with the cell wall) and the cytosol, as well as the PM electrical potential gradient (inside approximately -150 mV), it would be expected that passive Ca^{2+} conduction through channels would dominate the overall Ca^{2+} uptake pathway into the plant cell, and this appears to be the case (White *et al.*, 2002). The plant genome does not encode any channel proteins with deduced sequences corresponding to the Ca^{2+}-selective channels of animals. Patch clamp studies indicate that the major inward Ca^{2+} current across the PM occurs through non-selective weakly voltage-gated cation channels (Demidchik *et al.*, 2002).

13.6 Channels Involved in Cytosolic Ca^{2+} Elevation Associated With Pathogen Defence Signalling

13.6.1 Endomembrane channels

Peiter *et al.* (2005) characterized TPC1 (a non-selective cation channel) and demonstrated that it is responsible for the major conductance of

Ca²⁺ across the tonoplast from the vacuole to the cytosol in *Arabidopsis*. TPC1 conductance is strongly gated by both voltage and Ca^{2+}; Ca^{2+} rise from homeostatic levels in the cytosol could dramatically activate this Ca^{2+}-activated Ca^{2+}-conducting channel. Thus this tonoplast protein provides a mechanism for amplification of an initial pathogen-induced rise in cytosolic Ca^{2+} into an even greater (either larger and/or more sustained) Ca^{2+} signal in the cytosol. Peiter *et al.* (2005) have demonstrated that TPC1-dependent Ca^{2+} flux into the cytosol is involved in (hormone-based) Ca^{2+} signalling pathways and that *tpc1* loss-of-function mutant plants lack the ability to translate a cytosolic Ca^{2+} increase arising from influx across the PM into a Ca^{2+}-dependent physiological response. However, more recent work has suggested that TPC1 is not involved in pathogen-induced cytosolic Ca^{2+} elevation (Ranf *et al.*, 2008). It has recently been shown, however, that TPC1 is involved in long-distance systemic Ca^{2+} signalling in response to herbivore sensing (Kiep *et al.*, 2015). The studies of TPC1 involvement in pathogen responses (Ranf *et al.*, 2008) were not designed to consider the issue of systemic signalling.

There are up to 57 cation-conducting channels encoded in the plant genome (see Mäser *et al.*, 2001). Aside from K-selective channels and the tonoplast-localized TPC1, 40 of these cation channels belong to two large protein families. There are 20 members each in the glutamate receptor (GLR) channel and cyclic nucleotide-gated channel (CNGC) families of proteins.

13.6.2 Ligand-gated channels

Plant GLRs are orthologues of the non-selective cation (including Ca^{2+}) conducting ligand-gated channels found in neurons of the mammalian central nervous system; members of this channel family are activated by extracellular glutamate and other amino acids to varying degrees (Davenport, 2002; Qi *et al.*, 2006). Plant GLRs have been demonstrated to conduct Ca^{2+} into the plant cell (Qi *et al.*, 2006); however, there is little direct evidence associating any member of this channel family with the Ca^{2+} flux associated with pathogen defence signalling. An exception is the work of Manzoor *et al.* (2013). Another report provides a loose association between

GLR-mediated Ca^{2+} uptake and altered plant response to pathogens (Kang *et al.*, 2006); in this case GLR-mediated response to the pathogen did not involve HR. Recent reviews (Forde and Roberts, 2014) provide some new perspectives on GLR involvement in plant responses to biotic stress. In consideration of their possible activation by a wide range of amino acid ligands, they might be involved in signalling responding to pathogens or herbivory by sensing the contents (released into the apoplast) of cells damaged by these biotic assaults.

Plant cyclic nucleotide-gated channels (CNGC1-20; ion conductance by these channels is activated by binding of cAMP or cGMP to the cytosolic portion of the channel; Fig. 13.2 A, B and C) have been confirmed to conduct Ca^{2+} by expression of their coding sequences in heterologous systems as well as using native plant cell membranes (Ma and Berkowitz, 2007; Ma *et al.*, 2010). There is a substantial amount of evidence linking CNGCs with the Ca^{2+} signalling that occurs during plant immune responses. Mutations in several (*Arabidopsis*) CNGCs have been associated with altered plant responses to pathogens. Loss-of-function of CNGC2 (the 'defence-no-death' or *dnd1* mutant) and CNGC4 (the 'HR-like lesion mimic *hlm1*' or '*dnd2*' mutant) alters plant responses to avirulent pathogens (including *Pseudomonas syringae*). *Arabidopsis* plants with these mutations display impaired HR, constitutive expression of salicylic acid (SA), altered expression of pathogen defence-related genes, and (despite the lack of HR) increased resistance to pathogen growth unrelated to the HR. A recent study demonstrated that upon phytosulfokine (PSK) treatment, both *Arabidopsis* PSK receptor PSKR1 and CNGC17 physically interacted with BAK1 and the two H⁺-ATPase, AHA1 and AHA2, to regulate plant growth (Ladwig *et al.*, 2015). These results indicate that CNGC17 is involved in PSK signalling. However, no work has yet tested whether PSK-induced cytosolic Ca^{2+} elevation is critical to this signalling cascade. In addition to the evidence showing physical association of the CNGC with the PSKR1 coreceptor BAK1, PSKR1 has the putative guanylyl cyclase (GC) domain, and the mutation of one of the three putative functional amino acids, glycine, showed similar phenotypes to the PSKR double mutant *r1/r2* (Ladwig *et al.*, 2015). These findings further support the notion

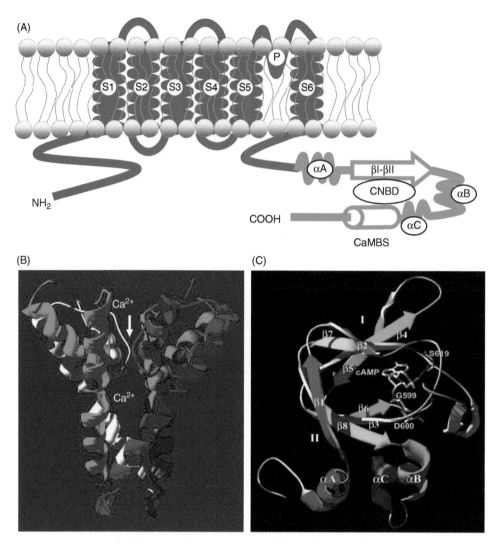

Fig. 13.2. Molecular structure of a plant cyclic nucleotide gated channel. CNGCs are thought to have the structure common to the superfamily of 'P-loop' ion channels; which are composed of four CNGC polypeptides. (A) A two-dimensional model of a single plant CNGC polypeptide. Each polypeptide has six membrane-spanning domains (S1–S6), a pore region (P) that dips into the membrane from the extracellular space, and amino- and carboxyl-termini extending into the cytosol. The carboxyl terminus has a cyclic nucleotide binding domain (CNBD) formed by three α helices (αA–C) and a β barrel formed by two β sheets. A CaM binding site (CaMBS) overlaps the CNBD. (B) The quaternary structure of the P loop region of a CNGC channel formed by four P loop regions of CNGC polypeptides (shown in different colours). The P loop regions traverse the membrane to form the ion (Ca^{2+}) conducting pathway. (C) A three-dimensional ribbon structure representation of CNBD on a single CNGC polypeptide, showing the α helices and β sheets (I and II) formed by eight β strands. Also shown is the activating ligand cAMP and specific amino acid residues that might contribute to ligand binding.

that the GC domain is functional and involved in the signalling, as asserted in work with other LRR-RLK receptors (Kwezi *et al.*, 2007; Meier *et al.*, 2010; Qi *et al.*, 2010; Ma *et al.*, 2012).

Evidence inconsistent with this model, however, can also be found in the literature. A crystal structural analysis of the brassinosteroid receptor, brassinosteroid insensitive1 (BRI1), showed that the putative GC domain could not be functional because it is in a hydrophobic environment and (theoretically) not exposed to the substrate GTP necessary for GC activity (Bojar et al., 2014).

The Arabidopsis mutant 'constitutive expresser of PR (pathogenesis-related) genes 22' (cpr22) was identified in a screen for mutations associated with altered activation of pathogen defence responses (Yoshioka et al., 2001). The cpr22 mutant displays different phenotypes from dnd1 and hlm1 (dnd2) plants, but does overlap with these CNGC loss-of-function mutants in having constitutively activated defence responses and enhanced resistance to P. syringae (cpr22 also was shown to have enhanced resistance to the oomycete pathogen Hyaloperonospora parasitica) (Yoshioka et al., 2001, 2006; Moeder and Yoshioka, 2008). The cpr22 mutation was identified as a 3-kb deletion that fuses two CNGC genes, CNGC11 and CNGC12, to generate a novel chimeric gene, CNGC11/12 (Yoshioka et al., 2006). Based on genetic and molecular analyses, it is suggested that the phenotype conferred by cpr22 is attributable to the expression of CNGC11/12. Analysis of the cngc11/12 mutant suggests that CNGC12 forms heteromeric channel protein complexes that are positive regulators of plant resistance to avirulent pathogens (Yoshioka et al., 2006). Urquhart et al. (2007) further investigated the nature of HR-like cell death induced by expression of CNGC11/12 using a transient expression system. In this study, cell death development was found to depend on Ca^{2+} influx into cells.

An increase in cAMP levels in plant leaves has been associated with the transient cytosolic Ca^{2+} spike occurring within minutes of pathogen perception (Ma et al., 2009). At a later time after pathogen inoculation (i.e. developing over several hours), cGMP levels rise (Meier et al., 2009). Other studies identify specific gene products as putative GCs that may play a role (i.e. activating CNGCs) in pathogen defence signalling. The translation product of the wall-associated kinase-like 10 gene has GC activity in vitro and is expressed along with pathogen defence-related genes (Meier et al., 2010). Some evidence suggests that a leucine-rich repeat receptor-like kinase (AtPEPR1) binds endogenous peptide signals in Arabidopsis, and is involved in pathogen defence signalling. This receptor may be a GC and acts upstream from CNGC2 in pathogen-induced

cytosolic Ca^{2+} elevations (Qi et al., 2010; Krol et al., 2010). Inhibitors of nucleotidyl cyclases block pathogen-induced cAMP elevation and prevent the coincident Ca^{2+} elevation and downstream pathogen defence signalling (Ma et al., 2009). Inhibitors of cyclic nucleotide breakdown (i.e. cyclic nucleotide phosphodiesterase inhibitors) lead to a greater elevation in cAMP and a faster onset of pathogen defence responses, presumably due to a stronger Ca^{2+} signal (Ma et al., 2009). These pharmacological studies, along with the work mentioned above with plants that have mutations in CNGC genes, and the study of cyclic nucleotide elevations during pathogen responses, provide evidence that the Ca^{2+} signal generated during pathogen defence responses in plants is dependent on CNGC function. Other than the studies cited above regarding AtPEPR1, there is no direct evidence identifying specific proteins involved in transducing the extracellular signal of pathogen perception to a cytosolic rise of cyclic nucleotide that could activate CNGCs during pathogen defence responses. An alternative possible mechanism for transduction of pathogen perception by a PRR protein in the PM, to a cytosolic signalling cascade which involves elevation of Ca^{2+} in the cytosol, could involve protein phosphorylation. A number of studies have demonstrated that pathogen and PAMP signalling are impaired by kinase inhibitors. Phosphorylation is known to be one mechanism of ion channel regulation. However, at present, no evidence links ion channel activation and Ca^{2+} conductance during pathogen signalling to any specific phosphorylation event involving an ion channel. Currently known Ca^{2+} conducting channel proteins are listed in Table 13.1.

13.6.3 Transcriptional regulation of CNGCs

As mentioned previously, although HR- and R gene-mediated immune resistance is suppressed in dnd1 and dnd2 mutants, both of the mutants are able to show enhanced resistance to pathogens, which could be partially due to the dramatic elevation of both free and glucoside-conjugated SA in both mutants. At the molecular level, the regulation of transcripts of these two genes

Table 13.1. Channel proteins involved in Ca²⁺ signalling and plant–pathogen interaction.

Proteins	Responses	References
CNGC2 and CNGC4	HR	Clough et al. (2000)
		Balagué et al. (2003)
		Jurkowski et al. (2004)
CNGC11 and CNGC12	Positively regulate plant defence	Yoshioka et al. (2006)
	HR-like cell death	Urquhart et al. (2007)
CNGC19	Significant transcript upregulation by flg22 or during	Moeder et al. (2011)
	bacterial pathogen infection	
CNGC19 and CNGC20	T-DNA mutants are more susceptible to pathogens	Moeder et al. (2011)
Radish GLR	No strong evidence shown to be involved in plant	Kang et al. (2006)
	immunity	
GLR3.3	glr3.3 showed defective defence responses to	Li et al. (2013)
	Pseudomonas syringae	
	Required for basal resistance to oomycete	Manzoor et al. (2013)
	pathogen, Hyaloperonospora arabidopsidis	
TPC1	Not involved in biotic stress in Arabidopsis.	Ranf et al. (2008)
	Regulate elicitor-induced hypersensitive cell	Kurusu et al. (2005)
	death in rice	
Calcium ATPase	HR	Zhu et al. (2010)
		Boursiac et al. (2010)

Note: If not indicated, the genes are from Arabidopsis. HR, hypersensitive response.

upon pathogen infection was investigated. Balagué et al. (2003) showed that HLM1 (CNGC4) transcription was slightly upregulated in the Columbia ecotype but not in the Ws ecotype between 8 and 24 h after the inoculation of an avirulent strain of Xanthomonas campestris. However, this finding is opposite to the later report that CNGC4 expression was significantly repressed 12 h after inoculation with Pseudomonas syringae pv. tomato (P.s.t.) DC3000 AvrRps4 (Zhu et al., 2010). Balagué et al. (2003) also demonstrated that CNGC2 is constitutively expressed at a low level and slightly reduced at the sites of infection in the Ws background. In addition, CNGC2 is upregulated in the hlm1(dnd2) mutant background, and seems to have no significant change in the transcript level after Xanthomonas infection (Balagué et al., 2003), suggesting that the two genes may be complementary to each other in response to pathogen infection when one is not functioning. Recently, it was shown that CNGC2 and CNGC4 physically interact with each other or themselves in vivo using bimolecular fluorescence complementation. This indicates that they could form heteromeric channels in the PM (Chin et al., 2013; also unpublished results by the present authors). In addition, two groups both reported that CNGC2

(García et al., 2010; Zhu et al., 2010) and CNGC4 (Zhu et al., 2010) transcription levels were highly repressed after Pseudomonas challenge, and the suppression depends on functioning EDS1. Zhu et al. (2010) further indicated that the suppression of the two transcripts is due to the negative transcriptional regulation of a transcriptional co-repressor, Topless-related1 (TPR1), which was identified from a screening for suppressors of suppressor of npr1, constitutive1 (snc1), which is a gain of function mutant of a TIR-NB-LRR type R gene (Zhang et al., 2003; Zhu et al., 2010).

Based on the above revealed correlation between the transcriptional regulation of CNGC2 and CNGC4 and plant innate immunity, it has been proposed that CNGC2 and CNGC4 are negative regulators of SA-dependent plant defence. However, the defective HR in both mutants indicates that CNGC2 and CNGC4 do not suppress plant defence to pathogens exclusively, but still act positively in plant innate immunity. CNGC2 and CNGC4 have been implicated to mediate Ca²⁺ fluxes (Hua et al., 2003; Ali et al., 2007), and Ca²⁺ has been demonstrated to play critical roles in plant innate immunity. Ali et al. (2007) reported that the NO generation induced by the PAMP lipopolysaccharide (LPS) is impaired in the *dnd1* mutant, due to the blockage of Ca^{2+} influx, and consequently diminishes HR response. As discussed above, *dnd1* and *dnd2* mutants hyperaccumulate SA. It has been reported that NO production can be induced by SA in Columbia and is Ca^{2+} dependent (Zottini *et al.*, 2007). This is consistent with the observation that even though there is an increased amount of SA in *dnd1* mutant, no excess accumulation of NO was detected because of the lack of cytosolic Ca^{2+} in *dnd1* mutant (Ali *et al.*, 2007). The counter reaction of *R* gene-mediated immune response and SA-dependent resistance in *dnd1* mutant suggests that Ca^{2+} could be a key switch of different plant innate immunity signalling pathways.

13.7 Downstream Effects of Cytosolic Ca^{2+} Elevation During Pathogen Signalling

Cytosolic Ca^{2+} elevation leads to a network of signalling events during plant immune response to pathogens; as mentioned above, one aspect of this resistance response that limits infection is HR, involving PCD (Hofius *et al.*, 2007). This signalling cascade is illustrated in Fig. 13.3. Downstream events from Ca^{2+} elevation in this signalling cascade (i.e. their effect is Ca^{2+}-dependent) which have been demonstrated to positively regulate PCD are: activation of CPKs, mitogen activated protein kinase (MAPK) cascades, NO and H_2O_2 (ROS) generation and increase in Ca^{2+} bound to calmodulin (CaM) or CaM-like proteins (CMLs). With several of these signal cascade events, it is clear that one way they facilitate HR is through effects on gene expression. ROS and NO can act independently and together to alter gene expression related to PCD during HR, and through actions independent of gene expression (Clarke *et al.*, 2000; Wendehenne *et al.*, 2001, 2004; Neill *et al.*, 2002; Delledonne, 2005; Torres *et al.*, 2006; van Breusegem and Dat, 2006; Zaninotto *et al.*, 2006; Hofius *et al.*, 2007).

MAPK signalling cascades (including the SA- and wound-induced MAPKs) downstream from Ca^{2+} elevation during pathogen response signalling activate plant pathogen defence-related gene expression including some transcription factors related to HR (Kim and Zhang, 2004; Pedley and Martin, 2005; Ren *et al.*, 2006; Hofius *et al.*,

2007; Stulemeijer *et al.*, 2007; Suarez-Rodriguez *et al.*, 2007; Pitzschke *et al.*, 2009). It is unclear how cytosolic Ca^{2+} elevation activates MAPK cascades but the Ca^{2+} dependency of the initial phosphorylations suggests that CPKs may be involved (Lecourieux-Ouaked *et al.*, 2000; Garcia-Brugger *et al.*, 2006). However, there is at least some degree of independence of signalling that acts through MAPKs and CPKs during plant pathogen defence cascades leading to transcriptional reprogramming (Boudsocq *et al.*, 2010). CPKs 4, 5, 6 and 11 act downstream from cytosolic Ca^{2+} elevation to translate PAMP perception to defence responses. MPK3 and MPK6 were found to have roles that were partially overlapping, but clearly independent of CPK, in the signalling pathway.

CaM (and/or calmodulin-like proteins (CMLs)) also affects pathogen response-signalling downstream from the cytosolic Ca^{2+} elevation by interacting with enzymatic proteins and transcriptional regulators. Overexpression of soybean (*Glycine max*) CaMs in tobacco (*Nicotiana tabacum*) leads to enhanced resistance to a range of pathogens (Heo *et al.*, 1999). Tobacco CaM silencing lines display increased susceptibility to virulent bacterial and fungal pathogens (Takabatake *et al.*, 2007). Silencing expression of the CML *APR134* in tomato (*Solanum lycopersicum*) impaired HR formation while overexpression of the *Arabidopsis* orthologue of APR134 (*AtCML43*) hastened HR development (Bouché *et al.*, 2005). LPS-induced NO generation in plant cells can be quenched by the addition of CaM antagonist (Ali *et al.*, 2007). AtCML42 and 37 have also been linked to biotic stress signalling (see multiple citations in Kiep *et al.*, 2015).

An arginine-dependent nitric oxide synthase- (NOS-)type enzyme is responsible for NO generation during pathogen response-signalling cascades (Delledonne *et al.*, 1998). A gene encoding a NOS-type enzyme has not yet been identified in plants; however, a recent breakthrough in the field has identified a NOS in a green alga (Foresi *et al.*, 2010). This newly identified plant NOS does have a deduced CaM binding domain and therefore supports a model whereby NOS-dependent NO generation could be activated by cytosolic Ca^{2+} elevation. A number of studies indicate that NOS-dependent NO generation is Ca^{2+}/CaM dependent and that this Ca^{2+} dependency is a component of pathogen defence signalling leading to NO generation

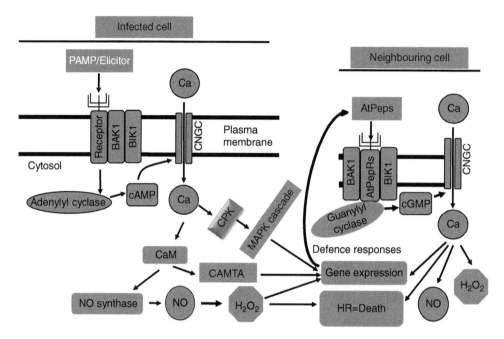

Fig. 13.3. Model of early steps in the plant innate immunity signal transduction pathway. In the infected cell, a PAMP binds to a PRR receptor leading to cAMP generation and activation of CNGC dependent Ca^{2+} conduction. Cytosolic Ca^{2+} elevation activates: (i) CPKs, leading to MAPK phosphorylation signalling, and (ii) CaM activation of CAMTA (CAM-binding transcription activator); both (i) and (ii) could alter gene expression. Ca^{2+}/CaM could activate NO synthase, leading to NO and downstream H_2O_2 production, and programmed cell death (HR). Defence gene expression leads to AtPep production and release as cells die. AtPeps could move to neighbouring cells and, upon binding to AtPEPR receptors, generate cGMP and activate CNGCs to amplify the pathogen defence signal.

(Corpas *et al.*, 2004; Ma *et al.*, 2008; Choi *et al.*, 2009). Thus, CNGC-mediated cytosolic Ca^{2+} elevation upon pathogen perception could lead to NO generation due to Ca^{2+} binding to CaM (or a CML) which then interacts either directly or indirectly with NOS.

During plant–pathogen interactions, Ca^{2+}/ CaM can activate NAD kinase, which would result in increased NADPH, the substrate for the oxidative burst responsible for ROS production during the HR (Harding and Roberts, 1998). A rise in cytosolic Ca^{2+} could affect ROS production during HR in other ways, such as through CPK signalling. Some studies suggest involvement of CPKs in HR. Two CPK isoforms in potato (*Solanum tuberosum*) have been shown to phosphorylate NADPH oxidase, which generates the ROS product H_2O_2 during HR (Kobayashi *et al.*, 2007). Other studies have suggested that this ROS-generating enzyme may be affected by direct binding of Ca^{2+} (Torres *et al.*, 2006).

Recent work has begun to provide insights into how CNGC-dependent conductance of Ca^{2+} into the cytosol could impact on expression of defence-related genes during pathogen response signalling. The *Arabidopsis CBP60g* gene product is a CaM-responsive protein involved in pathogen responses; it is required for PAMP-mediated SA generation and other defence responses (Wang *et al.*, 2009). The absence of CaM interaction with this protein results in enhanced disease progression. The expression product of the barley (*Hordeum vulgare*) *MLO* (mildew resistance locus *o*) gene is another CaM target involved in plant responses to pathogens (Kim *et al.*, 2002). MLO may be a negative regulator in plant defence responses. The *Arabidopsis* signal-responsive protein SR1, a member of the CaM binding transcription activator protein family, is also a negative regulator of plant pathogen signalling (Galon *et al.*, 2008; Du *et al.*, 2009), which could act downstream from cytosolic Ca^{2+} elevation through

CaM/CML signalling to regulate expression of defence-related genes (including *PR1* and regulators of SA synthesis).

ROS production is also a component of pathogen signalling downstream from cytosolic Ca^{2+} elevation. Plasma membrane-localized NADPH oxidase generates ROS (O_2^-) that causes the oxidative burst leading to HR during pathogen defence responses. The O_2^- is converted to H_2O_2 by superoxide dismutase (SOD) (Lecourieux *et al.*, 2006). Kobayashi *et al.* (2007) identified two CPKs as phosphorylating NADPH oxidase and thereby upregulating ROS production during pathogen signalling. In this recent work, however, prevention of NADPH oxidase activation in CPK loss-of-function mutants was not investigated, leaving open the possibility that other mechanisms may be involved as well. Recently, Dubiella *et al.* (2013) showed that *Arabidopsis* CPK5 directly phosphorylated RBOHD, the NADPH oxidase, to activate systemic defence signalling. NADPH oxidase has its own Ca^{2+}-binding domains, and other work has shown that Ca^{2+} is required for maximal NADPH oxidase activity *in vitro* (Sagi and Flurh, 2006), suggesting that Ca^{2+} elevations during pathogen signalling could directly activate ROS production.

13.8 Concluding Perspectives: Plant Immunity and the Generation of a Ca^{2+} Signal

The information provided above has presented a relatively detailed analysis of how the perception of non-self is translated to plant immune responses to pathogens, specifically focusing on the role Ca^{2+} plays in the stress response pathways and signal transduction. For nearly two decades, the linkage between pathogen recognition and plant immune responses at the cellular level has been known to involve influx into the cell of Ca^{2+} as an early and paramount step of pathogen defence response signal transduction cascades (Nürnberger *et al.*, 1994). However, it is only until recently that specific molecular steps in this Ca^{2+} signalling cascade have been identified, and many more await discovery. Nonetheless, a few closing points can provide a proper context for consideration of the significance of the signalling network that we have focused on here.

One point to consider is the recent identification of pathogen 'cloaking devices' akin to those of some imaginary alien invaders portrayed in popular science fiction literature. Some pathogens (of both animals and plants), such as *P. syringae*, *Xanthomonas campestris* pv. *campestris*, *Pseudomonas aeruginosa* and *Erwinia amylovora*, secrete high molecular extracellular polysaccharides (EPS) that have been linked with maximal pathogenicity and virulence (Aslam *et al.*, 2008). Interestingly, a recent study (Aslam *et al.*, 2008) has demonstrated that these EPS, 'surface mucos'-generating pathogens, take advantage of these polymers as Ca^{2+}-chelating agents. These EPS films bind to the apoplastic Ca^{2+} outside of plant cells that is required for the plant cell to initiate a cytosolic Ca^{2+} elevation. Thus, these pathogens use EPS to prevent plant cells from generating the 'signalling alarm' that transmits the message that non-self has been recognized at the cell PM into the Ca^{2+}-dependent immune signalling, which leads to the invocation of the plant cell's pathogen defence responses.

Another point emerging in the current literature that underscores the linkage between Ca^{2+} signalling and plant immune responses regards the ubiquity of the system. The generation of a cytosolic Ca^{2+} elevation appears to be a ubiquitous means of 'sounding the alarm' when it comes to signalling the perception of impending trouble due to pathogen invasion. The Ca^{2+} signal has been shown to be essential for PTI (Ali *et al.*, 2007; Aslam *et al.*, 2008; Ma *et al.*, 2009). A similar Ca^{2+} also underlies the HR, or ETI responses to avirulent pathogens (Grant *et al.*, 2000; Ali *et al.*, 2007). In addition to PAMPs and the HR responses that occur to limit the disease symptoms and spread of avirulent pathogens, a similar Ca^{2+} is also necessary for signalling leading to defence responses evoked by microbial pathogenic toxins (i.e. elicitors). It has been shown that plant immune responses to the bacterial toxin thaxtomin A and the fungal toxin cryptogein also require a Ca^{2+} signal (Garcia-Brugger *et al.*, 2006; Errakhi *et al.*, 2008). Work from this laboratory has linked signalling downstream from the plant-generated DAMP signal of AtPep peptides as requiring a cytosolic Ca^{2+} elevation (Qi *et al.*, 2010). The plant cell wall breakdown product oligogalacturonide, a HAMP, also evokes a similar Ca^{2+}

signal (Lecourieux *et al.*, 2006). These pathogen-associated signalling pathways clearly involve a wide variety of 'non-self'-recognizing receptors (i.e. beyond the LRR-RLK PRRs mentioned above) as initial proteins involved in generating the warning signal. The immune response pathways that occur downstream from this variety of pathogenic cues also vary; yet they all involve the Ca^{2+} signal as a critical component of the response to the biotic stress.

Acknowledgement

This work was supported by National Science Foundation award IOS-1146827 to GAB.

References

Ali, R., Ma, W., Lemtiri-Chlieh, F., Tsaltas, D., Leng, Q., von Bodman, S. and Berkowitz, G.A. (2007) Death don't have no mercy and neither does calcium: *Arabidopsis* cyclic nucleotide gated channel2 and innate immunity. *Plant Cell* 19, 1081–1095.

Arimura, G.I. and Maffei, M.E. (2010) Calcium and secondary CPK signaling in plants in response to herbivore attack. *Biochemical and Biophysical Research Communications* 400, 455–460.

Aslam, S.N., Newman, M.-A., Erbs, G., Morrissey, K.L., Chinchilla, D., Boller, T., Jensen, T.T., De Castro, C., Ierano, T., Molinaro, A., Jackson, R.W., Knight, M.R. and Cooper, R.M. (2008) Bacterial polysaccharides suppress induced innate immunity by calcium chelation. *Current Biology* 18, 1078–1083.

Balagué, C., Lin, B., Alcon, C., Flottes, G., Malmström, S., Köhler, C., Neuhaus, G., Pelletier, G., Gaymard, F. and Roby, D. (2003) HLM1, and essential signaling component in the hypersensitive response, is a member of the cyclic nucleotide-gated channel ion channel family. *Plant Cell* 15(2), 365–379.

Bigeard, J., Colcombet, J. and Hirt, H. (2015) Signaling mechanisms in pattern-triggered immunity (PTI). *Molecular Plant* 8, 521–539.

Bojar, D., Martinez, J., Santiago, J., Rybin, V., Bayliss, R. and Hothorn, M. (2014) Crystal structures of the phosphorylated BRI1 kinase domain and implications for brassinosteroid signal initiation. *Plant Journal* 1, 2–5.

Boller, T. and Felix, G. (2009) A renaissance of elicitors: perception of microbe-associated molecular patterns and danger signals by pattern-recognition receptors. *Annual Review of Plant Biology* 60, 379–406.

Boller, T. and He, S.Y. (2009) Innate immunity in plants: an arms race between effectors in microbial pathogens and pattern recognition receptors in plants. *Science* 324, 742–744.

Bouché, N., Yellin, A., Snedden, W.A. and Fromm, H. (2005) Plant-specific calmodulin-binding proteins. *Annual Review of Plant Biology* 56, 435–366.

Boudsocq, M., Willmann, M.R., McCormack, M., Lee, H., Shan, L., He, P., Bush, J., Cheng, S.H. and Sheen, J. (2010) Differential innate immune signaling via Ca^{2+} sensor protein kinases. *Nature* 464, 418–422.

Boursiac, Y., Lee, S.M., Romanowsky, S., Blank, R., Slakek, C., Chung, W.S. and Harper, J.F. (2010) Disruption of the vacuolar calcium-ATPases in Arabidopsis results in the activation of a salicylic acid-dependent programmed cell death pathway. *Plant Physiology* 154, 1158–1171.

Bricchi, I., Leitner, M., Foti, M., Mithöfer, A., Boland, W. and Maffei, M.E. (2010) Robotic mechanical wounding (MecWorm) versus herbivore-induced responses: early signaling and volatile emission in Lima bean (*Phaseolus lunatus* L.). *Planta* 232, 719–729.

Case, R.M., Eisner, D., Gurney, A., Jones, O., Muallemd, S. and Verkhratsky, A. (2007) Evolution of calcium homeostasis: from birth of the first cell to an omnipresent signalling system. *Cell Calcium* 42, 345–350.

Chin, K., DeFalco, T.A., Moeder, W. and Yoshioka, K (2013) The Arabidopsis cyclic nucleotide-gated ion channels AtCNGC2 and AtCNGC4 work in the same signaling pathway to regulate pathogen defense and floral transition. *Plant Physiology* 163, 611–624.

Choi, H.W., Lee, D.H. and Hwang, B.K. (2009) The pepper calmodulin gene *CaCaM1* is involved in reactive oxygen species and nitric oxide generation required for cell death and the defense response. *Molecule Plant-Microbe Interaction* 22, 1389–1400.

Clarke, A., Desikan, R., Hurst, R.D., Hancock, J.T. and Neill, S.J. (2000) NO way back: nitric oxide and programmed cell death in *Arabidopsis thaliana* suspension cultures. *Plant Journal* 24, 667–677.

Clough, S.J., Fengler, K.A., Yu, I.C., Lippok, B., Smith Jr, R.K. and Bent, A.F. (2000) The *Arabidopsis dnd1* "defense, no death" gene encodes a mutated cyclic nucleotide-gated ion channel. *Proceedings of the National Academy of Sciences USA* 97, 9323–9328.

Corpas, F.J., Barroso, J.B., Carreras, A., Quirós, M., León, A.M., Romero-Puertas, M.C., Esteban, F.J., Valderrama, R., Palma, J.M., Sandalio, L.M., Gómez, M. and del Río, L.A. (2004) Cellular and subcellular localization of endogenous NO in young and senescent pea plants. *Plant Physiology* 136, 2722–2733.

Davenport, R. (2002) Glutamate receptors in plants. *Annals of Botany* 90, 549–557.

Delledonne, M. (2005) NO news is good news for plants. *Current Opinion in Plant Biology* 8, 390–396.

Delledonne, M., Xia, Y., Dixon, R.A. and Lamb, C. (1998) NO functions as a signal in plant disease resistance. *Nature* 394, 585–588.

Demidchik, V., Bowen, H.C., Maathuis, F.J.M., Shabala, S.N., Tester, M.A., White, P.J. and Davies, J.M. (2002) *Arabidopsis thaliana* root nonselective cation channels mediate calcium uptake and are involved in growth. *Plant Journal* 32, 799–808.

Dodd, A.N., Kudla, J. and Sanders, D. (2010) The language of calcium signaling. *Annual Review of Plant Biology* 61, 593–620.

Du, L., Ali, G.S., Simons, K.A., Hou, J., Yang, T., Reddy, A.S. and Poovaiah, B.W. (2009) Ca^{2+}/calmodulin regulates salicylic-acid-mediated plant immunity. *Nature* 457, 1154–1158.

Dubiella, U., Seybold, H., Durian, G., Komander, E., Lassig, R. and Witte, C. (2013) Calcium-dependent protein kinase/NADPH oxidase activation circuit is required for rapid defense signal propagation. *Proceedings of the National Academy of Sciences USA* 110, 8744–8749.

Errakhi, R., Dauphin, A., Meimoun, P., Lehner, A., Reboutier, D., Vatsa, P., Briand, P.J., Madiona, K., Rona, J.P., Barakate, M., Wendehenne, D., Beaulieu, C. and Bouteau, F. (2008) An early Ca^{2+} influx is a prerequisite to thaxtomin A-induced cell death in *Arabidopsis thaliana* cells. *Journal of Experimental Botany* 59, 4259–4270.

Forde, B.G. and Roberts, M.R. (2014) Glutamate receptor-like channels in plants: a role as amino acid sensors in plant defence? *F1000 Prime Reports* 6, 37.

Foresi, N., Correa-Aragunde, N., Parisi, G., Calo, G., Salerno, G. and Lamattina, L. (2010) Characterization of a nitric oxide synthase from the plant kingdom: NO generation from the green alga *Ostreococcus tauri* is light irradiance and growth phase dependent. *Plant Cell* 22, 3816–3830.

Galletti, R., De Lorenzo, G. and Ferrari, S. (2009) Host-derived signals activate plant innate immunity. *Plant Signaling and Behavior* 4, 33–34.

Galon, Y., Nave, R., Boyce, J.M., Nachmias, D., Knight, M.R. and Fromm, H. (2008) Calmodulin-binding transcription activator (CAMTA) 3 mediates biotic defense responses in *Arabidopsis*. *FEBS Letter* 582, 943–948.

Galon, Y., Finkler, A. and Fromm, H. (2010) Calcium mediated transcription in plants. *Molecular Plant* 3, 653–669.

García, A.V., Blanvillain-Baufumé, S., Huibers, R.P., Wiermer, M., Li, G., Gobbato, E., Rietz, S. and Parker, J.E. (2010) Balanced nuclear and cytoplasmic activities of EDS1 are required for a complete plant innate immune response. *PLoS Pathogens* 6, e1000970.

Garcia-Brugger, A., Lamotte, O., Vandelle, E., Bourque, S., Lecourieux, D., Poinssot, B., Wendehenne, D. and Pugin, A. (2006) Early signaling events induced by elicitors of plant defenses. *Molecular Plant-Microbe Interaction* 19, 711–724.

Grant, M. and Lamb, C. (2006) Systemic immunity. *Current Opinion in Plant Biology* 9, 414–420.

Grant, M., Brown, I., Adams, S., Knight, M., Ainslie, A. and Mansfield, J. (2000) The *RPM1* plant disease resistance gene facilitates a rapid and sustained increase in cytosolic calcium that is necessary for the oxidative burst and hypersensitive cell death. *Plant Journal* 23, 441–450.

Grün, S., Lindermayr, C., Sell, S. and Durner, J. (2006) Nitric oxide and gene regulation in plants. *Journal of Experimental Botany* 57, 507–516.

Harding, S.A. and Roberts, D.M. (1998) Incompatible pathogen infection results in enhanced reactive oxygen and cell death responses in transgenic tobacco expressing a hyperactive mutant calmodulin. *Planta* 206, 253–258.

He, P., Shan, L. and Sheen, J. (2007) Elicitation and suppression of microbe-associated molecular pattern-triggered immunity in plant-microbe interactions. *Cell Microbiology* 9, 1385–1396.

Heo, W.D., Lee, S.H., Kim, M.C., Kim, J.C., Chung, W.S., Chun, H.J., Lee, K.J., Park, C.Y., Park, H.C., Choi, J.Y. and Cho, M.J. (1999) Involvement of specific calmodulin isoforms in salicylic acid-independent activation of plant disease resistance responses. *Proceedings of the National Academy of Sciences USA* 96, 766–771.

Hetherington, A.M. and Brownlee, C. (2004) The generation of Ca^{2+} signals in plants. *Annual Review of Plant Biology* 5, 401–427.

Hofius, D., Tsitsigiannis, D.I., Jones, J.D. and Mundy, J. (2007) Inducible cell death in plant immunity. *Seminars in Cancer Biology* 17, 166–187.

Hu, X., Neill, S.J., Yang, Y. and Cai, W. (2009) Fungal elicitor Pep-25 increases cytosolic calcium ions, H$_2$O$_2$ production and activates the octadecanoid pathway in *Arabidopsis thaliana*. *Planta* 229, 1201–1208.

Hua, B.G., Mercier, R.W., Zielinski, R.E. and Berkowitz, G.A. (2003) Functional interaction of calmodulin with a plant cyclic nucleotide gated cation channel. *Plant Physiology and Biochemistry* 41, 945–954.

Huffaker, A., Pearce, G. and Ryan, C.A. (2006) An endogenous peptide signal in *Arabidopsis* activates components of the innate immune response. *Proceedings of the National Academy of Sciences USA* 103, 10098–100103.

Huffaker, A., Pearce, G., Veyrat, N., Erb, M., Turlingsc, T.C.J., Sartor, R., Shen, Z., Briggs, S.P., Vaughan, M.M., Alborn, H.T., Teala, P.E.A. and Schmelz, E.A. (2013) Plant elicitor peptides are conserved signals regulating direct and indirect antiherbivore defense. *Proceedings of the National Academy of Sciences USA* 110, 5707–5712.

Jaiswal, J.K. (2001) Calcium—how and why? *Journal of Biosciences* 26, 357–363.

Jones, J.D. and Dangl, J.L. (2006) The plant immune system. *Nature* 444, 323–329.

Jurkowski, G.I., Smith Jr, R.K., Yu, I.C., Ham, J.H., Sharma, S.B., Klessig, D.F., Fengler, K.A. and Bent, A.F. (2004) *Arabidopsis DND2*, a second cyclic nucleotide-gated ion channel gene for which mutation causes the "defense, no death" phenotype. *Molecular Plant-Microbe Interaction* 17, 511–520.

Kadota, Y., Sklenar, J., Derbyshire, P., Stransfeld, L., Asai, S., Ntoukakis, V., Jones, D.G.J., Shirasu, K., Menke, F., Jones, A. and Zipfel, C. (2014) Direct regulation of the NADPH oxidase RBOHD by the PRR-associated kinase BIK1 during plant immunity. *Molecular Cell* 54, 43–55.

Kanchiswamy, C.N., Takahashi, H., Quadro, S., Maffei, M.E., Bossi, S., Bertea, C., Zebelo, S.A., Muroi, A., Ishihama, N., Yoshioka, H., Boland, W., Takabayashi, J., Endo, Y., Sawasaki, T. and Arimura, G. (2010) Regulation of Arabidopsis defense responses against *Spodoptera littoralis* by CPK-mediated calcium signaling. *BMC Plant Biology* 10, 97.

Kang, S., Kim, H.B., Lee, H., Choi, J.Y., Heu, S., Oh, C.J., Kwon, S.I. and An, C.S. (2006) Over expression in *Arabidopsis* of a plasma membrane-targeting glutamate receptor from small radish increases glutamate-mediated Ca²⁺ influx and delays fungal infection. *Molecular Cells* 21, 418–427.

Kiep, V., Vadassery, J., Lattke, J., Maaß, J.-P., Boland, W., Peiter, E. and Mithöfer, A. (2015) Systemic cytosolic Ca²⁺ elevation is activated upon wounding and herbivory in Arabidopsis. *New Phytologist* 207, 996–1004.

Kim, C.Y. and Zhang, S. (2004) Activation of a mitogen-activated protein kinase cascade induces WRKY family of transcription factors and defense genes in tobacco. *Plant Journal* 38, 142–151.

Kim, M.C., Panstruga, R., Elliott, C., Müller, J., Devoto, A., Yoon, H.W., Park, H.C., Cho, M.J. and Schulze-Lefert, P. (2002) Calmodulin interacts with MLO protein to regulate defense against mildew in barley. *Nature* 416, 447–451.

Kim, M.C., Chung, W.S., Yun, D.-J. and Cho, M.J. (2009) Calcium and calmodulin-mediated regulation of gene expression in plants. *Molecular Plant* 2, 13–21.

Klauser, D., Desurmont, G.A., Glauser, G., Vallat, A., Flury, P., Boller, T., Turlings, T.C.J. and Bartels, S. (2015) The *Arabidopsis* Pep-PEPR system is induced by herbivore feeding and contributes to JA-mediated plant defence against herbivory. *Journal of Experimental Botany* 66, 1–10.

Kobayashi, M., Ohura, I., Kawakita, K., Yokota, N., Fujiwara, M., Shimamoto, K., Doke, N. and Yoshioka, H. (2007) Calcium-dependent protein kinases regulate the production of reactive oxygen species by potato NADPH oxidase. *Plant Cell* 19, 1065–1080.

Krol, E., Mentzel, T., Chinchilla, D., Boller, T., Felix, G., Kemmerling, B., Postel, S., Arents, M., Jeworutzki, E., Al-Rasheid, K.A., Becker, D. and Hedrich, R. (2010) Perception of the *Arabidopsis* danger signal peptide 1 involves the pattern recognition receptor AtPEPR1 and its close homologue AtPEPR2. *Journal of Biological Chemistry* 285, 13471–13479.

Kudla, J., Batistic, O. and Hashimoto, K. (2010) Calcium signals: the lead currency of plant information processing. *Plant Cell* 22, 541–563.

Kurusu, T., Yagala, T., Miyao, A., Hirochika, H. and Kuchitsu, K. (2005) Identification of a putative voltage-gated Ca²⁺ channel as a key regulator of elicitor-induced hypersensitive cell death and mitogen-activated protein kinase activation in rice. *Plant Journal* 42, 798–809.

Kwezi, L., Meier, S., Mungur, L., Ruzvidzo, O., Irving, H. and Gehring, C. (2007) The *Arabidopsis thaliana* brassinosteroid receptor (AtBRI1) contains a domain that functions as a guanylyl cyclase *in vitro*. *PLoS ONE* 2, e449.

Ladwig, F., Dahlke, R.I., Stührwohldt, N., Hartmann, J., Harter, K. and Sauter, M. (2015) Phytosulfokine regulates growth in Arabidopsis through a response module at the plasma membrane that includes cyclic nucleotide-gated channel17, H⁺-ATPase, and BAK1. *Plant Cell* 27, 1718–1729.

Lecourieux, D., Ranjeva, R. and Pugin, A. (2006) Calcium in plant defence-signalling pathways. *New Phytologist* 171, 249–269.

Lecourieux-Ouaked, F., Pugin, A. and Lebrun-Garcia, A. (2000) Phosphoproteins involved in the signal transduction of cryptogein, an elicitor of defense reactions in tobacco. *Molecular Plant-Microbe Interaction* 13, 821–829.

Li, F., Wang, J., Ma, C., Zhao, Y., Wang, Y., Hasi, A. and Qi, Z. (2013) Glutamate receptor-like channel 3.3 is involved in mediating glutathione-triggered cytosolic calcium transients, transcriptional changes, and innate immunity responses in Arabidopsis. *Plant Physiology* 162, 1497–1509.

Lin, W., Li, B., Lu, D., Chen, S., Zhu, N., He, P. and Shan, L. (2014) Tyrosine phosphorylation of protein kinase complex BAK1/BIK1 mediates *Arabidopsis* innate immunity. *Proceedings of the National Academy of Sciences USA* 111, 3632–3637.

Liu, Z., Wu, Y., Yang, F., Zhang, Y., Chen, S., Xie, Q., Tian, X. and Zhou, J.-M. (2013) BIK1 interacts with PEPRs to mediate ethylene-induced immunity. *Proceedings of the National Academy of Sciences USA* 110, 6205–6210.

Lu, D., Wu, S., Gao, X., Zhang, Y., Shan, L. and He, P. (2010) A receptor-like cytoplasmic kinase, BIK1, associates with a flagellin receptor complex to initiate plant innate immunity. *Proceedings of the National Academy of Sciences USA* 107, 496–501.

Ma, W. and Berkowitz, G.A. (2007) The grateful dead: calcium and cell death in plant innate immunity. *Cellular Microbiology* 9, 2571–2585.

Ma, W., Smigel, A., Tsai, Y.C., Braam, J. and Berkowitz, G.A. (2008) Innate immunity signaling: cytosolic Ca^{2+} elevation is linked to downstream nitric oxide generation through the action of calmodulin or a calmodulin-like protein. *Plant Physiology* 148, 818–828.

Ma, W., Qi, Z., Smigel, A., Walker, R.K., Verma, R. and Berkowitz, G.A. (2009) Ca^{2+}, cAMP, and transduction of non-self perception during plant immune responses. *Proceedings of the National Academy of Sciences USA* 106, 20995–21000.

Ma, W., Yoshioka, K., Gehring, C.A. and Berkowitz, G.A. (2010) The function of cyclic nucleotide gated channels in biotic stress. In: Demidchik, V. and Maathuis, F.J.M. (eds) *Ion Channels and Plant Stress Responses*. Springer, Berlin/Heidelberg, Germany, pp. 159–174.

Ma, Y., Walker, K., Zhao, Y. and Berkowitz, G.A. (2012) Linking ligand perception by PEPR pattern recognition receptors to cytosolic Ca^{2+} elevation and downstream immune signaling in plants. *Proceedings of the National Academy of Sciences USA* 109, 19852–19857.

Ma, Y., Zhao, Y., Walker, R.K. and Berkowitz, G.A. (2013) Molecular steps in the immune signaling pathway evoked by plant elicitor peptides: Ca^{2+}-dependent protein kinases, nitric oxide, and reactive oxygen species are downstream from the early Ca^{2+} signal. *Plant Physiology* 163, 1459–1471.

Maffei, M., Bossi, S., Spiteller, D., Mithöfer, A. and Boland, W. (2004) Effects of feeding *Spodoptera littoralis* on lima bean leaves. I. Membrane potentials, intracellular calcium variations, oral secretions, and regurgitate components. *Plant Physiology* 134, 1752–1762.

Manzoor, H., Kelloniemi, J., Chiltz, A., Wendehenne, D., Pugin, A., Poinssot, B. and Garcia-Brugger, A. (2013) Involvement of the glutamate receptor AtGLR3.3 in plant defense signaling and resistance to *Hyaloperonospora arabidopsidis*. *Plant Journal* 76, 466–480.

Mäser, P., Thomine, S., Schroeder, J.I., Ward, J.M., Hirschi, K., Sze, H., Talke, I.N., Amtmann, A., Maathuis, F.J., Sanders, D., Harper, J.F., Tchieu, J., Gribskov, M., Persans, M.W., Salt, D.E., Kim, S.A. and Guerinot, M.L. (2001) Phylogenetic relationships within cation transporter families of *Arabidopsis*. *Plant Physiology* 126, 1646–1667.

McAinsh, M.R. and Pittman, J.K. (2009) Shaping the calcium signature. *New Phytologist* 181, 275–294.

Meier, S., Madeo, L., Ederli, L., Donaldson, L., Pasqualini, S. and Gehring, C. (2009) Deciphering cGMP signatures and cGMP-dependent pathways in plant defence. *Plant Signaling and Behavior* 4, 307–309.

Meier, S., Ruzvidzo, O., Morse, M., Donaldson, L., Kwezi, L. and Gehring, C. (2010) The *Arabidopsis* wall associated kinase-like 10 gene encodes a functional guanylyl cyclase and is co-expressed with pathogen defense related genes. *PLoS ONE* 5, e8904.

Mészáros, T., Helfer, A., Hatzimasoura, E., Magyar, Z., Serazetdinova, L., Rios, G., Bardóczy, V., Teige, M., Koncz, C., Peck, S. and Bögre, L. (2006) The *Arabidopsis* MAP kinase kinase MKK1 participates in defense responses to the bacterial elicitor flagellin. *Plant Journal* 48, 485–498.

Mithöfer, A. and Boland, W. (2008) Recognition of herbivory-associated molecular patterns. *Plant Physiology* 146, 825–831.

Moeder, W. and Yoshioka, K. (2008) Lesion mimic mutants. *Plant Signaling and Behavior* 3, 764–767.

Moeder, W., Urquharta, W., Unga, H. and Yoshioka, K. (2011) The role of cyclic nucleotide-gated channels in plant immunity. *Molecular Plant* 4, 442–452.

Munemasa, S., Oda, K., Watanabe-Sugimoto, M., Nakamura, Y., Shimoishi, Y. and Murata, Y. (2007) The coronatine-insensitive 1 mutation reveals the hormonal signaling interaction between abscisic acid and methyl jasmonate in *Arabidopsis* guard cells. Specific impairment of ion channel activation and second messenger production. *Plant Physiology* 143, 1398–1407.

Naito, K., Taguchi, F., Suzuki, T., Inagaki, Y., Toyoda, K., Shiraishi, T. and Ichinose, Y. (2008) Amino acid sequence of bacterial microbe-associated molecular pattern flg22 is required for virulence. *Molecular Plant-Microbe Interaction* 21, 1165–1174.

Neill, S.J., Desikan, R., Clarke, A., Hurst, R.D. and Hancock, J.T. (2002) Hydrogen peroxide and nitric oxide as signalling molecules in plants. *Journal of Experimental Botany* 53, 1237–1247.

Newman, M.-A., Dow, J.M., Molinaro, A. and Parrilli, M. (2007) Priming, induction and modulation of plant defense responses by bacterial lipopolysaccharides. *Journal of Endotoxin Research* 13, 69–84.

Nicaise, V., Roux, M. and Zipfel, C. (2009) Recent advances in PAMP-triggered immunity against bacteria: pattern recognition receptors watch over and raise the alarm. *Plant Physiology* 150, 1638–1647.

Nürnberger, T., Nennstiel, D., Jabs, T., Sacks, W.R., Hahlbrock, K. and Scheel, D. (1994) High affinity binding of a fungal oligopeptide elicitor to parsley plasma membranes triggers multiple defense responses. *Cell* 78, 449–460.

Pedley, K.F. and Martin, G.B. (2005) Role of mitogen-activated protein kinases in plant immunity. *Current Opinion in Plant Biology* 8, 541–547.

Peiter, E., Maathuis, F.J.M., Mills, L.N., Knight, H., Pelloux, J., Hetherington, A.M. and Sanders, D. (2005) The vacuolar Ca^{2+}-activated channel TPC1 regulates germination and stomatal movement. *Nature* 434, 404–408.

Pitzschke, A., Schikora, A. and Hirt, H. (2009) MAPK cascade signaling networks in plant defence. *Current Opinion in Plant Biology* 12, 421–426.

Postel, S., Kufner, I., Beuter, C., Mazzotta, S., Schwedt, A., Borlotti, A., Halter, T., Kennerling, B. and Nurnberger, T. (2010) The multifunctional leucine-rich repeat kinase BAK1 is implicated in *Arabidopsis* development and immunity. *European Journal of Cell Biology* 89, 169–174.

Qi, Z., Stephens, N.R. and Spalding, E.P. (2006) Calcium entry mediated by GLR3.3, an *Arabidopsis* glutamate receptor with a broad agonist profile. *Plant Physiology* 142, 963–971.

Qi, Z., Verma, R., Gehring, C., Yamaguchi, Y., Zhao, Y., Ryan, C.A. and Berkowitz, G.A. (2010) Ca^{2+} signaling by plant *Arabidopsis thaliana* Pep peptides depends on *At*PepR1, a receptor with guanylyl cyclase activity, and cGMP-activated Ca^{2+} channels. *Proceedings of the National Academy of Sciences USA* 107, 21193–21198.

Qutob, D., Kemmerling, B., Brunner, F., Küfner, I., Engelhardt, S., Gust, A.A., Luberacki, B., Seitz, H.U., Stahl, D., Rauhut, T., Glawischnig, E., Schween, G., Lacombe, B., Watanabe, N., Lam, E., Schlichting, R., Scheel, D., Nau, K., Dodt, G., Hubert, D., Gijzen, M. and Nürnberger, T. (2006) Phytotoxicity and innate immune responses induced by Nep1-like proteins. *Plant Cell* 18, 3721–3744.

Ranf, S., Wunnenberg, P., Lee, J., Becker, D., Dunkel, M., Hedrich, R., Scheel, D. and Dietrich, P. (2008) Loss of the vacuolar cation channel *At*TPC1, does not impair Ca^{2+} signals induced by abiotic and biotic stresses. *Plant Journal* 53, 287–299.

Ren, D., Yang, K.Y., Li, G.J., Liu, Y. and Zhang, S. (2006) Activation of Ntf4, a tobacco mitogen-activated protein kinase, during plant defense response and its involvement in hypersensitive response-like cell death. *Plant Physiology* 141, 1482–1493.

Robatzek, S., Chinchilla, D. and Boller, T. (2006) Ligand-induced endocytosis of the pattern recognition receptor FLS2 in *Arabidopsis*. *Genes & Development* 20, 537–542.

Sagi, M. and Fluhr, R. (2006) Production of reactive oxygen species by plant NADPH oxidases. *Plant Physiology* 141, 336–340.

Schulze, B., Mentzel, T., Jehle, A.K., Mueller, K., Beeler, S., Boller, T., Felix, G. and Chinchilla, D. (2010) Rapid heteromerization and phosphorylation of ligand-activated plant transmembrane receptors and their associated kinase BAK1. *Journal of Biological Chemistry* 285, 9444–9451.

Seybold, H., Trempel, F., Ranf, S., Scheel, D., Romeis, T. and Lee, J. (2014) Ca^{2+} signalling in plant immune response: from pattern recognition receptors to Ca^{2+} decoding mechanisms. *New Phytologist* 204, 782–790.

Stulemeijer, I.J., Stratmann, J.W. and Joosten, M.H. (2007) Tomato mitogen-activated protein kinases LeMPK1, LeMPK2, and LeMPK3 are activated during the Cf-4/Avr4-induced hypersensitive response and have distinct phosphorylation specificities. *Plant Physiology* 144, 1481–1494.

Suarez-Rodriguez, M.C., Adams-Phillips, L., Liu, Y., Wang, H., Su, S.H., Jester, P.J., Zhang, S., Bent, A.F. and Krysan, P.J. (2007) MEKK1 is required for flg22-induced MPK4 activation in *Arabidopsis* plants. *Plant Physiology* 143, 661–669.

Takabatake, R., Karita, E., Seo, S., Mitsuhara, I., Kuchitsu, K. and Ohashi, Y. (2007) Pathogen-induced calmodulin isoforms in basal resistance against bacterial and fungal pathogens in tobacco. *Plant Cell and Physiology* 48, 414–423.

Thomma, B.P.H.J., Nürnberger, T. and Joosten, M.H.A.J. (2011) Of PAMPs and effectors: the blurred PTI-ETI dichotomy. *Plant Cell* 23, 4–15.

Torres, M.A., Jones, J.D. and Dangl, J.L. (2006) Reactive oxygen species signaling in response to pathogens. *Plant Physiology* 141, 373–378.

Urquhart, W., Gunawardena, A.H., Moeder, W., Ali, R., Berkowitz, G.A. and Yoshioka, K. (2007) The chimeric cyclic nucleotide-gated ion channel ATCNGC11/12 constitutively induces programmed cell death in a Ca^{2+} dependent manner. *Plant Molecular Biology* 65, 747–761.

van Breusegem, F. and Dat, J.F. (2006) Reactive oxygen species in plant cell death. *Plant Physiology* 141, 384–390.

Verret, F., Wheeler, G., Taylor, A.R., Farnham, G. and Brownlee, C. (2010) Calcium channels in photosynthetic eukaryotes: implications for evolution of calcium-based signaling. *New Phytologist* 187, 23–43.

Wang, L., Tsuda, K., Sato, M., Cohen, J.D., Katagiri, F. and Glazebrook, J. (2009) Arabidopsis CaM binding protein CBP60g contributes to MAMP-induced SA accumulation and is involved in disease resistance against *Pseudomonas syringae*. *PLoS Pathology* 5, e1000301.

Wendehenne, D., Pugin, A., Klessig, D.F. and Durner, J. (2001) Nitric oxide: comparative synthesis and signaling in animal and plant cells. *Trends in Plant Science* 6, 177–183.

Wendehenne, D., Durner, J. and Klessig, D.F. (2004) Nitric oxide: a new player in plant signalling and defense responses. *Current Opinion in Plant Biology* 7, 449–455.

White, P.J. and Broadley, M.R. (2003) Calcium in plants. *Annals of Botany* 92, 487–511.

White, P.J., Bowen, H.C., Demidchik, V., Nichols, C. and Davies, J.M. (2002) Genes for calcium-permeable channels in the plasma membrane of plant root cells. *Biochimica et Biophysica Acta* 1564, 299–309.

Yoshioka, K., Kachroo, P., Tsui, F., Sharma, S.B., Shah, J. and Klessig, D.F. (2001) Environmentally sensitive, SA-dependent defense responses in the *cpr22* mutant of Arabidopsis. *Plant Journal* 26, 447–459.

Yoshioka, K., Moeder, W., Kang, H.G., Kachroo, P., Masmoudi, K., Berkowitz, G. and Klessig, D.F. (2006) The chimeric *Arabidopsis* cyclic nucleotide-gated ion channel11/12 activates multiple pathogen resistance responses. *Plant Cell* 18, 747–763.

Zaninotto, F., La Camera, S., Polverari, A. and Delledonne, M. (2006) Cross talk between reactive nitrogen and oxygen species during the hypersensitive disease resistance response. *Plant Physiology* 141, 379–383.

Zhang, Y., Goritschnig, S., Dong, X. and Li, X. (2003) A gain-of-function mutation in a plant disease resistance gene leads to constitutive activation of downstream signal transduction pathways in *suppressor of npr1-1, constitutive 1*. *Plant Cell* 15, 2636–2646.

Zhu, X., Caplan, J., Mamillapalli, P., Czymmek, K. and Dinesh-Kumar, S.P. (2010) Function of endoplasmic reticulum calcium ATPase in innate immunity-mediated programmed cell death. *EMBO Journal* 29, 1007–1018.

Zipfel, C. (2009) Early molecular events in PAMP-triggered immunity. *Current Opinion in Plant Biology* 12, 414–420.

Zipfel, C., Robatzek, S., Navarro, L., Oakeley, E.J., Jones, J.D.G., Felix, G. and Boller, T. (2004) Bacterial disease resistance in *Arabidopsis* through flagellin perception. *Nature* 428, 764–767.

Zottini, M., Costa, A., De Michele, R., Ruzzene, M., Carimi, F. and Lo Schiavo, F. (2007) Salicylic acid activates nitric oxide synthesis in *Arabidopsis*. *Journal of Experimental Botany* 58, 1397–1405

Index

Note: Page numbers in **bold** refer to figures; those in *italic* refer to tables.